3A	4A	5A	6A	7A	NOBLE ELEMENTS
								HELIUM 4.0026 0.179 **He** 2 $1s^2$ 3.57 HEX ~1.0 (26 Atm) 26^{LT}
			BORON 10.81 2.34 **B** 5 $1s^22s^22p^1$ 8.73 TET 2600 1250	CARBON 12.01 2.26 **C** 6 $1s^22s^22p^2$ 3.57 DIA (4300) 1860	NITROGEN 14.007 1.03 **N** 7 $1s^22s^22p^3$ 4.039 HEX 63.3 (β) 79^{LT}	OXYGEN 15.999 1.43 **O** 8 $1s^22s^22p^4$ 6.83 CUB 54.7 (γ) 46^{LT}	FLUORINE 18.998 1.97 (α) **F** 9 $1s^22s^22p^5$ MCL 53.5	NEON 20.18 1.56 **Ne** 10 $1s^22s^22p^6$ 4.43 FCC 24.5 63
⟶	1B	2B	ALUMINUM 26.982 2.70 **Al** 13 [Ne] $3s^23p^1$ 4.05 FCC 933 394	SILICON 28.086 2.33 **Si** 14 [Ne] $3s^23p^2$ 5.43 DIA 1683 625	PHOSPHORUS 30.974 1.82 (white) **P** 15 [Ne] $3s^23p^3$ 7.17 CUB 317.3	SULFUR 32.064 2.07 **S** 16 [Ne] $3s^23p^4$ 10.47 ORC 2.339 1.229 386	CHLORINE 35.453 2.09 **Cl** 17 [Ne] $3s^23p^5$ 6.24 ORC 1.324 0.718 172.2	ARGON 39.948 1.78 **Ar** 18 [Ne] $3s^23p^6$ 5.26 FCC 83.9 85
CKEL 58.71 9 **Ni** 28 [Ar] $3d^84s^2$ 52 FCC 726 375	COPPER 63.55 8.96 **Cu** 29 [Ar] $3d^{10}4s^1$ 3.61 FCC 1356 315	ZINC 65.38 7.14 **Zn** 30 [Ar] $3d^{10}4s^2$ 2.66 HEX 693 234	GALLIUM 69.72 5.91 **Ga** 31 [Ar] $3d^{10}4s^24p^1$ 4.51 ORC 303 240	GERMANIUM 72.59 5.32 **Ge** 32 [Ar] $3d^{10}4s^24p^2$ 5.66 DIA 1211 360	ARSENIC 74.922 5.72 **As** 33 [Ar] $3d^{10}4s^24p^3$ 4.13 RHL 1090 285	SELENIUM 78.96 4.79 **Se** 34 [Ar] $3d^{10}4s^24p^4$ 4.36 HEX 490 150^{LT}	BROMINE 79.91 4.10 **Br** 35 [Ar] $3d^{10}4s^24p^5$ 6.67 ORC 266	KRYPTON 83.80 3.07 **Kr** 36 [Ar] $3d^{10}4s^24p^6$ 5.72 FCC 116.5 73^{LT}
ALLADIUM 106.40 .0 **Pd** 46 [Kr] $4d^{10}5s^0$ 89 FCC 825 275	SILVER 107.87 10.5 **Ag** 47 [Kr] $4d^{10}5s^1$ 4.09 FCC 1234 215	CADMIUM 112.40 8.65 **Cd** 48 [Kr] $4d^{10}5s^2$ 2.98 HEX 594 120	INDIUM 114.82 7.31 **In** 49 [Kr] $4d^{10}5s^25p^1$ 4.59 TET 429.8 129	TIN 118.69 7.30 **Sn** 50 [Kr] $4d^{10}5s^25p^2$ 5.82 TET 505 170	ANTIMONY 121.75 6.62 **Sb** 51 [Kr] $4d^{10}5s^25p^3$ 4.51 RHL 904 200	TELLURIUM 127.60 6.24 **Te** 52 [Kr] $4d^{10}5s^25p^4$ 4.45 HEX 723 139^{LT}	IODINE 126.90 4.94 **I** 53 [Kr] $4d^{10}5s^25p^5$ 7.27 ORC 387	XENON 131.30 3.77 **Xe** 54 [Kr] $4d^{10}5s^25p^6$ 6.20 FCC 161.3 55^{LT}
ATINUM 195.09 .4 **Pt** 78 [Xe] $4f^{14}5d^{10}6s^0$ 92 FCC 45 230	GOLD 196.97 19.3 **Au** 79 [Xe] $4f^{14}5d^{10}6s^1$ 4.08 FCC 1337 170	MERCURY 200.59 13.6 **Hg** 80 [Xe] $4f^{14}5d^{10}6s^2$ 2.99 RHL 234.3 100	THALLIUM 204.37 11.85 **Tl** 81 [Xe] $4f^{14}5d^{10}6s^26p^1$ 3.46 HEX 577 96	LEAD 207.19 11.4 **Pb** 82 [Xe] $4f^{14}5d^{10}6s^26p^2$ 4.95 FCC 601 88	BISMUTH 208.98 9.8 **Bi** 83 [Xe] $4f^{14}5d^{10}6s^26p^3$ 4.75 RHL 544.5 120	POLONIUM 210 9.4 **Po** 84 [Xe] $4f^{14}5d^{10}6s^26p^4$ 3.35 SC 527	ASTATINE 210 **At** 85 [Xe] $4f^{14}5d^{10}6s^26p^5$ (575)	RADON 222 (4.4) **Rn** 86 [Xe] $4f^{14}5d^{10}6s^26p^6$ (FCC) (202)

JROPIUM 151.96 90 **Eu** 63 [Xe] $4f^75d^06s^2$ 61 BCC 095 107^{LT}	GADOLINIUM 157.25 8.23 **Gd** 64 [Xe] $4f^75d^16s^2$ 3.64 HEX 1585 176^{LT}	TERBIUM 158.92 8.54 **Tb** 65 [Xe] $4f^95d^06s^2$ 3.60 HEX 1633 188^{LT}	DYSPROSIUM 162.50 8.78 **Dy** 66 [Xe] $4f^{10}5d^06s^2$ 3.59 HEX 1680 186^{LT}	HOLMIUM 164.93 9.05 **Ho** 67 [Xe] $4f^{11}5d^06s^2$ 3.58 HEX 1743 191^{LT}	ERBIUM 167.26 9.37 **Er** 68 [Xe] $4f^{12}5d^06s^2$ 3.56 HEX 1795 195^{LT}	THULIUM 168.93 9.31 **Tm** 69 [Xe] $4f^{13}5d^06s^2$ 3.54 HEX 1818 200^{LT}	YTTERBIUM 173.04 6.97 **Yb** 70 [Xe] $4f^{14}5d^06s^2$ 5.49 FCC 1097 118^{LT}	LUTETIUM 174.97 9.84 **Lu** 71 [Xe] $4f^{14}5d^16s^2$ 3.51 HEX 1929 207^{LT}
MERICIUM 243 .8 **Am** 95 [Rn] $5f^76d^07s^2$ 267	CURIUM 247 **Cm** 96 [Rn] $5f^76d^17s^2$ 1600	BERKELIUM 247 **Bk** 97 [Rn] $5f^76d^27s^2$	CALIFORNIUM 251 **Cf** 98 [Rn] $5f^96d^17s^2$	EINSTEINIUM 254 **Es** 99	FERMIUM 257 **Fm** 100	MENDELEVIUM 256 **Md** 101	NOBELIUM 254 **No** 102	LAWRENCIUM 257 **Lw** 103

Introductory
QUANTUM
MECHANICS

THIRD EDITION

RICHARD L. LIBOFF
CORNELL UNIVERSITY

 ADDISON-WESLEY

An imprint of Addison Wesley Longman, Inc.

Reading, Massachusetts • Menlo Park, California • New York • Harlow, England
Don Mills, Ontario • Sydney • Mexico City • Madrid • Amsterdam

Library of Congress Cataloging-in-Publication Data

Liboff, Richard L.
    Introductory quantum mechanics / Richard L. Liboff. -- 3rd ed.
        p.   cm.
    Includes bibliographical references and index.
    ISBN 0-201-87879-8 (hardcover)
    1. Quantum theory.   I. Title.
QC174.12.L52   1997
530.1'2--dc20

96-23160
CIP

ISBN 0-201-87879

    4 5 6 7 8 9 10 — MA — 99

To Myra

"She openeth her mouth with wisdom; and in her tongue is the law of kindness. . . ."

PREFACE

It remains true, since the first edition of this work appeared, that physics continues to evolve in both esoteric and pragmatic directions. The present edition follows this trend and the more extensive list of subjects included here should provide instructors with a broader base from which to choose topics for their particular course.

In Section 10.3, descriptions of the spherical quantum well and cylindrical quantum well are presented, including a discussion of the relation of eigenfunction angular momentum quantum numbers to spherical and cylindrical symmetry. These quantum-well configurations are relevant to the "quantum dot" and the "quantum wire," respectively, which are pertinent to semiconductor technology. In Section 11.14, employing an assortment of symmetry principles, the *transfer matrix method* is developed, which is relevant to the problem of transmission of particles through a periodic potential. The transfer matrix method is employed to regain the Kronig–Penny dispersion relation (described earlier in Section 8.2). In present-day technology, the formalism of the transfer-matrix method finds application in the examination of quantum-well microdevices and in the more fundamental analysis of random matrices relevant to the study of disordered systems and the closely allied phenomenon of localization.

Section 12.9, addressing impurity semiconductors, was revised and a subsection was added on the p-n junction. A description was inserted in Section 13.3 on the phenomenon of symmetry breaking and removal of degeneracy. Section 13.10 was added to describe the Hartree–Fock model, which affords a method for obtaining approximate atomic wavefunctions and eigenenergies. This description complements the Thomas–Fermi model (Section 10.8) in which an effective atomic potential is constructed.

Beside this assortment of additions and expansions, a new chapter (Chapter 15) is included in the present edition in which relativistic wave equations are developed. The chapter begins with a review of basic relativistic concepts and continues with derivation of the Klein–Gordon equation appropriate to relativistic bosons. Components of this equation are employed to obtain the Dirac equation for relativistic fermions. In the Dirac theory of the electron, a gap appears in the energy spectrum of an electron, which is noted to imply electron-positron creation and annihilation. This equation is further employed to develop a relativistic formulation of the magnetic moment of the electron. The Dirac formulation of the four-dimensional spin operator is described, and the chapter concludes with a brief introduction to the covariant formulation of relativistic quantum mechanics.

There are additional entries in Appendix E (Physical Constants and Equivalence), including a list of relations of constants in Coulomb's law in cgs and MKS units. A new appendix (Appendix D) is included, which lists vector differential relations in the three primary orthogonal coordinate frames. The historical list in Section 2.1, addressing the early days of quantum mechanics, has been extended with a number of intriguing footnotes.

A number of interesting problems have been added to this edition, many of which carry solutions. These include, for example, problems addressing Ehrenfest's principle in quantum mechanics; the momentum operator in curvilinear coordinates; the Heisenberg picture; the number-phase uncertainty relation and the notion of "squeezed states"; the 21 cm line of atomic hydrogen; surpersymmetry (revised); normalization of associated Legendre polynomials (revised); Lorentz invariance of proper time; and relativistic γ and Γ matrices. (For specific problem-number citations, see "Topical Problems," which appears after the Table of Contents.)

Acknowledgments

A number of colleagues have been helpful in the development of this text. I remain indebted to these individuals and express my appreciation to Norman Ramsey, Ilya Prigogine, Norman Rostoker, Kurt Gottfried, Chung Tang, David Clark, Terence Fine, Jack Freed, Greg Ezra, Paul McIsaac, Geoffrey Canright, Robert Fay, Sidney Leibovich, Vaclav Kostroun, Richard Lovelace, Richard Crandall, Donald Scarl, Donald Yennie, Stanley Bashkin, Danny Heffernan, Michael Guillen, Brian Jones, Kenneth Gardner, Lloyd Hillman, David Faulconer, George George, Yervant Terzian, Joseph Saltzman, Gregory Schenter, Joseph DiGioia, Volkan Kaman, Yu-Hwa Lo, John Wendler, Steven Seidman, Mark Vaughn, Frank Goodwin, Matthew Angyal, Yann-Ting Chen, Ravindra Sudan, Johnathan Getty, Michael Parker, Dan Fekete, Stanley Lau, Kenneth Andrews, Michael Wong, Mark Coffey, Yosi Shacham, Hercules Nevis, Gregory Dionne, and Clifford Pollock.

It will be noted that a new, informative periodic chart appears on the inside flap of the front cover. This chart is a copy of that which appears in the classic work *Solid State Physics* by Neil Ashcroft and N. David Mermin. I am grateful to these colleagues for granting me permission to use this chart in the present work.

Once again I would like to express my appreciation to the many individuals who have taught from the prior editions of this text and the many who have learned from it. I sincerely hope that these teachers and students find this new edition equally valuable.

Ithaca, 1996   R. L. Liboff

‎.תושלב״ע

"I do not know what I may appear to the world; but to myself I seem to have been only like a boy playing on the seashore, and diverting myself in now and then finding a smoother pebble or a prettier shell than ordinary, whilst the great ocean of truth lay all undiscovered before me."

This statement by Isaac Newton shortly before his death in 1727 eloquently reflects the sentiments of all mature scientists from the ancient past to the present.

Piping down the valleys wild,
Piping songs of pleasant glee,
On a cloud I saw a child,
And he laughing said to me;

"Pipe a song about a Lamb!"
So I piped with merry chear.
"Piper, pipe that song again;"
So I piped; he wept to hear.

"Drop thy pipe, thy happy pipe;
Sing thy songs of happy chear;"
So I sung the same again,
While he wept with joy to hear.

"Piper, sit thee down and write
In a book, that all may read."
So he vanish'd from my sight,
And I pluck'd a hollow reed,

And I made a rural pen,
And I stain'd the water clear,
And I wrote my happy songs
Every child may joy to hear.

Introduction to Songs of Innocence
–William Blake (1757–1827)

CONTENTS

LIST OF TABLES

TOPICAL PROBLEMS

I

ELEMENTARY PRINCIPLES
AND APPLICATIONS TO PROBLEMS
IN ONE DIMENSION

1

REVIEW OF CONCEPTS OF CLASSICAL MECHANICS

This is a preparatory chapter in which we review fundamental concepts of classical mechanics important to the development and understanding of quantum mechanics. Hamilton's equations are introduced and the relevance of cyclic coordinates and constants of the motion is noted. In discussing the state of a system, we briefly encounter our first distinction between classical and quantum descriptions. The notions of forbidden domains and turning points relevant to classical motion, which find application in quantum mechanics as well, are also described. The experimental motivation and historical background of quantum mechanics are described in Chapter 2.

1.1 GENERALIZED OR "GOOD" COORDINATES

Our discussion begins with the concept of *generalized* or *good* coordinates.

A bead (idealized to a point particle) constrained to move on a straight rigid wire has *one degree of freedom* (Fig. 1.1). This means that only one variable (or parameter) is needed to uniquely specify the location of the bead in space. For the problem under

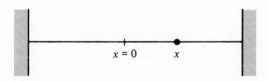

$x = 0 \qquad x$

FIGURE 1.1 A bead constrained to move on a straight wire has one degree of freedom.

discussion, the variable may be displacement from an arbitrary but specified origin along the wire.

A particle constrained to move on a flat plane has two degrees of freedom. Two independent variables suffice to uniquely determine the location of the particle in space. With respect to an arbitrary, but specified origin in the plane, such variables might be the Cartesian coordinates (x, y) or the polar coordinates (r, θ) of the particle (Fig. 1.2).

Two beads constrained to move on the same straight rigid wire have two degrees of freedom. A set of appropriate coordinates are the displacements of the individual particles (x_1, x_2) (Fig. 1.3).

A rigid rod (or dumbbell) constrained to move in a plane has three degrees of freedom. Appropriate coordinates are the location of its center (x, y) and the angular displacement of the rod from the horizontal, θ (Fig. 1.4).

Independent coordinates that serve to uniquely determine the orientation and location of a system in physical space are called *generalized* or *canonical* or *good* coordinates. *A system with N generalized coordinates has N degrees of freedom.* The orientation and location of a system with, say, three degrees of freedom are not specified until all three generalized coordinates are specified. The fact that *good*

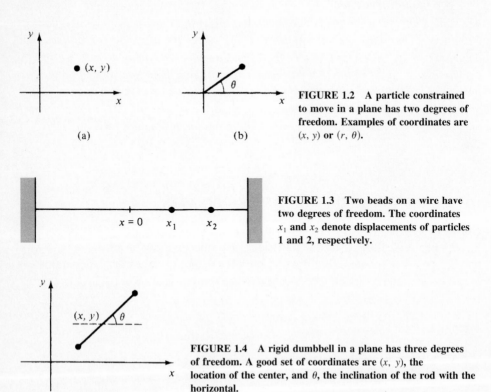

FIGURE 1.2 A particle constrained to move in a plane has two degrees of freedom. Examples of coordinates are (x, y) or (r, θ).

FIGURE 1.3 Two beads on a wire have two degrees of freedom. The coordinates x_1 and x_2 denote displacements of particles 1 and 2, respectively.

FIGURE 1.4 A rigid dumbbell in a plane has three degrees of freedom. A good set of coordinates are (x, y), the location of the center, and θ, the inclination of the rod with the horizontal.

coordinates may be specified independently of one another means that, given the values of all but one of the coordinates, the last coordinate remains arbitrary. Having specified (x, y) for a point particle in 3-space, one is still free to choose z independently of the assigned values of x and y.

PROBLEMS

1.1   For each of the following systems, specify the number of degrees of freedom and a set of good coordinates.

(a)   A bead constrained to move on a closed circular hoop that is fixed in space.
(b)   A bead constrained to move on a helix of constant pitch and constant radius.
(c)   A particle on a right circular cylinder.
(d)   A pair of scissors on a plane.
(e)   A rigid rod in 3-space.
(f)   A rigid cross in 3-space.
(g)   A linear spring in 3-space.
(h)   Any rigid body with one point fixed.
(i)   A hydrogen atom.
(j)   A lithium atom.
(k)   A compound pendulum (two pendulums attached end to end).

1.2   Show that a particle constrained to move on a curve of any shape has one degree of freedom.

Answer

A curve is a one-dimensional locus and may be generated by the parameterized equations

$$x = x(\eta), \qquad y = y(\eta), \qquad z = z(\eta)$$

Once the independent variable η (e.g., length along the curve) is given, x, y, and z are specified.

1.3   Show that a particle constrained to move on a surface of arbitrary shape has two degrees of freedom.

Answer

A surface is a two-dimensional locus. It is generated by the equation

$$u(x, y, z) = 0$$

Any two of the three variables x, y, z determine the third. For instance, we may solve for z in the equation above to obtain the more familiar equation for a surface (height z at the point x, y).

$$z = z(x, y)$$

In this case, x and y may serve as generalized coordinates.

1.4   How many degrees of freedom does a classical gas composed of 10^{23} point particles have?

1.2   ENERGY, THE HAMILTONIAN, AND ANGULAR MOMENTUM

These three elements of classical mechanics have been singled out because they have direct counterparts in quantum mechanics. Furthermore, as in classical mechanics, their role in quantum mechanics is very important.

Consider that a particle of mass m in the potential field $V(x, y, z)$ moves on the trajectory

$$
\begin{aligned}
x &= x(t) \\
y &= y(t) \\
z &= z(t)
\end{aligned}
$$

(1.1)

At any instant t, the energy of the particle is

(1.2) $$E = \frac{1}{2}mv^2 + V(x, y, z) = \frac{1}{2}m(\dot{x}^2 + \dot{y}^2 + \dot{z}^2) + V(x, y, z)$$

The velocity of the particle is $\mathbf{v}$. Dots denote time derivatives. The force on the particle $\mathbf{F}$ is the negative gradient of the potential.

(1.3) $$\mathbf{F} = -\mathbf{\nabla}V = -\left(\mathbf{e}_x \frac{\partial}{\partial x}V + \mathbf{e}_y \frac{\partial}{\partial y}V + \mathbf{e}_z \frac{\partial}{\partial z}V\right)$$

The three unit vectors $(\mathbf{e}_x, \mathbf{e}_y, \mathbf{e}_z)$ lie along the three Cartesian axes.

Here are two examples of potential. The energy of a particle in the gravitational force field,

$$\mathbf{F} = -\mathbf{e}_z mg = -\mathbf{\nabla}mgz$$

is

(1.4) $$E = \frac{1}{2}m(\dot{x}^2 + \dot{y}^2 + \dot{z}^2) + mgz$$

The particle is at the height z above sea level. For this example,

$$V = mgz$$

An electron of charge q and mass m, between capacitor plates that are maintained at the potential difference Φ_0 and separated by the distance d (Fig. 1.5), has potential

$$V = \frac{q\Phi_0}{d}z$$

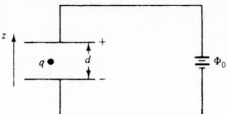

FIGURE 1.5   Electron in a uniform capacitor field.

The displacement of the electron from the bottom plate is z. The electron's energy is

$$(1.5) \qquad E = \frac{1}{2}m(\dot{x}^2 + \dot{y}^2 + \dot{z}^2) + \frac{q\Phi_0}{d}z$$

In both examples above, the system (particle) has three degrees of freedom. The Cartesian coordinates (x, y, z) of the particle are by no means the only "good" coordinates for these cases. For instance, in the last example, we may express the energy of the electron in spherical coordinates (Fig. 1.6):

$$(1.6) \qquad E = \frac{1}{2}m(\dot{r}^2 + r^2\dot{\theta}^2 + r^2\dot{\phi}^2 \sin^2 \theta) + \frac{q\Phi_0}{d} r \cos \theta$$

In cylindrical coordinates (Fig. 1.7) the energy is

$$(1.7) \qquad E = \frac{1}{2}m(\dot{\rho}^2 + \rho^2\dot{\phi}^2 + \dot{z}^2) + \frac{q\Phi_0}{d}z$$

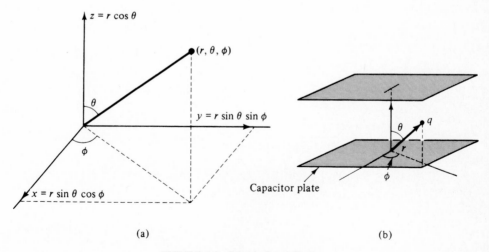

(a)   (b)

FIGURE 1.6   Spherical coordinates.

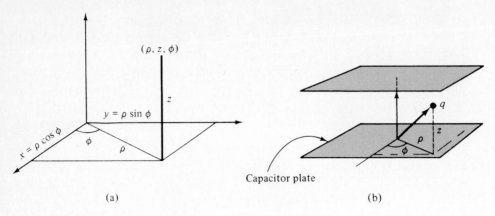

FIGURE 1.7   Cylindrical coordinates.

The hydrogen atom has six degrees of freedom. If (x_1, y_1, z_1) are the coordinates of the proton and (x_2, y_2, z_2) are the coordinates of the electron, the energy of the hydrogen atom appears as

(1.8)
$$E = \frac{1}{2}M(\dot{x}_1^2 + \dot{y}_1^2 + \dot{z}_1^2) + \frac{1}{2}m(\dot{x}_2^2 + \dot{y}_2^2 + \dot{z}_2^2)$$

$$-\frac{q^2}{\sqrt{(x_1 - x_2)^2 + (y_1 - y_2)^2 + (z_1 - z_2)^2}}$$

(Fig. 1.8). The mass of the proton is M and that of the electron is m. In all the cases above, the energy is a *constant of the motion*. A constant of the motion is a dynamical function that is constant as the system unfolds in time. For each of these cases,

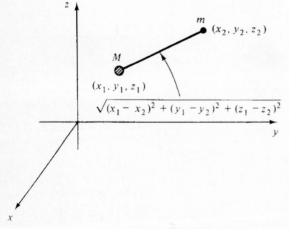

FIGURE 1.8   The hydrogen atom has six degrees of freedom. The Cartesian coordinates of the proton and electron serve as good generalized coordinates.

whatever E is initially, it maintains that value, no matter how complicated the subsequent motion is. Constants of the motion are extremely useful in classical mechanics and often serve to facilitate calculation of the trajectory.

A system that in no way interacts with any other object in the universe is called an *isolated system*. The total energy, linear momentum, and angular momentum of an isolated system are constant. Let us recall the definition of linear and angular momentum for a particle. A particle of mass m moving with velocity $\mathbf{v}$ has linear momentum

(1.9)
$$\mathbf{p} = m\mathbf{v}$$

The angular momentum of this particle, measured about a specific origin, is

(1.10)
$$\mathbf{L} = \mathbf{r} \times \mathbf{p}$$

where $\mathbf{r}$ is the radius vector from the origin to the particle (Fig. 1.9).

If there is no component of force on a particle in a given (constant) direction, the component of momentum in that direction is constant. For example, for a particle in a gravitational field that is in the z direction, p_x and p_y are constant.

If there is no component of torque $\mathbf{N}$ in a given direction, the component of angular momentum in that direction is constant. This follows directly from Newton's second law of angular momentum.

(1.11)
$$\mathbf{N} = \frac{d\mathbf{L}}{dt}$$

For a particle in a gravitational field that is in the minus z direction, the torque on the particle is

$$\mathbf{N} = \mathbf{r} \times \mathbf{F} = -\mathbf{r} \times \mathbf{e}_z mg$$

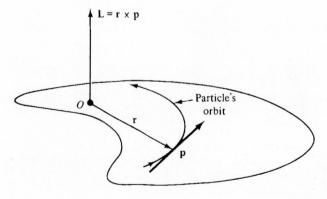

FIGURE 1.9 Angular momentum of a particle with momentum p about the origin O.

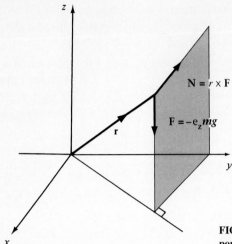

FIGURE 1.10 **The torque r × F has no component in the z direction.**

The radius vector from the origin to the particle is **r** (Fig. 1.10). Since $\mathbf{e}_z \times \mathbf{r}$ has no component in the z direction ($\mathbf{e}_z \cdot \mathbf{e}_z \times \mathbf{r} = 0$), it follows that

(1.12) $$L_z = xp_y - yp_x = \text{constant})$$

Since p_x and p_y are also constants, this equation tells us that the projected orbit in the xy plane is a straight line (Fig. 1.11).

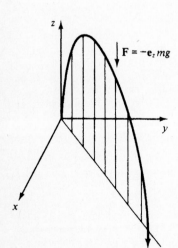

FIGURE 1.11 **The projected motion in the xy plane is a straight line. Its equation is given by the constant z component of angular momentum: $L_z = xp_y - yp_x$.**

Hamilton's Equations

The constants of motion for more complicated systems are not so easily found. However, there is a formalism that treats this problem directly. It is Hamiltonian mechanics. Consider the energy expression for an electron between capacitor plates (1.5). Rewriting this expression in terms of the linear momentum **p** (as opposed to velocity) gives

$$(1.13) \quad E(x, y, z, \dot{x}, \dot{y}, \dot{z}) \rightarrow H(x, y, z, p_x, p_y, p_z) = \frac{1}{2m}(p_x^2 + p_y^2 + p_z^2) + \frac{q\Phi_0}{d}z$$

The energy, written in this manner, as a function of coordinates and momenta is called the *Hamiltonian, H.* One speaks of p_x as being the momentum *conjugate* to x; p_y is the momentum conjugate to y; and so on.

The equations of motion (i.e., the equations that replace Newton's second law) in Hamiltonian theory are (for a point particle moving in three-dimensional space)

$$(1.14)$$

$$\frac{\partial H}{\partial x} = -\dot{p}_x \qquad \frac{\partial H}{\partial p_x} = \dot{x}$$

$$\frac{\partial H}{\partial y} = -\dot{p}_y \qquad \frac{\partial H}{\partial p_y} = \dot{y}$$

$$\frac{\partial H}{\partial z} = -\dot{p}_z \qquad \frac{\partial H}{\partial p_z} = \dot{z}$$

Cyclic Coordinates

For the Hamiltonian (1.13) corresponding to an electron between capacitor plates, one obtains

$$(1.15) \qquad \frac{\partial H}{\partial x} = \frac{\partial H}{\partial y} = 0$$

The Hamiltonian does not contain x or y. When coordinates are missing from the Hamiltonian, they are called *cyclic* or *ignorable.* The momentum conjugate to a cyclic coordinate is a constant of the motion. This important property follows directly from Hamilton's equations, (1.14). For example, for the case at hand, we see that $\partial H/\partial x = 0$ implies that $\dot{p}_x = 0$, so p_x is constant; similarly for p_y. (Note that there is no component of force in the x or y directions.) The remaining four Hamilton's equations give

$$(1.16) \qquad \dot{p}_z = -\frac{q\Phi_0}{d}, \qquad p_x = m\dot{x}, \qquad p_y = m\dot{y}, \qquad p_z = m\dot{z}$$

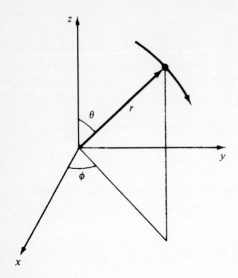

FIGURE 1.12   **Motion of a particle in spherical coordinates with r and ϕ fixed: $v_\theta = r\dot\theta$, $p_\theta = rmv_\theta = mr^2\dot\theta$. The moment arm is r.**

The last three equations return the definitions of momenta in terms of velocities. The first equation is the z component of Newton's second law. (For an electron, $q = -|q|$. It is attracted to the positive plate.)

Consider next the Hamiltonian for this same electron but expressed in terms of spherical coordinates. We must transform E as given by (1.5) to an expression involving r, θ, ϕ, and the momenta conjugate to these coordinates. The momentum conjugate to r is the component of linear momentum in the direction of $\mathbf{r}$. If $\mathbf{e}_r$ is a unit vector in the $\mathbf{r}$ direction, then

$$(1.17) \qquad p_r = \frac{\mathbf{r} \cdot \mathbf{p}}{r} = \mathbf{e}_r \cdot \mathbf{p} = m\mathbf{e}_r \cdot \mathbf{v} = m\dot r$$

The momentum conjugate to the angular displacement θ is the component of angular momentum corresponding to a displacement in θ (with r and ϕ fixed). The moment arm for this motion is r. The velocity is $r\dot\theta$. It follows that

$$(1.18) \qquad p_\theta = mr(r\dot\theta) = mr^2\dot\theta$$

(Fig. 1.12).

The momentum conjugate to ϕ is the angular momentum corresponding to a displacement in ϕ (with r and θ fixed). The moment arm for this motion is $r\sin\theta$. The velocity is $r\dot\phi\sin\theta$ (Fig. 1.13). The angular momentum of this motion is

$$(1.19) \qquad p_\phi = mr^2\dot\phi\sin^2\theta$$

Since such motion is confined to a plane normal to the z axis, p_ϕ is the z component of angular momentum. This was previously denoted as L_z in (1.12).

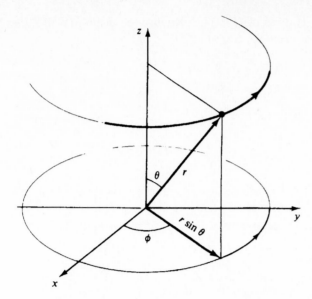

FIGURE 1.13 **Motion of a particle with r and θ fixed: $v_\phi = r \sin \theta \, \dot{\phi}$.**
The moment arm is $r \sin \theta$,
$p_\phi = (r \sin \theta) m v_\phi = mr^2 \, \dot{\phi} \sin^2 \theta.$

In terms of these coordinates and momenta, the energy expression (1.6) becomes

$$(1.20) \quad H(r, \theta, \phi, p_r, p_\theta, p_\phi) = \frac{p_r^2}{2m} + \frac{p_\theta^2}{2mr^2} + \frac{p_\phi^2}{2mr^2 \sin^2 \theta} + \frac{q\Phi_0}{d} r \cos \theta$$

Hamilton's equations for a point particle, in spherical coordinates, become

$$(1.21) \quad \begin{array}{cc} \dfrac{\partial H}{\partial \theta} = -\dot{p}_\theta & \dfrac{\partial H}{\partial p_\theta} = \dot{\theta} \\[3mm] \dfrac{\partial H}{\partial \phi} = -\dot{p}_\phi & \dfrac{\partial H}{\partial p_\phi} = \dot{\phi} \\[3mm] \dfrac{\partial H}{\partial r} = -\dot{p}_r & \dfrac{\partial H}{\partial p_r} = \dot{r} \end{array}$$

From the form of the Hamiltonian (1.20) we see that ϕ is a cyclic coordinate. That is,

$$(1.22) \quad \frac{\partial H}{\partial \phi} = 0 = -\dot{p}_\phi$$

It follows that p_ϕ, as given by (1.19), is constant. Thus, the component of angular momentum in the z direction is conserved. The torque on the particle has no component in this direction.

Again the momentum derivatives of H in (1.20) return the definitions of momenta in terms of velocities. For example, from (1.20),

$$(1.23) \qquad \frac{\partial H}{\partial p_\theta} = \dot{\theta} = \frac{p_\theta}{mr^2}$$

which is (1.18). Hamilton's equation for $\dot{p}_r$ is

$$(1.24) \qquad -\frac{\partial H}{\partial r} = \dot{p}_r = \frac{p_\theta^2}{mr^3} + \frac{p_\phi^2}{mr^3 \sin^2 \theta} - \frac{q\Phi_0}{d} \cos \theta$$

The first two terms on the right-hand side of this equation are the components of centripetal force in the radial direction, due to θ and ϕ displacements, respectively. The last term is the component of electric force $-\mathbf{e}_z q\Phi_0/d$ in the radial direction. Hamilton's equation for $\dot{p}_\theta$ is

$$(1.25) \qquad -\frac{\partial H}{\partial \theta} = \dot{p}_\theta = \frac{p_\phi^2 \cos \theta}{mr^2 \sin^3 \theta} + \frac{q\Phi_0}{d} r \sin \theta$$

The right-hand side is a component of torque. It contains the centripetal force factor due to the ϕ motion ($p_\phi^2/mr^3 \sin^3 \theta$) and a moment arm factor, $r \cos \theta$. At any instant of time this component of torque is normal to the plane swept out by r due to θ motion alone.

A very instructive example concerns the motion of a free particle. A free particle is one that does not interact with any other particle or field. It is free of all interactions and is an isolated system. A particle moving by itself in an otherwise empty universe is a free particle. In Cartesian coordinates the Hamiltonian for a free particle is

$$(1.26) \qquad H = \frac{1}{2m} p^2 = \frac{1}{2m} (p_x^2 + p_y^2 + p_z^2)$$

All coordinates (x, y, z) are cyclic. Therefore, the three components of momenta are constant and may be equated to their respective initial values at time $t = 0$.

$$(1.27) \qquad \begin{aligned} p_x &= p_x(0) \\ p_y &= p_y(0) \\ p_z &= p_z(0) \end{aligned}$$

Combining these with the remaining three Hamilton's equations gives

$$(1.28) \qquad \begin{aligned} m\dot{x} &= p_x(0) \\ m\dot{y} &= p_y(0) \\ m\dot{z} &= p_z(0) \end{aligned}$$

These are simply integrated to obtain

$$x(t) = \frac{p_x(0)}{m} t + x(0)$$

(1.29)
$$y(t) = \frac{p_y(0)}{m} t + y(0)$$

$$z(t) = \frac{p_z(0)}{m} t + z(0)$$

which are parametric equations for a straight line.

Let us calculate the y component of angular momentum of the (free) particle.

$$(1.30) \quad L_y = zp_x - xp_z = \left[z(0) + \frac{p_z(0)}{m} t \right] p_x(0) - \left[x(0) + \frac{p_x(0)}{m} t \right] p_z(0)$$

Canceling terms, we obtain

$$(1.31) \qquad L_y = z(0)p_x(0) - x(0)p_z(0) = L_y(0)$$

and similarly for L_x and L_z. It follows that

$$(1.32) \qquad \mathbf{L} = (L_x, L_y, L_z) = \text{constant}$$

for a free particle.

Investigating the dynamics of a free particle in Cartesian coordinates has given us immediate and extensive results. We know that $\mathbf{p}$ and $\mathbf{L}$ are both constant. The orbit is rectilinear.

We may also consider the dynamics of a free particle in spherical coordinates. The Hamiltonian is

$$(1.33) \qquad H = \frac{p_r^2}{2m} + \frac{p_\theta^2}{2mr^2} + \frac{p_\phi^2}{2mr^2 \sin^2 \theta}$$

Only ϕ is cyclic, and we immediately conclude that p_ϕ (or equivalently, L_z) is constant. However, p_r and p_θ are not constant. From Hamilton's equations, we obtain

$$\dot{p}_r = \frac{p_\theta^2}{mr^3} + \frac{p_\phi^2}{mr^3 \sin^2 \theta}$$

(1.34)

$$\dot{p}_\theta = \frac{p_\phi^2 \cos \theta}{mr^2 \sin^3 \theta}$$

These centripetal terms were interpreted above. In this manner we find that the rectilinear, constant-velocity motion of a free particle, when cast in a spherical coordinate frame, involves accelerations in the r and θ components of motion. These accelerations arise from an inappropriate choice of coordinates. In simple language: Fitting a straight line to spherical coordinates gives peculiar results.

A comparison of the Hamiltonian for a free particle in Cartesian, spherical, and cylindrical coordinates is shown in Table 1.1.

Canonical Coordinates and Momenta

While the reader may feel some familiarity with the components of linear momentum (p_x, p_y, p_z) and angular momentum (p_θ, p_ϕ), it is clear that these intuitive notions are exhausted for a system with, say, 17 degrees of freedom. If we call the seventeenth coordinate q_{17}, what is the momentum p_{17} conjugate to q_{17}? There is a formal procedure for determining the momentum conjugate to a given generalized coordinate. For example, it gives $p_\theta = mr^2\dot{\theta}$ as the momentum conjugate to θ for a particle in spherical coordinates. This procedure is described in any book on graduate mechanics.[1]

The coordinates of a system with N degrees of freedom, $(q_1, q_2, q_3, \ldots, q_N)$, and conjugate momenta $(p_1, p_2, p_3, \ldots, p_N)$ are also called *canonical* coordinates and momenta. A set of coordinates and momenta are canonical if with the Hamiltonian, $H(q_1, \ldots, q_N, p_1, \ldots, p_N, t)$, Hamilton's equations

$$(1.35) \qquad \frac{\partial H}{\partial q_l} = -\dot{p}_l, \qquad \frac{\partial H}{\partial p_l} = \dot{q}_l \qquad (l = 1, \ldots, N)$$

are entirely consistent with Newton's laws of motion. We have seen this to be the case for all the problems considered above. (Time-dependent Hamiltonians are considered in Chapter 13.)

Other important functions and concepts of classical mechanics include the *Lagrangian, action integral, and Hamilton's principle.* These topics are discussed in Section 7.11, which addresses the Feynman path integral.

PROBLEMS

1.5 Show that the z component of angular momentum for a point particle

$$L_z = xp_y - yp_x$$

when expressed in spherical coordinates, becomes

$$L_z = p_\phi = mr^2\dot{\phi}\sin^2\theta$$

(*Hint:* Recall the transformation equations

$$z = r\cos\theta$$

$$y = r\sin\theta\sin\phi$$

$$x = r\sin\theta\cos\phi.)$$

---

[1] See, for example, H. Goldstein, *Classical Mechanics,* 2d ed., Addison-Wesley, Reading, Mass., 1980.

TABLE 1.1   Hamiltonian of a free particle in three coordinate frames

	Cartesian Coordinates	Spherical Coordinates	Cylindrical Coordinates
Frames	(x, y, z)	(r, θ, ϕ)	(ρ, z, ϕ)
Hamiltonian	$H(x, y, z, p_x, p_y, p_z)$ $= \dfrac{1}{2m}\left(p_x^2 + p_y^2 + p_z^2\right)$	$H(r, \theta, \phi, p_r, p_\theta, p_\phi)$ $= \dfrac{1}{2m}\left[p_r^2 + \dfrac{1}{r^2}\left(p_\theta^2 + \dfrac{p_\phi^2}{\sin^2\theta}\right)\right]$ $= \dfrac{1}{2m}\left(p_r^2 + \dfrac{L^2}{r^2}\right)$	$H(\rho, z, \phi, p_\rho, p_z, p_\phi)$ $= \dfrac{1}{2m}\left(p_\rho^2 + p_z^2 + \dfrac{p_\phi^2}{\rho^2}\right)$
Momenta	$p_x = m\dot{x}$ $p_y = m\dot{y}$ $p_z = m\dot{z}$	$p_r = m\dot{r}$ $p_\theta = mr^2\dot{\theta}$ $p_\phi = mr^2\dot{\phi}\sin^2\theta$	$p_\rho = m\dot{\rho}$ $p_z = m\dot{z}$ $p_\phi = m\rho^2\dot{\phi}$
Cyclic coordinates	x, y, z	ϕ	z, ϕ
Constant momenta	p_x, p_y, p_z	$p_\phi = L_z$	p_z, p_ϕ

1.6   (a)   Calculate $\dot{p}_r$, $\dot{p}_\theta$, and $\dot{p}_\phi$ as explicit functions of time for the following motion of a particle.

$$y = y_0, \qquad z = z_0, \qquad x = v_0 t$$

(b)   For what type of free-particle orbit are the following conditions obeyed?
  (1) $\dot{p}_r = 0$
  (2) $\dot{p}_\theta = 0$
  (3) $\dot{p}_\phi = 0$
  (4) $\dot{p}_r = \dot{p}_\theta = \dot{p}_\phi = 0$

(c)   Describe an experiment to measure p_r, at a given instant, for the motion of part (a).

1.7   Show that the energy of a free particle may be written

$$H = \frac{p_r^2}{2m} + \frac{L^2}{2mr^2}$$

where $\mathbf{L} = \mathbf{r} \times \mathbf{p}$. [*Hint:* Use the vector relation

$$L^2 = (\mathbf{r} \times \mathbf{p})^2 = r^2 p^2 - (\mathbf{r} \cdot \mathbf{p})^2$$

together with the definition $p_r = (\mathbf{r} \cdot \mathbf{p})/r$.]

1.8   Show that angular momentum of a free particle obeys the relation

$$L^2 = L_x^2 + L_y^2 + L_z^2 = p_\theta^2 + \frac{p_\phi^2}{\sin^2 \theta}$$

(*Hint:* Employ the results of Problem 1.7.)

1.9   A particle of mass m is in the environment of a force field with components

$$F_z = -Kz, \qquad F_x = 0, \qquad F_y = 0$$

with K constant.

  (a)   Write down the Hamiltonian of the particle in Cartesian coordinates. What are the constants of motion?

  (b)   Use the fact that the Hamiltonian itself is also constant to obtain the orbit.

  (c)   What is the Hamiltonian in cylindrical coordinates? What are the constants of motion?

1.10   Suppose that one calculates the Hamiltonian for a given system and finds a coordinate missing. What can be said about the symmetry of the system?

1.11   A particle of mass m is attracted to the origin by the force

$$\mathbf{F} = -K\mathbf{r}$$

Write the Hamiltonian for this system in spherical and Cartesian coordinates. What are the cyclic coordinates in each of these frames? [*Hint:* The potential for this force, $V(r)$, is given by $\mathbf{F} = -K\mathbf{r} = -\nabla V(r)$.]

1.12   A "spherical pendulum" consists of a particle of mass m attached to one end of a weightless rod of length a. The other end of the rod is fixed in space (the origin). The rod is free to rotate about this point. If at any instant the angular velocity of the particle about the origin is ω, its energy is

$$E = \frac{1}{2}ma^2\omega^2 = \frac{1}{2}I\omega^2$$

The motion of inertia is I.) What is the Hamiltonian of this system in spherical coordinates? (*Hint:* Recall the relation $L = I\omega$.) What is the Hamiltonian of the pendulum in a gravity field? (Let g denote the acceleration of gravity.)

1.3   THE STATE OF A SYSTEM

To know the values of the generalized coordinates of a system at a given instant is to know the location and orientation of the system at that instant. In classical physics we can ask for more information about the system at any given instant. We may ask for its motion as well. The location, orientation, and motion of the system at a given instant specify the state of the system at that instant. For a point particle in 3-space, the classical state Γ is given by the six quantities (Fig. 1.14)

(1.36) $\Gamma = (x,\ y,\ z,\ \dot{x},\ \dot{y},\ \dot{z})$

In terms of momenta,

(1.37) $\Gamma = (x,\ y,\ z,\ p_x,\ p_y,\ p_z)$

More generally, the state of a system is a minimal aggregate of information about the system which is maximally informative. A set of good coordinates and their corresponding time derivatives (generalized velocities) or corresponding momenta (canonical momenta) always serves as such a minimal aggregate which is maximally informative and serves to specify the state of a system in classical physics.

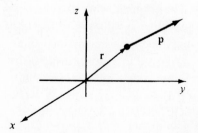

FIGURE 1.14   The classical state of a free particle is given by six scalar quantities $(x,\ y,\ z,\ p_x,\ p_y,\ p_z)$.

The state of the system composed of two point particles moving in a plane is given by the eight parameters

$$(1.38) \qquad \Gamma = (x_1,\ y_1,\ x_2,\ y_2,\ p_{x_1},\ p_{y_1},\ p_{x_2},\ p_{y_2})$$

Just as the set of generalized coordinates one assigns to a given system is not unique, neither is the description of the state Γ. For instance, the state of a point particle moving in a plane in Cartesian representation is

$$(1.39) \qquad \Gamma = (x,\ y,\ p_x,\ p_y)$$

In polar representation it is

$$(1.40) \qquad \Gamma = (r,\ \theta,\ p_r,\ p_\theta)$$

All representations of the state of a given system in classical mechanics contain an equal number of variables. If we think of Γ as a vector, then for a system with N degrees of freedom, Γ is $2N$-dimensional. In classical mechanics change of representation is effected by a change from one set of canonical coordinates and momenta (q, p) to another valid set of canonical coordinates and momenta (q', p').

$$\Gamma(q_1, \ldots, q_N, p_1, \ldots, p_N) \rightarrow \Gamma(q_1', \ldots, q_N', p_1', \ldots, p_N')$$

One form of canonical transformation results simply from a change in coordinates. For example, the transformation from Cartesian to polar coordinates for a particle moving in a plane effects the following change in representation:

$$\Gamma(x,\ y,\ p_x,\ p_y) \rightarrow \Gamma(r,\ \theta,\ p_r,\ p_\theta)$$

Representations in Quantum Mechanics

Next, we turn briefly to the form these concepts take in quantum mechanics. The specification of parameters that determines the state of a system in quantum mechanics is more subtle than in classical mechanics. As will emerge in the course of development of this text, in quantum mechanics one is not free to simultaneously specify certain sets of variables relating to a system. For example, while the classical state of a free particle moving in the x direction is given by the values of its position x, and momentum p_x, in quantum mechanics such simultaneous specification cannot be made. Thus, if the position x of the particle is measured at a given instant, the particle is left in a state wherein the particle's momentum is maximally uncertain. If on the other hand the momentum p_x is measured, the particle is left in a state in which its position is maximally uncertain. Suppose it is known that the particle has a specific value of momentum. One may then ask if there are any other variables whose values may be ascertained without destroying the established value of momentum. For a free particle one may further specify the energy E; that is, in quantum mechanics it is possible for the particle to be in a state such that measurement of momentum definitely finds the value p_x and measurement of energy definitely finds the value E. Suppose there are no further observable properties of the free particle that may be specified simultaneously with those two variables. Consequently, values of p_x and E comprise the most informative statement one can make about the particle and these values may be taken to comprise the state of the system of the particle

$$\Gamma = \Gamma(p_x, E)$$

As remarked above, if the particle is in this state, it is certain that measurement of momentum finds p_x and measurement of energy finds E. Such values of p_x and E are sometimes called *good quantum numbers*. As with their classical counterpart, good quantum numbers are an independent set of parameters which may be simultaneously specified and which are maximally informative.

   For some problems in quantum mechanics it will prove convenient to give the state in terms of the Cartesian components of angular momentum: L_x, L_y, and L_z. We will find that specifying the value of L_z, say, induces an uncertainty in the accompanying components of L_x and L_y, so that, for example, it is impossible to simultaneously specify L_z and L_x for a given system. One may, however, simultaneously specify L_z together with the square of the magnitude of the total momentum, L^2. For a particle moving in a spherically symmetric environment, one may also simultaneously specify the energy of the particle. This is the most informative[1] statement one can make about such a particle, and the values of energy, L^2 and L_z, comprise a quantum state of the system.

(1.41) $$\Gamma = (E, L^2, L_z)$$

The values of E, L^2, and L_z are then good quantum numbers. That is, they are an independent set of parameters which may be simultaneously specified and which are maximally informative.

   Just as change in representation, as discussed above, plays an important role in classical physics, so does its counterpart in quantum mechanics. A representation in quantum mechanics relates to the observables that one can precisely specify in a given state. In transforming to a new representation, new observables are specified in the state. For a free point particle moving in 3-space, in one representation the three components of linear momentum p_x, p_y, and p_z are specified while in another representation the energy $p^2/2m$, the square of the angular momentum L^2, and any component

---

[1] More precisely, Γ includes the *parity* of the system. This is a purely quantum mechanical notion and will be discussed more fully in Chapter 6.

of angular momentum, say L_z, are specified. In this change of representation,

(1.42) $$\Gamma(p_x, p_y, p_z) \rightarrow \Gamma(E, L^2, L_z)$$

When treating the problem of the angular momentum of two particles ($\mathbf{L}_1$ and $\mathbf{L}_2$, respectively) in one representation, ($L_1{}^2, L_2{}^2, L_{z_1}, L_{z_2}$) are specified while in another representation ($L_1{}^2, L_2{}^2, L^2, L_z$) are specified. Here we are writing $\mathbf{L}$ for the *total angular momentum of the system* $\mathbf{L} = \mathbf{L}_1 + \mathbf{L}_2$. In this change of representation,

(1.43) $$\Gamma(L_1{}^2, L_2{}^2, L_{z_1}, L_{z_2}) \rightarrow \Gamma(L_1{}^2, L_2{}^2, L^2, L_z{}^2)$$

Finally, in this very brief introductory description, we turn to the concept of the change of the quantum state in time. In classical mechanics, Newton's laws of motion determine the change of the state of the system in time. In quantum mechanics, the evolution in time of the state of the system is incorporated in the *wave* (or *state*) *function* and its equation of motion, the *Schrödinger equation*. Through the wavefunction, one may calculate (expected) values of observable properties of the system, including the time development of the state of the system.

These concepts of the quantum state—its evolution in time and change in representation—comprise principal themes in quantum mechanics. Their understanding and application are important and are fully developed later in the text.

PROBLEMS

1.13 Write down a set of variables that may be used to prescribe the classical state for each of the 11 systems listed in Problem 1.1.

Answer (partial)

(e) A rigid rod in 3-space: Since the system has five degrees of freedom, the classical state of the system is given by 10 parameters. For example,

$$\Gamma = \{x, y, z, \theta, \phi, \dot{x}, \dot{y}, \dot{z}, \dot{\theta}, \dot{\phi}\}$$

[*Note:* The quantum state is less informative. For example, such a state is prescribed by five variables (x, y, z, θ, ϕ). Another specification of the quantum state is given by five momenta $(p_x, p_y, p_z, p_\theta, p_\phi)$. However, simultaneous specification of, say, x and p_x is not possible in quantum mechanics.]

1.14 (a) Use Hamilton's equations for a system with N degrees of freedom to show that H is constant in time if H does not contain the time explicitly. [*Hint:* Write

$$\frac{dH}{dt} = \frac{\partial H}{\partial t} + \sum_{l=1}^{N} \left(\frac{\partial H}{\partial q_l} \dot{q}_l + \frac{\partial H}{\partial p_l} \dot{p}_l \right).]$$

(b) Construct a simple system for which H is an explicit function of the time.

1.15   For a system with N degrees of freedom, the Poisson bracket of two dynamical functions A and B is defined as

$$\{A, B\} = \sum_{l=1}^{N} \left(\frac{\partial A}{\partial q_l} \frac{\partial B}{\partial p_l} - \frac{\partial B}{\partial q_l} \frac{\partial A}{\partial p_l} \right)$$

(a)   Use Hamilton's equations to show that the total time rate of change of a dynamical function A may be written

$$\frac{dA}{dt} = \frac{\partial A}{\partial t} + \{A, H\}$$

where H is the Hamiltonian of the system.

(b)   Prove the following: (1) If $a(q, p)$ does not contain the time explicitly and $\{A, H\} = 0$, then A is a constant of the motion. (2) If A does contain the time explicitly, it is constant if $\partial A/\partial t = \{H, A\}$.

(c)   For a free particle moving in one dimension, show that

$$A = x - \frac{pt}{m}$$

satisfies the equation

$$\frac{\partial A}{\partial t} = -\{A, H\}$$

so that it is a constant of the motion. What does this constant correspond to physically?

1.16   How many degrees of freedom does the compound pendulum depicted in Fig. 1.15 have? Choose a set of generalized coordinates (be certain they are independent). What is the Hamiltonian for this system in terms of the coordinates you have chosen? What are the immediate constants of motion?

1.17   How many constants of the motion does a system with N degrees of freedom have?

Answer

Each of the coordinates $\{q_l\}$ and momenta $\{p_l\}$ satisfies a first-order differential equation in time (i.e., Hamilton's equations). Every such equation has one constant of integration. These comprise $2N$ constants of the motion.

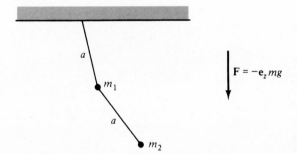

FIGURE 1.15   Compound pendulum composed of two masses connected by weightless rods of length a. The motion is in the plane of the paper. (See Problem 1.16.)

1.4 PROPERTIES OF THE ONE-DIMENSIONAL POTENTIAL FUNCTION

Consider a particle that is constrained to move in one dimension, x. The particle is in the potential field $V(x)$ depicted in Fig. 1.16. What is the direction of force at the point $x = A$? We can calculate the gradient (in the x direction) and conclude that the direction of force at A is in the $+x$ direction. There is a simpler technique. Imagine that the curve drawn is the contour of a range of mountain peaks. If a ball is placed at A, it rolls down the hill. The force is in the $+x$ direction. If placed at B (or C), it remains there. If placed at D, it rolls back toward the origin; the force is in the $-x$ direction. This technique always works (even for three-dimensional potential surfaces) because the gravity potential is proportional to height z, so the potential surface for a particle constrained to move on the surface of a mountain is that same surface.

The one-dimensional spring potential, $V = Kx^2/2$, is depicted in Fig. 1.17. If the particle is started from rest at $x = A$, it oscillates back and forth in the potential well between $x = +A$ and $x = -A$.

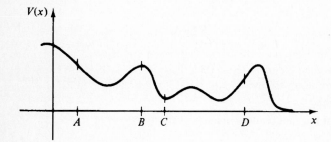

FIGURE 1.16 Arbitrary potential function.

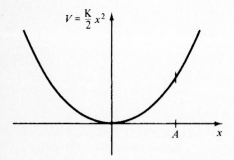

FIGURE 1.17 Spring potential.

Motion described by a potential function is said to be *conservative*. For such motion, the energy

(1.44) $E = T + V$

is constant. In terms of the kinetic energy T,

(1.45) $T = \dfrac{mv^2}{2} = E - V$

Forbidden Domains

From (1.45) we see that if $V > E$, then $T < 0$ and the velocity becomes imaginary. In classical physics, particles are excluded from such domains. They are called *forbidden regions*. Again consider a one-dimensional problem with potential $V(x)$ shown in Fig. 1.18. The constant energy E is superimposed on this diagram. Segments AB and CD are forbidden regions. Points A, B, C, and D are *stationary* or *turning* points. Since $E = V$ at these points, $T = 0$ and $\dot{x} = 0$. Suppose that a particle is started from rest from the point C. What is the subsequent motion? The particle is trapped in the potential well between B and C. It accelerates down the hill, slows down in climbing the middle peak, then slows down further in climbing to B, where it comes to rest and turns around. This periodic motion continues without end.

The one-dimensional potential depicted in Fig. 1.18 can be effected by appropriately charging and spacing a linear array of plates with holes bored along the axis. The potential depicted in Fig. 1.18 is seen by an electron constrained to move along this (x) axis.

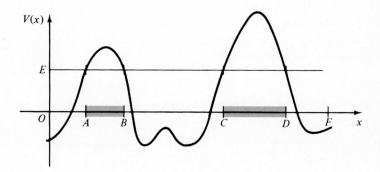

FIGURE 1.18   Forbidden domains at energy E.

PROBLEMS

1.18   A particle constrained to move in one dimension (x) is in the potential field

$$V(x) = \frac{V_0(x-a)(x-b)}{(x-c)^2} \qquad (0 < a < b < c < \infty)$$

(a)   Make a sketch of V.

(b)   Discuss the possible motions, forbidden domains and turning points. Specifically, if the particle is known to be at $x = -\infty$ with

$$E = \frac{3V_0}{c-b}(b - 4a + 3c)$$

at which value of x does it reflect?

1.19   A particle of mass m moves in a "central potential," $V(r)$, where r denotes the radial displacement of the particle from a fixed origin.

(a)   What is the (vector) force on the particle? Recall here the components of the ∇ operator in spherical coordinates.

(b)   Show that the angular momentum $\mathbf{L}$ of the particle about the origin is constant. (*Hint:* Calculate the time derivative of $\mathbf{L} = \mathbf{r} \times \mathbf{p}$ and recall that $\mathbf{p} = m\dot{\mathbf{r}}$.)

(c)   Show that the energy of the particle may be written

$$E = \frac{p_r^2}{2m} + \frac{L^2}{2mr^2} + V(r)$$

(d)   From Hamilton's equations obtain a "one-dimensional" equation for $\dot{p}_r$, in the form

$$\dot{p}_r = -\frac{\partial}{\partial r} V_{\text{eff}}(r)$$

where V_{eff} denotes an "effective" potential that is a function of r only.

(e)   For the case of gravitational attraction between two masses (M, m), $V = -GmM/r$, where G is the gravitational constant. Make a sketch of V_{eff} versus r for this case. Use this sketch to establish the conditions for circular motion (assume that M is fixed in space) for a given value of L^2.

1.20   Complex variables play an important role in quantum mechanics. The following two problems are intended as a short review.

If

$$\psi = |\psi| \exp(i\alpha_1)$$
$$\chi = |\chi| \exp(i\alpha_2)$$

show that

$$|\psi + \chi|^2 = |\psi|^2 + |\chi|^2 + 2|\psi\chi| \cos(\alpha_1 - \alpha_2)$$

1.21   Use the expression

$$e^{i\theta} = \cos\theta + i\sin\theta$$

to derive the following relations

(a) $\cos(\theta_1 + \theta_2) = \cos\theta_1 \cos\theta_2 - \sin\theta_1 \sin\theta_2$

(b) $\sin(\theta_1 + \theta_2) = \cos\theta_1 \sin\theta_2 + \sin\theta_1 \cos\theta_2$

(c) $2\sin\theta_1 \cos\theta_2 = \sin(\theta_1 - \theta_2) + \sin(\theta_1 + \theta_2)$

(d) $2\cos\theta_1 \cos\theta_2 = \cos(\theta_1 + \theta_2) + \cos(\theta_1 - \theta_2)$

(e) $2\sin\theta_1 \sin\theta_2 = \cos(\theta_1 - \theta_2) - \cos(\theta_1 + \theta_2)$

(f) $2\cos^2\theta = 1 + \cos 2\theta$

(g) $2\sin^2\theta = 1 - \cos 2\theta$

(h) $e^{i\theta} - 1 = 2ie^{i\theta/2}\sin(\theta/2)$

(i) $\frac{1}{2}|e^{i\theta_1} + e^{i\theta_2}|^2 = \frac{1}{2}(e^{i\theta_1} + e^{i\theta_2})(e^{i\theta_1} + e^{i\theta_2})* = 1 + \cos(\theta_1 - \theta_2)$

(j) $2\operatorname{Re} z = z + z*$

(k) $2i\operatorname{Im} z = z - z*$

(l) $(\exp z)* = \exp z*$

(m) $|\exp z|^2 = \exp(2\operatorname{Re} z)$

(n) $i^i = e^{-\pi/2},\ e^{-5\pi/2},\ e^{-9\pi/2}\ \ldots$ $\left(i = \exp\left[i\pi\left(\dfrac{1 + 4n}{2}\right)\right]\right)$

(o) $\displaystyle\sum_{n=1}^{l} \exp(-i2\pi n/l) = 0$

where l is an integer. [*Hint:* Rewrite the given relation as a summation over the complex unit vector $z(\theta) = \exp(i\theta)$.]

1.22   Vector calculus plays an important role in quantum mechanics. The following is intended as a short review.

(a)   Write down $\nabla^2\varphi$ in Cartesian (x, y, z), spherical (r, θ, ϕ), and cylindrical (ρ, ϕ, z) coordinates.

(b)   If (dA, dV) represents differential of (area, volume), working in spherical coordinates, show that

$$dA = r^2\, d\Omega, \qquad dV = r^2\, dr\, d\Omega$$

where $d\Omega$ is the differential of *solid angle*

$$d\Omega = \sin\theta\, d\theta\, d\phi$$

(c)   Calculate the value of

$$\Lambda \equiv \int_S d\Omega$$

where the integral is over a closed surface, S, about the origin. The units of solid angle are *steradians*. Note that, in two dimensions, the corresponding expression is

$$\Lambda_L = \oint d\theta = 2\pi \ \text{(radians)}$$

(d)   Show that

$$\nabla \times (\nabla \varphi) = 0$$

(e)   Show that

$$\nabla \cdot (\nabla \times \mathbf{A}) = 0$$

Hint: The latter two properties are most easily established with Gauss's and Stokes's integral laws.

(f)   What is the value of the line integral

$$I \equiv \int_{\mathbf{r}_1}^{\mathbf{r}_2} \nabla \varphi \cdot d C ?$$

The integral is along a curve C which connects the end points $\mathbf{r}_1, \mathbf{r}_2$.

1.23   The dumbbell molecule plays an important role in later discussions of angular momentum. Consider a dumbbell molecule of reduced mass μ and atomic separation a. Let I denote the moment of inertia of the molecule with respect to rotation of the dumbbell about its center of mass. If one writes $I = \mu r_{\text{eff}}^2$, obtain an expression for r_{eff} in terms of m_1, m_2, and the separation, a.

Answer

Respective displacements of masses from the center of mass of the dumbbell are

$$r_1 = \frac{a m_2}{M}, \qquad r_2 = \frac{a m_1}{M}$$

where $M \equiv m_1 + m_2$. Thus,

$$I = \frac{m_1 a^2 m_2^2 + m_2 a^2 m_1^2}{M^2}$$

$$= \frac{m_1 m_2 a^2 M}{M^2} = \mu a^2$$

so that $r_{\text{eff}} = a$.

1.24   Time reversibility in classical physics is described by the invariance of Hamilton's equations (1.35) to the transformation $(t \rightarrow -t, \ p \rightarrow -p)$. What implication does reversibility have on:

(a)   The Hamiltonian of the system?

(b)   The dynamical solutions $x(t)$, $p(t)$?

1.25   A transformation of coordinates and momenta

$$(q_1, \ldots, q_N; p_1, \ldots, p_N) \to (Q_1, \ldots, Q_N; P_1, \ldots, P_N)$$

is said to be canonical, providing these variables satisfy the *fundamental Poisson-bracket relations*

$$\{Q_i, Q_j\} = \{P_i, P_j\} = 0, \qquad \{Q_i, P_j\} = \delta_{ij}$$

where $\{,\}$ denotes Poisson brackets (Problem 1.15) evaluated with respect to $(q_1, \ldots, q_N; p_1, \ldots, p_N)$ variables.

(a)   Consider a rigid body that rotates about a fixed interior point in the absence of external forces, with energy given by

$$E = \frac{L_x^{\,2}}{2I_1} + \frac{L_y^{\,2}}{2I_2} + \frac{L_z^{\,2}}{2I_3}$$

where I denotes moment of inertia. Is this energy a valid Hamiltonian form?

(b)   Now consider a one-dimensional rigid rod that rotates about a fixed point on the rod in the absence of external forces, with energy $L^2/2I$. Is this a valid Hamiltonian form?

Answer (partial)

(a)   This Hamiltonian is not valid as the angular momenta (L_x, L_y, L_z) satisfy the relations, $\{L_x, L_y\} = L_z$, etc., and consequently do not comprise a canonical set of momenta.

2

HISTORICAL REVIEW: EXPERIMENTS AND THEORIES

The following sections summarize experiments and theories formulated during the early decades of the century. These observations and theories comprise the genesis of quantum mechanics. The important concept of the wavefunction is introduced and the Born interpretation of this function in terms of probability density is described. A more formal presentation of the postulates of quantum mechanics appears in Chapter 3.

2.1 DATES

Physics at the turn of the century was in a state of turmoil. There was a Pandora's box of experimental observations which, on the grounds of otherwise firmly established classical theory, was totally inexplicable. One by one all these perplexing questions were answered—with the drama and flair of a story told by a masterful raconteur. Out of the turmoil came a new philosophy of science. A new way of thinking was called for. At the very core of natural law lay subjective probability—not objective determinism.

What were some of these perplexing observations? Light exhibits interference and therefore may be assumed to be a wave phenomenon. However, if we try to explain the photoelectric effect (light hitting a metal surface ejects electrons) on the basis of the wave nature of light, we obtain erroneous results. It is found that the energy of an emitted electron is dependent only on the frequency of the incident radiation, not on the intensity as might be expected from the classical theory of light.

In 1911 it was established by Rutherford that an atom has a positive central core and satellite electrons. Hydrogen, for instance, has a proton at its center and one outer electron. But such a circulating (and therefore accelerating) electron radiates and soon should collapse into the nucleus. So why do we not see a burst of ultraviolet radiation emitted as the electron spirals into the nucleus? Why is the frequency spectrum of light emitted from an atom a discrete line spectrum and not a continuous spectrum?

Another dilemma lay in the observations of the spectrum of radiant energy in a cavity whose walls are maintained at a fixed temperature. Theory (on the basis of the wave nature of light) was unable to account for the observed frequency distribution of radiant energy.

Outlined below are the discoveries and events that occurred near the turn of the century that removed the enigmas and led naturally to the development of quantum mechanics.

1898	Mme. Curie[1]	Radioactive polonium and radium
1901	Planck	Blackbody radiation
1905	Einstein	Photoelectric effect
1911	Rutherford[2]	Model of the atom
1913	Bohr	Quantum theory of spectra
1922	Compton	Scattering photons off electrons
1924	Pauli	Exclusion principle
1925	de Broglie	Matter waves
1926	Schrödinger	Wave equation
1927	Heisenberg	Uncertainty principle
1927	Davisson and Germer	Experiment on wave properties of electrons
1927	Born	Interpretation of the wavefunction
1928	Dirac	Relativistic wave equation; prediction of existence of the positron

In the remainder of this chapter we will outline these topics in more detail, except for the work of Schrödinger, which is formally presented in Chapter 3, and the

---

[1] Marja Sklodowska (1867–1934), born in Poland.

[2] Ernest Rutherford (1871–1937), born in poverty in New Zealand to parents who were potato farmers. Scholarship of the Great Exhibition of 1851 awards him study at Cambridge.

work of Pauli, which is presented in Chapter 12. The Compton effect is discussed in Problem 2.28, and the Dirac equation is studied in Chapter 15.

2.2   THE WORK OF PLANCK. BLACKBODY RADIATION

Place a closed, evacuated container (with a small window in the wall) in an oven of uniform temperature. Wait until all components of the experiment reach the same temperature (thermal equilibrium). At a sufficiently high temperature, visible light emerges from the window of the container cavity. The cavity contains radiant energy, which is in thermal equilibrium with the cavity walls. Suppose that the total radiant energy per unit volume in the cavity (at any instant) is U. How much of this energy is in electromagnetic waves with frequency between ν and $\nu + d\nu$? Let us call the answer $u(\nu)\, d\nu$. The function $u(\nu)$ then gives the energy per frequency interval per unit volume. The total energy per unit volume in the radiation field in the cavity is

$$(2.1) \qquad U = \int_0^\infty u(\nu)\, d\nu$$

The radiation is called *blackbody radiation* because it is assumed that any light falling on the window is totally absorbed. The window acts as a perfect radiator and a perfect absorber. This property is characteristic of ideal black surfaces. At any given temperature, no object emits or absorbs radiation more efficiently than does an (ideal) blackbody.

The experimentally observed curve of $u(\nu)$ is shown in Fig. 2.1. Classical electrodynamic and thermodynamic theory give two properties of the spectral distribution of a radiation field in equilibrium at the temperature T. The Rayleigh—Jeans (1900) approximation

$$u_{RJ}(\nu) = \frac{8\pi\nu^2}{c^3} k_B T$$

is appropriate for low frequencies. In this formula k_B is Boltzmann's constant,

$$k_B = 1.381 \times 10^{-16} \text{ erg/K}$$

and c is the speed of light. While this approximation is valid at low frequencies, it is seen to diverge at larger frequencies, where as shown in Fig. 2.1, the correct spectral distribution falls off to zero. Wien's law (1893) specifies that u, as a function of

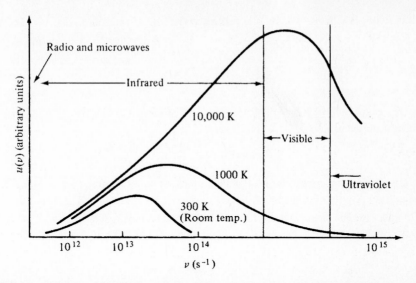

FIGURE 2.1   Spectrum of blackbody radiation. The curves have been distorted to bring out some important features. In reality the curve at 10,000 K is about 37,000 times higher than the curve at 300 K. Also, the radio and microwave domain is only about 1/30,000 of the ν axis depicted.

wavelength $\lambda = c/\nu$, is of the form

$$u_W(\lambda) = \frac{W(\lambda T)}{\lambda^5}$$

where W is an arbitrary continuous function of the product of wavelength λ and temperature T. Although this formula is valid over the whole spectrum of wavelengths, it is incomplete in that $W(\lambda T)$ is undetermined. The complete explicit form for the spectral distribution u cannot be obtained from classical physics. A quantum hypothesis must be invoked. Such was the assumption made by Planck to obtain a uniformly valid formula for $u(\nu)$. It implied that energy of radiation with frequency ν exists only in multiples of $h\nu$, where h is a constant of nature (Planck's constant). A quantum of radiation of energy $h\nu$ is called a *photon*.

(2.2) $E = h\nu$

The correct formula for $u(\nu)$ which results is (see Problems 2.36 and 2.37)

(2.3) $$u(\nu) = \frac{8\pi h\nu^3}{c^3} \frac{1}{e^{h\nu/k_B T} - 1}$$

$$h = 6.626 \times 10^{-27} \text{ erg-s}$$

This expression precisely matches the experimental curves shown in Fig. 2.1.

PROBLEMS

2.1 (a)   Show that for photons of frequency ν and wavelength λ:

(1) $d\nu = -c\, d\lambda/\lambda^2$

(2) $u(\lambda)\, d\lambda = -u(\nu)\, d\nu$

(3) $u(\lambda)\, d\lambda = u(\nu)c\, d\lambda/\lambda^2$

(b)   Show that the Rayleigh–Jeans spectral distribution of blackbody radiation, $u_{RJ}(\nu)$, is of the form required by Wien's law,

$$u_W(\lambda) = \frac{W(\lambda T)}{\lambda^5}$$

(c)   Obtain the correct form of Wien's undetermined function $W(\lambda T)$ from Planck's formula.

2.2   A spherical enclosure is in equilibrium at the temperature T with a radiation field that it contains. Show that the power emitted through a hole of unit area in the wall of enclosure is

$$P = \frac{1}{4} cU$$

Answer

Let the cavity be very large, so that its walls can be considered to be flat. The energy that flows through a hole in the wall, of unit area, in 1 s is the power radiated. This energy is due to photons that lie in a hemisphere of radius c, centered at the hole (Fig. 2.2). The energy in the volume

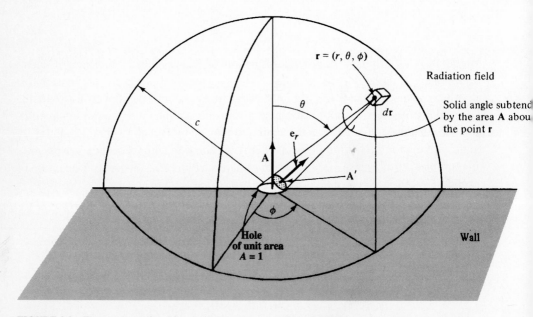

FIGURE 2.2   The power radiated by an electromagnetic field in equilibrium at temperature T is due to photons that lie in a hemisphere of radius c, centered at the hole. (See Problem 2.2.)

element $d\mathbf{r}$ about the point $\mathbf{r}$ is $U\,d\mathbf{r}$. Owing to the isotropy of the radiation field, the amount of this energy that passes through the hole is $U\,d\mathbf{r}$ times the ratio of solid angle Ω subtended by the area of the hole about the point $\mathbf{r}$, to 4π, the total solid angle about the point $\mathbf{r}$.

$$dP = \frac{\Omega}{4\pi}\,U\,d\mathbf{r} = \frac{A'}{4\pi r^2}\,U\,d\mathbf{r} = \frac{\mathbf{e}_r \cdot A U\,d\mathbf{r}}{4\pi r^2} = \frac{U\cos\theta}{4\pi r^2}\,d\mathbf{r}$$

$$= -\frac{U\,d\phi\,\cos\theta\,d\cos\theta\,r^2\,dr}{4\pi r^2}$$

The radiation energy that passes through the hole in 1 s from all volume elements in the hemisphere is the total power radiated per unit area.

$$P = \int_{\text{hemisphere}} dP = \frac{U}{4\pi}\int_0^{2\pi} d\phi \int_0^1 \cos\theta\,d\cos\theta \int_0^c dr = \frac{1}{4}cU$$

2.3   Show that the energy density $U(T)$ of a radiation field in equilibrium at the temperature T is directly proportional to T^4. The corresponding expression for the emitted power is

$$P = \sigma T^4$$

where σ is the Stefan–Boltzmann constant

$$\sigma = \frac{\pi^2}{60}\frac{k_B{}^4}{\hbar^3 c^2} = 0.567 \times 10^{-4}\ \text{erg/s-cm}^2\text{-K}^4$$

[*Hint:* Nondimensionalize the integration over (2.3) through the variable $x \equiv h\nu/k_B T$.]

2.4   Use (2.3) to prove Wien's displacement law

$$\lambda_{\max} T = \text{constant} = 0.290\ \text{cm K}$$

The wavelength $\lambda_{\max}$ is such that $u(\lambda_{\max})$ is maximum. [*Hint:* Differentiate $u(\lambda)$ with respect to the variable $x \equiv hc/kT\lambda$ and set equal to zero.]

2.5   From the sketch of u versus ν given in Fig. 2.1, make a sketch of u versus λ, where $\nu\lambda = c$.

2.6   What is the photon flux (photons/cm^2 s) at a distance of 1 km from a light source emitting 50 W of radiation in the visible domain, with wavelength 6000 Å?

2.7   The average energy in a unit volume in the ν frequency mode of a blackbody radiation field is

$$\langle U \rangle = \frac{h\nu}{e^{h\nu/k_B T} - 1}$$

What does $\langle U \rangle$ reduce to in the limit (a) $\nu \to 0$? (b) $T \to \infty$?

2.8   As discussed above, the radiation field interior to a closed cavity whose walls are in thermal equilibrium (i.e., at the same temperature) with the radiation field is called blackbody radiation. Prove that blackbody radiation has the following properties by showing that if any of these properties are not true, a device can be constructed which violates the second law of thermodynamics.

   (a)   The flux of radiation is the same in all directions. (The radiation field is *isotropic*.)

   (b)   The energy density is the same at all points inside the cavity. (The radiation field is *homogeneous*.)

   (c)   The energy density interior to the cavity is the same (function of frequency) at a given temperature, regardless of the material of the cavity wall.

2.9   Prove that the radiation emitted by the surface of an ideal blackbody at the temperature T is the same as that which travels in one direction inside a closed isothermal cavity at the same temperature.

Answer

Immerse an ideal black cube inside the isothermal container. The radiation that falls on any face of the cube is completely absorbed. For equilibrium to be maintained, the radiation emitted must be balanced by that absorbed, so that the radiation emitted is precisely that which flows into the face.

   If, on the other hand, the cube is not ideally black, equilibrium is maintained by balancing the absorbed radiation by the reflected plus emitted radiation. Since energy density in the cavity is the same as in the case above (both experiments are at the same temperature), the radiation emitted by the nonblack surface is less intense than that emitted by the ideally black surface.

2.10   One of the theories of the origin of the universe is that it was contained in a primeval fireball which began its expansion about 10^{10} years ago. As it expanded, it cooled. Measurements of the energy spectrum of cosmic photons suggest a (blackbody) temperature of 3 K. At what frequency is maximum energy observed?

2.11   Suppose that you are inside a blackbody radiation cavity which is at temperature T. Your job is to measure the energy in the radiation field in the frequency interval 10^{14} to 89×10^{14} Hz. You have a detector that will do the job. For best results, should the temperature of the detector T' be $T' > T$, $T' = T$, $T' < T$, or $T' = 0$; or is the temperature of the detector irrelevant to the measurement?

2.3   THE WORK OF EINSTEIN. THE PHOTOELECTRIC EFFECT

The experimental setup that exhibits the photoelectric effect is depicted in Fig. 2.3. The observation is as follows. A metal plate (e.g., copper) is irradiated with light of a given frequency. Electrons are ejected from the photo cathode and current is registered in the ammeter A. As the potential on the collecting plate is made more negative, the current diminishes, until finally at the potential V_{stop}, current ceases. The

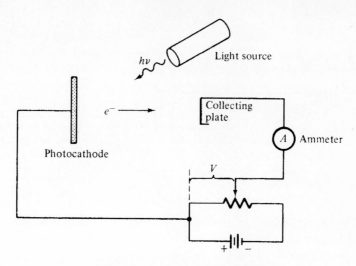

FIGURE 2.3 Photoelectric experiment.

energy that an electron must have in order to climb the potential hill imposed by the negative bias V is eV. Only the most energetic electrons reach the plate near V_{stop}. At V_{stop} the electrons with maximum kinetic energy $\mathbb{E}$ have been repelled. Then

$$(2.4) \qquad \mathbb{E} = eV_{stop}$$

At a given frequency ν, one makes a measurement of $\mathbb{E}$ and plots a point on an $\mathbb{E}$ versus ν graph (Fig. 2.4). If the intensity of light is increased while ν is held fixed, $\mathbb{E}$ remains constant. On the other hand, when ν is increased, $\mathbb{E}$ increases. A typical collection of data is shown in Fig. 2.4.

To explain this effect Einstein hypothesized that light is composed of localized bundles of electromagnetic energy called photons. At frequency ν, the energy of a photon is $h\nu$. When striking the metal surface the photon interacts with an electron and ejects it from the metal. Let us consider the *Sommerfeld model* of a conductor (Fig. 2.5). The conductor is composed of fixed positive sites (e.g., Cu^{2+} ions in copper) and

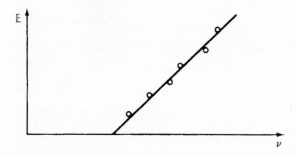

FIGURE 2.4 Typical data showing energy of most energetic electrons as a function of frequency ν in the photoelectric experiment.

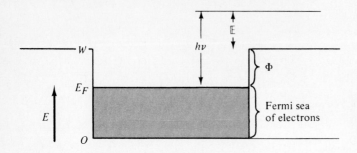

FIGURE 2.5   Sommerfeld model for energy distribution of electrons in a metal.

free electrons. The positive ions generate a potential well in which the electrons are trapped. The electrons have energy from 0 to E_F, the *Fermi energy*. The minimum work required to remove an electron from the metal is $W - E_F$, which is called the *work function*, Φ. The depth of the well is W.

Electrons distribute themselves in accordance with the *Pauli exclusion principle*. This principle precludes more than one electron existing in the same quantum state. For example, the distribution of electron energies shown in Fig. 2.5 is maintained at 0 K. At this temperature electrons fall to lowest allowable energies. They cannot all fall to the single lowest level, owing to the Pauli principle. Once this level is occupied, the next electron must seek the next higher level. The maximum value of energy so reached is the Fermi energy E_F.

Suppose that a photon of energy $h\nu$ hits an electron and ejects it with kinetic energy $\mathbb{E}$. The most energetic electrons come from the top of the *Fermi sea*. The energy $\mathbb{E}$ of such an electron ejected by a photon of energy $h\nu$ is given by

(2.5) $$\mathbb{E} = h\nu - \Phi$$

If we plot $\mathbb{E}$ versus ν from this equation, we obtain the curve shown in Fig. 2.4. Note that the slope of the curve is Planck's constant h, and the ν intercept gives the work function (of the photocathode metal). If $\Phi \equiv h\nu_{th}$, ν_{th} is called the *threshold frequency*. A few typical values are:

Metal	ν_{th}(Hz)	E_F (eV)
Silver	1.14×10^{15}	5.5
Potassium	0.51×10^{15}	2.1
Sodium	0.56×10^{15}	3.1

Millikan in 1916 used the photoelectric experiment to obtain a value of Planck's constant, h [see (2.3)].

Contact Potential

The preceding description may also be used to explain the phenomenon of *contact potential,* the finite potential that develops between two dissimilar metals which are brought into contact with each other. To describe this effect we consider a parallel-plate capacitor with one plate made of metal A and the other made of metal B. When the plates are isolated and displaced far from each other, the common zero in potential of both metals corresponds to zero free-particle kinetic energy (Fig. 2.6a).

Now let the metals be brought into contact with each other. Electrons then "fall" from the Fermi level of metal A, which has the smaller work function, to the deeper-lying Fermi level of metal B, until the tops of the two electron energy distributions are equalized. Having lost electrons, metal A is left electropositive with respect to metal B and a potential difference exists between the plates (Fig. 2.6b).

This description leads to the conclusion that the contact potential difference V_C between two metals should be well approximated by the difference in work functions:

$$eV_C = \Phi_B - \Phi_A$$

The validity of this relation is borne out by experiment.

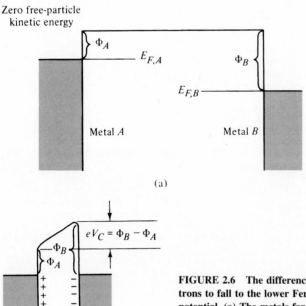

(a)

(b)

FIGURE 2.6   The difference in work functions causes electrons to fall to the lower Fermi level thereby creating a contact potential. (a) The metals far removed from each other. (b) The metals in contact with each other. The sloping curve represents the potential seen by an electron.

PROBLEMS

2.12 (a) A monochromatic point source of light radiates 25 W at a wavelength of 5000 Å. A plate of metal is placed 100 cm from the source. Atoms in the metal have a radius of 1 Å. Assume that the atom can continually absorb light. The work function of the metal is 4 eV. How long is it before an electron is emitted from the metal?

(b) Is there sufficient energy in a single photon in the radiation field to eject an electron from the metal?

2.13 The photoelectric threshold of tungsten is 2300 Å. Determine the energy of the electrons ejected from the surface by ultraviolet light of wavelength 1900 Å.

2.14 The work function of zinc is 3.6 eV. What is the energy of the most energetic photoelectron emitted by ultraviolet light wavelength 2500Å?

2.15 Photoelectrons emitted from a cesium plate illuminated with ultraviolet light of wavelength 2000 Å are stopped by a potential of 4.21 V. What is the work function of cesium?

2.4 THE WORK OF BOHR. A QUANTUM THEORY OF ATOMIC STATES

Consider a discharge tube filled with hydrogen gas. At sufficient voltage the gas glows. If the light is examined in a spectroscope, it is seen that only a discrete set of frequencies—a line spectrum—is emitted. Bohr was able to account for the discrete emission spectra in an analysis based on two postulates:

1. Hydrogen exists in discrete energy states. These states are characterized by discrete values of the angular momentum as given by the relation

$$(2.6) \qquad \oint p_\theta \, d\theta = nh$$

   with n an integer greater than zero. In these states the atom does not radiate. The line integral follows the electron in one complete orbit about the nucleus.

2. When an atom undergoes a change in energy from E_n to E_m, electromagnetic radiation (a photon) is emitted at a frequency ν given by

$$(2.7) \qquad h\nu = E_n - E_m$$

Let us recall how condition (2.6) leads to a discrete set of energies $\{E_n\}$. The energy of a (stationary) hydrogen atom whose electron is moving in circular motion is

$$(2.8) \qquad E = \frac{1}{2}mv^2 - \frac{e^2}{r} = \frac{p_\theta^2}{2mr^2} - \frac{e^2}{r}$$

The radius r obeys the centripetal condition

(2.9)
$$\frac{mv^2}{r} = \frac{p_\theta^2}{mr^3} = \frac{e^2}{r^2}$$

so that, with (2.6),

(2.10)
$$\frac{e^2}{r} = \frac{p_\theta^2}{mr^2} = \frac{n^2\hbar^2}{mr^2} \qquad \left(\hbar \equiv \frac{h}{2\pi}\right)$$

(2.11)
$$r_n = \frac{n^2\hbar^2}{me^2}$$

These are the quantized values of r at which the electron persists without radiating. The values of the energy at these radii are

(2.12)
$$E_n = -\frac{p_\theta^2}{2mr^2} = -\frac{n^2\hbar^2}{2m}\left(\frac{me^2}{n^2\hbar^2}\right)^2$$
$$= -\frac{\mathbb{R}}{n^2}$$

where $\mathbb{R}$ is the Rydberg constant:

(2.13)
$$\mathbb{R} = \frac{me^4}{2\hbar^2} = 2.18 \times 10^{-11} \text{ erg} = 13.6 \text{ eV}$$

The negative quality of the energy reflects the fact that we are dealing with *bound states*. When $n = 1$, the atom is in the *ground state* and has energy, $-\mathbb{R}$. To ionize the atom when it is in this state takes $+\mathbb{R}$ ergs of energy. The value of r when the atom is in the ground state is

(2.14)
$$r_1 \equiv a_0 = \frac{\hbar^2}{me^2} = 5.29 \times 10^{-9} \text{ cm}$$

This is a fundamental length in physics. It is called the *Bohr radius*.

When the electron and proton are infinitely far removed and at rest, $r_n = \infty$. From (2.11) we see that this corresponds to $n = \infty$. In this state $E_n = 0$; there is no kinetic energy and no potential energy. If the electron is given a tap, it becomes a free particle. The composite system of proton plus electron then has positive energy (kinetic only), with all (unquantized) positive values of energy allowed (Fig. 2.7).

The emission spectra of hydrogen is generated by the values for E_n (2.12) and the second postulate (2.7). The frequencies so generated (with some minor refinements, e.g., accounting for the motion of the proton) agree to a high degree of accuracy with the data. Characteristically, the spectrum divides into various series of lines. The Lyman series is comprised of frequencies generated by transitions to the ground state:

(2.15)
$$h\nu_L = E_n - E_1 \qquad (n > 1) \qquad \text{(UV)}$$

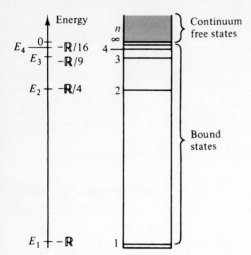

FIGURE 2.7 Bohr spectrum.

The Balmer series is generated by transitions to the second excited state:

(2.16) $\qquad h\nu_B = E_n - E_2 \qquad (n - 2) \qquad$ (UV and visible)

and so forth (Fig. 2.8). Remaining series are Paschen, Brackett, and Pfund (all IR).

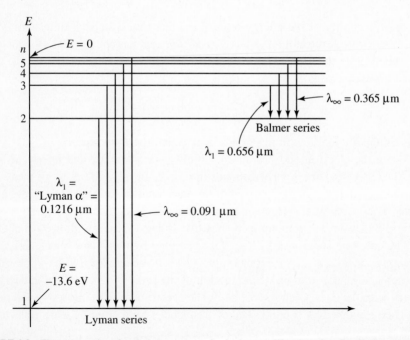

FIGURE 2.8 First two series of emission spectrum for hydrogen. Wavelengths of radiation are given in units of microns (10^{-4} cm).

PROBLEMS

2.16 (a) Consider the spherical pendulum described in Problem 1.12. Use the Bohr formula (2.6) to obtain the quantum energies of this system. Note the identity $p_\theta = L$.

(b) Suppose that the pendulum is comprised of a proton attached to a weightless rod of length $a = 2$ Å. What is the ground rotational state of this system (in eV)? (See Problem 10.40.)

2.17 (a) What is the formula for the frequency ν of radiation emitted when the hydrogen atom decays from state n to state n'? Give your answer in terms of $\mathbb{R}$, n, n', and h only.

(b) What is the corresponding formula for the wavelength λ emitted in the same transition? Now your formula will contain the additional constant c, the speed of light.

2.18 The angular momentum of an isolated system is constant (when referred to any origin). Derive an expression for the angular momentum p_θ carried away by a photon emitted when a hydrogen atom decays from the state n to the state n' (in the Bohr model).

2.19 In classical electromagnetic theory an accelerating charge e radiates energy at the rate

$$W = \frac{2}{3} \frac{a^2 e^2}{c^3} \text{ ergs/s}$$

The acceleration is a and c is the speed of light. At time $t = 0$, a hydrogen atom has a radius 1 Å. Assuming classical circular motion:

(a) What is a initially?

(b) What is the initial frequency of radiation that the atom emits?

(c) How long does it take for the radius to collapse from 1 Å to 0.5 Å? (Assume that a is constant.)

(d) What is the frequency of radiation at the radius 0.5 Å?

2.20 The dimensionless number

$$\alpha \equiv \frac{e^2}{\hbar c} = \frac{1}{137.037}$$

is called the *fine-structure constant*.

(a) Show that the Rydberg constant may be written $\mathbb{R} = \frac{1}{2}\alpha^2 mc^2$.

(b) If the rest-mass energy of the electron is $mc^2 = 0.511$ MeV, calculate $\mathbb{R}$ in eV.

(c) Obtain an expression for the Bohr energies E_n in terms of α and mc^2.

2.5   WAVES VERSUS PARTICLES

Suppose that a disturbance propagates from one point in space to another point in space. What is propagating, waves or particles? A principal distinguishing characteristic is that waves exhibit interference, particles do not.

Consider the two-slit experiment shown in Fig. 2.9. A continuous spray of particles is fired from the source S. They strike the wall or pass through the two slits A and B. An intensity I_1 (number/unit area · second) emerges from A and an intensity I_2

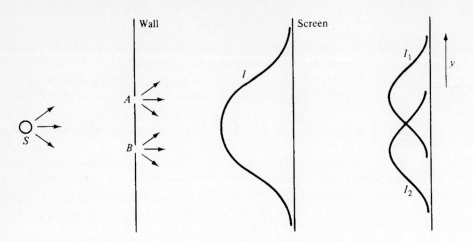

FIGURE 2.9 Particle double-slit experiment. Particle intensities add.

emerges from B. When striking the screen, the two streams of particles superimpose and the net intensity measured is

$$(2.17) \qquad\qquad I = I_1 + I_2$$

This is nothing more than the statement that numbers of particles add.

Now consider the same experimental setup, but instead of a source of particles, let S represent a source of waves, say water waves (Fig. 2.10). Waves are characterized

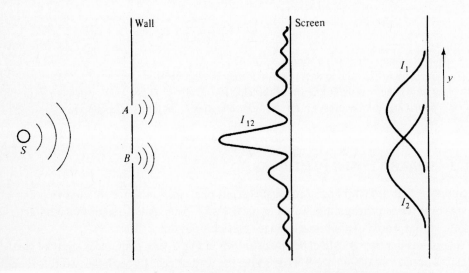

FIGURE 2.10 Wave double-slit experiment. Amplitudes add.

by an amplitude function ψ such that the absolute square of this function gives the intensity I:

$$(2.18) \qquad I = |\psi|^2$$

(Absolute values are taken for complex amplitudes.) Let the two propagating wave disturbances have (complex) scalar (as opposed to vector) amplitudes $\psi_1(\mathbf{r}, t)$ and $\psi_2(\mathbf{r}, t)$, respectively. These functions have the representations

$$(2.19) \qquad \psi_1 = |\psi_1| e^{i\alpha_1}, \qquad \psi_2 = |\psi_2| e^{i\alpha_2}$$

where α is the phase of the wave, which in general is also a function of $(\mathbf{r}, t)$. The intensities of these waves are

$$(2.20) \qquad I_1 = |\psi_1|^2, \qquad I_2 = |\psi_2|^2$$

At a common value of $\mathbf{r}$ and t, the two wave amplitudes superimpose to give the resultant amplitude:

$$(2.21) \qquad \psi = \psi_1 + \psi_2$$

The corresponding resultant intensity is

$$(2.22) \qquad \begin{aligned} I = |\psi|^2 = |\psi_1 + \psi_2|^2 &= (\psi_1 + \psi_2)(\psi_1 + \psi_2)^* \\ &= |\psi_1|^2 + |\psi_2|^2 + |\psi_1\psi_2|[e^{i(\alpha_1-\alpha_2)} + e^{-i(\alpha_1-\alpha_2)}] \\ &= I_1 + I_2 + 2\sqrt{I_1 I_2} \cos(\alpha_1 - \alpha_2) \end{aligned}$$

Comparing this with the resultant intensity I for the particle case (2.17), we note that the wave intensity carries the additional term

$$(2.23) \qquad \Delta \equiv 2\sqrt{I_1 I_2} \cos(\alpha_1 - \alpha_2)$$

This is an interference term. As the y component of r traverses the screen in Fig. 2.10, Δ oscillates and gives a pattern of the form depicted.

Hence, we have uncovered an operational, distinguishing characteristic between particles and waves. Waves exhibit interference, particles do not. Consider the example of a propagating electric field $\mathcal{E}(\mathbf{r}, t)$. The intensity of the wave (energy flux) is proportional to the time average of $|\mathcal{E}|^2$. If two electric waves $\mathcal{E}_1$ and $\mathcal{E}_2$ are superimposed, the new value of the electric field becomes

$$(2.24) \qquad \mathcal{E} = \mathcal{E}_1 + \mathcal{E}_2$$

The intensity is proportional to the time average of squared amplitude, $|\mathcal{E}_1 + \mathcal{E}_2|^2$.

So we have the following important rule: *when two noninteracting beams of particles combine in the same region of space, intensities add; when waves interact, amplitudes add.* The intensity is then proportional to the time average of the absolute square of the resultant amplitude.

PROBLEMS

2.21   In a given wave double-slit experiment, a detector traces across a screen along a straight line whose coordinate we label y. If one slit is closed, the amplitude

$$\psi_1 = \sqrt{\frac{1}{2}} e^{-y^2/2} e^{i(\omega t - ay)}$$

is measured. If the other slit is closed, the amplitude

$$\psi_2 = \sqrt{\frac{1}{2}} e^{-y^2/2} e^{i(\omega t - ay - by)}$$

is measured. What is the intensity pattern along the y axis if both slits are open?

2.6   THE DE BROGLIE HYPOTHESIS AND THE DAVISSON–GERMER EXPERIMENT

In preceding sections we have seen that for a consistent explanation of certain experiments it is necessary to ascribe particle (photon) behavior to light. The energy of such a photon of frequency ν is $E = h\nu$. Its momentum is

$$(2.25) \qquad\qquad p = \frac{E}{c} = \frac{h\nu}{c}$$

This formula can also be written in terms of wavelength λ. The relation between λ and ν for light is particularly simple. It is

$$(2.26) \qquad\qquad \lambda\nu = c$$

In terms of wavenumber k (cm^{-1}) and angular frequency ω,

$$(2.27) \qquad\qquad \omega = 2\pi\nu, \qquad k = \frac{2\pi}{\lambda}$$

Equations (2.25) appear as

$$(2.28) \qquad E = \hbar\omega, \qquad p = \hbar k, \qquad \omega = ck \qquad \left(\hbar \equiv \frac{h}{2\pi}\right)$$

The last of these three equations is called a *dispersion relation*. It reveals a linear dependence between ω and k. The significance of this is that the phase velocity (ω/k) of a monochromatic wave of frequency ω is independent of ω or k. It is the constant

c (speed of light). If a *wave packet* composed of a collection of waves of different wavelengths (or, equivalently, different wavenumbers) is constructed, it propagates with no distortion (dispersion). All component waves have the same speed, c.

The first two equations of (2.28) reveal that photons, which are in essence particles, are identified by two wave parameters: wavenumber k and frequency ω. Now in what sense is a photon different from other more familiar particles (e.g., electrons, protons, etc.)? A photon is special in that it has zero rest mass and travels only at the speed of light. The more familiar particles with finite rest mass also have wave properties. For a (nonrelativistic) particle of kinetic energy

$$(2.29) \qquad\qquad E = \frac{p^2}{2m}$$

the wavelength for the corresponding ("matter") wave is

$$(2.30) \qquad\qquad \lambda = \frac{h}{p} \qquad \text{or} \qquad p = \hbar k$$

which we see from (2.25) and (2.26) is equally relevant to photons. Equation (2.30) is, in essence, the *de Broglie hypothesis*. It ascribes a wave property to particles. While the Planck hypothesis, which assigned a particle quality to electromagnetic waves, had strong experimental motivation, the de Broglie hypothesis, when first introduced in 1925, had little. Such motivation lay to a large degree in the mystery that surrounded the Bohr recipe for the hydrogen atom. What was the physical basis of the first rule for stationary orbits (2.6)? For circular orbits of radius r, with electron momentum p, this rule gives

$$(2.31) \qquad\qquad 2\pi r p = nh$$

In terms of the de Broglie wavelength λ, the last equation reads

$$(2.32) \qquad\qquad 2\pi r = n\lambda$$

The stationary orbits in the Bohr model have an integral number of wavelengths precisely fitting the circumference (Fig. 2.11). This is the classical criterion for the existence of (standing) waves on a circle.

Thus, we see that the de Broglie hypothesis returns the stationary orbit radii of the Bohr theory. This result lends support to the idea that the electron has something "wavy" associated with it, this property being characterized by the de Broglie wavelength (2.30). It was not until two years later (1927) that M. Born suggested what is believed today to be the correct interpretation of this wave property (see Section 2.8).

If electrons (in some respect) propagate as waves, they should exhibit interference. This is the essence of the *Davisson–Germer experiment* (1927). Reflect a beam

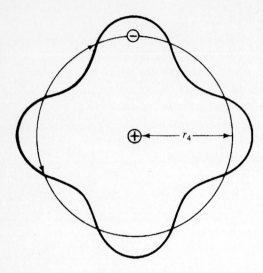

FIGURE 2.11 De Broglie wavelength λ and the $n = 4$ Bohr orbit of the hydrogen atom.

of electrons with well-defined momentum (therefore, wavelength) off a crystal surface whose ion sites are separated by a distance a (the lattice constant) which is of the order of the de Broglie wavelength of the electrons. In the actual experiment, low-energy ($\sim$200-eV) electrons were reflected from the face of a nickel crystal ($a = 3.52 \times 10^{-8}$ cm). An interference pattern was observed which could most consistently be interpreted as the diffraction of plane waves (with de Broglie wavelength) by the regularly spaced atoms of the crystal (Fig. 2.12).

To bring out the full physical interpretation of these results we will consider a simpler experiment in which the same principles are involved. We revert to the two-slit configuration depicted in Fig. 2.10. The source S is able to eject single electrons with well-defined momenta[1] $\mathbf{p} = \hbar\mathbf{k}$. This is a vector normal to the diffracting wall. The

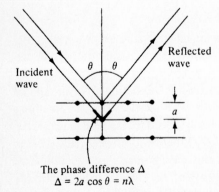

The phase difference Δ
$\Delta = 2a \cos \theta = n\lambda$

FIGURE 2.12 Reflection of plane waves from a lattice. Conditions stated are for constructive interference, with n an integer.

[1] The consistency of this arrangement with the *uncertainty principle* is discussed in the next section.

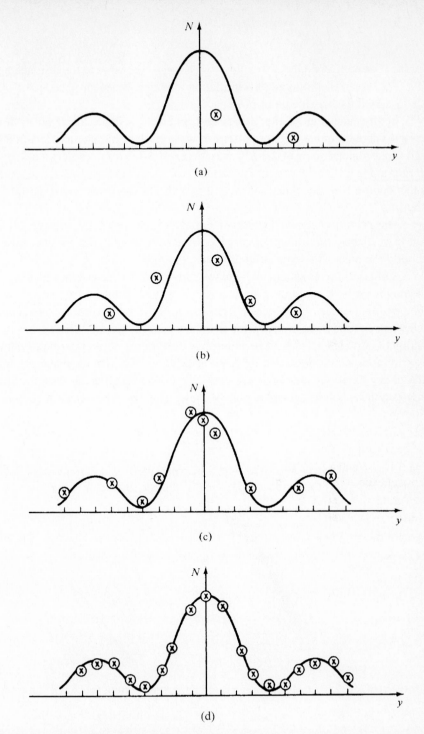

FIGURE 2.13 Number and location of flashes in electron double-slit experiment depicted in Fig. 2.10. Each point is an average of flashes over a unit interval. The solid curve is the theoretical interference pattern corresponding to the de Broglie wavelength.

distance between S and the diffracting wall is large compared to the distance between slits. The screen is composed of scintillation material. When an electron hits it, there is a localized flash at the point of impact.

In any single run of this experiment, one sees a single localized flash on the screen. There is no interference pattern. If we record the number and location (idealized to one dimension, y) of these flashes, the results of 5 runs are shown in Fig. 2.13a; 10 runs in Fig. 2.13b; 50 runs in Fig. 2.13c; 10,000 runs in Fig. 2.13d. The solid curve is the theoretically calculated diffraction pattern obtained with the de Broglie wavelength.

The electrons begin to distribute themselves in an interference pattern. It follows that if we change the source to eject a *current pulse* containing many electrons, the scintillation plate will show an interference pattern.

Similarly, for a source of light we can use a detection plate made of many photomultipliers. If the source emits a single photon, a single pulse from one of the photomultiplier tubes is registered. There is no diffraction pattern. A single particle under any circumstance always gives a single localized "flash." Wherein lies the wave quality of particles? Clearly, it is centered in a *statistical* interpretation of data. Such description was first presented by Born in 1927. Before turning to this analysis, we will give a brief account of a discovery by Heisenberg, which was to throw the well-established philosophical dogma of the seventeenth to nineteenth centuries into disarray.

PROBLEMS

2.22 A photon of energy $h\nu$ collides with a stationary electron of rest mass m. Show that it is not physically possible for the photon to impart all its energy to the electron.

Answer

We must do this problem relativistically. Let us assume that the photon does give up all its energy to the electron. Conservation of energy and momentum then give

$$h\nu + mc^2 = m\gamma c^2$$

$$\frac{h\nu}{c} = m\gamma\beta c$$

where

$$\beta \equiv \frac{v}{c}, \qquad \gamma \equiv (1 - \beta^2)^{-1/2}, \qquad m = \text{rest mass}$$

The speed of the electron after collision is v. Eliminating $h\nu$ from the conservation equations gives

$$\gamma\beta = \gamma - 1$$

whose only (real) solution is $\beta = 0$ ($\gamma = 1$). This is a contradiction.

2.23 Show that the de Broglie wavelength of an electron of kinetic energy $E(eV)$ is

$$\lambda_e = \frac{12.3 \times 10^{-8}}{E^{1/2}} \text{ cm}$$

and that of a proton is

$$\lambda_p = \frac{0.29 \times 10^{-8}}{E^{1/2}} \text{ cm}$$

2.24 At what speed is the de Broglie wavelength of an α particle equal to that of a 10-keV photon?

2.25 Show that in order to associate a de Broglie wave with the propagation of photons (electromagnetic radiation), photons must travel with the speed of light c and their rest mass must be zero. (Do relativistically.)

Answer

For a de Broglie wave associated with a particle of rest mass m,

$$\lambda = \frac{h}{p} = \frac{h}{m\gamma v}$$

For a photon with rest mass m,

$$\lambda = \frac{c}{\nu} = \frac{hc}{h\nu} = \frac{hc}{m\gamma c^2}$$

Equating these relations gives

$$v = c$$

which gives a noninfinite mass, γm, only for $m = 0$.

2.26 The relativistic kinetic energy T of a particle of rest mass m and momentum $p = \gamma m v$ is

$$T = \sqrt{p^2 c^2 + m^2 c^4} - mc^2$$

(a) Show that

$$T = mc^2(\gamma - 1)$$

(b) Show that in the limit $\beta \ll 1$,

$$T = \frac{1}{2}mv^2 + O(\beta^4)$$

(c) Show that the relativistic expression for T above gives the correct energy–momentum relation for a photon if $m = 0$.

(d) What is the total relativistic energy, E (i.e., including rest-mass energy) of a particle of mass m?

(e) What is the total relativistic energy of a particle moving in a potential field $V(x)$? What is the corresponding Hamiltonian, $H(p, x)$?

Answers (partial)

(d) $E = \gamma mc^2 = \sqrt{p^2 c^2 + m^2 c^4}$.

(e) $E = \gamma mc^2 + V(x)$; $H = \sqrt{p^2 c^2 + m^2 c^4} + V(x)$.

(An extensive discussion of relativistic quantum mechanics is presented in Chapter 15.)

2.27 Assuming the sun to be a blackbody with a surface temperature of 6000 K, (a) calculate the rate at which energy is radiated from it. (b) Determine the loss in solar mass per day due to this radiation.

2.28 In 1922, A. H. Compton applied the photon concept of electromagnetic radiation to explain the scattering of x rays from electrons. In the analysis it is assumed that a photon of energy $h\nu$ and momentum $h\nu/c = h/\lambda$ is incident on a stationary but otherwise free electron of rest mass m. The photon scatters from the electron. Its new momentum, h/λ', makes an angle θ with the incident (old) momentum. The momentum of the recoiling electron makes an angle ϕ with the incident momentum (Fig. 2.14). If the system of electron and photon is an isolated system, its energy and total momentum are constant. Conservation of energy reads

$$h\nu + mc^2 = h\nu' + m\gamma c^2$$

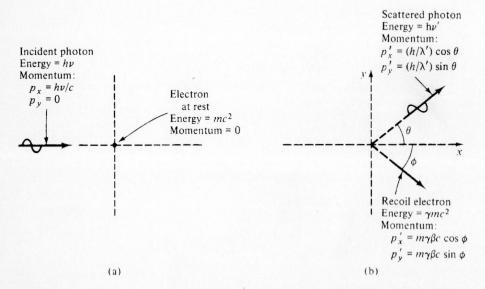

Incident photon
Energy = $h\nu$
Momentum:
$p_x = h\nu/c$
$p_y = 0$

Electron
at rest
Energy = mc^2
Momentum = 0

Scattered photon
Energy = $h\nu'$
Momentum:
$p'_x = (h/\lambda') \cos \theta$
$p'_y = (h/\lambda') \sin \theta$

Recoil electron
Energy = γmc^2
Momentum:
$p'_x = m\gamma\beta c \cos \phi$
$p'_y = m\gamma\beta c \sin \phi$

(a)   (b)

FIGURE 2.14 Angles θ and ϕ in the Compton scattering of photons from electrons. (See Problem 2.28.)

Conservation of momentum (the whole collision occurs in a plane) gives

$$\frac{h}{\lambda} = m\gamma\beta c \cos\phi + \left(\frac{h}{\lambda'}\right) \cos\theta$$

$$0 = -m\gamma\beta c \sin\phi + \left(\frac{h}{\lambda'}\right) \sin\theta$$

Using these three conservation equations, derive the *Compton effect equation* for the difference in wavelengths:

$$\lambda' - \lambda = \lambda_C(1 - \cos\theta)$$

$$\lambda_C = \text{the Compton wavelength} = \frac{h}{mc} = 2.43 \times 10^{-10} \text{ cm}$$

2.29   The "classical radius of the electron," r_0, is obtained by setting the potential e^2/r_0 equal to the rest-mass energy of the electron, mc^2.

$$\frac{e^2}{r_0} = mc^2$$

$$r_0 = \frac{e^2}{mc^2}$$

Show that the successive powers of the fine-structure constant

$$\alpha \equiv \frac{e^2}{\hbar c}$$

are measures of the Bohr radius to the Compton wavelength; and of the Compton wavelength to the classical radius of the electron. That is, show that

$$a_0 : \lambdabar_C : r_0 = 1 : \alpha : \alpha^2$$

2.7   THE WORK OF HEISENBERG. UNCERTAINTY AS A CORNERSTONE OF NATURAL LAW

It is an essential feature of Newton's second law that, given the initial coordinates and velocity of a particle, $\mathbf{r}(0)$ and $\dot{\mathbf{r}}(0)$, respectively, and knowing all the forces on the particle, the orbit $\mathbf{r}(t)$ is uniquely determined. The same holds true for a system of particles. This is the essence of *determinism*. Laplace, in the eighteenth century, took the implications of the latter statements to their extreme: the entire universe consists of bodies moving through space and obeying Newton's laws. Once the interaction between these bodies is precisely known and the position and velocities of all the bodies at any given instant are known, these coordinates and velocities are determined (through Newton's second law) for all time.

Quantum mechanics was to bring down the walls of this deterministic philosophy. The instrument of destruction was the *Heisenberg uncertainty principle*. What Heisenberg put forth in 1927 implied the following: if the momentum of a particle is known precisely, it follows that the position (location) of that same particle is completely unknown. Quantitatively, if an identical experiment involving an electron is performed many times, and in each run of the experiment the position (x) of the electron is measured, then although the experimental setup is identical (same electron momentum) in each run, measurement of the position of the electron does not give the same result. Let the average of these measurements be $\langle x \rangle$. Then we can form the mean-square deviation

$$(2.33) \qquad (\Delta x)^2 \equiv \langle (x - \langle x \rangle)^2 \rangle$$

The standard deviation is labeled Δx. If Δx is small compared to some typical length in the experiment, one is more certain to find the value $x = \langle x \rangle$ in any given run. If Δx is large, it is not certain what the measurement of x will yield (Fig. 2.15). For this reason Δx is also called the *uncertainty in* x.

Similarly, one may speak of an uncertainty in any physically observable quantity: magnetic field $\mathscr{B}$, energy E, momentum $\mathbf{p}$, and so forth.

$$(2.34) \qquad \begin{aligned} \Delta \mathscr{B}_x &= \sqrt{\langle (\mathscr{B}_x - \langle \mathscr{B}_x \rangle)^2 \rangle}, & \Delta \mathscr{B}_y &= \cdots \\ \Delta E &= \sqrt{\langle (E - \langle E \rangle)^2 \rangle} \\ \Delta p_x &= \sqrt{\langle (p_x - \langle p_x \rangle)^2 \rangle}, & \Delta p_y &= \cdots \end{aligned}$$

Heisenberg's uncertainty relation for momentum p_x and position x (parallel components) appears as

$$(2.35) \qquad \Delta x \, \Delta p_x \geq \hbar/2$$

If it can be said with certainty what the position of a particle is ($\Delta x = 0$), then there is total uncertainty regarding the momentum of the particle ($\Delta p_x = \infty$). Observable

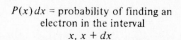

$P(x)\,dx$ = probability of finding an
electron in the interval
x, $x + dx$

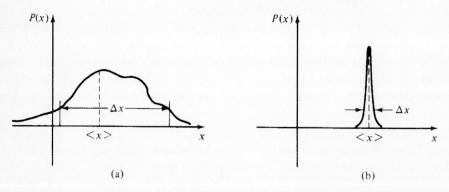

(a) (b)

FIGURE 2.15 (a) **Large uncertainty in** x: $(\Delta x)^2 = \langle x^2 \rangle - \langle x \rangle^2$. (b) **Small uncertainty in** x.

parameters obeying a relation such as (2.35) are called *complementary variables.*
Examples include (1) coordinates and momenta (x, p_x); (2) energy and time (E, t); and
(3) any two Cartesian components of angular momentum (L_x, L_y). Later in the text a
formal technique is presented to determine if two observables are complementary (as
opposed to *compatible*).

When an electron (or photon) exists within a well-defined locality of space
(momentum is ill defined), it acts very much like a particle. This is the case since in
a double-slit experiment such a localized disturbance would only go through one slit;
and we can therefore follow it in time, so it is very much like a true particle. When the
electron does not exist in a well-defined locality of space, its momentum can be
defined more precisely. Under such circumstances the wave character of the electron
manifests itself. We cannot follow *it*. A whole wave is propagating. Nevertheless, it
should be borne in mind that when a scintillation screen is put across its path one gets
a single flash—although one has little idea of when or where the event will occur.

PROBLEMS

2.30   Consider a particle with energy $E = p^2/2m$ moving in one dimension (x). The uncer-
tainty in its location is Δx. Show that if $\Delta x \, \Delta p > \hbar$, then $\Delta E \, \Delta t > \hbar$, where $(p/m) \, \Delta t = \Delta x$.

2.31   The size of an atom is approximately 10^{-8} cm. To locate an electron within the atom,
one should use electromagnetic radiation of wavelength not longer than, say, 10^{-9} cm. (a) What
is the energy of a photon with such a wavelength (in eV)? (b) What is the uncertainty in the
electron's momentum if we are uncertain about its position by 10^{-9} cm?

2.8   THE WORK OF BORN. PROBABILITY WAVES

When discussing the double-slit wave diffraction experiment and the Davisson–
Germer experiment, we found it appropriate to introduce an amplitude function ψ, the
square of whose modulus, $|\psi|^2$, was set equal to the intensity (2.18) of the wave.

Born suggested in 1927 that, when referred to the propagation of particles, $|\psi|^2$
is more appropriately termed a *probability density.* The function ψ is called the
wavefunction (also the *state function* or *state vector*) of the particle. Quantitatively,
the Born postulate states the following (in Cartesian space). The wavefunction for a
particle $\psi(x, y, z, t)$ is such that

$$(2.36) \qquad\qquad |\psi|^2 \, dx \, dy \, dz = P \, dx \, dy \, dz$$

where $P \, dx \, dy \, dz$ is the probability that measurement of the particle's position at the
time t finds it in the volume element $dx \, dy \, dz$ about the point x, y, z.

This statement is quite consistent with the discussions above relating to the
interference of photons or electrons. In all cases an interference pattern exhibits itself
when an abundance of particles is present. The wavefunction ψ generates the interfer-

ence pattern. Where $|\psi|^2$ is large, the probability that a particle is found there is large. When enough particles are present, they distribute themselves in the probability pattern outlined by the density function $|\psi|^2$.

The rules of quantum mechanics (Chapter 3) give a technique for calculating the wavefunction ψ to within an arbitrary multiplicative constant. The equation one solves to find ψ is called the *Schrödinger equation*. This is a homogeneous linear equation. Suppose that we solve it and obtain a function ψ. Then $A\psi$ is also a solution, where A is any constant. The Born postulate specifies[1] A. For problems where it can be said with certainty that the particle is somewhere in a given volume V,

$$(2.37) \qquad \int_V |\psi|^2 \, dx \, dy \, dz = 1$$

This is a standard property that probability densities satisfy. It is the mathematical expression of the certainty that the particle is in the volume V.

As an example, consider the following one-dimensional problem. A particle that is known to be somewhere on the x axis has the wavefunction

$$(2.38) \qquad \psi = A e^{i\omega t} e^{-x^2/2a^2}$$

The frequency ω and length a are known constants. The (real) constant A is to be determined. Since it is certain that the particle is somewhere in the interval $-\infty < x < +\infty$, it follows that

$$(2.39) \qquad 1 = \int_{-\infty}^{\infty} \psi^* \psi \, dx = A^2 \int_{-\infty}^{\infty} e^{-i\omega t} e^{+i\omega t} e^{-x^2/a^2} \, dx$$

$$= A^2 a \int_{-\infty}^{\infty} e^{-\eta^2} \, d\eta = A^2 a \sqrt{\pi}$$

The nondimensional variable $\eta \equiv x/a$. This calculation gives

$$(2.40) \qquad A = \frac{1}{a^{1/2} \pi^{1/4}}$$

The normalized wavefunction is therefore

$$(2.41) \qquad \psi = \frac{1}{a^{1/2} \pi^{1/4}} e^{i\omega t} e^{-x^2/2a^2}$$

For the stated problem, $|\psi|^2$ as obtained from (2.41) is the correct probability density.

---

[1] If A is complex it may be determined only to within an arbitrary phase factor, $e^{i\alpha}$, where α is a real number.

PROBLEMS

2.32   The wavefunction for a particle in one dimension is given by

$$|\psi_1| = A_1 e^{-y^2/4}$$

Another state that the particle may be in is

$$|\psi_2| = A_2 y e^{-y^2/8}$$

A third state the particle may be in is

$$|\psi_3| = A_3 (e^{-y^2/4} + y e^{-y^2/8})$$

Normalize all three states in the interval $-\infty < y < +\infty$ (i.e., find A_1, A_2, and A_3). Is the probability of finding the particle in the interval $0 < y < 1$ when the particle is in the state ψ_3 the same as the sum of the separate probabilities for the states ψ_1 and ψ_2? Answer the same question for the interval $-1 < y < +1$.

2.33   The energy density (ergs/cm^3) of electromagnetic radiation is proportional to $\mathscr{E}^2$, where $\mathscr{E}$ is the electric field. Present an argument to demonstrate that $\mathscr{E}^2 \, d\mathbf{r}$ is a measure of the probability of finding a photon in the volume element $d\mathbf{r}$. Assume a monochromatic radiation field.

2.34   Suppose that in a sample of 1000 electrons, each has a wavefunction

$$\psi = e^{-|x|} e^{-i\omega t} \cos \pi x$$

Measurements are made (at a specific time, $t = t'$) to determine the locations of electrons in the sample. Approximately how many electrons will be found in the interval $-\frac{1}{2} \le x \le \frac{1}{2}$? A graphical approximation is adequate.

2.35   A beam of monochromatic electromagnetic radiation incident normally on a totally absorbing surface exerts a pressure on the surface of

$$P = U = \frac{\mathscr{E}^2}{8\pi}$$

where $\mathscr{E}$ is the amplitude of the electric field vector. If $P = 3 \times 10^{-6}$ dyne/cm^2 and the wavelength of radiation is $\lambda = 8000$ Å, what is the photon flux (cm^{-2}/s) striking the surface?

2.9   SEMIPHILOSOPHICAL EPILOGUE TO CHAPTER 2

The wavefunction ψ affords information related to experiments on, say, an electron. Consider once again the double-slit experiment of Fig. 2.10. Again, we suppose that the source is able to fire single electrons with well-defined momenta. There is a corresponding (propagating) wavefunction $\psi(\mathbf{r}, t)$ which is diffracted by the slits. When measurements are made, the scintillation screen gives a single flash (for a single electron). If we calculate $|\psi|^2$ at the screen, we find an interference pattern. What is the significance of this pattern? Suppose that the electron is a bullet. We can play a quantum mechanical Russian roulette. The game is to stand at the screen so that the bullet misses you. The first thing to do is solve the Schrödinger equation and calculate $|\psi|^2$ at the screen. Stand where it is minimum. But this, of course, does not guarantee that the bullet does not find its mark. The laws of nature do not provide a more definite knowledge of the electron's trajectory.

Now when a pulse of electrons (assume that they are all independent of one another) is fired at the slits, the scintillation screen registers an interference pattern. Eventually (i.e., when a sufficient number are fired), the electrons begin to follow the dictates of $|\psi|^2$ and fall into place (Fig. 2.13). It is interesting that all the information in ψ cannot be extracted from an experiment on one electron. To get this information one has to do many experiments, each of which involves many more than one electron.

At this point the reader may well ask: If electrons are particles, why not follow their trajectories through the slits (an electron can only go through one slit at a time) and onto the screen? One could then add the intensities of particles stemming from each individual slit and obtain the (noninterference) pattern depicted in Fig. 2.9. Well, we can do exactly that, and the interference pattern does vanish. In the process of "watching" the electrons, the interference is destroyed.

This is seen as follows. Let us see if we can discern which slit the electron goes through. The uncertainty in measurement of its y coordinate must obey the inequality

$$(2.42) \qquad\qquad \Delta y \ll \frac{d}{2}$$

If the interference pattern is not to be destroyed, the uncertainty in an electron's y momentum Δp_y, induced by encounter with a photon, must be substantially smaller than that which would displace the electron from a maximum in the interference pattern to a neighboring minimum. With the aid of Fig. 2.16 this condition is

$$(2.43) \qquad\qquad \Delta p_y \ll \frac{\theta}{2} p_x = \frac{h}{2d}$$

In the latter equality we have recalled the de Broglie relation (2.30).

The first inequality for Δy, (2.42), enables one to observe which slit the electron goes through. The second inequality for Δp_y, (2.43), guarantees the preservation of the interference pattern. Combining these two inequalities gives the relation

$$(2.44) \qquad\qquad \Delta y \, \Delta p_y \ll \frac{h}{4}$$

which is in contradiction to the Heisenberg uncertainty principle.

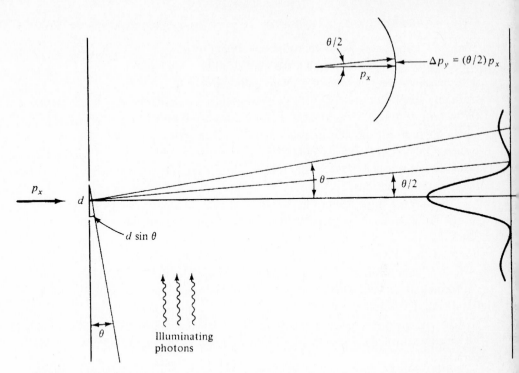

FIGURE 2.16   First maximum at $\theta = 0$; second maximum at $\sin \theta \simeq \theta = \lambda/d$. **The angle between the first minimum and the second maximum** $= \theta/2 = \lambda/2d$.

We conclude that if it is possible to observe which slit electrons go through, their interference pattern is destroyed. In observing the positions of the electrons, their wave quality (e.g., interference-producing mechanism) diminishes. When the light (whose photons are illuminating the electrons' path) is switched off, the interference pattern reappears.

In general, we may note the fundamental rule that quantum mechanics does not delineate the trajectory of a single particle. One may calculate the probability that an electron is in some region of space, but this is again a probability and not a guarantee that the electron will be found there. To realize this probability, one must in principle observe many experiments on the same system with identical initial conditions obeyed in each experiment. Average results then fall to the dictates of quantum mechanics. In this regard Einstein[1] has remarked that quantum mechanics is incapable of describing the behavior of a single system (such as an electron).

[1] P. A. Schlipp, ed., *Albert Einstein: Philosopher–Scientist*. Harper & Row, New York, 1959.

Hidden Variables

There is another, somewhat philosophical school of thought (due primarily to Bohm and de Broglie[1]) which holds that the impossibility of quantum mechanics to predict with certainty the outcome of a given measurement on an individual system stems from one's inability to know the exact values of certain *hidden variables* relating to the system. In this description, the wavefunction is viewed as a mathematical object which contains all the information one possesses regarding an incompletely known system. Quantum formulas should emerge as averages over the hidden parameters in much the same way as the laws of classical physics do in fact follow in averaging over the quantum equations.[2]

Bethe[3] has argued that the existence of such hidden variables for an electron would imply electronic degrees of freedom which are not specified in atomic physics. However, the success of the present theory in formulation of the periodic table indicates that this is not the case (i.e., no further degrees of freedom exist).[4]

In Chapter 3 we discuss the postulates of quantum mechanics. These are clear-cut formal statements whose mastery enables the student to treat many problems in the quantum mechanical domain. In addition, a deeper understanding is gained of some of the questions raised in this semiphilosophical epilogue. The notion of hidden variables is returned to in Section 11.13 in discussion of the Einstein–Padolsky–Rosen paradox.

PROBLEMS

2.36   In deriving the Planck radiation formula (2.3), one first sets

$$u(\nu) = h\nu n(\nu)$$

where $n(\nu)\,d\nu$ is the density of photons in the frequency interval ν, $\nu + d\nu$. An expression for $n(\nu)$ is obtained through the relation

$$n(\nu) = g(\nu) f_{BE}(\nu)$$

---

[1] D. Bohm, *Phys, Rev.* **85**, 166, 180, (1952): *L. de Broglie, Physicien et Penseur*, Albin Michel, Paris, 1953. For further discussion and reference, see J. S. Bell. *Rev. Mod. Phys.* **38**, 447 (1966); F. J. Belinfante, *A Survey of Hidden-Variable Theories*, Pergamon Press. New York, 1973.

[2] This relation between quantum and classical mechanics is called *Ehrenfest's principle* and is discussed fully in Chapter 6.

[3] H. A. Bethe and R. W. Jackiw, *Intermediate Quantum Mechanics*, 2d ed., W. A. Benjamin, New York, 1968.

[4] Experimental evidence obtained by S. Freedman and J. Clauser, *Phys. Rev. Lett.* **28**, 938 (1972), also appears to point against a hidden-variable theory.

where $g(\nu)\, d\nu$ is the density of modes (i.e., vibrational states) in the said frequency interval and f_{BE}, the Bose–Einstein factor, gives the average number of photons per mode at the frequency ν. Use Planck's hypothesis to obtain the expression

$$f_{BE}(\nu) = \frac{1}{e^{\,h\nu/k_B T} - 1}$$

[The calculation of $g(\nu)$ is considered in Problem 2.37.]

Answer

We seek the average number of photons per mode at the frequency ν. At this frequency the modes of excitation of the radiation field have energies $h\nu$, $2h\nu$, $3h\nu$, ... Let us assume that the probability that the Nth energy mode is excited is given by the *Boltzmann distribution*

$$p(N) = e^{-Nx} \bigg/ \sum_{N=0}^{\infty} e^{-Nx}$$

$$x \equiv \frac{h\nu}{k_B T}$$

There are N photons of frequency ν in the Nth mode. Averaging over N gives

$$f_{BE} = \langle N \rangle = \sum_N N P(N)$$

$$= \sum_N N e^{-nx} \bigg/ \sum_N e^{-Nx} = -\frac{\partial}{\partial x} \ln \sum_N e^{-Nx}$$

$$= -\frac{\partial}{\partial x} \ln \sum_N (e^{-x})^N$$

$$= -\frac{\partial}{\partial x} \ln \frac{1}{1 - e^{-x}} = \frac{\partial}{\partial x} \ln (1 - e^{-x})$$

$$= \frac{1}{e^x - 1}$$

2.37   In Problem 2.36 we noted that the number of photons per unit volume in the frequency interval ν, $\nu + d\nu$, is given by

$$n(\nu) = g(\nu) f_{BE}\!\left(\frac{h\nu}{k_B T}\right)$$

(a)   Calculate the density of states $g(\nu)$, assuming that the blackbody radiation field consists of standing waves in a cubical box with perfectly reflecting walls and edge length a.

(b)   Obtain the Rayleigh–Jeans law for the radiant energy density $u_{RJ}(\nu)$, assuming the classical *equipartition hypothesis* for the electromagnetic field: that is, each mode of vibration contains k_BT ergs of energy.

(c)   Make a sketch of $u_{RJ}(\nu)$ and compare it to the Planck formula for $u(\nu)$. In what frequency domain do the two theories agree?

(d)   What property of the vibrational energy levels of the radiation field (at a given frequency) allows the classical description (i.e., u_{RJ}) to be valid?

Answers (partial)

(a)   The spatial components of a standing electric field in a cubical box of volume $V = a^3$, with perfectly reflecting walls, are

$$\mathcal{E}_x = A \cos k_x x \sin k_y y \sin k_z z$$
$$\mathcal{E}_y = B \sin k_x x \cos k_y y \sin k_z z$$
$$\mathcal{E}_z = C \sin k_x x \sin k_y y \cos k_z z$$

These fields have the required property that the tangential component of $\mathcal{E}$ vanishes at all six walls provided that

$$k_x a = n_x \pi, \qquad k_y a = n_y \pi, \qquad k_z a = n_z \pi$$

where n_x, n_y, and n_z assume positive integer values. There is a mode of vibration for each triplet of values (n_x, n_y, n_z). We seek the number of such modes in the frequency interval ν, $\nu + d\nu$. First note that for each mode the square sum

$$n^2 = n_x{}^2 + n_y{}^2 + n_z{}^2 = \left(\frac{a}{\pi}\right)^2 (k_x{}^2 + k_y{}^2 + k_z{}^2)$$
$$= \left(\frac{a}{\pi}\right)^2 k^2 = \left(\frac{2a}{c}\right)^2 \nu^2$$

is proportional to the square of ν, the frequency of vibration $(2\pi\nu = ck)$. Next consider Cartesian n space with axes n_x, n_y, and n_z. Each point in this space corresponds to a mode of vibration. It is clear that all points which fall on a spherical surface of radius $(2a\nu/c)$ correspond to modes at the frequency ν. It follows that the number of modes in the frequency interval ν, $\nu + d\nu$, is given by the volume in n space of a spherical shell of thickness dn and radius n (Fig. 2.17):

$$Vg(\nu)\,d\nu = 2 \times \frac{1}{8} \times 4\pi n^2 \, dn = \pi n^2 \, dn$$

The factor 2 enters because of the two possible polarizations of an electric field in a given mode. The factor $\frac{1}{8}$ is due to the fact we wish only to consider positive frequencies so that only that

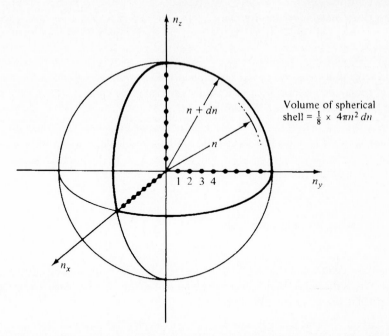

Volume of spherical shell $= \frac{1}{8} \times 4\pi n^2\, dn$

FIGURE 2.17   Cartesian n space for the enumeration of standing electromagnetic wave states in a box of edge length a. All points that fall in the shell of thickness dn and radius $(2a\nu/c)$ correspond to modes with frequencies in the interval ν, $\nu + d\nu$. (See Problem 2.37.)

portion of the shell in the first octant is counted. This gives the desired result:

$$g(\nu) = \frac{8\pi \nu^2}{c^3}$$

(b)   If there is $k_B T$ energy per mode, the spectral energy density is

$$u_{RJ} = \frac{8\pi \nu^2}{c^3}\, k_B T$$

2.38   For a gas of N noninteracting particles (an *ideal gas*) in thermodynamic equilibrium at temperature T and confined to volume V, the pressure P is given by

$$PV = N\langle p_x v_x \rangle$$

The momentum of a particle is $\mathbf{p}$ and $\mathbf{v}$ is its velocity. The average is taken over all particles in the gas. Show that for a *gas* of photons, this relation gives

$$PV = \frac{1}{3}E$$

while for a gas of mass points it gives

$$PV = \frac{2}{3}E$$

The total energy of the gas is E. (*Note:* The principle of *equipartition of energy* ascribes equal portions of energy, on the average, to each degree of freedom of a particle. Thus, if the average energy of a mass particle is ε, then $\langle \frac{1}{2}mv_x^2 \rangle = \varepsilon/3$.)

2.39   The energy density $U = E/V$ of a blackbody radiation field is a function only of temperature, $U = U(T)$. Using this fact together with the result of the last problem, $E = 3PV$, show that Stefan's law, $U = (4\sigma/c)T^4$, follows from purely thermodynamic arguments, thereby establishing the law as a classical result.

Answer

The first two laws of thermodynamics give

$$T\,dS = dE + P\,dV$$

The second law defines the entropy S, while the first law gives the conservation of energy statement in the form: heat added = increase in internal energy + work done. Using the given relations permits this equation to be rewritten

$$dS = \frac{V}{T}\,dU + \frac{4}{3}\frac{U}{T}\,dV$$

We recognize this equation to be in the form

$$dS = \left.\frac{\partial S}{\partial U}\right|_V \,dU + \left.\frac{\partial S}{\partial V}\right|_U \,dV$$

It follows that

$$\left.\frac{\partial V/T}{\partial V}\right|_{U(T)} = \frac{4}{3}\left.\frac{\partial U(T)/T}{\partial U}\right|_V$$

which integrates to the desired result.

2.40   An *adiabatic process* is one in which a system exchanges no heat with its environment.

   (a)   What form does the first law of thermodynamics assume for an adiabatic process?

   (b)   Using the form obtained in part (a), show that for an adiabatic expansion of a blackbody radiation field,

$$VT^3 = \text{constant}$$

   (c)   Consider that the primeval fireball described in Problem 2.10 contains mass $M = \rho V$ and radiation energy $E = UV$, where V is volume. Show that in an expanding universe, the radiation density U decreases faster than the mass density ρ. Thus, although it is believed that radiation density of the primeval fireball far exceeded its mass density, the fact that in our present universe mass density dominates over radiation density is seen to be consistent with an adiabatically expanding universe in which Stefan's law holds.

Answers (partial)

(a)   Since $dS = 0$ for an adiabatic process, the first law becomes

$$dE + P\,dV = 0$$

(b)   *Hint:* To find $P = P(T)$, use $E = 3PV$ in conjunction with Stefan's law.

(c)   *Hint:* Compare $\rho(V)$ at constant M with $U(V)_{adb}$.

2.41   Two plates of an ideal parallel-plate capacitor are made of platinum and silver, respectively. When the plates are brought to the displacement 10^{-3} cm, electrons "tunnel" through[1] the potential barrier, thereby creating a contact potential. What is the electric field (V/cm) between the plates? Which metal is left positive?

2.42   As described in Section 2.3, in a sample of metal at absolute zero, electrons completely fill the lowest levels such that no more than one electron occupies each state. All levels are filled from zero energy to E_F, the Fermi energy (see Fig. 2.5). Let $g(E)\,dE$ represent the number of energy states that are available for occupation in the interval $E, E + dE$, per unit volume. Since each state from 0 to E_F is occupied, the number of free electrons, n, per unit volume is given by the number of available states per unit volume in this interval. That is,

$$n = \int_0^{E_F} g(E)\,dE$$

For free electrons, *the density of states* $g(E)$ (see Section 8.8) is given by

$$g = \frac{8\sqrt{2}\pi m^{3/2} E^{1/2}}{h^3}$$

(a)   Using the expressions above, obtain an explicit formula for E_F for a metal with n free electrons per unit volume.

(b)   Given that $E_F(Cu) = 7.0$ eV and $E_F(Na) = 3.1$ eV, use your formula to obtain the density (cm^{-3}) of Cu and Na nuclei, respectively, in samples at absolute zero. (The periodic chart is given in Table 12.4.)

Answer (partial)

(a) $E_F = (h^2/2m)(3n/8\pi)^{2/3}$.

2.43   (a)   Show that when expressed in terms of *angular* frequency ω, the radiant energy density relevant to a blackbody radiation field is given by

$$u(\omega) = \frac{\hbar\omega^3/\pi^2 c^3}{e^{\hbar\omega/k_B T} - 1}$$

(b)   Show that the corresponding density of states is given by

$$g(\omega) = \frac{\omega^2}{\pi^2 c^3}$$

[1] The mechanism of quantum mechanical tunneling is described in Section 7.7.

(c)   Answers to the preceding questions may be found by setting

$$u(\omega)\, d\omega = u(\nu)\, d\nu$$

What is the physical meaning of this equality? (See also Problem 2.1.)

2.44 (a)   What is the wavelength of an electromagnetic wave of frequency ν propagating in vacuum?

(b)   What is the de Broglie wavelength of a photon of frequency ν?

(c)   How are the wavelengths of parts (a) and (b) related?

2.45   An idealized model of a *plasma* (i.e., a fluid of electrons and ions) is given by a so-called one-component plasma (OCP). In this configuration electrons move in a uniformly distributed charge-neutralizing background. This background is sometimes referred to as "jellium." It was originally suggested by E. P. Wigner that a high-density OCP will crystallize at some critical value of electron density as this parameter is relaxed from its extreme high value.

(a)   What is the potential energy of an electron situated at the center of a sphere of radius a which is uniformly filled with positive neutralizing charge?

(b)   What is the uncertainty in momentum of an electron confined to a region of dimension a? What is the corresponding kinetic energy E_k?

(c)   Assuming that crystallization will occur when potential energy per particle exceeds uncertainty energy, obtain a criterion for crystallization in terms of the distance between electrons a and the Bohr radius a_0.

Answers

(a)   The charge density of the neutralizing medium is

$$\rho = \frac{e}{\frac{4}{3}\pi a^3}$$

The potential of interaction between electron and positive jellium in a shell of thickness dr at r is

$$dV = \frac{e\rho 4\pi r^2\, dr}{r}$$

Thus

$$V = \int_0^a \frac{e\rho 4\pi r^2\, dr}{r} = \frac{3e^2}{2a}$$

(b)   From the uncertainty relation (2.35) we have $\Delta p \simeq \hbar/a$ so that

$$E_k \simeq \frac{(\Delta p)^2}{2m} = \frac{\hbar^2}{2ma^2}$$

(c)   The relation $V > E_k$ gives the order-of-magnitude criterion,

$$a > a_0$$

where a_0 is the Bohr radius. Thus, Wigner crystallization can be expected to occur when interparticle spacing exceeds the Bohr radius.

2.46 In a nuclear reactor, uranium atoms break apart and release several high-energy neutrons. These neutrons have an initial energy of 2 MeV, but after many collisions with other atoms they slow down to an average energy of 0.025 eV. Thus there will be a continuous distribution of neutrons with energies ranging from 2 MeV to *less* than 0.025 eV.

(a) What are the de Broglie wavelengths of the 2-MeV neutrons and the 0.025-eV neutrons? Give your answer in Å.

(b) A "low-pass" energy filter can be constructed by passing the neutrons through a long piece of polycrystalline graphite. Explain briefly why low-energy "particles" can travel straight through the block but high-energy "particles" are *reflected* out the sides.

(c) What is the shortest de Broglie wavelength of the neutrons that are totally transmitted through the crystal? Give your answer in Å. Spacing between lattice planes in graphite is $a \simeq 2$ Å.

Answer (partial)

(c) The condition for constructive interference from adjacent planes (Fig. 2.12) is

$$n\lambda = 2a \cos \theta \leq 2a$$

$$\lambda \leq \frac{2a}{n} \leq 2a$$

Thus for reflection, $\lambda \leq 2a = \lambda_{max}$.

2.47 De Broglie waves are incident on a crystal with lattice planes separated by a. Show that constructive interference of waves reflected from planes separated by $2a$ is ruled out by parallel reflection from planes in between.

2.48 As described in Section 7.10, a criterion which discerns if a given configuration is classical or quantum mechanical may be stated in terms of the de Broglie wavelength λ. Namely, if L is a scale length characteristic to the configuration at hand, then one has the following criteria:

$\lambda \ll L$:   Classical

$\lambda \gtrsim L$:   Quantum mechanical

Use these criteria to describe which physics is relevant to the following configurations:

(a) An atomic electron. For the typical length choose the Bohr radius. For typical energy choose the Rydberg.

(b) A proton in a nucleus. For nuclear size choose $\simeq 10^{-13}$ cm (1 fermi). For energy choose $\simeq 10$ MeV.

(c) An electron in a vacuum tube operating at 10 kV.

(d) An electron gas of density 10^{20} cm^{-3} and temperature 300 K.

2.49 A particle constrained to move on the x axis has a probability of $\frac{1}{5}$ for being in the interval $(-d - a, -d + a)$ and $\frac{4}{5}$ for being in the interval $(d - a, d + a)$, where $d \gg a$.

(a) Sketch the wavefunction that describes this situation.

(b) Call the normalized wavefunction for the "left" interval $\varphi_-(x)$ and that for the "right" interval $\varphi_+(x)$. What is the normalized wavefunction $\varphi(x)$ for the particle?

(c) Again, with $d \gg a$, what is the probability density $P(x)$ for the particle? What does the integral over all x of $P(x)$ give?

Answers (partial)

(b) $\varphi_\pm = \sqrt{\frac{1}{5}}\varphi_- \pm \sqrt{\frac{4}{5}}\varphi_+$

(c) $P(x) = |\varphi_\pm|^2 = \frac{1}{5}|\varphi_-|^2 + \frac{4}{5}|\varphi_+|^2$

Note that for $d \gg a$, $\varphi_+\varphi_- = 0$.

$$\int_{-\infty}^{\infty} P(x)\,dx = \frac{1}{5} + \frac{4}{5} = 1$$

Note: For mutually exclusive events a and b, the summational probability $P_1(a) + P_2(b)$ gives the probability of a or b occurring; whereas for independent events, the product probability $P_1(a)P_2(b)$ gives the probability of a and b occurring (and is called the *joint probability*). Note further that mutually exclusive events are not independent.

2.50   Explain the meaning of the following statement: Interference between photons does not occur. A photon can only interfere with itself.[1]

---

[1] For further discussion see P. A. M. Dirac, *The Principles of Quantum Mechanics* 4th ed., Section 3, Oxford University Press, New York, 1958.

3

THE POSTULATES OF QUANTUM MECHANICS. OPERATORS, EIGENFUNCTIONS, AND EIGENVALUES

3.1 *Observables and Operators*
3.2 *Measurement in Quantum Mechanics*
3.3 *The State Function and Expectation Values*
3.4 *Time Development of the State Function*
3.5 *Solution to the Initial-Value Problem in Quantum Mechanics*

In this chapter we consider four basic postulates of quantum mechanics, which when taken with the Born postulate described in Section 2.8, serve to formalize the rules of quantum mechanics. Mathematical concepts material to these postulates are developed along with the physics. The postulates are applied throughout the text. We choose the simplest problems first to exhibit their significance and method of application— that is, problems in one dimension.

3.1 OBSERVABLES AND OPERATORS

Postulate I[1]

This postulate states the following: To any self-consistently and well-defined observable in physics (call it A), such as linear momentum, energy, mass, angular momentum, or number of particles, there corresponds an operator (call it $\hat{A}$) such that

---

[1] The order in which these postulates appear is by no means conventional.

TABLE 3.1   Examples of operators

$\hat{D} = \partial/\partial x$	$\hat{D}\varphi(x) = \partial\varphi(x)/\partial x$
$\hat{\Delta} = -\partial^2/\partial x^2 = -\hat{D}^2$	$\hat{\Delta}\varphi(x) = -\partial^2\varphi(x)/\partial x^2$
$\hat{M} = \partial^2/\partial x\,\partial y$	$\hat{M}\varphi(x, y) = \partial^2\varphi(x, y)/\partial x\,\partial y$
$\hat{I}$ = operation that leaves φ unchanged	$\hat{I}\varphi = \varphi$
$\hat{Q} = \int_0^1 dx'$	$\hat{Q}\varphi(x) = \int_0^1 dx'\varphi(x')$
$\hat{F}$ = multiplication by $F(x)$	$\hat{F}\varphi(x) = F(x)\varphi(x)$
$\hat{B}$ = division by the number 3	$\hat{B}\varphi(x) = \frac{1}{3}\varphi(x)$
$\hat{\Theta}$ = operator that annihilates φ	$\hat{\Theta}\varphi = 0$
$\hat{P}$ = operator that changes φ to a specific polynomial of φ	$\hat{P}\varphi = \varphi^3 - 3\varphi^2 - 4$
$\hat{G}$ = operator that changes φ to the number 8	$\hat{G}\varphi = 8$

measurement of A yields values (call these measured values a) which are eigenvalues of $\hat{A}$. That is, the values, a, are those values for which the equation

(3.1) $$\hat{A}\varphi = a\varphi \qquad \boxed{\text{an eigenvalue equation}}$$

has a solution φ. The function φ is called the *eigenfunction* of $\hat{A}$ corresponding to the eigenvalue a.

Examples of mathematical operators, which are not necessarily connected to physics, are offered in Table 3.1. (Labels such as D, G, and M are of no special significance.) An operator operates on a function and makes it something else (except for the identity operator $\hat{I}$).

Let us now turn to operators that correspond to physical observables. Two very important such observables are the momentum and the energy.

The Momentum Operator $\hat{p}$

The operator that corresponds to the observable linear momentum is

(3.2) $$\hat{p} = -i\hbar\nabla$$

What are the eigenfunctions and eigenvalues of the momentum operator? Consider that the particle (whose momentum is in question) is constrained to move in one dimension (x). Then the momentum has only one nonvanishing component, p_x. The corresponding operator is

(3.3) $$\hat{p}_x = -i\hbar\frac{\partial}{\partial x}$$

The eigenvalue equation for this operator is

$$(3.4) \qquad -i\hbar \frac{\partial}{\partial x} \varphi = p_x \varphi$$

The values p_x represent the possible values that measurement of the x component of momentum will yield. The eigenfunction $\varphi(x)$ corresponding to a specific value of momentum (p_x) is such that $|\varphi|^2 \, dx$ is the probability of finding the particle (with momentum p_x) in the interval $x, x + dx$. Suppose we stipulate that the particle is a *free* particle. It is unconfined (along the x axis). For this case there is no boundary condition on φ and the solution to (3.4) is

$$(3.5) \qquad \varphi = A \exp\left(\frac{ip_x x}{\hbar}\right) = Ae^{ikx}$$

where we have labeled the wavenumber k and have deleted the subscript x.

$$(3.6) \qquad k = \frac{p}{\hbar}$$

The eigenfunction given by (3.5) is a periodic function (in x). To find its wavelength λ, we set

$$(3.7) \qquad e^{ikx} = e^{ik(x+\lambda)}$$
$$1 = e^{ik\lambda} = \cos k\lambda + i \sin k\lambda$$

which is satisfied if

$$(3.8) \qquad \cos k\lambda = 1$$
$$\sin k\lambda = 0$$

The first nonvanishing solution to these equations is

$$(3.9) \qquad k\lambda = 2\pi$$

which (with 3.6) is equivalent to the de Broglie relation

$$(3.10) \qquad p = \frac{h}{\lambda}$$

We conclude that the eigenfunction of the momentum operator corresponding to the eigenvalue p has a wavelength that is the de Broglie wavelength h/p.

   In quantum mechanics it is convenient to speak in terms of wavenumber k instead of momentum p. In this notation one says that the eigenfunctions and eigenvalues of the momentum operator are

$$(3.11) \qquad \varphi_k = Ae^{ikx}, \qquad p = \hbar k$$

The subscript k on φ_k denotes that there is a continuum of eigenfunction and eigenvalues, $\hbar k$, which yield nontrivial solutions to the eigenvalue equation, (3.4).

The Energy Operator $\hat{H}$

The operator corresponding to the energy is the Hamiltonian $\hat{H}$, with the momentum **p** replaced by its operator counterpart, $\hat{\mathbf{p}}$. For a single particle of mass m, in a potential field $V(\mathbf{r})$,

(3.12)
$$\hat{H} = \frac{\hat{p}^2}{2m} + V(\mathbf{r}) = -\frac{\hbar^2}{2m}\nabla^2 + V(\mathbf{r})$$

The eigenvalue equation for $\hat{H}$,

(3.13)
$$\hat{H}\varphi(\mathbf{r}) = E\varphi(\mathbf{r})$$

is called the *time-independent Schrödinger equation*. It yields the possible energies E which the particle may have. Again consider the free particle. The energy of a free particle is purely kinetic, so

(3.14)
$$\hat{H} = \frac{\hat{p}^2}{2m} = -\frac{\hbar^2}{2m}\nabla^2$$

Constraining the particle to move in one dimension, the time-independent Schrödinger equation becomes

(3.15)
$$-\frac{\hbar^2}{2m}\frac{\partial^2}{\partial x^2}\varphi = E\varphi$$

In terms of the wave vector

(3.16)
$$k^2 = \frac{2mE}{\hbar^2}$$

(3.15) appears as

(3.17)
$$\varphi_{xx} + k^2\varphi = 0$$

The subscript x denotes differentiation. For a free particle there are no boundary conditions and we obtain[1]

(3.18)
$$\varphi = Ae^{ikx} + Be^{-ikx}$$

This is the eigenfunction of $\hat{H}$ which corresponds to the energy eigenvalue

(3.19)
$$E = \frac{\hbar^2k^2}{2m}$$

---

[1] The solution to (3.17) with boundary conditions imposed is discussed in Section 4.1.

We have found above (3.11) that the momentum of a free particle is $\hbar k$. This is clearly the same $\hbar k$ that appears in (3.19), since for a free particle

$$(3.20) \qquad E = \frac{p^2}{2m} = \frac{\hbar^2 k^2}{2m}$$

Note also that the eigenfunction of $\hat{H}$ (3.18), with $B = 0$, is also an eigenfunction of $\hat{p}$ (3.11). That $\hat{H}$ and $\hat{p}$ for a free particle have common eigenfunctions is a special case of a more general theorem to be discussed later.[1] The following simple argument demonstrates this fact. Let

$$(3.21) \qquad \hat{p}\varphi = \hbar k \varphi$$

Let us see if φ is also an eigenfunction of $\hat{H}$ (for a free particle).

$$(3.22) \qquad \hat{H}\varphi = \frac{\hat{p}}{2m}(\hat{p}\varphi) = \frac{\hat{p}(\hbar k \varphi)}{2m} = \frac{\hbar k}{2m}\hat{p}\varphi$$
$$= \frac{(\hbar k)^2}{2m}\varphi$$

It follows that φ is also an eigenfunction of $\hat{H}$.

Both the energy and momentum eigenvalues for the free particle comprise a continuum of values:

$$(3.23) \qquad \boxed{E = \frac{\hbar^2 k^2}{2m} \qquad p = \hbar k}$$

That is, these are valid eigenvalues for *any* wavenumber k. The eigenfunction (of both $\hat{H}$ and $\hat{p}$) corresponding to these eigenvalues is

$$(3.24) \qquad \varphi_k = A e^{ikx}$$

If the free particle is in this state, measurement of its momentum will definitely yield $\hbar k$, and measurement of its energy will definitely yield $(\hbar^2 k^2/2m)$.

Suppose that we measure its position x; what do we find? Well, where is the particle most likely to be? Again we call on the Born postulate. If the particle is in the state φ_k, the probability density relating to the probability of finding the particle in the interval $x, x + dx$, is

$$(3.25) \qquad |\varphi_k|^2 = |A|^2 = \text{constant}$$

[1] The commutator theorem, Chapter 5.

The probability density is the same constant value for all x. That means we would be equally likely to find the particle at any point from $x = -\infty$ to $x = +\infty$. This is a statement of maximum uncertainty which is in agreement with the Heisenberg uncertainty principle. In the state φ_k, it is known with absolute certainty that measurement of momentum yields $\hbar k$. Therefore, for the state φ_k, $\Delta p = 0$, whence $\Delta x = \infty$.

We mentioned in Section 2.7 that E and t are complementary variables; that is, they obey the relation $\Delta E \, \Delta t \geq \hbar$. Specifically, this means that if the energy is uncertain by amount ΔE, *the time it takes to measure E* is uncertain by $\Delta t \geq \hbar/\Delta E$. Now for the problem at hand, in the state φ_k, it is certain that measurement of E yields $\hbar^2 k^2/2m$. Therefore, $\Delta E = 0$. To measure E we have to let the particle interact with some sort of energy-measuring apparatus, say a plate with a spring attached to measure the momentum imparted to the plate when the particle hits it head on. Well, if the plate with attached spring is placed in the path of the particle, how long must we wait before we detect something? We can wait 10^{-8} s—or we can wait 10^{10} yr. The uncertainty Δt is infinite in the present case, since there is an infinite uncertainty in x.

PROBLEMS

3.1   For each of the operators listed in Table 3.1 ($\hat{D}$, $\hat{A}$, $\hat{M}$, etc.), construct the square, that is, $\hat{D}^2$, $\hat{A}^2$, ...

Answer (*partial*)

$$\hat{I}^2\varphi = \hat{I}\varphi = \varphi$$

$$\hat{Q}\varphi = \hat{Q}\int_0^1 dx'\varphi(x') = \int_0^1 dx'' \int_0^1 dx'\varphi(x')$$

$$\hat{F}^2\varphi = F^2\varphi$$

$$\hat{B}^2\varphi = \frac{1}{9}\varphi$$

$$\hat{P}^2\varphi = \hat{P}(\hat{P}\varphi) = (\varphi^3 - 3\varphi^2 - 4)^3 - 3(\varphi^3 - 3\varphi^2 - 4)^2 - 4$$

3.2   The inverse of an operator $\hat{A}$ is written $\hat{A}^{-1}$. It is such that

$$\hat{A}^{-1}\hat{A}\varphi = \hat{I}\varphi = \varphi$$

Construct the inverses of $\hat{D}$, $\hat{I}$, $\hat{F}$, $\hat{B}$, $\hat{\Theta}$, $\hat{G}$, provided that such inverses exist.

3.3   An operator $\hat{O}$ is *linear* if

$$\hat{O}(a\varphi_1 + b\varphi_2) = a\hat{O}\varphi_1 + b\hat{O}\varphi_2$$

where a and b are arbitrary constants. Which of the operators in Table 3.1 are linear and which are nonlinear?

3.4   The displacement operator $\hat{\mathcal{D}}$ is defined by the equation

$$\hat{\mathcal{D}} f(x) = f(x + \zeta)$$

Show that the eigenfunctions of $\hat{\mathcal{D}}$ are of the form

$$\varphi_\beta = e^{\beta x} g(x)$$

where

$$g(x + \zeta) = g(x)$$

and β is any complex number. What is the eigenvalue corresponding to φ_β?

3.5   An electron moves in the x direction with de Broglie wavelength 10^{-8} cm.
   (a)   What is the energy of the electron (in eV)?
   (b)   What is the time-independent wavefunction of the electron?

3.2   MEASUREMENT IN QUANTUM MECHANICS

Postulate II

The second postulate[1] of quantum mechanics is, measurement of the observable A that yields the value a leaves the system in the state φ_a, where φ_a is the eigenfunction of $\hat{A}$ that corresponds to the eigenvalue a.

   As an example, suppose that a free particle is moving in one dimension. We do not know which state the particle is in. At a given instant we measure the particle's momentum and find the value $p = \hbar k$ (with k a specific value, say 1.3×10^{10} cm^{-1}). This measurement[2] leaves the particle in the state φ_k, so immediate subsequent measurement of p is certain to yield $\hbar k$.

   Suppose that one measures the position of a free particle and the position $x = x'$ is measured. The first two postulates tell us the following. (1) There is an operator corresponding to the measurement of position, call it $\hat{x}$. (2) Measurement of x that yields the value x' leaves the particle in the eigenfunction of $\hat{x}$ corresponding to the eigenvalue x'.

   The operator equation appears as

(3.26)
$$\hat{x}\delta(x - x') = x'\delta(x - x')$$

---

[1] This postulate has been the source of some discussion among physicists. For further reference, see B. S. DeWitt, *Phys. Today* **23**, 30 (September 1970).

[2] Measurement is taken in the idealized sense. More formal discussion on the theory of measurement may be found in K. Gottfried, *Quantum Mechanics,* W. A. Benjamin, New York, 1966; J. Jauch, *Foundations of Quantum Mechanics,* Addison-Wesley, Reading, Mass., 1968; and E. C. Kemble, *The Fundamental Principles of Quantum Mechanics with Elementary Applications,* Dover, New York, 1958.

Dirac Delta Function

The eigenfunction of $\hat{x}$ has been written[1] $\delta(x - x')$ and is called the *Dirac delta function*. It is defined in terms of the following two properties. The first are the integral properties

$$\int_{-\infty}^{\infty} f(x')\delta(x - x')\,dx' = f(x)$$

(3.27)
$$\int_{-\infty}^{\infty} \delta(x - x')\,dx' = 1$$

or equivalently, in terms of the single variable y

$$\int_{-\infty}^{\infty} f(y)\delta(y)\,dy = f(0)$$

(3.28)
$$\int_{-\infty}^{\infty} \delta(y)\,dy = 1$$

The second defining property is the value

(3.29) $\delta(y) = 0$ (for $y \neq 0$)

A sketch of $\delta(y)$ is given in Fig. 3.1. Properties of $\delta(y)$ are usually proved with the aid of the defining integral (3.27). For instance, consider the relation

(3.30) $y\delta'(y) = -\delta(y)$

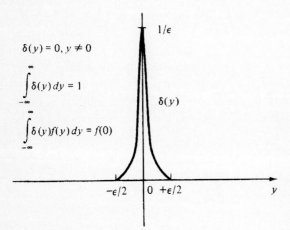

$\delta(y) = 0, y \neq 0$

$\int_{-\infty}^{\infty} \delta(y)\,dy = 1$

$\int_{-\infty}^{\infty} \delta(y)f(y)\,dy = f(0)$

$1/\epsilon$

$\delta(y)$

$-\epsilon/2$ 0 $+\epsilon/2$ y

FIGURE 3.1   Dirac delta function $\delta(y)$. The curve is distorted to bring out essential features. A more accurate picture is obtained in the limit $\epsilon \to 0$.

[1] More accurately one says that $\delta(x - x')$ is an eigenfunction of $\hat{x}$ in the coordinate representation. This topic is returned to in Section 7.4 and in Appendix A.

To establish this relation we employ a *test function* $f(y)$ and perform the following integration by parts.

$$(3.31) \quad \int_{-\infty}^{\infty} f(y) y \delta'(y) \, dy = \int_{-\infty}^{\infty} \frac{d}{dy} (fy\delta) \, dy - \int_{-\infty}^{\infty} \delta \frac{d}{dy} (yf) \, dy$$

$$= -\int_{-\infty}^{\infty} \delta(y) \left(y \frac{df}{dy} + f \right) dy$$

$$= -\int_{-\infty}^{\infty} \delta(y) f(y) \, dy$$

which establishes (3.30).

The student should not lose sight of the fact that $\hat{x}$, when operating on a function $f(x)$, merely represents multiplication by x. For example, $\hat{x}f(x) = xf(x)$. These topics will be returned to in Chapter 11 and discussed further in Appendix A.

PROBLEMS

3.6 Establish the following properties of $\delta(y)$:

(a) $\delta(y) = \delta(-y)$

(b) $\delta'(y) = -\delta'(-y)$

(c) $y\delta(y) = 0$

(d) $\delta(ay) = |a|^{-1}\delta(y)$

(e) $\delta(y^2 - a^2) = |2a|^{-1}[\delta(y - a) + \delta(y + a)]$

(f) $\displaystyle\int_{-\infty}^{\infty} \delta(a - y)\delta(y - b) \, dy = \delta(a - b)$

(g) $f(y)\delta(y - a) = f(a)\delta(y - a)$

(h) $y\,\delta'(y) = -\delta(y)$

(i) $\displaystyle\int g(y)\delta[f(y) - a] \, dy = \left. \frac{g(y)}{|df/dy|} \right|_{\substack{y=y_0 \\ f(y_0)=a}}$

3.7 Show that the following are valid representations of $\delta(y)$:

(a) $2\pi\delta(y) = \displaystyle\int_{-\infty}^{\infty} e^{iky} \, dk$

(b) $\pi\delta(y) = \displaystyle\lim_{\eta \to \infty} \frac{\sin \eta y}{y}$

Note: In mathematics, an object such as $\delta(y)$, which is defined in terms of its integral properties, is called a *distribution*. Consider all $\chi(y)$ defined on the interval $(-\infty, \infty)$ for which

$$\int_{-\infty}^{\infty} |\chi(y)|^2 \, dy < \infty$$

Then two distributions, δ_1 and δ_2, are *equivalent* if for all $\chi(y)$.

$$\int_{-\infty}^{\infty} \chi \, \delta_1 \, dy = \int_{-\infty}^{\infty} \chi \, \delta_2 \, dy$$

When one establishes that a mathematical form such as $\int_{-\infty}^{\infty} \exp{(iky)} \, dy$ is a representation of $\delta(y)$, one is in effect demonstrating that these two objects are *equivalent as distributions*.

3.8   Show that the continuous set of eigenfunctions $\{\delta(x - x')\}$ obeys the "orthonormality" condition

$$\int_{-\infty}^{\infty} \delta(x - x')\delta(x - x'') \, dx = \delta(x' - x'')$$

3.9   (a)   Show that $\delta(\sqrt{x}) = 0$.
      (b)   Evaluate $\delta(\sqrt{x^2 - a^2})$.

3.3   THE STATE FUNCTION AND EXPECTATION VALUES

Postulate III

The third postulate of quantum mechanics establishes the existence of the state function and its relevance to the properties of a system: The state of a system at any instant of time may be represented by a state or wave function ψ which is continuous and differentiable. All information regarding the state of the system is contained in the wavefunction. Specifically, if a system is in the state $\psi(\mathbf{r}, t)$, the average of any physical observable C relevant to that system at time t is

$$(3.32) \qquad \langle C \rangle = \int \psi^* \hat{C} \psi \, d\mathbf{r}$$

(The differential of volume is written $d\mathbf{r}$.) The average, $\langle C \rangle$, is called the *expectation value* of C.

The physical meaning of the average of an observable C involves the following type of (conceptual) measurements. The observable C is measured in a specific experiment, X. One prepares a very large number (N) of identical replicas of X. The initial states $\psi(\mathbf{r}, 0)$ in each such replica are all identical. At the time t, one measures

C in all these replica experiments and obtains the set of values $C_1, C_2, \ldots, C_N$. The average of C is then given by the rule

$$(3.33) \qquad \langle C \rangle = \frac{1}{N} \sum_{i=1}^{N} C_i \qquad (N \gg 1)$$

The postulate stated above claims that this experimentally calculated average (3.33) is the same as that given by the integral in (3.32). Another way of defining $\langle C \rangle$ is in terms of the probability $P(C_i)$. This function gives the probability that measurement of C finds the value C_i. For $\langle C \rangle$, we then have

$$(3.34) \qquad \langle C \rangle = \sum_{\text{all } C} C_i P(C_i)$$

This is a consistent formula if all the values C may assume comprise a discrete set (e.g., the number of marbles in a box). In the event that the values that C may assume comprise a continuous set (e.g., the values of momentum of a free particle), $\langle C \rangle$ becomes

$$(3.35) \qquad \langle C \rangle = \int C P(C) \, dC$$

The integration is over all values of C. Here $P(C) \, dC$ is the probability of finding C in the interval $C, C + dC$.

The quantity $\langle C \rangle$ is also called the *expectation value* of C because it is representative of the value one expects to obtain in any given measurement of C. This will be especially true if the deviation of values of C from the mean value $\langle C \rangle$ is not large. As discussed in Section 2.7, a measure of this spread of values about the value $\langle C \rangle$ is given by the mean-square deviation ΔC, defined through

$$(3.36) \qquad (\Delta C)^2 = \langle (C - \langle C \rangle)^2 \rangle = \langle C^2 \rangle - \langle C \rangle^2$$

In order to become familiar with the operational use of postulate III, we work out the following one-dimensional problem. A particle is known to be in the state

$$(3.37) \qquad \psi(x, t) = A \exp \left[\frac{-(x - x_0)^2}{4a^2} \right] \exp \left(\frac{ip_0 x}{\hbar} \right) \exp \left(-i\omega_0 t \right)$$

The lengths x_0 and a are constants, as are the momentum p_0 and frequency ω_0. The (real) constant A is determined through normalization. This then ensures that $\psi^* \psi$ is a numerically correct probability density.

$$\int_{-\infty}^{\infty} |\psi|^2 \, dx = A^2 a \int_{-\infty}^{\infty} e^{-\eta^2/2} \, d\eta = \sqrt{2\pi} \, A^2 a = 1$$

(3.38)
$$A^2 = \frac{1}{a\sqrt{2\pi}}$$

The nondimensional "dummy" variable η and constant η_0 are such that

$$\eta = \frac{x - x_0}{a}$$

(3.39)
$$x = a(\eta + \eta_0)$$

$$\eta_0 = \frac{x_0}{a}$$

Having obtained A, we may now calculate the expectation of x:

(3.40) $$\langle x \rangle = \int_{-\infty}^{\infty} \psi^* \hat{x} \psi \, dx = \int_{-\infty}^{\infty} \psi^* x \psi \, dx$$

$$= A^2 a^2 \int_{-\infty}^{\infty} e^{-\eta^2/2}(\eta + \eta_0) \, d\eta = a\eta_0 \left(aA^2 \int_{-\infty}^{\infty} e^{-\eta^2/2} \, d\eta \right)$$

which, with the normalization condition (3.38), gives

(3.41) $$\langle x \rangle = a\eta_0 = x_0$$

[Note that integration of the odd integrand $\eta \exp(-\eta^2)$ in (3.40) vanishes.] That x_0 is the proper value for $\langle x \rangle$ is evident from the sketch of $|\psi|^2$ shown in Fig. 3.2.
     If we call

(3.42) $$|\psi|^2 \, dx = P(x) \, dx$$

the probability of finding the particle in the interval $x, x + dx$, then

(3.43) $$\langle x \rangle = \int_{-\infty}^{\infty} x P(x) \, dx$$

This is consistent with definition (3.35).
     The probability density

$$P(x) = \frac{1}{a\sqrt{2\pi}} \exp \left[\frac{-(x - x_0)^2}{2a^2} \right]$$

is called the *Gaussian* or *normal distribution*, and a^2 is called the *variance* of x. It is a measure of the spread of $P(x)$ about the mean value

$$\langle x \rangle = x_0$$

$$|\psi|^2 \, dx = P(x) \, dx$$

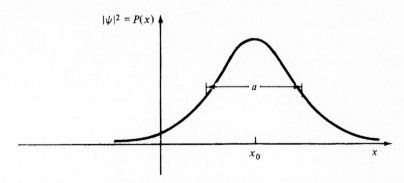

FIGURE 3.2   Gaussian probability density with variance a^2. The variance measures the spread of $P(x)$ about the mean, $\langle x \rangle = x_0$. In quantum mechanics the square root of variance is called the uncertainty in x and is denoted as Δx, so for the case under discussion,

$$a = \Delta x = \sqrt{\langle x^2 \rangle - \langle x \rangle^2}$$

As shown in Problem 3.10, the variance of x is the same as the mean-square deviation, $(\Delta x)^2$.

$$(\Delta x)^2 = \langle x^2 \rangle - \langle x \rangle^2 = a^2 + x_0{}^2 - x_0{}^2 = a^2$$

If it is known that a particle is in the state $\psi(x)$ at a given instant of time, and that in this state $\langle x \rangle = x_0$, one may then ask: With what certainty will measurement of x find the value x_0? A measure of the relative *uncertainty* is given by the square root of the variance, Δx. If this value is large (compared to $\langle x \rangle$), one may say with little certainty that measurement will find the particle at x_0. If, on the other hand, Δx is small, one is more certain that measurement will find the particle at $x = x_0$. In quantum mechanics Δx is called the *uncertainty in x,* introduced previously in Section 2.7.

Next, we calculate the expectation of the momentum for a particle in the state ψ, (3.37).

$$(3.44) \quad \langle p \rangle = \int_{-\infty}^{\infty} \psi^* \hat{p} \psi \, dx = \int_{-\infty}^{\infty} \psi^* \left(-i\hbar \frac{\partial}{\partial x} \right) \psi \, dx$$

$$= A^2 a \int_{-\infty}^{\infty} \left(p_0 + \frac{i\hbar}{2a} \eta \right) e^{-\eta^2/2} \, d\eta = p_0 \left(A^2 a \int_{-\infty}^{\infty} e^{-\eta^2/2} \, d\eta \right)$$

$$= p_0$$

It follows that the parameter p_0 which appears in the state function ψ is the average value of p. Note in particular that the values, $\langle x \rangle = x_0$ and $\langle p \rangle = p_0$, refer to a specific time, t. As time increases beyond t, the "Gaussian" wavefunction (3.37) propagates and, because of various momentum components of this wavefunction, disperses. For further discussion, see (6.40).

PROBLEMS

3.10   For the state ψ, given by (3.37), show that

$$(\Delta x)^2 = a^2$$

Argue the consistency of this conclusion with the change in shape that $|\psi|^2$ suffers with a change in the parameter a.

3.11   Calculate the uncertainty Δp for a particle in the state ψ given by (3.37). Do you find your answer to be consistent with the uncertainty principle? (In this problem one must calculate $\langle \hat{p}^2 \rangle$. The operator $\hat{p}^2 = -\hbar^2 \, \partial^2/\partial x^2$.)

3.12   Let s be the number of spots shown by a die thrown at random.
   (a)   Calculate $\langle s \rangle$.
   (b)   Calculate Δs.

3.13   The number of hairs (N_1) on a certain rare species can only be the number 2^l $(l = 0, 1, 2, \ldots)$. The probability of finding such an animal with 2^l hairs is $e^{-1}/l!$ What is the expectation, $\langle N \rangle$? What is ΔN?

3.4   TIME DEVELOPMENT OF THE STATE FUNCTION

Postulate IV

The fourth postulate of quantum mechanics specifies the time development of the state function $\psi(\mathbf{r}, t)$: the state function for a system (e.g., a single particle) develops in time according to the equation

$$(3.45) \qquad i\hbar \, \frac{\partial}{\partial t} \, \psi(\mathbf{r}, t) = \hat{H}\psi(\mathbf{r}, t)$$

This equation is called the *time-dependent Schrödinger* equation.[1] The operator $\hat{H}$ is the Hamiltonian operator. For a single particle of mass m, in a potential field $V(\mathbf{r})$, it is given by (3.12). If $\hat{H}$ is assumed to be independent of time, we may write

$$(3.46) \qquad \hat{H} = \hat{H}(\mathbf{r})$$

Under these circumstances, one is able to construct a solution to the time-dependent Schrödinger equation through the technique of separation of variables. We assume a solution of the form

$$(3.47) \qquad \psi(\mathbf{r}, t) = \varphi(\mathbf{r})T(t)$$

---

[1] A formulation of the Schrödinger equation that has its origin in the classical principle of least action has been offered by R. P. Feynman, *Rev. Mod. Phys.* **60**. 367 (1948). An elementary description of this derivation may be found in S. Borowitz, *Quantum Mechanics,* W. A. Benjamin. New York, 1967.

Substitution into (3.45) gives

$$(3.48) \qquad i\hbar \frac{T_t}{T} = \frac{\hat{H}\varphi}{\varphi}$$

The subscript t denotes differentiation with respect to t. Equation (3.48) is such that the left-hand side is a function of t only, while the right-hand side is a function of $\mathbf{r}$ only. Such an equation can be satisfied only if both sides are equal to the same constant, call it E (we do not yet know that E is the energy).

$$(3.49) \qquad \hat{H}\varphi(\mathbf{r}) = E\varphi(\mathbf{r})$$

$$(3.50) \qquad \left(\frac{\partial}{\partial t} + \frac{iE}{\hbar} \right) T(t) = 0$$

The first of these equations is the time-independent Schrödinger equation (3.13). This identification serves to label E, in (3.49), the energy of the system. That is, E, as it appears in this equation, is an eigenvalue of $\hat{H}$. But the eigenvalues of $\hat{H}$ are the allowed energies a system may assume, and we again conclude that E is the energy of the system.

The second equation (3.50) is simply solved to give the oscillating form

$$(3.51) \qquad T(t) = A \exp\left(-\frac{iEt}{\hbar} \right)$$

Suppose that we solve the time-independent Schrödinger equation and obtain the eigenfunctions and eigenvalues

$$(3.52) \qquad \hat{H}\varphi_n = E_n\varphi_n$$

For each such eigensolution, there is a corresponding eigensolution to the time-dependent Schrödinger equation

$$(3.53) \qquad \psi_n(\mathbf{r}, t) = A\varphi_n(\mathbf{r}) \exp\left(-\frac{iE_n t}{\hbar} \right)$$

In equations (3.52) and (3.53) the index n denotes the set of integers $n = 1, 2, \ldots$. This notation is appropriate to the case where the solution to the time-independent Schrödinger equation gives a discrete set of eigenfunctions, $\{\varphi_n\}$. Such is the case for problems that pertain to a finite system, such as a particle confined to a finite domain of space. We will encounter this property in Chapter 4 when we solve the problem of a bead constrained to move on a straight wire strung between two impenetrable walls.

In the one-dimensional free-particle case treated in Section 3.2, one obtains a continuum of eigenfunctions $\varphi_k(x)$ and, correspondingly, a continuum of eigenvalues, E_k. To repeat, these values are

$$(3.54) \qquad\qquad \hat{H}\varphi_k = E_k\varphi_k$$

$$(3.55) \qquad\qquad \varphi_k = A \exp(ikx), \qquad E_k = \frac{\hbar^2 k^2}{2m}$$

For each such time-independent solution, there is a solution to the time-dependent Schrödinger equation

$$(3.56) \qquad\qquad \psi_k(x, t) = A e^{i(kx - \omega t)}$$

where we have labeled

$$(3.57) \qquad\qquad \hbar\omega = E_k$$

The structure of the solution (3.56) is characteristic of a propagating wave. More generally, any function of x and t of the form

$$(3.58) \qquad\qquad f(x, t) = f(x - vt)$$

represents a wave propagating in the positive x direction with velocity v. To see this, we note the following property of f:

$$(3.59) \qquad\qquad f(x + v\,\Delta t, t + \Delta t) = f(x, t)$$

At any given instant t, one may plot the x dependence of f (Fig. 3.3). If t increases to $t + \Delta t$, this curve is displaced to the right (as a rigid body) by the amount $v\,\Delta t$. We conclude from these arguments that the disturbance f (3.58) propagates with the wave speed v.

Now let us return to the free-particle eigenstate, (3.56), and rewrite it in the form

$$(3.60) \qquad\qquad \psi_k(x, t) = A \exp\left[ik\left(x - \frac{\omega}{k}t\right)\right]$$

Comparison with the waveform (3.58) indicates that (1) ψ_k is a propagating wave (moving to the right), and (2) the speed of this wave is

$$(3.61) \qquad\qquad v = \frac{\omega}{k} = \frac{\hbar\omega}{\hbar k} = \frac{p^2/2m}{p} = \frac{p}{2m} = \frac{v_{CL}}{2}$$

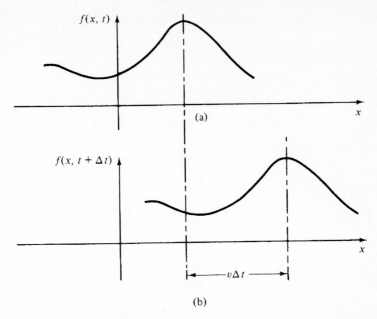

FIGURE 3.3   **Propagating wave,** $f(x, t) = f(x - vt)$**: (a) at time** t**; (b) at time** $t + \Delta t$**.**

The velocity v_{CL} represents the classical velocity of a particle of mass m and momentum p. Thus we find that the wave speed of the state function of a particle with well-defined momentum, $p = \hbar k$, is half the classical speed, $v_{CL} = p/m$.

This discrepancy is due to the following fact. Suppose that we calculate the probability density corresponding to the state given in (3.56). We obtain the result that it is uniformly probable to find the particle anywhere along the x axis. This is not a typical classical property of a particle. The state function that better represents a classical (localized) particle is a wave packet. The shape of such a function is sketched in Fig. 3.4. Such a state may be constructed as a sum of eigenstates of the form given

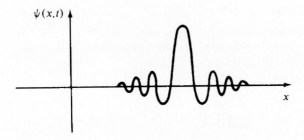

FIGURE 3.4   Wave packet at a given instant of time t**.**

in (3.56) (a Fourier series). The velocity with which the packet moves is called the *group velocity*,[1]

(3.62)
$$v_g = \frac{\partial \omega}{\partial k}$$

For a wave packet composed of free-particle eigenstates, v_g takes the value

(3.63)
$$v_g = \frac{\partial \hbar \omega}{\hbar \, \partial k} = \frac{\partial (\hbar^2 k^2 / 2m)}{\hbar \, \partial k} = \frac{\hbar k}{m} = \frac{p}{m}$$
$$= v_{CL}$$

The value of k that enters the formula for v_g is the value about which there is a superabundance of ψ_k component waves. These topics will be more fully developed in Chapter 4. For the moment we are concerned only with the identification given in (3.63).

PROBLEMS

3.14   Describe the evolution in time of the following wavefunctions:

$$\psi_1 = A \sin \omega t \cos k(x + ct)$$
$$\psi_2 = A \sin (10^{-5} \, kx) \cos k(x - ct)$$
$$\psi_3 = A \cos k(x - ct) \sin [10^{-5} k(x - ct)]$$

3.15   What is the expectation of momentum $\langle p \rangle$ for a particle in the state

$$\psi(x, t) = A e^{-(x/a)^2} e^{-i\omega t} \sin kx?$$

3.5   SOLUTION TO THE INITIAL-VALUE PROBLEM IN QUANTUM MECHANICS

Functions of Operators

The time-dependent Schrödinger equation permits solution of the initial-value problem: given the initial value of the state function $\psi(\mathbf{r}, 0)$, determine $\psi(\mathbf{r}, t)$. We will formulate the solution to the problem for a time-independent Hamiltonian. The more general case is given as an exercise (Problem 3.18).

First we rewrite (3.45) in the form

(3.64)
$$\frac{\partial}{\partial t} \psi(\mathbf{r}, t) + \frac{i\hat{H}}{\hbar} \psi(\mathbf{r}, t) = 0$$

---

[1] The concepts of phase and group velocities are returned to in Section 6.1.

Next, we multiply this equation (from the left) by the integrating factor $\hat{U}^{-1}$

$$(3.65) \qquad \hat{U}^{-1} = \exp\left(\frac{it\hat{H}}{\hbar}\right)$$

which is the inverse of

$$(3.66) \qquad \hat{U} \equiv \exp\left(-\frac{it\hat{H}}{\hbar}\right)$$

This function of the operator, $\hat{H}$, is itself an operator. It is defined in terms of its Taylor series expansion.

$$(3.67) \qquad \hat{U}^{-1} = \exp\left(\frac{it\hat{H}}{\hbar}\right) = 1 + \frac{it\hat{H}}{\hbar} + \frac{1}{2!}\left(\frac{it\hat{H}}{\hbar}\right)^2 + \cdots$$

More generally for any operator $\hat{A}$, the function operator $f(\hat{A})$ is defined in terms of a series in powers of $\hat{A}$. A few examples are provided in the problems.

Let us return to the problem under discussion. Multiplying the time-dependent Schrödinger equation through by the integrating factor (3.65), one obtains the equation

$$(3.68) \qquad \frac{\partial}{\partial t}\left[\exp\left(\frac{it\hat{H}}{\hbar}\right)\psi(\mathbf{r}, t)\right] = 0$$

Integrating over the time interval $(0, t)$ gives

$$(3.69) \qquad \exp\left(\frac{it\hat{H}}{\hbar}\right)\psi(\mathbf{r}, t) - \psi(\mathbf{r}, 0) = 0$$

Multiplying this equation through by $\hat{U}$ gives the desired result:

$$(3.70) \qquad \boxed{\psi(\mathbf{r}, t) = \exp\left(-\frac{it\hat{H}}{\hbar}\right)\psi(\mathbf{r}, 0) = \hat{U}\psi(\mathbf{r}, 0)}$$

Here we have used the fact that

$$(3.71) \qquad \hat{U}\hat{U}^{-1} = \exp\left(-\frac{it\hat{H}}{\hbar}\right)\exp\left(\frac{it\hat{H}}{\hbar}\right) = \hat{I}$$

where $\hat{I}$ is the identity operator.

Suppose that in solution (3.70) we choose the initial state to be an eigenstate of $\hat{H}$. Call it φ_n, so that

$$(3.72) \qquad \psi_n(\mathbf{r}, 0) = \varphi_n(\mathbf{r})$$
$$\hat{H}\varphi_n = E_n\varphi_n$$

By virtue of the theorem presented in Problem 3.16,

$$\psi_n(\mathbf{r}, t) = \exp\left(-\frac{it\hat{H}}{\hbar}\right)\varphi_n = \exp\left(-\frac{iE_n t}{\hbar}\right)\varphi_n$$

(3.73)
$$= e^{-i\omega_n t}\varphi_n(\mathbf{r})$$

$$\hbar\omega_n = E_n$$

This is the solution of the time-dependent Schrödinger equation, derived in Section 3.4 by the technique of separation of variables. The solution given in (3.70) is more general. It exhibits the development of an arbitrary initial state $\psi(\mathbf{r}, 0)$ in time. It will be used extensively in the chapters to follow, where the student will gain a more workable understanding of the equation.

As a final topic of discussion in this chapter we note the following. Suppose that a system is in an eigenstate of the Hamiltonian at $t = 0$, described by (3.72). At this (initial) time the expectation of an observable A is

(3.74)
$$\langle A \rangle_{t=0} = \int \psi^*(\mathbf{r}, 0)\hat{A}\psi(\mathbf{r}, 0)\, d\mathbf{r} = \int \varphi_n^* \hat{A}\varphi_n\, d\mathbf{r}$$

What is $\langle A \rangle$ at a later time, $t > 0$? The state of the system at $t > 0$ is given by (3.73):

(3.75)
$$\psi_n(\mathbf{r}, t) = e^{-i\omega_n t}\varphi_n(\mathbf{r})$$

so that at $t > 0$ (assuming that $\partial\hat{A}/\partial t = 0$),

$$\langle A \rangle_t = \int \psi^*(\mathbf{r}, t)\hat{A}\psi(\mathbf{r}, t)\, d\mathbf{r} = e^{+i\omega_n t}e^{-i\omega_n t}\int \varphi_n^* \hat{A}\varphi_n\, d\mathbf{r}$$

(3.76)
$$= \int \varphi_n^* \hat{A}\varphi_n\, d\mathbf{r} = \langle A \rangle_{t=0}$$

$$\boxed{\langle A \rangle_{t>0} = \langle A \rangle_{t=0} \qquad \text{in a stationary state}}$$

The expectation of *any* observable is constant in time, if at any instant in time the system is in an eigenstate of the Hamiltonian. For this reason eigenstates of the Hamiltonian are called *stationary states*.

(3.77)
$$\boxed{\psi_n(\mathbf{r}, t) = e^{-i\omega_n t}\varphi_n(\mathbf{r}) \qquad \text{a stationary state}}$$

In the first three sections of this chapter we encountered functions relevant to a system which are eigenfunctions of operators corresponding to observable properties of that same system. In what sense are these eigenfunctions related to the *state function* of the system? From postulate II we know that ideal measurement of A leaves the system in the eigenstate of $\hat{A}$ corresponding to the value of A that was found in measurement. Thus, the state function of the system immediately after measurement is this same eigenstate of $\hat{A}$. The state function then evolves in time according to (3.70).

PROBLEMS

3.16 Let the eigenfunctions and eigenvalues of an operator $\hat{A}$ be $\{\varphi_n\}$ and $\{a_n\}$, respectively, so that

$$A\varphi_n = a_n\varphi_n$$

Let the function $f(x)$ have the expansion

$$f(x) = \sum_{t=0}^{\infty} b_t x^t$$

Show that φ_n is an eigenfunction of $f(\hat{A})$ with eigenvalue $f(a_n)$. That is,

$$f(\hat{A})\varphi_n = f(a_n)\varphi_n$$

3.17 If $\hat{p}$ is the momentum operator in the x direction, and $f(x)$ is an arbitrary "well-behaved" function, show that

$$\exp\left(\frac{i\zeta\hat{p}}{\hbar}\right) f(x) = f(x + \zeta)$$

The constant ζ represents a small displacement. In this problem the student must demonstrate that the left-hand side of the equation above is the Taylor series expansion of the right-hand side about $\zeta = 0$.

3.18 If $\hat{H}$ is an explicit function of time, show that the solution to the initial-value problem (by direct differentiation) is

$$\psi(\mathbf{r}, t) = \exp\left[-\frac{i}{h}\int_0^t dt' \hat{H}(t')\right]\psi(r, 0)$$

You may assume that $\hat{H}(t)\hat{H}(t') = \hat{H}(t')\hat{H}(t)$.

3.19 What is the effect of operating on an arbitrary function $f(x)$ with the following two operators?

(a) $\hat{O}_1 \equiv (\partial^2/\partial x^2) - 1 + \sin^2(\partial^3/\partial x^3) + \cos^2(\partial^3/\partial x^3)$

(b) $\hat{O}_2 \equiv \cos(2\partial/\partial x) + 2\sin^2(\partial/\partial x) + \int_a^b dx$

3.20 (a) The time-dependent Schrödinger equation is of the form

$$a\frac{\partial\psi}{\partial t} = \hat{H}\psi$$

Consider that a is an unspecified constant. Show that this equation has the following property. Let $\hat{H}$ be the Hamiltonian of a system composed of two independent parts, so that

$$\hat{H}(x_1, x_2) = \hat{H}_1(x_1) + \hat{H}_2(x_2)$$

and let the stationary states of system 1 be $\psi_1(x_1, t)$ and those of system 2 be $\psi_2(x_2, t)$. Then the stationary states of the composite system are

$$\psi(x_1, x_2) = \psi_1(x_1, t)\psi_2(x_2, t)$$

That is, show that this product form is a solution to the preceding equation for the given composite Hamiltonian.

Such a system might be two beads that are invisible to each other and move on the same straight wire. The coordinate of bead 1 is x_1 and the coordinate of bead 2 is x_2.

(b)   Show that this property is not obeyed by a wave equation that is second order in time, such as

$$a^2 \frac{\partial^2 \psi}{\partial t^2} = \hat{H}\psi$$

(c)   Arguing from the Born postulate, show that the wavefunction for a system composed of two independent components must be in the preceding product form, thereby disqualifying the wave equation in part (b) as a valid equation of motion for the wavefunction ψ.

Answer (partial)

(c)   If the two components are independent of each other, the joint probability density describing the state of the system is given by

$$P_{12} = P_1 P_2$$

This, in turn, guarantees that the probability density associated with component 1,

$$P_1(x_1) = \int P_{12}(x_1, x_2)\, dx_2$$

is independent of the form of $P_2(x_2)$ (and vice versa). The product form for P_{12} is guaranteed by the product structure for the wavefunction $\psi(x_1, x_2)$.

3.21   It is established in Problem 3.20 that for the joint probability for two independent systems to be consistently described by the time-dependent Schrödinger equation, this equation must be of the form

$$a \frac{\partial \psi}{\partial t} = \hat{H}\psi$$

where a is some number. Show that for this equation to imply wave motion, a must be complex. You may assume that $\hat{H}$ has only real eigenvalues.

Answer

Following development of the general solution (3.70), we find that the given equation implies the solution

$$\psi(\mathbf{r}, t) = \exp\left(\frac{t\hat{H}}{a}\right)\psi(\mathbf{r}, 0)$$

Since $\hat{H}$ has only real eigenvalues, the time dependence of $\psi(\mathbf{r}, t)$ is nonoscillating. It modulates $\psi(\mathbf{r}, 0)$ in time and does not give propagation. Thus, if a is real, ψ cannot represent a propagating wave. (*Note:* The fact that a is complex implies that ψ is complex. These last two problems illustrate the necessity of complex wavefunctions in quantum mechanics.)

3.22   Consider the wavefunction

$$\psi = Ae^{i(kx+\omega t)}$$

where k is real and $\omega > 0$ and is real. Is this wavefunction an admissible quantum state for a free particle? Justify your answer. If your answer is no, in what manner would you change the given function to describe a free particle moving in the $-x$ direction?

3.23   (a)   A free particle of mass m moves in one-dimensional space in the interval $0 \le x$, with energy E. There is a rigid wall at $x = 0$. Write down a time-independent wavefunction, $\varphi(x)$, which satisfies these conditions, in terms of x and k, where k is the wave vector of the motion. State the relation between k and E for this wavefunction.

(b)   Show explicitly that the wavefunction you have found in part (a) is an eigenfunction of the Hamiltonian for this system.

(c)   What is the time-dependent state, $\varphi(x, t)$, corresponding to the wavefunction, $\varphi(x)$?

4

PREPARATORY CONCEPTS. FUNCTION SPACES AND HERMITIAN OPERATORS

In this and the following two chapters, we continue development of physical principles and mathematical groundwork important to quantum mechanical descriptions. Included in the present chapter are the notions of Hilbert space and Hermitian operators. First, we obtain wavefunctions relevant to a particle in a one-dimensional box. These, together with previously derived free-particle wavefunctions, then serve as simply understood references for subsequent descriptions of Hilbert space and Hermitian operators.

4.1 PARTICLE IN A BOX AND FURTHER REMARKS ON NORMALIZATION

In Chapter 3 we solved the quantum mechanical free-particle problem. We recall that the free-particle Hamiltonian generates a continuous spectrum of eigenvalues, $\hbar^2 k^2 / 2m$, and eigenfunctions, $\varphi_k = A \exp (ikx)$, as given in (3.55).

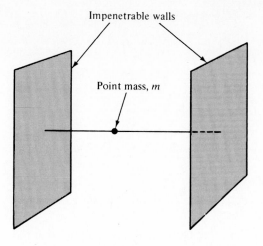

Impenetrable walls

Point mass, m

FIGURE 4.1   One-dimensional "box."

The second one-dimensional problem we wish to treat is that of a point mass m, constrained to move on an infinitely thin, frictionless wire which is strung tightly between two impenetrable walls a distance a apart (see Fig. 4.1). The corresponding potential has the values

(4.1)
$$V(x) = \infty \qquad (x \leq 0, \quad x \geq a) \qquad \text{(domain 1)}$$
$$V(x) = 0 \qquad (0 < x < a) \qquad \text{(domain 2)}$$

and is depicted in Fig. 4.2. This configuration is known as the *one-dimensional box.*[1]

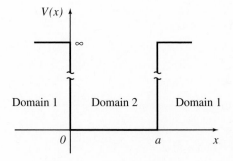

FIGURE 4.2   Potential corresponding to the one-dimensional box.

---

[1] A mathematically more accurate description of the one-dimensional box would characterize it as an infinitesimally thin, flat sheet of infinite extent and finite mass which moves between two walls of infinite extent. The two walls and sheet are all parallel and the velocity of the sheet is normal to the walls. Every point in space is then characterized by one coordinate, the normal displacement of the sheet from either of the walls.

The Hamiltonian for this problem is the following operator:

(4.2) $$\hat{H}_1 = \frac{\hat{p}^2}{2m} + \infty = \infty \qquad (x \le 0, \quad x \ge a) \qquad \text{(domain 1)}$$

(4.3) $$\hat{H}_2 = \frac{\hat{p}^2}{2m} \qquad (0 < x < a) \qquad \text{(domain 2)}$$

In domain 1 the time-independent Schrödinger equation gives $\varphi = 0$. For any finite eigenenergy E, in this domain the time-independent Schrödinger equation reads

(4.4) $$\hat{H}_1 \varphi = E\varphi$$

Since φ and E are finite, the right-hand side is finite. Therefore, the left-hand side is finite and φ must vanish in this domain.

The fact that $\varphi = 0$ in domain 1 implies that there is zero probability that the particle is found there ($|\varphi|^2 = 0$). This is in agreement with the discussion in Chapter 1 on "forbidden domains." These, we recall, are domains where $E < V$. Certainly, this is the case in domain 1 for any finite energy E.

In domain 2 the time-independent Schrödinger equation is

(4.5) $$-\frac{\hbar^2}{2m} \frac{\partial^2}{\partial x^2} \varphi_n = E_n \varphi_n$$

The subscript n is in anticipation of a discrete spectrum of energies E_n and eigenfunctions φ_n.

Since φ_n is a continuous function, it must have the values

(4.6) $$\varphi_n(0) = \varphi_n(a) = 0$$

First we rewrite (4.5) in the form

(4.7) $$\frac{\partial^2 \varphi_n}{\partial x^2} + k_n{}^2 \varphi_n = 0$$

(4.8) $$k_n{}^2 = \frac{2mE_n}{\hbar^2}$$

This is merely a change of variables from energy E_n to wavenumber k_n. The solution to (4.7) appears as

(4.9) $$\varphi_n = A \sin k_n x + B \cos k_n x$$

The boundary conditions (4.6) give

(4.10) $$B = 0$$

(4.11) $A \sin k_n a = 0$

The second of these equations serves to determine the eigenvalues k_n.

(4.12) $k_n a = n\pi,$ $n = 0, 1, 2, \ldots$

This is seen to be equivalent to the requirement that an integral number of half-wavelengths, $n(\lambda/2)$, fit into the width a.

The spectrum of eigenvalues and eigenfunctions is discrete. To find the constant A in (4.11), we normalize φ_n.

$$\int_0^a \varphi_n{}^2 \, dx = 1 = A^2 \int_0^a \sin^2\left(\frac{n\pi x}{a}\right) dx$$

(4.13)

$$1 = \frac{A^2 a}{n\pi} \int_0^{n\pi} \sin^2 \theta \, d\theta = \frac{A^2 a}{2}$$

The dummy variable $\theta = n\pi x/a$.

It follows that the eigenenergies E_n and normalized eigenfunctions φ_n for the one-dimensional box problem are

(4.14)
$$\boxed{E_n = n^2 E_1 \qquad E_1 = \frac{\hbar^2 k_1{}^2}{2m} = \frac{\hbar^2 \pi^2}{2ma^2} = \pi^2 \mathbb{R}\left(\frac{a_0}{a}\right)^2}$$

(4.15)
$$\boxed{\varphi_n = \sqrt{\frac{2}{a}} \sin\left(\frac{n\pi x}{a}\right)}$$

The eigenstate corresponding to $n = 0$ is $\varphi = 0$. This, together with the solution in domain 1, gives $\varphi = 0$ over the whole x axis. There is zero probability of finding the particle anywhere. This is equivalent to the statement that the particle does not exist in the $n = 0$ state. Another argument that disallows the $n = 0$ state follows from the uncertainty principle. The energy corresponding to $n = 0$ is $E = 0$. Since the energy in domain 2 is entirely kinetic, this, in turn, implies that the particle is in a state of

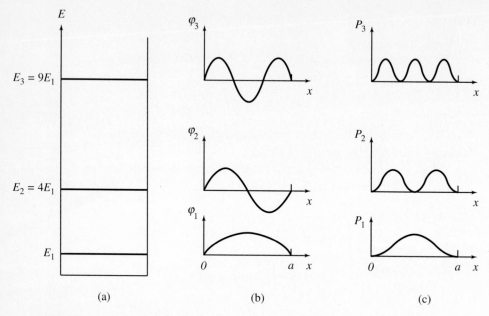

FIGURE 4.3 **(a) Eigenenergies for the one-dimensional box problem. (b) Eigenstates for the one-dimensional box problem:**

$$\varphi_n = \sqrt{\frac{2}{a}} \sin\left(\frac{n\pi x}{a}\right)$$

(c) Probability densities for the one-dimensional box problem:

$$P_n = |\varphi_n|^2 = \frac{2}{a} \sin^2\left(\frac{n\pi x}{a}\right)$$

absolute rest ($\Delta p = 0$), an illegitimate state of affairs for a particle constrained to move in a finite domain.

The eigenenergies and eigenfunctions given by (4.14) and (4.15), together with the corresponding probability densities $|\varphi_n|^2$, are sketched in Fig. 4.3.

The Arbitrary Phase Factor

In concluding this section we note the following important fact. As described in Section 3.3, the wavefunction ψ gives information about a system through calculation of averages of observable properties of that system, according to the rule

$$\langle C \rangle = \int \psi^* \hat{C} \psi \, dx$$

This equation, as well as the normalization condition

$$\int \psi^* \psi \, dx = 1$$

are invariant under the transformation $\psi \rightarrow e^{i\alpha}\psi$, where α is any real number. That is, a wavefunction is determined only to within a constant *phase factor* of the form $e^{i\alpha}$. Although associated with all wavefunctions, this arbitrary property has no effect upon any physical results.[1]

PROBLEMS

4.1 What are the energy eigenfunctions and eigenvalues for the one-dimensional box problem described above if the ends of the box are at $-a/2$ and $+a/2$? [Check your answer with (6.100).]

4.2 For what values of the real angle θ will the constant $C = \frac{1}{2}(e^{i\theta} - 1)$ have no effect in calculations involving the modulus $|C\psi|$?

4.2 THE BOHR CORRESPONDENCE PRINCIPLE

Let us now consider the *classical* motion of a particle in a one-dimensional box. As described previously, this configuration is effected by a bead sliding with no friction on a taut wire strung between two impenetrable walls a distance a apart. If the particle is a given a velocity v, its motion (between collisions with the wall) is

$$x = x_0 + vt$$

Now suppose that the initial position x_0 is completely unknown. What is the probability $P \, dx$ of finding the particle in the interval $x, x + dx$, at a subsequent time? The answer is, the fraction of time dt/T it spends in this interval.

(4.16)
$$P \, dx = \frac{dt}{T} = \frac{v \, dt}{a} = \frac{dx}{a}$$

so that

(4.17)
$$P = \frac{1}{a} = \text{constant}$$

It is uniformly probable to find the particle at any position on the wire. If we make a large number of replicas of this one-dimensional system, measurement (at random

---

[1] On the other hand, component phase factors for a composite wavefunction such as that discussed in Section 2.5 do contribute to measurable effects, such as interference.

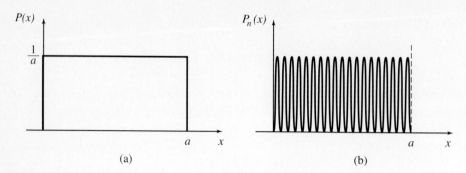

FIGURE 4.4   **(a) Classical probability density for the one-dimensional box. (b) Quantum mechanical probability density**

$$P_n = |\varphi_n|^2$$

for the one-dimensional box problem, for the case $n \gg 1$. The probability P_n vanishes $n + 1$ times in the interval $(0, a)$.

times) of the coordinate x of the bead will find all values $(0 \leq x \leq a)$ occurring equally often (Fig. 4.4).

On the other hand, in the quantum mechanical case, if the particle is in the state φ_3, say, the probability density P is peaked at $x = (a/6, a/2, 5a/6)$; see Fig. 4.3. In this case, measurement on an abundant number of replica systems finds the particle spending much of its time in the neighborhood of these three values of x. This situation is quite different from the classical case described above. Suppose we move to higher quantum states. At what values of x is the probability density P peaked? The solution is left to Problem 4.3, where one obtains that $|\varphi_n|^2$ is peaked at the values

(4.18)
$$x_j = \frac{2j + 1}{2n} a, \qquad j = 0, 1, 2, \ldots$$

As n becomes very large, the probability density oscillates with so large a frequency that it begins to assume a uniform quality. For any n, one can divide the interval $(0, a)$ into n strips of equal width Δx such that in each strip the probability $|\varphi_n|^2 \Delta x$ of finding the particle is equal. For the classical case the number of such strips is infinite. In the quantum mechanical case, the same situation is approached in the limit $n \rightarrow \infty$.

One encounters this transition to classical physics from the quantum mechanical domain in many problems. Bohr was the first to analyze this transition and offered the general rule that a quantum mechanical result must reduce to its classical counterpart in the classical domain. Since classical formulas do not contain $\hbar$, such a transition should be realized in the limit that $\hbar$ becomes small. For many problems this limit is attained in passing to high quantum numbers $(n \rightarrow \infty)$. This rule is called the *Bohr correspondence principle*.

Classical physics includes the dynamics of macroscopic bodies. An aggregate of particles (e.g., a gas) obeys classical laws when the de Broglie wavelength, λ, of a typical particle is small compared to all relevant lengths. For example, if the density of particles (number/cm^3) is n, the gas obeys classical statistics if $\lambda \ll n^{-1/3}$ (the mean distance between particles is $n^{-1/3}$). In the classical limit, fluctuations about the average become small and the probabilities indigenous to quantum mechanics reduce to certainties.

A rule of thumb in this area is that any quantum mechanical result that does not contain $\hbar$ is in essence a classical result. The first (fortuitous) example of this rule was Rutherford's classical calculation of the Coulomb cross section, relevant to the scattering of charged particles. The correct quantum mechanical calculation of this parameter is found not to contain $\hbar$. Rutherford's classical calculation yields the same result.

More examples of the correspondence principle will arise in the course of development of the text. Coulomb scattering is further described in Section 14.4.

PROBLEMS

4.3   For the one-dimensional box problem, show that $P = |\varphi_n|^2$ is maximum at the values $x = x_j$ given by

$$x_j = \frac{2j + 1}{2n} a, \qquad j = 0, 1, 2, \ldots, n - 1$$

4.3   DIRAC NOTATION

In this section we introduce a notation that proves to be an invaluable tool in calculation, called the *Dirac notation*. It gives a monogram to the integral of the product of two state functions, $\psi(x)$ and $\varphi(x)$, which appears as

$$(4.19) \qquad \langle \psi | \varphi \rangle = \int_{-\infty}^{\infty} \psi^*(x)\varphi(x) \, dx$$

In Dirac notation, the integral on the right is written in the form shown on the left.

More generally, the integral operation $\langle \psi | \varphi \rangle$ denotes: (1) take the complex conjugate of the object in the first slot ($\psi \rightarrow \psi^*$) and then, (2) integrate the product ($\psi^*\varphi$). This operation has the following simple properties. If a is any complex number and the functions ψ and φ are such that

$$(4.20) \qquad \int_{-\infty}^{\infty} \psi^*\varphi \, dx < \infty$$

the following rules hold:

(4.21) $$\langle \psi | a\varphi \rangle = a \langle \psi | \varphi \rangle$$

(4.22) $$\langle a\psi | \varphi \rangle = a^* \langle \psi | \varphi \rangle$$

(4.23) $$\langle \psi | \varphi \rangle^* = \langle \varphi | \psi \rangle$$

(4.24) $$\langle \varphi + \psi | = \langle \psi | + \langle \varphi |$$

(4.25) $$\int (\psi_1 + \psi_2)^* (\varphi_1 + \varphi_2) \, dx$$

$$= \langle \psi_1 + \psi_2 | \varphi_1 + \varphi_2 \rangle = ((\langle \psi_1 | + \langle \psi_2 |)(| \varphi_1 \rangle + | \varphi_2 \rangle)$$
$$= \langle \psi_1 | \varphi_1 \rangle + \langle \psi_1 | \varphi_2 \rangle + \langle \psi_2 | \varphi_1 \rangle + \langle \psi_2 | \varphi_2 \rangle$$

The object $\langle \psi |$ (called a "bra vector"). It joins in a product form with a ("ket vector") $| \varphi \rangle$, to form the "bra-ket," $\langle \psi | \varphi \rangle$.

A more fundamental description of Dirac notation in quantum mechanics is given in Appendix A.

PROBLEMS

4.4   Write the following equations for the state vectors f, g, and so on, in Dirac notation.

(a) $f(x) = g(x)$

(b) $c = \displaystyle\int g^*(x') h(x') \, dx'$

(c) $f(x) = \displaystyle\sum_n \varphi_n(x) \int \varphi_n^*(x') f(x') \, dx'$

(d) $\hat{O} \equiv \psi(x) \displaystyle\int dx' \varphi^*(x')$

(e) $\dfrac{\partial}{\partial x} f(x) = h(x) \displaystyle\int h^*(x') g(x') \, dx'$

4.5   Consider the operator $\hat{O} = | \varphi \rangle \langle \psi |$ and the arbitrary state function $f(x)$. Describe the following forms.

(a) $\langle f | \hat{O}$

(b) $\hat{O} | f \rangle$

(c) $\langle f | \hat{O} | f \rangle$

(d) $\langle f | \hat{O} | \psi \rangle$

Answer (partial)

(a) $\langle f | \hat{O}$ is the bra vector $C \langle \psi |$, where the constant $C \equiv \langle f | \varphi \rangle = \int_{-\infty}^{\infty} f^* \varphi \, dx$.

4.4   HILBERT SPACE

In this section we introduce the concept of a space of functions. Specifically we will deal with a Hilbert space. This serves the purpose of giving a geometrical quality to some of the abstract concepts of quantum mechanics.

We recall that in Cartesian 3-space a vector $\mathbf{V}$ is a set of three numbers, called components (V_x, V_y, V_z). Any vector in this space can be expanded in terms of the three unit vectors $\mathbf{e}_x$, $\mathbf{e}_y$, $\mathbf{e}_z$ (Fig. 4.5). Under such conditions one terms the triad $\mathbf{e}_x$, $\mathbf{e}_y$, $\mathbf{e}_z$, a *basis*.

$$(4.26) \qquad \mathbf{V} = \mathbf{e}_x V_x + \mathbf{e}_y V_y + \mathbf{e}_z V_z$$

The vectors $\mathbf{e}_x$, $\mathbf{e}_y$, $\mathbf{e}_z$ are said to *span* the vector space.

The inner ("dot") product of two vectors ($\mathbf{U}$ and $\mathbf{V}$) in the space is defined as

$$(4.27) \qquad \mathbf{V} \cdot \mathbf{U} = V_x U_x + V_y U_y + V_z U_z$$

The length of the vector $\mathbf{V}$ is $\sqrt{\mathbf{V} \cdot \mathbf{V}}$.

A Hilbert space is much the same type of object. Its elements are functions instead of three-dimensional vectors. The similarity is so close that the functions are sometimes called vectors. A Hilbert space $\mathfrak{H}$ has the following properties.

1. The space is linear. A function space is linear under the following two conditions: (a) If a is a constant and φ is any element of the space, then $a\varphi$ is also an element of the space. (b) If φ and ψ are any two elements of the space, then $\varphi + \psi$ is also an element of the space.

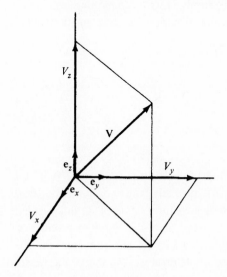

FIGURE 4.5   Vector V in Cartesian 3-space and its components (V_x, V_y, V_z). **The orthogonal triad** $(\mathbf{e}_x, \mathbf{e}_y, \mathbf{e}_z)$ **spans the space.**

2.   There is an *inner product,* $\langle \psi | \varphi \rangle$, for any two elements in the space. For functions defined in the interval $a \le x \le b$ (in one dimension), we may take

(4.28)
$$\langle \varphi | \psi \rangle = \int_a^b \varphi^* \psi \, dx$$

3.   Any element of $\mathfrak{H}$ has a norm ("length") that is related to the inner product as follows:

(4.29)
$$(\text{norm of } \varphi)^2 = \| \varphi \|^2 = \langle \varphi | \varphi \rangle$$

4. $\mathfrak{H}$ is complete. Every Cauchy sequence of functions in $\mathfrak{H}$ converges to an element of $\mathfrak{H}$. A Cauchy sequence $\{\varphi_n\}$ is such that $\| \varphi_n - \varphi_l \| \to 0$ as n and l approach infinity. (See Problem 4.24.) Loosely speaking, a Hilbert space contains all its limit points.

An example of a Hilbert space is given by the set of functions defined on the interval $(0 \le x \le a)$ with finite norm

(4.30)
$$\| \varphi \|^2 = \int_0^a \varphi^* \varphi \, dx < \infty \qquad \mathfrak{H}_1$$

Another example is the space of functions commonly referred to by mathematicians as "L^2 space." This is the set of square-integrable functions defined on the whole x interval.

(4.31)
$$\| \varphi \|^2 = \int_{-\infty}^{\infty} \varphi^* \varphi \, dx < \infty \qquad \mathfrak{H}_2$$

Let us see how the preceding concept of inner product (4.28) is similar to the definition of the inner product between two finite-dimensional vectors (4.27). To see this we interpret the function $\varphi(x)$ as a vector with infinitely many components. These components are the values that φ assumes at each distinct value of its independent variable x. Just as the inner product between $\mathbf{U}$ and $\mathbf{V}$ is a sum over the products of parallel components, so is the inner product between φ and ψ a sum over parallel components. This sum is nothing but the integral of the product of φ and ψ. The reason we complex-conjugate the first "vector" is to ensure that the "length" (square root of the inner product between a "vector" φ and itself) of a vector φ is real.

Thus we see that Hilbert space is closely akin to a vector space. Mathematicians[1] call it that—an infinite-dimensional vector space (also, a complete, normed, linear vector space). Elements of this space have length and one can form an inner product between any two elements. The vector quality of Hilbert space can be pushed a bit further. We recall that if two vectors **U** and **V** in three-dimensional vector space are orthogonal to each other, their inner product vanishes. In a similar vein two vectors in Hilbert space, φ and ψ, are said to be orthogonal if

(4.32) $$\langle \psi \,|\, \varphi \rangle = 0$$

Furthermore, we recall that the three unit vectors $\mathbf{e}_x$, $\mathbf{e}_y$, and $\mathbf{e}_z$ "span" 3-space. Similarly, there is a set of vectors that "spans" Hilbert space. For instance, the Hilbert space whose elements all have the property given by (4.30) is spanned by the sequence of functions $\{\varphi_n\}$, which are the eigenfunctions of the Hamiltonian relevant to the one-dimensional box problem (4.15). This means that any function φ in this Hilbert space may be expanded in a series of the sequence $\{\varphi_n\}$.

(4.33) $$\varphi(x) = \sum_{n=1}^{\infty} a_n \varphi_n(x)$$

The geometrical interpretation of this relation is depicted in Fig. 4.6. The coefficient a_n is the projection of φ onto the vector φ_n. To see this, first we state a fact to

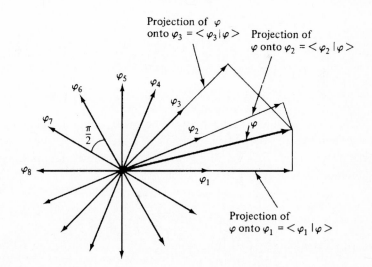

FIGURE 4.6 Projection of φ onto an orthonormal set of eigenfunctions in Hilbert space.

[1] A more mathematically accurate presentation of function spaces may be found in C. Goffman and G. Pedrick, *First Course in Functional Analysis,* Prentice-Hall, Englewood Cliffs, N.J., 1965. Another book in this area, but more directly related to quantum mechanics, is T. F. Jordan, *Linear Operators for Quantum Mechanics,* Wiley, New York, 1969.

be illustrated in the next section. The *basis* vectors $\{\varphi_n\}$ comprise an orthogonal set. That is,

(4.34) $$\langle \varphi_n | \varphi_{n'} \rangle = 0 \qquad (n \neq n')$$

Furthermore, φ_n is a unit vector; that is, it has unit "length"

(4.35) $$\langle \varphi_n | \varphi_n \rangle = \| \varphi_n \|^2 = 1$$

These latter two statements may be combined into the single equation

(4.36) $$\langle \varphi_n | \varphi_{n'} \rangle = \delta_{n,n'}$$

The symbol $\delta_{n,n'}$ is called the *Kronecker delta* and is defined by

(4.37) $$\delta_{n,n'} = 0 \quad \text{for } n \neq n', \qquad \delta_{n,n'} = 1 \quad \text{for } n = n'$$

Any sequence of functions that obeys (4.36) is called an *orthonormal set*.

To show that a_n is the projection of φ into φ_n, we first rewrite (4.33) in Dirac notation.

(4.38) $$| \varphi \rangle = \sum_n | a_n \varphi_n \rangle$$

Then we multiply from the left by $\langle \varphi_{n'} |$ and use the relation (4.36).

$$\langle \varphi_{n'} | \varphi \rangle = \sum_n \langle \varphi_{n'} | a_n \varphi_n \rangle$$

(4.39) $$= \sum_n a_n \langle \varphi_{n'} | \varphi_n \rangle = \sum_n a_n \delta_{n,n'} = a_{n'}$$

$$a_{n'} = \langle \varphi_{n'} | \varphi \rangle$$

The coefficient $a_{n'}$ is the inner product between the basis vector $\varphi_{n'}$ and the vector φ. Since $\varphi_{n'}$ is a "unit" vector, $a_{n'}$ is the projection of φ onto $\varphi_{n'}$ (Fig. 4.6). The student should recognize (4.33) to be a discrete Fourier series representation of φ, in terms of the trigonometric sequence (4.15).

Delta-Function Orthogonality

We will continue with the use of the labels $\mathfrak{H}_1$ and $\mathfrak{H}_2$ to denote the two Hilbert spaces defined by (4.30) and (4.31), respectively. As stated previously, the sequence $\{\varphi_n\}$ given by (4.15) "spans" $\mathfrak{H}_1$. The sequence $\{\varphi_n\}$ is a basis of $\mathfrak{H}_1$. What are the vectors which span $\mathfrak{H}_2$? The answer is, the eigenfunctions of the momentum operator $\hat{p}$,

(4.40)
$$\varphi_k(x) = \frac{1}{\sqrt{2\pi}} e^{ikx}$$

Let us see if this (continuous) set of functions is an orthogonal set. Toward these ends we form the inner product

(4.41)
$$\langle \varphi_k | \varphi_{k'} \rangle = \frac{1}{2\pi} \int_{-\infty}^{\infty} e^{ix(k'-k)} \, dx = \delta(k' - k)$$

It follows that the inner product between any two distinct eigenvectors of the operator $\hat{p}$ vanishes.

Any function in $\mathfrak{H}_2$ may be expanded in terms of the eigenvectors $\{\varphi_k\}$. Since this sequence comprises a continuous set, the expansion is not a discrete sum as in (4.33), but an integral. If $\varphi(x)$ is any element of $\mathfrak{H}_2$, then since $\{\varphi_k\}$ spans this space, one may write

(4.42)
$$\varphi(x) = \int_{-\infty}^{\infty} b(k)\varphi_k(x) \, dk$$

This is the Fourier integral representation of $\varphi(x)$. Again, the coefficient of expansion $b(k)$ is the projection of $\varphi(x)$ onto φ_k. To exhibit this fact, we first rewrite the last integral in the form

(4.43)
$$|\varphi\rangle = \int_{-\infty}^{\infty} dk \, |b(k)\varphi_k\rangle$$

Again, if this equation is compared to (4.38), we see how the sum over discrete a_n values is replaced by an integration over the continuum of $b(k)$ values. If we now multiply (4.43) from the left with $\langle \varphi_{k'} |$, there results

(4.44)
$$\langle \varphi_{k'} | \varphi \rangle = \int_{-\infty}^{\infty} dk \langle \varphi_{k'} | b(k)\varphi_k \rangle = \int_{-\infty}^{\infty} dk b(k) \langle \varphi_{k'} | \varphi_k \rangle$$
$$= \int_{-\infty}^{\infty} dk b(k) \delta(k' - k) = b(k')$$

The coefficient of expansion $b(k')$ is the inner product between $\varphi_{k'}$ and φ; hence it may be termed a projection of φ onto $\varphi_{k'}$. But $\varphi_{k'}$ does not appear to be a "unit" vector. Indeed, the vector φ_k is infinitely long.

(4.45)
$$\|\varphi_k\|^2 = \langle \varphi_k | \varphi_k \rangle = \delta(0) = \frac{1}{2\pi} \int_{-\infty}^{\infty} dx = \infty$$

Although this disqualifies the set $\{\varphi_k\}$ for membership in $\mathfrak{H}_2$, they nevertheless span the space. They comprise a valid set of basis vectors and the projection of any function in $\mathfrak{H}_2$ onto any member of the basis $\{\varphi_k\}$ gives a finite result. If φ is any function in $\mathfrak{H}_2$, then

(4.46) $$\langle \varphi_k \,|\, \varphi \rangle < \infty$$

The functions $\{\varphi_k\}$ may, through proper renormalization, be cast in a form which allows them to be members of $\mathfrak{H}_2$. (See Problem 4.6.)

PROBLEMS

4.6 Consider the functions

$$\varphi_k = \frac{1}{\sqrt{a}}\, e^{ikx}$$

defined over the interval $(-a/2, +a/2)$.

    (a) Show that these functions are all normalized to unity and maintain this normalization in the limit $a \to \infty$.

    (b) Show that these functions comprise an orthogonal set in the limit $a \to \infty$.

4.7 State to which space each of the functions listed belongs, $\mathfrak{H}_1$ or $\mathfrak{H}_2$.

    (a) $f_1 = (x^5 - x^4 - ax^4 + ax^3)/(x - 2a)$

    (b) $f_2 = (\sin x)e^{-x^2}$

    (c) $f_3 = \sqrt{\ln[x(x - a) + 1]}$

    (d) $f_4 = \sin 2n\pi[x(x - a) + 1]$, $n = 0, 1, 2, \ldots$

    (e) $f_5 = e^{i\alpha x}(x^2 + a^2)^{-1}$

    (f) $f_6 = x^{10}e^{-x^2}$

    (g) $f_7 = 1/\sin kx$

4.8 The function

$$g(x) = x(x - a)e^{ikx}$$

is in $\mathfrak{H}_1$. Calculate the coefficients of expansion, a_n, of this function, in the series representation (4.33), in terms of the constants a and k. Use the basis functions (4.15).

4.9 Two vectors ψ and φ in a Hilbert space are orthogonal. Show that their lengths obey the Pythagorean theorem,

$$\| \psi + \varphi \|^2 = \| \psi \|^2 + \| \varphi \|^2$$

4.10 Consider a free particle moving in one dimension. The state functions for this particle are all elements of $\mathfrak{H}_2$. Show that the expectation of the momentum $\langle p_x \rangle$ vanishes in any state that is purely real $(\psi^* = \psi)$. Does this property hold for $\langle H \rangle$? Does it hold for $\langle x \rangle$?

4.5   HERMITIAN OPERATORS

The average of an observable A for a system in the state $\psi(x, t)$ is given by (3.32). In Dirac notation this equation appears as (in one dimension)

$$(4.47) \qquad \langle A \rangle = \int \psi^*(x, t)\hat{A}\psi(x, t) \, dx = \langle \psi \, | \, \hat{A}\psi \rangle$$

Since t is a fixed parameter in this equation, we may conclude that the formula gives the expectation of A at the time t. Now one may ask: What are the possible state functions for a particle moving in one dimension at a given instant of time? The answer is: any function in $\mathfrak{H}_2$. For example, the particle could be in any of the following states at some specified time:

$$(4.48) \qquad \psi_1 = Be^{-x^2/a^2}, \qquad \psi_2 = \frac{Ce^{ikx}}{x}, \qquad \psi_3 = \frac{iD}{\sqrt{x^2 + a^2}}$$

where B, C, and D are normalization constants. Again consider the observable A. If the average of this observable is calculated in any of these states (that is, any member of $\mathfrak{H}_2$), the result must be a real number. This is a property that we demand an operator have if it is to qualify as the operator corresponding to a physical observable. The object $\langle \psi \, | \, \hat{A}\psi \rangle$ must be real for all ψ in $\mathfrak{H}_2$. When working with the one-dimensional box problem, $\langle \psi \, | \, \hat{A}\psi \rangle$ must be real for all ψ in $\mathfrak{H}_1$. For example, if $\hat{H}$ is the operator corresponding to energy, then

$$(4.49) \qquad \langle E \rangle = \langle \psi \, | \, \hat{H}\psi \rangle = -\int_0^a \frac{\psi^*\hbar^2}{2m} \frac{\partial^2}{\partial x^2} \psi \, dx$$

must be real for any state function ψ in $\mathfrak{H}_1$.

These observations give rise to the following rule: In quantum mechanics one requires that the eigenvalues of an operator corresponding to a physical observable be real numbers. In this section we discuss the class of operators that have this property. They are called *Hermitian operators* and are a cornerstone in the theory of quantum mechanics.

The Hermitian Adjoint

To understand what a Hermitian operator is, we must first understand what the *Hermitian adjoint* of an operator is. Consider the operator $\hat{A}$. The Hermitian adjoint of $\hat{A}$ is written $\hat{A}^\dagger$. Under most circumstances, it is an entirely different operator from $\hat{A}$.

For instance, the Hermitian adjoint of the complex number c is the complex conjugate of c. That is,

(4.50) $$c^\dagger = c*$$

How is the Hermitian adjoint defined? First, let us agree that an operator is defined over a specific Hilbert space, $\mathfrak{H}$. Also if $\hat{A}$ is the operator and ψ is any element of $\mathfrak{H}$, then $\hat{A}\psi$ is also in $\mathfrak{H}$. For any two elements of this space, say ψ_l and ψ_n, we can form the inner product

(4.51) $$\langle \psi_l | \hat{A}\psi_n \rangle$$

Suppose there is another operator, $\hat{A}^\dagger$, also defined over $\mathfrak{H}$, for which

(4.52) $$\langle \hat{A}^\dagger \psi_l | \psi_n \rangle = \langle \psi_l | \hat{A}\psi_n \rangle$$

Suppose further that this equality holds for *all* ψ_l and ψ_n in $\mathfrak{H}$. Then $\hat{A}^\dagger$ is called the *Hermitian adjoint* of $\hat{A}$. To find the Hermitian adjoint of an operator $\hat{A}$, we have to find the object $\hat{A}^\dagger$ that fits (4.52) for all ψ_l and ψ_n. Consider $\hat{A} = a$, a complex number. Then

(4.53) $$\langle a^\dagger \psi_l | \psi_n \rangle = \langle \psi_l | a\psi_n \rangle = a\langle \psi_l | \psi_n \rangle = \langle a*\psi_l | \psi_n \rangle$$

Equating the first and the last terms, we see that $a^\dagger = a*$. As a second example, consider the operator

(4.54) $$\hat{D} = \frac{\partial}{\partial x}$$

defined in $\mathfrak{H}_2$. Then

(4.55) $$\langle \psi_l | \hat{D}\psi_n \rangle = \int_{-\infty}^{\infty} dx\, \psi_l* \frac{\partial}{\partial x} \psi_n = [\psi_l*\psi_n]_{-\infty}^{+\infty} - \int_{-\infty}^{\infty} dx \left(\frac{\partial}{\partial x} \psi_l* \right) \psi_n$$

$$= \langle -\hat{D}\psi_l | \psi_n \rangle$$

The "surface" term is zero since ψ_l and ψ_n are elements of $\mathfrak{H}_2$. Thus we find

(4.56) $$\hat{D}^\dagger = -\hat{D}$$

For some cases we will find that the Hermitian adjoint of an operator is the operator itself. For such an operator $\hat{A}$, we may write

(4.57) $$\hat{A}^\dagger = \hat{A}$$

In terms of the defining equation (4.52), this implies that for all ψ_l and ψ_n in $\mathfrak{H}$ (over which $\hat{A}$ is defined),

$$(4.58) \qquad \langle \psi_l | \hat{A}\psi_n \rangle = \langle \hat{A}\psi_l | \psi_n \rangle$$

Operators that have this property are called *Hermitian operators*. The simplest example of a Hermitian operator is any real number a, since

$$(4.59) \qquad \langle \psi_l | a\psi_n \rangle = \langle a\psi_l | \psi_n \rangle$$

If $\hat{A}$ and $\hat{B}$ are two Hermitian operators, is the product operator $\hat{A}\hat{B}$ Hermitian? This is most simply answered with the aid of Problem 4.11(b), according to which

$$(4.60) \qquad (\hat{A}\hat{B})^\dagger = \hat{B}^\dagger \hat{A}^\dagger$$

If $\hat{A}$ and $\hat{B}$ are Hermitian, then

$$(4.61) \qquad (\hat{A}\hat{B})^\dagger = \hat{B}\hat{A}$$

and $\hat{A}\hat{B}$ is not (necessarily) Hermitian. What about $\hat{A}\hat{B} + \hat{B}\hat{A}$?

$$(4.62) \qquad (\hat{A}\hat{B} + \hat{B}\hat{A})^\dagger = \hat{B}^\dagger\hat{A}^\dagger + \hat{A}^\dagger\hat{B}^\dagger = \hat{B}\hat{A} + \hat{A}\hat{B}$$
$$= \hat{A}\hat{B} + \hat{B}\hat{A}$$

It follows that if $\hat{A}$ and $\hat{B}$ are both Hermitian, so is the bilinear form $(\hat{A}\hat{B} + \hat{B}\hat{A})$.

Is the square of a Hermitian operator Hermitian?

$$(4.63) \qquad (\hat{A}^2)^\dagger = (\hat{A}\hat{A})^\dagger = \hat{A}^\dagger\hat{A}^\dagger = \hat{A}\hat{A} = (\hat{A})^2$$

The answer is yes. Another way of doing this problem is as follows. Look at the inner product.

$$(4.64) \qquad \langle \psi_l | \hat{A}\hat{A}\psi_n \rangle = \langle \hat{A}\psi_l | \hat{A}\psi_n \rangle = \langle \hat{A}\hat{A}\psi_l | \psi_n \rangle$$

The first equality follows because $\hat{A}\psi_n$ is in $\mathfrak{H}$ and $\hat{A}$ is Hermitian, while the second equality follows simply because $\hat{A}$ is Hermitian. Comparing the first and third terms shows that $\hat{A}^2$ is Hermitian.

The Momentum and Energy Operators

Let us test the momentum operator $\hat{p}$ and see if it is Hermitian. For the free-particle case, $\hat{p}$ is Hermitian if for all ψ_l and ψ_n in $\mathfrak{H}_2$,

$$(4.65) \qquad \langle \psi_l | \hat{p}\psi_n \rangle = \langle \hat{p}\psi_l | \psi_n \rangle$$

Developing the left-hand side, we have

$$(4.66) \quad \int_{-\infty}^{\infty} \psi_l^* \left(-i\hbar \frac{\partial}{\partial x} \psi_n \right) dx = -i\hbar [\psi_l^* \psi_n]_{-\infty}^{\infty} + i\hbar \int_{-\infty}^{\infty} \left(\frac{\partial}{\partial x} \psi_l^* \right) \psi_n \, dx$$

$$= \int_{-\infty}^{\infty} \left(-i\hbar \frac{\partial}{\partial x} \psi_l \right)^* \psi_n \, dx = \langle \hat{p} \psi_l | \psi_n \rangle$$

This technique is, by and large, the principal method by which a specific operator is shown to be Hermitian.

Having shown that $\hat{p}$ is Hermitian, it follows that the free-particle Hamiltonian, $\hat{H}$, is Hermitian.

$$(4.67) \quad \hat{H} = \frac{\hat{p}^2}{2m}$$

$$(4.68) \quad \hat{H}^\dagger = \left(\frac{\hat{p}^2}{2m} \right)^\dagger = \frac{\hat{p}^2}{2m} = \hat{H}$$

[Recall (4.63).] For a particle in a potential field $V(x)$,

$$(4.69) \quad \hat{H} = \frac{\hat{p}^2}{2m} + V(x)$$

Since $V(x)$ is a real function that merely multiplies (say in $\mathfrak{H}_2$), it is Hermitian.

$$(4.70) \quad \langle \psi_l | V\psi_n \rangle = \int_{-\infty}^{\infty} \psi_l^* V\psi_n \, dx = \int_{-\infty}^{\infty} V\psi_l^* \psi_n \, dx$$

$$= \int (V\psi_l)^* \psi_n \, dx = \langle V\psi_l | \psi_n \rangle$$

It follows that $\hat{H}$ as given by (4.69) is Hermitian.

PROBLEMS

4.11 (a) Show that $(a\hat{A} + b\hat{B})^\dagger = a^* \hat{A}^\dagger + b^* \hat{B}^\dagger$.
(b) Show that $(\hat{A}\hat{B})^\dagger = \hat{B}^\dagger \hat{A}^\dagger$.
(c) What is the Hermitian adjoint of the real number a?
(d) What is the Hermitian adjoint of $\hat{D}^2$? [See (4.54).]
(e) What is the Hermitian adjoint of $(\hat{A}\hat{B} - \hat{B}\hat{A})$?
(f) What is the Hermitian adjoint of $(\hat{A}\hat{B} + \hat{B}\hat{A})$?
(g) What is the Hermitian adjoint of $i(\hat{A}\hat{B} - \hat{B}\hat{A})$?
(h) What is $(\hat{A}^\dagger)^\dagger$?
(i) What is $(\hat{A}^\dagger \hat{A})^\dagger$?

4.12   If $\hat{A}$ and $\hat{B}$ are both Hermitian, which of the following three operators are Hermitian?

(a) $i(\hat{A}\hat{B} - \hat{B}\hat{A})$

(b) $(\hat{A}\hat{B} - \hat{B}\hat{A})$

(c) $\left(\dfrac{\hat{A}\hat{B} + \hat{B}\hat{A}}{2}\right)$

(d)   If $\hat{A}$ is not Hermitian, is the product $\hat{A}^\dagger\hat{A}$ Hermitian?

(e)   If $\hat{A}$ corresponds to the observable A, and $\hat{B}$ corresponds to B, what is a "good" (i.e., Hermitian) operator that corresponds to the physically observable product AB?

4.13   If $\hat{A}$ is Hermitian, show that

$$\langle\hat{A}^2\rangle \geq 0$$

Answer (in $\mathfrak{H}_2$)

$$\langle\hat{A}^2\rangle = \int_{-\infty}^{\infty} \psi^* \hat{A}^2 \psi \, dx = \int_{-\infty}^{\infty} (\hat{A}\psi)^* \hat{A}\psi \, dx$$

$$= \int_{-\infty}^{\infty} |\hat{A}\psi|^2 \, dx \geq 0$$

4.14   If $\hat{A}$ is Hermitian, show that $\langle A\rangle$ is real; that is, show that $\langle A\rangle^* = \langle A\rangle$.

4.15   For a particle moving in one dimension, show that the operator $\hat{x}\hat{p}$ is not Hermitian. Construct an operator which corresponds to this physically observable product that is Hermitian.

4.6   PROPERTIES OF HERMITIAN OPERATORS

The first property of Hermitian operators we wish to establish is that their eigenvalues are real. Let $\hat{A}$ be a Hermitian operator. Let $\{\varphi_n\}$ and $\{a_n\}$ represent, respectively, the eigenfunctions and eigenvalues of the operator $\hat{A}$.

(4.71) $$\hat{A}\varphi_n = a_n\varphi_n$$

In Dirac notation

(4.72) $|\hat{A}\varphi_n\rangle = |a_n\varphi_n\rangle$ or equivalently $\hat{A}|\varphi_n\rangle = a_n|\varphi_n\rangle$

Multiplying from the left with $\langle\varphi_n|$ gives

(4.73) $$\langle\varphi_n|\hat{A}\varphi_n\rangle = \langle\varphi_n|a_n\varphi_n\rangle = a_n\langle\varphi_n|\varphi_n\rangle$$

Since $\hat{A}$ is Hermitian, we can write the left-hand side as

(4.74) $$\langle\hat{A}\varphi_n|\varphi_n\rangle = \langle a_n\varphi_n|\varphi_n\rangle = a_n^*\langle\varphi_n|\varphi_n\rangle$$

Equating the last terms in the latter two equations gives

$$(4.75) \qquad a_n{}^* = a_n$$

and a_n is real.

The second property of Hermitian operators we wish to establish is that *their eigenfunctions are orthogonal*. Again consider (4.72). Now multiply from the left with another eigenvector of $\hat{A}$, $\langle \varphi_l |$. There results

$$(4.76) \qquad \langle \varphi_l | \hat{A} \varphi_n \rangle = a_n \langle \varphi_l | \varphi_n \rangle$$

Since $\hat{A}$ is Hermitian, the left-hand side of this equation can be rewritten

$$(4.77) \qquad \langle \hat{A} \varphi_l | \varphi_n \rangle = a_l{}^* \langle \varphi_l | \varphi_n \rangle = a_l \langle \varphi_l | \varphi_n \rangle$$

The eigenvalue a_l is real because it is an eigenvalue of a Hermitian operator (i.e., $\hat{A}$). Subtracting the two equations above gives

$$(4.78) \qquad (a_l - a_n)\langle \varphi_l | \varphi_n \rangle = 0$$

If $a_l \neq a_n$ (which is the case if a_l and a_n are nondegenerate: see Section 5.3), this equation says that

$$(4.79) \qquad \langle \varphi_l | \varphi_n \rangle = 0$$

which is the expression of the orthogonality of the set of functions $\{\varphi_n\}$. If these functions are all normalized, then (4.79) may be generalized to read

$$(4.80) \qquad \langle \varphi_l | \varphi_n \rangle = \delta_{ln}$$

Thus, the eigenvalues of a Hermitian operator are real, and its eigenfunctions are orthogonal.

PROBLEMS

4.16 Show that if an operator $\hat{B}$ has an eigenvalue $b_1 \neq b_1{}^*$, then $\hat{B}$ is not Hermitian.

4.17 Consider the operator $\hat{C}$,

$$\hat{C}\varphi(x) = \varphi^*(x)$$

(a) Is $\hat{C}$ Hermitian?
(b) What are the eigenfunctions of $\hat{C}$?
(c) What are the eigenvalues of $\hat{C}$?

4.18 Given that the operator $\hat{O}$ annihilates the ket vector $| f \rangle$, that is, $\hat{O} | f \rangle = 0$, what is the value of the bra vector $\langle f | \hat{O}^\dagger$? Interpret the meaning of your answer.

4.19   The parallelogram law of geometry states that the sum of the squares of the diagonals of a parallelogram equals twice the sum of the squares of the sides. Show that this is also true in Hilbert space; that is, if ψ and φ are any two elements of a Hilbert space, then

$$\|\psi + \varphi\|^2 + \|\psi - \varphi\|^2 = 2\|\psi\|^2 + 2\|\varphi\|^2$$

4.20   Show that the standard properties of $\cos \theta$, together with the definition of the inner product between two vectors φ and ψ, in $\mathfrak{H}$, with respective lengths, $\|\varphi\|$ and $\|\psi\|$, imply the Cauchy–Schwartz inequality

$$|\langle \varphi | \psi \rangle| \leq \|\varphi\| \|\psi\|$$

4.21   Use the Cauchy–Schwartz inequality to prove the triangle inequality

$$\|\varphi + \psi\|^2 \leq (\|\varphi\| + \|\psi\|)^2$$

4.22   Construct the squared length of $(\psi - \varphi)$ to show that

$$\|\psi\|^2 + \|\varphi\|^2 \geq 2 \operatorname{Re} \langle \psi | \varphi \rangle$$

4.23   Let the sequence $\{\varphi_n\}$ be an orthononormal basis in $\mathfrak{H}$. Let the sequence $\{\cos \theta_n\}$ represent the angles between the vectors $\{\varphi_n\}$ and an arbitrary element ψ in $\mathfrak{H}$. Using Bessel's inequality,

$$\sum_{n=1}^{\infty} |\langle \varphi_n | \psi \rangle|^2 \leq \|\psi\|^2$$

show that

$$\sum_{n=1}^{\infty} \cos^2 \theta_n \leq 1$$

Under what circumstances does the equality hold?

4.24   Every convergent sequence is also a *Cauchy sequence*. A sequence $\{\varphi_n(x)\}$ is a Cauchy sequence if

$$\lim_{\substack{n \to \infty \\ l \to \infty}} \|\varphi_n - \varphi_l\| = 0$$

A function space $\mathfrak{H}$ is a *complete space* if every Cauchy sequence in $\mathfrak{H}$ converges to an element of $\mathfrak{H}$. This is a requirement that a function space must satisfy in order that it be termed a Hilbert space. [See property 4 after (4.27).] Show that the space of functions on the unit interval with the property $\varphi(0) = \varphi(1) = 0$ is not a Hilbert space.

4.25   In addition to a complete space, one also defines a *complete sequence*. An orthonormal sequence $\{\varphi_n\}$ is complete in $\mathfrak{H}$ if there is no vector ψ, in $\mathfrak{H}$ of nonzero length ($\|\psi\| > 0$), which is perpendicular to all the elements in the sequence $\{\varphi_n\}$. Show that if $\{\varphi_n\}$ is an orthonormal basis of $\mathfrak{H}$, it is complete in $\mathfrak{H}$.

Answer

Let $\{\varphi_n\}$ be an orthonormal basis of $\mathfrak{H}$. Let ψ be an element of $\mathfrak{H}$ with nonzero length, which is normal to all the elements of $\{\varphi_n\}$. If $\{\varphi_n\}$ is a basis, then we may expand ψ.

$$\psi = \sum a_n \varphi_n = \sum \langle \varphi_n | \psi \rangle \varphi_n$$

But ψ is normal to all φ_n. Therefore, $\langle \varphi_n | \psi \rangle = 0$, which gives $\psi = 0$, so the hypothesis leads to a contradiction, hence the hypothesis is an incorrect statement and there is no such ψ in $\mathfrak{H}$.

4.26 Show that any operator $\hat{A}$ may be expressed as the linear combination of a Hermitian and an anti-Hermitian ($\hat{B}^\dagger = -\hat{B}$) operator.

Answer

$$\hat{A} = \left(\frac{\hat{A} + \hat{A}^\dagger}{2} \right) + i \left(\frac{\hat{A} - \hat{A}^\dagger}{2i} \right)$$

[*Note:* $\hat{A} + \hat{A}^\dagger$ and $i(\hat{A} - \hat{A}^\dagger)$ are both Hermitian.]

4.27 Show that the wavefunctions for a particle in a one-dimensional box with walls at $x = 0$ and a satisfy the equality

$$\int_0^a \psi^* \psi_{xx} \, dx = -\int_0^a |\psi_x|^2 \, dx$$

The subscript x denotes differentiation.

4.28 Use the equality proved in Problem 4.27 to establish the following *variational principle*. If the expectation $\int \psi^* \hat{H} \psi \, dx$ is minimum, the normalized wavefunction ψ is the ground state. Specifically, establish the theorem for a particle in a one-dimensional box, assuming real wavefunctions.

Answer

Apart from a constant factor and with the results of Problem 4.27, we may write

$$\langle H \rangle = -\int_0^a \psi^* \psi_{xx} \, dx = -\int_0^a \psi_x^2 \, dx$$

Let ψ minimize $\langle H \rangle$. Then infinitesimal variation of ψ causes no change in $\langle H \rangle$. Let $\psi \to \psi + \delta\psi$. The variation $\delta\psi$ is an arbitrary infinitesimal function of x that vanishes at $x = 0$ and a. Then

$$\langle H \rangle = \int \psi_x^2 \, dx \to \int (\psi_x + \delta\psi_x)^2 \, dx = \langle H \rangle + \delta\langle H \rangle$$

$$\delta\langle H \rangle = 2\int \psi_x \, \delta\psi_x \, dx = 2\int \psi_x \frac{d}{dx} \delta\psi \, dx = 0$$

Integrating the last term by parts and dropping the "surface" terms gives

$$\int \psi_{xx} \, \delta\psi \, dx = 0$$

Variation of the normalization statement (both ψ and $\psi + \delta\psi$ are normalized) gives

$$\lambda \int \psi \delta\psi \, dx = 0$$

whence λ is an arbitrary undetermined multiplier. Combining the last two equations yields

$$\int_0^a \delta\psi(\psi_{xx} - \lambda\psi) \, dx = 0$$

If this equation is to be satisfied for arbitrary variation of ψ about the minimizing value, we may conclude

$$\psi_{xx} = \lambda\psi$$

where ψ is an eigenstate of $\hat{H}$. It follows that, in any eigenstate, $\langle H \rangle$ is stationary and is minimum in the ground state.

4.29   Let

$$A_{nl} \equiv \langle \varphi_n | \hat{A}\varphi_l \rangle$$

Show that

$$(\hat{A}^\dagger)_{ln} = (A_{nl})^*$$

Answer

$$A_{nl} = \langle \varphi_n | \hat{A}\varphi_l \rangle$$
$$= \langle \hat{A}\varphi_l | \varphi_n \rangle^* = \langle \varphi_l | \hat{A}^\dagger \varphi_n \rangle^*$$

Taking the complex conjugate of the last and first terms in this equality gives the desired result.

4.30   Employing Hermitian properties of $\hat{H}$, show that in general, extreme values of $\langle \hat{H} \rangle$ yield the eigenfunctions of $\hat{H}$, and that any eigenfunction of $\hat{H}$ makes $\langle \hat{H} \rangle$ an extremum. Your derivation should be independent of specific boundary conditions.

Answer

To incorporate wavefunction normalization in the analysis we write

$$\langle \hat{H} \rangle = \frac{\langle \psi | \hat{H}\psi \rangle}{\langle \psi | \psi \rangle}$$

Taking the variation of this form we obtain

$$(\langle \psi | \psi \rangle)^2 \, \delta \langle \hat{H} \rangle = -\langle \psi | \hat{H}\psi \rangle [\langle \delta\psi | \psi \rangle + \langle \psi | \delta\psi \rangle]$$
$$+ \langle \psi | \psi \rangle [\langle \delta\psi | \hat{H}\psi \rangle + \langle \psi | \hat{H}\delta\psi \rangle] + O[(\delta\psi)^2]$$

Neglecting terms of $O[(\delta\psi)^2]$, and recalling the hermiticity of $\hat{H}$, the preceding equation becomes

$$(\langle \psi | \psi \rangle)^2 \, \delta \langle \hat{H} \rangle = -\langle \psi | \hat{H}\psi \rangle [\langle \delta\psi | \psi \rangle + \langle \psi | \delta\psi \rangle]$$
$$+ \langle \psi | \psi \rangle [\langle \delta\psi | \hat{H}\psi \rangle + \langle \hat{H}\psi | \delta\psi \rangle]$$

With $\langle \psi | \hat{H}\psi \rangle = \lambda$, and $\langle \psi | \psi \rangle = 1$, it follows that any function ψ which makes $\delta \langle \hat{H} \rangle = 0$, to first order in $\delta\psi$, gives

$$0 = 2 \, \text{Re} \, \langle \delta\psi | (-\lambda + \hat{H})\psi \rangle$$

If we label $\langle \delta\psi | (-\lambda + \hat{H})\psi \rangle \equiv z$, where z is a complex number, then the preceding statement remains valid if $\delta\psi$ is multiplied by z, in which case the preceding gives

$$\langle \delta\psi | (-\lambda + \hat{H})\psi \rangle = 0$$

If this relation is to be satisfied for any arbitrary variation, $\delta\psi$, it must be the case that, $\hat{H}\psi = \lambda\psi$. Furthermore, it is clear from the equation above for $\delta \langle \hat{H} \rangle$, that any eigenfunction of $\hat{H}$ makes $\langle \hat{H} \rangle$ an extremum.

4.31 (a) An electron propagates through a periodic potential with period a. What is the maximum energy, E_e, for which electron dynamics is quantum-mechanical?

(b) Repeat the preceding problem for the case of a proton. Call your answer E_p.

(c) What can you conclude regarding classical versus quantum behavior for these two particles from the ratio E_p/E_e? Call electron and proton masses m and M, respectively.

Answer (partial)

(c) The motion is quantum-mechanical if $\lambda \gtrsim a$, where λ is the de Broglie wavelength. There results $E_p/E_e = \sqrt{m/M} \ll 1$. Thus, there is a wide range of energies for which electron motion is quantum-mechanical, but proton motion is classical.

4.32 Our definition of Hermitian adjoint (4.52) is that $\langle \hat{A}^\dagger \varphi_n | \varphi_m \rangle = \langle \varphi_n | \hat{A}\varphi_m \rangle$ is satisfied for all φ_n and φ_m in the relevant Hilbert space. Show that an equivalent definition is that $\langle \varphi_n | \hat{A}^\dagger \varphi_m \rangle = \langle \hat{A}\varphi_n | \varphi_m \rangle$ for all φ_n and φ_m in the same Hilbert space.

4.33 (a) A particle in a one-dimensional box is in the superposition state

$$\varphi(x) = b_1 e^{i\lambda_1} \varphi_1(x) + b_3 e^{i\lambda_3} \varphi_3(x)$$

where λ_1 and λ_3 are arbitrary real phase factors, b_1 and b_3 are real coefficients, and

$$b_1^2 + b_3^2 = 1$$

What is the functional dependence of the probability density, $\varphi^*\varphi$, on λ_1, λ_3?

(b)   Repeat the preceding problem for the superposition state

$$\varphi(x) = b_1 e^{i\lambda_1}\varphi_1(x) + b_3 e^{i\lambda_3}\varphi_3(x) + b_7 e^{i\lambda_7}\varphi_7(x)$$

where

$$b_1^2 + b_3^2 + b_7^2 = 1$$

Answers

(a)   The probability density is a function of $|\lambda_1 - \lambda_3|$.

(b)   The probability density is a function of $|\lambda_1 - \lambda_7|, |\lambda_1 - \lambda_3|, |\lambda_7 - \lambda_3|$.

4.34   Given that the inverses $\hat{A}^{-1}$ and $\hat{B}^{-1}$ are known:

(a)   What is the inverse of $\hat{S} = \hat{A}\hat{B}$?

(b)   Does your answer to part (a) depend on the Hermiticity of $\hat{A}$ or $\hat{B}$?

(c)   Employ your answer to part (a) to establish that if $\hat{A}^{-1}\hat{A} = \hat{I}$ then $\hat{A}\hat{A}^{-1} = \hat{I}$.

(An operator $\hat{A}$ and its inverse, $\hat{A}^{-1}$, satisfy the relation $\hat{A}^{-1}\hat{A} = \hat{I}$. For further discussion see Section 11.2.)

4.35   An electron in a one-dimensional box with walls at $x = (0, a)$ is in the quantum state

$$\psi(x) = A, \qquad 0 < x < a/2$$

$$\psi(x) = -A, \qquad a/2 < x < a$$

(a)   Obtain an expression for the normalization constant, A.

(b)   What is the lowest energy of the electron that will be measured in this state?

4.36   (a) Is the following wavefunction an energy eigenstate for a free particle moving in one dimension?

$$\psi(x) = A e^{ikx} + \frac{A}{\sqrt{2}} e^{-ikx}$$

(b)   Is this wavefunction an eigenstate of the one-dimensional momentum operator?

(c)   If your answer to part (b) is no, what values of momentum will be found in measurement and with what probabilities will these values occur?

5

SUPERPOSITION AND COMPATIBLE OBSERVABLES

5.1 *The Superposition Principle*
5.2 *Commutator Relations in Quantum Mechanics*
5.3 *More on the Commutator Theorem*
5.4 *Commutator Relations and the Uncertainty Principle*
5.5 *"Complete" Sets of Commuting Observables*

In this chapter we encounter the superposition principle, which is considered by many to be one of the more fundamental concepts of quantum mechanics. This principle represents one of the basic differences between classical and quantum mechanics and also provides a deeper understanding of the uncertainty principle. Closely related to the superposition principle are the commutator theorem and the notions of compatible observables and simultaneous eigenfunctions.

5.1 THE SUPERPOSITION PRINCIPLE

Ensemble Average

Consider again a particle in a one-dimensional box. Let us imagine a large number of identical replicas of the system (called an *ensemble* in statistical mechanics), such as described in Section 3.3. If each such box is in the *same* initial state $\psi(x, 0)$, after an interval of time t, each box will again be in a common state $\psi(x, t)$, as shown in Fig. 5.1. Suppose that we ask what the energy of the particle is in each box, at the time t. The laws of nature are such that the energy measured in each of the identical boxes, which are all in the identically same state $\psi(x, t)$, are not the same [save for the case that $\psi(x, 0)$ is an eigenstate of $\hat{H}$].

How does one answer the question above: What will the energy be? Since the energy measured at the time t in each box of the ensemble will most likely not be the same, more appropriate questions are: (1) What is the average of the energies measured in all the boxes of the ensemble? (2) If we measure the energy in one box, with what probability will the value, say, E_3 be found? To answer these questions, we first

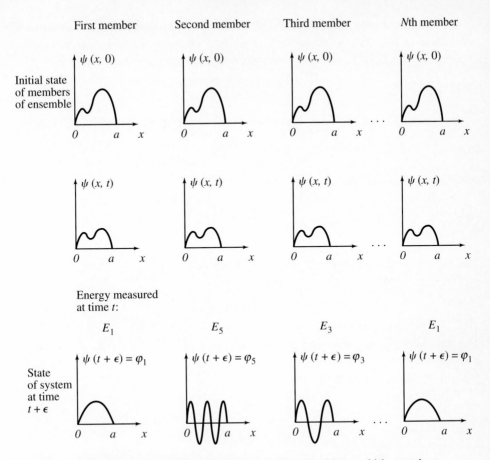

FIGURE 5.1 **Measurement of energy of N identical one-dimensional boxes which comprise an "ensemble." All boxes are in the same state at $t = 0$.**

recall that if the probability of finding the value E_n in a given measurement of energy is $P(E_n)$, then the average energy over measurements of all members of the ensemble in the limit as this number becomes large is given by the expression

$$(5.1) \qquad \langle E \rangle = \sum_{\text{all } E_n} P(E_n)E_n$$

[Recall (3.34)]. This formula holds for all physical observables. For example, the average particle position is given by

$$(5.2) \qquad \langle x \rangle = \int_0^a xP(x) \, dx$$

In this case the integral is a sum over the continuum of values x may assume.

The quantum mechanical prescription for calculating the average of a dynamical observable in the state ψ is given by the third postulate of quantum mechanics [Section 3.3, (3.32)]. Specifically, for the energy we have (in Dirac notation)

$$(5.3) \qquad \langle E \rangle = \langle \psi | \hat{H} \psi \rangle$$

Let us expand the state ψ in the eigenstates of $\hat{H}$. These eigenstates obey the equation

$$(5.4) \qquad \hat{H}\varphi_n = E_n \varphi_n$$

For the box problem they are explicitly (4.15)

$$(5.5) \qquad \varphi_n = \sqrt{\frac{2}{a}} \sin\left(\frac{n\pi x}{a}\right)$$

The expansion of ψ in these eigenstates appears as

$$(5.6) \qquad \psi(x, t) = \sum_{n=1}^{\infty} b_n(t)\varphi_n(x)$$

The state ψ is that of the system at the time t, so that it is, in general, a function of x and t. Since φ_n is a function of x only, the coefficients of expansion b_n may, in general, be functions of time.

In Dirac notation, (5.6) appears as

$$(5.7) \qquad |\psi\rangle = \sum_{n=1}^{\infty} |b_n\varphi_n\rangle$$

Substituting this series into (5.3) gives

$$
\begin{aligned}
(5.8) \qquad \langle E \rangle &= \left\langle \sum_n b_n\varphi_n \middle| \hat{H} \sum_l b_l\varphi_l \right\rangle \\
&= \sum_n \sum_l b_n{}^* b_l \langle \varphi_n | \hat{H}\varphi_l \rangle \\
&= \sum_n \sum_l b_n{}^* b_l E_l \langle \varphi_n | \varphi_l \rangle \\
&= \sum_n \sum_l b_n{}^* b_l E_l \delta_{nl} \\
&= \sum_{n=1}^{\infty} |b_n|^2 E_n
\end{aligned}
$$

Equating this average to that given by (5.1) gives

$$\sum_n |b_n|^2 E_n = \sum_n P(E_n)E_n \tag{5.9}$$

This equation dictates the following interpretation of the square of the modulus of b_n. It is the probability that at the time t, measurement of the energy of the particle which is in the state $\psi(x, t)$ yields the value E_n.

$$P(E_n) = |b_n|^2 \tag{5.10}$$

These coefficients have the correct normalization, provided that the states ψ and φ_n are normalized. In this case we have

$$1 = \langle \psi | \psi \rangle = \left\langle \sum_n b_n \varphi_n \middle| \sum_l b_l \varphi_l \right\rangle \tag{5.11}$$

$$= \sum_n \sum_l b_n{}^* b_l \langle \varphi_n | \varphi_l \rangle$$

$$= \sum_n \sum_l b_n{}^* b_l \delta_{nl}$$

$$= \sum_n |b_n|^2 = 1$$

When this is the case the coefficient $|b_n|^2$ is an *absolute* probability. If not, the correct expression for the probability that measurement finds E_n is

$$P(E_n) = \frac{|b_n|^2 |C_n|^2}{\sum |b_n|^2 |C_n|^2} = \frac{|b_n|^2 |C_n|^2}{\langle \psi | \psi \rangle} \tag{5.12}$$

where

$$|C_n|^2 = \langle \varphi_n | \varphi_n \rangle$$

Let us return to the expansion (5.7). The coefficients b_n are calculated in the following manner. Multiply this equation from the left with the bra vector $\langle \varphi_{n'} |$. Owing to the orthonormality of the set $\{\varphi_n\}$, one obtains

$$b_n = \langle \varphi_n | \psi_n \rangle \tag{5.13}$$

The coefficient b_n is the projection of ψ onto the eigenvector φ_n. The physical interpretation of b_n is that $|b_n|^2$ is the probability that measuring E finds the value E_n when the system is in the state ψ. This prescription is true for *any* dynamical observable. Consider the symbolic operator $\hat{F}$

$$(5.14) \qquad \hat{F}\varphi_n = f_n\varphi_n$$

At a given time t, the system is in the state $\psi(x, t)$. What is the probability that measurement of F at this time finds the value f_3? The state ψ is a superposition state. It is composed of the eigenstates of $\hat{F}$. Here we are assuming that the eigenstates of $\hat{F}$ are a basis for the Hilbert space that ψ is in. So we may write

$$(5.15) \qquad \psi = \sum b_n\varphi_n$$

$$(5.16) \qquad b_n = \langle \varphi_n | \psi \rangle$$

This assumption that an arbitrary state ψ may be represented as a superposition of the eigenstates of a physical observable is the essence of the superposition principle. With $\{\varphi_n\}$ and ψ normalized to unity, the probability that measurement finds the value f_3 is $|b_3|^2$. This procedure is depicted in Fig. 5.2.

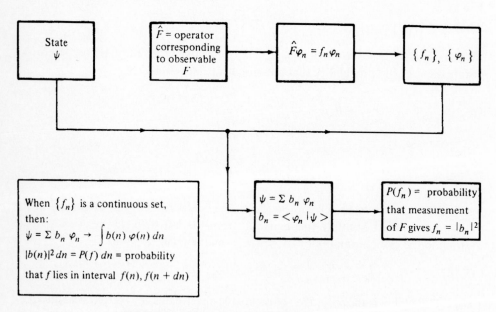

FIGURE 5.2   Elements of the superposition principle.

Hilbert-Space Interpretation

When we look in Hilbert space, $\{\varphi_n\}$ is one set of vectors and ψ is another vector. The system is in the state ψ. Measurement of F causes the state ψ to fall to one of the φ_n vectors. Chances are that it goes to the φ_n vector to which it is most inclined (in the geometrical sense; see Fig. 5.3).

   Consider the following illustrative example. A particle of mass m is in a one-dimensional box of width a. At $t = 0$ the particle is in the state

$$(5.17) \qquad \psi(x, 0) = \frac{3\varphi_2 + 4\varphi_9}{\sqrt{25}}$$

The φ_n functions are the orthonormal eigenstates of $\hat{H}$:

$$(5.18) \qquad \varphi_n = \sqrt{\frac{2}{a}} \sin\left(\frac{n\pi x}{a}\right)$$

What will measurements of E yield at $t = 0$ and what is the probability of finding this value? First let us see if ψ is normalized. In Dirac notation we have, for the state (5.17),

$$(5.19) \qquad |\psi\rangle = \frac{3\,|\,\varphi_2\rangle + 4\,|\,\varphi_9\rangle}{\sqrt{25}}$$

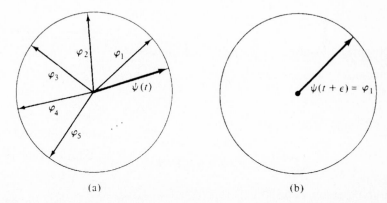

(a)   (b)

FIGURE 5.3   **(a) State of the system before measurement at t, superimposed on the basis $\{\varphi_n\}$, which are the eigenvectors of the operator $\hat{F}$. The probability that measurement of F finds the value f_n is proportional to the projection of ψ on φ_n. (b) State of the system immediately after measurement has found the value f_1. Measurement acts as a "wave filter." It filters out all components of the superposition $\psi(x, t) = \Sigma\, b_n(t)\varphi_n(x)$, passing only the φ_1 wave.**

so that

(5.20) $\langle \psi | \psi \rangle = \dfrac{1}{25}\{(3\langle \varphi_2| + 4\langle \varphi_9|)(3|\varphi_2\rangle + 4|\varphi_9\rangle)\}$

$= \dfrac{1}{25}\{9\langle \varphi_2|\varphi_2\rangle + 12\langle \varphi_2|\varphi_9\rangle + 12\langle \varphi_9|\varphi_2\rangle + 16\langle \varphi_9|\varphi_9\rangle\}$

$= 1$

and ψ is normalized. The inner products $\langle \varphi_2|\varphi_2\rangle = \langle \varphi_9|\varphi_9\rangle = 1$ while the other two are zero, owing to the orthogonality of the set $\{\varphi_n\}$.

The superposition principle stipulates the following. If we want the probability that measurement finds the value E_n, we must expand ψ in the eigenstates of $\hat{H}$. The square of the magnitude of the coefficient of φ_n is the said probability.

(5.21) $\psi = \sum b_n \varphi_n = \dfrac{3\varphi_2 + 4\varphi_9}{\sqrt{25}}$

In this simplified problem, by inspection we find that

$$b_2 = \frac{3}{\sqrt{25}}$$

(5.22) $$b_9 = \frac{4}{\sqrt{25}}$$

$$b_n = 0 \qquad (n \neq 2 \text{ or } 9)$$

Therefore, the probability $P(E_n)$ that measurement of E at $t = 0$ finds the value E_n is

$$P(E_2) = \frac{9}{25}$$

(5.23) $$P(E_9) = \frac{16}{25}$$

$$P(E_n) = 0 \qquad (n \neq 2 \text{ or } 9)$$

In an ensemble of 2500 identical one-dimensional boxes, each containing an identical particle in the same state $\psi(x, 0)$ given by (5.17), measurement of E at $t = 0$ finds about 900 particles to have energy $E_2 = 4E_1$ and about 1600 particles to have energy $E_9 = 81E_1$.

Is there a chance that, in an ensemble of 10^{17} boxes, measurement of E finds E_2 in all 10^{17} boxes? Yes. This remarkable response carries the philosophical impact of the superposition principle. Although the state $\psi(x, 0)$ is a precise superposition of well-defined eigenstates of the observables being measured, one is not *certain* what measurement will yield. There is nothing in classical physics that is similar to this concept. Any uncertainty in classical physics arises from uncertain initial data. In quantum mechanics, although the initial state $\psi(x, 0)$ is prescribed with perfect accuracy, one is never certain in which eigenstate, φ_n, measurement will leave the system.

However, once E is measured and, say, the value E_9 is found, then one knows with absolute certainty that the state of the system immediately after this measurement is φ_9.

The Initial Square Wave

As a second illustrative example, we consider the following free-particle problem in one dimension. Suppose that at $t = 0$ the system is in the state (Fig. 5.4)

$$(5.24) \qquad \psi(x, 0) \begin{cases} \sqrt{\dfrac{1}{a}} & |x| < \dfrac{a}{2} \\ 0 & \text{elsewhere} \end{cases}$$

If at this same instant, the momentum of the particle is measured, what are the possible values that will be found, and with what probability will these values occur?

To answer these questions we must first expand $\psi(x, 0)$ in a superposition of the eigenstates of $\hat{p}$:

$$(5.25) \qquad \varphi_k = \frac{1}{\sqrt{2\pi}} e^{ikx}$$

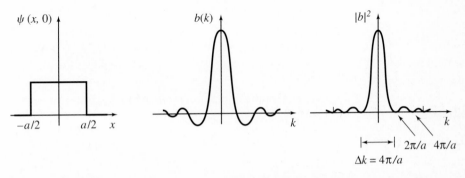

FIGURE 5.4 Square wave packet at $t = 0$ and corresponding momentum eigenstate amplitudes $b(k)$. The interval over which momentum values are most likely to be found is $\Delta p = \hbar\, \Delta k = 4\pi\hbar/a.$

Since these states comprise a continuum, the corresponding superposition of eigenstates of $\hat{p}$ is an integral.

$$(5.26) \qquad \psi(x, 0) = \int_{-\infty}^{\infty} b(k)\varphi_k \, dk$$

Inverting this equation (see 4.42 et seq.) gives the coefficient $b(k)$.

$$(5.27) \qquad b(k) = \int_{-\infty}^{\infty} \psi(x, 0)\varphi_k{}^* \, dx = \frac{1}{\sqrt{2\pi}} \int_{-\infty}^{\infty} \psi(x, 0)e^{-ikx} \, dx$$

$$= \frac{1}{\sqrt{2\pi a}} \int_{-a/2}^{+a/2} e^{-ikx} \, dx = \frac{1}{\sqrt{2\pi a}} \frac{2}{k}\left(\frac{e^{ika/2} - e^{-ika/2}}{2i}\right)$$

$$= \sqrt{\frac{2}{\pi a}} \frac{\sin(ka/2)}{k}$$

Again, this coefficient is the projection of the state $\psi(x, 0)$ onto the eigenstate φ_k. Its square times the differential dk is the probability that measurement of momentum yields $p = \hbar k$, in the interval $\hbar k$, $\hbar(k + dk)$. The corresponding probability density (in momentum space) is

$$(5.28) \qquad |b|^2 = \frac{2}{\pi a} \frac{\sin^2(ka/2)}{k^2}$$

This function has its maximum at $k = 0$. It drops to zero at

$$(5.29) \qquad \frac{ka}{2} = \pi$$

or equivalently at

$$(5.30) \qquad p = \hbar k = \frac{2\pi\hbar}{a}$$

It is most probable that measurement of momentum finds the value $p = 0$. The momentum values $(\pm n2\pi\hbar/a)$ with n an integer greater than 1 are never found, for at these values, $b(k) = 0$.

Referring to Fig. 5.4, we see that the interval of momentum values that measurements are most likely to uncover has the approximate width

$$\Delta k = \frac{4\pi}{a}$$

$$(5.31)$$

$$\Delta p = \hbar \, \Delta k = \frac{4\pi\hbar}{a}$$

On the other hand, from (5.24), it is uniformly probable that measurement of x finds the particle anywhere in the interval $(-a/2, +a/2)$, of width

$$\text{(5.32)} \qquad \Delta x = a$$

Combining these two latter uncertainties (5.31 and 5.32) gives

$$\text{(5.33)} \qquad \Delta x \, \Delta p \simeq \hbar$$

The approximation sign is used because of the qualitative manner in which Δp was calculated. The result (5.33) is another example of the Heisenberg uncertainty principle at work.

The Chopped Beam

To further exhibit the significance of the probability density $|b(k)|^2$, we consider the following problem. Suppose that the free-particle system above is composed of N noninteracting electrons. Every electron is in the state $\psi(x, 0)$ given by (5.24). The density ρ (number/length) is related to ψ through

$$\text{(5.34)} \qquad \text{Number of particles in } dx = \rho \, dx = N \, |\psi|^2 \, dx$$

The total number in the whole "beam" is

$$\text{(5.35)} \qquad N = \int_{-\infty}^{\infty} \rho(x) \, dx = N \int_{-a/2}^{a/2} |\psi|^2 \, dx = N$$

Suppose that we now ask how many electrons have momentum in the interval $(-2\pi\hbar/a, +2\pi\hbar/a)$, or equivalently, how many have wavenumber in the interval $(-2\pi/a, 2\pi/a)$. For a single electron, the probability of finding an electron with momentum in the interval $\hbar k$ to $\hbar k + \hbar dk$ is

$$\text{(5.36)} \qquad P(k) \, dk = |b(k)|^2 \, dk$$

This is a correct statement provided that

$$\text{(5.37)} \qquad \int_{-\infty}^{\infty} |b(k)|^2 \, dk = 1$$

If this is not the case, one must divide $|b(k)|^2$ in (5.36) by the last integral.

For a totality of N electrons in the beam, the number of them that have momentum in the interval $\hbar k$, $\hbar k + \hbar dk$ is

(5.38)
$$\rho(k)\,dk = N\,|b(k)|^2\,dk$$

The total number in the whole beam is

(5.39)
$$N = \int_{-\infty}^{\infty} \rho(k)\,dk = N \int_{-\infty}^{\infty} |b(k)|^2\,dk$$

For the example at hand

(5.40)
$$\int_{-\infty}^{\infty} |b(k)|^2\,dk = \frac{2}{\pi a} \int_{-\infty}^{\infty} \frac{\sin^2 (ka/2)}{k^2}\,dk$$

$$= \frac{1}{\pi} \int_{-\infty}^{\infty} \frac{\sin^2 \eta\,d\eta}{\eta^2} = 1$$

The dummy variable $\eta \equiv ka/2$. To return to the original question, the number of electrons ΔN in the beam with momentum in the interval $(-2\pi\hbar/a, +2\pi\hbar/a)$ is given by the integral

(5.41)
$$\Delta N = N \int_{-2\pi/a}^{+2\pi/a} \frac{2}{\pi a} \frac{\sin^2 (ka/2)}{k^2}\,dk$$

$$= \frac{N}{\pi} \int_{-\pi}^{+\pi} \frac{\sin^2 \eta}{\eta^2}\,d\eta = 0.903N$$

Thus, we find a majority of the electrons in this momentum interval.

Superposition and Uncertainty

Let us return to the case of a single electron in the state $\psi(x, 0)$ given by (5.24). Suppose at this time, $t = 0$, we measure the electron's momentum. What value do we find? The answer is (a) the values $p = \pm n2\pi\hbar/a$ are never found; (b) any other value may occur with corresponding probability density $|b(k)|^2$. Let the measurement find the electron to have the momentum

(5.42)
$$p = \frac{\pi\hbar}{a}$$

Immediately after this measurement, what is the state of the particle? The answer is

$$(5.43) \qquad \psi = \frac{1}{\sqrt{2\pi}} \exp\left(\frac{i\pi x}{a}\right)$$

The electron is now in the state (5.43). Suppose that we measure the energy of the particle. What value is found? Since this state is also an eigenstate of $\hat{H}$, it is a certainty that measurement yields

$$(5.44) \qquad E = \frac{(\pi\hbar/a)^2}{2m}$$

The system is still left in the eigenstate (5.43). Suppose that we now measure the position of the particle. What values may occur? The probability density is

$$(5.45) \qquad P = |\psi|^2 = \frac{1}{2\pi}$$

which is a constant. It is uniformly probable to find the electron anywhere along the whole x axis. The uncertainty in x is $\Delta x = \infty$. For this same state it is certain that measurement of momentum finds the value $\pi\hbar/a$, so that $\Delta p = 0$. Again we find corroborating evidence for the Heisenberg uncertainty principle.

Now we place a uniform array of scintillation detectors along the x axis. One of them scintillates at $x = x'$. What is the state of the electron immediately after measurement? The answer is the eigenstate of the position operator corresponding to the eigenvalue x' (Fig. 5.5).

$$(5.46) \qquad \psi = \delta(x - x')$$

Now we measure momentum again. What values can be found? To answer this question, we again call on the superposition recipe: expand ψ in the eigenstates of $\hat{p}$.

$$(5.47) \qquad \delta(x - x') = \frac{1}{\sqrt{2\pi}} \int_{-\infty}^{\infty} b(k)e^{ikx} \, dk$$

$$b(k) = \frac{1}{\sqrt{2\pi}} \int \delta(x - x')e^{-ikx} \, dx = \frac{1}{\sqrt{2\pi}} e^{-ikx'}$$

The corresponding momentum probability density is

$$(5.48) \qquad P(k) = |b(k)|^2 = \frac{1}{2\pi}$$

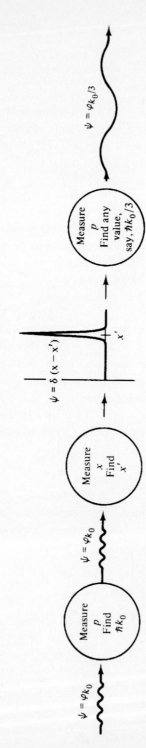

FIGURE 5.5 Measuring x destroys the momentum eigenstate φ_{k_0}.

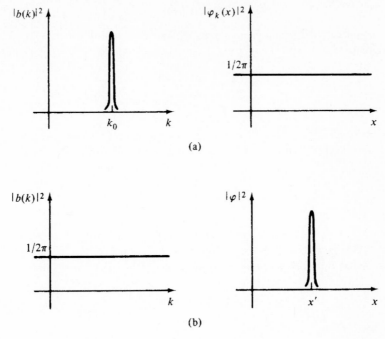

FIGURE 5.6   (a) **In the state** $\psi = \varphi_{k_0} = (1/\sqrt{2\pi})e^{ik_0 x}$, $\Delta p = 0$ **and** $\Delta x = \infty$. (b) **In the state** $\psi = \delta(x - x')$, $\Delta x = 0$ **and** $\Delta p = \infty$.

It is uniformly probable to find the electron with any momentum along the whole k axis. The uncertainty in momentum is $\Delta p = \infty$ for the state (5.46), for which $\Delta x = 0$, and the uncertainty principle holds firm (Fig. 5.6).

We have been using the phrase superposition principle, but have not given a concise statement of this principle. P. A. M. Dirac, one of the early investigators of quantum mechanics, was first to grasp the full significance of this principle. His description[1] is perhaps the most succinct. The superposition principle "requires us to assume that between . . . states there exist peculiar relationships such that whenever the system is definitely in one state we can consider it as being partly in each of two or more other states. The original state must be regarded as the result of a kind of superposition of the two or more new states, in a way that cannot be conceived on classical ideas."

The superposition principle is a cornerstone of quantum mechanics. We have used it previously in some elementary one-dimensional problems. We will return to it in the remainder of the text in relation to more extensive one-dimensional problems as well as more practical problems in two and three dimensions. A sound understanding of this principle is prerequisite to a working knowledge of quantum mechanics.

[1] P. A. M. Dirac, *The Principles of Quantum Mechanics,* 4th ed., Oxford University Press, New York, 1958.

PROBLEMS

5.1 If an arbitrary initial state function for a particle in a one-dimensional box is expanded in the discrete series of eigenstates of the Hamiltonian relevant to the box configuration, one obtains (5.6)

$$\psi(x, 0) = \sum_{n=1}^{\infty} b_n(0)\varphi_n(x)$$

On the other hand, if the particle is free, its Hamiltonian has a continuous spectrum of eigen-energies and the superposition of an arbitrary initial state in the eigenstates φ_k of $\hat{H}$ becomes an integral (5.26):

$$\psi(x, 0) = \int_{-\infty}^{\infty} b(k)\varphi_k \, dk$$

(a) What are the dimensions of $|b_n|^2$ and $|b(k)|^2$, respectively?
(b) What is the source of the difference to dimensionality?
(c) What are the dimensions and physical interpretation of the integral

$$\int_{-\infty}^{\infty} |b(k)|^2 \, dk?$$

Answer (partial)

(b) The term $|b_n|^2$ represents a probability, whereas $|b(k)|^2$ represents a probability density.

5.2 One thousand neutrons are in a one-dimensional box, with walls at $x = 0$, $x = a$. At $t = 0$, the state of each particle is

$$\psi(x, 0) = Ax(x - a)$$

(a) Normalize ψ and find the value of the constant A.
(b) How many particles are in the interval $(0, a/2)$ at $t = 0$?
(c) How many particles have energy E_5 at $t = 0$?
(d) What is $\langle E \rangle$ at $t = 0$?

5.3 Using the expressions for φ_k and ψ given by (5.25) and (5.26), respectively, show that

$$\langle \psi | \psi \rangle = 1 \rightarrow \int_{-\infty}^{\infty} |b(k)|^2 \, dk = 1$$

5.4 A pulse 1 m long contains 1000 α particles. At $t = 0$, each α particle is in the state

$$\psi(x, 0) = \begin{cases} \dfrac{1}{10}e^{ik_\sigma x}, & |x| \leq 50 \text{ cm}, \ k_0 = \pi/50 \\ 0 & \text{elsewhere} \end{cases}$$

(a)   At $t = 0$, how many α particles have momentum in the interval $(0 < \hbar k < \hbar k_0)$?

(b)   At which values of momentum will α particles not be found at $t = 0$?

(c)   Describe an experiment to "prepare" such a state.

(d)   Construct Δx and Δp for this state, formally. What is $\Delta x \, \Delta p$? [*Hint:* To calculate Δp, use $|b(k)|^2$.]

5.5   At $t = 0$ it is known that of 1000 neutrons in a one-dimensional box of width 10^{-5} cm, 100 have energy $4E_1$, and 900 have energy $225E_1$.

(a)   Construct a state function that has these properties. (Coefficients may be complex.)

(b)   Use the state you have constructed to calculate the density $\rho(x)$ of neutrons per unit length. [Note that $\rho(x)$ is a real function.]

(c)   How many neutrons are in the left half of the "box"?

5.6   Over a very long interval of the x axis, a uniform distribution of 10,000 electrons is moving to the right with velocity 10^8 cm/s and 10,000 electrons are moving to the left with velocity 10^8 cm/s. Assuming that the electrons do not interact with one another, construct a state function that yields the preceding properties for the combined beam. Calculate $\langle p \rangle$ for this state.

5.7   Given an argument in support of the conjecture that one cannot measure the momentum of a particle in a one-dimensional box with absolute accuracy. Support the theoretical argument with an argument involving an experiment.

5.8   A one-dimensional box containing an electron suffers an infinitesimal perturbation and emits a photon of frequency

$$h\nu = 3E_1$$

where E_1 denotes the ground state of the particle. A student concludes that the electron was in the state φ_2 prior to perturbation. Is he correct?

Answer

What the student has in mind is that the photon corresponds to the decay

$$h\nu = E_2 - E_1 = 3E_1$$

However, suppose that the electron was in the superposition state $(3\varphi_2 + 8\varphi_6)/\sqrt{73}$. Then it is still possible that a photon of frequency $h\nu = 3E_1$ is emitted. So the student is incorrect.

5.9   Measurement of the position of a particle in a one-dimensional box with walls at $x = 0$ and $x = a$ finds the value $x = a/2$.

(a)   Show that in the subsequent measurement, it is equally probable to find the particle in any odd-energy eigenstate.

(b)   Show that the probability of finding the particle in any even eigenstate is zero. (An eigenstate φ_n is even if n is even and odd if n is odd.)

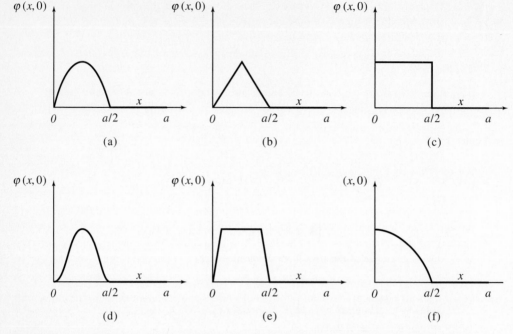

FIGURE 5.7 Six initial states for a particle in a one-dimensional box, with the property that $|\psi|^2 = 0$ in the right half of the box. (See Problem 5.10.)

5.10 It is known that at time $t = 0$, a particle in a box (described in Problem 5.9) is not in the right half of the box. The particle is in one of an infinite number of states. Six such states are depicted in Fig. 5.7.

(a) Write down an approximate wavefunction for each of these states.

(b) Calculate $\langle E \rangle$ for each of these states.

(c) Argue that the state depicted in Fig. 5.7a is the state of minimum $\langle E \rangle$ (assuming that $\varphi = A \sin 2\pi x/a$, $x < a/2$).

5.11 A particle in the one-dimensional box described in Problem 5.9 is in the ground state. One of the walls of the box is moved to the position $x = 2a$, in a time short compared to the natural period $2\pi/\omega_1$, where $\hbar\omega_1 = E_1$. If the energy of the particle is measured soon after this expansion, what value of energy is most likely to be found? How does this energy compare to the particle's initial energy (E_1)?

5.2 COMMUTATOR RELATIONS IN QUANTUM MECHANICS

An important operation in quantum mechanics is the *commutator* between two operators, $\hat{A}$ and $\hat{B}$. It is written $[\hat{A}, \hat{B}]$ and is defined as

(5.49) $$[\hat{A}, \hat{B}] = \hat{A}\hat{B} - \hat{B}\hat{A}$$

An immediate property of the commutator is that

(5.50)
$$[\hat{A}, \hat{B}] = -[\hat{B}, \hat{A}]$$

If

(5.51)
$$[\hat{A}, \hat{B}] = 0$$

the two operators are said to *commute* (*A* and *B* are *compatible*) with each other. That is,

(5.52)
$$\hat{A}\hat{B} = \hat{B}\hat{A}$$

Any operator $\hat{A}$ commutes with any constant a.

(5.53)
$$[\hat{A}, a] = 0$$

(5.54)
$$[\hat{A}, a\hat{B}] = [a\hat{A}, \hat{B}] = a[\hat{A}, \hat{B}]$$

Any operator $\hat{A}$ commutes with its own square, $\hat{A}^2$.

(5.55)
$$[\hat{A}, \hat{A}^2] = (\hat{A}\hat{A}^2 - \hat{A}^2\hat{A}) = (\hat{A}\hat{A}\hat{A} - \hat{A}\hat{A}\hat{A}) = 0$$

The meaning of this relation is that, no matter what $\hat{A}$ is, when $[\hat{A}, \hat{A}^2]$ operates on any function $g(x)$, one gets zero,

(5.56)
$$[\hat{A}, \hat{A}^2]g(x) = 0$$

More generally, $\hat{A}$ commutes with any function of $\hat{A}$, $f(\hat{A})$.

(5.57)
$$[f(\hat{A}), \hat{A}] = 0$$

As an example of this rule, consider the following commutator involving the momentum operator $\hat{p}$.

(5.58)
$$[e^{\hat{p}}, \hat{p}] = \left[\sum_{n=0}^{\infty} \frac{\hat{p}^n}{n!}, \hat{p} \right]$$

$$= \sum \frac{1}{n!} [\hat{p}^n, \hat{p}]$$

$$= [1, \hat{p}] + [\hat{p}, \hat{p}] + \frac{1}{2!} [\hat{p}^2, \hat{p}] + \cdots = 0$$

It follows that

(5.59)
$$[e^{\hat{p}}, \hat{p}]g(x) = \left[\exp\left(-\frac{i\hbar\partial}{\partial x} \right), -\frac{i\hbar\partial}{\partial x} \right] g(x) = 0$$

where $g(x)$ represents any function of x.

One of the most important commutators in physics is that between the coordinate, $\hat{x}$, and the momentum, $\hat{p}$. Let us calculate it.

(5.60)
$$[\hat{x}, \hat{p}]g(x) = i\hbar\left(-x\frac{\partial}{\partial x} + \frac{\partial}{\partial x}x\right)g(x)$$
$$= i\hbar\left(-x\frac{\partial g}{\partial x} + x\frac{\partial g}{\partial x} + g\right) = i\hbar g(x)$$

It follows that

(5.61)
$$\boxed{[\hat{x}, \hat{p}] = i\hbar}$$

In other words, the operator $[\hat{x}, \hat{p}]$ has the sole effect of a simple multiplication by the constant $i\hbar$. As an immediate consequence (using Problem 5.12)

(5.62)
$$[\hat{x}, \hat{p}^2] = [\hat{x}, \hat{p}]\hat{p} + \hat{p}[\hat{x}, \hat{p}]$$
$$= 2i\hbar\hat{p}$$

so that

(5.63)
$$[\hat{x}, \hat{p}^2]g(x) = 2\hbar^2 \frac{\partial g}{\partial x}$$

In a similar vein,

(5.64)
$$[\hat{x}^2, \hat{p}] = \hat{x}[\hat{x}, \hat{p}] + [\hat{x}, \hat{p}]\hat{x}$$
$$= 2i\hbar\hat{x} = 2i\hbar x$$

The operator $[\hat{x}^2, \hat{p}]$ multiplies by $2i\hbar x$.

We now prove an important theorem in quantum mechanics which is related to the commutator between two operators. It states: if $\hat{A}$ and $\hat{B}$ commute

(5.65)
$$[\hat{A}, \hat{B}] = 0$$

then $\hat{A}$ and $\hat{B}$ have a set of nontrivial (i.e., other than a constant) common eigenfunction. The proof is as follows.

Let φ_a be the eigenfunction of $\hat{A}$ that corresponds to the eigenvalue a.

(5.66)
$$\hat{A}\varphi_a = a\varphi_a$$

Then

(5.67) $$\hat{B}\hat{A}\varphi_a = a\hat{B}\varphi_a$$

Since $\hat{A}$ and $\hat{B}$ commute, the left-hand side of this last equation may be rewritten

(5.68) $$\hat{A}(\hat{B}\varphi_a) = a(\hat{B}\varphi_a)$$

Inspection of this equation reveals that $\hat{B}\varphi_a$ is also an eigenfunction of $\hat{A}$ corresponding to the eigenvalue a. If φ_a is the *only* linearly independent (defined below) eigenfunction of $\hat{A}$ that corresponds to the eigenvalue a, the function $\hat{B}\varphi_a$ can differ from φ_a by, at most, a multiplicative constant μ. That is,

(5.69) $$\hat{B}\varphi_a = \mu\varphi_a$$

($\hat{B}\varphi_a$ and $\mu\varphi_a$ are in the same *direction* in Hilbert space: see Fig. 5.8.) But this is the eigenvalue equation for the operator $\hat{B}$. It follows that φ_a is also an eigenfunction of $\hat{B}$.

   We have already encountered the implication of this theorem for the problem of the free particle moving in one dimension. For this case

(5.70) $$[\hat{p}, \hat{H}] = 0$$

It follows by the theorem above that $\hat{p}$ and $\hat{H}$ have common eigenfunctions. They do. We recall that

(5.71)
$$\hat{p}e^{ikx} = \hbar k e^{ikx}$$

$$\hat{H}e^{ikx} = \frac{\hbar^2 k^2}{2m} e^{ikx}$$

Before pursuing the case when φ_a is not the only linearly independent eigenfunction of $\hat{A}$ corresponding to the eigenvalue a, we consider the definition of linearly independent functions.

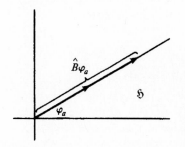

FIGURE 5.8 If $\hat{B}\varphi_a$ is eigenvector of $\hat{A}$ corresponding to the eigenvalue a, $\hat{B}\varphi_a$ and φ_a are in the same direction in Hilbert space $\mathfrak{H}$.

Linearly Independent Functions

When is a set of functions a linearly independent set? The N functions of the set $\{\varphi_n\}$ are linearly independent if the linear combination

(5.72)
$$\sum_{n=1}^{N} \lambda_n \varphi_n = 0$$

for all x is *only* satisfied when

(5.73)
$$\lambda_1 = \lambda_2 = \cdots = \lambda_n = 0$$

For example, the two functions e^x and $\sin x$ are linearly independent since

(5.74)
$$\lambda_1 e^x + \lambda_2 \sin x = 0$$

for all x is only satisfied by

(5.75)
$$\lambda_1 = \lambda_2 = 0$$

The two functions e^x and $3e^x$ are not linearly independent since

(5.76)
$$\lambda_1 e^x + 3\lambda_2 e^x = 0$$

is true for all x if

(5.77)
$$\lambda_1 = -3\lambda_2 \neq 0$$

The concept of linearly independent functions has an interesting geometrical interpretation in Hilbert space. If two "vectors" φ_1 and φ_2 in a Hilbert space $\mathfrak{H}$ are linearly independent, they do not lie along the same axis (line) in $\mathfrak{H}$ (Fig. 5.9). Similarly, if the set of N vectors $\{\varphi_n\}$ is such that all members are linearly independent, no two elements of this set lie on the same axis. If φ_1 and φ_2 are linearly independent, one must "rotate" φ_1 to align it with φ_2.

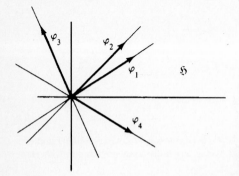

FIGURE 5.9 If $\{\varphi_1, \varphi_2, \varphi_3, \varphi_4\}$ are a linearly independent set, no two lie along the same axis in Hilbert space $\mathfrak{H}$.

If φ_a is the only linearly independent eigenfunction of $\hat{A}$ corresponding to the eigenvalue a, all eigenfunctions of $\hat{A}$ corresponding to a must be of the form $\mu\varphi_a$. The functions φ_a and $\mu\varphi_a$ are two linearly dependent eigenfunctions of $\hat{A}$ corresponding to the eigenvalue a.

(5.78)
$$\hat{A}(\mu\varphi_a) = \mu\hat{A}\varphi_a = \mu a\varphi_a = a(\mu\varphi_a)$$

How many such vectors are there? Since μ can be any constant, there is a continuum of such linearly dependent eigenfunctions of $\hat{A}$ corresponding to the eigenvalue a. In any given problem only one of these states is relevant. For bound states $(|\psi|^2 \rightarrow 0, |x| \rightarrow \infty)$, ψ is fixed (and therefore μ) by normalization. For an unbound state $(|\psi|^2 \nrightarrow 0, |x| \rightarrow \infty)$, ψ is fixed through an appropriate boundary condition. The latter case is appropriate to beam or scattering problems, where the boundary conditions usually involve stipulations on particle current or number density at $|x| = \infty$. These concepts are discussed in greater detail in Section 7.6, which concerns   one-dimensional barrier problems.

PROBLEMS

5.12 If $\hat{A}$, $\hat{B}$, and $\hat{C}$ are three distinct operators, show that:
  (a) $[\hat{A} + \hat{B}, \hat{C}] = [\hat{A}, \hat{C}] + [\hat{B}, \hat{C}]$
  (b) $[\hat{A}\hat{B}, \hat{C}] = \hat{A}[\hat{B}, \hat{C}] + [\hat{A}, \hat{C}]\hat{B}$

5.13 If $\hat{A}$ and $\hat{B}$ are both Hermitian, show that $\hat{A}\hat{B}$ is Hermitian if $[\hat{A}, \hat{B}] = 0$.

5.14 Show that the solution to the time-dependent Schrödinger equation given in Problem 3.18, that is,

$$\psi(\mathbf{r}, t) = \exp\left[-\frac{i}{\hbar} \int_0^t dt' \hat{H}(t') \right] \psi(\mathbf{r}, 0)$$

is correct, provided that

$$[\hat{H}(t), \hat{H}(t')] = 0 \qquad (t \neq t')$$

Answer

For $\psi(\mathbf{r}, t)$ as given above to be a solution, the expansion

$$\frac{\partial}{\partial t} e^{\hat{W}} \psi = e^{\hat{W}} \frac{\partial \psi}{\partial t} + e^{\hat{W}} \hat{H}\psi$$

must be valid in order to obtain the Schrödinger equation. For this to be so, $e^{\hat{W}}$ in the second term must precede $\hat{H}$. Here we have set

$$\hat{W} \equiv \int_0^t \hat{H}(t') \, dt'$$

and have absorbed the constant $-i/\hbar$ into $\hat{H}$. We must show that

$$\frac{\partial}{\partial t} e^{\hat{W}} = e^{\hat{W}} \frac{\partial \hat{W}}{\partial t}$$

In general

$$\frac{\partial}{\partial t} e^{\hat{W}} = \frac{\partial}{\partial t} \left(1 + \hat{W} + \frac{1}{2} \hat{W}^2 + \frac{1}{6} \hat{W}^3 + \cdots \right)$$

$$= \frac{\partial \hat{W}}{\partial t} + \frac{1}{2} \left(\hat{W} \frac{\partial \hat{W}}{\partial t} + \frac{\partial \hat{W}}{\partial t} \hat{W} \right) + \cdots$$

Thus the equality above holds if we are able to set

$$\left[\hat{W}, \frac{\partial \hat{W}}{\partial t} \right] = 0$$

In this case

$$\frac{\partial}{\partial t} e^{\hat{W}} = \left(\frac{\partial \hat{W}}{\partial t} + \hat{W} \frac{\partial \hat{W}}{\partial t} + \frac{1}{2} \hat{W}^2 \frac{\partial \hat{W}}{\partial t} + \cdots \right)$$

$$= e^{\hat{W}} \frac{\partial \hat{W}}{\partial t}$$

In terms of the integral definition of $\hat{W}$, the commutation criterion above becomes

$$\hat{H}(t) \int_0^t \hat{H}(t')\, dt' = \left(\int_0^t \hat{H}(t')\, dt' \right) \hat{H}(t)$$

which is guaranteed if $[\hat{H}(t), \hat{H}(t')] = 0$.

5.15 Discuss the linear independence of the following sets of functions.
   (a) $\{x, 3x, e^x\}$
   (b) $\{e^{ix}, \sin x, \cos x\}$
   (c) $\{x^2, x^3, x^5\}$
   (d) $\{x, 3, \sin^2 x, 4\cos^2 x, \ln x\}$

5.16 If μ is an arbitrary constant, the two vectors φ and $\mu\varphi$ in $\mathfrak{H}$ are linearly dependent. Show that the cosine of the angle between these two vectors has modulus 1.

$$|\cos \theta| = 1$$

5.17 From Problem 5.16 we conclude that φ and $\mu\varphi$ lie along the same axis in $\mathfrak{H}$. Show also that $\mu\varphi$ is $|\mu|$ times longer than φ, that is, that (see Fig. 5.10)

$$\|\mu\varphi\| = |\mu| \|\varphi\|$$

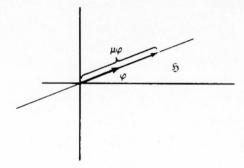

FIGURE 5.10   The vectors φ and $\mu\varphi$ in Hilbert space $\mathfrak{H}$ lie along the same axis and $\|\mu\varphi\| = |\mu|\,\|\varphi\|$. (See Problem 5.17.)

5.18 Show that if $\hat{A}\varphi_n = a_n\varphi_n$ and $\hat{B}\varphi_n = b_n\varphi_n$ for all eigenvalues $\{a_n\}$ and $\{b_n\}$ of $\hat{A}$ and $\hat{B}$, respectively (i.e., $\hat{A}$ and $\hat{B}$ have completely common eigenstates), then $[A, \ B] = 0$ on the space of functions spanned by the basis $\{\varphi_n\}$. (*Hint:* Any element of this space may be written

$$\psi = \sum c_n\varphi_n$$

and one need merely show that

$$[\hat{A}, \hat{B}] \sum c_n\varphi_n = 0.)$$

Note: In a more general vein one may say the following: let the eigenstates common to $\hat{A}$ and $\hat{B}$ span a subspace $\mathcal{G}$ of a Hilbert space $\mathfrak{H}$. Then $[\hat{A}, \hat{B}]\psi = 0$, where ψ is any element of $\mathcal{G}$.

5.3   MORE ON THE COMMUTATOR THEOREM

The Concept of Degeneracy

Suppose there are two (and *only* two) linearly independent eigenfunctions of the operator $\hat{A}$ which both correspond to the eigenvalue a. Call them φ_1 and φ_2.

(5.79)
$$\hat{A}\varphi_1 = a\varphi_1$$
$$\hat{A}\varphi_2 = a\varphi_2$$

Under such circumstances one says that *the eigenvalue a is doubly degenerate*. The eigenfunctions φ_1 and φ_2 are degenerate. Now we ask, what is the most general eigenfunction of $\hat{A}$ that corresponds to the eigenvalue a? The answer is, any function of the form

(5.80)
$$\varphi_a = \alpha\varphi_1 + \beta\varphi_2$$

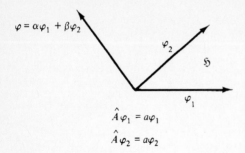

$\varphi = \alpha\varphi_1 + \beta\varphi_2$

φ_2

$\mathfrak{H}$

φ_1

$\hat{A}\varphi_1 = a\varphi_1$

$\hat{A}\varphi_2 = a\varphi_2$

FIGURE 5.11 If φ_1 and φ_2 are two linearly independent degenerate eigenvectors of $\hat{A}$, they span "plane" (two-dimensional subspace) in Hilbert space $\mathfrak{H}$. Any vector in this plane is an eigenvector of $\hat{A}$ corresponding to the eigenvalue a.

with α and β arbitrary constants. Let us test that this is the case.

$$(5.81) \qquad \hat{A}\varphi_a = \hat{A}(\alpha\varphi_1 + \beta\varphi_2) = \alpha a\varphi_1 + \beta a\varphi_2$$
$$= a(\alpha\varphi_1 + \beta\varphi_2)$$

In Hilbert space the two functions φ_1 and φ_2 span a plane (two-dimensional subspace). Equation (5.80) indicates that any vector φ_a in this plane is an eigenfunction of $\hat{A}$ corresponding to the eigenvalue a (Fig. 5.11).

Let us return to the commutator theorem discussed in Section 5.2. The operators $\hat{A}$ and $\hat{B}$ commute. If we operate on the first of equations (5.79) with $\hat{B}$ and use the commuting property of $\hat{A}$ and $\hat{B}$, there results

$$(5.82) \qquad \hat{B}\hat{A}\varphi_1 = a(\hat{B}\varphi_1) = \hat{A}(\hat{B}\varphi_1)$$

We conclude that $\hat{B}\varphi_1$ is an eigenstate of $\hat{A}$ that corresponds to the eigenvalue a. But there is a continuum of such eigenstates, all of the form (5.80). All we can say is that there are some α and β such that

$$(5.83) \qquad \hat{B}\varphi_1 = \mu(\alpha\varphi_1 + \beta\varphi_2)$$

Inspection of this equation [compare with (5.69)] reveals that φ_1 need not be an eigenfunction of $\hat{B}$.

So we have the following rule: If $[\hat{A}, \hat{B}] = 0$, and a is a degenerate eigenvalue of $\hat{A}$, the corresponding eigenfunctions of $\hat{A}$ (which all have the same eigenvalue, a) are not necessarily eigenfunctions of $\hat{B}$. Loosely speaking, degenerate operators have "more" eigenstates than nondegenerate operators. This concept may be illustrated in terms of the Venn diagrams depicted in Fig. 5.12.

A very simple physical example of this situation is provided by the problem of the free particle moving in one direction. The eigenvalue

$$(5.84) \qquad E_k = \frac{\hbar^2 k^2}{2m}$$

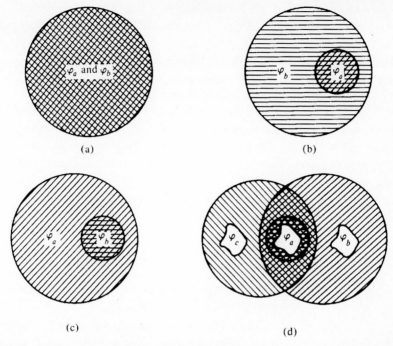

FIGURE 5.12   **Various cases pertaining to the sets of eigenfunctions of two compatible operators, $\hat{A}$ and $\hat{B}$. $[\hat{A}, \hat{B}] = 0$. (a) Eigenfunctions of $\hat{A}$ = all eigenfunctions of $\hat{B}$. (b) $\hat{A}$ has only nondegenerate eigenfunctions. (c) $\hat{B}$ has only nondegenerate eigenfunctions. (d) $[\hat{A}, \hat{B}] = [\hat{B}, \hat{C}] = [\hat{A}, \hat{C}] = 0$. $\hat{A}$ has only nondegenerate eigenfunctions.**

of the Hamiltonian [see (3.17)]

$$(5.85) \qquad\qquad \hat{H} = \frac{\hat{p}^2}{2m}$$

is doubly degenerate. All of the following functions are eigenfunctions of $\hat{H}$ corresponding to this eigenvalue.

$$(5.86) \qquad\qquad (\varphi_1, \varphi_2, \varphi_3) = \{\cos kx, \sin kx, \exp (ikx)\}$$

This is not a linearly independent set. However, any two are, so that for the free particle the eigenvalue (5.84) is doubly degenerate. For example, the two linearly independent functions, say,

$$(5.87) \qquad\qquad \{\varphi_1, \varphi_2\} = \{\cos kx, \sin kx\}$$

both have the eigenvalue $\hbar^2 k^2/2m$. Although $[\hat{p}, \hat{H}] = 0$, for the free particle, the set of functions (5.87), being degenerate eigenfunctions of energy, need not be eigenfunctions of $\hat{p}$. In fact, they are not.

Another linearly independent set of degenerate eigenstates corresponding to the eigenenergy $\hbar^2 k^2/2m$ is $\{\varphi_2, \varphi_3\}$. Of these, φ_3 is an eigenstate of $\hat{p}$ and φ_2 is not. Of the set $\{\varphi_1, \varphi_3\}$, φ_1 is not an eigenstate of $\hat{p}$, and again φ_3 is.

When there are n (and only n) linearly independent eigenstates of an operator $\hat{A}$ that all correspond to the same eigenvalue, the eigenvalue is *n-fold degenerate*. Suppose that $[\hat{A}, \hat{B}] = 0$. What can then be said is that from these n degenerate eigenstates of $\hat{A}$, one can form n linear combinations which are n linearly independent eigenstates of both $\hat{A}$ and $\hat{B}$.

For instance, from the two degenerate eigenstates (5.87) in the free-particle problem above, we can form

(5.88) $\varphi_+ = \varphi_1 + i\varphi_2 = \cos kx + i \sin kx = e^{ikx}$

(5.89) $\varphi_- = \varphi_1 - i\varphi_2 = \cos kx - i \sin kx = e^{-ikx}$

These two functions are common eigenstates of $\hat{H}$ and $\hat{p}$. They remain degenerate eigenstates of $\hat{H}$ but are nondegenerate eigenstates of $\hat{p}$.

PROBLEMS

5.19   Construct two linearly independent linear combinations of φ_2 and φ_3 given in (5.86) which are common eigenfunctions of $\hat{H}$ and $\hat{p}$.

5.20   Given that $\hat{x}$ and $\hat{p}$ operate on functions in $\mathfrak{H}_2$ and the relation $[\hat{x}, \hat{p}] = i\hbar$, show that if $\hat{x} = x$ (i.e., multiplication by x), $\hat{p}$ has the representation

$$\hat{p} = -i\hbar \frac{\partial}{\partial x} + f(x)$$

where $f(x)$ is an arbitrary function of x.

[*Note:* Dirac[1] has shown that through proper choice of *phase factor* (Section 2.4), the arbitrary function f may always be made to vanish. Thus, the basic commutator relation between $\hat{x}$ and $\hat{p}$ is equivalent to explicit operator forms for these variables. Listing of such commutator relations may serve in place of postulate I (Section 3.1).]

5.21   The operators $\hat{A}$ and $\hat{B}$ both have a denumerable number of eigenstates. Of these, the single eigenstate φ is known to be common to both. That is,

$$\hat{A}\varphi = a\varphi, \qquad \hat{B}\varphi = b\varphi$$

(a)   What can be said about the commutability of $\hat{A}$ and $\hat{B}$?

(b)   Suppose that it is further known that all the eigenstates of $\hat{A}$ and $\hat{B}$ are degenerate. Does this additional information in any way change your answer to part (a)?

---

[1] Dirac, *The Principles of Quantum Mechanics*.

5.4   COMMUTATOR RELATIONS AND THE UNCERTAINTY PRINCIPLE

As we have seen above, owing to the fact that for a free particle, $\hat{p}$ and $\hat{H}$ commute, they have a set of simultaneous eigenfunctions. Namely, any function of the form

$$(5.90) \qquad\qquad \varphi = Ae^{ikx}$$

is a common eigenstate of both $\hat{p}$ and $\hat{H}$. If the system (particle) is in this state, it is certain that measurement of p gives $\hbar k$ and measurement of energy gives $\hbar^2 k^2 / 2m$. Since (5.90) is a common eigenstate of $\hat{p}$ and $\hat{H}$, measurement of p, which (absolutely) gives $\hbar k$, leaves the particle in the state (5.90). Subsequent measurement of E gives $\hbar^2 k^2 / 2m$ and also leaves the particle in the state (5.90).[1] The operators $\hat{H}$ and $\hat{p}$ are *compatible;* that is, they commute. Quantum mechanics allows p and E to be simultaneously specified (for a free particle). Furthermore, there is only one (unique) state which gives these two values, the state (5.90).

Although there exists a state in which both energy and momentum may be specified simultaneously, the same is not true for the observables $\hat{x}$ and $\hat{p}$. There is no state in which measurement is certain to yield definite values of x and p. Measurement of p leaves the system in an eigenstate of $\hat{p}$ (5.90). Subsequent measurement of x is infinitely uncertain. The state (5.90) is not an eigenstate of $\hat{x}$. Conversely, measurement of x that finds x' leaves the system in the eigenstate of $\hat{x}$,

$$(5.91) \qquad\qquad \psi = \delta(x - x')$$

When the particle is in this state, measurement of momentum is infinitely uncertain.

For the free particle, there are states in which the uncertainty in energy and momentum obeys the relation

$$(5.92) \qquad\qquad \Delta E \, \Delta p = 0$$

On the other hand, in any state, the uncertainties in observation of p and x are such that the product $\Delta p \, \Delta x$ is always greater than a fixed magnitude.

$$(5.93) \qquad\qquad \Delta x \, \Delta p \geq \frac{\hbar}{2}$$

---

[1] Here we mean an *ideal* measurement. This is a measurement which *least* perturbs the system. Any real measurement causes the system to suffer a greater perturbation. After the energy of the particle in the state (5.90) is measured, *ideal* measurement maintains that state. However, it is also possible that after finding $\hbar^2 k^2 / 2m$, the particle is in any linear combination of the independent degenerate energy eigenfunctions of $\hat{H}$ which correspond to this eigenvalue (e.g., $\alpha \cos kx + \beta \sin kx$). However, measurement that leaves the particle in this state must have interfered with the momentum, since this state is a superposition of momentum eigenstates. Measurement that leaves the system in the original state (5.90) does not perturb the momentum. It is the *ideal* measurement of energy.

It is quite clear at this point that these uncertainty relations have their origin in the compatibility properties (5.51) of the operators that correspond to the observables being measured.

Suppose that two observables $\hat{A}$ and $\hat{B}$ are not compatible:

$$(5.94) \qquad\qquad [\hat{A}, \hat{B}] = \hat{C} \neq 0$$

For example, such is the case for displacement and kinetic energy. Then one can show the following[1]: If measurement of A, in the state ψ, is uncertain by the amount ΔA, then measurement of B is uncertain by the amount ΔB, such that[2]

$$(5.95) \qquad\qquad \Delta A \, \Delta B \geq \frac{1}{2}|\langle C \rangle|$$

We recall (Section 3.3) that the uncertainty of an observable A in the state ψ is the root mean square of the deviation of A away from the mean $\langle A \rangle$.

$$(\Delta A)^2 \equiv \langle (A - \langle A \rangle)^2 \rangle = \langle A^2 \rangle - \langle A \rangle^2$$

Expectation values in (5.95) are calculated in the state ψ. For example,

$$(5.96) \qquad \langle C \rangle = \langle \psi | \hat{C} \psi \rangle, \qquad (\Delta A)^2 = \langle (\hat{A} - \langle A \rangle)\psi | (\hat{A} - \langle A \rangle)\psi \rangle$$

The mechanism at work behind these uncertainty relations is as follows. If $\hat{A}$ and $\hat{B}$ do not commute, then the eigenstate φ_a of $\hat{A}$ which the system goes into on measurement of A is not necessarily an eigenstate of $\hat{B}$. Subsequent measurement of B will give any of the spectrum of eigenvalues of $\hat{B}$ with a corresponding probability distribution $P(b)$. This probability distribution is obtained from the coefficients in the expansion of φ_a in the eigenstates φ_b of $\hat{B}$.

$$(5.97) \qquad\qquad P(b) = |\langle \varphi_b | \varphi_a \rangle|^2$$

(with $\{\varphi_a\}$ and $\{\varphi_b\}$ normalized). Remeasurement of A is then in no way certain of finding the system in the state φ_a.

We note that the commutator-uncertainty relation, (5.94) and (5.95), is among the more fundamental relations in quantum mechanics. In addition to its important practical significance, it stands as an immutable barrier separating quantum and classical physics.

---

[1] See Problem 5.42.

[2] This generalization of the uncertainty principle is sometimes called the Robertson–Schrödinger relation [H. P. Robertson, *Phys. Rev.* **35**, 667A (1930); E. Schrödinger. *Sitzungsber. Preuss. Akad. Wiss.* (1930), p. 296].

PROBLEMS

5.22   How do the states for a free particle

$$\varphi_1 = Ae^{ikx}$$

$$\varphi_2 = B \cos kx$$

differ with regard to measurements of momentum and energy?

5.23   For a particle in a one-dimensional potential field $V(x)$, show that

$$\Delta E \, \Delta x \geq \frac{\hbar}{2m} \langle p_x \rangle$$

5.24   Consider three observables, $\hat{A}$, $\hat{B}$, and $\hat{C}$. If it is known that

$$[\hat{B}, \hat{C}] = \hat{A}$$

$$[\hat{A}, \hat{C}] = \hat{B}$$

show that

$$\Delta(AB) \, \Delta C \geq \frac{1}{2} \langle A^2 + B^2 \rangle$$

5.25   Obtain uncertainty relations for the following products
(a) $\Delta x \, \Delta E$
(b) $\Delta p_x \, \Delta E$
(c) $\Delta x \, \Delta T$
(d) $\Delta p_x \, \Delta T$

relevant to a particle whose kinetic energy is T and whose total energy is E. (A closely related example is discussed in Problem 2.30.)

5.26   If $g(x)$ is an arbitrary function of x, show that

$$[\hat{p}_x, g] = -i\hbar \frac{dg}{dx}$$

5.27   If $g(x)$ and $f(x)$ are both analytic functions, show that

$$g(\hat{A}) f(\varphi) = g(a) f(\varphi), \qquad \text{where } \hat{A}\varphi = a\varphi$$

5.28   The time-dependent Schrödinger equation permits the identification

$$\hat{E} = i\hbar \frac{\partial}{\partial t}$$

Using this identification together with the rule (5.95), give a formal derivation of the uncertainty relation

$$\Delta E \, \Delta t \geq \frac{1}{2}\hbar$$

Note that in a stationary state (eigenstate of $\hat{H}$), $\Delta E = 0$. The implication for this case is that a stationary state may last indefinitely.

5.29   Can the total energy and linear momentum of a particle moving in one dimension in a constant potential field be measured consecutively with no uncertainty in the values obtained?

5.30   If

$$[\hat{A}, \hat{B}] = i\hat{C}$$

and $\hat{A}$ and $\hat{B}$ are both Hermitian, show that $\hat{C}$ is also Hermitian.

5.31   Prove that if $\hat{A}$ and $\hat{B}$ are Hermitian, $[\hat{A}, \hat{B}]$ is Hermitian if and only if $[\hat{A}, \hat{B}] = 0$.

Answer (partial)

Set $[\hat{A}, \hat{B}] \equiv \hat{K}$. Then $\hat{K} = -\hat{K}^{\dagger}$. But $\hat{K}^{\dagger} = \hat{K}$, hence $\hat{K} = -\hat{K}$.

5.32   (a)   Obtain an uncertainty relation for mass and time from the relativistic mass-energy equivalency formula.

(b)   A free neutron has a mean lifetime of $\simeq 10^3$ s. Apply the uncertainty relation found in part (a) to find the uncertainty in the neutron's mass.

Answer (partial)

(b) $\Delta m \simeq 10^{-27}$ amu $(M_p \simeq 1$ amu$)$

5.5   "COMPLETE" SETS OF COMMUTING OBSERVABLES

We have already seen that for the free particle in one dimension, the eigenvalues of $\hat{H}$ are doubly degenerate. The two eigenfunctions of $\hat{H}$ corresponding to the eigenvalue $\hbar^2 k^2 / 2m$ are $\exp(+ikx)$ and $\exp(-ikx)$. However, once we specify what p is (say, $+\hbar k$), in addition to E, then one can say that the system is in one and only one state, $\exp(+ikx)$ (to within a multiplicative constant). Merely prescribing the energy of the particle does not uniquely determine the state of the particle. Further specifying the momentum removes this ambiguity and the state of the particle is uniquely determined.

Suppose that an operator $\hat{A}$ has degenerate eigenvalues. If a is one of these values, specifying a does not uniquely determine which state the system is in. Let $\hat{B}$ be another operator which is compatible with $\hat{A}$. Consider all the eigenstates $\{\varphi_{ab}\}$ which are common to $\hat{A}$ and $\hat{B}$. Of the degenerate eigenstates of $\hat{A}$, only a subset of these are also eigenfunctions of $\hat{B}$. Under such conditions, if we specify the eigenvalue b and the eigenvalue a, then the state that the system can be in is a smaller set than that determined by specification of a alone. Suppose further that there is only one other operator $\hat{C}$ which is compatible with both $\hat{A}$ and $\hat{B}$. Then they all share a set of common eigenstates. Call these states φ_{abc}. Then

(5.98)
$$\hat{A}\varphi_{abc} = a\varphi_{abc}$$
$$\hat{B}\varphi_{abc} = b\varphi_{abc}$$
$$\hat{C}\varphi_{abc} = c\varphi_{abc}$$

These functions are still a smaller set than the set $\{\varphi_a\}$ or $\{\varphi_{ab}\}$. Indeed, let us consider that φ_{abc} is *uniquely* determined by the values a, b, and c. This means that having measured a, b, and c: (1) Since φ_{abc} is a common eigenstate of $\hat{A}$, $\hat{B}$, and $\hat{C}$, simultaneous measurement (or a succession of three immediately repeated "ideal" measurements) of A, B, and C will definitely find the values a, b, and c. (2) The state φ_{abc} cannot be further resolved by more measurement. This state contains a maximum of information which is permitted by the laws of quantum mechanics. (3) There are no other operators independent of $\hat{A}$, $\hat{B}$, and $\hat{C}$ which are compatible with these. If there were, the state φ_{abc} could be further resolved. An exhaustive set (in the sense that there are no other independent operators compatible with $\hat{A}$, $\hat{B}$, and $\hat{C}$) of commuting operators such as $\hat{A}$, $\hat{B}$, and $\hat{C}$ above, whose common eigenstates are uniquely determined by the eigenvalues a, b, and c and are a basis of Hilbert space, is called a *complete set of commuting operators*.

Maximally Informative States

The values, a, b, and c, which may be so specified in the state φ_{abc}, are sometimes referred to as *good quantum numbers*. These are analogous to the generalized coordinates whose values determine the state of a system classically. As discussed in Section 1.1, such classical coordinates are also labeled *good variables*.

  Suppose that there are, in all, five independent operators that specify the properties of a system: $\hat{A}$, $\hat{B}$, $\hat{C}$, $\hat{D}$, and $\hat{F}$. Of these, $\hat{A}$, $\hat{B}$, and $\hat{C}$ are compatible with one another and $\hat{D}$ and $\hat{F}$ are compatible. However, these two sets are incompatible with one another, so that, for example,

(5.99)
$$[\hat{A}, \hat{D}] \neq 0$$

One can simultaneously specify either the eigenvalues a, b, and c or the eigenvalues d and f. One cannot, for instance, say that the system is in a state for which measurement of A definitely gives a and measurement of D definitely gives d. For this case there are two sets of states that are maximally informative: $\{\varphi_{abc}\}$ and $\{\varphi_{df}\}$.

Suppose that $\hat{A}$ has degenerate eigenvalue a. What is the state of the system after one has measured and found a? The state lies in a subspace of Hilbert space which is spanned by the degenerate eigenfunctions that correspond to a. This subspace $\mathfrak{H}_a$ has dimensionality $\mathcal{N}_a$ (a is an $\mathcal{N}_a$-fold degenerate eigenvalue). After measurement of $\hat{B}$, the state of the system lies in the space $\mathfrak{H}_{ab}$, which is a subspace of $\mathfrak{H}_a$ and is spanned by the eigenfunction common to $\hat{A}$ and $\hat{B}$. This subspace has dimensionality $\mathcal{N}_{ab}$, which is not greater than $\mathcal{N}_a$.

$$(5.100) \qquad \mathcal{N}_{ab} \leq \mathcal{N}_a$$

Subsequent measurement of $\hat{C}$ (mutually compatible with $\hat{A}$ and $\hat{B}$) leaves the state of the system in a space $\mathfrak{H}_{abc}$ that is a subspace of $\mathfrak{H}_{ab}$ and whose dimensionality does not exceed that of $\mathfrak{H}_{ab}$.

$$(5.101) \qquad \mathcal{N}_{abc} \leq \mathcal{N}_{ab}$$

In this manner we can proceed to measure more and more mutually compatible observables. At each step of the way the eigenstate is forced into subspaces of lesser and lesser dimensionality, until finally after the successive measurement of A, B, C, D, ... the state of the system is forced into a subspace of dimensionality $N = 1$. This is a space spanned by only one function. It is the eigenstate common to the complete set of observables $(\hat{A}, \hat{B}, \hat{C}, \hat{D}, \ldots)$: namely, $\varphi_{abcd} \ldots$ This state cannot be further resolved by additional measurements. Measurement of any of the observables $(A, B, C, D, \ldots)$ in this state is certain to find the respective values $(a, b, c, d, \ldots)$.

PROBLEMS

5.33 (a) Show that for a particle in a one-dimensional box, in an arbitrary state $\psi(x, 0)$,

$$\langle E \rangle \geq E_1$$

(b) Under what conditions does the equality maintain?

5.34 A free particle at a given instant of time is in the state

$$\psi = \frac{A}{(xk_0)^2 + 4}$$

At this same instant, (ideal) measurement of the energy finds that

$$E = \frac{\hbar^2 k_0^2}{2m}$$

The measurement leaves the momentum uncertain. Under such circumstances, what is the state $\tilde{\psi}$ of the particle immediately after measurement?

Answer

Since the momentum is uncertain after measurement, we know that the state is not one of the eigenstates of momentum $\varphi_{\pm k_0}$. Instead, one may say that the state vector lies in a subspace of $\mathfrak{H}$ spanned by the vectors $\cos k_0 x$ and $\sin k_0 x$.

$$\tilde{\psi} = \alpha \cos k_0 x + \beta \sin k_0 x$$

The coefficients α and β are proportional to the projections of ψ on $\cos k_0 x$ and $\sin k_0 x$, respectively. Since

$$\langle \psi \,|\, \sin k_0 x \rangle = 0$$

it follows that after measurement, the state of the particle is

$$\tilde{\psi} = \alpha \cos k_0 x$$

5.35 Show that

$$e^{\hat{A}} \hat{B} e^{-\hat{A}} = \hat{B} + [\hat{A}, \hat{B}] + \frac{1}{2!} [\hat{A}, [\hat{A}, \hat{B}]] + \frac{1}{3!} [\hat{A}, [\hat{A}, [\hat{A}, \hat{B}]]] + \cdots$$

(*Hint:* Taylor-series expand $\tilde{f}(\eta) \equiv e^{\eta \hat{A}} \hat{B} e^{-\eta \hat{A}}$ about $\eta = 0$. Also note the derivative property of $\hat{f}(\eta)$: $d\hat{f}/d\eta = [\hat{A}, \hat{f}]$.)

5.36 Show that [Baker–Hausdorf lemma]

$$e^{\hat{A}} e^{\hat{B}} = e^{\hat{A}+\hat{B}} e^{(1/2)[\hat{A}, \hat{B}]}$$

given that $\hat{A}$ and $\hat{B}$ each commutes with $[\hat{A}, \hat{B}]$. (*Hint:* First show that $[e^{\eta \hat{A}}, \hat{B}] = \eta e^{\eta \hat{A}}[\hat{A}, \hat{B}]$. Then establish that the derivative of

$$\hat{g}(\eta) \equiv e^{\eta \hat{A}} e^{\eta \hat{B}} e^{-\eta(\hat{A}+\hat{B})}$$

is

$$\frac{d\hat{g}}{d\eta} = \eta [\hat{A}, \hat{B}] \hat{g}$$

and integrate.) Note that for $\beta \ll 1$, one may always write

$$e^{\beta(\hat{A}+\hat{B})} \simeq e^{\beta \hat{A}} e^{\beta \hat{B}} e^{-(1/2)\beta^2 [\hat{A}, \hat{B}]}$$

This relation is important in statistical mechanics, where β plays the role of inverse temperature and $\hat{A} + \hat{B}$ *is the Hamiltonian.*

5.37 The operator $\hat{A}$ has only nondegenerate eigenvectors and eigenvalues, $\{\varphi_n\}$ and $\{a_n\}$. What are the eigenvectors and eigenvalues of the inverse operator, $\hat{A}^{-1}$? Is your answer consistent with the commutator theorem?

5.38 (a) Construct a one-dimensional wave packet that has zero probability density outside a domain of length a at time $t = 0$ and which has average momentum $\langle p \rangle = \hbar k_0$. That is, it is propagating to the right.

(b)   The wave packet collides with a mass m. Estimate the probability that the mass is deflected to the left with momentum $\hbar k_0/10 \pm \hbar k_0/100$. Take $k_0 a = 10\pi$. (Assume that complete momentum exchange occurs simultaneously at $t = 0$. The mass is located at the origin.)

5.39   Show that the expectation of an observable A of a system that is in the superposition state

$$\psi(x, t) = \sum_n b_n \varphi_n e^{i\omega_n t}$$

may be written in the form

$$\langle A \rangle = 2 \sum_{n>l} b_n^* b_l \langle n|A|l \rangle \cos(\omega_n - \omega_l)t + \sum_n |b_n|^2 \langle n|A|n \rangle$$

for $\{b_n^* b_l\}$ and $\langle n|A|l \rangle$ real. The states $\varphi_n \exp(i\omega_n t)$ are eigenstates of the Hamiltonian of the system. Here we are writing $|n\rangle$ for φ_n.

5.40   What is the average $\langle x \rangle$ and square root of variance Δx for the following probability densities?

(a) $P(x) = A[a^2 + (x - x_0)^2]^{-1}$

(b) $P(x) = Ax^2 e^{-x^2/2a^2}$

(c) $P(x) = A \sin^2\left(\dfrac{x - x_0}{\sqrt{2}a} - 8\pi\right) \exp\left\{-\left[\dfrac{(x - x_0)^2}{2a^2}\right]\right\}$

5.41   (a)   Show that for a particle in a one-dimensional box with walls at $(-a/2, a/2)$

$$\Delta p_{\min} = \sqrt{\langle p^2 \rangle_{\min}} = \frac{h}{2a}$$

(b)   Show for this same configuration that

$$\Delta x_{\max} = \sqrt{\langle x^2 \rangle_{\max}} = \frac{a}{2\sqrt{3}}$$

(c)   In which states are $\Delta p_{\min}$ and $\Delta x_{\max}$ realized?

(d)   From part (a) obtain the following momentum uncertainty relation for this configuration:

$$a \, \Delta p \geq \frac{h}{2}$$

5.42   Given that $\hat{A}$ and $\hat{B}$ are Hermitian operators and that

$$[\hat{A}, \hat{B}] = i\hat{C}$$

show that

$$\Delta A \, \Delta B \geq \frac{1}{2}|\langle C \rangle|$$

Answer

The uncertainties in $\hat{A}$ and $\hat{B}$, when written in terms of the operators

$$\hat{\delta}_A \equiv \hat{A} - \langle A \rangle$$

$$\hat{\delta}_B \equiv \hat{B} - \langle B \rangle$$

appear as

$$(\Delta A)^2 = \langle \hat{\delta}_A \psi | \hat{\delta}_A \psi \rangle = \|\hat{\delta}_A \psi\|^2$$

$$(\Delta B)^2 = \|\hat{\delta}_B \psi\|^2$$

These expressions may be incorporated into the Cauchy–Schwartz inequality (Problem 4.20). There results

$$\|\hat{\delta}_A \psi\|^2 \|\hat{\delta}_B \psi\|^2 \geq |\langle \hat{\delta}_A \psi | \hat{\delta}_B \psi \rangle|^2$$

$$(\Delta A)^2 (\Delta B)^2 \geq |\langle \hat{\delta}_A \psi | \hat{\delta}_B \psi \rangle|^2 = |\langle \psi | \hat{\delta}_A \hat{\delta}_B \psi \rangle|^2$$

The latter equality is due to the Hermiticity of $\hat{\delta}_A$. We now recall that any operator can be written as a linear combination of two Hermitian operators:

$$\hat{\delta}_A \hat{\delta}_B = \frac{1}{2}(\hat{\delta}_A \hat{\delta}_B + \hat{\delta}_B \hat{\delta}_A) + \frac{1}{2}[\hat{\delta}_A, \hat{\delta}_B] \equiv \hat{G} + \frac{i}{2}\hat{C}$$

Here we have used the fact that $[\hat{\delta}_A, \hat{\delta}_B] = [\hat{A}, \hat{B}]$. Substituting the expression above into the preceding inequality gives

$$(\Delta A)^2 (\Delta B)^2 \geq \left| \left\langle \psi \left| \left(\hat{G} + \frac{i}{2}\hat{C} \right) \psi \right\rangle \right|^2 = \left| \langle G \rangle + \frac{i}{2}\langle C \rangle \right|^2$$

Owing to the Hermiticity of $\hat{G}$ and $\hat{C}$, their expectation values are both real. It follows that

$$(\Delta A)^2 (\Delta B)^2 \geq |\langle G \rangle|^2 + \frac{1}{4}|\langle C \rangle|^2 \geq \frac{1}{4}|\langle C \rangle|^2$$

5.43 The linear independence of two functions $u(x)$ and $v(x)$ may be specified in terms of their *Wronskian*,

$$W(u, v) = \begin{vmatrix} u & v \\ u' & v' \end{vmatrix}$$

Thus, if u and v are solutions to a linear, second-order differential equation, and $W(u, v) \neq 0$ in some interval, then u and v are independent solutions in this interval. Employing this criterion, establish that the two functions given by (5.87) are independent over the entire x axis. What value of W do you find for this case?

5.44 Determine an expression for the Bohr radius a_0 from the following crude approximation. The electron moves to the nucleus to lower its potential energy,

$$V(r) = -\frac{e^2}{r}$$

If the electron is in the domain $0 \leq r \leq \bar{r}$, then we may write $\Delta p \simeq \hbar/\bar{r}$, with corresponding kinetic energy $\hbar^2/2m\bar{r}^2$. With this information estimate a_0 by minimizing the total energy. How does your answer compare with the actual expression for a_0 given by (2.14)?

5.45   Show that in three dimensions, the coordinate-momentum commutation relation (5.61) may be written

$$[\mathbf{r}, \mathbf{p}] = i\hbar \hat{I}$$

In this expression, the dyadic commutator has nine components and may be written as a 3×3 matrix and $\hat{I}$ is the identity operator.

5.46   Show that in three dimensions, the coordinate-momentum uncertainty relation (5.93) may be written

$$(\Delta r)^2 (\Delta p)^2 \geq \frac{9}{4} h^2$$

where

$$(\Delta r)^2 = \langle (\mathbf{r} - \langle \mathbf{r} \rangle)^2 \rangle$$
$$(\Delta p)^2 = \langle (\mathbf{p} - \langle \mathbf{p} \rangle)^2 \rangle$$

Answer

Working in a frame where $\langle \mathbf{p} \rangle = \langle \mathbf{r} \rangle = 0$, we have $(\Delta \mathbf{r})^2 = \langle r^2 \rangle$, $\langle \Delta \mathbf{p} \rangle^2 = \langle p^2 \rangle$. The resulting nine Cartesian products separate into two groups as follows;

$$(\Delta r)^2 (\Delta p)^2 = [(\Delta x \Delta p_x)^2 + \cdots] + \{(\Delta x \Delta p_y)^2 + (\Delta y \Delta p_x)^2 + \cdots\}$$

The first three bracketed terms give

$$[\quad] \geq \frac{3}{4} \hbar^2$$

In the second bracketed six terms, with $\Delta p_y \geq \hbar/\Delta y$, etc., we write

$$\{\quad\} \geq \left(\frac{\hbar}{2}\right)^2 \left[\left(\frac{\Delta x}{\Delta y}\right)^2 + \left(\frac{\Delta y}{\Delta x}\right)^2\right] + \cdots$$

As $(z + z^{-1}) \geq 2$, for all positive z, we find

$$\{\quad\} \geq 6 \left(\frac{\hbar}{2}\right)^2$$

Adding both contributions gives the desired result.

5.47   (a) A particle of mass m moves in one dimension (x). It is known that the momentum of the particle is $p_x = \hbar k_0$, where k_0 is a known constant. What is the *time-independent* (unnormalized) wavefunction of this particle, $\psi_a(x)$?

(b) The particle interacts with a system. After interaction it is known that the probability of measuring the momentum of the particle is $\frac{1}{5}$ for $p_x = 2\hbar k_0$ and $\frac{4}{5}$ for $p_x = 8\hbar k_0$. What is the time-independent (unnormalized) wavefunction of the particle in this state, $\psi_b(x)$?

(c) What is the average momentum for the particle, $\langle p_x \rangle$, in the state $\psi_b(x)$?

(d) What is the particle's average kinetic energy $\langle T \rangle$ in the state $\psi_b(x)$? Express your answer in terms of the constant $E_0 \equiv \hbar^2 k_0^2 / 2m$.

5.48 Consider a situation where it is equally likely that an electron has momentum $\pm \mathbf{p}_0$. Measurement at a given instant of time finds the value $+\mathbf{p}_0$. A student concludes that the electron must have had this value of momentum prior to measurement. Is the student correct?

Answer

The given information indicates that the electron was in a superposition state prior to measurement. In quantum mechanics one cannot rely on the premise of inference. The student is incorrect.

5.49 (a) Show that in one dimension, the energy spectrum of bound states is always *nondegenerate*. That is, to an eigenenergy there corresponds only one linearly independent eigenstate.

(b) In what step in your derivation does the Wronskian (Problem 5.43) come into play?

(c) In what manner does your proof depend on the given bound-state property?

(d) What is the nature of the potential you have included in your proof?

[*Hint:* See Problem 10.68. (The notion of degeneracy is discussed in detail in Chapter 8 et seq.)]

5.50 Relevant to one-dimensional motion, what is the value of the commutator

$$[\hat{p}_x, x \sin (\hat{p}_x)]?$$

5.51 A free particle of mass m, moving in one-dimensional space, has the following momentum probability amplitude.

$$b(k) = b_0, \qquad |ka| < \pi$$
$$b(k) = 0, \qquad |ka| \geq \pi$$

(a) Obtain the value of b_0 from the normalization of the wavefunction, $\psi(x)$. What are the dimensions of $b(k)^* \, b(k) \, dk$? Does your answer agree with these dimensions?

(b) At which values of x will the particle not be found?

$$\text{Data:} \qquad \int_{-\infty}^{\infty} \frac{\sin^2 cy}{y^2} \, dy = c\pi$$

5.52 A student argues the following. If a particle is in an eigenstate of a one-dimensional box of width a, then we know its energy exactly. But the energy in the box is purely kinetic.

Therefore we know the particle's momentum as well. This conclusion is a counter example to the uncertainty relation as the uncertainty in the particle's location is finite (a). Punch a hole in the student's argument.

5.53 A free particle moving in one dimension is in the state

$$\psi(x) = \int_{-\infty}^{\infty} \exp\left[-(ak)^2/2\right](i \sin ak)e^{ikx}\,dk$$

(a) What values of momentum, p_x, of the particle will not be found?

(b) If momentum of the particle in this state is measured, in which momentum state is the particle most likely to be found?

(c) If $a = 2.1$ Å, and the particle is an electron, what value of energy (in eV) will measurement find in the state described in part (b)?

6

TIME DEVELOPMENT, CONSERVATION THEOREMS, AND PARITY

In this chapter we pursue the study of time development of the state function in greater generality than we did in our previous discussion in Chapter 3. This description leads naturally to the concept of constants of the motion in quantum mechanics and again to the notion of stationary states. The distortion of a wave packet in time is obtained with the aid of the free-particle propagator. Classical motion of the packet is obtained in the limit $\hbar \rightarrow 0$. The significance to physics of constants of the motion was described in Chapter 1. We now find that such constants stem from related fundamental symmetries in nature. In the two chapters to follow, the principles and mathematical formalism developed to this point are applied to some practical one-dimensional problems.

6.1 TIME DEVELOPMENT OF STATE FUNCTIONS

The Discrete Case

Let us recall the recipe for solution to the *initial-value problem* in quantum mechanics (Section 3.5). The initial-value problem poses the question: Given the state $\psi(x, 0)$, at time $t = 0$, what is the state at $t > 0$, $\psi(x, t)$? The answer is (3.70):

$$(6.1) \qquad \psi(x, t) = \exp\left(\frac{-i\hat{H}t}{\hbar}\right) \psi(x, 0)$$

We recall that the exponential operation is written for its series representation,

$$(6.2) \qquad \exp\left(\frac{-i\hat{H}t}{\hbar}\right) = 1 - \frac{i\hat{H}t}{\hbar} - \frac{\hat{H}^2 t^2}{2!\,\hbar^2} + \cdots$$

Suppose that this exponential operator operates on an eigenfunction φ_n of $\hat{H}$. Then $\hat{H}$ as it appears in the exponential is simply replaced by E_n; that is,

$$(6.3) \qquad \exp\left(\frac{-i\hat{H}t}{\hbar}\right)\varphi_n = \exp\left(\frac{-iE_n t}{\hbar}\right)\varphi_n$$

As an application of this property we consider the problem of a particle in a one-dimensional box with walls at $(0, a)$, which is initially in an eigenstate of the Hamiltonian of this system.

$$(6.4) \qquad \psi_n(x, 0) = \varphi_n(x)$$

Then the state at time t is

$$\psi_n(x, t) = \exp\left(\frac{-i\hat{H}t}{\hbar}\right)\varphi_n(x) = e^{-i\omega_n t}\varphi_n(x)$$

$$(6.5) \qquad \psi_n(x, t) = e^{-i\omega_n t}\varphi_n(x)$$

$$\hbar\omega_n = E_n = n^2 E_1$$

As described in Section 3.5, the time-dependent eigenstates, $\psi_n(x, t)$ of $\hat{H}$, are called *stationary states*. We recall a very important property of a stationary state (3.76)— that the expectation of any operator (which does not contain the time explicitly) is constant in a stationary state. As an example of a stationary state, consider the $n = 5$ eigenstate of the problem at hand,

$$(6.6) \qquad \psi_5(x, t) = e^{-i25E_1 t/\hbar}\sqrt{\frac{2}{a}}\sin\left(\frac{5\pi x}{a}\right)$$

The eigenstate ψ_5 oscillates with the frequency $25E_1/h$. Both real and imaginary parts of $\psi_5(x, t)$ are *standing waves*. The expectation of energy in this state is constant and equal to $25E_1$.

Suppose, on the other hand, that $\psi(x, 0)$ is not an eigenstate of $\hat{H}$. Under such circumstances, to determine the time development of $\psi(x, 0)$ one calls on the superposition principle and writes $\psi(x, 0)$ as a linear superposition of the eigenstates of $\hat{H}$.

$$(6.7) \qquad \psi(x, 0) = \sum b_n \varphi_n(x)$$

$$b_n = \langle \varphi_n | \psi(x, 0) \rangle$$

If we now invoke (6.1), the calculation of $\psi(x, t)$ becomes tractable.

$$\psi(x, t) = \exp\left(\frac{-i\hat{H}t}{\hbar}\right) \sum b_n \varphi_n(x)$$

(6.8)
$$= \sum b_n \exp\left(\frac{-i\hat{H}t}{\hbar}\right)\varphi_n(x)$$

$$= \sum b_n e^{-i\omega_n t}\varphi_n(x)$$

$$\hbar\omega_n = E_n = n^2 E_1$$

This solution indicates that each component amplitude $b_n\varphi_n$ oscillates with the corresponding angular eigenfrequency ω_n.

Consider the specific example in which the initial state is

(6.9)
$$\psi(x, 0) = \sqrt{\frac{2}{a}}\,\frac{\sin(2\pi x/a) + 2\sin(\pi x/a)}{\sqrt{5}}$$

This state is depicted in Fig. 6.1 and is simply the superposition of the two eigenstates φ_2 and φ_1. That is, in the expansion (6.7), one obtains

$$b_1 = \frac{2}{\sqrt{5}}, \qquad b_2 = \frac{1}{\sqrt{5}}$$

(6.10)
$$b_n = 0 \qquad\qquad \text{(for all other } n)$$

The state of the system at $t > 0$ is given by (6.8).

(6.11)
$$\psi(x, t) = \sqrt{\frac{2}{a}}\left(\frac{e^{-i\omega_2 t}\sin(2\pi x/a) + 2e^{-i\omega_1 t}\sin(\pi x/a)}{\sqrt{5}}\right)$$

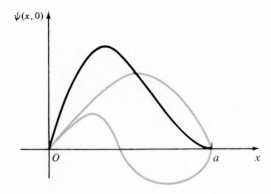

FIGURE 6.1   Initial state

$$\psi(x, 0) = \sqrt{\frac{2}{a}}\left(\frac{\sin(2\pi x/a) + 2\sin(\pi x/a)}{\sqrt{5}}\right)$$

How are these time-dependent solutions related to experimental observations? Let us rewrite (6.8) in the form

$$\psi(x, t) = \sum \bar{b}_n(t) \varphi_n(x) \tag{6.12}$$

so that $\bar{b}_n(t)$ now includes the exponential time factor

$$\bar{b}_n(t) \equiv e^{-i\omega_n t} b_n \tag{6.13}$$

Suppose that the energy is measured at $t > 0$. What values will result, and with what probabilities will these values occur? As in Section 5.1, calculation of the expectation of E yields

$$\langle E \rangle = \sum |\bar{b}_n(t)|^2 E_n \tag{6.14}$$

Again, we find that the square of the coefficient of expansion $b_n(t)$ gives the probability that measurement of E at the time t finds the value E_n.

$$P(E_n) = |\bar{b}_n(t)|^2 \tag{6.15}$$

For the state (6.9) this probability distribution is

$$P(E_1) = \frac{4}{5}$$

$$P(E_2) = \frac{1}{5} \tag{6.16}$$

$$P(E_n) = 0 \qquad \text{(for all other } n\text{)}$$

For the initial state (6.9), at *any time* $t > 0$, the probability that measurement of energy finds the value E_1 is $\frac{4}{5}$. Similarly, the probability that measurement finds the value $4E_1$ is $\frac{1}{5}$.

What is the expectation of E at $t > 0$ for the initial state (6.9)?

$$\langle E \rangle_{t>0} = \frac{(\langle e^{-i\omega_2 t} \varphi_2| + \langle 2e^{-i\omega_1 t} \varphi_1|)(\hat{H}| e^{-i\omega_2 t} \varphi_2\rangle + \hat{H}|2e^{-i\omega_1 t} \varphi_1\rangle)}{5} \tag{6.17}$$

$$= \frac{E_2 + 4E_1}{5} = \langle E \rangle_{t=0} = \frac{8}{5} E_1$$

The "cross terms" vanish due to orthogonality of the eigenstates of $\hat{H}$, and one finds that the expectation of energy is constant in time. More generally, for any isolated system, in any initial state: (1) the probability of finding a specific energy E_n is constant in time; (2) the expectation $\langle E \rangle$ is constant in time.

These rules follow directly from (6.13) through (6.15).

$$P(E_n) = |\bar{b}_n(t)|^2 = e^{+i\omega_n t}e^{-i\omega_n t}b_n{}^*b_n$$

(6.18)
$$P(E_n) = |b_n|^2 = \text{constant in time}$$

(6.19)
$$\langle E \rangle = \sum |b_n(t)|^2 E_n = \sum |b_n|^2 E_n = \text{constant in time}$$

The Continuous Case. Wave Packets

Next, consider the problem of a free particle moving in one dimension. Let the particle be initially in a localized state $\psi(x, 0)$ such as that depicted in Fig. 6.2.

Since the eigenstates of the Hamiltonian for a free particle comprise a continuum, the representation of $\psi(x, 0)$ as a superposition of energy eigenstates is an integral [see (5.26) et seq.].

(6.20)
$$\psi(x, 0) = \frac{1}{\sqrt{2\pi}} \int_{-\infty}^{\infty} b(k)e^{ikx}\, dk$$

$$b(k) = \frac{1}{\sqrt{2\pi}} \int_{-\infty}^{\infty} \psi(x, 0)e^{-ikx}\, dx$$

The state of the particle at $t > 0$ follows from (6.1).

(6.21)
$$\psi(x, t) = \exp\left(\frac{-i\hat{H}t}{\hbar}\right)\frac{1}{\sqrt{2\pi}} \int_{-\infty}^{\infty} b(k)e^{ikx}\, dk$$

$$\psi(x, t) = \frac{1}{\sqrt{2\pi}} \int_{-\infty}^{\infty} b(k)e^{i(kx-\omega t)}\, dk$$

(6.22)
$$\hbar\omega = \frac{\hbar^2 k^2}{2m} = E_k$$

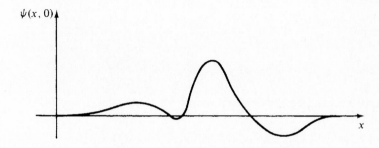

FIGURE 6.2   Initial state for a free particle.

While the component amplitudes of the state function of a particle in a box oscillate as standing waves, the k-component amplitudes of the free-particle state function propagate. For each value of k, the integrand of (6.22) appears as

$$(6.23) \qquad b(k) \exp\left[ik\left(x - \frac{\omega}{k}t\right)\right]$$

The phase of this component, $[x - (\omega/k)t]$, is constant on the propagating "surface,"

$$(6.24) \qquad x = \frac{\omega}{k}t$$

This is a surface of constant phase. It propagates with the phase velocity

$$(6.25) \qquad v = \frac{\omega}{k} = \frac{\hbar k}{2m}$$

The components with larger wavenumbers (shorter wavelengths) propagate with larger speeds. The long-wavelength components propagate more slowly.

Suppose that at $t = 0$, the state $\psi(x, 0)$ is a tight bundle of eigenstates of $\hat{H}$. When the clocks begin to move, each k component propagates with a distinct phase velocity. The initial state begins to distort. It may be that the initial state remains somewhat intact and moves. In this case one speaks of a *propagating wave packet*. To have a wave packet propagate, it is necessary that the average momentum of the particle in the initial state does not vanish.

$$(6.26) \qquad \langle p \rangle_{t=0} = \langle \psi(x, 0) | \hat{p}\psi(x, 0) \rangle \neq 0$$

Furthermore, since the packet is localized in space,

$$(6.27) \qquad |\psi(x, 0)|^2 \neq 0 \qquad \text{only over a small domain}$$

The velocity with which such a packet moves is called the *group velocity*.

$$(6.28) \qquad v_g = \frac{\partial \omega}{\partial k}\bigg|_{k_{max}}$$

The meaning of k_{max} is that the amplitude $|b(k)|^2$ is maximum $k = k_{max}$.

$$(6.29) \qquad \hbar k_{max} \simeq \langle p \rangle = \int_{-\infty}^{\infty} |b(k)|^2 \hbar k \, dk$$

This approximation becomes more accurate the more peaked is[1] $|b(k)|^2$.

---

[1] However, if $|b(k)|^2$ becomes too peaked, condition (6.27) is violated; that is, $\psi(x)$ spreads out too much.

Combining (6.28) and (6.29) gives

$$(6.30) \qquad v_g = \left.\frac{\partial \omega}{\partial k}\right|_{k_{max}} = \left.\frac{\partial \hbar \omega}{\hbar \partial k}\right|_{k_{max}} = \left.\frac{\partial (\hbar^2 k^2 / 2m)}{\hbar \partial k}\right|_{k_{max}}$$

$$= \frac{\hbar k_{max}}{m} = \frac{\langle p \rangle}{m} = v_{CL}$$

The packet moves with the classical velocity $\langle p \rangle / m$.

As an example of these concepts, consider a beam of neutrons each of which has momentum $\hbar k_0$. The beam is "chopped," producing a pulse a cm long and containing N neutrons (Fig. 6.3). The state function for each neutron at the instant after the pulse is produced is

$$(6.31) \qquad \psi(x, 0) = \begin{cases} \dfrac{1}{\sqrt{a}} e^{ik_0 x} & -\dfrac{a}{2} \leq x \leq +\dfrac{a}{2} \\[2ex] 0 & \text{elsewhere} \end{cases}$$

If the momentum of any one of the neutrons is measured at $t > 0$, what values may be found and with what probability do these values occur? To answer this question, we need calculate only the expansion coefficients $b(k)$ of (6.20).

$$(6.32) \qquad b(k) = \frac{1}{\sqrt{2\pi a}} \int_{-a/2}^{+a/2} e^{ik_0 x} e^{-ikx} \, dx$$

$$= \sqrt{\frac{2}{\pi a}} \frac{\sin \left[(k - k_0) a / 2 \right]}{k - k_0}$$

The state at time $t > 0$ is

$$(6.33) \qquad \psi(x, t) = \frac{1}{\pi \sqrt{a}} \int_{-\infty}^{\infty} \frac{\sin \left[(k - k_0) a / 2 \right]}{k - k_0} e^{i(kx - \omega t)} \, dk$$

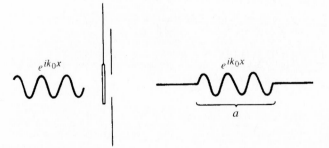

FIGURE 6.3   Chopped wave of length a.

with

(6.34)
$$\hbar\omega = \frac{\hbar^2 k^2}{2m}$$

The amplitude $b(k)$ is sketched in Fig. 6.4.

The momentum probability density $P(k)$ gives the probability that measurement of momentum of any of the neutrons yields a value in the interval $\hbar k$ to $\hbar(k + dk)$. It is given by

(6.35)
$$P(k) = \frac{|b(k)|^2}{\int_{-\infty}^{\infty} |b(k)|^2 \, dk} = |b(k)|^2$$

$$= \frac{2}{\pi a} \frac{\sin^2 [(k - k_0)a/2]}{(k - k_0)^2}$$

This probability density is constant in time.[1] At any time $t > 0$, it is most likely that measurement of momentum of any particle in the pulse finds the value

(6.36)
$$p = \hbar k_{max} = \hbar k_0$$

Recall that this was the only momentum the neutrons had before the beam was chopped.

At any time $t > 0$, the momentum values

(6.37)
$$\hbar k = \hbar k_0 + \frac{2n\pi\hbar}{a} \qquad (n = 1, 2, 3, \ldots)$$

have zero probability of being found. These momentum eigenstates do not enter into the superposition construction of $\psi(x, 0)$.

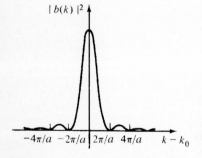

FIGURE 6.4   **Momentum probability density corresponding to the pulsed wave of Fig. 6.3.**

[1] This property of the free-particle momentum probability density is more fully developed in Section 7.4.

How many neutrons will be found with momentum in the interval $\hbar(k - k_0) - \hbar k_0$ to $\hbar(k - k_0) + \hbar k_0$? The answer is

(6.38)
$$\Delta N = N \int_{(k-2k_0)}^{k} |b(k)|^2 \, dk$$

This number is also constant in time.

Consider next the Fourier decomposition of a square wave packet as depicted in Fig. 5.4. There we see that the largest k component corresponds to $k = 0$ with

(6.39)
$$\varphi_0 = \frac{1}{\sqrt{2\pi}}$$

This is a "flat" wave. The other k components in the superposition of the square wave serve to taper the sides of the pulse. Since $p = 0$ for this packet, it does not propagate—it only *diffuses*.

The Gaussian Wave Packet

A more rewarding problem both from the pedagogical and physical points of view is that of the diffusion and propagation of a *Gaussian wave packet*, discussed previously in Section 3.3. The initial state is

(6.40)
$$\psi(x, 0) = \frac{1}{a^{1/2}(2\pi)^{1/4}} e^{ik_0 x} e^{-x^2/4a^2}$$

The corresponding initial probability density

(6.41)
$$P(x, 0) = \psi^*\psi = \frac{1}{a\sqrt{2\pi}} e^{-x^2/2a^2}$$

is properly normalized as

$$\int_{-\infty}^{\infty} P \, dx = 1$$

The initial uncertainty in position of a particle in the state (6.40) is the square root of the variance

(6.42)
$$\Delta x = a$$

The complex modulation $\exp(ik_0 x)$ in the state (6.40) serves to give the particle the average momentum

(6.43)
$$\langle p \rangle = \hbar k_0$$

It follows that the initial Gaussian state function (6.40) represents a particle localized within a spread of a about the origin and moving with an average momentum $\hbar k_0$.

The momentum amplitude corresponding to this initial state is

$$(6.44) \qquad b(k) = \frac{1}{a^{1/2}(2\pi)^{3/4}} \int_{-\infty}^{\infty} e^{-x'^2/4a^2} e^{ix'(k_0-k)} \, dx'$$

$$= \sqrt{\frac{2a}{\sqrt{2\pi}}} \, e^{-a^2(k_0-k)^2}$$

The Fourier transform of a Gaussian is itself Gaussian (see Fig. 6.5). The initial momentum probability density

$$(6.45) \qquad |b(k)|^2 = \frac{2a}{\sqrt{2\pi}} \, e^{-2a^2(k_0-k)^2}$$

is normalized, centered about the value $k = k_0$, and has a spread $\Delta k = (2a)^{-1}$. It follows that in the initial Gaussian state,

$$(6.46) \qquad \Delta x \, \Delta p \bigg|_{\text{Gauss}} = \Delta x \hbar \, \Delta k = \frac{\hbar}{2} = \Delta x \, \Delta p \bigg|_{\text{min}}$$

The product of uncertainties has its minimum value in a Gaussian packet.

Free-Particle Propagator

Next we turn to the construction of $\psi(x, t)$ from the initial state (6.40). The value of this function may be obtained from (6.21 et seq.)

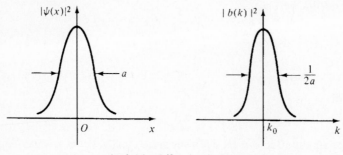

$$\Delta x \, \hbar \, \Delta k = \hbar/2 = \Delta x \, \Delta p|_{\text{min}}$$

FIGURE 6.5 The momentum probability density $|b|^2$ corresponding to a Gaussian position probability density, $|\psi|^2$, is Gaussian. In this state $\Delta p \, \Delta x$ has its minimum value at $\hbar/2$.

$$(6.47) \qquad \psi(x, t) = \frac{1}{2\pi} \int_{-\infty}^{\infty} \int_{-\infty}^{\infty} dx' \, dk e^{-ikx'} \psi(x', 0) e^{i(kx - \omega t)}$$

$$= \frac{1}{a^{1/2}} \frac{1}{(2\pi)^{5/4}} \int_{-\infty}^{\infty} dx' \exp\left(ik_0 x' - \frac{x'^2}{4a^2} \right)$$

$$\times \int_{-\infty}^{\infty} dk \exp\left\{ i\left[k(x - x') - \frac{k^2 a^2 t}{\tau} \right] \right\}$$

where the time constant τ is defined as

$$(6.48) \qquad \tau\omega = k^2 a^2 \qquad \omega = \frac{\hbar k^2}{2m}$$

Let us take advantage of our construction of $\psi(x, t)$ at this point of the analysis to introduce the free-particle propagator, $K(x', x; t)$. This function provides a formal solution to the free-particle, initial-value problem through the prescription

$$(6.49) \qquad \psi(x, t) = \int_{-\infty}^{\infty} dx' \, \psi(x', 0) K(x', x; t)$$

The explicit form of $K(x', x; t)$ is inferred from (6.47).

$$(6.50) \qquad K(x', x; t) = \frac{1}{2\pi} \int_{-\infty}^{\infty} dk \exp\left\{ i\left[k(x - x') - \frac{k^2 a^2 t}{\tau} \right] \right\}$$

With the aid of the integral (see Problem 6.5)

$$(6.51) \qquad \int_{-\infty}^{\infty} e^{-uy^2} e^{vy} \, dy = \sqrt{\frac{\pi}{u}} e^{v^2/4u} \qquad (\text{Re } u > 0)$$

there results[1]

$$(6.52) \qquad K(x', x; t) = \sqrt{\frac{\tau}{i4\pi a^2 t}} \exp\left[\frac{i(x - x')^2 \tau}{4a^2 t} \right]$$

$$= \sqrt{\frac{m}{2\pi i \hbar t}} \exp\left[\frac{im(x - x')^2}{2\hbar t} \right]$$

Having found this explicit form for the free-particle propagator, let us return to (6.49) and see its meaning. The wavefunction $\psi(x, t)$ gives the probability amplitude related to finding the particle at x at the instant t. If the particle was at x' at $t = 0$, then

---

[1] To obtain a convergent integral, first replace i by $\alpha \equiv i + \epsilon$, where ϵ is a small real positive number. After integrating, let $\epsilon \to 0$.

the probability that it is found at x at $t > 0$ depends on the probability that the particle propagated from x' to x in the interval t. This is what (6.49) says. The probability amplitude that the particle is at x at time t is equal to the initial amplitude that the particle is at x' multiplied by the probability amplitude of propagation from x' to x in the interval t, summed over all x'. Thus we may interpret $K(x, x'; t)$ as the probability amplitude that a particle initially at x' propagates to x in the interval t. It should be noted that the explicit form (6.52) is appropriate only for free-particle propagation. For more general problems involving interaction, the form of (6.49) still maintains, although the propagator function is more complicated (see Problem 6.26).

Distortion of the Gaussian State in Time

Let us return to the calculation of $\psi(x, t)$, given the initial Gaussian distribution (6.40). To complete the calculation one need merely complete the x' integration in (6.49).

$$\psi(x, t) = \frac{1}{a^{1/2}(2\pi)^{1/4}} \int_{-\infty}^{\infty} dx' \left[\exp\left(ik_0 x' - \frac{x'^2}{4a^2} \right) \right] K(x', x; t)$$

Employing the explicit form (6.52) for K and once again utilizing the integral formula (6.51) gives the desired result.

(6.53)

$$\psi(x, t) = \frac{1}{a^{1/2}(2\pi)^{1/4}(1 + it/\tau)^{1/2}} \exp\left[i\frac{\tau}{t}\left(\frac{x}{2a} \right)^2 \right] \exp\left[-\frac{(i\tau/4a^2 t)(x - \hbar k_0 t/m)^2}{1 + it/\tau} \right]$$

The corresponding probability density is

(6.54) $$P(x, t) = |\psi(x, t)|^2 = \frac{1}{a\sqrt{2\pi}(1 + t^2/\tau^2)^{1/2}} \exp\left[-\frac{(x - \hbar k_0 t/m)^2}{2a^2(1 + t^2/\tau^2)} \right]$$

If we compare this form with the initial probability density we see that the generic shape of $P(x, 0)$ (i.e., that of a bell) has remained intact with three modifications. It has become wider,

$$a \rightarrow a(1 + t^2/\tau^2)^{1/2}$$

Second, the center of symmetry of the packet is now at

$$x = v_0 t$$

where we have labeled

$$v_0 \equiv \frac{\hbar k_0}{m}$$

It follows that the probability density of a Gaussian wave packet propagates with a velocity that is directly related to the expectation of momentum of the particle in the Gaussian state. Finally, the height of the density function has diminished.

$$\frac{1}{a\sqrt{2\pi}} \rightarrow \frac{1}{a\sqrt{2\pi}\,(1 + t^2/\tau^2)^{1/2}}$$

The area under the curve P, at any time, remains unity.

A sequence of packet contours is shown in Fig. 6.6. It is quite clear that the packet begins to distort significantly after a time interval τ. If we represent a piece of chalk by a wave packet, $a \simeq 1$ cm, $m \simeq 1$ g, there results

$$\tau \simeq 10^{27}\text{ s} \simeq 10^{20}\text{ yr}$$

But the universe is only $\sim 10^{10}$ yr old. That is why classical objects are never observed to suffer a quantum mechanical spreading.

Flattening of the δ Function

There are two limits that can be taken on the probability density $P(x, t)$ related to the Gaussian wave packet which are very revealing. The first evolves from the initial state

(6.55) $$P(x, 0) = |\psi(x, 0)|^2 = \delta(x)$$

A valid representation of the delta function is given by the limit

(6.56) $$\delta(x) = \lim_{a \to 0} \frac{1}{a(2\pi)^{1/2}} e^{-x^2/2a^2}$$

This function has the correct delta function properties (3.27 et seq.).

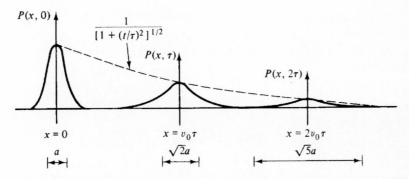

FIGURE 6.6   **Shrinkage and spreading of the probability distribution corresponding to a Gaussian wave packet. At any time** t**,**

$$\int_{-\infty}^{\infty} P(x, t)\, dx = 1$$

Measurement of the position of a particle which finds the value $x = 0$ leaves the particle in the state

(6.57) $$\psi = \delta(x)$$

This state is not normalizable. The state given by (6.55) is a little less sharply peaked than (6.57) and is normalizable.

To obtain the probability density $P(x, t)$ which follows from the initial value (6.55), we merely examine (6.54) in the limit, $a \to 0$. There results

(6.58) $$\lim P(x, t) = \lim \frac{2ma}{t\hbar\sqrt{2\pi}} \exp\left[-\frac{2a^2(x - \hbar k_0 t/m)^2}{t^2\hbar^2/m^2}\right]$$

$$= \lim \frac{2ma}{t\hbar\sqrt{2\pi}}[1 + O(a^2)]$$

The notation $O(a^2)$ denotes "order" of a^2. It stands for a group of terms, the sum of which goes to zero like a^2, with decreasing a.

From expression (6.58) we see that for all $t > 0$, P vanishes uniformly for all x, in the limit $a \to 0$. This instantaneous flattening of an infinitely peaked state (6.55) is due to the following circumstance. The momentum probability density $|b(k)|^2$ corresponding to such a state, depicted in Fig. 5.6b, is flat. This means that it is equally probable to find any k value, no matter how large k is, in this state. At any instant $t > 0$, at any point x, the components φ_k with k values which obey the inequality, $(\hbar k/m)t \geq x$, have overtaken that point. The initial infinitely peaked distribution assumes an (almost) instantaneous flattening.[1]

The Classical Particle

The second limit we wish to consider relative to the probability density $P(x, t)$, (6.54), changes $P(x, t)$ to the classical probability relating to a point particle of mass m moving with velocity $\hbar k_0/m$. This is accomplished by setting $\hbar \to 0$ in $P(x, t)$ (except where $\hbar$ appears in $p_0 = \hbar k_0$).

(6.59) $$\lim_{\hbar \to 0} P(x, t) = \frac{1}{a\sqrt{2\pi}} \exp\left[-\frac{(x - p_0 t/m)^2}{2a^2}\right]$$

$$\hbar k_0 = p_0 = \text{constant}$$

---

[1] Of course, these conclusions become erroneous for $x/t \geq c$. To obtain a completely physically valid solution for the infinitely peaked initial state, it is necessary to solve the relativistic form of the Schrödinger equation. See related discussions on the *Dirac equation* in A. Messiah, *Quantum Mechanics*, Wiley, New York, 1966. See also Chapter 15 of the present work.

For this probability to relate to a "point particle" we impose the additional constraint, $a \to 0$. This gives

(6.60)
$$\lim P(x, t) = \delta\left(x - \frac{p_0 t}{m}\right) = P_{CL}(x, t)$$

The probability of finding the particle at t is zero everywhere except on the classical trajectory

(6.61)
$$x = \frac{p_0 t}{m}$$

This is another example of the correspondence principle at work. In essence, the "leading term" (i.e., the term not containing $\hbar$) in the expansion of $P(x, t)$ about $\hbar = 0$ gives the classical result.

PROBLEMS

6.1 (a)   Find $\psi(x, t)$ and $P(E_n)$ at $t > 0$, relevant to a particle in a one-dimensional box with walls at $(0, a)$, for each of the following initial states.

(1) $\psi(x, 0) = A_1 \sin\left(\dfrac{3\pi x}{a}\right) \cos\left(\dfrac{\pi x}{a}\right)$

(2)[1] $\psi(x, 0) = A_2 x^2 (x - a)^2$

(3) $\psi(x, 0) = A_3[e^{i\pi(x-a)/a} - 1]$

(b)   If measurement of E finds that $E = 4E_1$ at 6 s, what is $\psi(x, t)$ at $t > 6$ s for each of the initial states in part (a)?

6.2   Consider the following three dispersion relations.

(1) $\omega^2 = gk$

(2) $\omega^2 = \dfrac{c^2 k^2}{1 - (\omega_0/\omega)^2}$

(3) $\omega^2 = \omega_p{}^2 + 3C^2 k^2$

The first relation obtains to deep-water surface waves (g is the acceleration due to gravity), the second to electromagnetic waves in a waveguide, and the third to longitudinal waves in a "warm" plasma (ω_p is the *plasma frequency* and C is the *thermal speed*). For all three cases find (1) the phase velocity and (2) the group velocity of a wave packet propagating in the respective medium.

6.3 (a)   Show that the free-particle propagator (6.52) has the following property and interpret the result physically.

$$K(x', x; 0) = \delta(x' - x)$$

[1] Faulty apparatus. E_2 cannot be measured in this state.

(b) Show that K satisfies the integral equation

$$K(x', x; t - t_0) = \int K(x', x''; t - t_1)K(x'', x; t_1 - t_0)\, dx''$$

and interpret this result physically in terms of the evolution in time of the state $\psi(x, t_0)$, first from t_0 to t_1 and then from t_1 to t.

Answer (partial)

(a) Set $it = \epsilon^2$ and compare with (C6) of Appendix C. The interpretation of this result is that for infinitesimally short time intervals, the probability amplitude for propagation away from the initial point x' is zero, except in a small neighborhood about the initial point.

6.4 At $t = 0$, 10^5 noninteracting protons are known to be on a line segment 10 cm long. It is equally probable to find any proton at any point on this segment. How many protons remain on the segment at $t = 10$ s? [*Hint:* Let the center of the segment be at $x = 0$. Then the formal answer to the problem with $\psi(x, t)$ normalized is

$$\Delta N = 10^5 \int_{-5}^{5} |\psi(x, 10)|^2\, dx$$

To construct $\psi(x, t)$, the initial square pulse must first be written as a superposition of φ_k states. With $b(k)$ calculated,

$$|\psi(x, t)|^2 = \frac{1}{2\pi} \int_{-\infty}^{\infty} \int_{-\infty}^{\infty} dk\, dk'\, b(k)b*(k')e^{i(\omega'-\omega)t} e^{i(k-k')x}$$

where ω' is written for $\omega(k')$.]

6.5 The integration involved in obtaining (6.51) is of the type

$$S = \int_{-\infty}^{\infty} e^{-uy^2} e^{vy}\, dy$$

where u and v are constants. Evaluate this integral.

Answer

The aim is to transform the exponent $-uy^2 + vy$ to a perfect square. First we set

$$uy^2 - vy \equiv \alpha^2 y^2 - 2\alpha\beta y$$

which gives

$$\alpha^2 = u, \qquad \beta^2 = \frac{v^2}{4u}$$

The exponent may now be written

$$uy^2 - vy = (\alpha y - \beta)^2 - \beta^2$$

and the integral S becomes (with $\eta = \alpha y - \beta$)

$$S = \frac{e^{\beta^2}}{\alpha} \int_{-\infty}^{\infty} e^{-\eta^2} \, d\eta = \frac{\sqrt{\pi} \, e^{\beta^2}}{\alpha} = \sqrt{\frac{\pi}{u}} \, e^{v^2/4u}$$

6.6 (a) An electron is in a Gaussian wave packet. If the packet is to remain intact for at least the time it takes light to move across 1 Bohr diameter, $2a_0 = 2\hbar^2/me^2$, what is the minimum width, a, that the Gaussian packet may have (in cm)?

(b) What is the diffusion time (τ) for an electron in a Gaussian wave packet of width e^2/mc^2 (in s)? This is the classical radius of the electron. How far does light travel in this time (in cm)?

6.7 A free particle of mass m moving in one dimension is known to be in the initial state

$$\psi(x, 0) = \sin(k_0 x)$$

(a) What is $\psi(x, t)$?

(b) What value of p will measurement yield at the time t, and with what probabilities will these values occur?

(c) Suppose that p is measured at $t = 3$ s and the value $\hbar k_0$ is found. What is $\psi(x, t)$ at $t > 3$ s?

6.8 A particle moving in one dimension has the wavefunction

$$\psi(x, t) = A \exp\left[i(ax - bt)\right]$$

where a and b are constants.

(a) What is the potential field $V(x)$ in which the particle is moving?

(b) If the momentum of the particle is measured, what value is found (in terms of a and b)?

(c) If the energy is measured, what value is found?

6.2 TIME DEVELOPMENT OF EXPECTATION VALUES

The law that covers the time development of the expectation of an observable, $\langle A \rangle$, follows from the time-dependent Schrödinger equation. We wish to calculate $d\langle A \rangle/dt$. Since $\langle A \rangle$ has all its spatial dependence integrated out, it is at most a function of time. We may therefore write

(6.62)
$$\frac{d\langle A \rangle}{dt} = \frac{\partial \langle A \rangle}{\partial t}$$

In the state $\psi(x, t)$, this expression becomes

(6.63)
$$\frac{d\langle \psi \,|\, \hat{A}\psi \rangle}{dt} = \int dx \, \frac{\partial}{\partial t} \left(\psi^* \hat{A} \psi \right)$$

The time derivative of the product is

(6.64)
$$\frac{\partial}{\partial t}(\psi^*\hat{A}\psi) = \left(\frac{\partial \psi^*}{\partial t}\right)\hat{A}\psi + \psi^*\hat{A}\,\frac{\partial \psi}{\partial t} + \psi^*\,\frac{\partial \hat{A}}{\partial t}\,\psi$$

Employing the time-dependent Schrödinger equation

(6.65)
$$\frac{\partial \psi}{\partial t} = \frac{-i\hat{H}}{\hbar}\,\psi, \qquad \frac{\partial \psi^*}{\partial t} = \frac{i\hat{H}\psi^*}{\hbar}$$

in (6.64) gives

(6.66)
$$\frac{\partial}{\partial t}(\psi^*\hat{A}\psi) = \frac{i}{\hbar}\left(\hat{H}\psi^*\hat{A}\psi - \psi^*\hat{A}\hat{H}\psi + \frac{\hbar}{i}\,\psi^*\,\frac{\partial \hat{A}}{\partial t}\,\psi\right)$$

Substituting this expansion in (6.63) gives

(6.67)
$$\frac{d\langle\hat{A}\rangle}{dt} = \frac{i}{\hbar}\left(\langle\hat{H}\psi\,|\,\hat{A}\psi\rangle - \langle\psi\,|\,\hat{A}\hat{H}\psi\rangle + \frac{\hbar}{i}\left\langle\psi\,\left|\,\frac{\partial \hat{A}}{\partial t}\,\psi\right.\right\rangle\right)$$

Since $\hat{H}$ is Hermitian, the first term on the right-hand side of (6.67) may be rewritten to yield the final result,

(6.68)
$$\boxed{\frac{d\langle A\rangle}{dt} = \left\langle\frac{i}{\hbar}\,[\hat{H},\,\hat{A}] + \frac{\partial \hat{A}}{\partial t}\right\rangle}$$

If $\hat{A}$ does not contain the time explicitly, then the last term on the right-hand side vanishes and

(6.69)
$$\frac{d\langle A\rangle}{dt} = \frac{i}{\hbar}\langle[\hat{H},\,\hat{A}]\rangle$$

In the event that $\hat{A}$ commutes with $\hat{H}$, the quantity $\langle A\rangle$ is constant in time and A is called a *constant of the motion*. For a free particle, $\hat{p}$ commutes with $\hat{H}$ and $\langle p\rangle$ is constant in time for any state (wave packet). Since $\hat{H}$ commutes with itself, $\langle H\rangle$, the expectation of the energy, is always constant in time.

Let a particle moving in one dimension be in the presence of the potential $V(x)$. The Hamiltonian of the particle is

(6.70)
$$\hat{H} = \frac{\hat{p}^2}{2m} + V(x)$$

How does $\langle x \rangle$ vary in time? Equation (6.69) gives

(6.71)
$$\frac{d\langle x \rangle}{dt} = \frac{i}{\hbar} \langle [\hat{H}, \hat{x}] \rangle$$

$$= \frac{i}{\hbar} \left\langle \left[\frac{\hat{p}^2}{2m}, \hat{x} \right] \right\rangle = \frac{i}{2m\hbar} \langle \hat{p}[\hat{p}, \hat{x}] + [\hat{p}, \hat{x}]\hat{p} \rangle$$

$$= \frac{i}{2m\hbar} \langle -2i\hbar p \rangle = \left\langle \frac{p}{m} \right\rangle$$

or, equivalently,

(6.72)
$$m \frac{d\langle x \rangle}{dt} = \langle p \rangle$$

This equation bears the same relation between expected values of displacement and momentum as in the classical case. Equation (6.72) cannot hold for the eigenvalues of $\hat{x}$ and $\hat{p}$, since such an equation implies that $x(t)$ and $p(t)$ are simultaneously known.

Ehrenfest's Principle

The reduction of quantum mechanical equations to classical forms when averages are taken, such as demonstrated above, is known as *Ehrenfest's principle*. Newton's second law follows from the commutator $[\hat{H}, \hat{p}]$, which for the Hamiltonian (6.70) is

(6.73)
$$[\hat{H}, \hat{p}] = i\hbar \frac{\partial V}{\partial x}$$

Again using (6.68), one obtains

(6.74)
$$\frac{d\langle p \rangle}{dt} = -\left\langle \frac{\partial V}{\partial x} \right\rangle$$

which is the x component of the vector relation

(6.75)
$$\frac{d\langle \mathbf{p} \rangle}{dt} = -\langle \boldsymbol{\nabla} V(x, y, z) \rangle = \langle \mathbf{F}(x, y, z) \rangle$$

where $\mathbf{F}$ is the force at (x, y, z). In any state $\psi(x, t)$, the time development of the averages of $\hat{x}$ and $\hat{p}$ follow the laws of classical dynamics, with the force at any given point replaced by its expectation in the state $\psi(x, t)$. (See Problem 6.31.)

PROBLEMS

6.9 Show that if $[\hat{H}, \hat{A}] = 0$ and $\partial \hat{A}/\partial t = 0$, then $\langle \Delta A \rangle$ is constant in time.

6.10 Show that

$$\frac{d}{dt} \langle A \rangle = 0$$

in a stationary state, provided that $\partial \hat{A}/\partial t = 0$, using the commutator relation (6.68).

Answer

$$\frac{d\langle A \rangle}{dt} = \frac{i}{\hbar} \langle \varphi_n | [\hat{H}, \hat{A}] \varphi_n \rangle = \frac{i}{\hbar} \langle \varphi_n | (\hat{H}\hat{A} - \hat{A}\hat{H}) \varphi_n \rangle$$

$$= \frac{i}{\hbar} (\langle \hat{H}\varphi_n | \hat{A}\varphi_n \rangle - \langle \varphi_n | \hat{A}\hat{H}\varphi_n \rangle)$$

$$= \frac{i}{\hbar} E_n (\langle \varphi_n | \hat{A}\varphi_n \rangle - \langle \varphi_n | \hat{A}\varphi_n \rangle) = 0$$

6.11 Show that for a wave packet propagating in one dimension,

$$m \frac{d\langle x^2 \rangle}{dt} = \langle xp \rangle + \langle px \rangle$$

6.12 A particle moving in one dimension interacts with a potential $V(x)$. In a stationary state of this system show that

$$\frac{1}{2} \left\langle x \frac{\partial}{\partial x} V \right\rangle = \langle T \rangle$$

where $T = p^2/2m$ is the kinetic energy of the particle.

Answer

In a stationary state,

$$\frac{d}{dt} \langle xp \rangle = \frac{i}{\hbar} \langle [\hat{H}, \hat{x}\hat{p}] \rangle = 0$$

Expanding the right-hand side, we obtain

$$0 = \langle \hat{x}[\hat{H}, \hat{p}] + [\hat{H}, \hat{x}]\hat{p} \rangle$$

$$= \langle x[\hat{V}, \hat{p}] + [\hat{T}, \hat{x}]\hat{p} \rangle$$

$$= i\hbar \left\langle x \frac{\partial V}{\partial x} - 2T \right\rangle$$

6.13 Consider an operator $\hat{A}$ whose commutator with the Hamiltonian $\hat{H}$ is the constant c.

$$[\hat{H}, \hat{A}] = c$$

Find $\langle A \rangle$ at $t > 0$, given that the system is in a normalized eigenstate of $\hat{A}$ at $t = 0$, corresponding to the eigenvalue a.

6.14   A system is in a superposition of the two energy eigenstates φ_1 and φ_2. Physical properties of the system characteristically depend on the probability density $\psi^*\psi$. Show that resolution of any such property involves measurements over an interval $\Delta t > \hbar / |E_1 - E_2|$.

Answer

The superposition state is

$$\psi(\mathbf{r}, t) = \varphi_1(\mathbf{r}) \exp\left(\frac{-iE_1 t}{\hbar}\right) + \varphi_2(\mathbf{r}) \exp\left(\frac{-iE_2 t}{\hbar}\right)$$

so that

$$\psi^*\psi = |\varphi_1|^2 + |\varphi_2|^2 + 2 \operatorname{Re} \varphi_1^*\varphi_2 \exp\left[\frac{i(E_1 - E_2)t}{\hbar}\right]$$

This function oscillates between the two extremes $(|\varphi_1| + |\varphi_2|)^2$ and $(|\varphi_1| - |\varphi_2|)^2$ with the period $\hbar / |E_1 - E_2|$. It follows that changes in related properties become discernible only after an interval greater than or of the same order as this period. The situation is similar to the process of tuning an oscillator to a frequency ω_0 by "listening" for beats. The period between beats varies as the inverse frequency $(\omega - \omega_0)^{-1}$. Thus one is certain that $\omega = \omega_0$ only after an infinite interval.

6.3   CONSERVATION OF ENERGY, LINEAR AND ANGULAR MOMENTUM

The principle of conservation of energy in classical physics states that the energy of an *isolated system* or a *conservative system* is constant in time. A conservative system is one whose dynamics are describable in terms of a potential function. A particle in a one-dimensional box is a conservative system. Suppose that at $t = 0$, the state of the particle is

(6.76)
$$\psi(x, 0) = \frac{3\varphi_1 + 4\varphi_5}{\sqrt{25}}$$

What can be said of the energy of the particle at the time $t > 0$? Measurement of the energy has a $\frac{9}{25}$ probability of finding the value E_1 and a $\frac{16}{25}$ probability of finding the value $25E_1$. At $t > 0$ the state (6.76) becomes

(6.77)
$$\psi(x, t) = \frac{3\varphi_1(x)e^{-iE_1 t/\hbar} + 4\varphi_5 e^{-iE_5 t/\hbar}}{\sqrt{25}}$$

The probability that measurement yields E_1 is

(6.78)
$$P(E_1) = \frac{(3e^{-iE_1 t/\hbar}) * (3e^{-iE_1 t/\hbar})}{25} = \frac{9}{25}$$

A similar calculation of $P(E_5)$ yields the constant value $\frac{16}{25}$. In other words, in the state given, one cannot say with certainty what the energy is at $t \geq 0$. In what sense is energy conserved? The answer is, in the average sense. It follows directly from (6.69) that

(6.79)
$$\langle H \rangle = \langle E \rangle = \text{constant}$$

For the example given, at any instant in time the expectation of the energy is

(6.80)
$$\langle E \rangle = \frac{9E_1 + 16E_5}{25} = 16.36E_1 = \text{constant}$$

For a free particle, $\hat{p}$ also commutes with $\hat{H}$; hence we can conclude from (6.68) that

(6.81)
$$\langle p \rangle = \text{constant}$$

The energy and total momentum of an isolated system are constants of the motion.

Conservation theorems in physics are closely related to symmetry principles. Consider, for example, the fact that the laws of physics do not depend on the time at which they are applied. Newton's second law, Maxwell's equations, and so on, do not change their structure with time. This symmetry of time (i.e., homogeneity) gives rise to the conservation of energy. Let H be the Hamiltonian of the whole universe. Homogeneity of time implies that H is not an explicit function of time. This together with (6.68) implies that $d\langle E \rangle / dt = 0$. We may reach the same conclusion for any *isolated system* (Fig. 6.7).

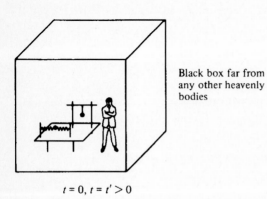

Black box far from any other heavenly bodies

$t = 0, t = t' > 0$

FIGURE 6.7   The laws of physics are the same at t and $t' \rightarrow \partial H / \partial t = 0 \rightarrow \langle E \rangle = $ constant.

Conservation of momentum for an isolated system depends on the homogeneity of space. Go out in space to a point far removed from other objects. Enclose yourself in a box with no windows and opaque walls. Let the box suffer a "virtual" displacement (Fig. 6.8). There is no experiment which will reveal that the box is at a new location. Consequently, for example, the dynamical laws of an isolated system of particles can only depend on the relative orientation of particles, not on the distances from these particles to some arbitrarily chosen origin. Equivalently, the Hamiltonian of the system can always be transformed so that it does not contain these variables (i.e., the coordinates of the center of mass).

To find the basis of the relation between the homogeneity of space and conservation of linear momentum, we turn back to Problem 3.17, where it was shown that the $\hat{p}$ operator effects the displacement

$$(6.82) \qquad \hat{\mathscr{D}}(\zeta) f(x) = \exp\left(\frac{i\zeta\hat{p}_x}{\hbar}\right) f(x) = f(x + \zeta)$$

In this expression f is any differentiable funtion of x. For infinitesimal displacement ($\zeta \to 0$), the displacement operator becomes

$$(6.83) \qquad \hat{\mathscr{D}}(\zeta) = \hat{I} + \frac{i\zeta\hat{p}_x}{\hbar}$$

or, equivalently,

$$\hat{p}_x = \frac{\hbar}{i\zeta}[\hat{\mathscr{D}}(\zeta) - \hat{I}]$$

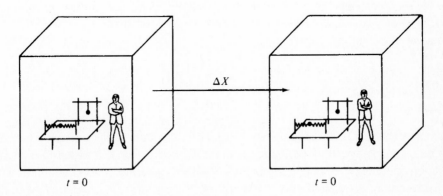

FIGURE 6.8   The laws of physics stay the same $\to \partial H/\partial x = 0 \to \langle p \rangle$ = constant. Note that in this "thought" experiment, translation occurs in zero time. This is called a "virtual" displacement. It is demanded by the details of the argument. To physically effect a virtual displacement one merely imagines two identical, noninterfering boxes a distance Δx apart.

where the identity operator is $\hat{I}$. As observed previously, the Hamiltonian of an isolated system cannot depend on displacement of the system from an origin at an arbitrary point in space. Therefore, the displacement operator $\hat{\mathscr{D}}$ commutes with $\hat{H}$, whence $\hat{p}_x$ does also. Again calling on (6.69), we recapture the constancy of $\langle p_x \rangle$. However, in the present argument we see how this conservation theorem finds its origin in the symmetry of the homogeneity of space.

In three dimensions the displacement operator becomes

$$(6.84) \qquad \hat{\mathscr{D}}f(\mathbf{r}) = \exp\left(\frac{i\boldsymbol{\zeta} \cdot \hat{\mathbf{p}}}{\hbar}\right) f(\mathbf{r}) = f(\mathbf{r} + \boldsymbol{\zeta})$$

Again, for an isolated system, one may conclude that $\hat{H}$ commutes with $\hat{\mathscr{D}}$ and therefore with $\hat{\mathbf{p}}$, the total linear momentum of the system. It follows that the vector $\langle \mathbf{p} \rangle$ is conserved.

Let us return to the experimental "black box" described above. The fact that experiments performed within the box are impervious to the box's location in space or time implies, respectively, conservation of linear momentum and energy. Suppose now that the box undergoes a rotation through the angle $\Delta\phi$ about an arbitrary fixed axis in space. Owing to the isotropy of space, experiments within the box cannot detect such rotational displacement. They are impervious to the box's orientation in space (Fig. 6.9). It follows that the Hamiltonian of the system cannot depend on ϕ, the rotational orientation with respect to some fixed axis, in the same way that it cannot depend on the displacement ζ from an arbitrary point in space. As a consequence of this rotational symmetry, the total angular momentum of the system is conserved.

Suppose that there is a property of the system which is dependent on the system's rotational orientation ϕ about a fixed axis, which we designate the z axis. Let the measure of this property be $f(\phi)$. After rotation of the system through the angle $\Delta\phi$, $\phi \rightarrow \phi + \Delta\phi$ and $f(\phi) \rightarrow f(\phi + \Delta\phi)$. This transformation of function is effected by the rotation operator $\hat{R}_{\Delta\phi}$:

$$(6.85) \qquad \begin{aligned} \hat{R}_{\Delta\phi}f(\phi) &= f(\phi + \Delta\phi) \\ \hat{R}_{\Delta\phi} &= \exp\left(\frac{i\Delta\phi\hat{L}_z}{\hbar}\right) \end{aligned}$$

Here $\hat{L}_z$ is the z component of the total angular momentum of the system.[1] Since the Hamiltonian of the (isolated) system cannot depend on ϕ, it is insensitive to the rotation operator, $\hat{R}_{\Delta\phi}$; that is, $\hat{H}$ commutes with $\hat{R}_{\Delta\phi}$. Hence it also commutes with $\hat{L}_z$ and we may conclude that $\langle L_z \rangle$ is constant. More generally, rotation through the vector angle $\Delta\boldsymbol{\phi}$ (the direction of $\Delta\boldsymbol{\phi}$ is parallel to the axis of rotation) is effected by

---

[1] This relation is derived in Problem 9.17.

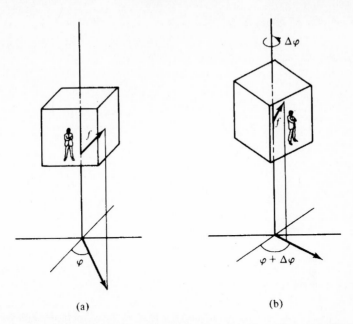

(a)

(b)

FIGURE 6.9 If $f(\phi)$ is a property of an isolated system that depends on the orientation ϕ of the system about an arbitrary fixed axis, rotation of the system through the angle $\Delta\phi$ causes the property to change to $f(\phi + \Delta\phi)$. Isotropy of space precludes the existence of such a property. This invariance with respect to rotation implies conservation of angular momentum.

the operator

$$(6.86) \qquad \hat{R}_{\Delta\phi} = \exp\left(\frac{i\mathbf{\Delta\phi} \cdot \hat{\mathbf{L}}}{\hbar}\right)$$

The argument demonstrating the constancy of $\hat{L}_z$ carries over to $\hat{\mathbf{L}}$, the total angular momentum of the system.

In summary, with $\mathbf{p}$ and $\mathbf{L}$ denoting, respectively, the total linear and angular momentum of an isolated system whose Hamiltonian is H, the following symmetry-conservation principles hold.

Homogeneity of Space

$$(6.87) \qquad [\hat{H}, \hat{\mathbf{p}}] = 0 \rightarrow \frac{d}{dt}\langle\mathbf{p}\rangle = 0$$

Isotropy of Space

$$(6.88) \qquad [\hat{H}, \hat{\mathbf{L}}] = 0 \rightarrow \frac{d}{dt}\langle\mathbf{L}\rangle = 0$$

Homogeneity of Time

(6.89)
$$\frac{\partial \hat{H}}{\partial t} = 0 \rightarrow \frac{d}{dt} \langle E \rangle = 0$$

PROBLEMS

6.15   Under what conditions is the expectation of an operator $\hat{A}$ (which does not contain the time explicitly) constant in time?

Answer

Under either of the following conditions:
   (a) $[\hat{A}, \hat{H}] = 0$.
   (b) $\langle A \rangle$ is calculated in a stationary state.

6.4   CONSERVATION OF PARITY

Consider an experiment and its mirror image (Fig. 6.10). Such an experiment might be the observation of the orbit of a missile fired in a uniform gravity field or two particles colliding. These phenomena obey certain physical laws. Suppose that we formulate the laws obeyed by the image orbits in the mirror. They are the same as the

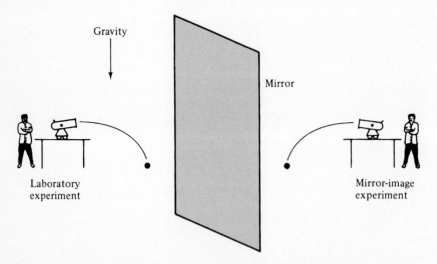

FIGURE 6.10   **The laws of physics are the same for the lab experiment and for the mirror-image experiment. This symmetry statement gives rise to the conservation of parity.**

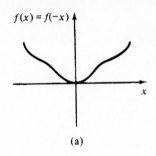

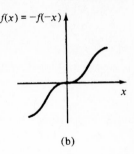

(a)   (b)

FIGURE 6.11 Any even function is an eigenfunction of $\hat{\mathbb{P}}$ with eigenvalue +1:

$$\hat{\mathbb{P}}f(x) = +1 f(x)$$

Any odd function is an eigenfunction of $\hat{\mathbb{P}}$ with eigenvalue −1:

$$\hat{\mathbb{P}}f(x) = -1 f(x)$$

laws that the orbits in the real world obey. This is a symmetry principle.[1] In quantum mechanics this principle is associated with a conservation law, *conservation of parity.*[2]

Parity is a property of a function. A function $f(x)$ has *odd parity* if

$$f(-x) = -f(x)$$

A function has *even parity* if (see Fig. 6.11)

$$f(-x) = f(x)$$

The parity operator $\hat{\mathbb{P}}$ is defined as[3]

(6.90) $$\hat{\mathbb{P}}f(x) = f(-x)$$

What are the eigenvalues of $\hat{\mathbb{P}}$? Let g be an eigenfunction of $\hat{\mathbb{P}}$ with eigenvalue α; then

(6.91) $$\hat{\mathbb{P}}g(x) = g(-x) = \alpha g(x)$$

To find α we operate again with $\hat{\mathbb{P}}$.

(6.92) $$\hat{\mathbb{P}}\hat{\mathbb{P}}g(x) = \hat{\mathbb{P}}g(-x) = g(x) = \alpha^2 g(x)$$

---

[1] At the molecular level, a striking example of mirror asymmetry is present in biological amino acids. All such molecules are "left-handed." (The plane of polarization of plane-polarized light, when passing through an aqueous solution of left-handed molecules at a given frequency, rotates in a direction opposite to that caused by a solution of mirror-image right-handed molecules.)

[2] "Broken symmetry" refers to a phenomenon in which a given symmetry law is not obeyed. For example, parity is not conserved in weak-interaction β-decay. For further discussion, see H. Frauenfelder and H. Henley, *Subatomic Physics,* Prentice-Hall, Englewood Cliffs, N. J., 1974.

[3] In three dimensions $\hat{\mathbb{P}}f(x, y, z) = f(-x, -y, -z)$. See Problem 6.23.

Hence

(6.93)
$$\alpha^2 = 1, \qquad \alpha = \pm 1$$

For $\alpha = +1$, from (6.91), we obtain

(6.94)
$$g(-x) = g(x)$$

Any even function is an eigenfunction of $\hat{\mathbb{P}}$ with eigenvalue $+1$. For $\alpha = -1$,

(6.95)
$$g(-x) = -g(x)$$

Any odd function is an eigenfunction of $\hat{\mathbb{P}}$ with eigenvalue -1. The order of degeneracy of $\alpha = \pm 1$ is infinite. There are no other eigenvalues of $\hat{\mathbb{P}}$.

   How is this parity property connected with the symmetry principle relating to mirror images mentioned above? Consider that a particle (m) moving in one dimension interacts with another stationary particle (M) which is at the position $x = 0$. The potential of interaction between the particles is $V(x)$. Suppose that the (moving) particle is at a position $x' > 0$. The image of the particle seen in a mirror which intersects the x axis normally at $x = 0$ is at $x = -x' < 0$ (Fig. 6.12). The temporal behavior of the image particle will be the same as that for the laboratory particle if $V(x) = V(-x)$. [The potential "seen" by the image particle is $V(-x)$.]

   The Hamiltonian for the particle in the laboratory system is

(6.96)
$$\hat{H} = \frac{\hat{p}^2}{2m} + V(x)$$

For $V(x)$ an even function, $\hat{H}$ commutes with $\hat{\mathbb{P}}$. To show that $\hat{\mathbb{P}}$ commutes with $V(x)$, let $g(x)$ be an arbitrary function of x. Then

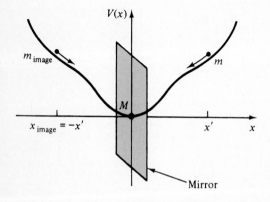

FIGURE 6.12 A mass m interacting with stationary mass M through potential $V(x)$. If image dynamics is to be the same as lab dynamics, $V(x) = V(-x)$.

$$(6.97) \qquad \hat{\mathbb{P}}V(x)g(x) = V(-x)g(-x) = V(x)\hat{\mathbb{P}}g(x)$$

The fact that $\hat{\mathbb{P}}$ commutes with the kinetic-energy part of $\hat{H}$ is shown in Problem 6.16. Since $\hat{\mathbb{P}}$ commutes with both parts of $\hat{H}$, it commutes with $\hat{H}$ itself.

$$(6.98) \qquad [\hat{H}, \hat{\mathbb{P}}] = 0$$

Together with (6.69) this gives the conservation principle

$$(6.99) \qquad \langle \mathbb{P} \rangle = \text{constant}$$

The parity of the state of a system is a constant of the motion.

As an example of this principle, consider a one-dimensional box centered at the origin, so that its walls are at $x = a/2$, $x = -a/2$ (see Problem 4.1). The eigenstates and eigenenergies of the Hamiltonian for this system are

$$(6.100) \qquad
\left.
\begin{aligned}
\tilde{\varphi}_n &= \sqrt{\frac{2}{a}} \sin\left(\frac{n\pi x}{a}\right) & (n = 2, 4, 6, \ldots, 2j, \ldots) \\
\varphi_n &= \sqrt{\frac{2}{a}} \cos\left(\frac{n\pi x}{a}\right) & (n = 1, 3, 5, \ldots, 2j + 1, \ldots)
\end{aligned}
\right\} \quad E_n = n^2 E_1
$$

The eigenstates $\tilde{\varphi}_n$ are odd while φ_n are even.

$$(6.101) \qquad
\begin{aligned}
\hat{\mathbb{P}}\tilde{\varphi}_n &= -\tilde{\varphi}_n \\
\hat{\mathbb{P}}\varphi_n &= \varphi_n
\end{aligned}
$$

Suppose that at $t = 0$ the particle is in the state

$$(6.102) \qquad
\begin{aligned}
\psi(x, 0) &= \sqrt{\frac{2}{45a}} \left[6 \sin\left(\frac{2\pi x}{a}\right) + 3 \cos\left(\frac{\pi x}{a}\right) \right] \\
&= \frac{1}{\sqrt{45}} (6\tilde{\varphi}_2 + 3\varphi_1)
\end{aligned}
$$

At $t > 0$,

$$(6.103) \quad \psi(x, t) = \sqrt{\frac{2}{45a}} \left[6 \sin\left(\frac{2\pi x}{a}\right) e^{-iE_2 t/\hbar} + 3 \cos\left(\frac{\pi x}{a}\right) e^{-iE_1 t/\hbar} \right]$$

The expectation of $\mathbb{P}$ at $t = 0$ is

(6.104)
$$\langle \mathbb{P} \rangle = \langle \psi(x, 0) | \hat{\mathbb{P}} \psi(x, 0) \rangle$$

$$\hat{\mathbb{P}} \psi(x, 0) = \frac{1}{\sqrt{45}} (-6\tilde{\varphi}_2 + 3\varphi_1)$$

$$\langle \mathbb{P} \rangle = \frac{1}{45} \{ -36 \langle \tilde{\varphi}_2 | \tilde{\varphi}_2 \rangle + 9 \langle \varphi_1 | \varphi_1 \rangle \} = -\frac{27}{45}$$

Since $[\hat{\mathbb{P}}, H] = 0$, this is the value of $\langle \mathbb{P} \rangle$ for all time. In that the initial state (6.102) is a superposition of the eigenstates of $\hat{\mathbb{P}}$, the squares of the coefficients of expansion give the probability that measurement finds the system in a state of even or odd parity. Measurements on an ensemble of 4500 boxes all of whose particles are in the initial state (6.102) at the time $t = 0$ would find approximately 3600 of the particles in the odd state $\tilde{\varphi}_2(t)$ and approximately 900 of the particles in the even state $\varphi_1(t)$ at the subsequent time $t > 0$.

PROBLEMS

6.16 (a) Show that $\hat{\mathbb{P}}$ *anticommutes* with the momentum operator $\hat{p}$. That is, show that

$$[\hat{\mathbb{P}}, \hat{p}]_+ \equiv \hat{\mathbb{P}}\hat{p} + \hat{p}\hat{\mathbb{P}} = 0$$

(b) Use your answer to part (a) to show that $\hat{\mathbb{P}}$ commutes with the kinetic-energy operator $\hat{T} = \hat{p}^2/2m$.

6.17 A particle in one dimension is in the energy eigenstate

$$\varphi_{k_0} = A \cos (k_0 x)$$

Ideal measurement of energy finds the value

$$E = \frac{\hbar^2 k_0^2}{2m}$$

What is the state of the particle after measurement?

Answer

If we recall postulate II of quantum mechanics (Section 3.2), the system is left in the eigenstate of $\hat{H}$ corresponding to the eigenenergy above. Any state in the two-dimensional subspace of Hilbert space spanned by $\sin (k_0 x)$ and $\cos (k_0 x)$ [or, equivalently, $\exp (i k_0 x)$ and $\exp (-i k_0 x)$] gives the eigenenergy above. However, an *ideal measurement* perturbs the system least. In the state before measurement the probability distribution relating to momentum is $1/2$ for $p = \pm \hbar k_0$. If we guess that the system is left in the state $\exp (i k_0 x)$ after measurement, the momentum distribution of the original state was disturbed. If we guess that the state of the measurement is $\sin(k_0 x)$, then measurement has not disturbed the momentum distribution; however, this measurement has disturbed the parity. In the original state the parity is $+1$ (with respect to the origin $x = 0$) while that of the hypothesized state after measurement is -1. This is still not the ideal measurement. We can find a measurement that perturbs the system even less. Consider that the particle is left in the state $\cos(k_0 x)$. The corresponding measurement did not perturb the momentum distribution or the parity of the original state. It is the ideal measurement.

6.18 (a) If $f(x)$ is any function, show that

$$f_+ = \frac{f(x) + f(-x)}{2} = \text{even function}$$

$$f_- = \frac{f(x) - f(-x)}{2} = \text{odd function}$$

(b) Show that

$$\hat{P}_+ \equiv \frac{\hat{I} + \hat{P}}{2}$$

is such that

$$\hat{P}_+ f(x) = f_+(x)$$

The identity operator is $\hat{I}$.

(c) Show that

$$\hat{P}_- \equiv \frac{\hat{I} - \hat{P}}{2}$$

is such that

$$\hat{P}_- f(x) = f_-(x)$$

The operator $\hat{P}_+$ "projects" f onto f_+ while $\hat{P}_-$ projects f onto f_-.

(d) Show that the *projection operators* $\hat{P}_+$ and $\hat{P}_-$ satisfy the following properties:

$$\hat{P}_\pm^2 = \hat{P}_\pm$$
$$[\hat{P}_+, \hat{P}_-] = 0$$
$$\hat{P}_+ + \hat{P}_- = \hat{I}$$

6.19 What is $\langle \mathbb{P} \rangle$ for a particle in a one-dimensional box with walls at $(-a/2, +a/2)$ in the initial state

$$\psi(x, 0) = \frac{1}{\sqrt{29}}(3\tilde{\varphi}_2 + 4\tilde{\varphi}_4 + 2\varphi_3)$$

6.20 For the same one-dimensional box as described in Problem 6.19, it is known that the particle is in a state with energy probabilities

$$P(E_1) = \frac{1}{3}, \qquad P(E_2) = \frac{1}{3}, \qquad P(E_3) = \frac{1}{3}$$

$$P(E_n) = 0, \qquad (n \neq 1, 2, 3)$$

The parity of the state is measured ideally and -1 is found. If some time later E is measured, what value is found? What is the answer if the original measurement found the parity to be $+1$?

6.21 For a free particle moving in one dimension, divide the following set of operators into subsets of commuting operators.

$$\{\hat{\mathbb{P}}, \hat{x}, \hat{H}, \hat{p}\}$$

6.22 A free particle moving in one dimension is in the initial state $\psi(x, 0)$. Prove that $\langle p \rangle$ is constant in time by direct calculation (i.e., without recourse to the commutator theorem regarding constants of the motion).

Answer

With

$$\bar{b}(k, t) \equiv b(k)e^{i\omega t}$$

$$\frac{\hbar^2 k^2}{2m} = \hbar\omega$$

$$b(k) = \frac{1}{\sqrt{2\pi}} \int_{-\infty}^{\infty} \psi(x, 0)e^{-ikx}\, dx$$

the expectation of p appears as

$$\langle p \rangle = \int_{-\infty}^{\infty} \hbar k\, |\bar{b}(k, t)|^2\, dk$$

which, given the structure of $\bar{b}(k, t)$ above, is time-independent. Alternatively, we may note that for a free-particle wave packet, $\hat{H} = \hat{p}^2/2m$, so that $d\langle p \rangle/dt = (1/i\hbar)\langle[H, p]\rangle = 0$.

6.23 In three dimensions, $\hat{\mathbb{P}}$ is defined as

$$\hat{\mathbb{P}}\psi(x, y, z) = \psi(-x, -y, -z)$$

(a) What does this definition become if ψ is measured in spherical coordinates: that is, $\psi = \psi(r, \theta, \phi)$?

(b) What does it become if ψ is measured in cylindrical coordinates: that is, $\psi = \psi(\rho, z, \phi)$?

(c) What are the parities of the following functions?

$$\psi_1 = A(x + y + z)e^{-(x^2+y^2+z^2)}$$

$$\psi_2 = Bre^{-r^2}\cos\theta$$

$$\psi_3 = C\frac{\sqrt{\rho^2 + z^2}\sin\phi}{z^5}$$

6.24 Discuss the consistency or inconsistency of the concept of the trajectory of a particle

$$\mathbf{r} = \mathbf{r}(t)$$

(whose mass is also known) in quantum mechanics.

Answer

The trajectory (through differentiation) implies the momentum, $\mathbf{p} = \mathbf{p}(t)$. Thus, at each instant, $\mathbf{r}$ and $\mathbf{p}$ are known, which is in violation of the uncertainty principle. On the other hand, for a wave packet one may construct the equation

$$\frac{d}{dt}\langle\mathbf{r}\rangle = \frac{1}{m}\langle\mathbf{p}\rangle$$

If $\langle\mathbf{p}\rangle$ is a known function of time, this equation may be integrated to obtain the trajectory, that is, $\langle\mathbf{r}\rangle$, as a function of time. The classical trajectory ensues when quantum fluctuations about the mean $\langle\mathbf{r}\rangle$ oscillate rapidly, thereby "averaging out." Equivalently, one may say that this limit is reached when the de Broglie wavelength is small compared to characteristic distances of the configuration.

6.25 Let a particle move in the potential

$$V = Ax^n$$

where A is constant and n is a finite integer. Show that Ehrenfest's equation (6.74) gives the classical relation

$$\frac{dp}{dt} = -Anx^{n-1}$$

only for $n = 2$ (the harmonic oscillator).

Answer

From (6.75) we obtain

$$\frac{d\langle p\rangle}{dt} = -An\langle x^{n-1}\rangle$$

To obtain the classical form we must equate $\langle x^{n-1}\rangle = \langle x\rangle^{n-1}$, which is only valid for $n = 2$.

6.26 (a) Consider that a particle is bound to the potential $V(x)$ and is initially in the state $\psi(x, 0)$. If eigenfunctions of the Hamiltonian $\hat{H} = \hat{p}^2/2m + V(x)$ are written $\varphi_n(x)$.

$$\hat{H}\varphi_n = E_n\varphi_n$$

obtain the state function $\psi(x, t)$ in terms of an integral over the propagator $K(x', x; t)$. That is, construct an explicit form for K as it appears in the integral

$$\psi(x, t) = \int \psi(x', 0)K(x', x; t) \, dx'$$

(b)   Show that the propagator you have constructed satisfies the integral equation

$$K(x', x; t - t_0) = \int K(x', x''; t - t_1)K(x'', x; t_1 - t_0) \, dx''$$

Answer (partial)

(a)   The formal solution to the time-dependent Schrödinger equation appears as [see (3.70)]

$$\psi(x, t) = \exp\left(-\frac{it\hat{H}}{\hbar}\right)\psi(x, 0)$$

Expanding the initial state in eigenstates of $\hat{H}$ [see (5.6)] gives

$$\psi(x, 0) = \sum_n b_n \varphi_n(x)$$

$$b_n = \int \psi(x', 0)\varphi_n*(x') \, dx'$$

Substituting in the above gives

$$\psi(x, t) = \sum_n b_n \exp\left(-\frac{iE_n t}{\hbar}\right)\varphi_n(x)$$

$$= \int \psi(x', 0)\left[\sum_n \varphi_n*(x')\varphi_n(x)e^{-iE_n t/\hbar}\right] dx'$$

$$= \int \psi(x', 0)K(x', x; t) \, dx'$$

which serves to identify the propagator

$$K(x', x; t) = \sum_n \varphi_n*(x')\varphi_n(x)e^{-iE_n t/\hbar}$$

In Dirac notation this equation appears as

$$\hat{K}(x', x; t) = \sum_n |\varphi_n(x)\rangle e^{-iE_n t/\hbar}\langle\varphi_n(x')|$$

which allows the solution to be written:

$$|\psi(x, t)\rangle = \hat{K}|\psi(x, 0)\rangle$$

6.27 The wavefunction of a particle of mass m which moves in one dimension is

$$\Psi(x,\ t) = A e^{i(kx-\omega t)}$$

where A, k, and ω are constants. Determine the potential $V(x)$ in which the particle moves in terms of m, $\hbar$, k, and ω.

6.28 Consider a particle of mass m moving in one dimension (x). Let $\varphi(x)$, $\overline{\varphi}(x)$ be eigenstates of the particle with corresponding eigenenergies $E > \overline{E}$. Let x_1, x_2 be two successive zeros of $\overline{\varphi}(x)$. [Values of x for which $\overline{\varphi}(x) = 0$ are called "zeros" of $\overline{\varphi}(x)$.]

(a) Show that $\varphi(x)$ has at least one zero in the interval (x_1, x_2).

(b) If energies are discrete and E is the next larger energy than $\overline{E}$, then show that $\varphi(x)$ has only one zero in the interval (x_1, x_2). (Zeros are then said to be *interlaced*. See, for example, Fig. 10.3.)

Answer (partial)

(a) The Schrödinger equation may be written [compare with (7.2)]

$$\varphi''(x) + k^2(x)\varphi(x) = 0$$

$$\overline{\varphi}''(x) + \overline{k}^2(x)\overline{\varphi}(x) = 0$$

$$k^2 > \overline{k}^2$$

where primes denote differentiation. Let x_1, x_2 be two successive zeros of $\overline{\varphi}(x)$. Assume that $\varphi(x) \neq 0$ on (x_1, x_2) and that (without loss of generality) $\varphi(x)$, $\overline{\varphi}(x) > 0$ on (x_1, x_2). Consider the Wronskian of these two solutions:

$$W(\varphi, \overline{\varphi}) = \varphi\overline{\varphi}' - \overline{\varphi}\varphi'$$

$$\frac{dW(\varphi, \overline{\varphi})}{dx} = \varphi\overline{\varphi}'' - \overline{\varphi}\varphi''$$

$$= \varphi\overline{\varphi}(k^2 - \overline{k}^2) > 0$$

In the second equation we employed the preceding Schrödinger equations. Thus $W(x)$ has positive slope, and we may conclude

$$W(x_2) - W(x_1) > 0$$

Since $\overline{\varphi}(x)$ vanishes at x_1, x_2, we obtain (from the first of the preceding equations)

$$W(x_1) = \varphi(x_1)\overline{\varphi}'(x_1) > 0$$

$$W(x_2) = \varphi(x_2)\overline{\varphi}'(x_2) < 0$$

so that

$$W(x_2) - W(x_1) < 0$$

which contradicts our previous result. Thus our assumptions are incorrect and we conclude that $\varphi(x)$ has at least one zero in (x_1, x_2). (*Note:* The preceding result is often referred to as the *Sturm comparison theorem* in the theory of differential equations.)

6.29 (a) What is the value of the spread a of the classical probability density, (6.60)?

(b) What is the spreading time τ of this distribution? Are your answers compatible with the picture one has of a classical particle?

6.30 An electron is initially in the Gaussian state, (6.40). If $a \approx \lambda_e$, the de Broglie wavelength, what is the spread time τ of the subsequent time development of the wave packet? Is this a rapid or a slow spread?

6.31 To complete Ehrenfest's correspondence principle, one must convert (6.74) to the classical relation

$$\frac{d\langle p \rangle}{dt} = -\frac{\partial V \langle x \rangle}{\partial \langle x \rangle}$$

For what class of potentials would this relation be valid?

Answer

If $V(x)$ is slowly varying. To show this, write

$$\frac{\partial V}{\partial x} \equiv G(x)$$

Then under the said condition we may expand

$$G(x) = G(\langle x \rangle) + (x - \langle x \rangle)G'(\langle x \rangle) + \frac{(x - \langle x \rangle)^2}{2!} G''(\langle x \rangle) + \cdots$$

Neglecting derivatives of G (second derivatives of V) gives the desired result:

$$\frac{\partial V(x)}{\partial x} \approx \frac{\partial V(\langle x \rangle)}{\partial \langle x \rangle}$$

6.32 Let $\hat{\Gamma}$ represent a symmetry operation such as translation, rotation, etc. Suppose a given system with Hamiltonian $\hat{H}$ is invariant under an operation represented by $\hat{\Gamma}$.

(a) What is the value of $[\hat{H}, \hat{\Gamma}]$?

(b) If the eigenstates of $\hat{H}$ are known, what can you say about the eigenstates of $\hat{\Gamma}$?

(c) What can you say about the expectation, $\langle \Gamma \rangle$?

(d) From descriptions given in this chapter, state four examples of $\hat{\Gamma}$ and respective systems on which $\hat{\Gamma}$ operates for which your answers to parts (a), (b), and (c) are valid.

6.33 In quantum mechanics, a process described by a wavefunction $\psi(x, t)$ is said to be reversible if $\psi(x, t) = \psi^*(x, -t)$. Show that solutions to the time-dependent Schrödinger equation are reversible. (Alternatively, one says that the Schrödinger equation is *invariant under time reversal.*)

Note: As described in Problem 1.24, time reversibility in classical physics includes the operation $(t \rightarrow -t, p \rightarrow -p)$. In quantum mechanics, momentum, p, is replaced by the operator, $-i\hbar\ \partial/\partial x$. It follows that the classical prescription, $p \to -p$, may be effected in quantum mechanics through complex conjugation.

6.34   Consider the time-dependent Hamiltonian

$$\hat{H}\ (x,\ t) = \hat{H}_0(x) + \hbar\ \frac{d\Theta}{dt}$$

where

$$\hat{H}_0\ \varphi_n(x) = E_n\varphi_n(x) \equiv \hbar\omega_n\varphi_n(x)$$

and θ is a dimensionless c-number, which depends only on time.

(a)   What are the time-dependent eigenfunctions, $\psi_n(x,\ t)$, of $\hat{H}\ (x,\ t)$?
(b)   What is $\langle E \rangle$ in the state $\psi_n(x,\ t)$?
(c)   What is the time dependence of energy for the related classical system?

Answers (*partial*)

(a) $\psi_n(x,\ t) = \exp\{-i\left[\omega_n t + \Theta(t)\right]\}\ \varphi_n(x)$

(b) $\langle E \rangle = E_n + \hbar\ \dfrac{d\Theta}{dt}$

7

ADDITIONAL ONE-DIMENSIONAL PROBLEMS. BOUND AND UNBOUND STATES

In this and the following chapters we examine some practical and fundamental problems in one dimension. Included are the very important examples of the harmonic oscillator and scattering configurations in one dimension. Creation and annihilation operators are introduced in algebraic construction of eigenenergies of the harmonic oscillator. The purely quantum mechanical effect of tunneling is encountered in a study of transmission through a barrier. The chapter continues with a description of the WKB technique of solution appropriate in the near-classical domain. This method of approximation finds application in still more realistic configurations, such as cold emission from a metal surface and α decay from a radioactive nucleus. A review of Hamilton's principal of least action precedes a description of Feynman's path integral formalism.

7.1   GENERAL PROPERTIES OF THE ONE-DIMENSIONAL SCHRÖDINGER EQUATION

The time-independent Schrödinger equation for a particle of mass m moving in one dimension in a potential field $V(x)$ appears as

$$(7.1) \qquad \left[-\frac{\hbar^2}{2m}\frac{\partial^2}{\partial x^2} + V(x) \right]\varphi(x) = E\varphi(x)$$

With subscripts denoting differentiation, this equation may be rewritten

$$\varphi_{xx} = -k^2(x)\varphi$$

$$(7.2) \qquad \frac{\hbar^2 k^2}{2m} = E - V$$

The partition of energy

$$(7.3) \qquad E = T + V$$

permits us to identify $\hbar^2 k^2/2m$ as the kinetic energy of the particle

$$(7.4) \qquad T = \frac{\hbar^2 k^2}{2m}$$

This identification is especially relevant if $E > V$. More generally, there are three distinct possibilities (Fig. 7.1). These are $E > V$, $E < V$, and $E = V$. In the first case the kinetic energy is positive and the corresponding classical motion is permitted. Classical motion is forbidden in the second domain, where the kinetic energy is negative. The points where $E = V$ are the classical *turning points*. (Recall Section 1.4.)

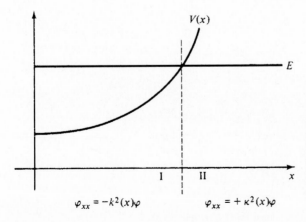

FIGURE 7.1   Domains relevant to a particle of energy E moving in a one-dimensional potential field $V(x)$. I: $E > V$. Kinetic energy is positive. II: $E < V$. Kinetic energy is negative ("forbidden domain"). III: $E = V$. This is a *turning point* of the corresponding classical motion.

In the domain where the kinetic energy is negative, the Schrödinger equation becomes

$$\varphi_{xx} = \kappa^2(x)\varphi$$

(7.5)
$$\frac{\hbar^2\kappa^2}{2m} = V - E > 0$$

$$\text{Kinetic energy} = -\frac{\hbar^2\kappa^2}{2m} = E - V < 0$$

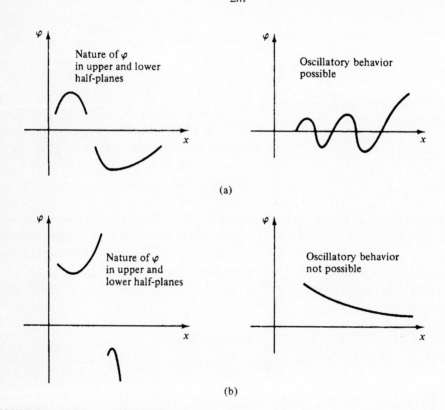

(a)

(b)

FIGURE 7.2   **(a) Kinetic energy positive:**

$$\varphi_{xx} = -k^2\varphi \qquad k^2 > 0$$

$$\varphi_{xx} < 0 \qquad \text{for} \quad \varphi > 0 \text{ (upper half-plane)}$$

$$\varphi_{xx} > 0 \qquad \text{for} \quad \varphi < 0 \text{ (lower half-plane)}$$

(b) Kinetic energy negative:

$$\varphi_{xx} = \kappa^2\varphi \qquad \kappa^2 > 0$$

$$\varphi_{xx} > 0 \qquad \text{for} \quad \varphi > 0 \text{ (upper half-plane)}$$

$$\varphi_{xx} < 0 \qquad \text{for} \quad \varphi < 0 \text{ (lower half-plane)}$$

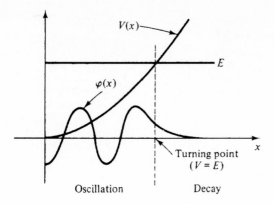

FIGURE 7.3   Characteristic behavior of wavefunction corresponding to the configuration shown in Fig. 7.1.

Recall now that in analytic geometry, φ_{xx} is related to the *curvature* of φ (at the point x). If $\varphi_{xx} > 0$, then φ is concave upward. If $\varphi_{xx} < 0$, then φ is concave downward. When the kinetic energy is positive, the Schrödinger equation takes the form (7.2) and φ has the following properties: φ_{xx} is less than zero in the upper half-plane so φ is concave downward; φ_{xx} is greater than zero in the lower half-plane, so φ is concave upward. As shown in Fig. 7.2, these conditions permit *oscillating solutions*. When the kinetic energy is negative, the Schrödinger equation takes the form (7.5) and the following properties pertain: φ_{xx} is greater than zero in the upper half-plane so φ is concave upward; φ_{xx} is less than zero in the lower half-plane and φ is concave downward. Again referring to Fig. 7.2, these conditions are seen to lead to growing or decaying solutions (as opposed to oscillating solutions). At a turning point, $\varphi_{xx} = 0$ and φ has a constant slope.

For the potential shown in Fig. 7.1, one might then expect an eigenfunction of the Hamiltonian to behave as depicted in Fig. 7.3.

PROBLEMS

7.1 (a)   Let a particle of mass m move in a one-dimensional potential field with energy E as sketched in Fig. 1.18. Write down the form of the time-independent Schrödinger equation [i.e., (7.2) or (7.5)] for the four domains that lie in the interval $0 \leq x \leq D$. In each case identify the wavenumber k or κ.

(b)   Given $\varphi(0) = \varphi_0 > 0$, make a rough sketch of $\varphi(x)$ in the interval $0 \leq x \leq F$.

7.2   THE HARMONIC OSCILLATOR

The configuration of a harmonic oscillator is depicted in Fig. 7.4. The classical equation of motion of a particle of mass m is given by Hooke's law,

(7.6)
$$m \frac{d^2 x}{dt^2} = -\mathrm{K}x$$

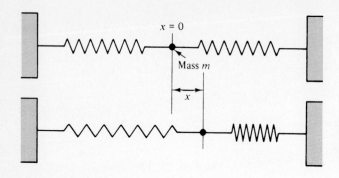

FIGURE 7.4   The one-dimensional harmonic oscillator. Displacement from equilibrium $(x = 0)$ is denoted by x.

The spring constant is K. In terms of the natural frequency ω_0,

(7.7)
$$\omega_0{}^2 = \frac{K}{m}$$

the above equation appears as

(7.8)
$$\frac{d^2x}{dt^2} + \omega_0{}^2 x = 0$$

Multiplying this equation by $\dot{x}$ gives

(7.9)
$$\frac{d}{dt}\left[\frac{1}{2}(\dot{x}^2 + \omega_0{}^2 x^2)\right] = 0$$

Integrating, one obtains the constant of motion

(7.10)
$$\frac{E}{m} = \frac{1}{2}(\dot{x}^2 + \omega_0{}^2 x^2)$$

$$E = \frac{1}{2}m\dot{x}^2 + \frac{K}{2}x^2$$

The potential energy is

(7.11)
$$V = \frac{K}{2}x^2$$

When the particle comes to rest, the energy is entirely potential.

(7.12)
$$E = \frac{K}{2}x_0{}^2$$

Such points (x_0) are turning points. For $x^2 > x_0^2$, the kinetic energy T is negative, so that classically this is a forbidden domain.

$$T = E - V = \frac{K}{2}(x_0^2 - x^2)$$

(7.13)

$$T < 0 \qquad (\text{for } x^2 > x_0^2)$$

See Fig. 7.5.

With these properties of the classical motion established, we turn next to the quantum mechanical formulation of the harmonic oscillator problem. The Hamiltonian for a particle of mass m in the potential (7.11) is

(7.14)
$$H = \frac{p^2}{2m} + \frac{K}{2}x^2$$

The corresponding Schrödinger equation appears as

(7.15)
$$-\frac{\hbar^2}{2m}\frac{\partial^2 \varphi}{\partial x^2} + \frac{K}{2}x^2\varphi = E\varphi$$

In the classically accessible domain, $E > Kx^2/2$, and this equation may be written

$$\varphi_{xx} = -k^2\varphi$$

(7.16)
$$\frac{\hbar^2 k^2(x)}{2m} = E - \frac{K}{2}x^2 > 0$$

The wavefunction φ is oscillatory in this domain.

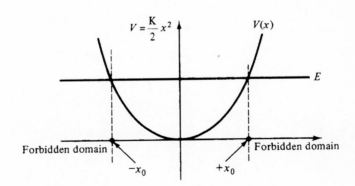

FIGURE 7.5   The turning points of the harmonic oscillator are at $x = \pm x_0$, where

$$\frac{Kx_0^2}{2} = E$$

In the classically forbidden domain where $x^2 > x_0{}^2$, $E < Kx^2/2$ and the Schrödinger equation becomes

$$\varphi_{xx} = \kappa^2 \varphi$$

(7.17)
$$\frac{\hbar^2 \kappa^2}{2m} = \frac{K}{2} x^2 - E > 0$$

so the wavefunction is nonoscillatory in this domain. In the asymptotic domain $Kx^2/2 \gg E$, the Schrödinger equation becomes

(7.18)
$$\varphi_{xx} = \frac{mK}{\hbar^2} x^2 \varphi = \beta^4 x^2 \varphi$$

where β is the characteristic wavenumber

(7.19)
$$\beta^2 \equiv \frac{m\omega_0}{\hbar}$$

In terms of the nondimensional displacement

(7.20)
$$\xi \equiv \beta x$$

(7.18) appears as

$$\varphi_{\xi\xi} = \xi^2 \varphi$$

In the domain under consideration, $\xi \gg 1$ and the solution to the latter equation appears as

$$\varphi \sim A \exp\left(\pm \frac{\xi^2}{2}\right) = A \exp\left[\pm \frac{(\beta x)^2}{2}\right]$$

The growing solution $(+)$ violates the normalization condition

(7.21)
$$\int_{-\infty}^{\infty} \varphi^* \varphi \, dx < \infty$$

and one is left with the exponentially decaying wavefunction

(7.22)
$$\varphi \sim A \exp\left(-\frac{\xi^2}{2}\right) = A \exp\left[-\frac{(\beta x)^2}{2}\right]$$

The character of the wavefunction changes from oscillatory for $x^2 < x_0{}^2$ to decaying

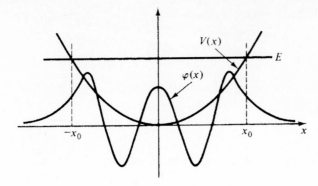

FIGURE 7.6 Typical behavior of energy eigenfunction for the simple harmonic oscillator.

for $x^2 > x_0^2$, so the turning points $x = \pm x_0$ are also physically relevant in quantum mechanics. These properties are depicted in Fig. 7.6.

Annihilation and Creation Operators

We turn to a general formulation of the solution to (7.15). The technique of solution we will develop is known as the *algebraic method*. It involves the operators

(7.23)
$$\hat{a} \equiv \frac{\beta}{\sqrt{2}} \left(\hat{x} + \frac{i\hat{p}}{m\omega_0} \right)$$

$$\hat{a}^\dagger = \frac{\beta}{\sqrt{2}} \left(\hat{x} - \frac{i\hat{p}}{m\omega_0} \right)$$

Inasmuch as $\hat{a} \neq \hat{a}^\dagger$, $\hat{a}$ is non-Hermitian. The properties that these operators have are determined through the fundamental commutator relation

(7.24)
$$[\hat{x}, \hat{p}] = i\hbar$$

For instance, it is readily shown that (see Problem 7.5)

$$[\hat{a}, \hat{a}^\dagger] = 1$$

(7.25)
$$\hat{a}\hat{a}^\dagger = 1 + \hat{a}^\dagger \hat{a}$$

With the aid of the inverse of (7.23),

(7.26)
$$\hat{x} = \frac{\hat{a} + \hat{a}^\dagger}{\sqrt{2}\beta}, \qquad \hat{p} = \frac{m\omega_0}{i} \frac{\hat{a} - \hat{a}^\dagger}{\sqrt{2}\beta}$$

the Hamiltonian for the harmonic oscillator becomes

(7.27) $$\hat{H} = \frac{\hat{p}^2}{2m} + \frac{K\hat{x}^2}{2} = \hbar\omega_0\left(\hat{a}^\dagger\hat{a} + \frac{1}{2}\right)$$

In this manner we see that the problem of finding the eigenvalues of $\hat{H}$ has been transformed to that of finding the eigenvalues of the operator

(7.28) $$\hat{N} \equiv \hat{a}^\dagger\hat{a}$$

Let φ_n be the eigenfunction of $\hat{N}$ corresponding to the eigenvalue n, so that

(7.29) $$\hat{N}\varphi_n = n\varphi_n$$

(We do not assume that n is an integer at this point. This property is established later.) Consider the effect of operating on $\hat{a}\varphi_n$ with $\hat{N}$.

(7.30)
$$\hat{N}\hat{a}\varphi_n = \hat{a}^\dagger\hat{a}\hat{a}\varphi_n = (\hat{a}\hat{a}^\dagger - 1)\hat{a}\varphi_n = \hat{a}(\hat{a}^\dagger\hat{a} - 1)\varphi_n$$
$$\hat{N}\hat{a}\varphi_n = \hat{a}(\hat{N} - 1)\varphi_n = \hat{a}(n - 1)\varphi_n = (n - 1)\hat{a}\varphi_n$$

It follows that $\hat{a}\varphi_n$ is the eigenfunction of $\hat{N}$ which corresponds to the eigenvalue $n - 1$. That is (apart from normalization factors),

(7.31) $$\hat{a}\varphi_n = \varphi_{n-1}$$

Similarly,

(7.32) $$\hat{a}\varphi_{n-1} = \varphi_{n-2}$$

and so forth. Because of this property, $\hat{a}$ is called an *annihilation* or *stepdown* or *demotion* operator.

In similar manner, if we consider the operation $\hat{N}\hat{a}^\dagger\varphi_n$, there results

(7.33) $$\hat{N}\hat{a}^\dagger\varphi_n = (n + 1)a^\dagger\varphi_n$$

This equation implies that $\hat{a}^\dagger\varphi_n$ is the eigenfunction of $\hat{N}$ corresponding to the eigenvalue $n + 1$.

(7.34) $$\hat{a}^\dagger\varphi_n = \varphi_{n+1}$$

Similarly,

(7.35) $$\hat{a}^\dagger\varphi_{n+1} = \varphi_{n+2}$$

and so forth. The operator $\hat{a}^\dagger$ is called a *creation* or *stepup* or *promotion* operator (Fig. 7.7).

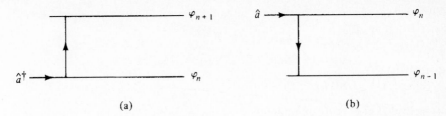

FIGURE 7.7   Schematic representation of the raising and lowering operators $\hat{a}^{\dagger}$ and $\hat{a}$.

Since the Hamiltonian for the harmonic oscillator is the sum of the squares of two Hermitian operators,

(7.36) $$\langle H \rangle \geq 0$$

(see Problem 4.13). In the eigenstate φ_n,

(7.37)
$$\hat{H}\varphi_n = \hbar\omega_0\left(\hat{N} + \frac{1}{2}\right)\varphi_n = \hbar\omega_0\left(n + \frac{1}{2}\right)\varphi_n$$

$$\langle\varphi_n|\hat{H}\varphi_n\rangle = \hbar\omega_0\left(n + \frac{1}{2}\right) \geq 0$$

This implies that the eigenvalues n must obey the condition

(7.38) $$n \geq -\frac{1}{2}$$

That is, all eigenstates of $\hat{H}$, or equivalently $\hat{N}$, corresponding to eigenvalues $n < -\frac{1}{2}$ must vanish identically. For the harmonic oscillator such states do not exist. This condition is guaranteed if we set

(7.39) $$\hat{a}\varphi_0 = 0$$

With (7.31) we obtain

(7.40)
$$\hat{a}\varphi_0 = \varphi_{-1} = 0$$

$$\hat{a}(\varphi_{-1}) = \varphi_{-2} = 0$$

As will be shown, (7.39) has a nontrivial (i.e., other than zero) solution for φ_0. Furthermore,

(7.41) $$\hat{N}\varphi_0 = \hat{a}^{\dagger}\hat{a}\varphi_0 = 0 = 0\varphi_0$$

and we may conclude that the eigenvalue of $\hat{N}$ corresponding to the eigenfunction φ_0 is zero. It follows that

$$\hat{N}\hat{a}^\dagger\varphi_0 = \hat{a}^\dagger\hat{a}\hat{a}^\dagger\varphi_0 = \hat{a}^\dagger(\hat{a}^\dagger\hat{a} + 1)\varphi_0 = \hat{a}^\dagger\varphi_0$$

(7.42)

$$\hat{N}\hat{a}^\dagger\varphi_0 = 1\hat{a}^\dagger\varphi_0 = \varphi_1$$

The eigenvalue of $\hat{N}$ corresponding to φ_1 is the integer 1. This construction [same as in (7.34) et seq.] allows one to conclude that the index n, which labels the eigenfunction φ_n, is indeed an integer.

Repeating (7.37),

(7.43)
$$\hat{H}\varphi_n = \hbar\omega_0\left(n + \frac{1}{2}\right)\varphi_n$$

one finds that the energy eigenvalues of the simple harmonic oscillator are

(7.44)
$$E_n = \hbar\omega_0\left(n + \frac{1}{2}\right) \qquad (n = 0, 1, 2, \ldots)$$

The energy levels are equally spaced by the interval $\hbar\omega_0$ (Fig. 7.8). If a molecule, for example HCl, which resembles a dumbbell, has vibrational modes of excitation (the

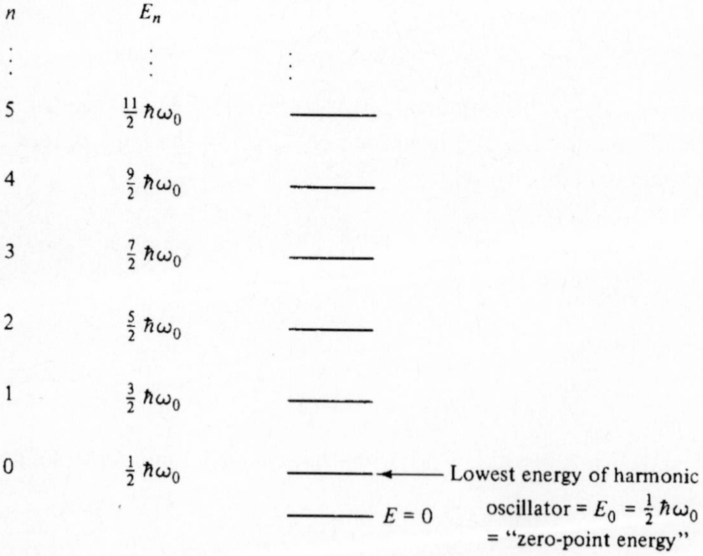

n	E_n	
$\vdots$	$\vdots$	$\vdots$
5	$\frac{11}{2}\hbar\omega_0$	———
4	$\frac{9}{2}\hbar\omega_0$	———
3	$\frac{7}{2}\hbar\omega_0$	———
2	$\frac{5}{2}\hbar\omega_0$	———
1	$\frac{3}{2}\hbar\omega_0$	———
0	$\frac{1}{2}\hbar\omega_0$	——— ←—— Lowest energy of harmonic

——— $E = 0$ oscillator $= E_0 = \frac{1}{2}\hbar\omega_0$
= "zero-point energy"

FIGURE 7.8 The energy levels of the simple harmonic oscillator are equally spaced.

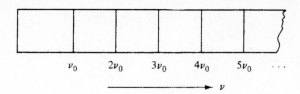

FIGURE 7.9 Spectrum of a vibrational diatomic molecule. (See Problem 10.59.)

arm of the dumbbell acts as a spring), the Bohr frequencies emitted by the molecule fall in the scheme

$$(7.45) \qquad h\nu = E_{n'} - E_n = \hbar\omega_0\left(n' + \frac{1}{2}\right) - \hbar\omega_0\left(n + \frac{1}{2}\right)$$

$$= \hbar\omega_0(n' - n) = \hbar\omega_0 s$$

$$\nu = s\nu_0, \qquad \omega_0 \equiv 2\pi\nu_0$$

In the latter sequence of equations, n and n' are integers, so their difference, s, is also an integer. It follows that the frequencies emitted by a vibrational diatomic molecule are integral multiples of the natural frequency of the molecule, ν_0 (Fig. 7.9).

PROBLEMS

7.2 A harmonic oscillator consists of a mass of 1 g on a spring. Its frequency is 1 Hz and the mass passes through the equilibrium position with a velocity of 10 cm/s. What is the order of magnitude of the quantum number associated with the energy of the system?

7.3 The spacing between vibrational levels of the CO molecule is 2170 cm^{-1}. Taking the mass of C to be 12 amu and O to be 16 amu, compute the effective spring constant K, which is a measure of the bond stiffness between the atoms of the molecule. [*Hint:* The mass that enters is the reduced mass, $mM/(m + M)$. The spacing between lines is given in terms of wavenumber $k = 2\pi/\lambda$, where $\omega = ck$ (c = the speed of light), so that $\Delta\omega = c\,\Delta k$.)

7.4 The derivation in the text of the eigenvalues of $\hat{N}$ is based on the constraint that there are no states corresponding to the eigenvalues $n < -\frac{1}{2}$. This constraint was guaranteed by setting $\hat{a}\varphi_0 = 0$. It would appear that it can also be guaranteed by setting $\hat{a}\varphi_{1/2} = 0$, for in this case

$$\hat{a}\varphi_{1/2} = \varphi_{-1/2} = 0$$

Show that $\varphi_{1/2}$ as defined is an eigenfunction of $\hat{N}$ with the eigenvalue zero; hence $\varphi_{1/2}$ is more properly termed φ_0.

7.5 Using the fundamental commutator relation

$$[\hat{x}, \hat{p}] = i\hbar$$

show that

$$[\hat{a}, \hat{a}^\dagger] = 1$$

7.3   EIGENFUNCTIONS OF THE HARMONIC OSCILLATOR HAMILTONIAN

When written in terms of the nondimensional displacement ξ (7.20),

$$(7.46) \qquad \xi^2 \equiv \frac{m\omega_0}{\hbar} x^2 \equiv \beta^2 x^2$$

the operators $\hat{a}$ and $\hat{a}^\dagger$ become

$$(7.47) \qquad \begin{aligned} \hat{a} &= \frac{\beta}{\sqrt{2}} \left(\hat{x} + \frac{i\hat{p}}{m\omega_0} \right) = \frac{\beta}{\sqrt{2}} \left(x + \frac{\hbar}{m\omega_0} \frac{\partial}{\partial x} \right) = \frac{1}{\sqrt{2}} \left(\xi + \frac{\partial}{\partial \xi} \right) \\ \hat{a}^\dagger &= \frac{\beta}{\sqrt{2}} \left(\hat{x} - \frac{i\hat{p}}{m\omega_0} \right) = \frac{\beta}{\sqrt{2}} \left(x - \frac{\hbar}{m\omega_0} \frac{\partial}{\partial x} \right) = \frac{1}{\sqrt{2}} \left(\xi - \frac{\partial}{\partial \xi} \right) \end{aligned}$$

The time-independent Schrödinger equation becomes

$$(7.48) \qquad \left(2\hat{a}^\dagger \hat{a} + 1 - \frac{2E}{\hbar\omega_0} \right) \varphi = \varphi_{\xi\xi} + \left(\frac{2E}{\hbar\omega_0} - \xi^2 \right) \varphi = 0$$

The ground-state wavefunction φ_0 of the simple harmonic oscillator Hamiltonian obeys (7.39)

$$\hat{a}\varphi_0 = 0$$

or, equivalently,

$$(7.49) \qquad \frac{1}{\sqrt{2}} \left(\xi + \frac{\partial}{\partial \xi} \right) \varphi_0 = 0$$

This has the solution

$$(7.50) \qquad \varphi_0 = A_0 e^{-\xi^2/2}$$

The requirement that $\varphi_0(\xi)$ be normalized implies that

$$(7.51) \qquad \begin{aligned} 1 &= \int_{-\infty}^{\infty} |\varphi_0|^2 \, d\xi = A_0{}^2 \int_{-\infty}^{\infty} e^{-\xi^2} \, d\xi = \sqrt{\pi A_0{}^2} \\ A_0 &= \pi^{-1/4} \end{aligned}$$

so

$$(7.52) \qquad \varphi_0(\xi) = \pi^{-1/4} e^{-\xi^2/2}$$

In terms of the dimensional displacement x, the normalized ground state is

$$(7.53) \qquad \varphi_0(x) = B_0 e^{-\xi^2/2} = B_0 e^{-(\beta x)^2/2}$$

Normalization (with respect to x) gives

$$
1 = \int_{-\infty}^{\infty} |\varphi_0(x)|^2 \, dx = \frac{B_0^{\,2}}{\beta} \int_{-\infty}^{\infty} e^{-\xi^2} \, d\xi = \frac{B_0^{\,2}\sqrt{\pi}}{\beta}
$$

(7.54)

$$
\varphi_0(x) = \left(\frac{\beta^2}{\pi}\right)^{1/4} e^{-(\beta x)^2/2}
$$

The ground state φ_0 is a purely exponentially decaying wavefunction. It has no oscillatory component. The higher-energy eigenstates, on the other hand, will be found to oscillate in the classically allowed domain and decay exponentially in the classically forbidden domain.

With φ_0 given by (7.52), the remaining normalized eigenstates of the harmonic oscillator Hamiltonian are generated with the aid of the creation operator $\hat{a}^\dagger$, in the following manner:

$$
\varphi_1 = \hat{a}^\dagger \varphi_0
$$

(7.55)

$$
\varphi_2 = \frac{1}{\sqrt{2}} \hat{a}^\dagger \varphi_1 = \frac{1}{\sqrt{2}} (\hat{a}^\dagger)^2 \varphi_0
$$

$$
\varphi_n = \frac{1}{\sqrt{n!}} (\hat{a}^\dagger)^n \varphi_0
$$

With $\hat{a}^\dagger$ written in terms of ξ, as in (7.47), the equation for φ_1 above becomes

$$
\varphi_1 = A_1\left(\xi - \frac{\partial}{\partial \xi}\right) e^{-\xi^2/2}
$$

(7.56)

$$
\varphi_1 = A_1 2\xi e^{-\xi^2/2}
$$

$$
A_1 = (2\sqrt{\pi})^{-1/2}
$$

where A_1 is the normalization constant of φ_1. The nth eigenstate is given by the formula

(7.57)

$$
\varphi_n = A_n\left(\xi - \frac{\partial}{\partial \xi}\right)^n e^{-\xi^2/2}
$$

The nth-order differential operator $(\hat{a}^\dagger)^n$, when acting on the exponential form $\exp(-\xi^2/2)$, reproduces the same exponential factor, multiplied by an nth-order polynomial in ξ.

(7.58)

$$
\left(\xi - \frac{\partial}{\partial \xi}\right)^n e^{-\xi^2/2} = \mathcal{H}_n(\xi) e^{-\xi^2/2}
$$

Thus the nth eigenstate of the simple harmonic oscillator Hamiltonian may be written together with its eigenvalue as

(7.59)

$$\varphi_n = A_n \mathcal{H}_n(\xi) e^{-\xi^2/2}$$

$$E_n = \hbar \omega_0 \left(n + \frac{1}{2} \right)$$

The nth-order polynomials $\mathcal{H}_n(\xi)$ are well-known functions in mathematical physics. They are called *Hermite polynomials*. From (7.56) we see that $\mathcal{H}_1 = 2\xi$. The first six Hermite polynomials are listed in Table 7.1.

The nth-order Hermite polynomial $\mathcal{H}_n$ enters in the eigenfunctions φ_n of the quantum mechanical harmonic oscillator as

$$\varphi_n(\xi) = A_n \mathcal{H}_n(\xi) e^{-\xi^2/2}$$

$\mathcal{H}_n$ is a solution to *Hermite's equation*,

$$\mathcal{H}_n'' - 2\xi \mathcal{H}_n' + 2n\mathcal{H}_n = 0$$

The formulas connecting φ_n and φ_{n+1} (see Problem 7.9) are very useful in many problems relating to the simple harmonic oscillator. In Dirac notation they appear as

(7.60)

$$\hat{a} | \varphi_n \rangle = n^{1/2} | \varphi_{n-1} \rangle$$

$$\hat{a}^\dagger | \varphi_n \rangle = (n + 1)^{1/2} | \varphi_{n+1} \rangle$$

In place of $| \varphi_n \rangle$, let us write the ket vector $| n \rangle$. In this notation the equations above appear as

(7.61)

$$\hat{a} | n \rangle = n^{1/2} | n - 1 \rangle$$

$$\hat{a}^\dagger | n \rangle = (n + 1)^{1/2} | n + 1 \rangle$$

TABLE 7.1 **The first six eigenenergies and eigenstates of the simple harmonic oscillator Hamiltonian**

n	E_n	φ_n
0	$\hbar\omega_0/2$	$A_0 e^{-\xi^2/2}$
1	$3\hbar\omega_0/2$	$A_1 2\xi e^{-\xi^2/2}$
2	$5\hbar\omega_0/2$	$A_2(4\xi^2 - 2)e^{-\xi^2/2}$
3	$7\hbar\omega_0/2$	$A_3(8\xi^3 - 12\xi)e^{-\xi^2/2}$
4	$9\hbar\omega_0/2$	$A_4(16\xi^4 - 48\xi^2 + 12)e^{-\xi^2/2}$
5	$11\hbar\omega_0/2$	$A_5(32\xi^5 - 160\xi^3 + 120\xi)e^{-\xi^2/2}$
		$A_n = (2^n n! \sqrt{\pi})^{-1/2}$

Let us check that

(7.62) $$\hat{N}|n\rangle = n|n\rangle$$

With the aid of (7.61), we obtain

(7.63)
$$\hat{a}^\dagger \hat{a}|n\rangle = \hat{a}^\dagger n^{1/2}|n-1\rangle = n^{1/2}n^{1/2}|n\rangle$$
$$\hat{a}^\dagger \hat{a}|n\rangle = \hat{N}|n\rangle = n|n\rangle$$

Inasmuch as $\{\varphi_n\}$ are normalized and are eigenstates of a Hermitian operator, they comprise an orthonormal sequence.

(7.64) $$\int_{-\infty}^{\infty} \varphi_n {}^*\varphi_l \, d\xi = \langle n|l\rangle = \delta_{nl}$$

To gain familiarity with the manner in which these concepts are used in problems, we will work out a few illustrative examples.

First, consider the question: What is $\langle x\rangle$ in the nth eigenstate φ_n? Here we must calculate

(7.65)
$$\langle x\rangle = \langle n|\hat{x}|n\rangle$$
$$= \frac{1}{\sqrt{2}\beta}\langle n|\hat{a} + \hat{a}^\dagger|n\rangle$$
$$= \frac{1}{\sqrt{2}\beta}\{n^{1/2}\langle n|n-1\rangle + (n+1)^{1/2}\langle n|n+1\rangle\}$$
$$= 0$$

The last step follows from the orthogonality relation (7.64). The fact that the average value of x in any eigenstate φ_n vanishes is a consequence of the symmetry of the probability density $P = |\varphi_n|^2$ about the origin (see Fig. 7.10).

The second example we consider is the expectation of momentum p, in the nth eigenstate φ_n.

(7.66)
$$\langle p\rangle = \langle n|\hat{p}|n\rangle = \frac{m\omega_0}{\sqrt{2}i\beta}\langle n|\hat{a} - \hat{a}^\dagger|n\rangle$$
$$= \frac{m\omega_0}{\sqrt{2}i\beta}\{n^{1/2}\langle n|n-1\rangle - (n+1)^{1/2}\langle n|n+1\rangle\}$$
$$= 0$$

In any eigenstate φ_n of the Hamiltonian of the simple harmonic oscillator, the probability of finding the particle with momentum $\hbar k$ is equal to that of finding the particle with momentum $-\hbar k$. Were we to express $\varphi_n(x)$ as a superposition of momentum

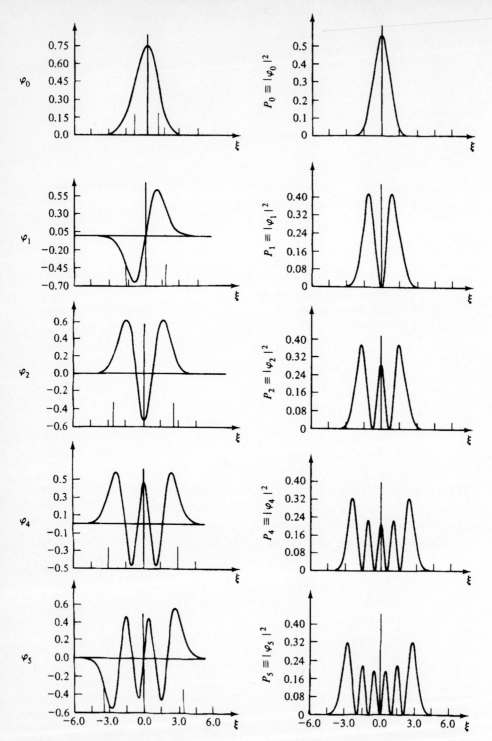

FIGURE 7.10 The first few eigenstates of the simple harmonic oscillator and corresponding probability densities. Turning points, $\xi_0^{(n)} = \sqrt{1 + 2n}$, are denoted by vertical marks.

eigenstates, exp (ikx), we would find the probability amplitude $b(k)$ to be an even (symmetric) function of k [i.e., $b(k) = b(-k)$].

Correspondence Principle

Next, we consider the manner in which the solution to the quantum mechanical harmonic oscillator problem obeys the *correspondence principle*. To these ends let us calculate the classical probability density P, corresponding to a one-dimensional spring with natural frequency ω_0. Let the particle be at the origin at $t = 0$ with velocity $x_0 \omega_0$. The displacement at the time t is then given by

(7.67)
$$x = x_0 \sin (\omega_0 t)$$
$$\dot{x} = x_0 \omega_0 \cos (\omega_0 t)$$

This gives the correct initial data

(7.68)
$$x(0) = 0$$
$$\dot{x}(0) = x_0 \omega_0$$

The product $P(x)\, dx$ is the probability of finding the particle in the interval dx about the point x at any time. If T_0 is the period of oscillation

(7.69)
$$T_0 = \frac{2\pi}{\omega_0}$$

then

(7.70)
$$P\, dx = \frac{dt}{T_0} = \frac{\omega_0\, dt}{2\pi}$$

where

(7.71)
$$dt = \frac{dx}{\dot{x}}$$

Using (7.67), one obtains

(7.72)
$$dt = \frac{dx}{\omega_0 \sqrt{x_0{}^2 - x^2}}$$

so that

(7.73)
$$P\, dx = \frac{\omega_0}{2\pi}\, dt = \frac{dx}{2\pi \sqrt{x_0{}^2 - x^2}}$$

The probability density so found is normalized with respect to the angular displacement $d\theta = \omega_0 \, dt, 0 \le \theta \le 2\pi$. The interval in displacement x is one-half as long, so the properly normalized P function, over the interval $-x_0 < x < +x_0$, is

(7.74)

$$P = \frac{1}{\pi \sqrt{x_0^2 - x^2}}$$

$$\int_{-x_0}^{+x_0} P(x) \, dx = 1$$

This function is sketched in Fig. 7.11, where it is superimposed on the quantum mechanical probability density corresponding to a state with $n \gg 1$. The singularities in P at the turning points $\pm x_0$ are due to the fact that the particle comes to rest at these points.

The correspondence which the quantum mechanical formulation displays in the present case is clearly exhibited in Fig. 7.11, where we see that

(7.75)

$$\lim_{n \to \infty} \langle P_n^{\text{QM}} \rangle = P^{\text{CL}}$$

$$\langle P_n^{\text{QM}} \rangle = \frac{1}{2\epsilon} \int_{x-\epsilon}^{x+\epsilon} \varphi_n^*(y) \varphi_n(y) \, dy$$

The superscripts QM and CL denote quantum mechanical and classical, respectively, and ϵ is an arbitrarily small interval. The integral above is called a *local average*. It represents the average of P^{QM} in a small interval centered at x.

The classical configuration that corresponds to the quantum state in which a set of commuting observables are specified is the configuration which includes these same parameters as constant and known. Thus, in the problem of a particle confined to a one-dimensional box, considered in Chapter 4, when one concludes that the classical probability density is uniform, it should be noted that this is the case provided that all one knows about the particle is its energy. The classical state of this system permits a more elaborate description. Unlike the quantum case, for the classical particle one may specify both its energy and position in time, $x(t)$. Given this maximally informative classical description, the configurational probability density becomes $\delta[x(t) - x]$. When one speaks of configurational correspondence in the limit of high quantum numbers, what is usually meant is that in this limit the quantum probability density goes to the classical probability density in which, consistent with the quantum description, not more than the energy is specified.

In our consideration of correspondence for the harmonic oscillator, this rule is again obeyed. The expression (7.74) for the classical probability density is relevant to the case where only the amplitude x_0, or equivalently the energy $E = Kx_0^2/2$, is known. The quantum density sketched in Fig. 7.11 is likewise connected to the energy eigenstate φ_n for which measurement of energy finds with certainty the value E_n.

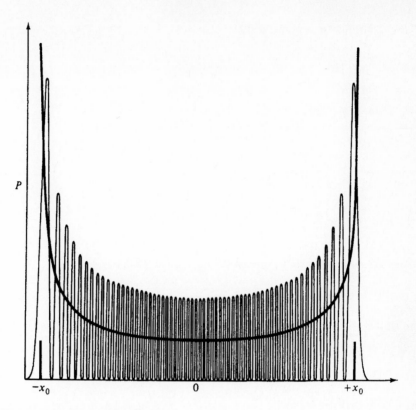

FIGURE 7.11   **Classical probability density**

$$P^{\text{CL}} = \frac{1}{\pi\sqrt{x_0^2 - x^2}}$$

superimposed on the quantum mechanical probability density

$$P_n^{\text{QM}} = |\varphi_n|^2$$

For the case $n \gg 1$,

$$\lim_{n \to \infty} \langle P_n^{\text{QM}} \rangle = P^{\text{CL}}$$

PROBLEMS

7.6   (a)   Show that the Hermite polynomials generated in the Taylor series expansion

$$\exp\left(2\xi t - t^2\right) = \sum_{n=0}^{\infty} \frac{\mathcal{H}_n(\xi)}{n!} \, t^n$$

are the same as generated in (7.58).

(b) Show that $\mathcal{H}_n$ as generated by

$$\mathcal{H}_n(\xi) = (-1)^n \left(e^{\xi^2} \frac{\partial^n}{\partial \xi^n} e^{-\xi^2} \right)$$

are equivalent to those given by (7.58).

(c)   Use any of the preceding relations to establish

$$\mathcal{H}_n' = 2n\mathcal{H}_{n-1}$$

(d)   and the recursion relation

$$\mathcal{H}_{n+1} = 2\xi\mathcal{H}_n - 2n\mathcal{H}_{n-1}$$

(e)   Use the generating formula of part (b) to find $\mathcal{H}_0(\xi)$, $\mathcal{H}_1(\xi)$, and $\mathcal{H}_2(\xi)$.

7.7   The general formula for the normalization constant of φ_n is

$$A_n = (2^n n! \sqrt{\pi})^{-1/2}$$

Show that this gives correct normalization for φ_4.

7.8   Show directly from the form of φ_n given by (7.57) that

$$\hat{\mathbb{P}}\varphi_n = (-)^n \varphi_n$$

where $\hat{\mathbb{P}}$ is the parity operator.

7.9   (a)   Show that the normalized nth eigenstate φ_n is generated from the normalized ground state φ_0 through

$$\varphi_n = \frac{1}{\sqrt{n!}} (\hat{a}^\dagger)^n \varphi_0$$

(b)   Show that part (a) implies the following relations.

$$\hat{a}\varphi_n = n^{1/2}\varphi_{n-1}$$
$$\hat{a}^\dagger \varphi_n = (n+1)^{1/2}\varphi_{n+1}$$

where φ_n, φ_{n-1}, and φ_{n+1} are normalized.

7.10   Show that in the nth eigenstate of the harmonic oscillator, the average kinetic energy $\langle T \rangle$ is equal to the average potential energy $\langle V \rangle$ (the virial theorem). That is,

$$\langle V \rangle = \frac{K}{2} \langle x^2 \rangle = \langle T \rangle = \frac{1}{2m} \langle p^2 \rangle = \frac{1}{2} \langle E \rangle = \frac{\hbar\omega_0}{2} \left(n + \frac{1}{2} \right)$$

Answer (partial)

$$\frac{K}{2}\langle x^2 \rangle = \left(\frac{K}{4\beta^2}\right)\langle n | (\hat{a} + \hat{a}^\dagger)^2 | n \rangle$$

$$= \left(\frac{K}{4\beta^2}\right)\langle n | \hat{a}^2 + \hat{a}^{\dagger 2} + (\hat{a}\hat{a}^\dagger + \hat{a}^\dagger \hat{a}) | n \rangle$$

$$= \left(\frac{K}{4\beta^2}\right)\{0 + 0 + \langle n | (1 + 2\hat{N}) | n \rangle\}$$

$$\langle V \rangle = \frac{\hbar\omega_0}{4}(1 + 2n) = \frac{\hbar\omega_0}{2}\left(n + \frac{1}{2}\right)$$

7.11   In Problem 7.2, what is the average spacing (in cm) between zeros of an eigenstate with such a quantum number?

7.12   A harmonic oscillator is in the initial state

$$\psi(x, 0) = \varphi_n(x)$$

that is, an eigenstate of $\hat{H}$. What is $\psi_n(x, t)$?

7.13   For a harmonic oscillator in the superposition state

$$\psi(x, t) = \frac{1}{\sqrt{2}}[\psi_0(x, t) + \psi_1(x, t)]$$

show that

$$\langle x \rangle = C \cos(\omega_0 t)$$

In the notation above,

$$\psi_n(x, t) \equiv \varphi_n(x) \exp\left(-\frac{iE_n t}{\hbar}\right)$$

7.14   Show that in the nth state of the harmonic oscillator,

$$\langle x^2 \rangle = (\Delta x)^2$$

$$\langle p^2 \rangle = (\Delta p)^2$$

7.15   Find $\langle x \rangle$ for a harmonic oscillator in the superposition state

$$\psi(x, t) = \frac{1}{\sqrt{2}}[\psi_0(x, t) + \psi_3(x, t)]$$

The harmonic oscillator has natural frequency ω_0.

7.16 A large dielectric cube with edge length a is uniformly charged throughout its volume so that it carries a total charge Q. It fills the space between condenser plates, which have a potential difference Φ_0 across them. An electron is free to move in a small canal drilled in the dielectric normal to the plates (Fig. 7.12).

The Hamiltonian for the electron is (with x measured from the center of the canal and e written for $-|e|$)

$$\hat{H} = \frac{\hat{p}^2}{2m} + \frac{Kx^2}{2} + \frac{e\Phi_0}{a}x$$

(a) What is the spring constant K in terms of the total charge Q?

(b) What are eigenenergies and eigenfunctions of $\hat{H}$? [*Hint:* Rewrite the potential energy of the electron as

$$V = \frac{K}{2}(x^2 + 2\gamma x) = \frac{K}{2}[(x + \gamma)^2 - \gamma^2]$$

$$\gamma \equiv \frac{e\Phi_0}{aK}$$

then change variables to $z \equiv x + \gamma$. To evaluate K, use Gauss's law (neglecting "edge effects").]

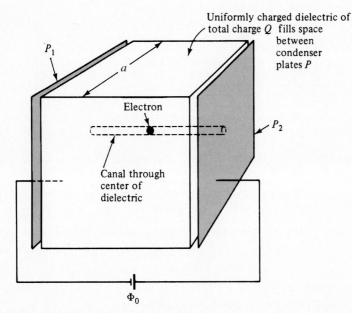

FIGURE 7.12   Configuration described in Problem 7.16.

7.17 (a) Show that the time-independent Schrödinger equation for the harmonic oscillator, with the energy eigenvalues (7.44), may be written

$$\varphi_{\xi\xi} + (2n + 1 - \xi^2)\varphi = 0$$

(b) Using the relations of Problem 7.6, show that

$$\varphi_n = \mathcal{H}_n(\xi)e^{-\xi^2/2}$$

is a solution to this equation.

(c) Obtain Hermite's equation

$$\mathcal{H}_n'' - 2\xi\mathcal{H}_n' + 2n\mathcal{H}_n = 0$$

7.18 Use the uncertainty principle between x and p to derive the "zero-point" energy

$$E_0 = \frac{1}{2}\hbar\omega_0$$

of a harmonic oscillator with natural frequency ω_0 (see Fig. 7.8).

7.19 Show that

$$\hat{a}^\dagger\hat{a} = \frac{1}{2}\left(\xi^2 - \frac{\partial^2}{\partial\xi^2} - 1\right)$$

in the nondimensional ξ notation.

7.20 Show that the asymptotic exponential behavior of $\varphi_n(\xi)$ agrees with that obtained directly from the Schrödinger equation, in the limit that $\xi \to \infty$.

7.21 Show that

$$\left(\xi + \frac{\partial}{\partial\xi}\right)^\dagger = \xi - \frac{\partial}{\partial\xi}$$

in $\mathfrak{H}_2$ [see (4.31)].

7.22 (a) What is the asymptotic solution φ_n to the Schrödinger equation (as given in Problem 7.17)

$$\varphi_{\xi\xi} + (2n + 1 - \xi^2)\varphi = 0$$

in the domain

$$\xi^2 \ll 1 + 2n \simeq 2n?$$

(b) Show that

$$\lim_{n \gg 1} \langle P_n\rangle = \langle|\tilde{\varphi}_n|^2\rangle = \text{constant}$$

Answer

In this domain the Schrödinger equation above becomes

$$\tilde{\varphi}_{\xi\xi} + 2n\tilde{\varphi} = 0$$

which has the (even) solution

$$\tilde{\varphi}_n = C \cos (\sqrt{2n}\ \xi)$$

It follows that the local average of $|\tilde{\varphi}_n|^2$ in this domain is given by

$$\langle |\tilde{\varphi}_n|^2\rangle = \frac{C^2}{2\epsilon} \int_{\xi-\epsilon}^{\xi+\epsilon} \cos (\sqrt{2n}\xi)\ d\xi$$

$$= \frac{C^2}{2\epsilon} \left\{ \epsilon + \frac{1}{2\sqrt{2n}} [\sin (2\sqrt{2n}\ \epsilon) \cos (2\sqrt{2n}\ \xi)] \right\}$$

$$\lim_{n\to\infty} \langle P_n\rangle = \frac{C^2}{2}$$

This result explains the flatness of $\langle P^{\mathrm{QM}}\rangle$ in the central domain $\xi^2 \ll 2n$, as seen in Fig. 7.11.

7.23   Estimate the length of interval about $x = 0$ which corresponds to the classically allowed domain for the ground state of the simple harmonic oscillator.

Answer

The turning points occur at

$$\xi = \pm 1 \quad \text{or equivalently at} \quad x = \pm \sqrt{\frac{\hbar}{m\omega_0}}$$

At this value, $|\varphi_0|^2$ is e^{-1} times smaller than its value at the origin (Fig. 7.10).

7.24   Show that in the nth stationary state $|n\rangle$ of a harmonic oscillator with fundamental frequency ω_0,

$$\Delta p\ \Delta x = \frac{E_n}{\omega_0} = \hbar \left(n + \frac{1}{2} \right)$$

7.4   THE HARMONIC OSCILLATOR IN MOMENTUM SPACE

Representations in Quantum Mechanics

Let us recall (4.42) et seq., which relate the wavefunction $\varphi(x)$ to the momentum coefficient $b(k)$.

$$\varphi(x) = \int_{-\infty}^{\infty} b(k)\varphi_k\ dk$$

(7.76)

$$b(k) = \int_{-\infty}^{\infty} \varphi(x)\varphi_k{}^*\ dx$$

The eigenfunction of momentum corresponding to the value $p = \hbar k$ is φ_k. The wave-function $\varphi(x)$ gives the probability density in coordinate space through the Born relation

$$(7.77) \qquad\qquad P(x) = |\varphi(x)|^2$$

The momentum coefficient $b(k)$ gives the probability density [probability of finding the particle to have momentum in the interval $\hbar k$ to $\hbar(k + dk)$] in momentum (k) space through the relation

$$(7.78) \qquad\qquad P(k) = |b(k)|^2$$

The integral formulas (7.76) serve to determine $\varphi(x)$ given $b(k)$, and vice versa. It follows that any information contained in $\varphi(x)$ can be obtained from knowledge of $b(k)$ and vice versa. Given the Hamiltonian of a system, $\varphi(x)$ is determined. Let us construct an equation which similarly determines $b(k)$ from the Hamiltonian [i.e., without first finding $\varphi(x)$]. To these ends we first recall the time-independent Schrödinger equation for the harmonic oscillator.

$$(7.79) \qquad \left(-\frac{\hbar^2}{2m}\frac{\partial^2}{\partial x^2} + \frac{Kx^2}{2} \right)\varphi(x) = E\varphi(x)$$

Substituting the Fourier decomposition of $\varphi(x)$ above and noting the equality.

$$(7.80) \qquad\qquad x\varphi_k = -i\frac{\partial \varphi_k}{\partial k}$$

gives

$$(7.81) \qquad \int_{-\infty}^{\infty} dk\, b(k)\left(\frac{\hbar^2 k^2}{2m} - \frac{K}{2}\frac{\partial^2}{\partial k^2} \right)\varphi_k = E\int_{-\infty}^{\infty} dk\, b(k)\varphi_k$$

Integrating the second term on the left-hand side by parts twice and setting

$$(7.82) \qquad\qquad b(k)\big|_{k=\pm\infty} = b'(k)\big|_{k=\pm\infty} = 0$$

gives

$$(7.83) \qquad \int_{-\infty}^{\infty} dk\, \varphi_k \left[\left(\frac{\hbar^2 k^2}{2m} - \frac{K}{2}\frac{\partial^2}{\partial k^2} - E \right)b(k) \right] = 0$$

It follows that the term in brackets is the Fourier transform of zero, which is zero. We conclude that $b(k)$ (appropriate to the harmonic oscillator) satisfies the *k-dependent Schrödinger equation*

$$(7.84) \qquad \left(\frac{\hbar^2 k^2}{2m} - \frac{K}{2}\frac{\partial^2}{\partial k^2}\right) b(k) = Eb(k)$$

This equation is also called the *Schrödinger equation in momentum representation*.
 We note that the Hamiltonian in momentum representation includes the simple multiplicative operator $\hbar k$ in place of p and the differential operator $+i\partial/\partial k$ in place of x. This rule for obtaining the structure of the Hamiltonian in momentum representation always holds providing the potential $V(x)$ is an analytic function[1] of x (i.e., has a well-defined power-series expansion). For such cases the Schrödinger equation in either coordinate or momentum space is obtained through the recipes:

$$\text{In } x\text{-space:} \quad \hat{H}(x, p) \rightarrow \hat{H}\left(x, -\frac{i\hbar\,\partial}{\partial x}\right)$$

$$\text{In } p\text{-space:} \quad \hat{H}(x, p) \rightarrow \hat{H}\left(-\frac{i\partial}{\partial k}, \hbar k\right)$$

The time-dependent Schrödinger equation in momentum representation appears as

$$(7.85) \qquad i\hbar\frac{\partial}{\partial t}b(k, t) = \hat{H}(k)b(k, t)$$

Paralleling the development of (3.70) permits the solution to (7.85) for the initial-value problem for $b(k, t)$ to be written

$$(7.86) \qquad b(k, t) = \exp\left(-\frac{it\hat{H}}{\hbar}\right)b(k, 0)$$

For free-particle motion with $\hat{H} = \hbar^2 k^2/2m$, the latter relation gives

$$(7.87) \qquad |b(k, t)|^2 = |b(k, 0)|^2$$

The momentum probability density for free-particle motion is constant in time.
 Geometrically, the function $b(k)$ is the projection of the state $\varphi(x)$ onto the momentum eigenstate φ_k [recall (4.44)].

$$(7.88) \qquad b(k) = \langle \varphi_k | \varphi \rangle$$

---

[1] In the more general case the Schrödinger equation in momentum space becomes an integral equation. These topics are discussed in greater detail in E. Merzbacher, *Quantum Mechanics*, 2d ed., Wiley, New York, 1970.

For any given state $\varphi(x)$, the function $b(k)$ represents a distribution of values, corresponding to the projections of $\varphi(x)$ onto the set of basis vectors $\{\varphi_k\}$. The functions $b(k)$ and $\varphi(x)$ are equally informative. In momentum representation a state of the system is represented by its projections onto the basis of Hilbert space $\{\varphi_k\}$. (See Fig. 4.6.)

This is analogous to the statement that a vector **B** in 3-space is represented by its projections onto the three unit vectors $\mathbf{e}_x$, $\mathbf{e}_y$, and $\mathbf{e}_z$, namely, B_x, B_y, and B_z. These are not the only basis vectors one can use to represent the vector **B**. For instance, one can employ the basis $\mathbf{e}_x'$, $\mathbf{e}_y'$, and $\mathbf{e}_z'$ given by

$$\mathbf{e}_x' = \frac{1}{\sqrt{2}}(\mathbf{e}_x + \mathbf{e}_y)$$

(7.89)
$$\mathbf{e}_y' = \frac{1}{\sqrt{2}}(\mathbf{e}_x - \mathbf{e}_y)$$

$$\mathbf{e}_z' = \mathbf{e}_z$$

(See Fig. 7.13.) In this basis **B** is represented by the three components

$$\mathbf{B} = (B_x', B_y', B_z') = \frac{1}{\sqrt{2}}(B_x + B_y, B_x - B_y, \sqrt{2}B_z)$$

There are countless other triads of unit vectors which are valid bases of 3-space (i.e., they all span 3-space). The three components of **B** in any one of these representations completely specify **B**.

Similarly, one may describe the state of a system in quantum mechanics in different representations. In each of these, a distinct set of vectors serves as a basis of

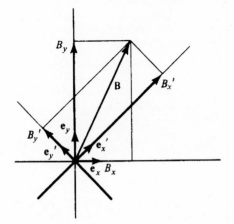

FIGURE 7.13 Projections onto two sets of basis vectors (in the xy plane) of the vector B. The two bases are related through (7.89). The z component is the same in both representations.

Hilbert space. Particularly important in the theory of representations is the concept of common eigenfunctions of some *complete set of commuting operators* relevant to a given system. Suppose that the complete set of commuting operators are $\hat{A}$, $\hat{B}$, and $\hat{C}$. In the state φ_{abc} (common eigenstates of $\hat{A}$, $\hat{B}$, and $\hat{C}$) one may specify the "good" quantum numbers a, b, and c. The state of the system cannot be further resolved. Such states may serve as a basis of Hilbert space. The representation in which all states are referred to the basis $\{\varphi_{abc}\}$ is called the *abc representation,* just as we call the representation in which states are referred to the eigenfunctions of momentum, the *momentum representation.*[1] One also speaks of the *abc* representation as the one in which $\hat{A}$, $\hat{B}$, and $\hat{C}$ are *diagonal.*

PROBLEMS

7.25   Show that $b(k)$ is even for any even potential $V(x) = V(-x)$. What can be concluded about the oddness or evenness of $b(k)$ if $V(x)$ is an odd function, $V(x) = -V(-x)$?

7.26   What are the eigenfunctions $b_n(k)$ of the harmonic oscillator Hamiltonian $\hat{H}(k)$ in momentum space [as given by (7.84)]? [*Hint:* Note the similarity between $\hat{H}(k)$ and $\hat{H}(x)$, and the boundary conditions of $\varphi_n(x)$ and $b_n(k)$.]

7.27   What is the Schrödinger equation in momentum representation for a free particle moving in one dimension? What are the eigenfunctions $b(k)$ of this equation?

7.28   Consider the Gaussian wave packet whose initial momentum probability density is given by (6.45).
   (a)   What is $|b(k, t)|^2$ at $t > 0$?
   (b)   What is $\Delta x \, \Delta p$ at $t > 0$?

7.29   Consider an arbitrary differentiable function of p, $\varphi(p)$. Show that with $\hat{p} = p$ and $\hat{x} = +i\hbar \, \partial/\partial p$,

$$[\hat{x}, \hat{p}]\varphi(p) = i\hbar\varphi(p)$$

7.30   What is the eigenfunction of the operator $\hat{x}$, in the momentum representation, corresponding to the eigenvalue x? That is, give the solution to the equation

$$\hat{x}\varphi_x(p) = x\varphi_x(p)$$

7.31   Let $|x'\rangle$ denote an eigenvector of the position operator $\hat{x}$ with eigenvalue x' and let $|k'\rangle$ denote an eigenvector of the momentum operator $\hat{p}$ with eigenvalue $\hbar k'$. Show that
   (a) $\langle k|k'\rangle = \delta(k - k')$
   (b) $\langle x|x'\rangle = \delta(x - x')$
   (c) $\langle x|k\rangle = \dfrac{1}{\sqrt{2\pi}} \exp{(ikx)}$

For each case, state in which representation you are working.

---

[1] For further discussion of the coordinate and momentum representations, see Appendix A.

7.32   Suppose that the operators $\hat{a}$ and $\hat{a}^\dagger$ in

$$\hat{H} = \hbar\omega_0\left(\hat{a}^\dagger\hat{a} + \frac{1}{2}\right)$$

obey the *anticommutation* relation

$$\{\hat{a}, \hat{a}^\dagger\} \equiv \hat{a}\hat{a}^\dagger + \hat{a}^\dagger\hat{a} = 1$$

(a)   What are the values of $\hat{a}|n\rangle$ and $\hat{a}^\dagger|n\rangle$ that follow from the anticommutation relation above?

(b)   Since $\langle H\rangle \geq 0$, for consistency we may again set

$$\hat{a}|0\rangle = 0$$

Combining this fact with your answer to part (a), which are the only nonvanishing states $|n\rangle$?

(c)   If, in addition to the anticommutation property above, $\hat{a}$ and $\hat{a}^\dagger$ also obey the relations, $\{\hat{a}, \hat{a}\} = \{\hat{a}^\dagger, \hat{a}^\dagger\} = 0$, show that $\hat{N}^2 = \hat{N}$.

Answers (partial)

(a) $\hat{a}|n\rangle = \sqrt{n}|1 - n\rangle$
   $\hat{a}^\dagger|n\rangle = \sqrt{1 - n}|1 - n\rangle$

(b)   The only nonvanishing states are $|0\rangle$ and $|1\rangle$. [*Note:* Anticommutation relations between $\hat{a}$ and $\hat{a}^\dagger$ are used to describe particles that obey the *Pauli exclusion principle*.[1] In this context the operator $\hat{N}$ denotes the number of particles in a given state so that (b) implies that there is no more than one particle in any state. The $|0\rangle$ state is called the *vacuum state*. The formalism is known as *second quantization*.[2]]

7.33   What is the lowest value of kinetic energy $\langle T\rangle$ a harmonic oscillator with frequency ω_0 can have?

Answer

In Problem 7.10 we found that in the *n*th eigenstate of the oscillator

$$\langle V\rangle = \langle T\rangle = \frac{\hbar\omega_0}{2}\left(n + \frac{1}{2}\right) \geq \frac{\hbar\omega_0}{4}$$

Thus, the lowest allowed energy of the oscillator is $\hbar\omega_0/2$. It is impossible to force the oscillator to a lower energy. In a solid, for example, whose nuclei are bound together by harmonic forces, this zero-point energy persists at $0\,\text{K}$. (See Problem 7.18.)

---

[1] A formal statement of this principle is given in Chapter 12. See also Appendix B.

[2] Second quantization is encountered again in Problem 13.37.

7.5 UNBOUND STATES

If a wavefunction ψ represents a bound state (in one dimension), then

$$(7.90) \qquad |\psi|^2 \to 0, \qquad |x| \to \infty$$

for all t. A wavefunction that does not obey this condition represents an *unbound state*. The square modulus of a *bound state* gives a finite integral over the infinite interval.

$$(7.91) \qquad \int_{-\infty}^{\infty} |\psi|^2 \, dx < \infty$$

The square modulus of an *unbound state* gives a finite integral over *any* finite interval.

$$(7.92) \qquad \int_{a}^{b} |\psi|^2 \, dx < \infty, \qquad |b - a| < \infty$$

The eigenstate of the momentum operator

$$(7.93) \qquad \varphi_k(x) = \frac{1}{\sqrt{2\pi}} e^{ikx}$$

represents an unbound state. The eigenfunction of the simple harmonic oscillator Hamiltonian

$$(7.94) \qquad \varphi_n(\xi) = A_n \mathcal{H}_n(\xi) e^{-\xi^2/2}$$

(see Section 7.3) represents a bound state. Unbound states are relevant to scattering problems. Such problems characteristically involve a beam of particles which is incident on a potential barrier (Fig. 7.14).

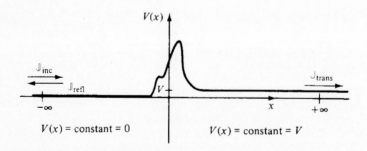

FIGURE 7.14 **One-dimensional scattering problem. Incident particle current $\mathbb{J}_{inc}$ initiated at $x = -\infty$ is partially transmitted ($\mathbb{J}_{trans}$) and partially reflected ($\mathbb{J}_{refl}$) by a potential barrier $V(x)$. The potential is constant outside the scattering domain.**

Since $\int_{-\infty}^{\infty} |\psi|^2 \, dx$ diverges for unbound states, it is convenient to normalize the wavefunction for scattering problems in terms of the particle density ρ. For one-dimensional scattering problems we take

$$|\psi|^2 \, dx = \rho \, dx = dN$$

$$= \text{number of particles in the interval } dx$$

(7.95)

$$\int_a^b |\psi|^2 \, dx = N$$

$$= \text{number of particles in the interval } (b - a)$$

For a one-dimensional beam of 10^3 neutrons/cm, all moving with momentum $p = \hbar k_0$, the wavefunction is written

$$\psi = 10^{3/2} \, e^{i(k_0 x - \omega t)}, \qquad |\psi|^2 = 10^3 \text{ cm}^{-1}$$

(7.96)

$$\frac{\hbar^2 k_0^2}{2m} = \hbar \omega$$

The sole difference between ψ so defined and a wavefunction whose square modulus is probability density is a multiplicative constant. It follows that $|\psi|^2$, when referred to particle density, is proportional to probability density also. For uniform beams, $|\psi|^2$ is constant, which in turn implies that it is uniformly probable to find particles anywhere along the beam. This is consistent with the uncertainty principle. For instance, for the wavefunction (7.96), the momentum of any neutron in the beam is $\hbar k_0$, whence its position is maximally uncertain.

Continuity Equation

One-dimensional barrier problems involve incident, reflected, and transmitted *current densities*, $\mathbb{J}_{\text{inc}}$, $\mathbb{J}_{\text{refl}}$, and $\mathbb{J}_{\text{trans}}$, respectively. In three dimensions the number density and current density $\mathbf{J}$ are related through the *continuity equation*

(7.97)

$$\frac{\partial \rho}{\partial t} + \mathbf{\nabla} \cdot \mathbf{J} = 0$$

To clarify the physical meaning of this equation, we integrate it over a volume V and obtain

(7.98)

$$\frac{\partial N}{\partial t} = -\int_S \mathbf{J} \cdot d\mathbf{S}$$

The total number of particles in the volume V is

(7.99)
$$N = \int_V \rho \, d\mathbf{r}$$

(Gauss's theorem was used to transform the divergence term.) The surface S encloses the volume V (Fig. 7.15). Equation (7.98) says that the number of particles in the volume V changes by virtue of a net flux of particles out of (or into) the volume V. It is a statement of the *conservation of matter* because it says that this is the *only* way the total number of particles in V can change. If particles are born spontaneously in V with no net flux of particles through the surface S, then $\partial N/\partial t > 0$, while $\int \mathbf{J} \cdot d\mathbf{S} = 0$ and (7.98) is violated.

If particles are moving only in the x direction,

(7.100)
$$\mathbf{J} = (\mathbb{J}_x, 0, 0)$$

and the continuity equation becomes

(7.101)
$$\frac{\partial \rho}{\partial t} + \frac{\partial \mathbb{J}_x}{\partial x} = 0$$

We already have identified ρ with $|\psi|^2$. To relate $\mathbb{J}_x$ to ψ, we must construct an equation that looks identical to (7.101) with $|\psi|^2$ in place of ρ. Then the functional of ψ which appears after $\partial/\partial x$ is $\mathbb{J}_x$.

The wavefunction for particles in the beam obeys the Schrödinger equation

(7.102)
$$\frac{\partial \psi}{\partial t} = \frac{i}{\hbar} \hat{H}\psi, \qquad \frac{\partial \psi^*}{\partial t} = +\frac{i}{\hbar} \hat{H}\psi^*$$

The time derivative of the particle density $\psi^*\psi$ is

(7.103)
$$\frac{\partial \psi^*\psi}{\partial t} = \psi^* \frac{\partial \psi}{\partial t} + \psi \frac{\partial \psi^*}{\partial t} = \psi^*\left(\frac{-i\hat{H}}{\hbar}\psi\right) + \psi\left(\frac{+i\hat{H}}{\hbar}\psi^*\right)$$

For the typical one-dimensional Hamiltonian

(7.104)
$$\hat{H} = \frac{\hat{p}^2}{2m} + V(x)$$

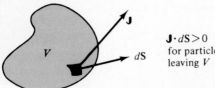

$\mathbf{J} \cdot d\mathbf{S} > 0$
for particles
leaving V

FIGURE 7.15   Geometry relevant to integration of the continuity equation.

the latter equation becomes

$$\frac{\partial \psi^* \psi}{\partial t} = \frac{i\hbar}{2m} (\psi^* \psi_{xx} - \psi \psi^*_{xx})$$

(7.105)

$$\frac{\partial \psi^* \psi}{\partial t} + \frac{\partial}{\partial x} \left[\frac{\hbar}{2mi} \left(\psi^* \frac{\partial \psi}{\partial x} - \psi \frac{\partial \psi^*}{\partial x} \right) \right] = 0$$

(The subscript x denotes differentiation.) Comparison of this equation with (7.101) permits the identification

(7.106)
$$\mathbb{J}_x = \frac{\hbar}{2mi} \left(\psi^* \frac{\partial \psi}{\partial x} - \psi \frac{\partial \psi^*}{\partial x} \right)$$

Note that the dimensions of $\mathbb{J}_x$ are number per second. In three dimensions the current density is written

(7.107)
$$\mathbf{J} = \frac{\hbar}{2mi} (\psi^* \boldsymbol{\nabla} \psi - \psi \boldsymbol{\nabla} \psi^*)$$

and has dimensions $\text{cm}^{-2}\,\text{s}^{-1}$.

Transmission and Reflection Coefficients

For one-dimensional scattering problems, the particles in the beam are in plane-wave states with definite momentum. Given the wavefunctions relevant to incident, reflected, and transmitted beams, one may calculate the corresponding current densities according to (7.106). The *transmission coefficient T* and *reflection coefficient R* are defined as

(7.108)
$$T \equiv \left| \frac{\mathbb{J}_\text{trans}}{\mathbb{J}_\text{inc}} \right|, \qquad R \equiv \left| \frac{\mathbb{J}_\text{refl}}{\mathbb{J}_\text{inc}} \right|$$

These one-dimensional barrier problems are closely akin to problems on the transmission and reflection of electromagnetic plane waves through media of varying index of refraction (see Fig. 7.16). In the quantum mechanical case, the scattering is also of waves.

For one-dimensional barrier problems there are three pertinent beams. Particles in the incident beam have momentum

(7.109)
$$p_\text{inc} = \hbar k_1$$

Particles in the reflected beam have the opposite momentum

(7.110)
$$p_\text{refl} = \hbar k_1$$

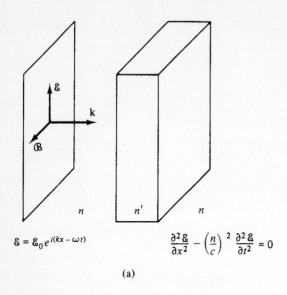

$$\mathcal{E} = \mathcal{E}_0 e^{i(kx - \omega t)}$$

$$\frac{\partial^2 \mathcal{E}}{\partial x^2} - \left(\frac{n}{c}\right)^2 \frac{\partial^2 \mathcal{E}}{\partial t^2} = 0$$

(a)

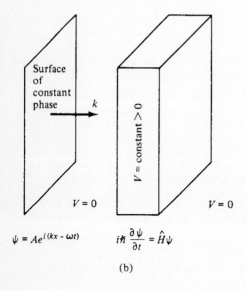

$$\psi = A e^{i(kx - \omega t)}$$

$$i\hbar \frac{\partial \psi}{\partial t} = \hat{H} \psi$$

(b)

FIGURE 7.16 (a) Scattering of plane electromagnetic waves through domains of different index of refraction n. (b) Scattering of plane, free-particle wave-functions through domains of different potential.

In the event that the environment (i.e., the potential) in the domain of the transmitted beam ($x = +\infty$) is different from that of the incident beam ($x = -\infty$), the momenta in these two domains will differ. Particles in the transmitted beam will have momentum $\hbar k_2 \neq \hbar k_1$,

(7.111)
$$p_{\text{trans}} = \hbar k_2$$

In all cases the potential is constant in the domains of the incident and transmitted beams (see Fig. 7.14), so the wavefunctions in these domains describe free particles, and we may write

(7.112)
$$\psi_{\text{inc}} = Ae^{i(k_1 x - \omega_1 t)}, \qquad \hbar\omega_1 = E_{\text{inc}} = \frac{\hbar^2 k_1^2}{2m}$$

$$\psi_{\text{refl}} = Be^{i(k_1 x + \omega_1 t)}, \qquad \hbar\omega_1 = E_{\text{refl}} = E_{\text{inc}}$$

$$\psi_{\text{trans}} = Ce^{i(k_2 x - \omega_2 t)}, \qquad \hbar\omega_2 = E_{\text{trans}} = \frac{\hbar^2 k_2^2}{2m} + V$$

$$= E_{\text{inc}} = \hbar\omega_1$$

Energy is conserved across the potential hill so that frequency remains constant ($\omega_1 = \omega_2$). The change in wavenumber k corresponds to changes in momentum and kinetic energy. Using (7.106) permits calculation of the currents

(7.113)
$$\mathbb{J}_{\text{inc}} = \frac{\hbar}{2mi} 2ik_1 |A|^2$$

$$\mathbb{J}_{\text{trans}} = \frac{\hbar}{2mi} 2ik_2 |C|^2$$

$$\mathbb{J}_{\text{refl}} = \frac{\hbar}{2mi} 2ik_1 |B|^2$$

It should be noted that these relations are equivalent to the classical prescription for particle current, $\mathbb{J} = \rho v$, with $\rho = |\psi|^2$ and $v = \hbar k/m$. These formulas, together with (7.108), give the T and R coefficients

(7.114)
$$T = \left|\frac{C}{A}\right|^2 \frac{k_2}{k_1}, \qquad R = \left|\frac{B}{A}\right|^2$$

In the event that the potentials in domains of incident and transmitted beams are equal, $k_1 = k_2$ and $T = |C/A|^2$. More generally, to calculate C/A and B/A as functions of the parameters of the scattering experiment (namely, incident energy, structure of potential barrier), one must solve the Schrödinger equation across the domain of the potential barrier.

PROBLEMS

7.34   Show that the current density $\mathbf{J}$ may be written

$$\mathbf{J} = \frac{1}{2m}[\psi^*\hat{\mathbf{p}}\psi + (\psi^*\hat{\mathbf{p}}\psi)^*]$$

where $\hat{\mathbf{p}}$ is the momentum operator.

7.35   Show that for a one-dimensional wavefunction of the form [where $\phi(x, t)$ is real]

$$\psi(x, t) = A \exp[i\phi(x, t)],$$

$$\mathbb{J} = \frac{\hbar}{m}|A|^2 \frac{\partial \phi}{\partial x}$$

7.36   Show that for a *wave packet* $\psi(x, t)$, one may write

$$\int_{-\infty}^{\infty} \mathbb{J}\, dx = \frac{1}{2m}(\langle p \rangle + \langle p \rangle^*) = \frac{\langle p \rangle}{m}$$

7.37   Show that a complex potential function, $V^*(x) \neq V(x)$, contradicts the continuity equation (7.97).

7.38   (a)   Show that if $\psi(x, t)$ is real, then

$$\mathbb{J} = 0$$

for all x.

(b)   What type of wave structure does a real state function correspond to?

7.6   ONE-DIMENSIONAL BARRIER PROBLEMS

In a one-dimensional scattering experiment, the intensity and energy of the particles in the incident beam are known in addition to the structure of the potential barrier $V(x)$. Three fundamental scattering configurations are depicted in Fig. 7.17. The energy of the particles in the beam is denoted by E.

The Simple Step

Let us first consider the simple step (Fig. 7.17a) for the case $E > V$. We wish to obtain the space-dependent wavefunction φ for all x. The potential function is zero for $x < 0$ and is the constant V, for $x \geq 0$. The incident beam comes from $x = -\infty$. To construct φ we divide the x axis into two domains: region I and region II, depicted in Fig. 7.18. In region I, $V = 0$, and the time-independent Schrödinger equation appears as

(7.115)
$$-\frac{\hbar^2}{2m}\varphi_{xx} = E\varphi$$

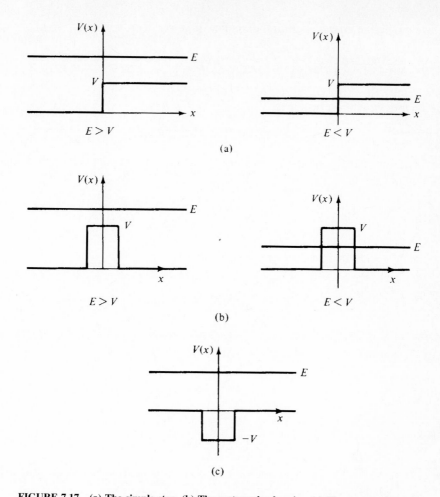

FIGURE 7.17   (a) The simple step. (b) The rectangular barrier. (c) The rectangular well.

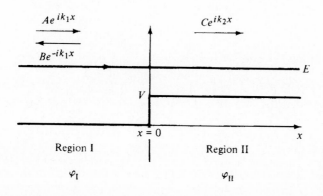

FIGURE 7.18   Domains relevant to the simple-step scattering problem for the case $E \geq V$.

In this domain the energy is entirely kinetic. If we set

(7.116)
$$\frac{\hbar^2 k_1^2}{2m} = E$$

then the latter equation becomes

(7.117) $\varphi_{xx} = -k_1^2 \varphi$ in region I

In region II the potential is constant V and the time-independent Schrödinger equation appears as

(7.118) $-\frac{\hbar^2}{2m} \varphi_{xx} = (E - V)\varphi$

The kinetic energy decreases by V and is given by

(7.119) $\frac{\hbar^2 k_2^2}{2m} = E - V$

In terms of k_2, (7.118) appears as

(7.120) $\varphi_{xx} = -k_2^2 \varphi$ in region II

Writing φ_I for the solution to (7.117) and φ_{II} for the solution to (7.120), one obtains

(7.121)
$$\varphi_I = A e^{ik_1 x} + B e^{-ik_1 x}$$
$$\varphi_{II} = C e^{ik_2 x} + D e^{-ik_2 x}$$

Since the term $D e^{-ik_2 x}$ (together with the time-dependent factor $e^{-i\omega_2 t}$) represents a wave emanating from the right ($x = +\infty$ in Fig. 7.18), and there is no such wave, we may conclude that $D = 0$. The interpretation of the remaining A, B, and C terms is given in (7.112). To repeat, $A \exp(ik_1 x)$ represents the incident wave; $B \exp(-ik_1 x)$, the reflected wave; and $C \exp(ik_2 x)$, the transmitted wave.

It is important at this time to realize that φ_I and φ_{II} (with $D \equiv 0$) represent a single solution to the Schrödinger equation for all x, for the potential curve depicted in Fig. 7.18. Since any wavefunction and its first derivative are continuous (see Section 3.3), at the point $x = 0$ where φ_I and φ_{II} join it is required that

$$\varphi_I(0) = \varphi_{II}(0)$$

(7.122)
$$\frac{\partial}{\partial x} \varphi_I(0) = \frac{\partial}{\partial x} \varphi_{II}(0)$$

These equalities give the relations

$$A + B = C$$

(7.123)

$$A - B = \frac{k_2}{k_1} C$$

Solving for C/A and B/A, one obtains

(7.124) $$\frac{C}{A} = \frac{2}{1 + k_2/k_1}, \qquad \frac{B}{A} = \frac{1 - k_2/k_1}{1 + k_2/k_1}$$

Substituting these values into (7.114) gives

(7.125) $$T = \frac{4k_2/k_1}{[1 + (k_2/k_1)]^2}, \qquad R = \left| \frac{1 - k_2/k_1}{1 + k_2/k_1} \right|^2$$

The ratio k_2/k_1 is obtained from (7.116) and (7.119).

(7.126) $$\left(\frac{k_2}{k_1} \right)^2 = 1 - \frac{V}{E}$$

In the present case $E \geq V$, so $0 \leq k_2/k_1 \leq 1$. For $E \gg V$, $k_2/k_1 \to 1$ and $T \to 1$, $R \to 0$. There is total transmission. For $E = V$, $k_2/k_1 = 0$ and $T = 0$, $R = 1$. There is total reflection and zero transmission. The T and R curves for the simple-step potential are sketched in Fig. 7.19. For all values of (k_2/k_1) we note that

(7.127) $$T + R = 1$$

The validity of this relation for all one-dimensional barrier problems is proved in Problem 7.39.

In the second configuration for the simple-step barrier, $E < V$ (see Fig. 7.17a). Again the x domain is divided into two regions, as shown in Fig. 7.20. In region I the

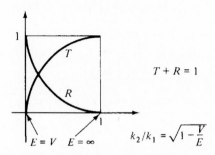

$$T + R = 1$$

$$k_2/k_1 = \sqrt{1 - \frac{V}{E}}$$

FIGURE 7.19 T and R versus k_2/k_1 for the simple-step scattering problem for $E \geq V$.

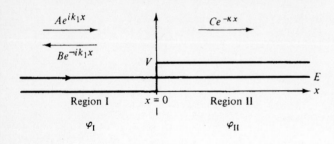

FIGURE 7.20   Domains relevant to the simple-step scattering problem for the case $E \leq V$.

Schrödinger equation becomes

(7.128) $$\varphi_{xx} = -k_1^2 \varphi \qquad \text{in region I}$$

where

(7.129) $$\frac{\hbar^2 k_1^2}{2m} = E$$

In region II the Schrödinger equation is

(7.130) $$\varphi_{xx} = \kappa^2 \varphi \qquad \text{in region II}$$

where

(7.131) $$\frac{\hbar^2 \kappa^2}{2m} = V - E > 0$$

The kinetic energy in this domain is negative $(-\hbar^2 \kappa^2/2m)$. In classical physics region II is a "forbidden" domain. In quantum mechanics, however, it is possible for particles to penetrate the barrier.

Again calling the solution to (7.128) φ_I and the solution to (7.130) φ_{II}, we obtain

$$\varphi_I = Ae^{ik_1 x} + Be^{-ik_1 x}$$

(7.132) $$\varphi_{II} = Ce^{-\kappa x}$$

Continuity of φ and φ_x at $x = 0$ gives

$$1 + \frac{B}{A} = \frac{C}{A}$$

(7.133)

$$1 - \frac{B}{A} = i\frac{\kappa}{k_1}\frac{C}{A}$$

Solving for (C/A) and (B/A) one obtains

(7.134)
$$\frac{C}{A} = \frac{2}{1 + i\kappa/k_1}$$
$$\frac{B}{A} = \frac{1 - i\kappa/k_1}{1 + i\kappa/k_1}$$

The coefficient B/A is of the form z^*/z, where z is a complex number. It follows that $|B/A| = 1$, so

(7.135)
$$R = \left|\frac{B}{A}\right|^2 = 1, \qquad T = 0$$

There is total reflection; hence the transmission must be zero.

To obtain the latter result analytically from our equations above, we must calculate the transmitted current. The function φ_{II} is of the form of a complex amplitude times a real function of x (7.132). Such wavefunctions do not represent propagating waves. They are sometimes called *evanescent waves*. That they carry no current is most simply seen by constructing $\mathbb{J}_{\text{trans}}$ (7.106).

(7.136)
$$\mathbb{J}_{\text{trans}} = \frac{\hbar}{2mi}|C|^2\left(e^{-\kappa x}\frac{\partial}{\partial x}e^{-\kappa x} - e^{-\kappa x}\frac{\partial}{\partial x}e^{-\kappa x}\right) = 0$$

We conclude that $T = 0$.

PROBLEMS

7.39   Show that

$$T + R = 1$$

for all one-dimensional barrier problems.

Answer

Since the scattering process is assumed to be steady-state, the continuity equation (7.101) becomes

$$\frac{\partial \mathbb{J}_x}{\partial x} = 0$$

Integrating this equation, one obtains

$$\int_{-\infty}^{\infty} \left(\frac{\partial \mathbb{J}_x}{\partial x} \right) dx = \mathbb{J}_{+\infty} - \mathbb{J}_{-\infty} = 0$$

But

$$\mathbb{J}_{-\infty} = \mathbb{J}_{inc} - \mathbb{J}_{refl}$$

$$\mathbb{J}_{+\infty} = \mathbb{J}_{trans}$$

so that the equation above becomes

$$\mathbb{J}_{trans} + \mathbb{J}_{refl} = \mathbb{J}_{inc}$$

Dividing through by $\mathbb{J}_{inc}$ gives the desired result.

7.40 Electrons in a beam of density $\rho = 10^{15}$ electrons/m are accelerated through a potential of 100 V. The resulting current then impinges on a potential step of height 50 V.

(a) What are the incident, reflected, and transmitted currents?

(b) Design an electrostatic configuration that gives a simple-step potential.

7.41 Show that the reflection coefficients for the two cases depicted in Fig. 7.21 are equal.

7.42 For the scattering configuration depicted in Fig. 7.20, given that $V = 2E$, at what value of x is the density in region II half the density of particles in the incident beam?

7.43 Equation (7.123) may be written in the matrix form

$$\begin{pmatrix} -1 & 1 \\ 1 & k_2/k_1 \end{pmatrix} \begin{pmatrix} B/A \\ C/A \end{pmatrix} = \begin{pmatrix} 1 \\ 1 \end{pmatrix}$$

Calling the 2×2 matrix $\mathcal{D}$, the left column vector $\mathcal{V}$, and the right column vector $\mathcal{U}$ permits this equation to be more simply written

$$\mathcal{D}\mathcal{V} = \mathcal{U}$$

This inhomogeneous matrix equation has the solution

$$\mathcal{V} = \mathcal{D}^{-1}\mathcal{U}$$

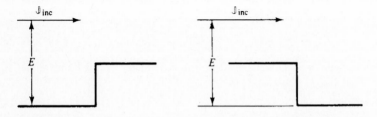

FIGURE 7.21   Reflection coefficients for these two configurations are equal. (See Problem 7.41.)

where $\mathcal{D}^{-1}$ is the inverse of $\mathcal{D}$, that is,

$$\mathcal{D}^{-1}\mathcal{D} = \begin{pmatrix} 1 & 0 \\ 0 & 1 \end{pmatrix}$$

(a) Find $\mathcal{D}^{-1}$ and then construct $\mathcal{V}$ using the technique above. Check your answer with (7.124).

(b) Do the same for (7.133) and (7.134).

7.7   THE RECTANGULAR BARRIER. TUNNELING

The scattering configuration we now wish to examine is depicted in Fig. 7.17b. The energy of the particles in the beam is greater than the height of the potential barrier, $E > V$. For the case at hand there are three relevant domains (see Fig. 7.22):

(7.137)

Region I: $x < -a$, $V = 0$.

Region II: $-a \le x \le +a$, $V > 0$,   and constant.

Region III: $a < x$, $V = 0$

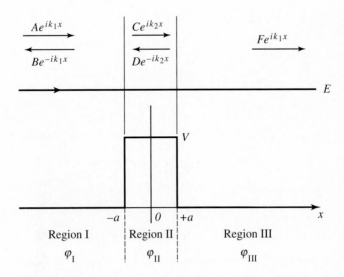

FIGURE 7.22   Domains relevant to the rectangular-barrier scattering problem for the case $E \ge V$.

The solutions to the time-independent Schrödinger equation in each of the three domains are

$$\varphi_{\text{I}} = Ae^{ik_1x} + Be^{-ik_1x}, \qquad \frac{\hbar^2 k_1^2}{2m} = E$$

$$\varphi_{\text{II}} = Ce^{ik_2x} + De^{-ik_2x}, \qquad \frac{\hbar^2 k_2^2}{2m} = E - V$$

(7.138)

$$\varphi_{\text{III}} = Fe^{ik_1x}, \qquad \frac{\hbar^2 k_1^2}{2m} = E$$

$$(ak_1)^2 - (ak_2)^2 = \frac{2ma^2 V}{\hbar^2} \equiv \frac{g^2}{4}$$

The parameter g contains all the barrier (or well) characteristics. The latter equation (conservation of energy) reveals the simple manner in which ak_1 and ak_2 are related. In Cartesian ak_1, ak_2 space they lie on a hyperbola (Fig. 7.23). The permitted values of k_1 (and therefore E) comprise a positive unbounded continuum. For each such eigen-k_1-value, there is a corresponding eigenstate (φ_{I}, φ_{II}, φ_{III}) which is determined in terms of the coefficients, (B/A, C/A, D/A, F/A). Knowledge of these coefficients gives the scattering parameters

$$T = \left| \frac{F}{A} \right|^2, \qquad R = \left| \frac{B}{A} \right|^2$$

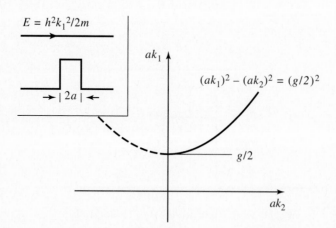

FIGURE 7.23   **For rectangular-barrier scattering with $E \geq V$, ak_1 and ak_2 lie on a hyperbola.**

$$ak_1 \geq ak_2 \geq 0$$

The energy spectrum $\hbar^2 k_1^2/2m$ comprises an unbounded continuum.

The coefficients are determined from the boundary conditions at $x = a$ and $x = -a$,

(7.139)

$$e^{-ik_1a} + \left(\frac{B}{A}\right)e^{ik_1a} = \left(\frac{C}{A}\right)e^{-ik_2a} + \left(\frac{D}{A}\right)e^{ik_2a}$$

$$k_1\left[e^{-ik_1a} - \left(\frac{B}{A}\right)e^{ik_1a}\right] = k_2\left[\left(\frac{C}{A}\right)e^{-ik_2a} - \left(\frac{D}{A}\right)e^{ik_2a}\right]$$

$$\left(\frac{C}{A}\right)e^{ik_1a} + \left(\frac{D}{A}\right)e^{-ik_2a} = \left(\frac{F}{A}\right)e^{ik_1a}$$

$$k_2\left[\left(\frac{C}{A}\right)e^{ik_2a} - \left(\frac{D}{A}\right)e^{-ik_2a}\right] = k_1\left(\frac{F}{A}\right)e^{ik_1a}$$

These are four linear, algebraic, inhomogeneous equations for the four unknowns: (B/A), (C/A), (D/A), and (F/A). Solving the last two for (D/A) and (C/A) as functions of (F/A) and substituting into the first two permits one to solve for (B/A) and (F/A). These appear as

(7.140)

$$\frac{F}{A} = e^{-2ik_1a}\left[\cos(2k_2a) - \frac{i}{2}\left(\frac{k_1^2 + k_2^2}{k_1k_2}\right)\sin(2k_2a)\right]^{-1}$$

$$2\left(\frac{B}{A}\right) = i\left(\frac{F}{A}\right)\frac{k_2^2 - k_1^2}{k_1k_2}\sin(2k_2a)$$

The transmission coefficient is most simply obtained from the second of these, together with the relation

(7.141)
$$T + R = \left|\frac{F}{A}\right|^2 + \left|\frac{B}{A}\right|^2 = 1$$

There results

(7.142)
$$\frac{1}{T} = \left|\frac{A}{F}\right|^2 = 1 + \frac{1}{4}\left(\frac{k_1^2 - k_2^2}{k_1k_2}\right)^2\sin^2(2k_2a)$$

Rewriting k_1 and k_2 in terms of E and V as given by (7.138), one obtains

(7.143)
$$\boxed{\frac{1}{T} = 1 + \frac{1}{4}\frac{V^2}{E(E-V)}\sin^2(2k_2a) \qquad E > V}$$

The reflection coefficient is $1 - T$.

For the case $E < V$, as depicted in Fig. 7.24a, we find that the structure of the solutions (7.138) are still appropriate, with the simple modification

(7.144)
$$ik_2 \to \kappa, \qquad \frac{\hbar^2\kappa^2}{2m} = V - E > 0$$

$$(ak_1)^2 + (a\kappa)^2 = \frac{2ma^2 V}{\hbar^2} = \frac{g^2}{4}$$

This latter conservation of energy statement indicates that the variables ak_1 and $a\kappa$ lie on a circle of radius $g/2$ (Fig. 7.25). The permitted eigen-k_1-values now comprise a positive, bounded continuum, so that the eigenenergies

$$E = \frac{\hbar^2 k_1^2}{2m}$$

also comprise a positive, bounded continuum.

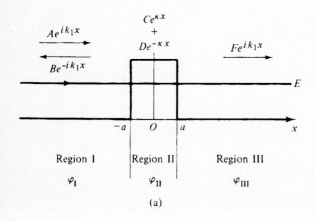

(a)

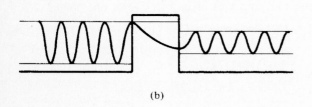

(b)

FIGURE 7.24 (a) Domains relevant to the rectangular-barrier scattering problem, for the case $E \leq V$. (b) Real part of φ for the case above, showing the hyperbolic decay in the barrier domain and decrease in amplitude of the transmitted wave.

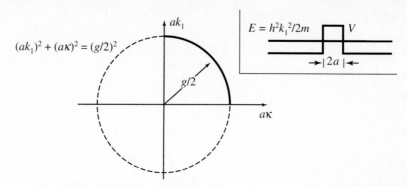

FIGURE 7.25 For rectangular-barrier scattering with $E \leq V$, ak_1 and $a\kappa$ lie on a circle $ak_1 \geq 0$, $a\kappa \geq 0$. The energy spectrum ($\hbar^2 k_1^2/2m$) comprises a bounded continuum.

The algebra leading to (7.140) remains unaltered so that the transmission coefficient for this case is obtained by making the substitution of (7.144) into (7.142). We also recall that $\sin (iz) = i \sinh z$. There results

(7.145)
$$\frac{1}{T} = 1 + \frac{1}{4}\left(\frac{k_1^2 - \kappa^2}{k_1 \kappa}\right)^2 \sinh^2 (2\kappa a)$$

which, with (7.144), gives

(7.146)
$$\frac{1}{T} = 1 + \frac{1}{4}\frac{V^2}{E(V - E)} \sinh^2 (2\kappa a)$$

Writing this equation in terms of T,

(7.147)
$$\boxed{T = \frac{1}{1 + \dfrac{1}{4}\dfrac{V^2}{E(V - E)} \sinh^2 (2\kappa a)} \qquad E < V}$$

indicates that in the domain $E < V$, $T < 1$. The limit that $E \to V$ deserves special attention. With

$$\frac{V - E}{V} = \frac{\hbar^2 \kappa^2}{2mV} \equiv \epsilon \to 0$$

one obtains

$$T = \frac{1}{1 + g^2/4} + O(\epsilon) < 1$$

(7.148)
$$g^2 \equiv \frac{2m(2a)^2 V}{\hbar^2}$$

The expression $O(\epsilon)$ represents a sum of terms whose value goes to zero with ϵ. We conclude that for scattering from a potential barrier, the transmission is less than unity at $E = V$ (Fig. 7.26).

Returning to the case $E > V$, (7.143) indicates that $T = 1$ when $\sin^2 (2k_2 a) = 0$, or equivalently when

(7.149) $2ak_2 = n\pi$ $(n = 1, 2, \ldots)$

Setting $k_2 = 2\pi/\lambda$, the latter statement is equivalent to

(7.150) $2a = n\left(\frac{\lambda}{2}\right)$

When the barrier width $2a$ is an integral number of half-wavelengths, $n(\lambda/2)$, the barrier becomes transparent to the incident beam; that is, $T = 1$. This is analogous to the case of total transmission of light through thin refracting layers.

Written in terms of E and V, the requirement for perfect transmission, (7.149), becomes

(7.151) $E - V = n^2\left(\frac{\pi^2\hbar^2}{8a^2m}\right) = n^2 E_1$

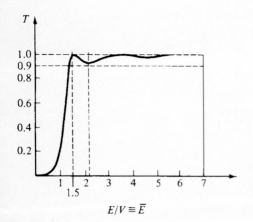

FIGURE 7.26   Transmission coefficient T versus E/V for scattering from a rectangular barrier with $2m(2a)^2V/\hbar^2 \equiv g^2 = 16$. The additional lines are in references to Problems 7.50 et seq.

where E_1 is the ground-state energy of a one-dimensional box of width $2a$ [see (4.14)].

Equations (7.143) and (7.146) give the transmission coefficient T, as a function of E, V, and the width of the well $2a$. The former of these indicates that $T \rightarrow 1$ with increasing energy of the incident beam. The transmission is unity for the values of E given by (7.151). Equation (7.146) gives T for $E \leq V$. The transmission is zero for $E = 0$ and is less than 1 for $E = V$. A sketch of T versus $E/V \equiv \bar{E}$ for the case $g^2 = 16$ is given in Fig. 7.26.

The fact that T does not vanish for $E < V$ is a purely quantum mechanical result. This phenomenon of particles passing through barriers higher than their own incident energy is known as *tunneling*. It allows emission of α particles from a nucleus and field emission of electrons from a metal surface in the presence of a strong electric field.

PROBLEMS

7.44   In terms of the new variables,

$$\alpha_{\pm} \equiv \frac{k_1^2 \pm k_2^2}{2k_1k_2}, \qquad \beta \equiv 2k_2a$$

$$\frac{F}{A} = \sqrt{T}\, e^{i\varphi_T}, \qquad \frac{B}{A} = \sqrt{R}\, e^{i\varphi_R}$$

(7.140) may be rewritten in the simpler form

$$\sqrt{T}\, e^{i\varphi_T} = \frac{e^{2iak_1}}{\cos \beta - i\alpha_+ \sin \beta}$$

$$\sqrt{R}\, e^{i\varphi_R} = i\alpha_- \sqrt{T} e^{i\varphi_T} \sin \beta$$

Use these expressions to show:
  (a) $T + R = 1$
  (b) $\phi_T = \phi_R - n(\pi/2)$, $n = 1, 2, 3, \ldots$.
  (c) $\tan (\phi_T - 2k_1a) = \alpha_+ \tan \beta$
  (d)   What is ϕ_R for the infinite potential step: $V(x) = \infty$, $x \geq 0$; $V(x) = 0$, $x < 0$?

Answers (partial)
  (a)   Solving for $T + R$ from (7.140) gives

$$T + R = \frac{1 + \alpha_-^2 \sin^2 \beta}{\cos^2 \beta + \alpha_+^2 \sin^2 B}$$

Substituting the definitions of $\alpha_{\pm}$ gives the desired result.

(c)   From the first of the two given equations above, we obtain

$$\sqrt{T}\, e^{i(\varphi_T - 2k_1 a)} = \frac{1}{\cos \beta + i\alpha_+ \sin \beta}$$

$$= \frac{e^{-i\varphi}}{\sqrt{\cos^2 \beta + \alpha_+{}^2 \sin^2 \beta}}$$

Equating the tangents of the phases of both sides gives the desired result.

7.45   An electron beam is sent through a potential barrier 4.5 Å long. The transmission coefficient exhibits a third maximum at $E = 100$ eV. What is the height of the barrier?

7.46   An electron beam is incident on a barrier of height 10 eV. At $E = 10$ eV, $T = 3.37 \times 10^{-3}$. What is the width of the barrier?

7.47   Use the correspondence principle with (7.147) to show that $T = 0$ for $E < V$, for the classical case of a beam of particles of energy E incident on a potential barrier of height V.

7.8   THE RAMSAUER EFFECT

The configuration for this case is depicted in Fig. 7.17c. The relevant domains are shown in Fig. 7.27. Once again eqs. 7.138 et seq. apply with the modification

(7.152)
$$\frac{\hbar^2 k_2{}^2}{2m} = E - V = E + |V|$$

The transmission coefficient (7.143) becomes, for $E \geq 0$,

(7.153)
$$\boxed{\frac{1}{T} = 1 + \frac{1}{4}\frac{V^2}{E(E + |V|)}\sin^2 (2k_2 a)}$$

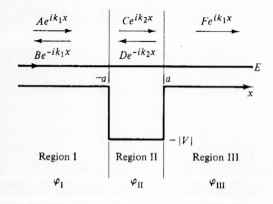

Region I   Region II   Region III

φ_I φ_II φ_III

FIGURE 7.27   Domains relevant to the rectangular-well scattering problem, $E > 0$.

Again there is perfect transmission when an integral number of half-wavelengths fit the barrier width.

(7.154) $$2ak_2 = n\pi \qquad (n = 1, 2, \ldots)$$

This condition may also be cast in terms of the eigenenergies of a one-dimensional box of width $2a$:

(7.155) $$E + |V| = n^2 E_1$$

From (7.153) we see that $T \to 1$ with increasing incident energy. At $E = 0$, $T = 0$. Thus we obtain an idea of the shape of T versus E. It is similar to the curve shown in Fig. 7.26. The transmission is zero for $E = 0$ and rises to the first maximum (unity) at $E = E_1 - |V|$. It has successive maxima of unity at the values given by (7.155), and approaches 1 with growing incident energy E.

The preceding theory of scattering of a beam of particles by a potential well has been used as a model for the scattering of low-energy electrons from atoms. The attractive well represents the field of the nucleus, whose positive charge becomes evident when the scattering electrons penetrate the shell structure of the atomic electrons. The reflection coefficient is a measure of the scattering cross section.[1] Experiments in which this cross section is measured (for rare gas atoms) detect a low-energy minimum which is consistent with the first maximum that T goes through for typical values of well depth and width according to the model above, (7.153). This transparency to low-energy electrons of rare gas atoms is known as the Ramsauer effect.

The student should not lose sight of the following fact. For any of the solutions to the scattering problems considered in these last few sections, we have in essence found the eigenfunctions and eigenenergies for the corresponding Hamiltonian. These Hamiltonians are of the form

(7.156) $$H = \frac{p^2}{2m} + V(x)$$

with the potential $V(x)$ depicted by any of the configurations of Fig. 7.17. In each case considered, the spectrum of energies is a continuum, $E = \hbar^2 k^2 / 2m$. For each value of k, a corresponding set of coefficient ratios (B/A, C/A for the simple step and B/A, C/A, D/A, F/A for the rectangular potential) are determined. The coefficient A is fixed by the data on the incident beam. These coefficients then determine the wavefunction,

[1] The notion of scattering cross section is discussed in Chapter 14.

which is an eigenfunction of the Hamiltonian above. All such scattering eigenstates are unbound states. A continuous spectrum is characteristic of unbound states, while a discrete spectrum is characteristic of bound states (e.g., particle in a box, harmonic oscillator).

The transmission coefficients corresponding to the one-dimensional potential configurations considered above are summarized in Table 7.2.

TABLE 7.2   Transmission coefficients for three elementary potential barriers

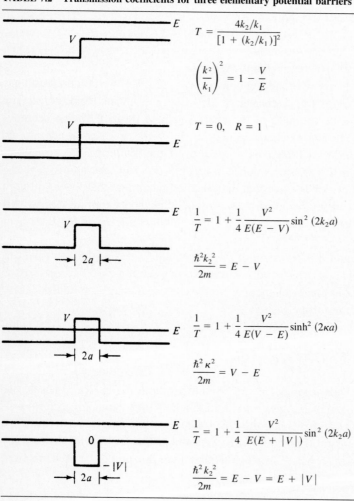

$$T = \frac{4k_2/k_1}{[1 + (k_2/k_1)]^2}$$

$$\left(\frac{k_2}{k_1}\right)^2 = 1 - \frac{V}{E}$$

$$T = 0, \quad R = 1$$

$$\frac{1}{T} = 1 + \frac{1}{4} \frac{V^2}{E(E - V)} \sin^2 (2k_2a)$$

$$\frac{\hbar^2 k_2^2}{2m} = E - V$$

$$\frac{1}{T} = 1 + \frac{1}{4} \frac{V^2}{E(V - E)} \sinh^2 (2\kappa a)$$

$$\frac{\hbar^2 \kappa^2}{2m} = V - E$$

$$\frac{1}{T} = 1 + \frac{1}{4} \frac{V^2}{E(E + |V|)} \sin^2 (2k_2a)$$

$$\frac{\hbar^2 k_2^2}{2m} = E - V = E + |V|$$

PROBLEMS

7.48   The scattering cross section for the scattering of electrons by a rare gas of krypton atoms exhibits a low-energy minimum at $E \simeq 0.9$ V. Assuming that the diameter of the atomic well seen by the electrons is 1 Bohr radius, calculate its depth.

7.49   Show that the transmission coefficient for the rectangular barrier may be written in the form

$$T = T(g, \bar{E})$$

where

$$g^2 \equiv \frac{2m(2a)^2 V}{\hbar^2}$$

$$\bar{E} \equiv \frac{E}{V}$$

Answer (partial)

For $\bar{E} \geq 1$.

$$T^{-1} = 1 + \frac{1}{4} \frac{1}{\bar{E}(\bar{E} - 1)} \sin^2 \sqrt{g^2(\bar{E} - 1)}$$

7.50   Using your answer to Problem 7.49, derive an equation for an approximation to the curve on which minimum values of T fall.

$$T_{\min} = T_{\min}(\bar{E})$$

Show that the values of T and $\bar{E}$ at the first minimum in the sketch of T versus $\bar{E}$ depicted in Fig. 7.26 ($g^2 = 16$) agree with your equation. [*Hint:* The minima of T fall at the values of $\bar{E}$ where T^{-1} is maximum. From Problem 7.49,

$$T^{-1} \leq 1 + \frac{1}{4} \frac{1}{\bar{E}(\bar{E} - 1)}.]$$

7.51   For the rectangular barrier:
  (a)   Write the values of $\bar{E}$ for which $T = 1$ as a function of g.
  (b)   Using your answer to part (a) and the two preceding problems, make a sketch of T versus $\bar{E}$ in the two limits $g \gg 1$, $g \ll 1$. Cite two physical situations to which these limits pertain.
  (c)   Show that for an electron, $g^2/V \equiv 2m(2a)^2/\hbar^2 = 0.26(2a)^2(eV)^{-1}$, where a is in Å.

7.52   For the case depicted in Fig. 7.26, show that the first maximum falls at a value consistent with your answer to part (a) of Problem 7.51.

7.53   Write the transmission coefficient for the rectangular well as a function of g and $\bar{E}$.

Answer

$$T^{-1} = 1 + \frac{1}{4}\frac{1}{\bar{E}(\bar{E} + 1)} \sin^2 \sqrt{g^2(\bar{E} + 1)}$$

7.54   In the limit $g^2 \gg 1$, show that the minima of T for the rectangular well fall on a curve which is well approximated by

$$T_{\min} = 4\bar{E}$$

Use this result together with (7.155) for the values of $\bar{E}$ where $T = 1$ to obtain a sketch of T versus $\bar{E}$ for the case $g^2 = 10^5$.

Answer

See Fig. 7.28.

7.55   Show that the spaces between resonances in T for the case of scattering from a potential well grow with decreasing g.

7.56   (a)   Calculate the transmission coefficient T for the double potential step shown in Fig. 7.29a.

(b)   If we call T_1 the transmission coefficient appropriate to the single potential step V_1, and T_2 that appropriate to the single potential step V_2, show that

$$T_2 \le T_1, \qquad T \ge T_2$$

Offer a physical explanation for these inequalities.

(c)   What are the three sets of conditions under which T is maximized? What do these conditions correspond to physically?

(d)   A student argues that T is the product $T_1 T_2$ on the following grounds. The particle current that penetrates the V_1 barrier is $T_1 J_{\text{inc}}$. This current is incident on the V_2 barrier so that

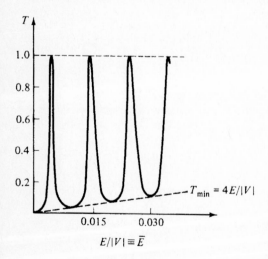

FIGURE 7.28   Resonances in the transmission coefficient for scattering by a potential well for $g^2 = 10^5$. (See Problems 7.54 et seq.)

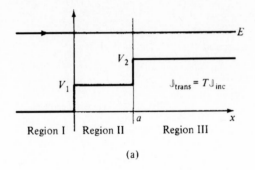

(a)

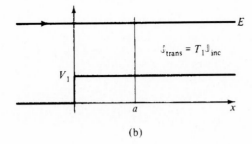

(b)

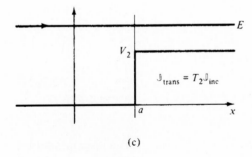

(c)

**FIGURE 7.29   (a) Double potential step show-
ing three regions discussed in Problem 7.56. (b)
and (c) Two related single potential steps:
$T_1 \geq T_2$ and $T_2 \leq T$.**

$T_2(T_1 \mathcal{J}_{inc})$ is the current transmitted through the second barrier. What is the incorrect assumption in his argument?

Answer (partial)

Applying boundary conditions to the wavefunctions

$$\varphi_I = Ae^{ik_1x} + Be^{-ik_1x} \qquad \text{(region I)}$$

$$\varphi_{II} = Ce^{ik_2x} + De^{-ik_2x} \qquad \text{(region II)}$$

$$\varphi_{III} = Fe^{ik_3x} \qquad \text{(region III)}$$

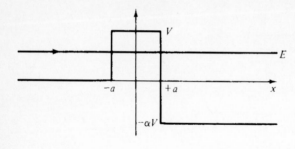

FIGURE 7.30   Tunneling configuration for Problem 7.57. The constant α
is real and greater than zero.

at $x = 0$ and $x = a$, respectively, and solving for $T = (k_3/k_1)|F/A|^2$ gives the desired result:

$$T = \frac{4k_1k_3k_2^2}{k_2^2(k_1 + k_3)^2 + (k_3^2 - k_2^2)(k_1^2 - k_2^2)\sin^2(k_2a)} \qquad (k_1 \geq k_2 \geq k_3)$$

Note that

$$T = \frac{4k_1k_3k_2^2}{k_2^2(k_1 + k_3)^2 - \Delta^2} \geq \frac{4k_1k_3k_2^2}{k_2^2(k_1 + k_3)^2} = T_2$$

where Δ^2 is as implied. With (7.125) we see that $T_1 \geq T_2$.

7.57   Calculate the transmission coefficient for the potential configuration and energy of incident particles depicted in Fig. 7.30. (*Note: T* is easily obtained from the answer given to Problem 7.56.)

7.9   KINETIC PROPERTIES OF A WAVE PACKET SCATTERED FROM A POTENTIAL BARRIER

The time-dependent one-dimensional scattering problem addresses itself primarily to the problem of a wave packet incident on a potential barrier. It seeks the shape of the reflected and transmitted pulses. We will restrict our discussion to the kinematic properties of these pulses.

To formulate this problem we first construct a wave packet whose center is at $x = -X$ at $t = 0$. In previous chapters we obtained such wave packets centered at $x = 0$ at $t = 0$. They are of the form

(7.157) $$\psi(x, t) = \frac{1}{\sqrt{2\pi}} \int_{-\infty}^{\infty} b(k)e^{i(kx - \omega t)}\, dk$$

For this same packet to be centered at $x = -X$ initially, one merely effects a translation in x so that

$$(7.158) \qquad \psi(x, t) = \frac{1}{\sqrt{2\pi}} \int_{-\infty}^{\infty} b(k) e^{ik(x+X)} e^{-i\omega t} \, dk$$

$$= \frac{1}{\sqrt{2\pi}} \int_{-\infty}^{\infty} b(k) e^{ikX} e^{i(kx-\omega t)} \, dk$$

For example, for a chopped pulse, L cm long, containing particles moving with momentum $\hbar k_0$, $b(k)$ is given by (6.32):

$$(7.159) \qquad b(k) = \sqrt{\frac{2}{\pi L}} \frac{\sin (k - k_0)L/2}{k - k_0}$$

See Fig. 7.31. The group velocity of this packet is $v_0 = \hbar k_0/m$. Let us call the wave packet (7.158), ψ_{inc}. This packet is a superposition of plane-wave states of the form (7.112). Each such incident k-component plane wave is reflected and transmitted. The corresponding reflected and transmitted waves are constructed from the amplitude ratios B/A and F/A given by (7.140), which are functions of $k[k_1$ in (7.140)]. Reassembling all of these waves, one obtains

$$x < -a \qquad \psi_{\text{inc}} = \frac{1}{\sqrt{2\pi}} \int_{-\infty}^{\infty} b(k) e^{ikX} e^{i(kx-\omega t)} \, dk$$

$$(7.160) \quad x < -a \qquad \psi_{\text{refl}} = \frac{1}{\sqrt{2\pi}} \int_{-\infty}^{\infty} \sqrt{R} e^{i\phi_R} b(k) e^{ikX} e^{-i(kx+\omega t)} \, dk$$

$$x > +a \qquad \psi_{\text{trans}} = \frac{1}{\sqrt{2\pi}} \int_{-\infty}^{\infty} \sqrt{T} e^{i\phi_T} b(k) e^{ikX} e^{i(kx-\omega t)} \, dk$$

Here we are using the notation of Problem 7.44.

To uncover the kinetic properties of these packets, we use the method of *stationary phase*. This relies on the fact that the major contribution in a Fourier integral is

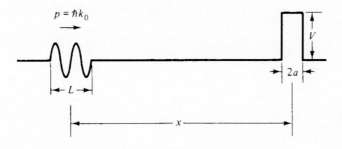

FIGURE 7.31   Wave packet incident on a potential barrier. $x \gg L \simeq 2a$.

due to the k component with stationary phase. If we call this component k_0, then the phase of the Fourier integral for ψ_{refl} vanishes when

(7.161)
$$\frac{\partial}{\partial k}(\phi_R + kX - kx - \omega t) = 0$$

This gives the trajectory of the reflected packet,

(7.162)
$$x = -\frac{\hbar k_0}{m}t + X + \left(\frac{\partial \phi_R}{\partial k}\right)_{k_0} \qquad x < -a$$

In like manner, for the incident and transmitted packets, one obtains

(7.163)
$$x = \frac{\hbar k_0}{m}t - X, \qquad\qquad x < -a$$

(7.164)
$$x = \frac{\hbar k_0}{m}t - X - \left(\frac{\partial \phi_T}{\partial k}\right)_{k_0}, \qquad x > a$$

The latter three equations illustrate the effect of a potential barrier on the trajectory of an incident wave packet. Were there no barrier, the packet would move freely in accordance with (7.163). However, there is a delay for both the transmitted and reflected packets. The transmitted pulse arrives at any plane $x > a$, $(\partial \phi_T / \partial k_0)v_0^{-1}$ seconds after the free pulse. The reflected pulse arrives at any plane $x < -a$, $(\partial \phi_R / \partial k_0)v_0^{-1}$ seconds after the free pulse would be reflected from an impenetrable wall at the $x = 0$ plane.[1]

PROBLEMS

7.58 For a pulse such as described in (7.158) and (7.159), containing 1.5-keV electrons, which scatters from a potential well of width 0.5×10^{-7} cm and of depth 25 keV, what is the delay in the transmitted beam (in s) imposed by the well?

7.59 Is there a delay in the scattering of a wave packet from a simple-step potential? Present an argument in support of your answer.

7.60 In the text we mentioned the method of stationary phase for evaluating Fourier integrals. Use this method to show that

$$\int_{-\infty}^{\infty} f(k)e^{is(k)}\,dk \simeq \sqrt{\frac{2\pi}{|s''(k_0)|}}\, f(k_0)e^{i[s(k_0)\pm\pi/4]}$$

$$s'(k_0) = 0$$

---

[1] The time development of a wave packet scattering from a potential barrier is graphically depicted in D. A. Saxon, *Elementary Quantum Mechanics*, Holden-Day, San Francisco, 1968.

The phase factor $+i\pi/4$ applies when $s''(k_0) > 0$ and $-i\pi/4$ applies when $s''(k_0) < 0$. Primes denote k differentiation. [*Hint:* Expand $s(k)$ in a Taylor series about $k = k_0$, keeping $O(k^2)$ terms.]

7.61   There is a tacit assumption in the construction of (7.160) that no interaction occurs between the incident wave packet and the potential barrier in the interval $0 \le t \le X/v_0$. Is this a valid assumption?

Answer

All k components in the distribution (7.159) with $k > k_0$ reach the barrier in a time less than X/v_0. The number of such components diminishes in the limit $X \gg 2a \simeq L$.

7.10   THE WKB APPROXIMATION[1]

Correspondence

In Section 7.3 we found that the quantum probability density goes over to the classical probability density in the limit of large quantum numbers. Such quantum states have many zeros and suffer rapid spatial oscillation. Equivalently, we may say that in this classical domain the local quantum (de Broglie) wavelength is small compared to characteristic distances of the problem. For the harmonic oscillator, such a characteristic distance is the maximum displacement or amplitude x_0 (7.12). More generally, this characteristic distance may be taken as the typical length over which the potential changes. Since the de Broglie wavelength changes only by virtue of a change in potential, the latter condition may be incorporated in the criterion (for classical behavior) that the quantum wavelength not change appreciably over the distance of one wavelength. Now the change in wavelength over the distance δx is

$$\delta\lambda = \frac{d\lambda}{dx}\,\delta x$$

In one wavelength this change is

$$\delta\lambda = \frac{d\lambda}{dx}\,\lambda$$

---

[1] Named for G. Wentzel, H. A. Kramers, and L. Brillouin.

In the classical domain, $\delta\lambda \ll \lambda$ (Fig. 7.32). This gives the criterion

(7.165)
$$\left|\frac{\delta\lambda}{\lambda}\right| = \left|\frac{d\lambda}{dx}\right| \ll 1$$

In terms of the momentum p, we find that

$$\left(\frac{h}{\lambda}\right)^2 = p^2 = 2m(E - V)$$

$$\frac{d(\lambda^2)}{dx} = -\frac{h^2}{p^4}\frac{d(p^2)}{dx} = -\frac{h^2}{p^4}\left(-2m\frac{dV}{dx}\right)$$

or, equivalently,

$$\frac{d\lambda}{dx} = \frac{mh}{p^3}\frac{dV}{dx}$$

Thus, the condition (7.165) for near-classical behavior becomes

(7.166)
$$\left|\frac{d\lambda}{\lambda}\right| = \left|\frac{mh}{p^3}\frac{dV}{dx}\right| \ll 1$$

The WKB Expansion

We seek solutions to the time-independent Schrödinger equation (7.1) which are valid in the near-classical domain (7.166).

If the potential V is slowly varying, one expects the wavefunction to closely approximate the free-particle state

$$\varphi(x) = Ae^{ikx} = Ae^{ipx/\hbar}$$

Thus we will look for solutions in the form

(7.167)
$$\varphi(x) = Ae^{iS(x)/\hbar}$$

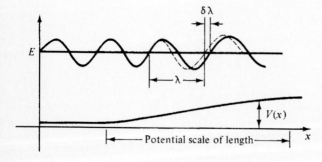

FIGURE 7.32 In the WKB analysis, the fractional change $\delta\lambda/\lambda \ll 1$. The potential scale of length is also large compared to wavelength.

Substitution of this function into (7.1) gives

$$(7.168) \qquad -i\hbar \frac{\partial^2 S}{\partial x^2} + \left(\frac{\partial S}{\partial x}\right)^2 = p^2(x)$$

$$p^2 = 2m[E - V(x)]$$

To further bias the solution (7.167) to the classical domain we examine the solutions to the nonlinear equation (7.168) in the limit $\hbar \rightarrow 0$. Recall (Section 6.1) that it is in this limit that the Gaussian packet reduces to the classical particle. To these ends we expand $S(x)$ in powers of $\hbar$ as follows:

$$(7.169) \qquad S(x) = S_0(x) + \hbar S_1(x) + \frac{\hbar^2}{2} S_2(x) + \cdots$$

Substituting this expansion into (7.168) gives

$$(7.170) \qquad 0 = \left[\left(\frac{\partial S_0}{\partial x}\right)^2 - p^2\right] + 2\hbar\left(\frac{\partial S_0}{\partial x}\frac{\partial S_1}{\partial x} - \frac{i}{2}\frac{\partial^2 S_0}{\partial x^2}\right)$$
$$+ \hbar^2\left[\frac{\partial S_0}{\partial x}\frac{\partial S_2}{\partial x} + \left(\frac{\partial S_1}{\partial x}\right)^2 - i\frac{\partial^2 S_1}{\partial x^2}\right] + O(\hbar^3)$$

Since this equation must be satisfied for small but otherwise arbitrary values of $\hbar$, it is necessary that the coefficient of each power of $\hbar$ vanish separately. In this manner we obtain the following series of coupled equations for the sequence $\{S_n\}$.

$$(7.171) \qquad \left(\frac{\partial S_0}{\partial x}\right)^2 = p^2$$

$$\frac{\partial S_0}{\partial x}\frac{\partial S_1}{\partial x} = \frac{i}{2}\frac{\partial^2 S_0}{\partial x^2}$$

$$\frac{\partial S_0}{\partial x}\frac{\partial S_2}{\partial x} + \left(\frac{\partial S_1}{\partial x}\right)^2 - i\frac{\partial^2 S_1}{\partial x^2} = 0$$

$$\vdots$$

Integrating the first of these equations gives

$$S_0(x) = \pm \int_{x_0}^{x} p(x)\, dx$$

or, equivalently, in terms of wavenumber $k = p/\hbar$,

$$(7.172) \qquad \frac{S_0}{\hbar} = \pm \int_{x_0}^{x} k(x)\, dx$$

Substituting this solution into the second equation in (7.171) and integrating gives

$$S_1 = \frac{i}{2} \ln \left(\frac{\partial S_0}{\partial x} \right) = \frac{i}{2} \ln \hbar k$$

or, equivalently,

(7.173)
$$\exp(iS_1) = \frac{1}{\hbar^{1/2} k^{1/2}}$$

Substituting (7.172) and (7.173) into the third equation in (7.171) and integrating gives

(7.174)
$$S_2 = \frac{1}{2} \frac{m(\partial V/\partial x)}{p^3} - \frac{1}{4} \int \frac{m^2(\partial V/\partial x)^2}{p^5} \, dx$$

In that S_1 is the log of the derivative of S_0, we cannot in general ignore S_1 compared to S_0, and both terms must be retained in the expansion (7.169). However, comparison of S_2 (7.174) with the criterion (7.166) shows that in the near-classical domain, the contribution of the second-order term $\hbar S_2/2$ to the phase of φ is small compared to unity. Higher-order contributions to $S(x)$ are likewise small. Thus, it is consistent to say that near the classical domain, φ is well described by the first two terms in the expansion (7.169). Inserting these solutions into (7.167) gives

(7.175)
$$\varphi(x) = \frac{A}{k^{1/2}} \exp \left(i \int k \, dx \right) + \frac{B}{k^{1/2}} \exp \left(-i \int k \, dx \right)$$

The Near-Classical Domain

In what sense does the solution (7.175) approximate classical behavior? To answer this question we consider the probability density $\varphi^*\varphi$. Specifically, consider that the momentum of the particle is specified so that it is known that the particle is moving to larger values of x. Then the corresponding WKB solution (7.175) reduces to

$$\varphi(x) = \frac{A}{k^{1/2}} \exp \left(i \int k \, dx \right)$$

The probability density for this state is

$$P(x) = \varphi^*\varphi = \frac{|A|^2}{k} = \frac{|A|^2 \hbar/m}{v}$$

where v is written for the classical velocity, $v = p/m$. The probability of finding the particle in the interval dx about x is

$$P \, dx = \left(\frac{|A|^2 \hbar}{m} \right) dt$$

This result, apart from a multiplicative constant, is the same as the classical probability, $P \, dx \sim dt$ [see (7.70)].

To obtain correspondence with the classical current, we renormalize φ so that it is relevant to a beam of N particles such as described in (7.95). Calculation of the current (7.106) gives

$$\mathbb{J} = \frac{N\hbar |A|^2}{m} = NP(x)v(x)$$

$$\mathbb{J} = \rho(x)v(x)$$

This is the classical expression for the current across a plane at the point x of a beam of particles with number density $\rho(x)$ moving with velocity $v(x)$.

Thus, the lowest-order WKB solution (7.175) reproduces the classical probability and current.

Application to Bound States

Consider the potential shown in Fig. 7.33. The WKB solution (7.175) invalid at the classical turning points x_1 and x_2, for at these points $E = V$ and $\hbar k = 0$, thereby violating the criterion (7.166). However, the WKB solution becomes valid in regions far removed from the turning points where $|E - V|$ is sufficiently large.

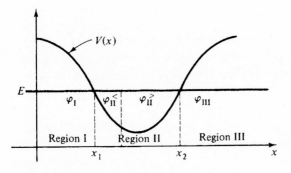

FIGURE 7.33   **Domains relevant to the WKB approximation of bound states.**

In region I, far to the left of x_1 ($x \to -\infty$), the solution is

$$\varphi_{\mathrm{I}} = \frac{1}{\sqrt{\kappa}} \exp \int_{x_1}^{x} \kappa \, dx$$

(7.176)
$$\frac{\hbar^2 \kappa^2}{2m} = V - E > 0$$

Far to the right of x_2, the wavefunction also decays exponentially.

(7.177)
$$\varphi_{\mathrm{III}} = \frac{A}{\sqrt{\kappa}} \exp \left(-\int_{x_2}^{x} \kappa \, dx \right)$$

In the classically allowed region II, the WKB solution is oscillatory. It is necessary in the WKB construction of φ to separate this component of the solution into two parts.

(7.178a)
$$\varphi_{\mathrm{II}}^{<}(x) = \frac{C}{\sqrt{k}} \sin \left(\int_{x_1}^{x} k \, dx + \delta \right), \qquad x_1 < x$$

(7.178b)
$$\varphi_{\mathrm{II}}^{>}(x) = \frac{B}{\sqrt{k}} \sin \left(\int_{x}^{x_2} k \, dx + \delta \right), \qquad x < x_2$$

$$\frac{\hbar^2 k^2}{2m} = E - V > 0$$

Through connection formulas obtained below, $\varphi_{\mathrm{II}}^{<}$ is matched to φ_{I} and $\varphi_{\mathrm{II}}^{>}$ is matched to φ_{III}. This connecting process will serve to determine all but one of the constants A, B, C, and δ. The remaining constant is determined in stipulating that $\varphi_{\mathrm{II}}^{<}$ join smoothly to $\varphi_{\mathrm{II}}^{>}$. This continuity condition will also be found to generate energy eigenvalues within the WKB approximation.

Connecting Formulas for Bound States

If φ_{I}, $\varphi_{\mathrm{II}}^{<}$, $\varphi_{\mathrm{II}}^{>}$, and φ_{III} were valid representations of φ throughout their respective domains, the constants of these functions could be obtained by simply matching these component solutions as was done in preceding sections of this chapter. This method clearly cannot be followed in the present analysis since the WKB solutions are invalid at the turning points.

The technique of matching φ_{I} to $\varphi_{\mathrm{II}}^{<}$ and $\varphi_{\mathrm{II}}^{>}$ to φ_{III} in the WKB approximation is as follows. The Schrödinger equation is solved exactly in the regions of the turning points for potentials that approximate $V(x)$ in these domains. The asymptotic forms of these exact solutions are then used to match φ_I to $\varphi_{II}^{<}$ and to match $\varphi_{II}^{>}$ to φ_{III}. Following this prescription, we approximate $V(x)$ in the neighborhood of x_1 with the linear potential $V_1(x)$.

$$(7.179) \qquad V(x) \simeq V_1(x) = E - F_1(x - x_1)$$

The constant F_1 is the slope of $V(x)$ at x_1. Similarly, in the neighborhood of x_2 we write

$$(7.180) \qquad V(x) \simeq V_2(x) \equiv E + F_2(x - x_2)$$

(See Fig. 7.34.) The Schrödinger equation then appears as

$$(7.181) \qquad \frac{d^2\varphi}{dx^2} + \frac{2mF_1}{\hbar^2}(x - x_1)\varphi = 0 \qquad x \text{ near } x_1$$

$$(7.182) \qquad \frac{d^2\varphi}{dx^2} - \frac{2mF_2}{\hbar^2}(x - x_2)\varphi = 0 \qquad x \text{ near } x_2$$

Further simplification of these equations is accomplished through the change in variable

$$y = -\left(\frac{2mF_1}{\hbar^2}\right)^{1/3}(x - x_1)$$

in (7.181) and

$$y = \left(\frac{2mF_2}{\hbar^2}\right)^{1/3}(x - x_2)$$

in (7.182). Both equations then reduce to the same equation (see Problem 7.93),

$$(7.182a) \qquad \frac{d^2\varphi}{dy^2} - y\varphi = 0$$

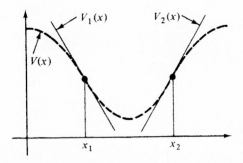

FIGURE 7.34 Approximate linear potentials $V_1(x)$ and $V_2(x)$ valid in the neighborhoods of the turning points x_1 and x_2, respectively.

The solutions to this equation are called[1] *Airy functions* and are denoted by the symbols $Ai(y)$ and $Bi(y)$ (see Table 7.3). For the problem at hand, the wavefunction $\varphi(x)$ must approach zero in the domains $x \ll x_1$ and $x \gg x_2$. Both these regions correspond to large positive values of $|y|$. The function with this property is $Ai(y)$, which has asymptotic forms

(7.183)

$$Ai(y) \sim \frac{1}{2\sqrt{\pi}\, y^{1/4}} \exp\left(-\frac{2}{3} y^{3/2}\right) \qquad (y > 0)$$

$$Ai(y) \sim \frac{1}{\sqrt{\pi}\,(-y)^{1/4}} \sin\left[\frac{2}{3}(-y)^{3/2} + \frac{\pi}{4}\right] \qquad (y < 0)$$

It is exponentially decaying for $y > 0$ and oscillatory for $y < 0$ and strongly resembles the behavior of a harmonic oscillator wavefunction across a turning point, such as shown in Fig. 7.6.

In the neighborhood of x_1, from (7.179), we obtain

$$p^2 = 2m(E - V_1) \simeq 2mF_1(x - x_1) = -(2mF_1\hbar)^{2/3} y$$
$$2mF_1\, dx = -(2mF_1\hbar)^{2/3}\, dy$$

To the left of x_1, $p^2 = -\hbar^2\kappa^2$, so

$$\hbar^2\kappa^2 = (2mF_1\hbar)^{2/3}\, y$$

and we may write

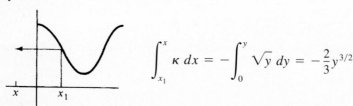

$$\int_{x_1}^x \kappa\, dx = -\int_0^y \sqrt{y}\, dy = -\frac{2}{3} y^{3/2}$$

To the right of x_1, in the oscillatory well domain, $p^2 = \hbar^2 k^2$ and

$$\hbar^2 k^2 = -(2mF_1\hbar)^{2/3}\, y$$

so y is negative in this domain The integral of k gives

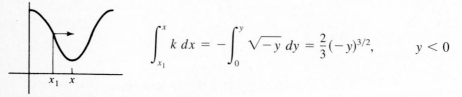

$$\int_{x_1}^x k\, dx = -\int_0^y \sqrt{-y}\, dy = \frac{2}{3}(-y)^{3/2}, \qquad y < 0$$

[1] Named for an English astronomer, G. B. Airy (1801–1892).

TABLE 7.3 Properties of Airy functions[a]

Differential equation

$$\frac{\partial^2 \varphi}{\partial x^2} - x\varphi = 0$$

Solutions

    (a) *Series representation*

$$Ai(x) = af(x) - bg(x)$$

$$Bi(x) = \sqrt{3}[af(x) + bg(x)]$$

       where

$$a = 3^{-2/3}/\Gamma(2/3) = 0.3550, \qquad b = 3^{-1/3}/\Gamma(1/3) = 0.2588$$

$$f(x) = 1 + \frac{1}{3!}x^3 + \frac{1 \cdot 4}{6!}x^6 + \frac{1 \cdot 4 \cdot 7}{9!}x^9 + \cdots$$

$$g(x) = x + \frac{2}{4!}x^4 + \frac{2 \cdot 5}{7!}x^7 + \frac{2 \cdot 5 \cdot 8}{10!}x^{10} + \cdots$$

    (b) *Integral representation*

$$Ai(x) = \frac{1}{\pi}\int_0^\infty \cos\left(\frac{s^3}{3} + sx\right)ds$$

$$Bi(x) = \frac{1}{\pi}\int_0^\infty \left[e^{sx-(1/3)s^3} + \sin\left(\frac{s^3}{3} + sx\right)\right]ds$$

Relations to Bessel functions of fractional order

    With $y \equiv \frac{2}{3}x^{3/2}$, the following relations hold.

$$Ai(x) = \frac{1}{\pi}\sqrt{x/3}\,K_{1/3}(y)$$

$$Ai(-x) = \frac{1}{3}\sqrt{x}\,[J_{1/3}(y) + J_{-1/3}(y)]$$

$$Bi(x) = \sqrt{x/3}\,[I_{-1/3}(y) + I_{1/3}(y)]$$

$$Bi(-x) = \sqrt{x/3}\,[J_{-1/3}(y) - J_{1/3}(y)]$$

    The I and K functions are modified Bessel functions of the first and second kind, respectively.

Asymptotic forms

    For large $|x|$, leading terms in asymptotic series are as follows:

$$Ai(x) \sim \frac{1}{2\sqrt{\pi}x^{1/4}}\exp\left(-\frac{2}{3}x^{3/2}\right), \qquad\qquad x > 0$$

$$Ai(x) \sim \frac{1}{\sqrt{\pi}(-x)^{1/4}}\sin\left[\frac{2}{3}(-x)^{3/2} + \frac{\pi}{4}\right], \qquad\qquad x < 0$$

$$Bi(x) \sim \frac{1}{\sqrt{\pi}x^{1/4}}\exp\left(\frac{2}{3}x^{3/2}\right), \qquad\qquad x > 0$$

$$Bi(x) \sim \frac{1}{\sqrt{\pi}(-x)^{1/4}}\cos\left[\frac{2}{3}(-x)^{3/2} + \frac{\pi}{4}\right], \qquad\qquad x < 0$$

[a]For further properties of these functions, see *Handbook of Mathematical Functions*, N. Abramowitz and I. A. Stegun, eds., Dover, New York, 1964; H. and B. S. Jeffries, *Methods of Mathematical Physics*, 3d ed., Cambridge University Press, New York, 1956.

In these same respective domains, the WKB functions φ_I (7.176) and $\varphi_{II}^<$ (7.178a), when written in terms of the variable y, appear as

$$\varphi_I = \frac{1}{y^{1/4}} \exp\left(-\frac{2}{3}y^{3/2}\right) \qquad\qquad y > 0$$

$$\varphi_{II}^< = \frac{C}{(-y)^{1/4}} \sin\left(\frac{2}{3}(-y)^{3/2} + \delta\right) \qquad\qquad y < 0$$

These agree with the asymptotic forms (7.183) for the exact Airy function solutions [corresponding to the approximate linear potential (7.179, 7.180)] provided that we set $C = 2$ and $\delta = \pi/4$.

In this manner we find that the WKB approximation in region I,

(7.184) $$\varphi_I(x) = \frac{1}{\sqrt{\kappa}} \exp\left(\int_{x_1}^x \kappa\, dx\right) \qquad (x < x_1)$$

matches (or "connects") with the WKB approximation

(7.185) $$\varphi_{II}^<(x) = \frac{2}{\sqrt{k}} \sin\left(\int_{x_1}^x k\, dx + \frac{\pi}{4}\right) \qquad (x_1 < x)$$

in region II.

In like manner we find that the WKB approximation in region III

(7.186) $$\varphi_{III} = \frac{A}{\sqrt{\kappa}} \exp\left(-\int_{x_2}^x \kappa\, dx\right) \qquad (x_2 < x)$$

matches with the WKB approximation

(7.187) $$\varphi_{II}^> = \frac{2A}{\sqrt{k}} \sin\left(\int_x^{x_2} k\, dx + \frac{\pi}{4}\right) \qquad (x < x_2)$$

in region II. The remaining constant A is determined in matching $\varphi_{II}^<$ to $\varphi_{II}^>$.

The Four Connection Formulas

There are in total four connection formulas which serve to relate WKB component wavefunctions across turning points. In the preceding analysis two of these relations were uncovered. Namely, these are given by the manner through which φ_I connects to $\varphi_{II}^<$ (7.184 and 7.185) and that by which $\varphi_{II}^>$ connects to φ_{III} (7.186 and 7.187). Carrying through a parallel analysis and employing the asymptotic forms for the Airy functions $Bi(y)$ gives the remaining two relations. The complete list of four connecting formulas[1] is given below with $x_{1,2}$ denoting either x_1 or x_2.

		$x < x_{1,2}$		$x_{1,2} < x$
(7.188a)		$\dfrac{2}{\sqrt{k}} \cos\left(\displaystyle\int_x^{x_2} k\,dx - \dfrac{\pi}{4}\right)$	$\rightleftharpoons$	$\dfrac{1}{\sqrt{\kappa}} \exp\left(-\displaystyle\int_{x_2}^x \kappa\,dx\right)$
(7.188b)		$\dfrac{1}{\sqrt{k}} \sin\left(\displaystyle\int_x^{x_2} k\,dx - \dfrac{\pi}{4}\right)$	$\rightleftharpoons$	$-\dfrac{1}{\sqrt{\kappa}} \exp\left(\displaystyle\int_{x_2}^x \kappa\,dx\right)$
(7.189a)		$\dfrac{1}{\sqrt{\kappa}} \exp\left(\displaystyle\int_{x_1}^x \kappa\,dx\right)$	$\rightleftharpoons$	$\dfrac{2}{\sqrt{k}} \cos\left(\displaystyle\int_{x_1}^x k\,dx - \dfrac{\pi}{4}\right)$
(7.189b)		$-\dfrac{1}{\sqrt{k}} \exp\left(-\displaystyle\int_{x_1}^x \kappa\,dx\right)$	$\rightleftharpoons$	$\dfrac{1}{\sqrt{k}} \sin\left(\displaystyle\int_{x_1}^x k\,dx - \dfrac{\pi}{4}\right)$

Bohr-Sommerfeld Quantization Rules

The energy levels of the finite well depicted in Fig. 7.33 may be obtained to within the accuracy of the WKB approximation by joining $\varphi_{II}^<$ and $\varphi_{II}^>$ smoothly within the well. This gives

$$\sin\left(\int_{x_1}^x k\,dx + \frac{\pi}{4}\right) = A \sin\left(\int_x^{x_2} k\,dx + \frac{\pi}{4}\right)$$

With

$$\eta \equiv \int_{x_1}^{x_2} k\,dx, \qquad a \equiv \int_x^{x_2} k\,dx + \frac{\pi}{4}$$

the continuity condition above becomes

$$\sin\left(\eta + \frac{\pi}{2} - a\right) = A \sin a$$

or, equivalently,

$$\sin\left(\eta + \frac{\pi}{2}\right)\cos a - \cos\left(\eta + \frac{\pi}{2}\right)\sin a = A \sin a$$

---

[1] Here we are assuming that no other linearly independent components of the wavefunction enter the analysis.

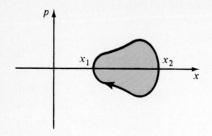

FIGURE 7.35   **Classical vibrational motion between the turning points (x_1, x_2) depicted in (x, p) "phase space." The enclosed surface has the area $\oint p \, dx$.**

The solution to this equation which gives A, constant and independent of the parameter a, is obtained by setting[1]

$$(7.190) \qquad \eta + \frac{\pi}{2} = (n + 1)\pi \qquad (n = 0, 1, 2, \ldots)$$

Corresponding values of A are $(-1)^n$. Thus, continuity of φ_{II} implies the condition

$$\eta = \int_{x_1}^{x_2} k \, dx = \left(n + \frac{1}{2}\right)\pi$$

When written in terms of momentum $p = \hbar k$, this criterion appears as

$$(7.191) \qquad \int_{x_1}^{x_2} p \, dx = \left(n + \frac{1}{2}\right)\frac{h}{2}$$

In the corresponding classical motion, the particle oscillates between the turning points x_1 and x_2. In Cartesian x, p space, this "orbit" is a closed loop, as depicted in Fig. 7.35, with area $\oint p \, dx$. With (7.191) we find that

$$(7.192) \qquad \oint p \, dx = \left(n + \frac{1}{2}\right)h$$

This equation is nearly the same as the Bohr-Sommerfeld quantization rule[2] (2.6). As discussed in Section 2.4, this rule prescribes that an integral number of wavelengths fit the orbit perimeter (see Fig. 2.11). When cast in terms of wavelength $\lambda = h/p$,

---

[1] Writing $n + 1$ instead of n ensures that η is nonnegative.

[2] The distinction between the loop integral in (2.6) and that in (7.192) is that $\oint p_\theta \, d\theta$ is relevant to rotational motion while $\oint p \, dx$ is relevant to vibrational motion. In either case such integrals play a major role in the study of periodic motion and are called *action* integrals. The Bohr-Sommerfeld quantization rule stipulates that these action integrals have only discrete values, nh. That is, in quantization of periodic systems, one quantizes the action variables $\oint p \, dx$.

(7.192) becomes

(7.193)
$$\oint \frac{dx}{\lambda} = n + \frac{1}{2}$$

The integral represents the number of wavelengths in the orbit perimeter. The distinction between this result and the Bohr-Sommerfeld prescription is that in the present WKB analysis, the wavefunction may extend into the classically forbidden region, or, equivalently, that the wavefunction need not vanish at the turning points. If the wavefunction must vanish at the turning points, such as is the case for a very sharply rising potential, a half-integer number of wavelengths are allowed between turning points, thereby returning the Bohr formula. On the other hand, leakage of the wavefunction into the classically forbidden domain in the WKB analysis is evident from the wavefunction (7.184) through (7.187) and matching value, $A = (-1)^n$.

WKB Eigenenergies

The continuity result (7.191) serves to determine eigenenergies within the WKB approximation. Since this analysis becomes more accurate for large energies, values so found will generally give better estimates for large quantum number n. In that this number is also a measure of the number of zeros of the wavefunction between turning points, we see that in this limit the wavelength becomes small compared to the distance between the turning points. As described in the first paragraph of this section, such is the domain of the classical WKB analysis.

As an example of the application of (7.191), let us consider calculation of the energies of the harmonic oscillator. For this configuration the momentum is given by

$$p = \sqrt{2m(E - m\omega_0^2 x^2/2)}$$

with turning points given by (7.12). Introducing the variable

$$\cos \theta \equiv x\sqrt{m\omega_0^2/2E}$$

permits the condition (7.191) to be written

$$\frac{4E}{\omega_0} \int_0^\pi \sin^2 \theta \, d\theta = \left(n + \frac{1}{2}\right)h$$

$$\frac{2\pi E_n}{\omega_0} = \left(n + \frac{1}{2}\right)h$$

These are seen to be, somewhat fortuitously, the exact eigenenergies of the harmonic oscillator (7.44), valid for all n.

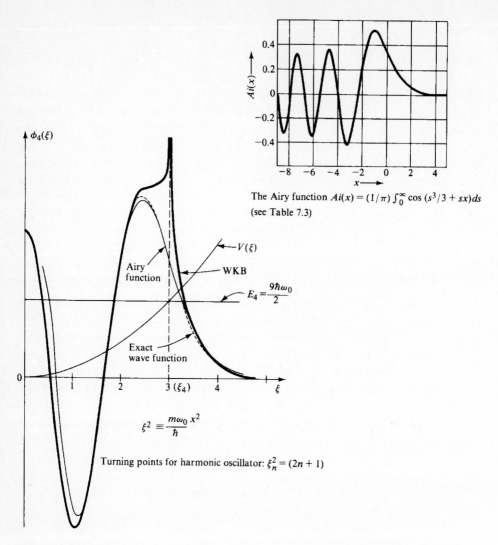

The Airy function $Ai(x) = (1/\pi) \int_0^\infty \cos{(s^3/3 + sx)}ds$
(see Table 7.3)

$V(\xi)$

Airy
function

WKB

$E_4 = \dfrac{9\hbar\omega_0}{2}$

Exact
wave function

$\xi^2 \equiv \dfrac{m\omega_0}{\hbar} x^2$

Turning points for harmonic oscillator: $\xi_n^2 = (2n + 1)$

FIGURE 7.36   WKB approximation for the fourth state of the harmonic oscillator, together with a graph of the Airy function. Also shown are the potential $V(\xi)$ and the fourth eigenenergy. Note the divergence of the WKB approximation at the turning point ξ_4. This calculation was performed previously by J. D. Powell and B. Craseman (*Quantum Mechanics*, Addison-Wesley, Reading, Mass., 1965). For extensive discussion of numerical techniques in the WKB analysis, see C. M. Bender and S. A. Orszag, *Advanced Mathematical Methods for Scientists and Engineers*, McGraw-Hill, New York, 1978.

An example in the opposite extreme is given by the one-dimensional box potential (4.1). There is no penetration of the particle wavefunction through the sharply rising potential wall, and the validity of the WKB formula (7.191) becomes very questionable, especially at low energies. Using this formula, we readily obtain the eigenenergies

$$E_n^{\text{WKB}} = \left(n + \frac{1}{2}\right)^2 \frac{h^2}{8ma^2} = E_n\left(1 + \frac{4n + 1}{4n^2}\right)$$

Here we have written E_n for the exact eigenenergies (4.14), $E_n = n^2E_1$. As expected, the estimate gives a large fractional error for small quantum numbers. However, in the high-quantum-number domain, $n \gg 1$, where the walls of the potential are many wavelengths apart, we find that the WKB estimate agrees with the exact result

$$E_n^{\text{WKB}} \simeq E_n \qquad (n \gg 1)$$

Note, however, that for this singular case where wavefunction penetration does not occur, the simpler Bohr-Sommerfeld rule (2.6) gives exact results.

WKB Wavefunctions

Wavefunctions in the WKB approximation incorporate Airy functions together with matching conditions (7.184) et seq. Results of this analysis to the fourth state of the harmonic oscillator are shown in Fig. 7.36. Here we may observe the dramatic disparity between the exact wavefunction and the WKB approximation in the vicinity of the turning points.

Application to Transmission Problems

In concluding this section we will obtain a very important formula for the transmission coefficient relevant to a potential barrier such as depicted in Fig. 7.37. As

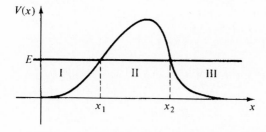

FIGURE 7.37   Domains relevant to the WKB approximation of the transmission through a potential barrier.

described previously in Section 7.5, the transmitted wave in region III has only one momentum component, so that to within the WKB approximation, we may write

$$(7.194) \qquad \varphi_{\text{III}} = \frac{A}{\sqrt{k}} \exp\left[i\left(\int_{x_2}^{x} k\,dx - \frac{\pi}{4}\right)\right]$$

The procedure we will follow to obtain the incident component of φ_{I} is as follows. Rewriting (7.194) as a combination of trigonometric functions permits application of the connection formulas (7.189), which allows calculation of φ_{II}. With φ_{II} so found, we again rewrite it in a manner that permits application of (7.188) to connect φ_{II} to φ_{I}. Finally, φ_{I} is decomposed into incident and reflected components. Comparison of the incident component with φ_{III} permits calculation of the transmission coefficient.

Rewriting (7.194) in the form

$$(7.195) \qquad \varphi_{\text{III}} = \frac{A}{\sqrt{k}}\left[\cos\left(\int_{x_2}^{x} k\,dx - \frac{\pi}{4}\right) + i\sin\left(\int_{x_2}^{x} k\,dx - \frac{\pi}{4}\right)\right]$$

permits application of (7.189) and we obtain

$$(7.196) \qquad \varphi_{\text{II}} = \frac{A}{2\sqrt{\kappa}} \exp\left(-\int_{x}^{x_2} \kappa\,dx\right) - \frac{iA}{\sqrt{\kappa}} \exp\left(\int_{x}^{x_2} \kappa\,dx\right)$$

Let r denote the integral (compare with η, p. 263)

$$(7.196\text{a}) \qquad r \equiv \exp\left(\int_{x_1}^{x_2} \kappa\,dx\right)$$

Appropriate division of the interval of integration gives the relations

$$\exp\left(-\int_{x}^{x_2} \kappa\,dx\right) = r^{-1} \exp\left(\int_{x_1}^{x} \kappa\,dx\right)$$

$$\exp\left(\int_{x}^{x_2} \kappa\,dx\right) = r \exp\left(-\int_{x_1}^{x} \kappa\,dx\right)$$

Substituting these expressions into (7.196) gives

$$(7.197) \qquad \varphi_{\text{II}} = \frac{A}{2r\sqrt{\kappa}} \exp\left(\int_{x_1}^{x} \kappa\,dx\right) - \frac{iAr}{\sqrt{\kappa}} \exp\left(-\int_{x_1}^{x} \kappa\,dx\right)$$

which allows application of the connection formulas (7.188). There results

$$(7.198) \qquad \varphi_{\text{I}} = -\frac{A}{2r\sqrt{k}} \sin\left(\int_{x}^{x_1} k\,dx - \frac{\pi}{4}\right) - \frac{i2Ar}{\sqrt{k}} \cos\left(\int_{x}^{x_1} k\,dx - \frac{\pi}{4}\right)$$

We are now at the point where we must extract the incident component of φ_I. If we label the argument $\int_x^{x_1} k \, dx - \pi/4 \equiv z$ and express both trigonometric terms as exponentials, (7.198) may be rewritten

(7.199)
$$\varphi_I = -\frac{A}{2r\sqrt{k}} \frac{e^{iz} - e^{-iz}}{2i} - \frac{i2Ar}{\sqrt{k}} \frac{e^{iz} + e^{-iz}}{2}$$

$$= i\frac{A}{\sqrt{k}}\left(\frac{1}{4r} - r\right)e^{iz} - i\frac{A}{\sqrt{k}}\left(\frac{1}{4r} + r\right)e^{-iz}$$

Far removed from the boundary, $k \simeq$ constant and $z \simeq -k(x - x_1) - \pi/4$. From this we may infer that the second term in the last equation for φ_I represents the incident component wavefunction. Employing the expression (7.108) for the transmission coefficient T, with the second term in (7.199) representing the incident wavefunction and (7.194) the transmitted wavefunction, we obtain

(7.200)
$$T = \frac{1}{(r + 1/4r)^2} = \frac{1}{r^2 + 1/2 + 1/16r^2}$$

It is consistent with the WKB criterion (7.166) to neglect all but the term r^2 in the denominator of (7.200), thereby obtaining (see Problem 7.89)

(7.201)
$$\boxed{T = r^{-2} = \exp\left(-2\int_{x_1}^{x_2} \kappa \, dx\right)}$$

The simplest application of this formula is in calculation of the transmission through a square potential barrier. Exact analysis gives the result (7.147). We should find that this expression reduces to the WKB formula in the limit

$$\kappa a = \sqrt{2ma^2(V - E)/\hbar^2} \gg 1.$$

In this limit (7.147) gives the transmission coefficient

$$T \simeq \frac{16E}{V} e^{-4\kappa a}$$

whereas (7.201) gives

$$T = e^{-4\kappa a}$$

which is seen to be in good order-of-magnitude agreement with the limiting form of the exact result given above. Further application of the exceedingly important result

(7.201) is left to the problems. A discussion on the Feynman path integral, closely allied to the WKB analysis, is given in Section 7.11.

PROBLEMS

7.62   In the phenomenon of *cold emission,* electrons are drawn from a metal (at room temperature) by an externally supported electric field. The potential well that the metal presents to the free electrons before the electric field is turned on is depicted in Fig. 2.5. After application of the constant electric field $\mathscr{E}$, the potential at the surface slopes down as shown in Fig. 7.38, thereby allowing electrons in the Fermi sea to "tunnel" through the potential barrier. If the surface of the metal is taken as the $x = 0$ plane, the new potential outside the surface is

$$V(x) = \Phi + E_F - e\mathscr{E}\, x$$

where E_F is the Fermi level and Φ is the work function of the metal.

(a)   Use the WKB approximation to calculate the transmission coefficient for cold emission.

(b)   Estimate the field strength $\mathscr{E}$, in V/cm, necessary to draw current density of the order of mA/cm^2 from a potassium surface. For $\mathbb{J}_{\text{inc}}$ [see (7.108)] use the expression $\mathbb{J}_{\text{inc}} = env$, where n is electron density and v is the speed of electrons at the top of the Fermi sea. The relevant expression for E_F may be found in Problem 2.42. Data for potassium is given in Section 2.3.

Answer(partial)

(a)   Using (7.201), and the form of the potential exterior to the metal given in the statement of the problem, we obtain, for transmission at the Fermi level ($V - E_F = \Phi - e\mathscr{E}x$),

$$T = \exp\left[-\frac{2}{\hbar} \int_0^{\Phi/e\mathscr{E}} \sqrt{2m(\Phi - e\mathscr{E}\, x)}\, dx \right]$$
$$= \exp\left(-\frac{4}{3} \frac{\sqrt{2m}}{\hbar} \frac{\Phi^{3/2}}{e\mathscr{E}} \right)$$

This equation is referred to as the *Fowler–Nordheim* equation.

7.63   An α particle is the nucleus of a helium atom. It is a tightly bound entity comprised of two protons and two neutrons, for which the approximate binding energy is 7 MeV. A primary mode of decay for radioactive nuclei is through the process of α decay. A consistent model for this process envisions the α particle bound to the nucleus by a spherical well

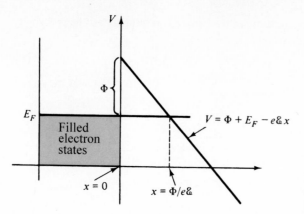

FIGURE 7.38 Potential configuration for the phenomenon of "cold emission." (See Problem 7.62.)

potential.[1] Outside the well the α particle is repelled from the residual nucleus by the potential barrier

$$V = \frac{2(Z - 2)e^2}{r} \equiv \frac{A}{r}$$

The original radioactive nucleus has charge Ze, while the α particle has charge $2e$ (Fig. 7.39).

(a) Use the WKB approximation to calculate the transmissivity T of the nuclear barrier to α decay in terms of the velocity $v = \sqrt{2E/m}$ and the dimensionless ratio $\sqrt{(r_0/r_1)} \equiv \cos W$. What form does T assume in the limit $r_0 \to 0$?

(b) Assuming that the α particle "bounces" freely between the walls presented by the spherical well potential with a speed $\sim 10^9$ cm/s and that the radius of the heavy radioactive nucleus (e.g., uranium) is $\sim 10^{-12}$ cm, one obtains that the α particle strikes the nuclear wall at the rate $\sim 10^{21}$ s^{-1}. In each collision the probability that the α particle penetrates the nuclear

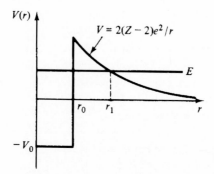

FIGURE 7.39 Nuclear α particle potential model for the process of α decay. (See Problem 7.63.)

[1] The rigid spherical well potential is described in Section 10.3. The effective one-dimensional Hamiltonian for the configuration at hand is given by (10.107) with angular momentum L set equal to zero. This corresponds to assuming in part (b) that the bounce motion of the α-particle is through the origin.

Coulomb barrier is equal to the transmissivity of the barrier T. It follows that the probability of tunneling through this barrier, per second, is

$$P = 10^{21} T$$

and that the mean lifetime of the nucleus is

$$\tau = \frac{1}{P} = \frac{10^{-21}}{T}$$

Use your answer to part (a) for T and the following expression for the nuclear radius

$$r_0 = 2 \times 10^{-13} Z^{1/3} \text{ cm}$$

to estimate the mean lifetime for uranium α decay.

Answer (partial)

(a)
$$T = \exp\left[-\frac{2}{\hbar} \int_{r_0}^{r_1} \sqrt{2m\left(\frac{A}{r} - E\right)}\, dr \right]$$

$$r_1 = \frac{A}{E}$$

Integrating, one obtains *Gamow's formula*,

$$T = \exp\left[-\frac{2A}{\hbar v}(2W - \sin 2W) \right]$$

As $r_0 \to 0$, $W \to \pi/2$ and $T \sim \exp(-2\pi A/\hbar v)$.

7.64 Use the WKB relation (7.191) to estimate the eigenenergies of the displaced spring potential

$$V = \frac{K}{2}(x^2 + 2\phi x)$$

How do these values compare with the exact values obtained in Problem 7.16?

7.65 An electron with charge $-e$ and mass m, constrained to move in the x direction, interacts with a uniform electric field $\mathcal{E}$, which points in the positive x direction.

(a) Show that energy eigenstates may be written as Airy functions.

(b) Do eigenenergies comprise a continuous or a discrete spectrum? *Hint for part (a):*
Set $e\mathcal{E}x - E = Kx'$ and then find the value of K that gives Airy's equation. (See Table 7.3.)

7.66 Use the WKB approximation to determine the bound-state energies of the potential well

$$V(x) = \frac{V_0}{a}|x|, \qquad |x| \le a$$

$$V(x) = V_0 = \frac{1}{m}\left(\frac{h}{a}\right)^2, \qquad |x| > a$$

Answer

Eigenenergies appear as

$$E_n = \bar{E}_0(n + \tfrac{1}{2})^{2/3}$$

$$\bar{E}_0^{3/2} = \frac{3V_0 h}{8a\sqrt{2m}}$$

With V_0 as given, we obtain $\bar{E}_0 = 0.41 V_0$, so there are four bound states: $E_0 = 0.63\bar{E}_0$, $E_1 = 1.31\bar{E}_0$, $E_2 = 1.85\bar{E}_0$, and $E_3 = 2.31\bar{E}_0$.

7.67 Show that for the singular potential

$$V = aV_0 \, \delta(x)$$

boundary conditions on $\varphi(x)$ become

(a) $\varphi(0)_- = \varphi(0)_+$

(b) $\dfrac{\hbar^2}{2m}[\varphi'(0)_+ - \varphi'(0)_-] = aV_0\varphi(0)$

7.68 Use the result of Problem 7.67 to construct the bound state of the potential well, $V = -aV_0\delta(x)$. (*Hint:* Set $E = -|E|$ and look for the solution for $x \neq 0$.)

7.69 Find T and R for the potential barrier $V = aV_0\delta(x)$.

7.70 The initial state for the harmonic oscillator

$$|\psi(0)\rangle = e^{-g^2/2} \sum_{n=0}^{\infty} \frac{g^n}{\sqrt{n!}} \, |n\rangle$$

represents a minimum uncertainty wave packet. The parameter g is real.

(a) Show that

$$|\psi(0)\rangle = e^{-g^2/2} \sum \frac{(g\hat{a}^\dagger)^n}{n!} \, |0\rangle = e^{-g^2/2}\, e^{g\hat{a}^\dagger} |0\rangle$$

where $|0\rangle$ is the $n = 0$ eigenstate.

(b) Show that for this state

$$\langle x \rangle = g\sqrt{\frac{2\hbar}{\omega_0 m}} \qquad\qquad \langle p \rangle = 0$$

$$\langle x^2 \rangle = \frac{\hbar}{2\omega_0 m}(4g^2 + 1) \qquad\qquad \langle p^2 \rangle = \frac{\hbar\omega_0 m}{2}$$

so that

$$(\Delta x)^2 = \frac{\hbar}{2\omega_0 m} \qquad\qquad (\Delta p)^2 = \frac{\hbar\omega_0 m}{2}$$

and

$$\Delta x \, \Delta p = \frac{\hbar}{2}$$

This property establishes the fact that the given superposition state represents a wave packet of minimum uncertainty.

(c)   Show that for this state

$$\langle H \rangle = \hbar\omega_0 \left(g^2 + \frac{1}{2} \right)$$

which gives a physical interpretation of the parameter g.

(d)   Show that $\psi(t)$ is

$$|\psi(t)\rangle = e^{-i\omega_0 t/2} e^{-g^2/2} \exp\left(g e^{-i\omega_0 t} \hat{a}^\dagger\right)|0\rangle$$

[*Hint:*

$$|\psi(t)\rangle = e^{-i\omega_0 t(\hat{a}^\dagger \hat{a}+1/2)}|\psi(0)\rangle$$

Also,

$$\exp\left(-i\omega_0 t \hat{a}^\dagger \hat{a}\right) \sum \frac{g^n}{\sqrt{n!}} |n\rangle = \sum \frac{(g e^{-i\omega_0 t})^n}{\sqrt{n!}} |n\rangle$$

Here we have recalled that $f(\hat{a}^\dagger \hat{a})|n\rangle = f(n)|n\rangle$.]

(e)   Show that in the state $\psi(t)$

$$\langle x \rangle = g\sqrt{\frac{2\hbar}{\omega_0 m}} \cos \omega_0 t \qquad \langle p \rangle = -g\sqrt{2\hbar\omega_0 m} \sin \omega_0 t$$

$$\langle x^2 \rangle = \langle x \rangle^2 + \frac{\hbar}{2\omega_0 m} \qquad \langle p^2 \rangle = \langle p \rangle^2 + \frac{\hbar\omega_0 m}{2}$$

so that

$$\Delta x \, \Delta p = \frac{\hbar}{2}$$

at any time t. The packet remains a packet of minimum uncertainty for all time and oscillates in classical simple harmonic motion. Note also that the probability density $\langle \psi(x, t)|\psi(x, t)\rangle$ is a Gaussian form, centered at $\langle x \rangle$.

7.71   In closed form the wave packet of minimum uncertainty for the harmonic oscillator appears (at time $t = 0$) as

$$\psi(x, 0) = \sqrt{\beta/\pi^{1/2}} \exp\left[\frac{ixp_0}{\hbar} - \frac{1}{2}\beta^2(x - x_0)^2\right]$$

(a)   Show that

$$\langle x \rangle = x_0 \qquad \Delta x = \frac{1}{\beta\sqrt{2}}$$

$$\langle p \rangle = p_0 \qquad \Delta p = \frac{\hbar\beta}{\sqrt{2}}$$

Hence $\Delta x\,\Delta p\,=\,\hbar/2$, and we are justified in calling ψ a packet of minimum uncertainty.

(b)   Show that in the initial state above,

$$\langle H\rangle = \frac{1}{2m}\,(p_0{}^2 + m^2\omega^2 x_0{}^2) + \frac{1}{2}\hbar\omega$$

and

$$\langle xp + px\rangle = 2x_0 p_0$$

(c)   In order to establish that $\psi(x)$ remains a packet of minimum uncertainty for all time, one must show that $\Delta x\,\Delta p$ is constant. Recalling the equation of motion for the average of an operator (6.68),

$$\frac{d}{dt}\langle A\rangle = \frac{l}{ih}\,\langle[\hat{A},\,\hat{H}]\rangle + \left\langle\frac{\partial\hat{A}}{\partial t}\right\rangle$$

and introducing the operator

$$\hat{\eta} = \hat{x}\hat{p} + \hat{p}\hat{x} - 2\langle x\rangle\langle p\rangle$$

show that

$$\frac{d}{dt}\,(\Delta x)^2 = \frac{\langle\eta\rangle}{m}$$

$$\frac{d}{dt}\,(\Delta p)^2 = -m\omega^2\langle\eta\rangle$$

$$\frac{d}{dt}\langle\eta\rangle = \frac{2(\Delta p)^2}{m} - 2m\omega^2(\Delta x)^2$$

Using these results, show that for the initial state above, $(\Delta x)^2$ and $(\Delta p)^2$ are both constants in time.

(d)   Show that $\psi(x, 0)$ is an eigenfunction of the annihilation operator $\hat{a}$. What is the eigenvalue of $\hat{a}$ in this state? [*Hint:* Employ the representation for $\psi(x, t)$ given in Problem 7.70.]

Note: In quantum optics the radiation field is viewed as a collection of harmonic oscillators. In this representation, the appropriate generalization of the state of minimum uncertainty is called the *coherent state*.

7.72   The Hamiltonian of a particle is

$$\hat{H} = A\hat{a}^{\dagger}\hat{a} + B(\hat{a} + \hat{a}^{\dagger})$$

where A and B are constants. What are the energy eigenvalues of the particle? (*Hint:* Introduce the operator $\hat{b} = \alpha\hat{a} + \beta; \hat{b}^{\dagger} = \alpha\hat{a}^{\dagger} + \beta$.)

7.73   What is the form of the potential that gives the Gaussian probability density with variance a^2 in the ground state?

7.74   The reflection coefficient for the smooth potential step

$$V(x) = \frac{V_0}{1 + e^{-\gamma x}}$$

for $E > V_0$ is[1]

$$R = \left(\frac{\sinh \left[\pi(k_1 - k_2)/\gamma \right]}{\sinh \left[\pi(k_1 + k_2)/\gamma \right]} \right)^2$$

The energy of incident particles at $x = -\infty$ is

$$E = \frac{\hbar^2 k_1^2}{2m}$$

while the kinetic energy of transmitted particles at $x = +\infty$ is

$$\frac{\hbar^2 k_2^2}{2m} = E - V_0$$

(a)   Make a sketch of the potential $V(x)$ and indicate roughly the length scale of potential and its relation to the wavenumber γ.

(b)   Show that in the limit that $V(x)$ approaches the simple step (Fig. 7.18), R goes to the value given by (7.125).

(c)   Show that the classical value of R emerges for wavelengths small compared to the potential scale of length.

7.75   The transmission coefficient for the symmetric potential hill

$$V(x) = \frac{V_0}{\cosh^2 (\gamma x)}$$

for $E < V$ is

$$T = \frac{\sinh^2 (\pi k/\gamma)}{\sinh^2 (\pi k/\gamma) + \cosh^2 \left[(\pi/2)\sqrt{\rho^2 - 1} \right]}$$

where

$$\rho^2 \equiv \frac{8mV_0}{\hbar^2 \gamma^2} > 1$$

Incident and transmitted particles at $x = -\infty$ and $x = +\infty$, respectively, have energy

$$E = \frac{\hbar^2 k^2}{2m}$$

---

[1] The coefficients R and T given in Problems 7.74 and 7.75, respectively, are calculated in L. Landau and E. Lifshitz, *Quantum Mechanics,* 2d ed., Addison-Wesley, Reading, Mass., 1965.

(a)   Sketch the potential and indicate roughly the length scale of potential and its relation to the wavenumber γ.

(b)   Show that the classical value of T emerges for values of γ appropriate to the classical domain.

(c)   Formulate an expression for the next-order approximation to the entirely classical result (b) for the transmission coefficient using the WKB analysis.

(d)   Obtain an explicit expression for the transmission coefficient in the near-classical domain that you have formulated in part (c) by expanding the exact formula for T given in the statement of this problem.

Answer (partial)

(d)   The classical limit is attained in the limit $\gamma \rightarrow 0$. From the given expression for T, we obtain

$$T \simeq \frac{e^{\pi k/\gamma}}{e^{\pi k/\gamma} + e^{\pi \rho/2}} = \frac{1}{1 + \exp{(\pi/\gamma\hbar)(\hbar\rho\gamma/2 - \hbar\kappa)}}$$

$$\simeq \exp\left[-\frac{\pi}{\gamma\hbar}(\sqrt{2mV_0} - \sqrt{2mE})\right]$$

7.76   A uniform homogeneous beam of electrons is incident on a rectangular potential barrier of height V. Each electron in the beam has energy $E > V$ and unit amplitude wavefunction

$$\varphi_{\text{inc}} = e^{ik_1 x}$$

If the transmitted electrons have wavefunction

$$\varphi_{\text{trans}} = \varphi_{\text{III}} = 0.97 e^{ik_1 x}$$

(a)   What is the total wavefunction φ_1, of electrons in region I?

(b)   If $E = 10$ eV and $V = 5$ eV, what is the minimum barrier width compatible with the information given above?

Answer

(a)   In general for unit amplitude incident waves,

$$\varphi_I = e^{ik_1 x} \pm i\sqrt{R}\, e^{-ik_1 x}$$

$$\varphi_{III} = \sqrt{T}\, e^{ik_1 x}$$

where the $\pm$ signs refer to the sign of $\sin{(2k_2 a)}$. (See Problem 7.44.) It follows that

$$T = (0.97)^2 = 0.94, \qquad R = 1 - T = 0.06, \quad \text{and} \quad \sqrt{R} = 0.24,$$

so that

$$\varphi_I = e^{ik_1 x} + i0.24 e^{-ik_1 x}$$

(b)   We then find that $\sin^2(k_2 2a) = 8R/T = 0.51$, $k_2 2a = 0.80 < \pi/2$, and

$$k_2 = \sqrt{\frac{2m(E - V)}{\hbar^2}} = 1.14 \times 10^8 \text{ cm}^{-1}$$

Therefore,

$$2a = 0.70 \text{ Å}$$

7.77   (a)   What are the values of k and κ at the "turning points" of a potential hill or potential barrier?

(b)   What are the values of the WKB wavefunctions, $|\varphi_I|$, $|\varphi_{II}|$, $|\varphi_{III}|$, at these points (for either bound or unbound states)?

(c)   A student argues the following: We see from part (b) of this problem that WKB wavefunctions blow up at the turning points and are therefore invalid. Such wavefunctions cannot be of any use. Is the student correct? Explain.

Answers

(a) $k = \kappa = 0$ at turning points.

(b) $|\varphi_I| = |\varphi_{II}| = |\varphi_{III}| = \infty$ at turning points.

(c)   The WKB wavefunctions are valid in domains removed from the turning points (where they were derived from the Schrödinger equation). For bound-state problems these solutions give an estimate for eigenenergies. For unbound-state problems they give an estimate for transmission and reflection coefficients.

7.11   PRINCIPLE OF LEAST ACTION AND FEYNMAN'S PATH INTEGRAL FORMULATION

Action Integral and the Lagrangian

Classical dynamics may be formulated in terms of *Hamilton's principle of least action*. This principle states the following: the classical trajectory between two fixed points $x_1(t_1)$ and $x_2(t_2)$ renders the integral

(7.202)
$$S = \int_{t_1}^{t_2} L[x(t), \dot{x}(t)] \, dt$$

a minimum. Here we have written

(7.203)
$$L(x, \dot{x}) \equiv T(x, \dot{x}) - V(x)$$

for the *Lagrangian,* where T represents kinetic energy and V potential energy and a dot represents differentiation with respect to time. (Whereas these expressions are written in one dimension, they are easily generalized to three dimensions.)

As noted above, Hamilton's principle states that of all possible paths between the fixed points x_1 and x_2, the path which minimizes the integral (7.202) is the actual physical path between these two points. Thus Hamilton's principle may be restated as follows: the physical path between the points x_1 and x_2 renders the integral (7.202) stationary. That is,

$$(7.204) \qquad \delta \int_{t_1}^{t_2} L(x, \dot{x}) \, dt = 0$$

where δ represents an arbitrary, infinitesimal variation about the true motion of the system. The variable S in (7.202) is called the *action*, and the rule (7.204) is alternatively called Hamilton's principle or the *principle of least action*.

Relation to the Hamiltonian

The relation of the Lagrangian (7.203) to the classical Hamiltonian (1.13) (for the present one-dimensional configuration) is given by

$$(7.205) \qquad H(x, p) = \dot{x} \frac{\partial L(x, \dot{x})}{\partial \dot{x}} - L(x, \dot{x})$$

Here we have written

$$(7.205a) \qquad p \equiv \frac{\partial L(x, \dot{x})}{\partial \dot{x}}$$

for the momentum *conjugate* to x. Employing the relation (7.205) in (7.202) and following a variational calculation similar to that described in Problem (4.28) leads to Hamilton's equations (1.35).[1] Thus Hamilton's principle (7.204) is an alternative description of classical mechanics.

Minimum Action

Let us examine the meaning of Hamilton's principle by way of a specific case. Thus, for example, consider the harmonic oscillator described previously in Section 7.2. Recalling (7.203) the corresponding Lagrangian is given by

$$(7.206) \qquad L = \frac{1}{2}m\dot{x}^2 - \frac{1}{2}Kx^2$$

---

[1] See, for example, H. Goldstein, *Classical Mechanics*, 2d ed., Addison-Wesley, Reading, Mass., 1980. Sec. 8.5.

The integral (7.202) then becomes

(7.207)
$$S = \int_{t_1}^{t_2} \frac{1}{2}(m\dot{x}^2 - Kx^2)\, dt$$

Consider the specific case that at $t = 0$ the particle is at $x = x_0$ with energy E. The motion for this problem is then given by

$$x(t) = x_0 \cos \omega t$$

(7.208)
$$E = \frac{1}{2}Kx_0^2, \qquad \omega^2 = \frac{K}{m}$$

Substituting these values into (7.207) with $t_1 = 0$ and t_2 related to time t, we find

(7.209)
$$S(t) = -\frac{E}{2\omega} \sin 2\omega t$$

Now in what sense is this value of the action minimum? To answer this question, we consider the varied motion

(7.210)
$$x(t) = x_0 \cos \omega t + \epsilon^2 x_0 \sin \frac{\omega t}{\epsilon}$$

where $\epsilon \ll 1$. The preceding function is seen to represent a high-frequency, small-amplitude oscillation, which follows the unperturbed $\cos \omega t$ motion. See Fig. 7.40.

Constructing the Lagrangian for the motion (7.210), we find [keeping terms to $0(\epsilon^2)$]

$$L = L_0 - 2\epsilon E\left[\sin \omega t \cos \frac{\omega t}{\epsilon} + \epsilon \cos \omega t \sin \frac{\omega t}{\epsilon} - \frac{\epsilon}{2}\cos^2 \frac{\omega t}{\epsilon} \right]$$

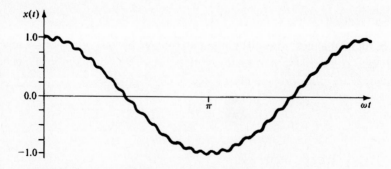

FIGURE 7.40   High-frequency, small-amplitude motion of (7.210).

where L_0 is the Lagrangian corresponding to the unperturbed motion (7.208). Integrating the preceding over time and passing to the limit $\epsilon \to 0$, we see that contributions from the first two of the bracketed terms "wash out," whereas the third term yields

$$(7.211) \quad S = S_0 + \frac{E\epsilon^3}{2\omega} \left(\frac{\omega t}{\epsilon} + \sin \frac{\omega t}{\epsilon} \cos \frac{\omega t}{\epsilon} \right) = S_0 + \frac{1}{2} \epsilon^2 E t + 0(\epsilon^3)$$

where S_0 is the action (7.209). Thus to $O(\epsilon^2)$ we may write

$$(7.212) \qquad\qquad S > S_0$$

We may conclude that the action S, (7.211), corresponding to the varied motion (7.210), is larger than S_0, (7.209), corresponding to the true motion (7.208), thereby corroborating Hamilton's principle.

Feynman Path Integral

In 1948, R. P. Feynman[1] presented a new formulation of quantum mechanics based in large part on the preceding classical concepts.[2]

Our discussion of this formalism begins with the solution of the initial-value problem (3.70) written in Dirac notation.

$$(7.213) \qquad\qquad |\psi(t)\rangle = e^{-it\hat{H}/\hbar} |\psi(0)\rangle$$

In coordinate representation (see Appendix A) this equation becomes

$$(7.214) \qquad \langle x | \psi(t) \rangle = \int \langle x | e^{-it\hat{H}/\hbar} | x' \rangle \langle x' | \psi(0) \rangle \, dx'$$

Note that in this expression $\hat{H} = \hat{H}(\hat{x}, \hat{p})$, whereas in (3.70) $\hat{H} = \hat{H}(x, -i\hbar\partial/\partial x)$. (See Problem 7.86.) We may write (7.214) in its equivalent form

$$(7.215) \qquad \psi(x, t) = \int \langle x | e^{-it\hat{H}/\hbar} | x' \rangle \psi(x', 0) \, dx'$$

In the Feynman description one assumes the following form for the propagation term in (7.215):

$$(7.216) \qquad \langle x | e^{-it\hat{H}/\hbar} | x' \rangle = \int_{all \atop paths} \exp\left[\frac{i}{\hbar} \int_{t'}^{t} L(x, \dot{x}) \, dt \right] \mathscr{D}[x', x]$$

[1] R. P. Feynman, *Rev. Mod. Phys.* **20**, 367 (1948).

[2] For further discussion, see R. P. Feynman and A. R. Hibbs, *Quantum Mechanics and Path Integrals,* McGraw-Hill, New York, 1965.

The integration on the right represents a sum of the exponential over all paths $x = x(t)$ between x' and x and $\mathscr{D}[x', x]$ is a differential-like measure related to the sum over paths. Furthermore, $L(x, \dot{x})$ is the classical Lagrangian. Over each path (7.216), $L(x, \dot{x})$ is evaluated from the given $x = x(t)$. Substituting (7.216) into (7.215) gives

$$(7.217) \qquad \psi(x, t) = \iint_{\substack{all \\ paths}} e^{iS/\hbar} \psi(x', 0) \, dx' \mathscr{D}[x', x]$$

which is Feynman's solution to the initial-value problem, where S is the action (7.202).

Consider that it is known that the particle is at x_0 at $t = 0$. Then

$$\psi(x', 0) = \delta(x' - x_0)$$

and (7.217) reduces to

$$(7.218) \qquad \psi(x, t) = \int_{\substack{all \\ paths}} \left[\exp \frac{i}{\hbar} \int_0^t L(x, \dot{x}) \, dt \right] \mathscr{D}[x, \dot{x}]$$

This expression represents an extension of the WKB approximation (7.167) to the full quantum domain.

Domains of Large and Small Contribution to the Wavefunction

In the integration over all paths in (7.218), most domains of integration do not contribute for the following reason. Away from the region where action is stationary, a small change in path causes a large change in action compared with $\hbar$. The corresponding rapid fluctuation in the exponential in (7.218) causes cancellation, thereby diminishing contribution to the wavefunction. On the other hand, for paths near the classical orbit, the action is stationary and the exponential term likewise does not vary, resulting in a net contribution to the wavefunction.

We may conclude that the main contribution to the wavefunction (7.218) stems from paths lying near the classical orbit. Thus, to lowest approximation, (7.218) returns the quasi-classical WKB expression (7.167):

$$(7.219) \qquad \psi(x) = A e^{(i/\hbar)S(x)}$$

In this expression the constant x_0 in (7.218) was absorbed in $S(x)$. Furthermore, the action $S(x)$ in (7.219) is evaluated on the classical orbit.

The reader will note that there is a difference between the action integral in (7.204) and that given by the result (7.172). The latter expression may be rewritten

$$S_0(x) = \pm \int_0^t 2T \, dt$$

Now note that

$$L = T - V = 2T - E$$

Thus, with the constraint $\delta E = 0$ in the variation (7.204), the principle of least action becomes

$$\delta \int 2T \, dt = 0$$

This relation contains the action relevant to (7.172).

PROBLEMS

7.78   The *scattering amplitude S,* in one dimension, may be defined by the equation

$$\psi_{\text{trans}} = S\psi_{\text{inc}}$$

For transmission above a potential well, one finds that the bound-state energies of the well are given by the negative real zeros of $S^{-1}(E)$.

(a)   What is the transmission coefficient T in terms of S corresponding to the configuration shown in Fig. 7.14? [See (7.112).]

(b)   Consider the transmission past the rectangular potential described in Section 7.8 (See Fig. 7.27.) Construct an expression for S for this configuration.

(c)   Using your answer to part (b) and the rule stated above, obtain the energy eigenvalues of the well (as discussed in Section 8.1).

Answers

(a)
$$T = \frac{k_2}{k_1} |S|^2$$

In the event that $V(+\infty) = V(-\infty)$, then

$$T = |S|^2$$

(b)
$$S = \frac{F}{A} = \frac{e^{2ik_2a}}{\cos 2k_2a - (i/2)[(k_1^2 + k_2^2)/k_1k_2] \sin 2k_2a}$$

(c)   Setting $k_1^2 \rightarrow -\kappa^2 = -2m|E|/\hbar^2$, and $k_2^2 \rightarrow k^2$, we find that the poles of S occur at

$$\tan 2ka = \frac{2\kappa k}{k^2 - \kappa^2}, \qquad (ak)^2 + (a\kappa)^2 = \frac{2ma^2V}{\hbar^2}$$

which are in the desired form (8.64).

7.79   (a)   Do the two integral expressions given in the connecting formula (7.188a) join smoothly at $x = x_1$?
   (b)   Explain your answer.

Answers

(a)   They do not join smoothly.
(b)   These two expressions represent asymptotic forms of the Airy function on either side of the turning point $x = x_1$. The (approximate) wavefunction at the turning point is the continuation of the Airy function from these asymptotic values.

7.80   Electrons in a beam have the wavefunction

$$\varphi_{\text{inc}}(x) = f(x)e^{ik_1x}$$

The beam passes through a potential barrier and electrons emerge with the wavefunction

$$\varphi_{\text{trans}} = g(x)e^{ik_2x}$$

In these expressions $f(x)$ and $g(x)$ are real functions and k_1 and k_2 are real constants. Show that

$$T = \frac{k_2}{k_1}\left|\frac{\varphi_{\text{trans}}}{\varphi_{\text{inc}}}\right|^2 = \frac{k_2}{k_1}\left|\frac{g(x)}{f(x)}\right|^2$$

7.81   A particle of mass m_0 confined to a potential well of width a_0 and depth V_0 is known to have $N_0 \gg 1$ bound states. How many bound states N does the well have if:
   (a) $a = 2a_0$ and $V = 25V_0$?
   (b) $a = a_0/2$ and $V = 4V_0$?
   (c) $m = 0.10m_0$, $V = 30V_0$, and $a = 0.577a_0$?

7.82   In physics, one requires that fundamental equations of motion be reversible in time. In quantum mechanics, the operation of *time reversibility* is given by $\psi(\mathbf{r}, t) \rightarrow \psi^*(\mathbf{r}, -t)$. Show that the quantum continuity equation (7.97) [with ρ given by (7.95) and $\mathbf{J}$ given by (7.107)] remains invariant under this operation.

7.83   Consider the plane wave

$$\psi(x, t) = A \exp i(kx - \omega t)$$

In the operation of time reversibility, what is the effect of the successive operations:
   (a) $t \rightarrow -t$?
   (b) $\psi \rightarrow \psi^*$?

Answers

(a)   This component of the transformation causes the wave to move in the reverse direction.

(b)   This component of the transformation causes momentum $\hbar k$ to reverse direction. Both of these properties come into play in classical dynamic reversibility.

7.84   (a)   Consider a particle of mass m in the complex potential field

$$\Phi = V(\mathbf{r}) + \frac{i\hbar}{2}\omega(\mathbf{r})$$

where $V(\mathbf{r})$ and $\omega(\mathbf{r})$ are real functions. What form does the continuity equation (7.97) assume for this potential?

(b)   Offer an interpretation for the field $\omega(\mathbf{r})$.

(c)   Is the new continuity equation found in part (a) time-reversible?

(d)   Is the related Hamiltonian Hermitian?

Answers (partial)

(a)   Repeating steps leading to (7.105), we obtain

$$\frac{\partial\rho}{\partial t} + \nabla \cdot \mathbf{J} = \omega\rho$$

(b)   The complex potential implies a source of particles (in violation of conservation of matter).

(c)   The equation does not obey time reversibility.

7.85   (a)   Show that the Gaussian wave packet given in Problem 7.71 satisfies the eigenvalue equation

$$\hat{a}\psi(x) = c\psi(x)$$

(b)   Under what conditions will the eigenvalue c be zero?

(c)   If conditions of part (b) are satisfied, what function does the Gaussian wave packet reduce to?

7.86   Given that [see (A.7)]

$$\langle x|\,\hat{p}\,|x'\rangle = -i\hbar\,\frac{\partial}{\partial x}\,\delta(x - x')$$

show that for free-particle motion

$$\langle x|\,e^{-it\hat{H}(\hat{x},\,\hat{p})/\hbar}\,|x'\rangle = \left[\exp-\frac{it}{\hbar}H\!\left(x, -i\hbar\frac{\partial}{\partial x}\right)\right]\delta(x - x')$$

Answer

For free-particle motion we label

$$\frac{it}{\hbar} H(\hat{x},\, \hat{p}) \equiv i\alpha\hat{p}^2$$

We may then write

$$\langle x | e^{-i\alpha\hat{p}^2} | x' \rangle = \left\langle x \left| \left[1 - i\alpha\hat{p}^2 + \frac{1}{2}(-i\alpha\hat{p}^2)^2 \right] + \cdots \right| x' \right\rangle$$

For the second term in the sum we obtain

$$\langle x | \hat{p}^2 | x' \rangle = \int dx''\, \langle x | \hat{p} | x'' \rangle \langle x'' | \hat{p} | x' \rangle$$

$$= \int dx'' \left[-i\hbar\, \frac{\partial}{\partial x}\, \delta(x - x'') \right]\left[-i\hbar\, \frac{\partial}{\partial x''}\, \delta(x'' - x') \right]$$

$$= -i\hbar\, \frac{\partial}{\partial x} \int dx''\, \delta(x - x'') \left[-i\hbar\, \frac{\partial}{\partial x''}\, \delta(x'' - x') \right]$$

$$= \left(-i\hbar\, \frac{\partial}{\partial x} \right)^2 \delta(x - x')$$

Following in this manner we obtain

$$\langle x | e^{-i\alpha\hat{p}^2} | x' \rangle = \exp\left[-i\alpha\left(-i\hbar\, \frac{\partial}{\partial x} \right)^2 \right] \delta(x - x')$$

which was to be shown.

7.87 (a) What is the value of the classical action $S_0(x_0, x)$ corresponding to free-particle motion of a particle of mass m which is at x_0 at time $t = 0$ and x at $t > 0$?

(b) Introduce an infinitesimal perturbation about the motion found in part (a) and find the new action $S(x_0, x)$. Following the steps (7.210 et seq.), show that $S > S_0$.

(c) What is the lowest-order wavefunction, $\psi(x, t)$, corresponding to the classical limit for this problem?

7.88 (a) Do the action functions $S_0, S_1, \ldots$ in the expansion (7.169) have the same dimensions?

(b) If not, what are their dimensions?

Answers

(a) They are not the same.

(b) From (7.167) we see that the dimensions of $S(x)$ are those of $\hbar$. Thus, from the series (7.169) we conclude that $[\hbar^n S_n] = \hbar$ or, equivalently, $[S_n] = \hbar^{1-n}$.

7.89 In developing the WKB expression for the transmission coefficient (7.201), it was assumed that $r \gg 1$, where r is given by (7.196a). Show that this assumption is valid within this approximation.

Answer

For slowly varying $V(x)$ of a potential barrier, one obtains $r \simeq (2a/\hbar)\sqrt{2m(V - E)}$, which, in the limit $V \gg E$, grows large. In this same limit $T \to \bar{0}$ [see (7.201)], which corresponds to the classical value. This, we recall, is the domain of validity of the WKB expansion.

7.90 An electron beam is incident on a rectangular barrier of width 5 Å and height 34.8 eV. The electrons have energy 2.9 eV. If the incident electron current is 10^{-2} cm^{-2} s^{-1}, what is the reflected current, within the WKB approximation? Is this a good approximation for the problem at hand? Why?

7.91 Again consider the configuration of Problem 7.62. However, now let the external potential be given by the following parabolic form:

$$V(x) = \left(\frac{E_F + \Phi}{x_0^2}\right)(x_0 - x)^2, \qquad 0 \le x \le x_0$$

$$V(x) = 0, \qquad\qquad\qquad x_0 \le x$$

(a) Draw the appropriate figure for this problem. Identify the Fermi energy.

(b) Calculate the transmission coefficient for emission through this potential barrier in the WKB approximation. Call your answer T_a. How does T_a compare to the transmission coefficient corresponding to the linear potential of Problem 7.26?

7.92 A particle of mass m is confined in a finite potential well of depth $|V|$ and width $2a$, as depicted in Fig. 7.27. (See also Fig. 8.1.)

(a) Employ the WKB method to obtain an expression for the bound-state eigenenergies of the particle.

(b) If

$$|V| = \left(\frac{13}{2}\right)\frac{h^2}{8m(2a)^2}$$

how many bound states $(E_n < |V|)$ does the system have?

(c) In what extreme do you expect your answer to agree with exact quantum mechanical values? (See Section 8.1)

7.93 Derive Airy's equation (7.182a) from either of the given transformations preceding (7.182a).

Answer

Let $\alpha \equiv 2\,mF/\hbar^2$, in which case the said transformation appears as

$$\alpha(x - x_1) = \alpha^{2/3}y$$

Furthermore,

$$\left(\frac{dy}{dx}\right)^2 = \alpha^{2/3}$$

Substituting these forms into (7.181) gives Airy's equation (7.182a).

7.94 Show that the transmission coefficient, $T = 1$, for the potential[1]

$$V = -V_0 \operatorname{sech}^2 (x/a)$$

where a is a scale length of the potential.

7.95 (a) Obtain an expression for the commutator $[\hat{p}, \hat{H}]$, for a one-dimensional harmonic oscillator.

   (b) The oscillator is in a state with uncertainty in momentum, $c \, \Delta p = 10^{-2}$ eV, where c is the speed of light. Given that the spring constant $K = 3.2 \times 10^8$ eV/cm^2 and $\langle x \rangle = 100$ Å, what is the related uncertainty in energy, ΔE (in eV), of the oscillator in this state?

7.96 (a) A one-dimensional current of electrons, each of energy $2\mathbb{R}$, is incident on a rectangular barrier of width $(\pi a_0/2)$ and height $\mathbb{R}$, where a_0 is the Bohr radius and $\mathbb{R}$ is the Rydberg. What is the value of the transmission coefficient, T, for this system? State units of your answer.

   (b) If particle current $\mathbb{J}_{\text{inc}} = 1.2 \times 10^4$ s^{-1} is incident on the barrier described above, what is the reflected current, $\mathbb{J}_{\text{refl}}$?

Hint: In part (a), recall $\hbar^2/2m = \mathbb{R}a_0^2$

---

[1] R. E. Crandall, *J. Phys.* **A16**, 3005 (1983).

8

FINITE POTENTIAL WELL, PERIODIC LATTICE, AND SOME SIMPLE PROBLEMS WITH TWO DEGREES OF FREEDOM

In this chapter we meet perhaps the most eminently successful application of quantum mechanics to a one-dimensional configuration. This is the problem of a charged particle in a periodic potential. When coupled with the exclusion principle for electrons, the analysis of this configuration provides a deep understanding of the process of conduction in solids. Some elementary problems in two dimensions are given, together with a discussion of degeneracy in quantum mechanics. The chapter continues with an approximation technique important to molecular and solid-state physics which carries the acronym LCAO. A concluding section describes density of states in various dimensions.

8.1   THE FINITE POTENTIAL WELL

Eigenstates

Scattering from a rectangular potential well was discussed previously in Section 7.8. The configuration is depicted again in Fig. 8.1. The scattering, unbound states correspond to a continuum of eigenenergies [e.g., (7.129)]:

$$E_k = \frac{\hbar^2 k^2}{2m}, \qquad E_k > 0$$

If we seek solutions to the Schrödinger equation for negative energies, $E < 0$, only a finite, discrete number of eigenstates are found. For the three regions depicted in Fig. 8.1, the Schrödinger equation and corresponding solutions are (for $|E| < |V|$, $E < 0$, $V < 0$):

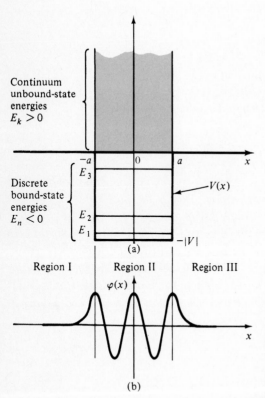

(a)

(b)

FIGURE 8.1   Finite rectangular potential well. (a) The potential function $V(x)$ and energy spectrum. (b) Typical structure of a bound eigenstate. Function oscillates in region II, where kinetic energy is positive, and decays in regions I and III, where kinetic energy is negative.

Region I: $x < -a$

$$-\frac{\hbar^2}{2m}\varphi_{xx} = -|E|\varphi, \qquad \varphi_{xx} = \kappa^2\varphi$$

(8.1)

$$\varphi_{\mathrm{I}} = Ae^{\kappa x}, \qquad \frac{\hbar^2\kappa^2}{2m} = |E| > 0$$

Region II: $-a \le x \le a$

$$-\frac{\hbar^2}{2m}\varphi_{xx} = (|V| - |E|)\varphi, \qquad \varphi_{xx} = -k^2\varphi$$

(8.2)

$$\varphi_{\mathrm{II}} = Be^{ikx} + Ce^{-ikx}, \qquad \frac{\hbar^2 k^2}{2m} = |V| - |E| > 0$$

Region III: $x > a$

$$-\frac{\hbar^2}{2m}\varphi_{xx} = -|E|\varphi, \qquad \varphi_{xx} = \kappa^2\varphi$$

(8.3)

$$\varphi_{\mathrm{III}} = De^{-\kappa x}, \qquad \frac{\hbar^2\kappa^2}{2m} = |E| > 0$$

First we note that k and κ obey the constraint

(8.4)

$$k^2 + \kappa^2 = \frac{2m|V|}{\hbar^2}$$

The coefficients A, B, C, and D determine the eigenstate corresponding to the eigenenergy $\hbar^2\kappa^2/2m$. These coefficients are determined by the continuity conditions at $x = a$, $x = -a$. Equating φ and its first derivative at these points gives

$$Ae^{-\kappa a} = Be^{-ika} + Ce^{ika}$$

$$Be^{ika} + Ce^{-ika} = De^{-\kappa a}$$

(8.5)

$$\kappa Ae^{-\kappa a} = ik(Be^{-ika} - Ce^{ika})$$

$$ik(Be^{ika} - Ce^{-ika}) = -\kappa De^{-\kappa a}$$

These are four linear, homogeneous equations for the four unknowns A, B, C, and D. They may be cast in the matrix form (where the right-hand side denotes the null column vector)

$$(8.6) \qquad \mathcal{D}\mathcal{V} \equiv \begin{pmatrix} e^{-\kappa a} & -e^{-ika} & -e^{ika} & 0 \\ 0 & e^{ika} & e^{-ika} & -e^{-\kappa a} \\ \kappa e^{-\kappa a} & -ike^{-ika} & ike^{ika} & 0 \\ 0 & ike^{ika} & -ike^{-ika} & \kappa e^{-\kappa a} \end{pmatrix} \begin{pmatrix} A \\ B \\ C \\ D \end{pmatrix} = 0$$

which serves to define the coefficient matrix $\mathcal{D}$ and the column vector $\mathcal{V}$. Cramer's rule tells us that this system has nontrivial solutions (i.e., other than $A = B = C = D = 0$) only if the determinant of the coefficient matrix vanishes.

$$(8.7) \qquad \det \mathcal{D} = 0$$

After a little manipulation (8.6) is rewritten

$$(8.8) \qquad \begin{pmatrix} G^* & G & 0 & 0 \\ G & G^* & 0 & 0 \\ e^{-ika} & -e^{ika} & -\dfrac{\kappa}{ik}e^{-\kappa a} & 0 \\ -e^{ika} & e^{-ika} & 0 & -\dfrac{\kappa}{ik}e^{-\kappa a} \end{pmatrix} \begin{pmatrix} B \\ C \\ A \\ D \end{pmatrix} = 0$$

where

$$(8.9) \qquad G \equiv (\kappa + ik)e^{ika}$$

(Note the rearrangement of the column vector $\mathcal{V}$.) Expanding about the fourth column, one obtains

$$(8.10) \quad \det \mathcal{D} = \det \begin{pmatrix} G^* & G & 0 & 0 \\ G & G^* & 0 & 0 \\ 0 & 0 & \dfrac{\kappa}{ik}e^{-\kappa a} & 0 \\ 0 & 0 & 0 & -\dfrac{\kappa}{ik}e^{-\kappa a} \end{pmatrix} = [G^2 - (G^*)^2]\left(\dfrac{\kappa}{ik}\right)^2 e^{-2\kappa a}$$

This is zero when

$$(8.11) \qquad G^2 = (G^*)^2$$

or, equivalently, when

$$(8.12) \qquad G = \pm G^*$$

Rewriting (8.9) as

$$G \equiv (\kappa + ik)e^{ika} \equiv |G|e^{i(ka+\phi)}$$

(8.13)
$$\tan \phi = \frac{k}{\kappa}$$

allows the conditions (8.12) to be recast in the form

(8.14)
$$e^{i(ka+\phi)} = \pm e^{-i(ka+\phi)}$$

The positive root gives $ka + \phi = 0$, or, equivalently,

(8.15)
$$\tan \phi = \frac{k}{\kappa} = -\tan ka$$

This may be put in the more normal form

(8.16)
$$k \cot ka = -\kappa, \qquad \frac{G}{G^*} = 1$$

The negative root gives $ka + \phi = \pi/2$ or, equivalently,

(8.17)
$$\tan \phi = \frac{k}{\kappa} = \tan\left(\frac{\pi}{2} - ka\right) = \cot ka$$

This may also be put in the more normal form

(8.18)
$$k \tan ka = \kappa, \qquad \frac{G}{G^*} = -1$$

The values of k that make det $\mathscr{D} = 0$ fall into two categories. These are the solutions to (8.16) and (8.18), respectively. From our starting matrix equation (8.8) we see that these values of k imply the relations

(8.19)
$$\frac{B}{C} = -\frac{G}{G^*} = \pm 1$$

The minus sign corresponds to the roots (8.16). Substituting this value ($B = -C$) into the last two equations of the set (8.5) gives [with (8.16)]

(8.20)
$$\frac{C}{B} = -1, \qquad \frac{A}{B} = -\frac{D}{B} = -2i \sin (ka)e^{\kappa a}$$

Substituting these values into (8.2) et seq. gives the eigenstate

(8.21)
$$\left.\begin{array}{l} \varphi_I = -2iB \sin (ka)e^{\kappa(x+a)} \\ \varphi_{II} = 2iB \sin kx \\ \varphi_{III} = 2iB \sin (ka)e^{-\kappa(x-a)} \end{array}\right\} \quad k \cot ka = -\kappa$$

This state has *odd parity;* that is,

(8.22)
$$\varphi(x) = -\varphi(-x)$$

The second class of solutions corresponds to the plus sign in (8.19) and stems from the roots (8.18). Substituting this value $(B = +C)$ into the last two equations of the set (8.5) gives

(8.23)
$$\frac{C}{B} = +1, \qquad \frac{A}{B} = \frac{D}{B} = 2 \cos (ka) e^{\kappa a}$$

The corresponding eigenstate is [with (8.18)]

(8.24)
$$\left.\begin{array}{l} \varphi_{\mathrm{I}} = 2B \cos (ka) e^{\kappa(x+a)} \\[4pt] \varphi_{\mathrm{II}} = 2B \cos kx \\[4pt] \varphi_{\mathrm{III}} = 2B \cos (ka) e^{-\kappa(x-a)} \end{array}\right\} \quad k \tan ka = \kappa$$

This state has *even parity.*

Since both eigenstates (8.21) and (8.24) are bound states, we may impose the normalization condition

(8.25)
$$\int_{-\infty}^{\infty} |\varphi|^2 \, dx = 1$$

This determines the remaining constant B. (See Problem 8.53.).

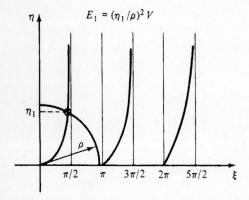

FIGURE 8.2   The curves $\eta = \xi \tan \xi$ and the circle $\xi^2 + \eta^2 = \rho^2$ for the case ρ slightly less than π. Intersection the first quadrant give bound-state eigenenergies for the pot well Hamiltonian which correspond to even eigenstates.

Next we turn to construction of the eigenenergies corresponding to the eigenstates (8.21) and (8.24). The energy is directly determined from κ. For the even eigenstates, the eigenenergies are determined from (8.4) and (8.24). Written in terms of nondimensional wavenumbers,

(8.26)
$$\xi = ka, \qquad \eta = \kappa a$$

these equations appear as

(8.27)
$$\xi \tan \xi = \eta \qquad\qquad\qquad \textit{even eigenstates}$$
$$\xi^2 + \eta^2 = \frac{2ma^2|V|}{\hbar^2} \equiv \frac{g^2}{4} \equiv \rho^2 \qquad (1 + \tan^2 \xi = 1/\cos^2 \xi)$$

For a given potential width $2a$, depth $|V|$, and particle mass m, (8.27) describes a circle of radius ρ, in Cartesian $\xi\eta$ space. The intersections of this circle (in the first quadrant) with the graph of the first equation of (8.27) determine the eigenenergies corresponding to the even eigenstates (8.24). This graphical technique is sketched in Fig. 8.2 for the case ρ slightly less than π. The sketch tells us that for this value of ρ, the finite potential well has only one bound even eigenstate.

The eigenenergies of the odd eigenstates (8.22) are the intersections of the two curves

(8.28)
$$\xi \cot \xi = -\eta \qquad \textit{odd eigenstates}$$
$$\xi^2 + \eta^2 = \rho^2 \qquad (1 + \cot^2 \xi = 1/\sin^2 \xi)$$

These curves are sketched in Fig. 8.3 for the case ρ slightly less than π. For this choice

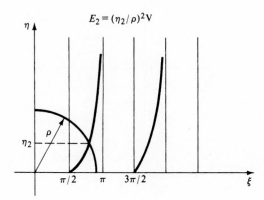

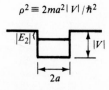

$$\rho^2 \equiv 2ma^2|V|/\hbar^2$$

FIGURE 8.3 The curves $\eta = -\xi \cot \xi$ and the circle $\xi^2 + \eta^2 = \rho^2$ again for the case ρ slightly less than π. Intersections in the first quadrant give bound-state eigenenergies for the potential well Hamiltonian which correspond to odd eigenstates. Note that E_2 lies higher than the ground state E_1.

of data, we see that there is only one bound odd eigenstate. These two lowest-energy eigenstates, (8.21) and (8.24), are sketched in Fig. 8.4.

At this point we wish to consider again the difference between the unbound scattering states of Chapter 7 and the bound states just encountered. The continuity conditions on the wavefunction φ and its derivative, together with the statement of conservation of energy, determine eigenenergies and eigenstates. For scattering states, the continuity conditions (7.139) are in the form of an *inhomogeneous* matrix equation,

$$(8.29) \qquad \mathscr{D}(k_1)\mathscr{V} = \mathscr{U}$$

where, for example, the column vector $\mathscr{V}$ is

$$(8.30) \qquad \mathscr{V} = \begin{pmatrix} B/A \\ C/A \\ D/A \\ F/A \end{pmatrix}$$

The solution to (8.29) is

$$(8.31) \qquad \mathscr{V} = \mathscr{D}^{-1}(k_1)\mathscr{U}$$

For these unbound scattering states, conservation of energy serves only to relate wavenumbers connected to distinct potential domains. The eigen-k_1-values comprise

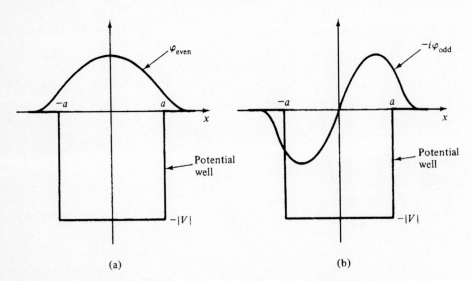

FIGURE 8.4 First two bound eigenstates for the potential well problem. For $2ma^2|V|/\hbar^2 < \pi^2$, these are the only bound states.

a continuum. For each such k_1 value, there corresponds an eigenstate of the form (7.138).

For bound states the continuity conditions (8.5) are in the form of a *homogeneous* matrix equation (8.6),

$$\mathscr{D}(\kappa)\mathscr{V} = 0$$

which has nontrivial solutions ($\mathscr{V} \neq 0$) only if

(8.32) $$\det \mathscr{D}(\kappa) = 0$$

This dispersion relation restricts the eigen-κ-values to values that obey certain transcendental relations [the first equation in (8.27) and in (8.28)]. In addition, κ is further restricted by the conservation of energy statement—namely, the second equation in (8.27). The intersections of this circle (depicted in Figs. 8.2 and 8.3) with the said transcendental curves generate a discrete spectrum of eigenenergies

$$E_n = -\frac{\hbar^2 \kappa_n^2}{2m}$$

Let us consider the time dependence of the eigenstates corresponding to the finite potential well. The bound time-dependent eigenstates appear as

(8.33) $$\psi_n(x, t) = \varphi_n(x)e^{-iE_n t/\hbar}, \qquad E_n < 0$$

with $\varphi_n(x)$ given by (8.1) et seq. For positive energy, the unbound time-dependent eigenstates form a continuum,

(8.34) $$\psi_{k_1}(x, t) = \varphi_{k_1}(x)e^{-iE_{k_1} t/\hbar}, \qquad E_{k_1} > 0$$

where $\varphi_{k_1}(x)$, for example, is of the form (7.138) with the modification $V \rightarrow -|V|$.

To employ the superposition principle in problems relating to the finite potential well, one must call on the finite number of bound states and infinite continuum of unbound states.[1]

The $E = 0$ Line

As stated above, energies relevant to the finite potential well are directly obtained from κ or, equivalently, η:

$$|E| = \frac{\hbar^2 \kappa^2}{2m} = \frac{\hbar^2 \eta^2}{2ma^2}$$

---

[1] Note the continuum of unbound states developed in this chapter excludes states with negative k in region III. The states discussed are appropriate to the superposition of a wave packet incident on a potential barrier from the left. For the superposition of a state with zero average momentum, one must include the negative k waves in region III.

These energies are measured with respect to the top of the well taken as the $E = 0$ line. It is sometimes convenient to measure energies with respect to the bottom of the well as the zero energy line. The energies, E', measured with respect to the bottom of the wall are directly obtained from k or, equivalently, ξ:

$$E' = |V| - |E| = \frac{\hbar^2 k^2}{2m} = \frac{\hbar^2 \xi^2}{2ma^2} > 0$$

See (8.2) and Fig. 8.5.

In either the modeling of one-electron atoms or the analysis of the present finite potential well, energy measured from the top of the well, $|E| \propto \eta^2$, represents binding energy. The energy $E' \propto \xi^2$ measures energy from the bottom of the well and, in accord with the infinite potential well (Section 4.1), is representative of the eigenenergy of the particle. This representation is also relevant to the modeling of quantum wells in superlattice structures where the $E' = 0$ line might, for instance, be set equal to the minimum energy of the conduction band. (See Sections 8.8 and 12.9.)

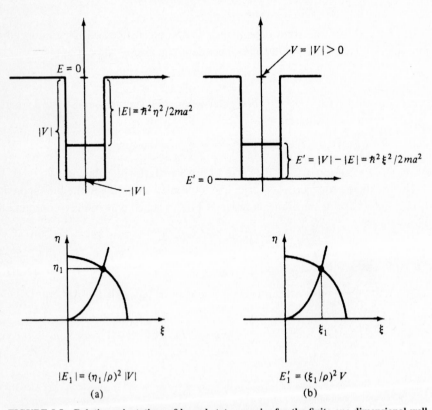

FIGURE 8.5   **Relative orientations of bound-state energies for the finite one-dimensional well.**

PROBLEMS

8.1   A deuteron, which is a neutron and a proton bound together, has only one bound state. Assume that the potential of interaction between the two particles may be described as a square well. The effective mass of the system is 0.84×10^{-24} g. The range of nuclear force is approximately 2.3×10^{-13} cm, while the ground state of the deuteron is 2.23 MeV below the zero-energy free-particle state. Assuming that only the odd-parity solutions are permitted for this case, estimate the depth of the potential well, $|V|$, which you may take to be large compared to the binding energy of the system.

8.2   An electron is confined to a potential well of finite depth and width, 10^{-9} cm. The eigenstate of highest energy of this system corresponds to the value $\xi = 3.2$.

   (a)   How many bound states does this system have?

   (b)   Estimate the energy of the ground state with respect to the zero energy line at the bottom of the well. Express your answer in eV.

8.3   Show that the graphical solutions of Figs. 8.2 and 8.3 give the eigenenergies of a one-dimensional box, in the limit that the well becomes infinitely deep.

Answer

In the said limit $\rho \to \infty$, the circles of constant ρ intersect the tan and cot curves on the vertical asymptotes

$$\xi = \frac{n\pi}{2} \qquad (n = 1, 2, \ldots)$$

[Compare (4.12). See also Problem 8.58.]

8.4   Given that

$$\frac{g^2}{4} = \frac{2ma^2|V|}{\hbar^2} = \left(\frac{7\pi}{4}\right)^2$$

for an electron in a potential well of depth $|V|$ and width $2a = 10^{-7}$ cm, if a 100-keV neutron is scattered by such a system, calculate the possible decrements in energy that the neutron may suffer.

8.5   For the potential well described in Problem 8.4, what is the parity of the eigenstate of maximum energy? How many zeros does this state have?

8.6   Consider a rectangular potential well of depth $|V|$ and width $2a$, such that $2ma^2|V|/\hbar^2 = (8\pi/18)^2$. The lowest-energy normalized bound state, φ_1, has wavenumber $k \simeq \pi/4a$. Let $\tilde{\varphi}$ be a wavefunction that is a square wave of height $1/\sqrt{4a}$ and width $4a$. The centers of $\tilde{\varphi}$ and the rectangular potential well are coincident. At $t = 0$ a particle of mass m is in the state

$$\psi(x, 0) = \frac{3\varphi_1 + 4\tilde{\varphi}}{5}$$

At time $t = 0$:

(a) What is the expectation of momentum of the particle?

(b) What is the expectation of energy?

(c) What is the parity of the state?

(d), (e), (f) Repeat parts (a), (b), and (c) for $t > 0$.

8.7 Consider the semi-infinite potential well

$$V(x) = \begin{cases} \infty, & x < 0 \\ -|V|, & 0 \le x \le a \\ 0, & a < x \end{cases}$$

(see Fig. 8.6).

(a) Using the solutions to the finite potential well (width $2a$) developed in the text, sketch the first three eigenfunctions of lowest energy for a particle in this well.

(b) Which ground-state energy is lower—that of the finite potential well (width $2a$) or that of the semi-infinite well (width a)?

(c) Are the eigenfunctions you have sketched eigenstates of the Hamiltonian appropriate to the finite potential well?

8.8 An electron is trapped in a rectangular potential well of width 3 Å and depth 1 eV. What are the possible frequencies of emission of this system (in hertz)?

8.9 Establish the following criteria for the number of bound states in a finite potential well:

(a) $(n\pi)^2 < \rho^2 < (n + 1)^2 \pi^2 (n + 1$ symmetric states$)$.

(b) $(n - \frac{1}{2})^2 \pi^2 < \rho^2 < (n + \frac{1}{2})^2 \pi^2 (n$ antisymmetric states$)$.

(c) Total number of bound states = maximum integer (ρ/π).

(d) Show that $\rho^2 = \bar{a}^2 |\bar{V}|$, where $\bar{a} \equiv a/a_0$, $\bar{V} \equiv V/\mathbb{R}$. (We recall that a_0 represents the Bohr radius and $\mathbb{R}$ represents the Rydberg.)

(e) Show that $E' = \hbar^2 \xi^2 / 2ma^2 = \mathbb{R}\xi^2 / \bar{a}^2$.

8.2 PERIODIC LATTICE. ENERGY GAPS

In this section we consider the problem of a particle in a periodic potential. This is of extreme practical importance in the theory of conduction and insulation in solids.

Consider the simple model of a solid (more precisely, a metal) in which the positive ions comprise a uniform array of fixed sites. The valence electrons are assumed to be free. They are the conduction electrons. For sodium, for instance, there is one free electron per ion. Each such electron finds itself in a periodic potential supported by the ions. Such a one-dimensional potential configuration is depicted in Fig. 8.7.

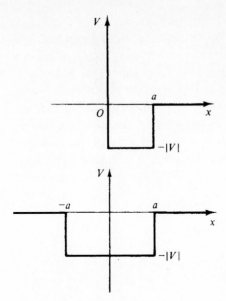

FIGURE 8.6   **Semi-infinite potential well and its companion finite potential well. (See Problem 8.7.)**

If the distance between sites is d, then inside the metal the potential is periodic in the distance d.

(8.35) $$V(x) = V(x + d)$$

A simple potential function that maintains this periodic quality and all the salient properties of the more realistic potential sketched in Fig. 8.7 is the *Kronig–Penney*

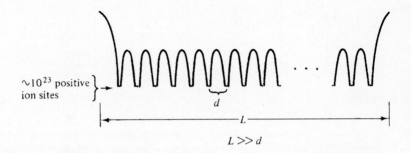

FIGURE 8.7   **Periodic potential that an electron sees in a one-dimensional crystalline solid.**

potential, depicted in Fig. 8.8. The periodic property of $V(x)$ as given by (8.35) fails at the ends of the lattice. To remove this difficulty the model is further simplified. This simplification derives from the fact that there are an overwhelmingly large number of ion sites in the length of the sample. The change in the character of the potential at the ends of the sample is therefore relatively unimportant to the transport properties of an interior electron. For this reason we change the ends of the sample to best facilitate analysis. It is assumed that when an electron leaves the end of the sample, it reenters the front of the sample. This idea is best realized if the one-dimensional potential function is assumed to lie on a circle of radius r which is very large compared to the distance between ion sites, d (see Fig. 8.9). The Hamiltonian for an electron in this potential is

$$H = \frac{p^2}{2m} + V(x)$$

(8.36)

$$V(x) = V(x + d)$$

Bloch Wavefunctions

To find the eigenfunctions of this Hamiltonian, we first recall the displacement operator $\hat{\mathcal{D}}$, introduced in Problem 3.4:

(8.37)
$$\hat{\mathcal{D}}f(x) = f(x + d)$$

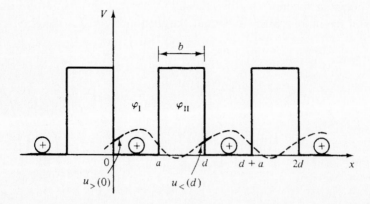

FIGURE 8.8   The Kronig–Penney model for a potential due to fixed ion sites separated by the distance d. The dashed curve represents a hypothetical periodic u component of the Bloch function $\varphi = u(x) \exp{(ikx)}$. The eigenfunction φ (8.48) et seq. and corresponding dispersion relations, (8.53) and (8.55), are obtained by matching u and u' at $x = 0 + \varepsilon$ to their respective values at $x = d - \varepsilon$ and matching φ and φ' across the potential barrier at $x = a$.

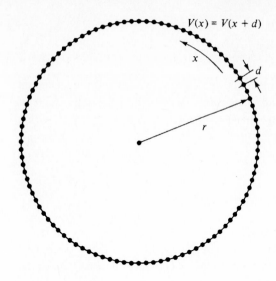

$$V(x) = V(x + d)$$

FIGURE 8.9   Ring model of a one-dimensional periodic potential. Black dots represent positive ion sites. For N sites in all, and $N \gg 1$, $Nd \simeq 2\pi r$.

The eigenfunctions of this operator are

(8.38)
$$\varphi = e^{ikx}u(x)$$

$$u(x) = u(x + d)$$

with k arbitrary. The eigenvalue of $\hat{\mathcal{D}}$ corresponding to φ is exp (ikd). Although both factors of φ, namely, exp (ikx) and $u(x)$, are periodic, φ need not be. The eigenfunction $\varphi(x)$ is periodic if d, the period of u, is commensurate with $2\pi/k$, the period of exp (ikx)—that is, if $2\pi/kd$ is a rational number.

Since $\hat{\mathcal{D}}$ commutes with $\hat{H}$

(8.39)
$$[\hat{\mathcal{D}}, \hat{H}] = 0$$

these two operators have common eigenfunctions. We conclude that the eigenfunctions of the Hamiltonian (8.36) are of the form (8.38). These functions are called *Bloch wavefunctions*. The related theorem that the eigenstates of a periodic Hamiltonian such as (8.36) are in the product form (8.38) is called *Bloch's theorem*.[1] We have obtained these functions using the displacement operator $\hat{\mathcal{D}}$. More simply, one may argue that, on the average, the density of an electron beam propagating through a crystal with a periodic potential should exhibit the same periodicity as the crystal. That is, one expects that

$$|\varphi(x)|^2 = |\varphi(x + d)|^2$$

[1] F. Bloch, Z. *Physik* **52** (1928).

This equation admits the solutions

$$\varphi(x) = u(x) \exp\left[i\alpha(x)\right]$$

where, again, $u(x)$ is periodic with period d and $\alpha(x)$ is any real function independent of d. In the limit that the periodic potential becomes constant, $V = $ constant, $d \to \infty$, and the wavefunction $\varphi(x)$ becomes the free-particle wavefunction $\exp\,(ikx)$, with k arbitrary but real. Since $\alpha(x)$ is independent of the period length (or *lattice constant*) d, this value of α (i.e., kx) is its value for all d and we again obtain the Bloch wavefunction

$$\varphi(x) = e^{ikx}u(x)$$

The shape of this wavefunction suggests the manner in which the crystal structure influences the wavefunctions of particles propagating through the crystal. This structure is primarily contained in the periodic factor $u(x)$, which in turn includes the lattice constant d and which modulates the free-particle form, *exp (ikx)*.

Another way of writing (8.38) is

(8.40)
$$\varphi(x + d) = e^{ikd}\varphi(x)$$
$$\varphi(x) = e^{ikd}\varphi(x - d)$$

If the eigenstate φ is known over any cell in the periodic lattice (more generally over any interval of length d), equations (8.40) generate the values of φ in all other cells.

For any value of k, the corresponding function φ, given by (8.38), is an eigenstate of $\hat{\mathcal{D}}$. When φ is also an eigenstate of $\hat{H}$, the values that k may assume become restricted. For example, the eigenstates of $\hat{H}$, with V defined over a ring, have the property

(8.41)
$$\varphi(x) = \varphi(x + Nd)$$

Substitution into (8.38) gives

(8.42) $e^{ikNd} = 1,$ $kNd = 2n\pi$ $(n = 0, \pm 1, \pm 2, \ldots)$

This implies that the allowed values of k form a discrete spectrum $[k_n = n(2\pi/L)]$. However, since N is very large (e.g., $N \simeq 10^8$), the difference between successive values of k is very small and the spectrum of the permitted values of k may be taken to comprise a continuum (see Fig. 8.10). With k restricted to the values given by (8.42), the ratio $2\pi/kd = N/n$, a rational number. It follows that for the closed-ring periodic potential, the eigenfunctions of $\hat{H}$ in the Bloch waveform (8.38) are periodic.

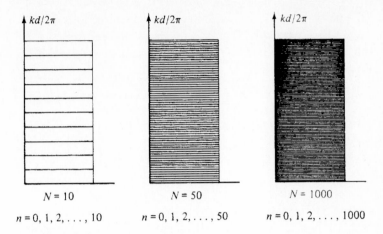

$$N = 10 \qquad\qquad N = 50 \qquad\qquad N = 1000$$
$$n = 0, 1, 2, \ldots, 10 \qquad n = 0, 1, 2, \ldots, 50 \qquad n = 0, 1, 2, \ldots, 1000$$

FIGURE 8.10 Permitted values of k for the periodic ring model depicted in Fig. 8.9. For $N \gg 1$ the spectrum of permitted k values approximates a continuum.

The Quasi-momentum

The variable $\hbar k$ is called the *quasi-momentum* of the particle. We list four of the properties of $\hbar k$ which motivate this name.

1. The eigenstates given in (8.38) resemble the form

    (8.43) $$\varphi_k = e^{ikx} \times \text{constant}$$

    This is the momentum eigenfunction of a free particle with momentum $\hbar k$. The momentum of an electron in a periodic lattice is, of course, not constant due to the lattice's space-dependent potential field. Nevertheless, there is a constant value of $\hbar k$ associated with every eigenenergy of the Hamiltonian (8.36).

2. The group velocity of an electron wave packet in a given band is given by

    (8.44) $$v_g = \frac{\partial E(k)}{\partial \hbar k}$$

    where $E(k)$ represents energy values in the band.[1] The relation above follows the classical recipe for obtaining the velocity of a free particle with energy E, provided that we associate $\hbar k$ with its momentum.

---

[1] These properties of the quasi-momentum $\hbar k$ are derived in L. D. Landau and E. M. Lifshitz, *Quantum Mechanics*, 2d ed., Addison-Wesley, Reading, Mass., 1965.

3. If a particle in a lattice is acted upon by an outside force **F**, its acceleration is not **F**/m, but **F**/m^*. The "effective mass" m^* may be less than m, greater than m, negative, and even infinite. In one dimension m^* is given by

(8.45)
$$m^* = \frac{\hbar^2}{\partial^2 E/\partial k^2}$$

which is suggestive of the classical relation for a free particle $E = p^2/2m^*$, again with $p = \hbar k$. (See Problem 8.57.)

4. Eigenenergies $E(k)$ are periodic in k with period $2\pi/d$, so

$$E(k) = E(k + 2\pi n/d),$$

where n is a positive or negative integer. The "central" $E(k)$ curve lies near the parabola $E = \hbar^2 k^2/2m$, which again suggests a free particle with momentum $\hbar k$.

Eigenstates

Next we turn to construction of the eigenstates and eigenenergies of the Kronig–Penney Hamiltonian. We know that eigenstates are in the Bloch form (8.38). The continuity conditions that apply to $\varphi(x)$ clearly apply also to the periodic component $u(x)$. It follows that $u(x)$ and $u'(x)$ must vary continuously from the right side of the point $x = 0$ to the left side of the point $x = d$, which is one periodic length displaced from $x = 0$ (see Fig. 8.8). With $u_>(0)$ denoting $u(x)$ evaluated at $x = 0 + \epsilon$, where ϵ is an infinitesimal, this condition on the periodic continuous quality of u and u' gives the two equations

(8.46a) $u_>(0) = u_<(d)$

(8.46b) $u_>'(0) = u_<'(d)$

Now

$$u = e^{-ikx}\varphi(x)$$

so that

$$u' = \varphi'e^{-ikx} - iku$$

and the continuity of u' (8.46b) across a periodic length becomes

(8.47a) $\varphi_>(0) = \varphi_<(d)e^{-ikd}$

(8.47b) $\varphi_>'(0) = \varphi_<'(d)e^{-ikd}$

In the well domain of the potential array

$$\varphi_{\mathrm{I}}(x) = Ae^{ik_1x} + Be^{-ik_1x} \qquad (0 \le x \le a)$$

(8.48)

$$\frac{\hbar^2 k_1^2}{2m} = E$$

In the barrier domain (with $E > V$)

$$\varphi_{\mathrm{II}}(x) = Ce^{ik_2x} + De^{-ik_2x} \qquad (a \le x \le a + b = d)$$

(8.49)

$$\frac{\hbar^2 k_2^2}{2m} = E - V$$

The periodicity conditions (8.46, 8.47) on $u(x)$ then become

(8.50)

$$A + B = e^{-ikd}(Ce^{ik_2d} + De^{-ik_2d})$$

$$k_1(A - B) = k_2 e^{-ikd}(Ce^{ik_2d} - De^{-ik_2d})$$

The remaining two equations for the four coefficients (A, B, C, D) are obtained from continuity of $\varphi(x)$ and $\varphi'(x)$ across the potential barrier at $x = a$. This gives

(8.51)

$$Ae^{ik_1a} + Be^{-ik_1a} = Ce^{ik_2a} + De^{-ik_2a}$$

$$k_1(Ae^{ik_1a} - Be^{-ik_1a}) = k_2(Ce^{ik_2a} - De^{-ik_2a})$$

The latter four equations may be rewritten in the matrix notation

$$\begin{pmatrix} 1 & 1 & -e^{id(k_2-k)} & -e^{-id(k_2+k)} \\ k_1 & -k_1 & -k_2 e^{id(k_2-k)} & k_2 e^{-id(k_2+k)} \\ e^{ik_1a} & e^{-ik_1a} & -e^{ik_2a} & -e^{-ik_2a} \\ k_1 e^{ik_1a} & -k_1 e^{-ik_1a} & -k_2 e^{ik_2a} & k_2 e^{-ik_2a} \end{pmatrix} \begin{pmatrix} A \\ B \\ C \\ D \end{pmatrix} = 0$$

With $\mathcal{D}$ representing the above 4×4 coefficient matrix and $\mathcal{V}$ the four-column vector, the preceding equation may be written

$$\mathcal{D}(k, k_1, k_2)\mathcal{V} = 0$$

This homogeneous equation has nontrivial solutions only if

(8.52) $$\det \mathcal{D} = 0$$

This is the desired dispersion relation which is seen to involve the propagation constant k and the wavenumbers k_1 and k_2. The latter two variables contain the energy (8.48, 8.49), so for a given value of k, the dispersion relation (8.52) determines the eigenenergy E. As will be shown, this dispersion relation also exhibits the band-gap quality of the energy spectrum attendant to all periodic potentials. The dispersion relation (8.52) is similar to (8.7), which gives the eigenenergies for the bound states of the potential well problem. As the domain of existence of the wavefunctions of $\hat{H}$ is over the finite interval $0 \leq x \leq Nd$, these wavefunctions are labeled "extended states." Our main goal is to obtain the energies of these states.

From (8.52) one obtains the dispersion relation (after a bit of algebra)

$$E > V$$

(8.53a) $$\cos k_1 a \cos k_2 b - \frac{k_1^2 + k_2^2}{2k_1 k_2} \sin k_1 a \sin k_2 b = \cos kd$$

(8.53b) $$k_1^2 - k_2^2 = \frac{2mV}{\hbar^2}$$

The related formula for the case $E < V$ is simply obtained from the latter relation through the substitution

(8.54) $$ik_2 \rightarrow \kappa, \qquad \frac{\hbar^2 \kappa^2}{2m} = V - E$$

There results

$$E < V$$

(8.55a) $$\cos k_1 a \cosh \kappa b - \frac{k_1^2 - \kappa^2}{2k_1 \kappa} \sin k_1 a \sinh \kappa b = \cos kd$$

(8.55b) $$k_1^2 + \kappa^2 = \frac{2mV}{\hbar^2}$$

Equations 8.53 and 8.55 are implicit equations for the eigenenergies E as a function of the propagation constant k, valid for all energies. Owing to the transcendental nature of these equations, one turns to a numerical technique for obtaining $E(k)$. For example, consider that $n = 2$, $N = 1000$. Then the right-hand side of (8.55a) is $\cos (4\pi/10^3) \simeq 1$. One then plots the left side of the same equation as a function of the dimensionless energy E/V. Superimposed on this same curve is the line RHS $= 1$ (Fig. 8.11). The values where these curves cross give the eigenenergies $E(k)$.

Energy Gaps

The fact that values of the right-hand sides of both (8.53a) and (8.55a) lie between $+1$ and -1 ($|\cos kd| \leq 1$) implies that the only solutions to these equations are values

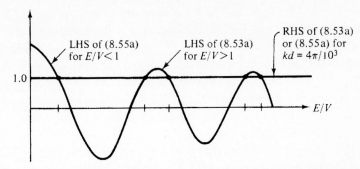

FIGURE 8.11 Graphical evaluation of eigenenergies of the Kronig–Penney Hamiltonian corresponding to $kd = 4\pi/10^3$. Eigenenergies are given by intersections of the horizontal line and the oscillating curve.

of E for which the left-hand sides of these respective equations fall in the same interval, that is, values of E for which

$$(8.56) \qquad -1 \leq [\text{left-hand sides of (8.53a) and (8.55a)}] \leq +1$$

Values of E that violate this condition are excluded from the energy spectrum.

The condition (8.56) gives rise to a "band" structure for the spectrum of eigenenergies. This is again well exhibited with a diagram. In Fig. 8.12, the left-hand sides of (8.53a) and (8.55a) are plotted versus E/V. On the same graph we draw the lines that represent the constant ordinates, $+1$ and -1. The values of E that qualify as eigenenergies are values for which the oscillating curve falls between the two horizontal lines, $+1$ and -1.

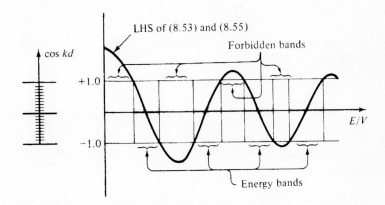

FIGURE 8.12 Band structure of the energy spectrum of the Kronig–Penney Hamiltonian. The only eigenenergies are those values for which the left-hand side of (8.53a) or (8.55a) falls between ±1.

This construction illustrates the band property of the energy spectrum of a particle in a periodic potential. This band feature is also illustrated in a plot of E versus k which may be inferred from the graph of Fig. 8.12. At the left of Fig. 8.12, values of cos kd are marked off. If a horizontal line is drawn from one of these values (e.g., cos $kd = 1/\sqrt{2}, kd = \pm\pi/4$), the intersections of this line with the oscillating curve give all the energies that correspond to the propagation-constant values, $k = \pm\pi/4d$. There are infinitely many of them. Continuing this process for all values of kd gives the curve of E versus k sketched in Fig. 8.13a.

If we look at any single band, the curve E versus k is periodic in k. This results from the fact that the right-hand sides of (8.53a) and (8.55a) maintain the same values if kd is replaced by $(kd + n2\pi)$, where n is an integer. The value of E that satisfies this equation [i.e., either (8.53) or (8.55)] for a given value of kd satisfies it for $(kd + n2\pi)$. It suffices then to draw all bands in the single interval $-\pi \leq kd \leq \pi$. This gives the *reduced-zone* description (Fig. 8.13b) of eigenstates. These bands consist of very closely packed discrete energies (recall Fig. 8.10) and constitute all the eigenenergies of the Hamiltonian (8.36). This discrete nature of the energy spectrum is a consequence of the boundedness of the system. The quasi-continuous quality (bands of closely packed levels) of the spectrum reflects the propagating nature of the eigenstates.

Superimposed on the E versus k curves in Fig. 8.13a are the free-particle energy curves

$$(8.57) \qquad E = \frac{\hbar^2(k + n2\pi d^{-1})^2}{2m} \qquad (n = \pm 0, 1, 2, \ldots)$$

This corresponds to a free-particle momentum

$$(8.58) \qquad p = \hbar(k + n2\pi d^{-1})$$

From Fig. 8.13 we see that (1) much of the locus of the E versus k curves falls near the free-particle energy curves, and (2) energy gaps occur at the values

$$(8.59) \qquad kd = q\pi$$

where q is a positive or negative integer. At these values of k an integral number of half-wavelengths span the distance d between ions.

Bragg Reflection

To understand the physical origin of energy gaps at these values of k, it is best to recall that the one-dimensional solution we have found is appropriate to propagation of plane waves through slabs of constant potential V, thickness b, and spaced a normal

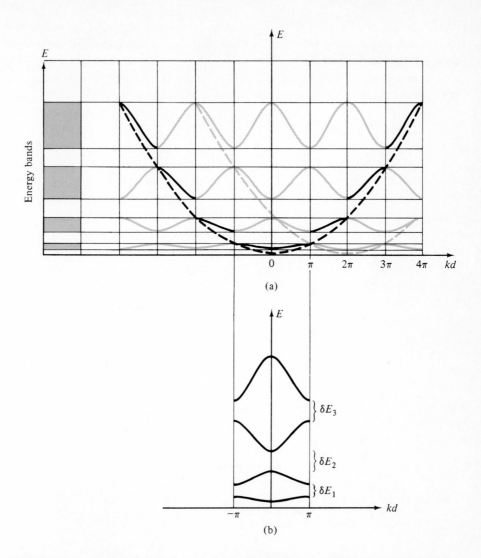

FIGURE 8.13   (a) Typical E versus k curves for the Kronig–Penney potential. The graininess of the curves stems from the fact that the k values in each band are discrete [i.e., $kd = (n/N)2\pi$, $N \gg 1$ or, equivalently, $(\Delta k)_{\min} = \pi/L$]. (b) The first four bands in the reduced-zone scheme. Also shown are the first three energy gaps, δE_1, δE_2, and δE_3. Consider that the second and third bands shown in (b) are identified as the valence and conduction bands, respectively. At $k = 0$, one finds that $m_e^* > 0$ and $m_h^* < 0$. This effective-mass property is typical in semiconductors. One notes further that effective mass is consistently defined in the neighborhood of extrema points in E versus k curves (i.e., points where $v_g = dE/dk = 0$). For further discussion see K. Hess, *Advanced Theory of Semiconductor Devices*, Prentice-Hall, Englewood Cliffs, N.J., 1988.

distance d from one another. This situation is depicted in Fig. 8.14, in which two typical wavefronts are also sketched. Suppose that these plane waves are incident at the angle θ, such as is drawn in Fig. 8.15. In the Bragg model of reflection, wavefronts scatter from ion-site lattice planes. We recall that the condition that reflected waves from adjacent planes add constructively is given by Bragg's formula,

$$(8.60) \qquad\qquad kd \cos \theta = q\pi$$

where θ is the angle that the incident $\mathbf{k}$ vector makes with the normal to the lattice plane. In the limit of normal incidence, $\theta \to 0$, and the one-dimensional model of the analysis above becomes relevant. Equation (8.60) reduces to the condition (8.59), which k satisfies at an energy gap. At these values of k, an integral number of wavelengths fit the distance $2d$ and the reflected waves constructively superpose. They are *Bragg-reflected*. Consider that a wave is moving in one direction with a critical k value (8.59). It is soon Bragg-reflected and propagates in the opposite direction. There is a similar reversal of direction of propagation on each reflection until finally the only steady-state solution is that which contains an equal number of waves traveling in either direction. As will be shown in the next section, at these critical values of k, the eigenstates of $\hat{H}$ are composed equally of waves moving to the right and left, so that, for example, in the well domain of the periodic potential, $\varphi \sim \exp{(ik_1 x)} + \exp{(-ik_1 x)} \sim \cos{(k_1 x_1)}$, which is the spatial component of a standing wave. A similar standing-wave structure prevails across the barrier domain. When these solutions are matched at the potential steps, a standing wave ensues over the whole periodic potential. In such states $\langle p \rangle = 0$. Electrons are trapped and lose their free-particle quality. Energy curves appropriate to a one-dimensional periodic potential are shown in Fig. 8.13b.

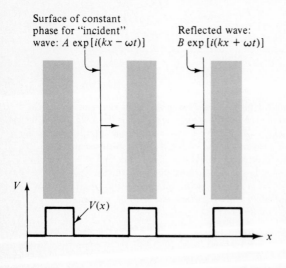

Surface of constant phase for "incident" wave: $A \exp{[i(kx - \omega t)]}$

Reflected wave: $B \exp{[i(kx + \omega t)]}$

V

$V(x)$

x

FIGURE 8.14 Shaded regions depict domains of constant potential. They are slabs that extend out of the paper. Surfaces of constant phase are also normal to the plane of the paper.

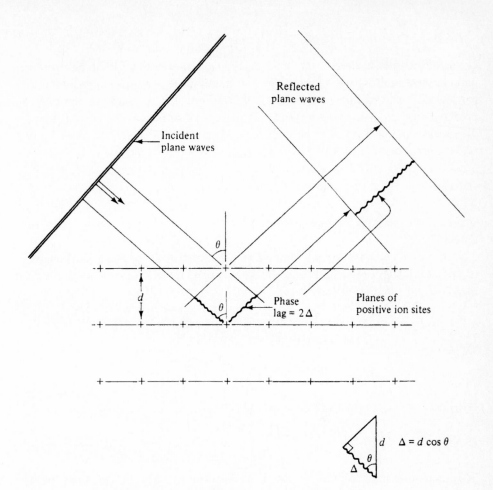

FIGURE 8.15 **Constructive interference between reflected waves from different planes occurs when**

$$2d \cos \theta = l\lambda = \frac{2\pi l}{k}$$

which gives

$$kd \cos \theta = l\pi$$

For normal incidence, $\theta = 0$, and this condition becomes

$$kd = l\pi$$

Spreading of the Bound States

Let us now demonstrate that the band-energy spectrum relevant to an electron in a periodic potential collapses to the discrete bound-state spectrum appropriate to a particle in a single finite potential well in the limit that the wells of the periodic potential grow far apart ($b \to \infty$). Toward these ends it suffices to demonstrate that the dispersion relation (8.55), for the states $E < V$ for a periodic potential, in the limit of $b \to \infty$, gives the relations (8.27) and (8.28):

$$(8.61) \qquad \tan \xi = \frac{\eta}{\xi}, \qquad \tan \xi = \frac{-\xi}{\eta}$$

These are the relations for the even and odd states, respectively, of a particle in a finite potential well.

Let us recall that the nondimensional parameters ξ, η, and ρ introduced in (8.26) and (8.27) contain half the well width, which for the Kronig–Penney potential is $a/2$, so for the present case

$$(8.62) \qquad \xi = \frac{k_1 a}{2}, \qquad \eta = \frac{\kappa a}{2}$$

$$\xi^2 + \eta = \rho^2 = \frac{2m(a/2)^2 V}{\hbar^2}$$

In terms of these variables, (8.55a) becomes

$$(8.63) \qquad \cos 2\xi \cosh \left(\frac{2b\eta}{a} \right) + \frac{\eta^2 - \xi^2}{2\eta\xi} \sin 2\xi \sinh \left(\frac{2b\eta}{a} \right) = \cos kd$$

This equation must reduce to the two equations (8.61) in the limit $b \to \infty$, $a = $ constant, $V = $ constant. First we note that in this limit

$$\cosh \left(\frac{2b\eta}{a} \right) \simeq \sinh \left(\frac{2b\eta}{a} \right) \simeq \frac{1}{2} \exp \left(\frac{2b\eta}{a} \right)$$

Dividing through by this exponential factor and allowing b to grow infinitely large reduces (8.63) to the form

$$(8.64) \qquad \tan 2\xi = \frac{2\xi\eta}{\xi^2 - \eta^2}$$

The double-angle formula for tangents permits this equation to be rewritten as

$$\frac{\tan \xi}{1 - \tan^2 \xi} = \frac{\xi\eta}{\xi^2 - \eta^2}$$

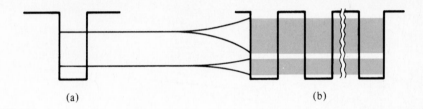

FIGURE 8.16 (a) Single isolated finite potential well with two bound states. (b) Corresponding periodic potential with two energy bands. For N wells each band contains N states.

which in turn may be rewritten as

$$\tan^2 \xi + \frac{\xi^2 - \eta^2}{\xi\eta} \tan \xi - 1 = 0$$

This is a quadratic equation for $\tan \xi$. Solving for the two roots gives

$$2 \tan \xi = \frac{\eta^2 - \xi^2}{\xi\eta} \pm \frac{\eta^2 + \xi^2}{\xi\eta}$$

These are the two relations (8.61) that give the discrete bound states of a single isolated finite potential well.

Thus we find that the band structure of the energy spectrum of a particle in a periodic potential collapses to the discrete energy spectrum of a particle in a finite potential well in the limit that the wells of the periodic array became far removed from one another. Consider, for instance, a finite potential well that has two bound states. Such, for example, in the case if $\rho = 5\pi/4$ (see Fig. 8.2). For a periodic array of such potentials, the relation (8.55) applies in the domain $E < V$. If the left-hand side of this equation is plotted versus E such as in Fig. 8.12, two bands will be found to fall in the domain $E < V$. This transition[1] from the discrete states of an isolated well to the band structure of a lattice is illustrated in Fig. 8.16.

PROBLEMS

8.10 (a) What is the expectation of momentum for an electron propagating in a Bloch wavefunction with spatial component

$$\varphi(x) = e^{ikx} u(x)?$$

(b) Show that if the periodic function $u(x)$ is real, $\langle p \rangle = \hbar k$.

[1] Details of a numerical analysis for this transition for the case of a well with four bound states may be found in V. Rojansky, *Introductory Quantum Mechanics,* Prentice-Hall, Englewood Cliffs, N.J., 1938.

Answer (partial)

(a) $\langle p \rangle = \hbar k + \langle u | \hat{p} | u \rangle$

8.11 What is the period of the Bloch wavefunction under the following conditions?

(a) $kd = 2l\pi$

(b) $kd = (2l + 1)\pi$

(c) $kd = n\pi/q$

Here l, n, and q are integers.

8.12 (a) Use the dispersion relation (8.55) to obtain the dispersion relation for the propagation of electrons through an infinite array of equally spaced delta-function potentials[1] separated by d cm (see Fig. 8.17). Note that the delta-function potential may be effected by constructing a potential barrier whose height is infinite and whose width is infinitesimal such that the area under the potential curve is fixed. This limit is easily constructed with the model at hand by setting

$$\lim_{\substack{\kappa \to \infty \\ b \to 0}} \kappa^2 bd = 2F = \text{constant}$$

(b) Make a plot of your dispersion function for the value $F = 3\pi/2$ and thereby illustrate the persistence of the band structure of the energy spectrum in this delta-function limit.

(c) How is it that electrons are able to propagate through the infinitely high potentials presented by the delta functions?

(d) Write down a formal expression for the potential you have considered.

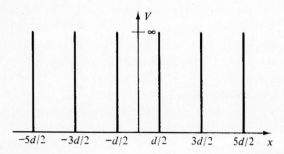

FIGURE 8.17 Periodic delta-function potential. The explicit form of the symmetric periodic delta-function potential is given by (see Problem 3.6e)

$$V(x) = V_0 \, d \left\{ \sum_{n=0}^{\infty} \delta \left[x - (2n + 1)\left(\frac{d}{2}\right) \right] + \sum_{n=0}^{\infty} \delta \left[x + (2n + 1)\left(\frac{d}{2}\right) \right] \right\}$$

$$= V_0 \, d^2 \sum_{n=0}^{\infty} (2n + 1)\delta \left[x^2 - (2n + 1)^2 \left(\frac{d}{2}\right)^2 \right]$$

(See Problem 8.12.)

[1] This limiting case was, in fact, the one treated by R. de L. Kronig and W. G. Penney in their original paper, *Proc. Roy. Soc.* **A130**, 499 (1931).

(e)   What are the eigenstates at the band edges $kd = n\pi$? Show that one of the energies at a band edge is the free-particle energy $\hbar^2 k^2/2m$, while the other energy is larger. In this manner obtain an expression for the width of the energy gap at $kd = n\pi$.

Answer (partial)

(a)   In a limit given above it follows that

$$\kappa b \to \frac{2F}{\kappa d} \to 0$$

Hence $\sinh \kappa b \to \kappa b$, $\cosh \kappa b \to 1$, and $d \to a$. The resulting dispersion relation appears as

$$\frac{F \sin k_1 d}{k_1 d} + \cos k_1 d = \cos kd$$

8.13   Show that the $E(k)$ spectrum for the arbitrary finite Kronig–Penney array draws close to the free-particle parabola $E = \hbar^2 k^2/2m$ in the limit $E \gg V$.

Answer

With $V/E = \epsilon \ll 1$, one obtains

$$k_2^2 = \frac{2m}{\hbar^2} E(1 - \epsilon) = k_1^2 + O(\epsilon)$$

Substitution into the dispersion relation (8.53) gives

$$\cos k_1 a \cos k_1 b - \sin k_1 a \sin k_1 b + O(\epsilon) = \cos k_1(a + b) + O(\epsilon) = \cos kd$$

Neglecting terms of $O(\epsilon)$ gives the spectrum $k_1 = k$ or, equivalently, $E = \hbar^2 k^2/2m$. Recalling further that k is discrete (8.42), one obtains $E_n = n^2(h^2/2mL^2)$.

8.14   The $E(k)$ spectrum for an electron in a periodic lattice, such as illustrated in Fig. 8.13, does not fall to zero at $k = 0$. Estimate this *zero-point* energy using results appropriate to a particle confined to a one-dimensional domain of length L. What value of k is implied by your answer? How does this value compare to the minimum value of k for a crystal of length L?

8.15   (a)   Show that the eigenenergies of the one-dimensional box of width a ($k_1 a = n\pi$) lie in the energy gaps of the Kronig–Penney potential of well width a (in the domain $E < V$). [*Hint:* Use (8.55).]

(b)   Show that these box energies become the lower energies of the band gaps for the periodic delta-function potential described in Problem 8.12.

8.16   (a)   Show that in the limit that the atomic sites of the Kronig–Penney model become far removed from each other ($b \to \infty$), energies of the more strongly bound electrons ($E \ll V$) become the eigenenergies $k_1 a = n\pi$ of a one-dimensional box of width a.

(b)   In this limit what do the lower-band $E(k)$ curves shown in Fig. 8.13 become? What is the functional form of $E(k)$ for these bands? (*Note:* The approximation in which one begins with electronic states of isolated atoms is called the *tight-binding approximation*.)

8.17   (a)   Construct an equation for the periodic component $u(k)$ of the Bloch wavefunction $\varphi = u \exp{(ikx)}$, from the Schrödinger equation with a periodic potential $V(x)$.

(b)   The periodic potential $V(x)$ may be expanded in a Fourier series as follows:

$$V(x) = \sum_{n=-\infty}^{\infty} V_n \exp\left[i2\pi n\left(\frac{x}{d}\right) \right]$$

Expand the periodic component $u(x)$ in a similar series, substitute in the equation obtained in part (a), and derive coupled equations for the coefficients of expansion u_n.

Answer

(a) $\dfrac{\hbar^2}{2m} (u'' + 2iku' - k^2 u) + [E - V(x)]u = 0$

(b) $[E - B_q(k)]u_q = \sum_{l=-\infty}^{\infty} V_{q-l}u_l, \quad 2mB_q(k)/\hbar^2 \equiv (2\pi q/d)^2 + 2k(2\pi q/d) + k^2$

8.18   Show that in the limit $V \to 0$, the equation for the periodic component u obtained in Problem 8.17 gives the free-particle eigenenergy

$$E = \frac{\hbar^2 k^2}{2m}$$

with $u = $ constant.

8.19   What is the number $\mathcal{N}_E$ of discrete energies in any of the energy bands depicted in Fig. 8.13?

Answer

The k values that enter the lowest energy band are given by the sequence

$$k = 0, \pm 1 \times \frac{2\pi}{L}, \pm 2 \times \frac{2\pi}{L}, \ldots, \pm \frac{N}{2}\frac{2\pi}{L}$$

$$L \equiv Nd$$

This series is cut off at $|kd| = \pi$ inasmuch as energy values begin to repeat beyond this value. Furthermore, as these k-values are at the zone boundary (see Fig. 8.13) they are counted only once. There is a distinct energy corresponding to each value of $|k|$ in the sequence above. This gives

$$\mathcal{N}_E = \left(\frac{N}{2} - 1\right) + 1 = \frac{N}{2}$$

8.20   What is the number $\mathcal{N}_k$ of independent eigenstates in a band for a one-dimensional crystal comprised of N uniformly spaced ions?

Answer

There is a distinct eigenstate (8.46 et seq.) corresponding to each value of k in the series in the example above. It follows that $\mathcal{N}_k = 2\mathcal{N}_E = N$. There are as many eigenstates in a sample as there are ion sites. There are approximately half as many eigenenergies.

This result may also be obtained geometrically. Referring to the reduced-zone energy diagram (Fig. 8.13b), each energy band has width $(\Delta kd)_b = 2\pi$. The minimum interval in each band is $(\Delta kd)_{min} = 2\pi/N$ or, equivalently, $(\Delta k)_{min} = 2\pi/Nd = 2\pi/L$. Thus the number of points (states) in each band is

$$\mathcal{N}_k = \frac{(\Delta kd)_b}{(\Delta kd)_{min}} = N$$

(*Note:* With the two spin orientations taken into account, one obtains $2N$ independent states in each band.[1] The concept of spin is described in Chapter 11.)

8.3   STANDING WAVES AT THE BAND EDGES

Let us return to the nature of the eigenstates of $\hat{H}$ at the band edges, that is, at $kd = n\pi$. We will demonstrate that these eigenstates are standing waves and illustrate the relation between the eigenenergies of these states and the energy gaps at the band edges.

The eigenstates of the Kronig–Penney Hamiltonian established above have components [see (8.48) and (8.49)]

(8.65)
$$\varphi_I = Ae^{ik_1x} + Be^{-ik_1x}$$
$$\varphi_{II} = Ce^{ik_2x} + De^{-ik_2x}$$

In order for these to be components of a standing wave, the magnitude of the amplitudes of the waves moving to right and left must be equal. That is, at the critical values $kd = n\pi$, one must have

$$\frac{|A|}{|B|} = 1, \qquad \frac{|C|}{|D|} = 1$$

We will establish the first equality and leave the second as a problem. At the values $kd = n\pi$, $\exp(ikd) = (-1)^n$. Consider that n is even so that $\exp(ikd) = +1$. With this value substituted into the equations of continuity, (8.50) and (8.51), one quickly obtains the following two equations for the expression $2Ck_2 \exp(ik_2a)$:

$$2Ck_2e^{ik_2a} = e^{-ik_2b}[A(k_1 + k_2) + B(k_2 - k_1)]$$
$$2Ck_2e^{ik_2a} = e^{ik_1a}A(k_1 + k_2) + e^{-ik_1a}B(k_2 - k_1)$$

[1] This result maintains in three dimensions, where N represents the number of *primitive cells* in the crystal. For further discussion, see C. Kittel, *Introduction to Solid State Physics*, 5th ed., Wiley, New York, 1976.

Setting these two expressions equal to each other and solving for A/B gives

(8.66)
$$\frac{A}{B} = \frac{k_2 - k_1}{k_2 + k_1} \left(\frac{e^{-ik_1 a} - e^{-ik_2 b}}{e^{-ik_2 b} - e^{ik_1 a}} \right)$$

Forming the square of the modulus $|A/B|^2 = (A/B)(A/B)*$ gives

(8.67)
$$\left| \frac{A}{B} \right|^2 = \left| \frac{k_2 - k_1}{k_2 + k_1} \right|^2 \frac{2 - 2\cos(k_1 a - k_2 b)}{2 - 2\cos(k_1 a + k_2 b)}$$

$$= \left| \frac{k_2 - k_1}{k_2 + k_1} \right|^2 \frac{1 - \cos k_1 a \cos k_2 b - \sin k_1 a \sin k_2 b}{1 - \cos k_1 a \cos k_2 b + \sin k_1 a \sin k_2 b}$$

We must show that this expression is unity for the allowable values of k_1 and k_2, that is, those values which are obtained from the dispersion relation (8.53). Again with exp $(ikd) = +1$, this relation reads

$$\cos k_1 a \cos k_2 b = 1 + \frac{k_1^2 + k_2^2}{2k_1 k_2} \sin k_1 a \sin k_2 b$$

When this formula for the cos product is substituted into (8.67), the desired result, $|A/B| = 1$, follows.

Let us proceed to construct such a standing-wave state. If the potential is placed in a symmetric position about the origin, such as in Fig. 8.18, then the Hamiltonian commutes with the parity operator $\hat{P}$ and these two operators share a set of common eigenfunctions; that is, $\hat{H}$ has even and odd eigenstates. It will be shown that eigenstates at the band edges exist in pairs, with each pair containing an even and an odd eigenstate. Very simply, one expects that this is the case since, in the steady-state situation, electron density $|\varphi|^2$ should enjoy the same symmetry as the periodic

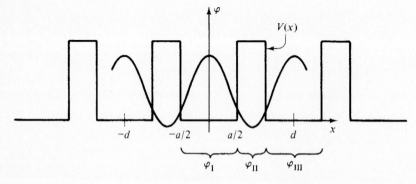

FIGURE 8.18   Standing-wave eigenfunction referred to in the construction $\varphi(x)$ as given by (8.68) to (8.70). Owing to the symmetric form of φ, it suffices to match φ_I to φ_{II} at $x = a/2$.

potential $V(x)$; that is, $|\varphi|^2$ is even, so φ is either even or odd. We will find that if kd/π is an even integer, eigenstates have period d. If kd/π is an odd integer, eigenstates have period $2d$. Let us consider the construction of the pair of eigenstates at a band edge corresponding to kd/π an even integer. Consider first the symmetric eigenstate. In the well domain about the origin

$$(8.68) \qquad \varphi_I(x) = \cos k_1 x \qquad (-a/2 \le x \le a/2)$$

To obtain φ_{III} in the second well domain, one uses Bloch's theorem (8.40) together with the value $\exp(ikd) = +1$.

$$(8.69) \quad \varphi_{III}(x) = e^{ikd} \varphi_I(x - d) = \cos k_1(x - d) \qquad (a/2 + b \le x \le a/2 + d)$$

The standing wave in the barrier region II which joins these waves is symmetric about the midpoint $d/2$ (see Fig. 8.18).

$$(8.70) \qquad \varphi_{II}(x) = D \cos k_2 \left(\frac{x - d}{2} \right) \qquad (a/2 \le x \le a/2 + b)$$

The coefficient D and energy $\hbar^2 k_1^2/2m$ are obtained by matching the components of φ and φ' at the potential interface at $x = a/2$. There results for the even eigenstates,

$$\cos\left(\frac{ak_1}{2}\right) = D \cos\left(\frac{k_2 b}{2}\right)$$

$$(8.71)$$

$$k_1 \sin\left(\frac{ak_1}{2}\right) = -Dk_2 \sin\left(\frac{k_2 b}{2}\right)$$

Identical equations are obtained by matching φ_{II} to φ_{III} at $a/2 + b$ (see Fig. 8.18). Equations 8.71, together with the energy statement

$$k_1^2 - k_2^2 = \frac{2mV}{\hbar^2}$$

are three equations for the three unknowns k_1, k_2, and D. They are reminiscent of (8.27) and (8.28) and appropriate to the bound states of a particle in a finite potential well. Here, as there, solution may be effected through a numerical procedure (see Problem 8.24). The companion odd eigenstate $\tilde{\varphi}$ may similarly be constructed with the modification that the standing wave in the barrier region II is odd about the midpoint $d/2$. One obtains

$$\tilde{\varphi}_I(x) = \sin \tilde{k}_1 x \qquad (-a/2 \le x \le a/2)$$

$$(8.72) \qquad \tilde{\varphi}_{II}(x) = \tilde{D} \sin \tilde{k}_2 \left(\frac{d}{2} - x \right) \qquad (a/2 \le x \le a/2 + b)$$

$$\tilde{\varphi}_{III}(x) = \sin \tilde{k}_1(x - d) \qquad (a/2 + b \le x \le a/2 + d)$$

Matching conditions at the interface position $a/2$ gives the following relations for the odd eigenstates:

$$\sin\left(\frac{\tilde{k}_1 a}{2}\right) = \tilde{D}\sin\left(\frac{\tilde{k}_2 b}{2}\right)$$

(8.73)
$$\tilde{k}_1\cos\left(\frac{\tilde{k}_1 a}{2}\right) = -\tilde{k}_2\tilde{D}\cos\left(\frac{\tilde{k}_2 b}{2}\right)$$

$$\tilde{k}_1^2 - \tilde{k}_2^2 = \frac{2mV}{\hbar^2}$$

Again numerical procedure yields values for the energy $\hbar^2\tilde{k}_1^2/2m$ and eigenstate parameter $\tilde{D}$.

The most significant result of such calculation is the width of the energy gap δE_n at the band edge $kd = n\pi$. This is the difference in energy between the even and odd standing-wave eigenstates.

$$(\delta E)_n = \frac{\hbar^2}{2m}(k_1^2 - \tilde{k}_1^2)$$

An analytic evaluation of this energy jump may be obtained in the "nearly free electron" model. This model is described in Section 13.4.

Parity Properties

Next we turn to a discussion of the parity properties of these standing-wave eigenstates at the band edges. These states are either even or odd in x. Again consider the case that kd/π is an even integer. Then the relation

(8.74)
$$\varphi(x + d) = e^{ikd}\varphi(x)$$

which is true for any eigenstate of the Kronig–Penney Hamiltonian, gives

(8.75)
$$\varphi(x + d) = \varphi(x)$$

It follows that for kd an even multiple of π, the period of φ is d. Setting $x = -d/2$ in (8.75) gives

(8.76)
$$\varphi\left(\frac{d}{2}\right) = \varphi\left(\frac{-d}{2}\right)$$

From this equation one concludes that φ can be an odd eigenfunction provided that

(8.77)
$$\varphi\left(\frac{d}{2}\right) = \varphi\left(\frac{-d}{2}\right) = 0$$

This property, taken together with the fact that φ has period d, gives

$$(8.78) \qquad \varphi\left(\pm n\,\frac{d}{2}\right) = 0 \qquad (\varphi \text{ is odd,}\quad kd = 2q\pi)$$

with n an integer. The only stipulation on the even eigenfunctions is that they are of period d.

In this manner we find that the eigenfunctions of the Kronig–Penney Hamiltonian at the band edges $kd = 2q\pi$ exist in pairs. Each pair contains an even eigenfunction and an odd eigenfunction. A typical pair of these functions is sketched in Fig. 8.19. The eigenenergies that accompany these eigenstates are the close-spaced pairs of values depicted in Fig. 8.13, where the vertical lines $kd = 2q\pi$ intersect the oscillating curves.

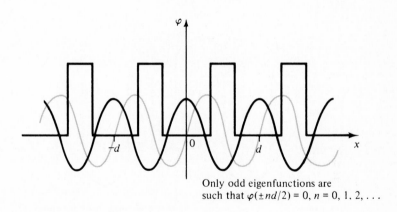

Only odd eigenfunctions are
such that $\varphi(\pm nd/2) = 0$, $n = 0, 1, 2, \ldots$

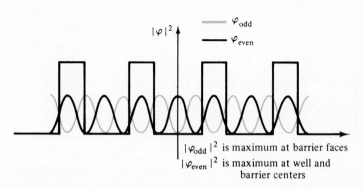

$|\varphi_{\text{odd}}|^2$ is maximum at barrier faces
$|\varphi_{\text{even}}|^2$ is maximum at well and
barrier centers

FIGURE 8.19   Typical pair of eigenfunctions for the Kronig–Penney Hamiltonian at the band edges: $kd = 2q\pi$. Periodicity of φ is d.

Having treated the case where kd is an even multiple of π, we next consider the case $kd = (2q + 1)\pi$, again with q an integer. From (8.74) one obtains

(8.79)
$$\varphi(x + d) = -\varphi(x)$$
$$\varphi(x + 2d) = -\varphi(x + d) = +\varphi(x)$$

It follows that for kd *an odd multiple of* π, *the period of* φ *is* $2d$. Setting $x = -d/2$ in the equation above gives

(8.80)
$$\varphi\left(\frac{d}{2}\right) = -\varphi\left(\frac{-d}{2}\right)$$

while $x = -d$ gives

(8.81)
$$\varphi(d) = \varphi(-d)$$

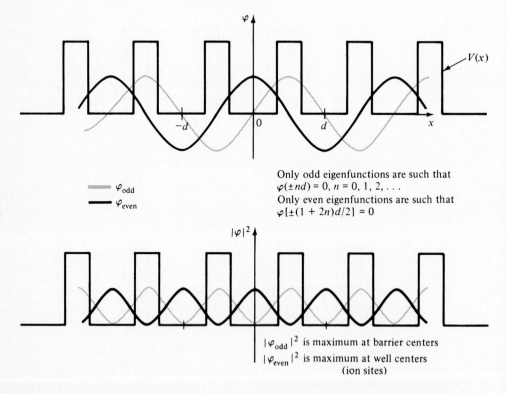

Only odd eigenfunctions are such that
$\varphi(\pm nd) = 0$, $n = 0, 1, 2, \ldots$
Only even eigenfunctions are such that
$\varphi[\pm(1 + 2n)d/2] = 0$

$|\varphi_{\text{odd}}|^2$ is maximum at barrier centers
$|\varphi_{\text{even}}|^2$ is maximum at well centers
(ion sites)

FIGURE 8.20 Typical pair of eigenfunctions for the Kronig–Penney Hamiltonian at the band edges: $kd = (2q + 1)\pi$. Periodicity of φ is $2d$.

Equation (8.80) indicates that φ can be an even eigenfunction provided that

(8.82)
$$\varphi\left[\pm(1 + 2n)\frac{d}{2}\right] = 0 \qquad [kd = (2q + 1)\pi]$$

with n an integer.

Equation (8.81) indicates that φ can be an odd eigenstate provided that

(8.83)
$$\varphi(\pm nd) = 0 \qquad [kd = (2q + 1)\pi]$$

again with n an integer.

A typical pair of eigenfunctions is sketched in Fig. 8.20, together with accompanying plots of electron density $|\varphi|^2$. From this sketch one notes that each pair of eigenstates, corresponding to kd an odd multiple of π, contains one eigenstate with density $|\varphi|^2$, maximum at the ion sites and minimum at the barrier centers, while the other eigenstate has its extremum values of $|\varphi|^2$ reversed.

Thus we conclude that at the band edges $[kd = 2q\pi$ or $kd = (2q + 1)\pi]$ eigenfunctions appropriate to the Kronig–Penney Hamiltonian are standing waves and that there are two such functions with opposite parity at each edge.

PROBLEMS

8.21 The standing-wave quality of the eigenstate $\varphi(x)$ at the band edges was demonstrated for the component of φ in the valley regions of the potential ($|A/B| = 1$) for the case $\exp(ikd) = +1$. Following this analysis, demonstrate that the component of φ in the barrier domain is also a standing wave (i.e., show that $|C/D| = 1$) for the case $\exp(ikd) = -1$. [See (8.65) et seq.]

8.22 Show that the expectation of momentum $\langle p \rangle$ vanishes for a particle in a standing-wave eigenstate.

8.23 Consider a typical pair of eigenstates appropriate to $kd = 2q\pi$ and the adjacent pair appropriate to $kd = (2q + 1)\pi$. By inspection only, conclude which pair of corresponding eigenenergies is of higher value.

8.24 (a) Introducing nondimensional variables

$$\xi \equiv \frac{k_1 a}{2}, \qquad \eta \equiv \frac{k_2 b}{2}, \qquad \rho^2 \equiv \frac{ma^2 V}{2\hbar^2}$$

show that (8.71) and (8.73) relevant to even and odd standing-wave band-edge solutions, respectively, give the dispersion relations

$$\xi \tan \xi = -\sqrt{\xi^2 - \rho^2} \tan\left(\frac{b}{a}\sqrt{\xi^2 - \rho^2}\right) \qquad \text{(even)}$$

$$\hat{\xi} \cot \hat{\xi} = -\sqrt{\hat{\xi}^2 - \rho^2} \cot\left(\frac{b}{a}\sqrt{\hat{\xi}^2 - \rho^2}\right) \qquad \text{(odd)}$$

(b)   Numerical solution of either equation may be effected by plotting the right-hand side and the left-hand side of the equation as functions of ξ (or $\tilde{\xi}$) on the same graph. Intersections then give the eigenenergies $E = (2\hbar^2/ma^2)\xi^2$. Use this procedure to estimate the lowest even and odd eigenstate energy corresponding to the barrier parameters, $\rho = \pi/2$, $(a/b)^2 = 15$.

(c)   Use your answer to part (b) to obtain the width of the energy gap δE at this band edge.

8.25   It was shown in Problem 8.13 that in the high-energy domain $E \gg V$, the $E(k)$ spectrum approaches the free-particle curve $E = \hbar^2 k^2/2m$. Show that the dispersion relation appropriate to the band edge $kd = 2n\pi$, (8.73), yields a free-particle standing wave in this limit.

Answer

With $k_1 \simeq k_2$, (8.73) gives (dropping the tilde notation)

$$D = \frac{\sin (k_1 a/2)}{\sin (k_1 b/2)} = -\frac{\cos (k_1 a/2)}{\cos (k_1 b/2)}$$

The second equality gives

$$\sin \left[\frac{k_1(a + b)}{2} \right] = 0$$

so

$$k_1 d = 2n\pi = kd$$

When substituted back into the expression above, one obtains $D = 1$, which is necessary in order that the eigenstate (8.72) with $k_1 = k_2$ be a free-particle standing wave.

8.4   BRIEF QUALITATIVE DESCRIPTION OF THE THEORY OF CONDUCTION IN SOLIDS

The spectrum of eigenenergies of electrons in an actual three-dimensional crystalline solid closely parallels that of the Kronig–Penney model described previously. In the three-dimensional case one also obtains a band structure for the allowed eigenenergies. The electrons in a solid occupy these bands. The properties of the two bands of highest energy for most practical cases determine whether the solid is an insulator, a conductor, or a semiconductor.

Suppose that the band structure of a solid is such that the band of highest energy is full (see Fig. 8.21). Furthermore, the gap between this filled band and the next completely unoccupied band is reasonably large. For example, for diamond this gap has a width of 6 eV. When an electric field is applied, electrons in the filled bands have no nearby unoccupied states to accelerate to. The sample remains nonconductive. It

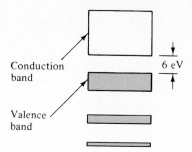

Conduction band

6 eV

Valence band

FIGURE 8.21 Energy bands of diamond, a good insulator.

is an insulator. Furthermore, photons that comprise the visible spectrum do not have sufficient energy (hv) to raise electrons from the *valence band* (last filled band) to the *conduction band*, so that diamond is transparent to light.

These statements are precisely true at absolute zero ($-273°C$). The student will recall that at absolute zero a system of particles falls to its lowest energy state, called the ground state of the system. When the temperature is raised, thermal agitation excites electrons to states of higher energy. For instance, for diamond at room temperature the characteristic energy of thermal agitation is ≈ 0.03 eV. The concentration of electrons which are raised to the conduction band is $\approx 1.1 \times 10^{-34}$ electron/cm^3. This gives rise to a conductivity which is lower than can be measured with present-day equipment, and diamond remains an insulator at room temperature.

In some crystalline solids, the conduction band is empty and the energy gap to the valence band is not prohibitively large. For instance, in silicon, this gap is 1.11 eV wide. In germanium it is 0.72 eV wide. At room temperature the concentration of electrons in the conduction band in silicon is 7×10^{10} electrons/cm^3. In germanium it is 2.5×10^{13} electrons/cm^3 (see Fig. 8.22). These densities give measurable conductivities. Such materials are called *intrinsic semiconductors*. The conduction of an *extrinsic semiconductor* is due to the presence of impurities in the sample.

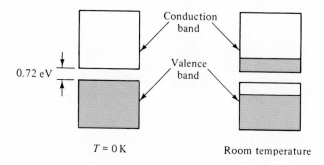

Conduction band

Valence band

0.72 eV

$T = 0$ K

Room temperature

FIGURE 8.22 Valence and conduction bands for germanium, a typical semiconductor, at absolute zero and room temperature.

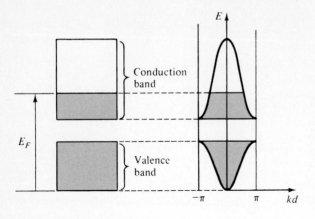

FIGURE 8.23   In a conductor the states in the conduction band are partially filled. The diagram to the right indicates the manner in which electrons fill the corresponding bands in the reduced-zone description for the idealized one-dimensional model. The Fermi energy E_F is also shown.

A semiconductor acts as an insulator at sufficiently low temperatures. It begins to conduct at higher temperatures. In a semiconductor, charge transfer in the valence band may also contribute to conduction. In this case it is simpler for, say, calculation of conductivity, to speak of *hole conduction*. A hole is an unfilled state, found usually in the valence band.

In a metal the band of highest energy is only partially filled and electrons are readily accelerated by an electric field, to states of higher energy (see Fig. 8.23). Photons also fall prey to these electrons, which explains the opacity of metals to light. Note that the Fermi energy, described in Chapter 2, appears in the conduction band.

The description we have presented for the band structure of energy levels in periodic structures is a vast simplification of that which occurs in actual solids. A more accurate description of the formation of such bands with shrinking interatomic distance is shown in Fig. 8.24 for the metal sodium. Note in particular the strong band overlap at very small interatomic distance. This property suggests why materials become electrically conductive under extreme compression. The theory of semiconductors is returned to in Section 12.9.

The problem of a particle propagating in a periodic potential is revisited in Section 11.14 where it is reformulated in terms of symmetry properties of the potential and the "transfer matrix" method.

PROBLEMS

8.26   What is the minimum frequency of radiation to which diamond is opaque? What kind of radiation is this (e.g., x rays, etc.)?

8.27   The mobility μ of an electron in an electric field **E** is defined by

$$\mathbf{v} = \mu \mathbf{E}$$

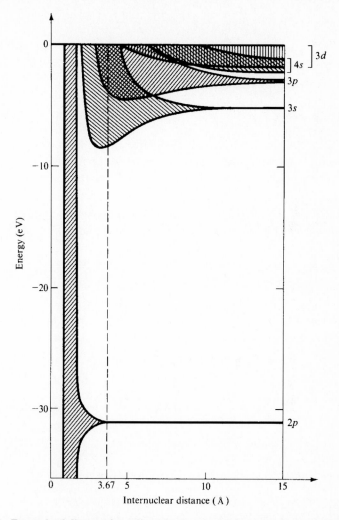

FIGURE 8.24 Energy-level diagram for sodium showing the development of a band structure as internucleon distance decreases. Labeling of states, $2p$, $3s$, etc., is fully described in Section 12.4. In the ground state of sodium, electrons fill up to the $3s$ level, which is half-empty. This property accounts for the conductivity of the metal.

where **v** is the drift velocity of the electron. In a given semiconductor the mobility of electrons is μ_n, while the mobility of holes is μ_p. If at a given temperature, the density of conduction electrons is n electrons/cm^3 and the density of holes is p holes/cm^3, obtain an expression for the current flow in the semiconductor if an electric field **E** is applied across it.

8.5 TWO BEADS ON A WIRE AND A PARTICLE IN A TWO-DIMENSIONAL BOX

Exchange Degeneracy

In this and the following section we discuss some simple examples of quantum mechanical systems with two degrees of freedom (see Section 1.2). The first such example is that of two beads constrained to move on a straight frictionless wire that is tightly stretched between two perfectly reflecting, rigid walls. The space between walls is a (see Fig. 8.25). We still assume that the particles do not interact with each other (they are "invisible" to each other). The Hamiltonian for this system is

$$(8.84) \qquad \hat{H}(x_1, x_2) = \frac{\hat{p}_1^{\,2}}{2m} + \frac{\hat{p}_2^{\,2}}{2m} + V(x_1) + V(x_2)$$

The two particles have the same mass, m. The potential functions $V(x_1)$ and $V(x_2)$ are relevant to a one-dimensional box. Their properties are given in Section 4.1.

This Hamiltonian may be partitioned into two independent terms,

$$\hat{H}(x_1, x_2) = \hat{H}_1(x_1) + \hat{H}_2(x_2)$$

$$(8.85) \qquad \hat{H}_1(x_1) = \frac{\hat{p}_1^{\,2}}{2m} + V(x_1)$$

$$\hat{H}_2(x_2) = \frac{\hat{p}_2^{\,2}}{2m} + V(x_2)$$

Under such circumstances, solution of the Schrödinger equation

$$(8.86) \qquad \hat{H}\varphi(x_1, x_2) = E\varphi(x_1, x_2)$$

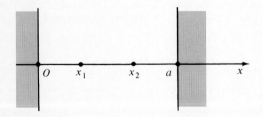

FIGURE 8.25 Coordinates of two beads on a wire stretched between two perfectly reflecting walls separated by the distance a.

is greatly simplified. It is given by the product

$$(8.87) \qquad \varphi_{n_1 n_2}(x_1, x_2) = \varphi_{n_1}(x_1)\varphi_{n_2}(x_2)$$

where

$$(8.88) \qquad \begin{aligned} \hat{H}_1 \varphi_{n_1}(x_1) &= E_{n_1}\varphi_{n_1}(x_1) \\ \hat{H}_2 \varphi_{n_2}(x_2) &= E_{n_2}\varphi_{n_2}(x_2) \end{aligned}$$

The function φ_{n_1} is the eigenfunction of $\hat{H}_1$ corresponding to the energy E_{n_1}, while φ_{n_2} is the eigenfunction of $\hat{H}_2$ corresponding to the eigenenergy E_{n_2}.

$$(8.89) \qquad \varphi_n = \sqrt{\frac{2}{a}} \sin\left(\frac{n\pi x}{a}\right), \qquad E_n = n^2 E_1$$

where n denotes either n_1 or n_2.

Let us test to see if $\varphi_{n_1 n_2}$, as given by (8.87), is an eigenstate of $\hat{H}(x_1, x_2)$.

$$(8.90) \qquad \begin{aligned} \hat{H}\varphi_{n_1 n_2}(x_1, x_2) &= (\hat{H}_1 + \hat{H}_2)\varphi_{n_1}(x_1)\varphi_{n_2}(x_2) \\ &= \varphi_{n_2}\hat{H}_1\varphi_{n_1} + \varphi_{n_1}\hat{H}_2\varphi_{n_2} \\ &= E_{n_1}\varphi_{n_1}\varphi_{n_2} + E_{n_2}\varphi_{n_1}\varphi_{n_2} \\ \hat{H}\varphi_{n_1 n_2} &= (E_{n_1} + E_{n_2})\varphi_{n_1}\varphi_{n_2} \end{aligned}$$

Thus we find that $\varphi_{n_1}\varphi_{n_2}$ is an eigenstate of $\hat{H}(x_1, x_2)$, and furthermore that the eigenenergy corresponding to this state is

$$(8.91) \qquad E_{n_1 n_2} = E_{n_1} + E_{n_2} = (n_1^2 + n_2^2)E_1$$

For example, the eigenstate

$$(8.92) \qquad \varphi_{2,3} = \frac{2}{a} \sin\left(\frac{2\pi x_1}{a}\right) \sin\left(\frac{3\pi x_2}{a}\right)$$

has corresponding eigenenergy

$$(8.93) \qquad E_{2,3} = E_1(4 + 9) = 13E_1$$

This energy is *doubly degenerate* since the eigenstate

$$(8.94) \qquad \varphi_{3,2} = \frac{2}{a} \sin\left(\frac{3\pi x_1}{a}\right) \sin\left(\frac{2\pi x_2}{a}\right)$$

also corresponds to the eigenenergy $E_{2,3}$. One may look upon the difference between $\varphi_{2,3}$ and $\varphi_{3,2}$ as being due to the exchange in the positions of particle 1 and particle 2. Such degeneracy is called *exchange degeneracy*.

Symmetric and Antisymmetric States

If two eigenstates correspond to the same eigenenergy, any linear combination of these eigenstates also corresponds to the same eigenenergy. Of all such linear combinations, two are of particular physical significance. These are of the form

(8.95)
$$\varphi_S = \frac{1}{\sqrt{2}} [\varphi_{n_1}(x_1)\varphi_{n_2}(x_2) + \varphi_{n_1}(x_2)\varphi_{n_2}(x_1)]$$

$$\varphi_A = \frac{1}{\sqrt{2}} [\varphi_{n_1}(x_1)\varphi_{n_2}(x_2) - \varphi_{n_1}(x_2)\varphi_{n_2}(x_1)]$$

The *symmetric state* φ_S has the property that

(8.96)
$$\varphi_S(x_1, x_2) = \varphi_S(x_2, x_1)$$

It is symmetric under the exchange of the particles. The *antisymmetric state* φ_A has the property that

(8.97)
$$\varphi_A(x_1, x_2) = -\varphi_A(x_2, x_1)$$

It is antisymmetric under the exchange of particles.

When referred to systems with two degrees of freedom, such as that of two particles in a one-dimensional box, the probability amplitude related to the system is given by (see Problem 3.20)

(8.98)
$$P_{12} \, dx_1 \, dx_2 = |\varphi(x_1, x_2)|^2 \, dx_1 \, dx_2$$

$P_{12} \, dx_1 \, dx_2$ is the probability of finding particle 1 in the interval dx_1 about the point x_1 *and* particle 2 in the interval dx_2 about the point x_2, in any given measurement.

When the two particles in the one-dimensional box are identical ($m_1 = m_2$), such as in the case considered, we note that for both classes of wavefunctions (symmetric and antisymmetric)

(8.99)
$$|\varphi_S(x_1, x_2)|^2 = |\varphi_S(x_2, x_1)|^2, \qquad |\varphi_A(x_1, x_2)|^2 = |\varphi_A(x_2, x_1)|^2$$

Physical properties of the system are not affected by an exchange of the position of the two particles. This is a manifestation of a quantum mechanical property attached to identical particles: that is, in quantum mechanics identical particles are also indistinguishable (they cannot be labeled). In the scattering of electrons off electrons, for example, the scattered beam contains both incident and target electrons. The indistinguishability of these particles must be taken into account in any consistent formulation of the theory of such scattering. It is the indistinguishability of identical particles which selects φ_A or φ_S (8.95) to be the physically relevant linear combination of eigenstates for the two-particle problem.

If the masses of the two particles in our one-dimensional box are different (m_1 and m_2), the Hamiltonian (8.84) becomes

$$(8.100) \qquad \hat{H}(x_1, x_2) = \frac{\hat{p}_1^2}{2m_1} + \frac{\hat{p}_2^2}{2m_2} + V(x_1) + V(x_2)$$

The particles are now distinguishable and the states of the system do not suffer exchange degeneracy. The eigenstate

$$(8.101) \qquad \varphi_{n_1 n_2} = \varphi_{n_1}(x_1)\varphi_{n_2}(x_2)$$

corresponds to the eigenenergy

$$(8.102) \qquad E_{n_1 n_2} = E_{n_1} + E_{n_2} = \left(\frac{n_1^2}{m_1} + \frac{n_2^2}{m_2}\right)\frac{\hbar^2 \pi^2}{2a^2}$$

The exchange state $\varphi_{n_2 n_1}$ corresponds to the eigenenergy

$$(8.103) \qquad E_{n_2 n_1} = \left(\frac{n_2^2}{m_1} + \frac{n_1^2}{m_2}\right)\frac{\hbar^2 \pi^2}{2a^2} \neq E_{n_1 n_2}$$

Thus the exchange degeneracy associated with systems containing identical particles is removed.

We now turn to the time-dependent Schrödinger equation for systems with two degrees of freedom.

$$(8.104) \qquad i\hbar \frac{\partial \psi}{\partial t} = \hat{H}\psi$$

The solution of this equation is (see Section 3.5)

$$(8.105) \qquad \psi_{n_1 n_2} = \phi_{n_1 n_2}(x_1, x_2) \exp\left(-\frac{iE_{n_1 n_2}t}{\hbar}\right)$$

Given the arbitrary initial state $\psi(x_1, x_2, 0)$, the state at time $t > 0$ is

$$\psi(x_1, x_2, t) = \exp\left(-\frac{it\hat{H}}{\hbar}\right)\psi(x_1, x_2, 0)$$

Examples of the use of this equation are given in the problems that follow the next subsection.

Symmetry and Accidental Denegeracy

Much of the preceding analysis may be carried over to the problem of a single particle moving in a two-dimensional box (see Fig. 8.26). This is another case of a system with two degrees of freedom. In the example of two beads on a wire, cited above, the good coordinates are (x_1, x_2). For the single particle in a two-dimensional box, good coordinates are (x, y). The Hamiltonian for this system appears as

$$(8.106) \qquad \hat{H}(x, y) = \frac{\hat{p}_x^{\ 2}}{2m} + \frac{\hat{p}_y^{\ 2}}{2m} + V(x) + V(y)$$

The potential $V(x)$ is the same as that of a one-dimensional box which lies between $x = 0$ and $x = a$ on the x axis, whereas $V(y)$ is the same as that of a one-dimensional box that lies between $y = 0$ and $y = a$ on the y axis. Eigenfunctions and eigenenergies are

$$(8.107) \qquad \varphi_{n_1 n_2}(x, y) = \frac{2}{a} \sin\left(\frac{n_1 \pi x}{a}\right) \sin\left(\frac{n_2 \pi y}{a}\right)$$

$$E_{n_1 n_2} = E_1(n_1^{\ 2} + n_2^{\ 2})$$

This eigenenergy also corresponds to the eigenstate

$$\varphi_{n_2 n_1}(x, y) = \frac{2}{a} \sin\left(\frac{n_2 \pi x}{a}\right) \sin\left(\frac{n_1 \pi y}{a}\right)$$

The probability density $|\varphi|^2$ may be plotted as a height above the xy plane. The distinction between $|\varphi_{n_1 n_2}|^2$ and $|\varphi_{n_2 n_1}|^2$ is then as follows. The surface $|\varphi_{n_1 n_2}|^2$ is

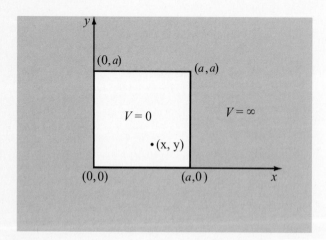

FIGURE 8.26   Particle in a two-dimensional box.

obtained from the surface $|\varphi_{n_2 n_1}|^2$ by reflecting this surface through the plane $x - y = 0$. The energy corresponding to both these distributions is the same. The degeneracy of these states is sometimes called *symmetry degeneracy* (as opposed to exchange degeneracy). Degeneracy that is neither symmetric nor exchange is often referred to as *accidental degeneracy* (see Problem 8.34).

PROBLEMS

8.28 At time $t = 0$, two particles of mass m_1 and m_2, respectively, in a one-dimensional box of width a are known to be in the state

$$\psi(x_1, x_2, 0) = \frac{3\varphi_5(x_1)\varphi_4(x_2) + 7\varphi_9(x_1)\varphi_8(x_2)}{\sqrt{58}}$$

(a) If the energy of the system is measured, what values will be found and with what probability will these values occur?

(b) Suppose the measurement finds the value $E_{5,4}$. What is the time-dependent state of the system subsequent to measurement?

(c) What is the probability of finding particle 1 (with mass m_1) in the interval $(0, a/2)$ at $t = 0$?

Answer (partial)

(c) If the state of the two-particle system is $\varphi(x_1, x_2)$, the probability of finding particle 1 in the interval dx_1 (independent of where particle 2 is) is

$$P(x_1)\, dx_1 = dx_1 \int_0^a |\varphi(x_1, x_2)|^2 \, dx_2$$

8.29 Show that $\varphi_A(x_1, x_2)$ in (8.95) may be written as a 2×2 determinant.

8.30 Show that φ_A and φ_S in (8.95) correspond to the same eigenenergy relevant to the Hamiltonian (8.84).

8.31 In the event that two particles in a one-dimensional box are identical one must ask: What is the probability of finding a particle in the interval dx about x? In this vein, show that for either $\varphi_S(x_1, x_2)$ or $\varphi_A(x_1, x_2)$, the two integrations

$$\int |\varphi|^2 \, dx_1 \quad \text{and} \quad \int |\varphi|^2 \, dx_2$$

give the same functional form.

8.32 Consider two identical particles in a one-dimensional box of length a. Calculate the expected value for the square of the interparticle displacement

$$d^2 \equiv (x_1 - x_2)^2$$

in the two states φ_S and φ_A. Show that

$$\langle d^2 \rangle_S \leq \langle d^2 \rangle_A$$

thus establishing that (in a statistical sense) particles in a symmetric state attract one another while particles in an antisymmetric state repel one another. Such attractions and repulsions are classified as exchange phenomena. They are discussed in further detail in Chapter 12.

Answer

With $\varphi_{n_1}(x_1)$ represented by $|n_1\rangle$, $\varphi_{n_1}(x_2)$ by $|\bar{n}_1\rangle$, and $\varphi_{n_1}(x_1)\varphi_{n_2}(x_2)$ by $|n_1\bar{n}_2\rangle$, the symmetric and antisymmetric states appear as

$$|\varphi_{S,A}\rangle = \frac{1}{\sqrt{2}} \left(|n_1\bar{n}_2\rangle \pm |\bar{n}_1 n_2\rangle \right)$$

Thus

$$2\langle d^2 \rangle_{S,A} = \left(\langle n_1\bar{n}_2| \pm \langle \bar{n}_1 n_2| \right) d^2 \left(|n_1\bar{n}_2\rangle \pm |\bar{n}_1 n_2\rangle \right)$$
$$= \langle n_1\bar{n}_2| d^2 |n_1\bar{n}_2\rangle + \langle \bar{n}_1 n_2| d^2 |\bar{n}_1 n_2\rangle \pm \langle n_1\bar{n}_2| d^2 |\bar{n}_1 n_2\rangle$$
$$\pm \langle \bar{n}_1 n_2| d^2 |n_1\bar{n}_2\rangle$$

In the last two $\pm$ contributions with $d^2 = x_1^2 + x_2^2 - 2x_1 x_2$, only the $-2x_1 x_2$ term is found to survive. Consider the term $\langle n_1\bar{n}_2| d^2 |\bar{n}_1 n_2\rangle = -2\langle n_1\bar{n}_2| x_1 x_2 |\bar{n}_1 n_2\rangle = -2\langle n_1|x_1|n_2\rangle\langle \bar{n}_2|x_2|\bar{n}_1\rangle$. In that $\langle n_1|x_1|n_2\rangle = \langle \bar{n}_1|x_2|\bar{n}_2\rangle \equiv x_{12}$ (write out the integrals and change variables), one obtains

$$\langle n_1\bar{n}_2| x_1 x_2 |\bar{n}_1 n_2\rangle = \langle n_1|x_1|n_2\rangle\langle n_2|x_1|n_1\rangle = |x_{12}|^2$$

There results

$$\langle d^2 \rangle_S = \langle d^2 \rangle_A - 4|x_{12}|^2 \leq \langle d^2 \rangle_A$$

8.33   For a single particle in a two-dimensional box such as described in the text, one may also construct symmetric and antisymmetric states. The symmetric state φ_S has the property

$$\varphi_S(x, y) = \varphi_S(y, x)$$

while the antisymmetric state has the property

$$\varphi_A(x, y) = -\varphi_A(y, x)$$

What are the eigenstates, φ_S and φ_A, that correspond to the energy $29E_1$? The symmetry of these eigenstates reflects the fact that there is no intrinsic distinction between the diagonal halves of the box depicted in Fig. 8.26.

8.34   Construct the eigenstates and eigenenergies of a particle in a two-dimensional rectangular box of edge lengths a and $2a$. Take the origin to be at a corner of the rectangle. Account geometrically for the removal of most of the degeneracy present in the case of the square, two-dimensional box described previously. The degeneracy present for this configuration (e.g., the energy $5E$ is doubly degenerate) is sometimes called *accidental degeneracy*, in that it is neither exchange- nor symmetry-degenerate.

8.6   TWO-DIMENSIONAL HARMONIC OSCILLATOR

The two-dimensional problem we consider now is that of a point particle of mass m, constrained by a set of four coplanar, orthogonal springs, all with the same spring constant K (see Fig. 8.27).

(8.108)

$$\hat{H}(x, y) = \frac{\hat{p}_x^{\,2}}{2m} + \frac{\hat{p}_y^{\,2}}{2m} + \frac{K}{2}x^2 + \frac{K}{2}y^2$$

$$\hat{H}(x, y) = \hat{H}(x) + \hat{H}(y)$$

Again, we find that the total Hamiltonian partitions into two independent parts, $\hat{H}(x)$ and $\hat{H}(y)$. These are the Hamiltonians relevant to one-dimensional harmonic oscillation in the x and y directions, respectively (see Sections 7.2 through 7.4). The eigenstates and eigenenergies of these Hamiltonians are

(8.109)

$$\varphi_{n_1}(\xi) = A_{n_1}\mathcal{H}_{n_1}(\xi)e^{-\xi^2/2}$$

$$E_{n1} = \hbar\omega_0\left(n_1 + \frac{1}{2}\right)$$

$$\varphi_{n_2}(\eta) = A_{n_2}\mathcal{H}_{n_2}(\eta)e^{-\eta^2/2}$$

$$E_{n_2} = \hbar\omega_0\left(n_2 + \frac{1}{2}\right)$$

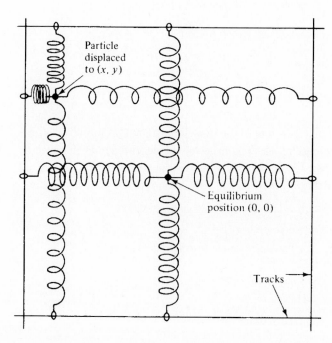

Particle displaced to (x, y)

Equilibrium position $(0, 0)$

Tracks

FIGURE 8.27   Two-dimensional harmonic oscillator. Springs are free to move on tracks but are otherwise constrained to displacements parallel to the coordinate axes. All springs have the same spring constant K.

The nondimensional displacements ξ and η are defined by (7.46)

(8.110)
$$\xi^2 \equiv \frac{m\omega_0 x^2}{\hbar} \equiv \beta^2 x^2$$

$$\eta^2 = \frac{m\omega_0 y^2}{\hbar} \equiv \beta^2 y^2$$

while $\mathcal{H}_n(\xi)$ is the nth-order Hermite polynomial (7.58) and A_n is a normalization constant (Problem 7.7).

Owing to the separability of $\hat{H}(x, y)$, it follows that its eigenstates are the product forms

(8.111)
$$\varphi_{n_1 n_2}(\xi, \eta) = \varphi_{n_1}(\xi)\varphi_{n_2}(\eta)$$

$$\varphi_{n_1 n_2} = A_{n_1 n_2}\mathcal{H}_{n_1}(\xi)\mathcal{H}_{n_2}(\eta)e^{-(\xi^2+\eta^2)/2}$$

while the eigenenergies of $\hat{H}(x, y)$ are the sums

(8.112)
$$E_{n_1 n_2} = E_{n_1} + E_{n_2}$$

$$E_{n_1 n_2} = \hbar\omega_0\left(n_1 + \frac{1}{2} + n_2 + \frac{1}{2}\right) = \hbar\omega_0(n_1 + n_2 + 1)$$

For example, the ground state of the two-dimensional harmonic oscillator is

(8.113)
$$\varphi_{0,0} = A_0 A_0 \mathcal{H}_0(\xi)\mathcal{H}_0(\eta)e^{-(\xi^2+\eta^2)/2}$$

$$\varphi_{0,0}(\xi, \eta) = \frac{1}{\sqrt{\pi}} e^{-(\xi^2+\eta^2)/2} = \frac{1}{\sqrt{\pi}} \exp\left[\frac{-\beta^2(x^2 + y^2)}{2}\right]$$

$$E_{0,0} = \hbar\omega_0$$

This is the only nondegenerate eigenstate of the two-dimensional harmonic oscillator. All the remaining states are degenerate. The order of the degeneracy of the eigenenergy $E_{n_1 n_2}$ is obtained from (8.112), from which we see that any eigenfunction $\varphi_{n_1'}\varphi_{n_2'}$ whose indices n_1', n_2' sum to the value $(n_1 + n_2)$ corresponds to the same eigenenergy, $E_{n_1 n_2}$.

(8.114)
$$\begin{bmatrix} \text{eigenfunctions corresponding to} \\ E_{n_1 n_2} \end{bmatrix} = \begin{bmatrix} \varphi_{n_1'}\varphi_{n_2'}, \text{ such that} \\ n_1' + n_2' = n_1 + n_2 \end{bmatrix}$$

For example, to find the eigenstates that correspond to the eigenenergy

(8.115)
$$E = 5\hbar\omega_0 = (4 + 1)\hbar\omega_0$$

one must find all pairs of integers n_1' and n_2' that sum to 4.

(8.116)
$$n_1' + n_2' = 4$$
$$(n_1', n_2') = (0, 4), (4, 0), (1, 3), (3, 1), (2, 2)$$

It follows that $E = 5\hbar\omega_0$ is a fivefold-degenerate eigenenergy. The five degenerate eigenstates are

(8.117)
$$\left.\begin{aligned}
\varphi_0(\xi)\varphi_4(\eta) &= \varphi_{04} \\
\varphi_4(\xi)\varphi_0(\eta) &= \varphi_{40} \\
\varphi_1(\xi)\varphi_3(\eta) &= \varphi_{13} \\
\varphi_3(\xi)\varphi_1(\eta) &= \varphi_{31} \\
\varphi_2(\xi)\varphi_2(\eta) &= \varphi_{22}
\end{aligned}\right\} = \text{eigenstates corresponding to } E = 5\hbar\omega_0$$

Of these five states, φ_{04} suffers symmetry degeneracy with φ_{40}, as does φ_{13} with φ_{31}. On the other hand, the three states φ_{04}, φ_{13}, and φ_{22} are accidentally degenerate with each other.

PROBLEMS

8.35   What is the order of degeneracy of the eigenstate

$$E_s = \hbar\omega_0(s + 1)$$

of the two-dimensional harmonic oscillator?

Answer

The degeneracy equals the number of ways of writing an integer s as the ordered sum of two numbers. There are $(s + 1)$ ways to do this.

8.36   (a)   Write down the Hamiltonians, eigenenergies, and eigenstates for a two-dimensional harmonic oscillator with distinct spring constants K_x and K_y.

(b)   If $K_y = 4K_x$, show that the eigenenergies may be written

$$E_{n_1 n_2} = \hbar\omega_0\left(n_1 + 2n_2 + \frac{3}{2}\right)$$

where n_1 corresponds to x motion and n_2 to y motion.

(c)   For part (b), what is the order of degeneracy of $E_{2,3}$? List the corresponding eigenstates. Account for the absence of symmetry degeneracy among these states.

8.37   A right circular cylinder of infinite height and large, but finite radius is uniformly, positively charged throughout its volume. The charge density is ρ_0 esu/cm^3. An electron moves in a plane normal to the cylinder. Its position is close to the central axis of the cylinder (see Fig. 8.28).

(a) What is the electrostatic potential Φ near the central axis of the cylinder?

(b) What are the eigenenergies of the electron? [*Hints:* For part (a), use Poisson's equation, $\nabla^2\Phi = -4\pi\rho = -4\pi\rho_0$. The radial operator in ∇^2, in cylindrical coordinates, is $r^{-1}\,\partial/\partial r(r\partial/\partial r)$. From symmetry you may assume $\Phi = \Phi(r)$. For part (b), note that the potential energy of the electron is $V(r) = -|e|\Phi(r)$, where $r^2 = x^2 + y^2$.]

8.38 A particle moves in the xy plane in the potential field

$$V = V(x) + V(y)$$
$$V(x) = V_1 = \text{constant}$$
$$V(y) = V_2 = \text{constant}$$

at constant energy E. Give the time-dependent wavefunction $\psi(x, y, t)$ of the particle corresponding to the initial data

$$\psi(x, 0, 0) = \psi(0, y, 0) = 0$$

8.39 A particle of mass m is confined to move on the two-dimensional strip

$$-a < x < a, \qquad -\infty < y < \infty$$

by two impenetrable parallel walls at $x = \pm a$.

(a) What is the minimum energy of the particle that measurement can find?

(b) Suppose that two additional walls are inserted at $y = \pm a$. Can measurement of the particle's energy find the value $3\pi^2\hbar^2/8ma^2$? Explain your answer.

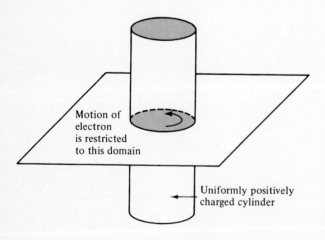

Motion of electron is restricted to this domain

Uniformly positively charged cylinder

FIGURE 8.28 Configuration for Problem 8.37.

8.40 Consider that three-dimensional space is divided into two semi-infinite domains of constant potential V_1 and V_2.

$$V = V_1, \qquad z > 0$$
$$V = V_2, \qquad z \leq 0$$

A beam of particles carrying the current $\hbar\mathbf{k}_1|A|^2/m$ particles/cm²-s is incident on the $z = 0$ interface and is in part reflected and transmitted. Particles in the reflected beam have momentum $\hbar\mathbf{k}_1'$, while those in the transmitted beam have momentum $\hbar\mathbf{k}_2$. The vectors $\mathbf{k}_1$, $\mathbf{k}_1'$, $\mathbf{k}_2$ are all parallel to the xz plane. The configuration is shown in Fig. 8.29.

    (a)   What is the wavefunction ψ_1 appropriate to a particle in the upper half-space $z > 0$? What is the wavefunction ψ_2 of a particle in the lower half-space (i.e., that of a particle in the transmitted beam)?

    (b)   Determine the relation between the angles α, α', and α'' through matching ψ_1 to ψ_2 and their derivatives across the $z = 0$ plane.

    (c)   Using the matching equations obtained in part (b) determine the transmission coefficient T and reflection coefficient R. Show that $T + R = 1$.

Answers (partial)

    (a)   The wavefunctions of particles in the upper and lower half-spaces are

$$\psi_1 = Ae^{i\phi_1} + Be^{i\phi_2}$$
$$\psi_2 = Ce^{i\phi_3}$$

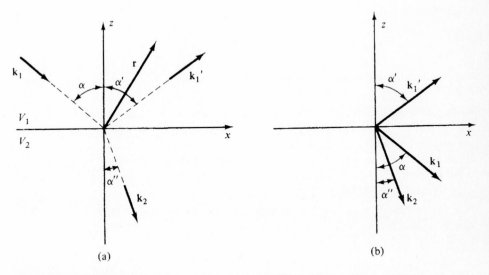

(a)   (b)

FIGURE 8.29 **Orientation of k vectors for a beam of particles incident on the $z = 0$ plane at the angle α. (See Problem 8.40.)**

The phases ϕ_1, ϕ_2, and ϕ_3 are

$$\phi_1 = \mathbf{k}_1 \cdot \mathbf{r} - \omega t = k_1 x \sin \alpha - k_1 z \cos \alpha - \omega t$$

$$\phi_2 = \mathbf{k}_1' \cdot \mathbf{r} - \omega t = k_1 x \sin \alpha' + k_1 z \cos \alpha' - \omega t$$

$$\phi_3 = \mathbf{k}_2 \cdot \mathbf{r} - \omega t = k_2 x \sin \alpha'' - k_2 z \cos \alpha'' - \omega t$$

$$\frac{\hbar^2 k_1^2}{2m} = E - V_1, \qquad \frac{\hbar^2 k_2^2}{2m} = E - V_2, \qquad \hbar \omega = E$$

(b) Matching ψ_1 to ψ_2 on $z = 0$ gives

$$A e^{ik_1 x \sin \alpha} + B e^{ik_1 x \sin \alpha'} = C e^{ik_2 x \sin \alpha''}$$

For this equation to be satisfied for all x, it is necessary that the phases be equal.

$$k_1 \sin \alpha = k_1 \sin \alpha' = k_2 \sin \alpha''$$

$$\alpha = \alpha' \quad \text{and} \quad \frac{\sin \alpha}{\sin \alpha''} = \frac{k_2}{k_1} = \sqrt{\frac{E - V_2}{E - V_1}} = \sqrt{\frac{1 - V_2/E}{1 - V_1/E}} = n$$

where n is the relative *index of refraction*. (Compare with Snell's laws of optical refraction.) These results, together with the matching equation $\partial \psi_1 / \partial z = \partial \psi_2 / \partial z$ on $z = 0$, give

$$A + B = C$$

$$A - B = \frac{k_2 \cos \alpha''}{k_1 \cos \alpha} C$$

which serve to determine T and R: namely,

$$R = \left| \frac{B}{A} \right|^2, \qquad T = \frac{k_2 \cos \alpha''}{k_1 \cos \alpha} \left| \frac{C}{A} \right|^2$$

The transmission coefficient is seen to involve only the normal components of incident and transmitted fluxes.

8.41   Calculate the reflection coefficient of sodium metal for low-energy electrons as a function of electron energy and angle of incidence. For electrons of sufficiently long wavelength, the potential barrier at the metal surface can be treated as discontinuous. Assume that the potential energy of an electron in the metal is -5 eV. Calculate the "index of refraction" of the metal for electrons. (See Problem 8.40.)

8.42   A beam of electrons of energy E in a potential-free region is incident on a potential step of 5 V at an incident angle of $45°$. Is there a threshold value of E below which all the electrons will be reflected? If so, what is this value?

Answer

Call the threshold (or "critical") incident energy E_c. Then at E_c, the angle of refraction $\alpha'' = \pi/2$. That is, at E_c the transmitted ray runs along the interface between the two media. If E is increased above E_c (at the same angle of incidence α), electrons penetrate the potential step and $R < 1$. On the other hand, if E is decreased below E_c, there is no transmitted ray at all. The analytic manifestation of this observation is that α'' becomes imaginary for $E < E_c$, while R maintains its value of unity for all such values of E. From Snell's law (for $\alpha = \pi/4$),

$$\sin \alpha = \frac{1}{\sqrt{2}} = n \sin \alpha''$$

where n is the index of refraction.

$$n = \sqrt{1 - \frac{V}{E}}$$

At E_c, $\sin \alpha'' = 1$ and one obtains $E_c = 2V = 10$ V. The reflection coefficient is given by

$$R = \left| \frac{\cos \alpha - n \cos \alpha''}{\cos \alpha + n \cos \alpha''} \right|$$

For the problem under discussion

$$\cos \alpha'' = \sqrt{1 - \sin^2 \alpha''} = \sqrt{1 - \frac{1}{2n^2}}$$

The critical value of $2n^2$ is $2n_c^2 = 1$. If $E < E_c$, then $2n^2 < 1$ and $\cos \alpha''$ becomes imaginary so that for these values of incident energy, R assumes the form $R = |\bar{z}/z|$, where z is a complex number and $\bar{z}$ is its conjugate. It follows that $R = 1$ for $E < E_c$.

8.43   This problem addresses the notion of *supersymmetry*[1] relevant to eigenenergies of a given Hamiltonian.

(a)   Working in one dimension, and in units $\hbar^2/2m = 1$, and deleting hats over operators, show that, apart from the ground state, the two Hamiltonians

$$H_\pm = \frac{p^2}{2m} + [W^2 \pm W']$$

$$W(x) = - (\psi_0'/\psi_0)$$

$$H_- \psi_0 = 0$$

have the same eigenenergies, where a prime denotes differentiation.

(b)   Obtain $H_\pm$ relevant to the harmonic oscillator (with $V = x^2$). Show explicitly that $H_- \psi_0 = 0$.

8.7   LINEAR COMBINATION OF ATOMIC ORBITALS (LCAO) APPROXIMATION

The LCAO approximation is employed to estimate the states of a molecule in terms of a linear combination s of quantum states ("orbitals") of isolated constituent atoms.

---

[1] For further discussion, see W. Keung, E. Kovacs and U. P. Sukhatme, *Phys. Rev. Letts.* **60**, 41 (1988).

This formalism finds wide application in solid-state physics in the tight-binding approximation.[1]

In the present analysis, the LCAO method is employed to estimate the ground-state energy and wavefunction of a given molecule.

The Molecular Ion

We will apply the LCAO method to the potential shown in Fig. 8.30, which, in the present discussion, is representative of a one-dimensional model of a homonuclear, diatomic molecular ion, such as H_2^+. For fixed internucleon spacing this example permits our analysis to remain purely one-dimensional, with the displacement of the valence electron as the only free variable.

The Hamiltonian for our system is given by

$$(8.118) \qquad \hat{H} = \frac{\hat{p}_x^2}{2m} + V_1(x) + V_2(x)$$

where $V_1(x)$ and $V_2(x)$ are identical square wells separated by the distance $2d$, as illustrated in Fig. 8.30.

Let us introduce the normalized ground-state wavefunctions $\varphi_1(x)$ and $\varphi_2(x)$ relevant to well 1 and well 2, respectively, given by the symmetric even wavefunction (8.24). See Fig. 8.31. For wells sufficiently separated, $d \gg a$, it is evident that

$$(8.119) \qquad \varphi_1(x)V_2(x) = \varphi_2(x)V_1(x) = 0$$

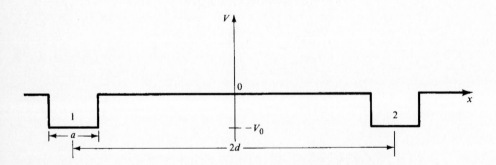

FIGURE 8.30   **Configuration of two identical wells (1 and 2) separated by the distance $2d$.**

[1] For further discussion, see C. W. N. Cumper, *Wave Mechanics for Chemists,* Heineman, London, 1966; W. A. Harrison, *Electronic Structure and the Properties of Solids,* Dover, New York, 1989.

FIGURE 8.31 Ground states of isolated atoms corresponding to the configuration depicted in Fig. 8.30.

With these equalities at hand, we may write

(8.120)
$$\hat{H}\varphi_1 = E_0\varphi_1$$
$$\hat{H}\varphi_2 = E_0\varphi_2$$

where E_0 is the ground state of an isolated well. It follows that the ground state of our composite decoupled molecule is doubly degenerate. Furthermore, as is evident from (8.120), this ground-state energy is the same as that of either one of the isolated atoms. See Fig. 8.32.

When atoms are brought together to form a molecular ion, it is assumed that the Hamiltonian maintains the summational form (8.118) but that φ_1 and φ_2 are no longer eigenstates of $\hat{H}$.

Ground-State Energy and Wavefunction

To estimate the ground state of our molecular ion in the LCAO approximation, one writes

(8.121)
$$\varphi(x) = c_1\varphi_1(x) + c_2\varphi_2(x)$$

where $\varphi_1(x)$ and $\varphi_2(x)$ remain centered about respective atomic sites.

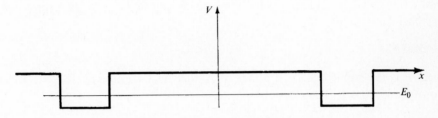

FIGURE 8.32 In the limit that atoms grow uncoupled, the ground-state energy of the composite molecule is the same as that of one of its isolated atoms.

To evaluate the coefficients c_1 and c_2, we recall Problems 4.28 and 4.30 according to which the ground state may be obtained through the variational statement

$$(8.122) \qquad\qquad \delta\langle E\rangle = 0$$

where

$$(8.123) \qquad\qquad \langle E\rangle = \frac{\langle\varphi|H|\varphi\rangle}{\langle\varphi|\varphi\rangle}$$

In developing the variational statement (8.122), we assume c_1, c_2, φ_1, and φ_2 to be real. There results

$$(8.124) \quad \langle E\rangle = \frac{\langle c_1\varphi_1 + c_2\varphi_2|\hat{H}|c_1\varphi_1 + c_2\varphi_2\rangle}{\langle c_1\varphi_1 + c_2\varphi_2|c_1\varphi_1 + c_2\varphi_2\rangle} = \frac{c_1{}^2 H_{11} + c_2{}^2 H_{22} + 2c_1 c_2 H_{12}}{c_1{}^2 + c_2{}^2 + 2c_1 c_2 S}$$

where

$$(8.124a) \qquad\qquad S = \langle\varphi_1|\varphi_2\rangle < 1$$

represents the overlap integral between φ_1 and φ_2. Due to the symmetry of our configuration,

$$H_{11} = \langle\varphi_1|\hat{H}\varphi_1\rangle = H_{22} = \langle\varphi_2|\hat{H}\varphi_2\rangle$$

As c_1 and c_2 are the only free variables in (8.124), to minimize $\langle E\rangle$ it suffices to set

$$(8.125) \qquad\qquad \frac{\partial\langle E\rangle}{\partial c_1} = \frac{\partial\langle E\rangle}{\partial c_2} = 0$$

There results

$$(8.126) \qquad \begin{aligned} c_1(H_{11} - \langle E\rangle) + c_2(H_{12} - \langle E\rangle S) &= 0 \\ c_1(H_{12} - \langle E\rangle S) + c_2(H_{22} - \langle E\rangle) &= 0 \end{aligned}$$

With c representing a column vector with components (c_1, c_2), the preceding equations may be written in the matrix form

$$(8.126a) \qquad\qquad Rc = 0$$

where R is the implied coefficient matrix. A nontrivial solution to (8.126a) occurs, providing $\det R = 0$. There results

$$(8.127) \qquad \begin{vmatrix} H_{11} - \langle E\rangle & H_{12} - \langle E\rangle S \\ H_{12} - \langle E\rangle S & H_{11} - \langle E\rangle \end{vmatrix} = 0$$

which gives

(8.128)
$$\langle E \rangle_{\pm} = \frac{H_{11} \pm H_{12}}{1 \pm S}$$

Substituting this finding into (8.126) reveals that $\langle E \rangle_-$ corresponds to the odd wavefunction $(c_1 = -c_2)$ and $\langle E \rangle_+$ to the even wavefunction $(c_1 = c_2)$. Consequently, we conclude that the even wavefunction, φ_+, corresponds to $\langle E \rangle_+$ and that the odd wavefunction, φ_-, corresponds $\langle E \rangle_-$. These states, with corresponding energies, are listed below (with $\langle E \rangle_{\pm}$ replaced by $E_{\pm}$).

(8.129)
$$E_- = \frac{H_{11} - H_{12}}{1 - S}, \qquad \varphi_- = \frac{A_-}{\sqrt{2}}(\varphi_1 - \varphi_2)$$

$$E_+ = \frac{H_{11} + H_{12}}{1 + S}, \qquad \varphi_+ = \frac{A_+}{\sqrt{2}}(\varphi_1 + \varphi_2)$$

where the normalization constants are given by

(8.129a)
$$A_{\pm} \equiv \frac{1}{\sqrt{1 \pm S}}$$

For the configuration at hand (see Fig. 8.30) we may set

(8.130)
$$H_{11} = -|H_{11}|$$
$$H_{12} = -|H_{12}|$$

Substituting these values into (8.129) reveals that for sufficiently small wavefunction overlap, the ground-state wavefunction and energy for the molecule at hand are given by

(8.131)
$$\varphi_G = \varphi_+$$

$$E_G = E_+ = -\frac{(|H_{11}| + |H_{12}|)}{1 + S}$$

The difference between new energies is

(8.132)
$$\Delta E = E_- - E_+ = 2\frac{(|H_{12}| - S|H_{11}|)}{1 + S^2}$$

For small wavefunction overlap, $S \ll 1$, we find

(8.133)
$$\Delta E = 2|H_{12}|$$

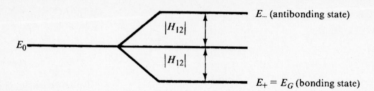

FIGURE 8.33 The unperturbed twofold degeneracy in energy of the uncoupled molecule is removed when atoms interact. For small wavefunction overlap, within the LCAO approximation, the spread in new energies is $2|H_{12}|$. In the bonding state the valence electron partially occupies the midregion between ions, thereby tending to bind them and lower the energy.

Thus we find that the interaction incurred by bringing the atoms in close proximity removes the twofold degeneracy of the uncoupled molecule. [Recall (8.120).] This situation is depicted in Fig. 8.33.

The lower energy of the symmetric state, φ_+, may be understood on the basis of the following. In this state the valence electron has a finite probability of occupying the mid-domain between positive ions, as opposed to the situation in the antisymmetric state, φ_-, for which the electron has zero probability of being found in this domain. Thus, the charge density associated with φ_+ tends to bind the positive ions, thereby lowering the energy of the system. A more realistic description of this situation is given in Section 12.7, which addresses the actual H_2 molecule.

Additional problems in this category of estimating properties of a system are encountered in Chapter 13 under the general topic of *perturbation theory*. In this context the configuration $d \gg a$ in the problem above is viewed as the *unperturbed system*. Perturbation is incurred due to the interaction between atoms, which comes into play with decrease in the separation d. Note in particular that the ground state in the present configuration is doubly degenerate and that the perturbation removes this degeneracy. See Fig. 8.33. That is, with the interaction between atoms "turned on," the twofold degenerate energy E_0 is split into two separate values, $E_\pm$. *Degenerate perturbation theory* is described in detail in Section 13.2.

PROBLEMS

8.44 Consider the following Gaussian form for the wavefunction of atom 1 corresponding to the configuration shown in Fig. 8.30.

$$\varphi_1(x) = A \exp\left[-\left(\frac{x + d}{\sqrt{2}a}\right)^2\right]$$

(a) What is the value of the normalization constant A?

(b) What is the form of the companion wavefunction $\varphi_2(x)$?

(c)   Obtain an expression for the overlap integral S (8.130a). In what manner does your expression vary with the parameter $\lambda \equiv d/a$?

8.45   Show that $E_{\pm}$, as given by (8.128), corresponds to $(c_1/c_2) = \pm 1$ in (8.121).

8.46   Establish the equality $H_{11} = H_{22}$. [See equation above (8.125).]

8.47   A linear molecule comprised of three identical, singly ionized atoms share a single electron. This molecular ion is modeled by three identical quantum wells, each of width a and depth V_0. The wells are separated from each other by the distance d and are symmetrically displayed about $x = 0$.

Let φ_1, φ_2, φ_3 be the ground-state wavefunctions of the three separate wells.

(a)   What is the net charge (in units of e) of the molecular ion?

(b)   Write down the Hamiltonian for the ionized triatomic molecule. What are the criteria for an uncoupled molecule?

(c)   What are the possible states φ of the uncoupled molecule in terms of φ_1, φ_2, φ_3 for the case of equal probabilities of occupation among the three wells?

(d)   What is the LCAO *form* of the ground state for the triatomic molecule? Call your constants c_1, c_2, c_3.

(e)   What is the form of $\langle E \rangle$ for this molecule?

(f)   Estimate the eigenenergies of this molecule corresponding to the wavefunction you have constructed, through the null variation of $\langle E \rangle$. Assume $H_{11} = H_{22} = H_{33} \equiv A$, $H_{12} = H_{21} = B$, and $H_{13} = H_{31} = S_{12} = S_{23} = S_{13} = 0$.

(g)   What are the relations among the constants c_1, c_2, c_3 corresponding to the energies you have found in part (f)? Sketch the corresponding wavefunctions and compare with corresponding sketches for part (c).

(h)   What is the ground-state wavefunction φ_G and ground-state energy E_G for this molecule, within the LCAO approximation?

Answers (partial)

(a) $+2|e|$

(b) $\varphi = \varphi_1 \pm (\varphi_2 \pm \varphi_3)$

(c) $\varphi = c_1\varphi_1 \pm c_2\varphi_2 + c_3\varphi_3$

8.48   A broad frequency band of photons is incident on a rare gas of homonuclear, singly ionized, diatomic molecules. It is noted that photons of frequency $\nu = 0.75 \times 10^{15}$ Hz are absorbed by the gas. Values of constants in the one-dimensional LCAO model of this molecule are $|H_{11}| = 10$ eV, $|H_{12}| = 3$ eV. What is the value of the overlap integral S? State units of your answer.

8.8   DENSITY OF STATES IN VARIOUS DIMENSIONS

We recall (see Problem 2.37) that the density of states $g(E)$ is defined so that $g(E)\,dE$ gives the number of energy states in the interval $E, E + dE$.

In this section we wish to obtain expressions for $g(E)$, relevant to a particle confined to boxes in one, two, and three dimensions. These results are then applied to a quantum-well defined in "slab geometry" important to electron-device physics.[1]

One Dimension

For a one-dimensional box of width L eigenenergies are given by (4.14):

$$(8.134) \qquad E_n = n^2 E_1, \qquad E_1 = \frac{h^2}{8mL^2}$$

It follows that there is an energy state for each value of the quantum number n. As is evident from (8.134), in the limit of large L, the separation between energy states diminishes and the energy spectrum grows quasi-continuous.[2] In this limit we may write

$$(8.135) \qquad g(E) = \frac{\Delta n}{\Delta E} \simeq \frac{dn}{dE}$$

which, with (8.134), gives

$$(8.136) \qquad g_1(E) = \frac{1}{2\sqrt{E_1 E}}$$

where the subscript 1 denotes one dimension. Thus, for a particle confined to a one-dimensional box, the density of states vanishes in the limit of large energy or, equivalently, large quantum number.

Two Dimensions

The two-dimensional square box was discussed in detail in Section 8.5. With the energy eigenvalue equation (8.107) we write

$$(8.137) \qquad E_{n_1 n_2} = E_1(n_1^2 + n_2^2)$$

There is an energy eigenvalue for every pair of quantum numbers (n_1, n_2). It follows that in Cartesian $n_1 n_2$ space, there is an energy eigenstate at every point (n_1, n_2) of

---

[1] For further discussion, see G. Burns, *Solid State Physics,* Academic Press, New York, 1985.

[2] Since energy increments grow with n, this quasi-continuous spectrum is realized for unbounded L.

this space. (Compare with Fig. 2.17.) Let

$$n = \sqrt{n_1^2 + n_2^2}$$

denote the radius vector in this space. Thus, all points that lie in the annular region $(n, n + dn)$ have the same energy $(E \propto n^2)$. Again, in the extreme of large L, eigenenergies form a quasi-continuum and the number of such points is given by the area of the annular region. There results

$$(8.138) \qquad g(E)\, dE = \frac{1}{4} \times 2\pi n\, dn = \frac{\pi}{4E_1}\, dE$$

The factor $\frac{1}{4}$ insures that only positive quantum numbers are included. It follows that in two dimensions, the density of states is given by

$$(8.139) \qquad g_2(E) = \frac{\pi}{4E_1}$$

which is constant.

Three Dimensions

The energy eigenvalue equation for a particle confined to a three-dimensional cubical box is given by (see Table 10.2)

$$(8.140) \qquad E_{n_1 n_2 n_3} = E_1(n_1^2 + n_2^2 + n_3^2)$$

In this case there is an energy eigenvalue corresponding to each point in Cartesian $n_1 n_2 n_3$ space. Again we define the radius vector

$$n = \sqrt{n_1^2 + n_2^2 + n_3^2}$$

and note once more that in the limit of large L all points within the spherical shell, n, $n + dn$, have the same energy. There results

$$g(E)\, dE = \frac{1}{8} \times 4\pi n^2\, dn = \frac{\pi}{4}\, \frac{\sqrt{E}}{E_1^{3/2}}\, dE$$

and

$$(8.141) \qquad g_3(E) = \frac{\pi}{4}\, \frac{\sqrt{E}}{E_1^{3/2}}$$

where the factor of $\frac{1}{8}$ in the preceding equation insures that only positive quantum numbers are included.

Here is a recapitulation of preceding results.

(8.142a)
$$g_1 = \frac{1}{2\sqrt{E_1 E}}$$

(8.142b)
$$g_2 = \frac{\pi}{4E_1}$$

(8.142c)
$$g_3 = \frac{\pi}{4} \frac{\sqrt{E}}{E_1^{3/2}}$$

If the particles we are considering have spin $\frac{1}{2}$, then each of the g-values above is multiplied by the factor 2. (The concept of spin is discussed in Chapter 11.)

Density of States per Unit Volume

With (8.135) we write

(8.143)
$$E_1 = \left(\frac{h^2}{8m}\right)\frac{1}{L^2} \equiv \frac{\alpha}{L^2}$$

where α is the inferred constant, which is seen to be independent of the box dimension L. Inserting this value into (8.142) gives the following expression for the density of states per unit volume for a particle confined to boxes in one-, two-, and three-dimensional Cartesian space, respectively.

(8.144a)
$$\bar{g}_1 = \frac{g_1}{L} = \frac{1}{2\sqrt{\alpha E}}$$

(8.144b)
$$\bar{g}_2 = \frac{g_2}{L^2} = \frac{\pi}{4\alpha}$$

(8.144c)
$$\bar{g}_3 = \frac{g_3}{L^3} = \frac{\pi}{4} \frac{\sqrt{E}}{\alpha^{3/2}}$$

(See Problem 8.49.) Note that in d dimensions, $\bar{g}_d \propto E^{(d-2)/2}$.

Slab Geometry

We wish to apply the preceding results to calculate the density of states of a particle of mass m confined to a well in slab geometry defined by the following potential:

(8.145a)
$$\left.\begin{array}{l} 0 \leq z \leq a \\ 0 \leq x \leq L \\ 0 \leq y \leq L \end{array}\right\} \quad V = 0$$

(8.145b) $V = \infty$, elsewhere

(8.145c) $a \ll L$

See Fig. 8.34.

The density of states for the particle so confined is an appropriate combination of $g_1^{(a)}$ (8.142a) relevant to the one-dimensional z confinement and $g_2^{(L)}$ (8.142b) relevant to the two-dimensional xy confinement.

We first note that the energy eigenfunctions of this particle may be written in the product form [compare with (8.111)]:

(8.146) $$\varphi(x, y, z) = \varphi_a(z)\varphi_L(x, y)$$

where

(8.146a) $$\varphi_a(z) = \sqrt{\frac{2}{a}} \sin \frac{n_1 \pi z}{a}$$

is relevant to the one-dimensional z box and

(8.146b) $$\varphi_L(x, y) = \frac{2}{L} \sin \frac{n_2 \pi x}{L} \sin \frac{n_3 \pi y}{L}$$

is relevant to the two-dimensional xy box.

Eigenenergies are given by the sum

(8.147) $$E_{n_1 n_2 n_3} = \frac{h^2}{8m} \left[\frac{n_1^2}{a^2} + \left(\frac{n_2^2 + n_3^2}{L^2} \right) \right] \equiv E_{n_1}^{(a)} + E_{n_2 n_3}^{(L)}$$

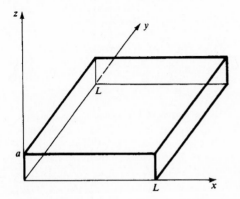

FIGURE 8.34   Sketch of the quantum well corresponding to the potential (8.145), with $L \gg a$.

As $L \gg a$, $E_1^{(L)} \ll E_1^{(a)}$ [recall (8.134)]. For example, if $a = L/10$, then $E_1^{(a)} = 50E_1^{(L)}$. See Fig. 8.35. However, as the wavefunction of the confined particle is the product (8.146), no motion exists for energy less than the ground-state energy of the z motion. Thus, $g(E < E_1^{(a)}) = 0$. Furthermore, the density of states for the xy motion is constant. [See (8.142).]

Thus, for the given configuration we have the following picture. The density of states $g(E)$ is zero for $E < E_1^{(a)}$. At $E = E_1^{(a)}$, two-dimensional motion is allowed and $g(E_1^{(a)}) = g_2^{(L)} = \pi/4E_1^{(L)}$. For larger energy, $g(E)$ maintains this constant value until the second excited state of the z motion is reached. At this value, $g(E_2^{(a)}) = 2g_2^{(L)}$ as now there are two allowed (z) modes allowing for an additional $g_2^{(L)}$ component.

This sequential development of $g(E)$ leads to the formula

(8.148)
$$g(E) = \frac{\pi}{4E_1^{(L)}} \sum_{n=1}^{\infty} \theta(E - E_n^{(a)})$$

where $\theta(x)$ is the step function

(8.148a)
$$\theta(x) = 1, \qquad x > 0$$
$$\theta(x) = 0, \qquad x < 0$$

See Fig. 8.36.

As a check on our calculation we should find that

$$g(E) \rightarrow g_3(E)$$

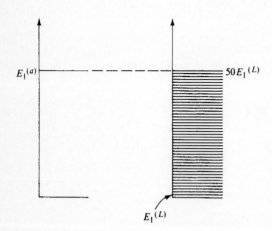

$E_1^{(a)}$

$50E_1^{(L)}$

$E_1^{(L)}$

FIGURE 8.35 Sketch of the eigenenergies corresponding to the z and xy motions in the quantum well (8.145) for the case $a = L/10$. States with energy $E < E_1^{(a)}$ do not exist.

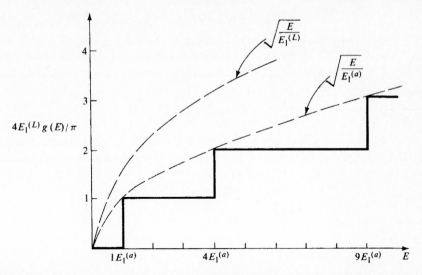

FIGURE 8.36   **At each step of the graph another quantum state (8.146) comes into play. The step function shown above applies to all values of $L \gg a$. In the limit $a \to L$, and L large, $g(E) \to g_3(E)$, appropriate to a large cubical box. In this same limit, the two dashed curves coalesce.**

in the limit

$$a \to L$$

As is evident from Fig. 8.36.

$$\text{Envelope of} \left[\sum_{n=1}^{\infty} \theta(E - E_n{}^{(a)}) \right] = \sqrt{\frac{E}{E_1{}^{(a)}}}$$

In the limit that $a = L$, which is large, the increments of the θ function grow closer and θ is well approximated by the envelope function. Thus

$$g(E) \to \frac{\pi}{4E_1{}^{(L)}} \sqrt{\frac{E}{E_1{}^{(L)}}} = g_3(E)$$

In device physics, the configuration in slab geometry considered above is called a "quantum well" and is realized by epitaxial growth of layers of semiconductors with different band gaps (see Section 12.9). Interfaces between different semiconductors are called *heterojunctions*. A multilayered structure of alternate semiconductors effects an array of quantum wells and is called a *superlattice*.

PROBLEMS

8.49   (a)   Obtain the density of states per unit volume for a free particle of mass m moving in 3-space. Call your answer $\overline{g}_0(E)$.

(b)   Compare your expression for $g_0(E)$ with $\overline{g}_3(E)$ as given by (8.144c). Explain the difference (or equality) between these results.

Answers

(a)   In six-dimensional Cartesian $\mathbf{x} - \mathbf{p}$ space there is an energy state at each point $(\mathbf{x}, \mathbf{p})$. Due to the uncertainty relation, the minimum volume for a state in phase space[1] is h^3. Thus,

$$g_0(E) \, dE = \frac{d\mathbf{x} \, d\mathbf{p}}{h^3}$$

For a free particle

$$E = \frac{p^2}{2m}$$

so that states in momentum space in the spherical shell $(p, p + dp)$ all have the same energy. As the volume of this shell is $4\pi p^2 \, dp$, we obtain

$$\overline{g}_0(E) \, dE = \frac{4\pi p^2 \, dp}{h^3}$$

With the preceding expression there results

$$\overline{g}_0(E) = 2\pi \frac{(2m)^{3/2}}{h^3} \sqrt{E} = \frac{\pi}{4} \frac{\sqrt{E}}{\alpha^{3/2}} = g_3(E)$$

(b)   We conclude that $\overline{g}_0(E)$ and $\overline{g}_3(E)$ are equal. However, it should be borne in mind that in summing states, $\overline{g}_0(E)$ is summed over the discrete energies of the box, whereas $\overline{g}_0(E)$ is integrated over the continuum of free-particle E-values. We may conclude that in any finite volume and energy interval, there are more free-particle energy states than box energy states.

8.50   Describe qualitatively how Fig. 8.36 changes if the well in the z direction of the quantum well is of finite potential height, V_a.

8.51   A particle of mass m is confined to a one-dimensional box of edge length L. Let the edge length be increased to $L + \delta L$. If $\Delta_0 E$ represents the difference between adjacent energies for the first box and $\Delta_1 E$ the difference of adjacent energies for the expanded box, show that

$$\Delta_1 E < \Delta_0 E$$

That is, show that energies grow closer with increase in L.

8.52   A particle of mass m is confined to move in two dimensions within an impenetrable rectangular barrier with edge lengths (a, b). Let the rectangle be situated in the first quadrant

---

[1] If the particle we are considering has, say, spin $\frac{1}{2}$, then this minimum volume is decreased by $\frac{1}{2}$. The concept of spin is discussed in detail in Chapter 11.

of Cartesian space with one corner at the origin and adjoining edges aligned with the (x, y) axes.

(a)   Working in this configuration, obtain the eigenstates, $\varphi_{nn'}(x, y)$, and eigenenergies, $E_{nn'}$, of the particle. Write eigenenergies in terms of E_0, the ground state of a particle in a one-dimensional box of edge length a.

(b)   Show that if a/b is a rational number, then states of the system are degenerate.

8.53   Derive expressions for the normalization constants $B_+ = 2B$, $B_- = 2iB$, appropriate to the even and odd wavefunctions, (8.24) and (8.21), respectively.

Answer

$$1/(2B_\pm)^2 = \frac{1}{k}[ka \pm \sin ka \cos ka] + \frac{1}{\kappa}\left[\begin{matrix} \cos^2 ka \\ \sin^2 ka \end{matrix}\right]$$

8.54   Consider the Schrödinger equation with a periodic potential of period a. Are the eigenfunctions of this equation necessarily periodic? Justify your answer.

8.55   What is the quantum mechanical current density (7.107) for a wavefunction that is (a) purely real, (b) purely imaginary, or (c) a complex constant times a real function?

8.56   (a)   Obtain an expression for the characteristic wavelength, λ, of emission due to an electron in an enclosure of diameter a.

(b)   At what value of a is $\lambda \simeq a$? What is the energy of photons at this value of λ (in eV)?

Answers

(a)   Working in one dimension we find,

$$f \simeq E_1/h = h/8ma^2$$

$$\lambda = \frac{c}{f} = \frac{8mca^2}{h}$$

(b)   With $\lambda_C \equiv h/mc$, we obtain,

$$\lambda = \frac{8a^2}{\lambda_C}$$

At $\lambda = a$, we find $\lambda = \lambda_C/8$. The Compton wavelength, λ_C, corresponds to the frequency, $f \simeq 10^{21}$ Hz. Photons at this wavelength have energy $E \simeq 1$ MeV (γ-radiation).

8.57   (a)   Starting with the relation for group velocity (8.44), derive the scalar expression for effective mass, m^*, given by (8.45).

(b)   In the preceding it was assumed that the matrix $\nabla_k \nabla_k E$ is diagonal so that applied force is collinear with acceleration. More generally, this is not the case. In this event the effective mass becomes a matrix of nine components (corresponding to components of effective mass in the directions $xx, xy, \ldots$). In "vector notation" this matrix is written as the *dyad*,

$\overline{\overline{m^*}}$, or, more precisely,

$$\overline{\overline{\left(\frac{1}{m^*}\right)}}$$

Derive an expression for this dyad. (Note that the inner product of a dyad and a vector is the product of a matrix and a column vector.) What is the expression for the xz component of the (inverse) density matrix?

Answer (partial)

   (a)   In vector notation, the starting equation becomes

$$\mathbf{v}_g = \frac{1}{\hbar}\frac{\partial E}{\partial \mathbf{k}}$$

$$\frac{d\mathbf{v}_g}{dt} = \frac{1}{\hbar}\frac{\partial^2 E}{\partial \mathbf{k}\,\partial t} = \frac{1}{\hbar}\overline{\overline{\left(\frac{\partial^2 E}{\partial \mathbf{k}\,\partial \mathbf{k}}\right)}}\cdot\frac{d\mathbf{k}}{dt}$$

The force equation is written

$$\hbar\frac{d\mathbf{k}}{dt} = \mathbf{F}$$

Inserting this expression in the preceding one obtains

$$\frac{d\mathbf{v}_g}{dt} = \frac{1}{\hbar^2}\overline{\overline{\left(\frac{\partial^2 E}{\partial \mathbf{k}\,\partial \mathbf{k}}\right)}}\cdot\mathbf{F}$$

which gives the desired expression

$$\overline{\overline{\left(\frac{1}{m^*}\right)}} = \frac{1}{\hbar^2}\overline{\overline{\left(\frac{\partial^2 E}{\partial \mathbf{k}\,\partial \mathbf{k}}\right)}}$$

8.58   A system composed of a particle of mass m confined to a quantum well of width $2a$ and depth V has $(N + 1)$ bound states, where $N \gg 1$. The eigenstate of largest energy of this system (with respect to $E = 0$ at the bottom of the well) corresponds to the value $\xi = N(\pi/2) + \epsilon$, where $\epsilon \ll 1$. Show that the ground-state energy measured from the bottom of the well corresponds to the value

$$\xi = \frac{\pi}{2} - \frac{1}{N}$$

Answer

We must solve the equation

$$\xi^2 (1 + \tan^2 \xi) = \rho^2 = \left[N\left(\frac{\pi}{2}\right) + \epsilon \right] \simeq \left[N\left(\frac{\pi}{2}\right) \right]^2$$

or, equivalently,

$$\frac{\xi^2}{\cos^2 \xi} \simeq \left[N\left(\frac{\pi}{2}\right) \right]^2$$

Consider the solution $\xi = (\pi/2) - \epsilon'$, where $\epsilon' \ll 1$. As $\cos[(\pi/2) - \epsilon'] = \sin \epsilon'$, we obtain

$$\frac{\frac{\pi}{2} - \epsilon'}{\sin \epsilon'} \simeq \frac{\pi}{2\epsilon'} = N\left(\frac{\pi}{2}\right)$$

which gives the desired result

$$\epsilon' = \frac{1}{N}$$

8.59 Show that a particle of mass m confined to the potential well

$$V(x) = -aV_0 \, \delta(x)$$

has one bound state. In the preceding expression, a is a length, V_0 is a constant potential, and $\delta(x)$ is the Dirac delta function. (Compare with Problem 7.68.) In the present version, this problem should be examined in the quantum-well formalism of the present chapter.

II

FURTHER DEVELOPMENT OF THE THEORY AND APPLICATIONS TO PROBLEMS IN THREE DIMENSIONS

9

ANGULAR MOMENTUM

Our study of the applications of quantum mechanics to three-dimensional problems begins with a description of the properties of angular momentum. We first consider orbital angular momentum, which is closely akin to angular momentum encountered in classical physics. Angular momentum in quantum mechanics, however, is a more general concept than its classical counterpart. In quantum mechanics, in addition to orbital angular momentum, one also encounters spin angular momentum. Spin angular momentum is an intrinsic, or internal, property of elementary particles such as electrons and photons, and has no classical counterpart. The operators corresponding to the Cartesian components of angular momentum in quantum mechanics obey a set of fixed, fundamental commutator relations. These relations are first derived for orbital angular momentum and then employed as the defining relations for angular momentum in general. Eigenvalues of angular momentum stemming from these commutator relations are obtained, and it is at this point that a distinction between orbital and spin angular momentum first emerges.

9.1 BASIC PROPERTIES

The significance of angular momentum in classical physics is that it is one of the fundamental constants of motion (together with linear momentum and energy) of an isolated system. As we will find, the counterpart of this statement also holds for isolated quantum mechanical systems. This conservation principle for angular momentum stems from the isotropy of space. That is, as described previously in Section 6.3, the physical laws relating to an isolated system are in no way dependent on the orientation of that system with respect to some fixed set of axes in space.

Classically, angular momentum of a particle is a property that depends on the particle's linear momentum **p** and its displacement **r** from some prescribed origin. It is given by (see Fig. 1.9)

(9.1)
$$\mathbf{L} = \mathbf{r} \times \mathbf{p}$$

One may also speak of the angular momentum of a system of particles or of a rigid body. For such extended aggregates, one must add the angular momentum of all particles in the system to obtain the total angular momentum of the system.

Cartesian Components

The classical Cartesian components of the orbital angular momentum **L** for a particle with momentum $\mathbf{p} = (p_x, p_y, p_z)$ at the displacement $\mathbf{r} = (x, y, z)$ are

(9.2)
$$L_x = yp_z - zp_y$$
$$L_y = zp_x - xp_z$$
$$L_z = xp_y - yp_x$$

The quantum mechanical operators $\hat{L}_x$, $\hat{L}_y$, and $\hat{L}_z$, corresponding to these observables, derive their definitions directly from the classical expressions above, with $\hat{\mathbf{p}}$ replaced by its corresponding gradient operator. There follows

(9.3)
$$\hat{L}_x = \hat{y}\hat{p}_z - \hat{z}\hat{p}_y = -i\hbar\left(y\frac{\partial}{\partial z} - z\frac{\partial}{\partial y}\right)$$
$$\hat{L}_y = \hat{z}\hat{p}_x - \hat{x}\hat{p}_z = -i\hbar\left(z\frac{\partial}{\partial x} - x\frac{\partial}{\partial z}\right)$$
$$\hat{L}_z = \hat{x}\hat{p}_y - \hat{y}\hat{p}_x = -i\hbar\left(x\frac{\partial}{\partial y} - y\frac{\partial}{\partial x}\right)$$

In terms of the three-dimensional vector linear momentum operator

(9.4)
$$\hat{\mathbf{p}} = (\hat{p}_x, \hat{p}_y, \hat{p}_z) = -i\hbar\left(\frac{\partial}{\partial x}, \frac{\partial}{\partial y}, \frac{\partial}{\partial z}\right) = -i\hbar\boldsymbol{\nabla}$$

the equations above may be written as the single vector equation

(9.5)
$$\mathbf{L} = -i\hbar\mathbf{r} \times \boldsymbol{\nabla}$$

Commutator Relations

Let us examine the commutation properties of these operators. If, for example, $\hat{L}_x$ does not commute with $\hat{L}_y$, then these components of angular momentum cannot be simultaneously specified in a single state, that is, these operators do not have common eigenfunctions.

To examine this specific question, we employ the basic commutator relation

(9.6) $$[\hat{x}, \hat{p}_x] = i\hbar$$

There follows

$$
\begin{aligned}
(9.7) \quad [\hat{L}_x, \hat{L}_y] &= \hat{L}_x\hat{L}_y - \hat{L}_y\hat{L}_x \\
&= (\hat{y}\hat{p}_z - \hat{z}\hat{p}_y)(\hat{z}\hat{p}_x - \hat{x}\hat{p}_z) - (\hat{z}\hat{p}_x - \hat{x}\hat{p}_z)(\hat{y}\hat{p}_z - \hat{z}\hat{p}_y) \\
&= \hat{x}\hat{p}_y(\hat{z}\hat{p}_z - \hat{p}_z\hat{z}) - \hat{y}\hat{p}_x(\hat{z}\hat{p}_z - \hat{p}_z\hat{z}) \\
&= i\hbar(\hat{x}\hat{p}_y - \hat{y}\hat{p}_x) \\
&= i\hbar\hat{L}_z
\end{aligned}
$$

In similar fashion we obtain

$$
\begin{aligned}
(9.8) \quad & \boxed{\begin{aligned} [\hat{L}_y, \hat{L}_z] &= i\hbar\hat{L}_x \\ [\hat{L}_z, \hat{L}_x] &= i\hbar\hat{L}_y \\ [\hat{L}_x, \hat{L}_y] &= i\hbar\hat{L}_z \end{aligned}}
\end{aligned}
$$

These commutator relations are sometimes combined in the single vector equation

(9.9) $$i\hbar\hat{\mathbf{L}} = \hat{\mathbf{L}} \times \hat{\mathbf{L}}$$

which in determinantal form appears as

(9.10) $$i\hbar(\mathbf{e}_x\hat{L}_x + \mathbf{e}_y\hat{L}_y + \mathbf{e}_z\hat{L}_z) = \begin{vmatrix} \mathbf{e}_x & \mathbf{e}_y & \mathbf{e}_z \\ \hat{L}_x & \hat{L}_y & \hat{L}_z \\ \hat{L}_x & \hat{L}_y & \hat{L}_z \end{vmatrix}$$

As illustrated in Problem 9.1, only one of the three Cartesian components of angular momentum may be specified in a quantum mechanical state. Suppose, for example, that φ is an eigenstate of $\hat{L}_z$. What will measurement of $\hat{L}_x$ find? To answer this question we must bring the superposition principle into play. Expand φ in the eigenstates of $\hat{L}_x$. The squares of the coefficients of expansion give the distribution of probabilities of finding different values of L_x.

Although no two values of the Cartesian components of angular momentum can be simultaneously specified in a quantum mechanical state, if one component, say the value of L_z, is specified, it is still possible to specify an additional property of angular momentum in that state. This additional property is the value of the square of the total angular momentum, L^2, or, equivalently, the magnitude of **L** ($L = \sqrt{\mathbf{L} \cdot \mathbf{L}} = \sqrt{L^2}$).

The total angular momentum operator is the vector operator

(9.11)
$$\hat{\mathbf{L}} = \mathbf{e}_x \hat{L}_x + \mathbf{e}_y \hat{L}_y + \mathbf{e}_z \hat{L}_z$$

from which we may form $\hat{L}^2$.

(9.12)
$$\hat{L}^2 = \hat{L}_x{}^2 + \hat{L}_y{}^2 + \hat{L}_z{}^2$$

To show that there are states of a system in which L_z and L^2 are simultaneously specified, one need merely show that $\hat{L}_z$ and $\hat{L}^2$ commute. Then we know that these operators have simultaneous eigenfunctions. That is, there are states that are eigenfunctions of both $\hat{L}_z$ and $\hat{L}^2$. Let us prove the commutability of $\hat{L}_z$ and $\hat{L}^2$.

$$\begin{aligned}
[\hat{L}_z, \hat{L}^2] &= [\hat{L}_z, \hat{L}_x{}^2 + \hat{L}_y{}^2 + \hat{L}_z{}^2] \\
&= [L_z, L_x{}^2] + [L_z, L_y{}^2] + 0 \\
&= L_x[L_z, L_x] + [L_z, L_x]L_x + L_y[L_z, L_y] + [L_z, L_y]L_y \\
&= i\hbar[L_x L_y + L_y L_x - L_y L_x - L_x L_y] \\
&= 0
\end{aligned}$$

In similar manner we find that $\hat{L}_x$ and $\hat{L}_y$ also commute with $\hat{L}^2$. This must be the case because we have in no way given any special significance to the z direction. In general

(9.13)
$$[\hat{L}_x, \hat{L}^2] = [\hat{L}_y, \hat{L}^2] = [\hat{L}_z, \hat{L}^2] = 0$$
$$[\hat{\mathbf{L}}, \hat{L}^2] = 0$$

It follows that the Cartesian components of $\hat{\mathbf{L}}$ have simultaneous eigenfunctions with $\hat{L}^2$. However, the individual components of $\hat{\mathbf{L}}$ do not have common eigenstates with one another (except for the special case of zero angular momentum). These properties are depicted in a Venn diagram in Fig. 9.1.

The preceding discussion tells us that $\hat{L}^2$ and $\hat{L}_z$, say, have common eigenfunctions. Let us call these eigenfunctions φ_{lm}. The integer indices l and m are related to the eigenvalues of $\hat{L}^2$ and $\hat{L}_z$ as in the following eigenfunction equations.

(9.14)
$$\hat{L}^2 \varphi_{lm} = \hbar^2 l(l+1)\varphi_{lm} \qquad (l = 0, 1, 2, \ldots)$$
$$\hat{L}_z \varphi_{lm} = \hbar m \varphi_{lm} \qquad (m = -l, \ldots, +l \text{ in integral steps})$$

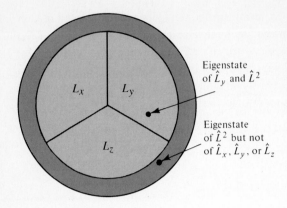

Eigenstate of $\hat{L}_y$ and $\hat{L}^2$

Eigenstate of $\hat{L}^2$ but not of $\hat{L}_x$, $\hat{L}_y$, or $\hat{L}_z$

FIGURE 9.1 Venn diagram for the eigenstates of $\hat{L}_x$, $\hat{L}_y$, $\hat{L}_z$, and $\hat{L}^2$. Every point represents an eigenfunction of $\hat{L}^2$. Depending on which sector the point is in, it is also an eigenfunction of $\hat{L}_x$, $\hat{L}_y$, or $\hat{L}_z$. The state at the center is the null eigenvector of $\hat{\mathbf{L}}$ and $\hat{L}^2$. It corresponds to the eigenvalues $L_x = L_y = L_z = 0$. Peripheral points depict states that are eigenstates of $\hat{L}^2$ only. Can you think of one such function? Note that the space of eigenstates of $\hat{L}^2$ is "bigger" than the space containing all the eigenstates of $\hat{L}_x$, $\hat{L}_y$, and $\hat{L}_z$. Compare with Fig. 5.12.

(These equations are derived in the next section.) The form of the first equation indicates the following. Suppose that a system (e.g., a wheel) is rotating somewhere in space, far removed from other objects. We measure the magnitude of its angular momentum. What possible values can be found? The values that experiment finds are only of the form $L = \hbar\sqrt{l(l + 1)}$, where l is some integer. For example, one would never measure the value $L = \hbar\sqrt{7}$, since it is not of the form $L^2 = \hbar^2 l(l + 1)$. There is no integer for which $l(l + 1) = 7$. This is similar to the fact that a particle in a one-dimensional box is never found to have the energy $E = 7E_1$. This value does not fit the energy eigenvalue recipe $E = n^2 E_1$.

Suppose that we measure the magnitude of angular momentum of the wheel and find the value $L^2 = 30\hbar^2$. This corresponds to the l value $l = 5$. Having measured L^2, the system is left in an eigenstate of $\hat{L}^2$. What value does subsequent measurement of L_z yield? The answer is given by the form of the eigenvalues of $\hat{L}_z$ given in (9.14). For the case in point, since $l = 5$, L_z can *only* be found to have one of the eleven values

$$L_z = 5\hbar, 4\hbar, 3\hbar, 2\hbar, \hbar, 0, -\hbar, -2\hbar, -3\hbar, -4\hbar, -5\hbar$$

Suppose that measurement finds $L_z = 3\hbar$. Then the wheel is left in the state $\varphi_{5,3}$.

The form of equations (9.14) indicates that the eigenvalues of $\hat{L}^2$ are $(2l + 1)$-fold degenerate. For the problem considered, all the eleven states $\varphi_{5,5}$; $\varphi_{5,4}$; . . . ; $\varphi_{5,-5}$ correspond to the same value of L^2 (i.e., $L^2 = 30\hbar^2$). (See Fig. 9.2.)

Uncertainty Relations

Angular momentum is a vector. The square magnitude of this vector is given by L^2. Having measured L^2, it is possible to measure any of the three Cartesian components of $\mathbf{L}$ and leave the system (such as a wheel, a particle, an atom, a rigid rod, etc.) with

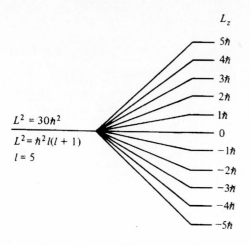

L_z

$5\hbar$
$4\hbar$
$3\hbar$
$2\hbar$
$1\hbar$
0
$-1\hbar$
$-2\hbar$
$-3\hbar$
$-4\hbar$
$-5\hbar$

$L^2 = 30\hbar^2$

$L^2 = \hbar^2 l(l+1)$

$l = 5$

FIGURE 9.2 The eigenvalue $L^2 = \hbar^2 l(l+1)$ is $(2l+1)$-**fold degenerate. For a fixed magnitude,** $L = \hbar\sqrt{l(l+1)}$, **there are only** $2l+1$ **possible projections of L onto a given axis. (See Fig. 9.3c.)**

the same value of L^2 that it had before measurement? Specifically, suppose that we measure L^2 and L_z and find the values $56\hbar^2$ and $3\hbar$, respectively ($l = 7$, $m = 3$). We know that the system is left in a simultaneous eigenstate of L^2 and L_z, namely, $\varphi_{7,3}$.

It is impossible further to resolve the state of the system. We cannot obtain more information on the vector **L** without destroying part of the information already known. Suppose that L_x is measured and the value $5\hbar$ is found. In measuring L_x, the information about L_z previously determined is destroyed.[1] The system is left in a simultaneous eigenstate of $\hat{L}^2$ and $\hat{L}_x$. Since this is not an eigenstate of $\hat{L}_z$, subsequent measurement of L_z is not certain to yield any specific value. Similarly for measurement of L_y. This conclusion is contained in the uncertainty relation

$$(9.15) \qquad \Delta L_y \, \Delta L_z \geq \frac{\hbar}{2}|\langle L_x \rangle| = \frac{\hbar L_x}{2} = \frac{5\hbar^2}{2}$$

Consider the case of a wheel whose center is fixed in space. L^2 and L_z are measured. What motion of the wheel will preserve these values but not preserve L_x and L_y? A very worthwhile model for such motion is given by a classical solution in which the angular momentum vector of constant magnitude precesses about the z axis at a constant inclination to that axis (see Fig. 9.3), thereby maintaining L_z. (Such motion is realized by a spinning top, with fixed vertex, in a gravity field.)

In the classical problem **L** is precisely determined as a function of time. At any instant **L** may be observed and completely specified. Not so for the quantum mechanical motion. If the wheel is in an eigenstate of $\hat{L}^2$ and $\hat{L}_z$, it is in a superposition state (i.e., a linear combination of the eigenstates) of $\hat{L}_x$ or $\hat{L}_y$. At best one can only speak

[1] That is, the outcome of subsequent measurement of L_z is rendered more uncertain.

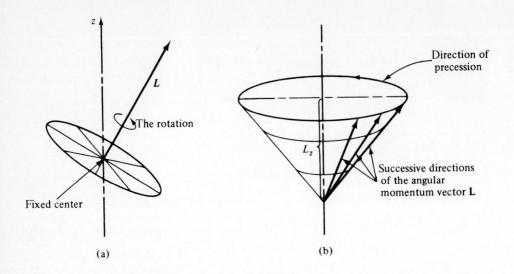

(a)   (b)

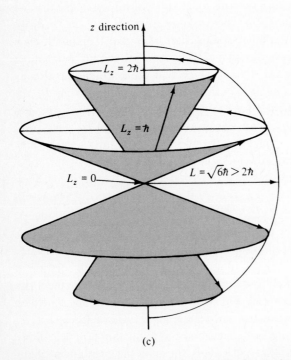

(c)

FIGURE 9.3   (a) The angular momentum vector **L** of a rotating wheel whose center is fixed in space. (b) Classical precession of **L** about the z axis with the constant projection L_z. (c) For $l = 2$, $L^2 = 6\hbar^2$. The only possible orientations of **L** onto the z axis are the five values shown. The precessional motion depicted preserves L^2 and L_z. $\theta = \cos^{-1} 2/\sqrt{6}$ is the smallest possible angle between **L** and the z axis.

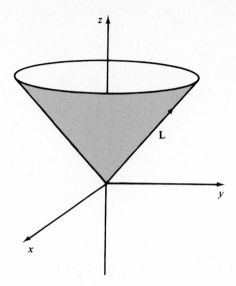

FIGURE 9.4 **For the quantum mechanical state in which L^2 and L_z are specified, L may be pictured as being uniformly distributed over the surface of a cone with half-apex-angle $\theta = \cos^{-1} m/\sqrt{l(l + 1)}$.**

of the *probability* of finding a certain value of L_x or L_y. If a system is in such a state with definite l and m values, it is therefore more consistent to view the related configuration as one in which the **L** vector is uniformly spread over a cone about the z axis with half apex angle $\theta = \cos^{-1} m/\sqrt{l(l + 1)}$ (see Fig. 9.4).

For a given value of L [i.e., $\hbar\sqrt{l(l + 1)}$] the maximum value of L_z is $\hbar l$. But $l < \sqrt{l(l + 1)}$. It follows that the angular momentum vector is never aligned with a given axis. Furthermore, there are only a discrete, finite $(2l + 1)$ number of inclinations that **L** makes with any given axis. This extraordinary property (classical physics permits a continuum of inclinations) is sometimes called the *quantization of space*. For reasons that will become clear in the following sections, l is often referred to as the *orbital quantum number* while m is often referred to as the *azimuthal* or *magnetic quantum number*.

Orbital Versus Spin Angular Momentum

The commutator relations (9.8) are the trademark of angular momentum in quantum mechanics. Although they are consistent with the differential and coordinate-momentum operator relations (9.3 et seq.), they may be taken independent of these and assumed to be the defining relations for quantum mechanical angular momentum. When such is the case, angular momentum need not refer to the space coordinates or linear momentum components of a particle, since the relations (9.8) by themselves do not. The first example that incorporates this concept is given in Section 9.2, where the eigenvalues of angular momentum are obtained using only the commutator relations (9.8). As will be shown in Section 9.3, only a subset of these eigenvalues are relevant to *orbital* angular momentum. Orbital angular momentum derives from the space and momentum coordinates of a particle and is akin to classical ($\mathbf{r} \times \mathbf{p}$) angular momentum. In contrast, *spin* angular momentum does not relate to a particle's coordinates or momenta, nor are the eigenstates of spin dependent on boundary conditions imposed in coordinate space. Spin, as mentioned previously, is an internal property of a particle, like mass or charge. It is an extra degree of freedom attached to a quantum mechanical particle and must be prescribed together with the values of all other compatible properties of a particle in order to designate the state of the particle. In particular, note that spin wavefunctions do not have a coordinate representation. The properties of spin are developed in detail in Chapter 11.

PROBLEMS

9.1 Show that, if a state exists which is a simultaneous eigenstate of $\hat{L}_x$ and $\hat{L}_y$, this state has the eigenvalues $L_x = L_y = L_z = 0$.

Answer

Let φ be the said state. Then

$$0 = [\hat{L}_x, \hat{L}_y]\varphi = i\hbar\hat{L}_z\varphi$$

It follows that φ is an eigenstate of $\hat{L}_z$ corresponding to the eigenvalue $L_z = 0$. (φ is a "null eigenfunction" of $\hat{L}_z$.) From the uncertainty principle, (5.94) and (5.95), and the fact that φ is an eigenstate of $\hat{L}_x$ and $\hat{L}_y$ we find that

$$0 = \Delta L_x \, \Delta L_z \geq \frac{\hbar}{2}|\langle L_y\rangle|$$

Since φ is an eigenstate of $\hat{L}_y$, there is no spread in the values obtained on measurement of L_y in this state. This fact, combined with the preceding equation, gives

$$\langle L_y\rangle = L_y = 0$$

Similarly, $L_x = 0$.

It follows that a state of a system corresponding to finite angular momentum cannot be a simultaneous eigenstate of any two of the Cartesian components of $\hat{\mathbf{L}}$. Furthermore, from the defining equations for $\hat{L}_x$, $\hat{L}_y$, and $\hat{L}_z$ (9.3), it follows that any constant is a simultaneous, null eigenfunction of $\hat{L}_x$, $\hat{L}_y$, and $\hat{L}_z$.

9.2 Show that $\hat{L}_x$ and $\hat{L}^2$ are Hermitian.

Answer (partial)

To prove the Hermiticity of $\hat{L}_x$, we must show that

$$\hat{L}_x = \hat{L}_x^{\dagger}$$

or, equivalently, that

$$(\hat{y}\hat{p}_z - \hat{z}\hat{p}_y)^\dagger = (\hat{y}\hat{p}_z - \hat{z}\hat{p}_y)$$

Look at the $\hat{y}\hat{p}_z$ term.

$$(\hat{y}\hat{p}_z)^\dagger = \hat{p}_z^\dagger \hat{y}^\dagger = \hat{p}_z \hat{y} = \hat{y}\hat{p}_z$$

The last two equalities follow from (a) $\hat{p}_z$ and $\hat{y}$ are Hermitian and (b) $[\hat{y}, \hat{p}_z] = 0$.

9.3 Measurements are made of the angle θ that $\mathbf{L}$ makes with the x axis of a collection of noninteracting rotators, all of which are known to have angular momentum $L = \hbar\sqrt{56}$. What is the minimum θ that will be measured?

9.4 If $[\hat{A}, \hat{L}_x] = [\hat{A}, \hat{L}_y] = [\hat{A}, \hat{L}_z] = 0$, what is $[\hat{A}^2, \hat{L}^2]$?

9.2 EIGENVALUES OF THE ANGULAR MOMENTUM OPERATORS

In this section we derive the eigenvalues of angular momentum that follow from the commutator relations (9.8). Eigenvalues relevant to two classes of angular momentum emerge: orbital and spin. In the remainder of the text $\hat{\mathbf{J}}$ will be used to denote angular momentum in general while $\hat{\mathbf{L}}$ will be reserved for orbital angular momentum and $\hat{\mathbf{S}}$ for spin. The operator $\hat{\mathbf{J}}$ may represent $\hat{\mathbf{L}}$, or $\hat{\mathbf{S}}$, or the combination $\hat{\mathbf{L}} + \hat{\mathbf{S}}$. The defining relations for the components of $\hat{\mathbf{J}}$ are

(9.16)
$$[\hat{J}_x, \hat{J}_y] = i\hbar\hat{J}_z$$
$$[\hat{J}_y, \hat{J}_z] = i\hbar\hat{J}_x$$
$$[\hat{J}_z, \hat{J}_x] = i\hbar\hat{J}_y$$

(9.17)
$$\hat{J}^2 = \hat{J}_x^2 + \hat{J}_y^2 + \hat{J}_z^2$$

The components of $\hat{\mathbf{J}}$ obey all rules obtained above from the commutator relations (9.8). These include:

(9.18)
$$[\hat{J}_x, \hat{J}^2] = [\hat{J}_y, \hat{J}^2] = [\hat{J}_z, \hat{J}^2] = 0$$
$$\Delta J_x \Delta J_y \geq \frac{\hbar}{2}|\langle J_z \rangle|$$

Ladder Operators

We seek the eigenvalues of $\hat{J}^2$ and $\hat{J}_z$. To facilitate the derivation we introduce the "ladder operators" $\hat{J}_+$ and $\hat{J}_-$. The reader will find these similar to the annihilation and creation operators $(\hat{a}, \hat{a}^\dagger)$ introduced in Section 7.2. The ladder operators are defined according to

(9.19)
$$\hat{J}_+ = \hat{J}_x + i\hat{J}_y$$
$$\hat{J}_- = \hat{J}_x - i\hat{J}_y = \hat{J}_+^\dagger$$

Some immediate properties of these operators are

$$[\hat{J}_z, \hat{J}_+] = \hbar\hat{J}_+$$
$$[\hat{J}_z, \hat{J}_-] = -\hbar\hat{J}_-$$

$$\boxed{[\hat{J}_z, \hat{J}_\pm] = \pm\hbar\hat{J}_\pm}$$

(9.20)

$$[\hat{J}^2, \hat{J}_+] = 0$$
$$[\hat{J}^2, \hat{J}_-] = 0$$

$$\boxed{[\hat{J}^2, \hat{J}_\pm] = 0}$$

The latter two equations follow from (9.8). To establish the first two relations one merely inserts the definitions of $\hat{J}_+$ and $\hat{J}_-$. For example,

(9.21)
$$[\hat{J}_z, \hat{J}_+] = [\hat{J}_z, \hat{J}_x + i\hat{J}_y] = [\hat{J}_z, \hat{J}_x] + i[\hat{J}_z, \hat{J}_y]$$
$$= i\hbar\hat{J}_y - i \cdot i\hbar\hat{J}_x = \hbar(\hat{J}_x + i\hat{J}_y) = \hbar\hat{J}_+$$

Other relations that $\hat{J}_+$ and $\hat{J}_-$ satisfy are

$$\hat{J}^2 = \hat{J}_-\hat{J}_+ + \hat{J}_z^2 + \hbar\hat{J}_z$$

(9.22)
$$= \hat{J}_+\hat{J}_- + \hat{J}_z^2 - \hbar\hat{J}_z$$

$$\boxed{\hat{J}^2 = \hat{J}_\mp\hat{J}_\pm + \hat{J}_z^2 \pm \hbar\hat{J}_z}$$

$$[\hat{J}_+, \hat{J}_-] = 2\hbar J_z, \qquad 2(\hat{J}^2 - \hat{J}_z^2) = \hat{J}_+\hat{J}_- + \hat{J}_-\hat{J}_+$$

Consider the relation

(9.23)
$$\hat{J}^2 = (\hat{J}_x - i\hat{J}_y)(\hat{J}_x + i\hat{J}_y) + \hat{J}_z^2 + \hbar\hat{J}_z$$
$$= \hat{J}_x^2 + \hat{J}_y^2 + \hat{J}_z^2 + i(\hat{J}_x\hat{J}_y - \hat{J}_y\hat{J}_x) + \hbar\hat{J}_z$$

With these relations between $\hat{J}^2$, $\hat{J}_z$, $\hat{J}_+$, and $\hat{J}_-$ established we turn to construction of the eigenvalues of $\hat{J}_z$ and $\hat{J}^2$.

Let

(9.24)
$$\hat{J}_z\varphi_m = \hbar m\varphi_m$$

We wish to show that m is either an integer or an odd multiple of one-half. Consider the operation

(9.25)
$$\hat{J}_z\hat{J}_+\varphi_m = (\hbar\hat{J}_+ + \hat{J}_+\hat{J}_z)\varphi_m = (\hbar\hat{J}_+ + \hat{J}_+\hbar m)\varphi_m$$
$$\hat{J}_z(\hat{J}_+\varphi_m) = \hbar(m + 1)(\hat{J}_+\varphi_m)$$

where we have employed (9.21). The latter equation (9.25) implies that $\hat{J}_+\varphi_m$ is an (unnormalized) eigenfunction of $\hat{J}_z$ corresponding to the eigenvalue $\hbar(m + 1)$. That is,

(9.26)
$$\hat{J}_+\varphi_m = \varphi_{m+1}$$

Applying $\hat{J}_+$ again gives

(9.27) $$\hat{J}_+(\hat{J}_+\varphi_m) = \hat{J}_+\varphi_{m+1} = \varphi_{m+2}$$

In a similar manner, we obtain

(9.28) $$\hat{J}_-\varphi_m = \varphi_{m-1}, \qquad \hat{J}_-\varphi_{m-1} = \varphi_{m-2}$$

Thus we have found a scheme of generating a sequence of (unnormalized) eigenfunctions of $\hat{J}_z$ from a single eigenfunction φ_m, with successive values of m in the sequence differing by unity.

(9.29) $$(\ldots, \varphi_{m-2}, \varphi_{m-1}, \varphi_m, \varphi_{m+1}, \varphi_{m+2}, \ldots)$$

Since $\hat{J}^2$ commutes with $\hat{J}_z$, these operators have common eigenfunctions. Let φ_m be a common eigenfunction with the eigenvalue $\hbar^2 K^2$, that is,

(9.30) $$\hat{J}^2\varphi_m = \hbar^2 K^2 \varphi_m$$

Operating on this equation with $\hat{J}_+$ gives [using the third equation in (9.20)]

(9.31) $$\hat{J}_+\hat{J}^2\varphi_m = \hbar^2 K^2(\hat{J}_+\varphi_m) = \hat{J}^2(\hat{J}_+\varphi_m)$$

The last equality asserts that $\hat{J}_+\varphi_m = \varphi_{m+1}$ is also an eigenfunction of $\hat{J}^2$ corresponding to the eigenvalue $\hbar^2 K^2$. It follows that the sequence of eigenfunctions of $\hat{J}_z$ found previously (9.29) are all eigenfunctions of $\hat{J}^2$ corresponding to the same eigenvalue $\hbar^2 K^2$. How many such eigenfunctions are there? From (9.30) one obtains

(9.32) $$\langle J^2 \rangle = \hbar^2 K^2 = \langle J_x^2 \rangle + \langle J_y^2 \rangle + \langle J_z^2 \rangle$$
$$\hbar^2 K^2 = \langle J_x^2 \rangle + \langle J_y^2 \rangle + \hbar^2 m^2$$

where the average has been taken in the φ_m state. If follows that

(9.33) $$\hbar^2 K^2 \geq \hbar^2 m^2$$

(recall $\langle J_x^2 \rangle \geq 0$; see Problem 4.13) or, equivalently,

(9.34) $$|K| \geq |m|$$

For a given value of $K > 0$, the possible values of m in the sequence (9.29) fall between $+K$ and $-K$. If m_{max} is the maximum value that m can assume for a given magnitude of angular momentum, $\hbar K$, then

(9.35) $$\hat{J}_+\varphi_{m_{max}} = 0$$

Similarly,

(9.36) $$\hat{J}_-\varphi_{m_{min}} = 0$$

From (9.22) and the last two equations, one obtains

(9.37)
$$\hat{J}^2 \varphi_{m_{max}} = \hbar^2 K^2 \varphi_{m_{max}} = \hat{J}_z^2 \varphi_{m_{max}} + \hbar \hat{J}_z \varphi_{m_{max}}$$
$$\hbar^2 K^2 = \hbar^2 m_{max}(m_{max} + 1)$$
$$\hat{J}^2 \varphi_{m_{min}} = \hbar^2 K^2 \varphi_{m_{min}} = \hat{J}_z^2 \varphi_{m_{min}} - \hbar \hat{J}_z \varphi_{m_{min}}$$
$$\hbar^2 K^2 = \hbar^2 m_{min}(m_{min} - 1)$$

It follows that

(9.38)
$$m_{max}(m_{max} + 1) = m_{min}(m_{min} - 1)$$

which is satisfied if

(9.39)
$$m_{max} = -m_{min}$$

The possible values that m may assume for a given value of J^2 form a symmetric sequence about $m = 0$ (see Fig. 9.5).

Let us call

(9.40)
$$m_{max} \equiv j$$

Since m runs from $-j$ to $+j$ in unit steps, one obtains

(9.41)

$j =$ an integer   if $m = 0$ is included in the sequence of m values

$j = \dfrac{1}{2} \times$ an odd integer   if $m = 0$ is not included in the sequence of m values

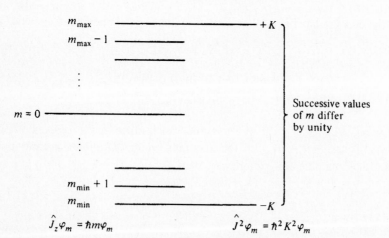

FIGURE 9.5 The possible values that m may assume, for a given value of $J^2 = \hbar^2 K^2$, form a symmetric sequence about $m = 0$.

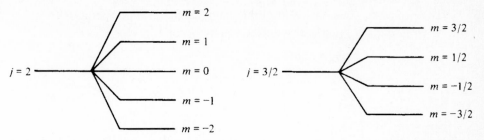

FIGURE 9.6   **The angular momentum quantum number j, which enters in the eigenvalue expression $J^2 = \hbar^2 j(j + 1)$, may be either integral or an odd multiple of one-half. In either case, for a given value of j, the azimuthal quantum number, m, runs from $-j$ to $+j$ in unit steps.**

Furthermore, if j is an integer, the related m values are integers. If j is an odd multiple of one-half, the related m values are odd multiples of one-half (Fig. 9.6).

In either case, inserting $j = m_{max} = -m_{min}$ into (9.37) gives the form of the eigenvalues of $\hat{J}^2$.

$$(9.42) \qquad J^2 = \hbar^2 K^2 = \hbar^2 j(j + 1)$$

Angular Momentum Eigenstates

In this manner we find that the eigenvalues of $\hat{J}^2$ and $\hat{J}_z$ take the form

$$(9.43) \qquad \begin{aligned} J^2 &= \hbar^2 j(j + 1) \\ J_z &= \hbar m_j \qquad (m_j = -j, \ldots, +j) \end{aligned}$$

with j an integer or half an odd integer. The structure of these eigenvalue equations is very significant and is another trademark of quantum mechanical angular momentum. In that they stem directly from the commutation relations (9.16), which in turn are obeyed by all quantum mechanical angular momenta, it follows that such eigenvalue relations are also appropriate to orbital angular momentum, $\hat{\mathbf{L}}$; spin angular momentum, $\hat{\mathbf{S}}$; or their sum, $\hat{\mathbf{L}} + \hat{\mathbf{S}}$. Such, for example, are the eigenvalues of $\hat{L}^2, \hat{L}_z$ as given in (9.14).

As will be shown in the following section, boundary conditions imposed on the common eigenstates of $\hat{L}^2, \hat{L}_z$ suggest that the related eigenvalues (l, m_l) be integral. Thus, of the entire spectrum of quantum angular momentum j values, only a subset $(l = j = \text{integer})$ correspond to orbital angular momentum. The complete j spectrum (integral and half-odd-integral values) will be found to correspond to either spin angular momentum or the combination of spin plus orbital angular momentum. An example of the latter case is given by atomic electrons which have both orbital and spin angular momentum and for which one must write $\hat{\mathbf{J}} = \hat{\mathbf{L}} + \hat{\mathbf{S}}$. For the present we will concentrate on orbital angular momentum.

The eigenvalue equations for the orbital angular momentum operators $\hat{L}^2$ and $\hat{L}_z$ (with m written for m_l), together with the equations for $\hat{L}_{\pm}$, appear as

(9.44)
$$\hat{L}^2 \varphi_{lm} = \hbar^2 l(l + 1)\varphi_{lm}$$
$$\hat{L}_z \varphi_{lm} = \hbar m \varphi_{lm} \qquad (m = -l, \ldots, +l)$$
$$\hat{L}_+ \varphi_{lm} = \varphi_{l,m+1} \qquad (\hat{L}_+ = \hat{L}_x + i\hat{L}_y)$$
$$\hat{L}_- \varphi_{lm} = \varphi_{l,m-1} \qquad (\hat{L}_- = \hat{L}_x - i\hat{L}_y)$$

Since $m = l$ is the maximum value of m and $m = -l$ is the minimum value of m,

(9.45)
$$L_+ \varphi_{ll} = 0$$
$$L_- \varphi_{l,-l} = 0$$

These equations will be used in the next section for the derivation of the φ_{lm} eigenfunctions.

The Rigid Rotator/Dumbbell Molecule

As an application of the preceding results relevant to the eigenvalues of $\hat{L}^2$ and $\hat{L}_z$, let us consider the problem of the energy spectrum of a rigid rotator, or, equivalently, a dumbbell molecule (at sufficiently low temperature[1]). The rotator has two particles each of mass M separated by a weightless rigid rod of length $2a$. The midpoint of the rotator is fixed in space (Fig. 9.7). The moment of inertia of the rotator, about an axis of rotation through this point, is

$$I = 2Ma^2$$

Let the rotator be far removed from any force fields so that its energy is purely kinetic.

(9.46)
$$E = \frac{L^2}{2I}$$

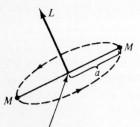

Midpoint fixed in
space

FIGURE 9.7   Rigid rotator with fixed midpoint. Moment of inertia about an axis of rotation through the midpoint is $I = 2Ma^2$.

[1] See footnote, p. 644.

The quantum mechanical Hamiltonian operator is

(9.47)
$$\hat{H} = \frac{\hat{L}^2}{2I}$$

and the time-independent Schrödinger equation for this system appears as

(9.48)
$$\hat{H}\varphi = \left(\frac{\hat{L}^2}{2I}\right)\varphi = E\varphi$$

The eigenvalues of $\hat{H}$ are the same as those of the square angular momentum operator $\hat{L}^2$. With the results obtained we may rewrite the equation above with the l, m indices.

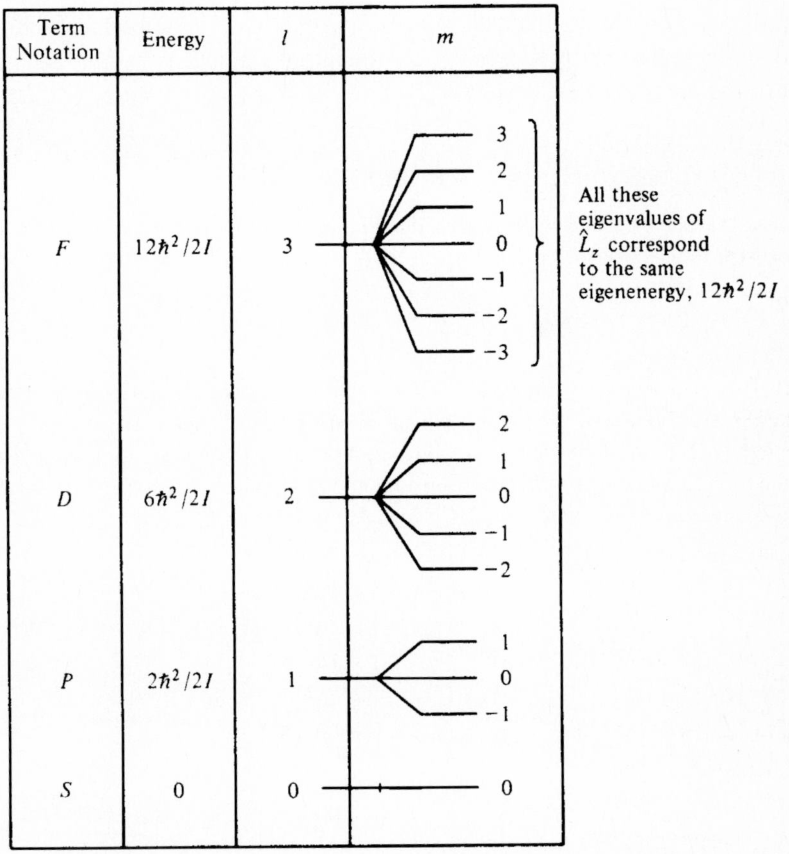

Term Notation	Energy	l	m
F	$12\hbar^2/2I$	3	3 2 1 0 −1 −2 −3
D	$6\hbar^2/2I$	2	2 1 0 −1 −2
P	$2\hbar^2/2I$	1	1 0 −1
S	0	0	0

All these eigenvalues of $\hat{L}_z$ correspond to the same eigenenergy, $12\hbar^2/2I$

FIGURE 9.8   Term diagram for the rigid rotator of moment of inertia, I. The lth eigenenergy, $\hbar^2 l(l+1)/2I$, is $(2l+1)$-fold degenerate.

$$\left(\frac{\hat{L}^2}{2I}\right)\varphi_{lm} = E_l\,\varphi_{lm}$$

(9.49)

$$E_l = \frac{\hbar^2 l(l+1)}{2I}$$

This energy is $(2l + 1)$-fold degenerate. For any value of l, there are $(2l + 1)$ eigenfunctions

(9.50) $\varphi_{l,l}, \ldots, \varphi_{l,-l} = \{\varphi_{lm}\}$

all corresponding to the same eigenenergy, (9.49). The energy of the rotator does not depend on the projection of **L** into the z axis or onto any other prescribed direction. The energy-level diagram for this system is sketched in Fig. 9.8, together with the "term notation" of levels. This notation is common to atomic spectroscopy and will be used in the next three chapters. When a particle is in a state of definite orbital angular momentum, characterized by the quantum number $l = 0, 1, 2, \ldots$, one speaks of the particle being, respectively, in an $S, P, D, F, \ldots$ state.

PROBLEMS

9.5   Show that the frequencies of photons due to energy decays between successive levels of a rotator with moment of inertia I are given by

$$\hbar\omega = \left(\frac{\hbar^2}{I}\right)(l+1), \quad \text{or} \quad \left(\frac{\hbar^2}{I}\right)l$$

9.6   An HCl molecule may rotate as well as vibrate. Discuss the difference in emission frequencies associated with these two modes of excitation. Assume that only $l \to l \pm 1$ transitions between rotational states are allowed. Assume the same for vibrational levels. For rotational levels assume $l \lesssim 50$. Spring constant and moment of inertia may be inferred from the equivalent temperature values for HCl: $\hbar\omega_0/k_B = 4150$ K; $\hbar^2/2Ik_B = 15.2$ K.

9.7   Show that

(a) $[\hat{L}_x, \hat{x}] = 0$ (f) $[\hat{p}_x, \hat{L}_x] = 0$

(b) $[\hat{L}_x, \hat{y}] = i\hbar\hat{z}$ (g) $[\hat{p}_x, \hat{L}_y] = i\hbar\hat{p}_z$

(c) $[\hat{L}_y, \hat{z}] = i\hbar\hat{x}$ (h) $[\hat{p}_y, \hat{L}_z] = i\hbar\hat{p}_x$

(d) $[\hat{L}_z, \hat{x}] = i\hbar\hat{y}$ (i) $[\hat{p}_z, \hat{L}_x] = i\hbar\hat{p}_y$

(e) $[\hat{L}_y, \hat{z}] = [\hat{y}, \hat{L}_z]$ (j) $[\hat{L}_y, \hat{p}_z] = [\hat{p}_y, \hat{L}_z]$

9.8   Calculate

(a) $\hat{L}_z kr$

(b) $\hat{L}_z \sin kr$

(c) $\hat{L}_z f(kr)$

explicitly in Cartesian coordinates, with $r^2 = x^2 + y^2 + z^2$. The function f is an arbitrary function of r, and k is a constant wavenumber.

9.9   (a)   Prove that

$$\hat{\mathbf{\Theta}} \equiv \hat{\mathbf{L}} \times \hat{\mathbf{r}} - i\hbar\hat{\mathbf{r}} = i\hbar\hat{\mathbf{r}} - \hat{\mathbf{r}} \times \hat{\mathbf{L}}$$

(b)   Show that this operator is Hermitian.

(c)   Show that

$$[\hat{L}^2, \hat{\mathbf{r}}] = -2i\hbar\hat{\mathbf{\Theta}}$$

9.10   Show that

$$[\hat{L}_x^2, \hat{L}_y^2] = [\hat{L}_y^2, \hat{L}_z^2] = [\hat{L}_z^2, \hat{L}_x^2]$$

9.11   Evaluate

(a) $[\hat{L}^2, \hat{\mathbf{p}}]$ (c) $[\hat{\mathbf{L}}, \hat{p}^2]$

(b) $[\hat{\mathbf{L}}, \hat{\mathbf{p}}]$ (d) $[\hat{\mathbf{L}}, \hat{\mathbf{L}} \times \hat{\mathbf{L}}]$

Note that parts (b) and (d) have nine components. They are called *dyadic operators*.

9.12   Show that

$$[\hat{L}_x, \hat{r}^2] = [\hat{L}_y, \hat{r}^2] = [\hat{L}_z, \hat{r}^2] = 0$$

9.13   Show that the expression

$$\langle J^2 \rangle = \hbar^2 j(j + 1)$$

is implied directly by the two assumptions:

(a)   The only possible values that the components of angular momentum can have on any axis are $\hbar(-j, \ldots, +j)$.

(b)   All these components are equally probable.

Answer

Because all axes are equivalent,

$$\langle J^2 \rangle = \langle J_x^2 + J_y^2 + J_z^2 \rangle = \langle J_x^2 \rangle + \langle J_y^2 \rangle + \langle J_z^2 \rangle = 3\langle J_x^2 \rangle$$

Since all values of J_x^2 are equally probable,

$$\langle J_x^2 \rangle = \hbar^2 \langle m^2 \rangle = \hbar^2 \frac{\Sigma_{m=-j}^j m^2}{2j + 1} = \frac{2\hbar^2 \Sigma_{m=1}^j m^2}{2j + 1}$$

Substituting the relation

$$\sum_{m=1}^{j} m^2 = \frac{j(j + 1)(2j + 1)}{6}$$

into the above gives

$$\langle J_x^2 \rangle = \frac{\hbar^2 j(j + 1)}{3} = \frac{1}{3}\langle J^2 \rangle$$

$$\langle J^2 \rangle = \hbar^2 j(j + 1)$$

9.3 EIGENFUNCTIONS OF THE ORBITAL ANGULAR MOMENTUM OPERATORS $\hat{L}^2$ AND $\hat{L}_z$

Spherical Harmonics

There are two techniques for obtaining the common eigenfunctions φ_{lm} of the orbital angular momentum operators $\hat{L}^2$ and $\hat{L}_z$. First, one may directly solve the eigenvalue equations

$$\begin{aligned} \hat{L}^2 \varphi_{lm} &= \hbar^2 l(l+1)\varphi_{lm} \\ \hat{L}_z \varphi_{lm} &= \hbar m \varphi_{lm} \end{aligned}$$

(9.51)

Second, one may seek solution to the equation

(9.52) $$\hat{L}_+ \varphi_{ll} = 0$$

Once having found φ_{ll}, the remaining eigenfunctions of $\hat{L}^2$ and $\hat{L}_z$, corresponding to the orbital quantum number l,

(9.53) $$\{\varphi_{lm}\} = (\varphi_{ll}, \varphi_{l,l-1}, \ldots, \varphi_{l,-l})$$

are obtained by applying $\hat{L}_-$ to φ_{ll}. That is,

$$\begin{aligned} \varphi_{l,l-1} &= \hat{L}_- \varphi_{ll} \\ \varphi_{l,l-2} &= \hat{L}_- \varphi_{l,l-1} \end{aligned}$$

(9.54)

In either technique for obtaining the eigenfunctions φ_{lm}, it proves both convenient and practical to work in spherical coordinates (r, θ, ϕ) (see Fig. 1.6). These coordinates are related to the Cartesian coordinates (x, y, z) through the transformation equations

$$\begin{aligned} x &= r \sin\theta \cos\phi \\ y &= r \sin\theta \sin\phi \\ z &= r \cos\theta \end{aligned}$$

(9.55)

With these equations, the Cartesian components of $\hat{\mathbf{L}}$, (9.3), are tranformed to (see Problem 9.14)

$$\begin{aligned} \hat{L}_x &= i\hbar\left(\sin\phi \frac{\partial}{\partial\theta} + \cot\theta \cos\phi \frac{\partial}{\partial\phi}\right) \\ \hat{L}_y &= i\hbar\left(-\cos\phi \frac{\partial}{\partial\theta} + \cot\theta \sin\phi \frac{\partial}{\partial\phi}\right) \\ \hat{L}_z &= -i\hbar \frac{\partial}{\partial\phi} \end{aligned}$$

(9.56)

Using expressions (9.56) we obtain first the ladder operators

(9.57)
$$\hat{L}_+ = \hat{L}_x + i\hat{L}_y = \hbar e^{i\phi}\left(i\cot\theta\,\frac{\partial}{\partial\phi} + \frac{\partial}{\partial\theta}\right)$$

$$\hat{L}_- = \hat{L}_x - i\hat{L}_y = \hbar e^{-i\phi}\left(i\cot\theta\,\frac{\partial}{\partial\phi} - \frac{\partial}{\partial\theta}\right)$$

and second, the operator $\hat{L}^2$

(9.58)
$$\hat{L}^2 = -\hbar^2\left[\frac{1}{\sin\theta}\frac{\partial}{\partial\theta}\left(\sin\theta\,\frac{\partial}{\partial\theta}\right) + \frac{1}{\sin^2\theta}\frac{\partial^2}{\partial\phi^2}\right]$$

We are now prepared to seek solutions to (9.51). This is the first technique, as mentioned above, for finding the eigenstates φ_{lm}. These solutions are quite common to many branches of physics. They are called *spherical harmonics* and are universally denoted by the symbol Y_l^m. Following this protocol we change notation: $\varphi_{lm} \to Y_l^m$.

Angular Momentum and Rotation

Before discussing these solutions we note two points. First, all the angular momentum operators, when expressed in spherical coordinates as listed above, are independent of r. They are functions only of the angular variables (θ, ϕ). This means that the eigenfunctions of $\hat{L}^2$ and $\hat{L}_z$ may be chosen independent of r, that is, $Y_l^m = Y_l^m(\theta, \phi)$. This property stems from the fact that angular momentum operators are related to rotation. For instance, the operator

(9.59)
$$\hat{R}_{\delta\phi} = 1 + i\delta\phi \cdot \frac{\mathbf{L}}{\hbar}$$

(described previously in Section 6.3) when acting on $f(\mathbf{r})$ rotates $\mathbf{r}$ through the azimuthal displacement $\delta\phi$, so that

(9.60)
$$\hat{R}_{\delta\phi} f(\mathbf{r}) = f(\mathbf{r} + \delta\mathbf{r})$$

$$\delta\mathbf{r} = \delta\phi \times \mathbf{r}$$

So the effect of the operation $\delta\phi \cdot \hat{\mathbf{L}}$ on a function $\mathbf{r}$ is to cause a rotational displacement of $\mathbf{r}$. If $\delta\phi$ is parallel to the z axis, $\delta\phi \cdot \hat{\mathbf{L}} = \delta\phi L_z$. This operator induces a rotation of $\mathbf{r}$ about the z axis, without changing the magnitude of $\mathbf{r}$. If we write $f(\mathbf{r}) = f(r, \theta, \phi)$, then $\hat{L}_z$ when operating on f affects only the variable ϕ. When L^2 operates on this function, θ and ϕ are both affected, but not r. So here we have the reason that the eigenstates of $\hat{L}^2$ and $\hat{L}_z$ may be chosen independent of r.

Normalization

The second point we wish to note relates to the normalization of the Y_l^m functions. This normalization is taken over the surface of a unit sphere. The differential element of area dS, on the surface of a sphere of radius a, is conveniently expressed in terms of the element of *solid angle* $d\Omega$.

$$(9.61) \qquad dS = a^2\, d\Omega = a^2 \sin\theta\, d\theta\, d\phi$$

(see Fig. 9.9). The solid angle subtended by dS about the origin is $dS/a^2 = d\Omega$. The solid angle subtended by a sphere (more generally any closed surface) about the origin is

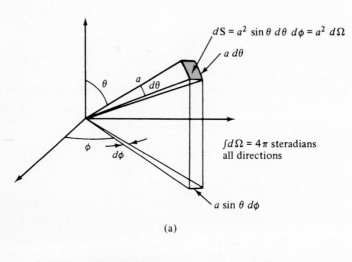

$dS = a^2 \sin\theta\, d\theta\, d\phi = a^2\, d\Omega$

$a\, d\theta$

$\int d\Omega = 4\pi$ steradians
all directions

$a \sin\theta\, d\phi$

(a)

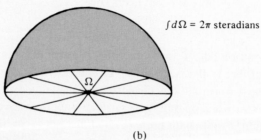

$\int d\Omega = 2\pi$ steradians

(b)

FIGURE 9.9   (a) Element of the solid angle $d\Omega = dS/a^2$. (b) Solid angle subtended by the hemisphere about the origin O is 2π.

(9.62)
$$\int_{\text{all directions}} d\Omega = \int_0^{2\pi} d\phi \int_0^{\pi} \sin\theta \, d\theta$$
$$= \int_0^{2\pi} d\phi \int_{-1}^{1} d\cos\theta = 4\pi \text{ steradians}$$

which is the same as the area of a unit sphere (a sphere of unit radius).

To normalize the eigenfunctions Y_l^m, set

(9.63)
$$\int_{4\pi} |Y_l^m|^2 \, d\Omega = 1$$

which we see is the same as requiring that $|Y_l^m|^2$ integrate to unity over the surface of a unit sphere.

We are now prepared to discuss the solutions to (9.51). The eigenfunction equation for $\hat{L}_z$ gives

(9.64)
$$\frac{\partial}{\partial\phi} Y_l^m = im Y_l^m$$

This equation determines only the ϕ dependence of Y_l^m. If we set

(9.65)
$$Y_l^m(\theta, \phi) = \Phi_m(\phi)\Theta_l^m(\theta)$$

the equation above gives

(9.66)
$$\Phi_m(\phi) = \frac{1}{\sqrt{2\pi}} e^{im\phi}$$

which satisfies the normalization

(9.67)
$$\int_0^{2\pi} d\phi \, |\Phi_m|^2 = 1$$

The index m can be determined from the single valuedness[1] of the wavefunction Φ. That is

(9.68)
$$\Phi(\phi) = \Phi(\phi + 2\pi)$$
$$e^{im\phi} = e^{im(\phi + 2\pi)}$$
$$e^{im2\pi} = 1$$

which is only satisfied for integral values of m:

(9.69)
$$m = 0, 1, 2, \ldots$$

---

[1] On physical grounds it is more appropriate to require that $|\Phi|^2$ be single-valued. However, this can be shown to be equivalent to the single valuedness of Φ for the case in point. For further discussion, see K. Gottfried, *Quantum Mechanics*, W. A. Benjamin, New York, 1966.

As demonstrated in Section 9.2 the values of m run from $-l$ to $+l$, whence l is also an integer. Thus we obtain the result stated previously that the l, m orbital angular momentum quantum numbers are integers only. We also see how this property follows directly from boundary conditions imposed on the wavefunctions Y_l^m. Spin, being an intrinsic property of a particle, is not so constrained, and the related quantum s numbers may assume half-odd-integral as well as integral values.

Legendre Polynomials

We have found that the eigenfunction Y_l^m has the structure

$$(9.70) \qquad Y_l^m = \frac{1}{\sqrt{2\pi}} \, e^{im\phi} \, \Theta_l^m(\theta)$$

Substituting this function into (9.51), together with the explicit expression for $\hat{L}^2$ as given by (9.58), gives the following equation for Θ_l^m (deleting l and m indices, for the moment):

$$(9.71) \qquad \frac{1}{\sin\theta} \frac{d}{d\theta} \left(\sin\theta \frac{d\Theta}{d\theta} \right) + \left[l(l+1) - \frac{m^2}{\sin^2\theta} \right] \Theta = 0$$

or equivalently, in terms of the variable,

$$\mu \equiv \cos\theta$$

$$(9.72) \qquad \frac{d}{d\mu} \left[(1 - \mu^2) \frac{d\Theta}{d\mu} \right] + \left[l(l+1) - \frac{m^2}{1 - \mu^2} \right] \Theta = 0$$

$$-1 \le \mu \le +1$$

Let us outline the method by which this equation is solved.[1] Setting $m = 0$ and $l(l+1) = \lambda$ in (9.72) gives *Legendre's equation*,

$$(9.73) \qquad \frac{d}{d\mu} \left[(1 - \mu^2) \frac{d\Theta_l}{d\mu} \right] + \lambda\Theta_l = 0$$

where we have set $\Theta_l^0 \equiv \Theta_l$. Referring to (9.58), we see that (9.73) is an eigenvalue equation for $\hat{L}^2/\hbar^2$ (corresponding to $L_z = 0$), with eigenvalue λ. A solution to (9.73) may be sought as a series[2] in powers of μ. The requirement that this series solution

[1] For a more detailed description of this method of solution, see E. Merzbacher, *Quantum Mechanics*, 2d ed., Wiley, New York, 1970. A closely related but more concise technique of solution is described in P. Stehle, *Quantum Mechanics*, Holden-Day, San Francisco, 1966.

[2] This method of series solution is explicitly demonstrated in Chapter 10 in the generation of Laguerre polynomials, which are components of the wavefunctions for the hydrogen atom.

remain bounded in the interval $-1 \leq \mu \leq +1$ implies that (1) the eigenvalue λ must be of the form $l(l + 1)$, where $l \geq 0$ and is an integer, and (2) the series solution for Θ_l contains, at most, a finite number of terms. The first conclusion returns the form of the eigenvalues of L^2 given previously by (9.14), namely, $L^2 = \hbar^2 l(l + 1)$. The second conclusion indicates that Θ_l is a polynomial of order l. These polynomials, called *Legendre polynomials,* are commonly denoted as $P_l(\mu)$, so that apart from a multiplicative constant, $\Theta_l(\mu) = P_l(\mu)$. The series summation for this solution may be expressed in the concise form, called the *formula of Rodrigues.*

$$(9.74) \qquad P_l(\mu) = \frac{1}{2^l l!} \frac{d^l}{d\mu^l} (\mu^2 - 1)^l$$

With this solution to (9.73) at hand, the solution to (9.72) is obtained by first constructing the *associated Legendre polynomials.* These are defined by[1] the following differential operation on $P_l(\mu)$:

$$(9.75) \qquad P_l^m(\mu) = (-1)^m (1 - \mu^2)^{m/2} \frac{d^m P_l(\mu)}{d\mu^m}$$

for positive integers $m \leq l$. Differentiating Legendre's equation (9.73) m times with $\lambda = l(l + 1)$, and Θ_l set equal to P_l, and employing the definition (9.75), one readily deduces the equation

$$(9.76) \qquad \frac{d}{d\mu}\left[(1 - \mu^2)\frac{dP_l^m}{d\mu}\right] + \left[l(l + 1) - \frac{m^2}{1 - \mu^2}\right]P_l^m = 0$$

Comparison with (9.72) indicates that $P_l^m(\mu)$ is a solution to this same equation. Furthermore, (9.72) remains unchanged if m is replaced by $-m$, and we may conclude that $P_l^{-m}(\mu)$ is also a solution to this equation, so that apart from a multiplicative constant, P_l^m is equal to P_l^{-m}.

In summary, we have found that the solutions $\Theta_l^m(\mu)$ to (9.72) are given by the associated Legendre polynomials $P_l^m(\mu)$. In addition, we see from the foregoing construction how the quantum conditions (9.14) emerge from the requirements that $Y_l^m(\Theta, \phi)$ remain nonsingular and single-valued in the intervals $-1 \leq \mu \leq +1$, $0 \leq \phi \leq 2\pi$.

---

[1] Another popular notation for these polynomials includes the $(-1)^m$ factor explicitly in the Y_l^m functions.

TABLE 9.1 The first few normalized spherical harmonics and corresponding associated Legendre polynomials[a]

$$Y_l^m(\theta, \phi) = \left[\frac{2l+1}{4\pi}\frac{(l-m)!}{(l+m)!}\right]^{1/2} P_l^m(\cos\theta)e^{im\phi}$$

$$\int_{-1}^{1} d\cos\theta \int_{0}^{2\pi} d\phi\, Y_l^m(Y_{l'}^{m'})^* = \delta_{mm'}\,\delta_{ll'}$$

$P_0 = 1$

$P_1^1 = -\sin\theta$

$P_1^0 = \cos\theta$

$P_1^{-1} = \dfrac{1}{2}\sin\theta$

$P_2^2 = 3\sin^2\theta$

$P_2^1 = -3\sin\theta\cos\theta$

$P_2^0 = \dfrac{1}{2}(3\cos^2\theta - 1)$

$P_2^{-1} = \dfrac{1}{2}\sin\theta\cos\theta$

$P_2^{-2} = \dfrac{1}{8}\sin^2\theta$

$P_3^3 = -15\sin^3\theta$

$P_3^2 = 15\sin^2\theta\cos\theta$

$P_3^1 = -\dfrac{3}{2}\sin\theta(5\cos^2\theta - 1)$

$P_3^0 = \dfrac{1}{2}(5\cos^3\theta - 3\cos\theta)$

$P_3^{-1} = \dfrac{1}{8}\sin\theta(5\cos^2\theta - 1)$

$P_3^{-2} = \dfrac{1}{8}\sin^2\theta\cos\theta$

$P_3^{-3} = \dfrac{1}{48}\sin^3\theta$

$$Y_l^{-l} = \frac{1}{2^l l!}\sqrt{\frac{(2l+1)!}{4\pi}}\,\sin^l\theta\, e^{-il\phi}$$

$$Y_l^0 = \sqrt{\frac{2l+1}{4\pi}}\,P_l(\cos\theta)$$

$$\sum_{m=-l}^{l}|Y_l^m(\theta,\phi)|^2 = \frac{2l+1}{4\pi}$$

$$Y_l^{-m} = (-1)^m(Y_l^m)^*$$

$$Y_0^0 = \left(\frac{1}{4\pi}\right)^{1/2}$$

$$Y_1^1 = -\frac{1}{2}\left(\frac{3}{2\pi}\right)^{1/2}\sin\theta\, e^{i\phi}$$

$$Y_1^0 = \frac{1}{2}\left(\frac{3}{\pi}\right)^{1/2}\cos\theta$$

$$Y_1^{-1} = \frac{1}{2}\left(\frac{3}{2\pi}\right)^{1/2}\sin\theta e^{-i\phi}$$

$$Y_2^2 = \frac{1}{4}\left(\frac{15}{2\pi}\right)^{1/2}\sin^2\theta e^{2i\phi}$$

$$Y_2^1 = -\frac{1}{2}\left(\frac{15}{2\pi}\right)^{1/2}\sin\theta\cos\theta e^{i\phi}$$

$$Y_2^0 = \frac{1}{4}\left(\frac{5}{\pi}\right)^{1/2}(3\cos^2\theta - 1)$$

$$Y_2^{-1} = \frac{1}{2}\left(\frac{15}{2\pi}\right)^{1/2}\sin\theta\cos\theta e^{-i\phi}$$

$$Y_2^{-2} = \frac{1}{4}\left(\frac{15}{2\pi}\right)^{1/2}\sin^2\theta e^{-2i\phi}$$

$$Y_3^3 = -\frac{1}{8}\left(\frac{35}{\pi}\right)^{1/2}\sin^3\theta\, e^{3i\phi}$$

$$Y_3^2 = \frac{1}{4}\left(\frac{105}{2\pi}\right)^{1/2}\sin^2\theta\cos\theta e^{2i\phi}$$

$$Y_3^1 = -\frac{1}{8}\left(\frac{21}{\pi}\right)^{1/2}\sin\theta(5\cos^2\theta - 1)e^{i\phi}$$

$$Y_3^0 = \frac{1}{4}\left(\frac{7}{\pi}\right)^{1/2}(5\cos^3\theta - 3\cos\theta)$$

$$Y_3^{-1} = \frac{1}{8}\left(\frac{21}{\pi}\right)^{1/2}\sin\theta(5\cos^2\theta - 1)e^{-i\phi}$$

$$Y_3^{-2} = \frac{1}{4}\left(\frac{105}{2\pi}\right)^{1/2}\sin^2\theta\cos\theta e^{-2i\phi}$$

$$Y_3^{-3} = \frac{1}{8}\left(\frac{35}{\pi}\right)^{1/2}\sin^3\theta e^{-3i\phi}$$

[a] Defining relations for $P_l(\mu)$ and $P_l^m(\mu)$ are given in Table 9.3. Comparison with other notations for the spherical harmonics and their related functions may be found in D. Park, *Introduction to the Quantum Theory*, 2d ed., McGraw-Hill, New York, 1974.

The precise relation between $\Theta_l{}^m(\mu)$ and $P_l{}^m(\mu)$ as defined by (9.75) follows from the normalization condition (9.63).

$$\int_{4\pi} |Y_l^m|^2 \, d\Omega = \int_0^{2\pi} d\phi \left| \frac{e^{im\phi}}{\sqrt{2\pi}} \right|^2 \int_{-1}^1 d\mu \, |\Theta_l{}^m(\mu)|^2 = 1$$

$$\int_{-1}^1 d\mu \, |\Theta_l{}^m(\mu)|^2 = 1$$

There results

(9.77) $$\Theta_l{}^m(\mu) = \left[\frac{2l+1}{2} \frac{(l-m)!}{(l+m)!} \right]^{1/2} P_l{}^m(\mu)$$

The first few spherical harmonics, Y_l^m, are listed in Table 9.1. Important properties of the Legendre polynomials, P_l, are listed in Table 9.2, while properties of the associated Legendre polynomials, P_l^m, are listed in Table 9.3.

TABLE 9.2   Properties of the Legendre polynomials

Generating formulas

$$(1 - 2\mu s + s^2)^{-1/2} = \sum_{l=0}^{\infty} P_l(\mu) s^l$$

$$P_l(\mu) = \frac{1}{2^l l!} \frac{d^l}{d\mu^l} (\mu^2 - 1)^l \begin{cases} -1 \le \mu \le 1 \\ l = 0, 1, 2, 3, \ldots \end{cases}$$

Legendre's Equation

$$(1 - \mu^2) \frac{d^2 P_l(\mu)}{d\mu^2} - 2\mu \frac{dP_l(\mu)}{d\mu} + l(l+1)P_l(\mu) = 0$$

Recurrence Relations

$$(l+1)P_{l+1}(\mu) = (2l+1)\mu P_l(\mu) - lP_{l-1}(\mu)$$

$$(1 - \mu^2) \frac{d}{d\mu} P_l(\mu) = -l\mu P_l(\mu) + lP_{l-1}(\mu)$$

Normalization and Orthogonality

$$\int_{-1}^1 P_l(\mu) P_m(\mu) \, d\mu = \frac{2}{2l+1} \qquad (l = m)$$

$$= 0 \qquad (l \ne m)$$

The First Few Polynomials

$$P_0 = 1 \qquad P_2 = \frac{1}{2}(3\mu^2 - 1) \qquad P_4 = \frac{1}{8}(35\mu^4 - 30\mu^2 + 3)$$

$$P_1 = \mu \qquad P_3 = \frac{1}{2}(5\mu^3 - 3\mu) \qquad P_5 = \frac{1}{8}(63\mu^5 - 70\mu^3 + 15\mu)$$

Special Values

$$P_l(\mu) = (-1)^l P_l(-\mu) \qquad P_l(1) = 1$$

TABLE 9.3   Properties of the associated Legendre polynomials

Definition

$$P_l^m(\mu) = (-1)^m (1 - \mu^2)^{m/2} \frac{d^m}{d\mu^m} P_l(\mu); \qquad P_l^0 = P_l$$

$$P_l^{-m}(\mu) = (-1)^m \frac{(l - m)!}{(l + m)!} P_l^m(\mu); \qquad P_l^{-1} = \frac{1}{2^l l!} \sin^l \theta$$

For these equations, m is taken as ≥ 0. In the formulas below, however, m may be < 0 also; $l = 0, 1, 2, 3, \ldots, |m| \leq l$.

Differential Equation

$$(1 - \mu^2) \frac{d^2 P_l^m(\mu)}{d\mu^2} - 2\mu \frac{d P_l^m(\mu)}{d\mu} + \left[l(l + 1) - \frac{m^2}{1 - \mu^2} \right] P_l^m(\mu) = 0$$

Recurrence Relations

$$(2l + 1)\mu P_l^m(\mu) = (l - m + 1)P_{l+1}^m(\mu) + (l + m)P_{l-1}^m(\mu)$$

$$(2l + 1)(1 - \mu^2)^{1/2} P_l^m(\mu) = P_{l-1}^{m+1}(\mu) - P_{l+1}^{m+1}(\mu)$$

$$(1 - \mu^2)\frac{d P_l^m(\mu)}{d\mu} = (l + 1)\mu P_l^m(\mu) - (l - m + 1)P_{l+1}^m(\mu)$$

$$= -l\mu P_l^m(\mu) + (l + m)P_{l-1}^m(\mu)$$

$$(1 - \mu^2)^{1/2} P_l^{m+1}(\mu) = (l - m)\mu P_l^m(\mu) - (l + m)P_{l-1}^m(\mu)$$

$$= -(l + m + 1)P_l^m(\mu) + (l - m + 1)P_{l+1}^m(\mu)$$

Normalization and Orthogonality

$$\int_{-1}^{1} P_l^m(\mu)P_k^m(\mu)\, d\mu = \frac{2}{2l + 1} \frac{(l + m)!}{(l - m)!} \qquad (l = k)$$

$$= 0 \qquad (l \neq k)$$

Polar Plots of Y_l^m and Spherical Harmonic Expansions

When a system such as a rigid rotator is in an eigenstate of $\hat{L}^2$ and $\hat{L}_z$, the z axis is said to be *preferred*. Namely, measurement of L_z is certain to find a specific value. However, in this state, it is still true that the x direction is in no way preferred over the y direction. Thus the probability density, $|Y_l^m|^2$, is rotationally symmetric about the z axis or, equivalently (from 9.70), $|Y_l^m|$ is independent of ϕ. The function $|Y_l^m|$ is a surface of revolution about the z axis.

$$(9.78) \quad |Y_l^m| = \frac{1}{\sqrt{2\pi}} |\Theta_l^m(\cos \theta)| = \left[\frac{2l + 1}{4\pi} \frac{(l - m)!}{(l + m)!} \right]^{1/2} |P_l^m(\cos \theta)|$$

Polar plots of these functions for $l = 0, 1, 2$, and all accompanying m values, in any plane through the z axis, are sketched in Fig. 9.10. These diagrams give the probability amplitude of rotational states for, say, a dumbbell molecule. Thus, in the diagram for $|Y_1^1|$, one notes that it is most probable that the molecule rotates in a plane normal to the z axis, corresponding to maximum projection of **L** onto the z axis.

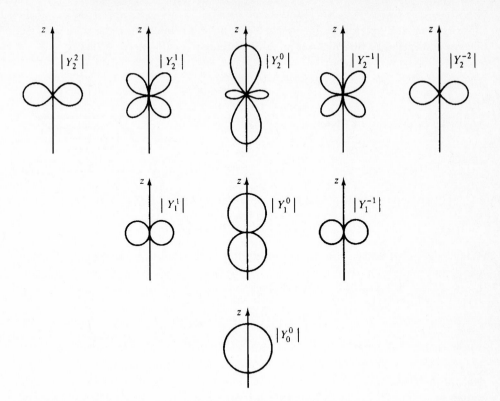

FIGURE 9.10 Polar plots of $|Y_l^m|$ versus θ in any plane through the z axis for $l = 0, 1, 2$. The equality $|Y_l^m| = |Y_l^{-m}|$ is exhibited.

The functions $Y_l^m(\theta, \phi)$ are a basis of the Hilbert space of square-integrable functions $\varphi(\theta, \phi)$ defined on the unit sphere. Such functions may be normalized as follows.

$$(9.79) \qquad \|\varphi(\theta, \phi)\|^2 = \langle \varphi | \varphi \rangle = \int_0^{2\pi} d\phi \int_{-1}^{1} d\cos\theta \; \varphi^*\varphi = 1$$

The expansion of φ in spherical harmonics is given by

$$(9.80) \qquad \varphi(\theta, \phi) = \sum_{l=0}^{\infty} \sum_{|m|\le l} a_{lm} Y_l^m(\theta, \phi)$$

The coefficient of expansion a_{lm} is given by the inner product,

$$(9.81) \qquad a_{lm} = \langle Y_l^m | \varphi \rangle = \int_0^{2\pi} d\phi \int_{-1}^{1} d\cos\theta [Y_l^m(\theta, \phi)]^*\varphi(\theta, \phi)$$

Suppose that at a given instant a system (e.g., a rigid rotator) is in the state $\varphi(\theta, \phi)$. Then the probability that measurement of L^2 finds the value $\hbar^2 l(l + 1)$ is

(9.82)
$$P[\hbar^2 l(l + 1)] = \sum_{m=-l}^{+l} |a_{lm}|^2$$

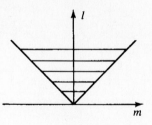

while the probability of finding L_z with the value $\hbar m$ is

(9.83)
$$P(\hbar m) = \sum_{l=|m|}^{\infty} |a_{lm}|^2$$

For example, consider that a rotator is in the state

Domain of (l, m) values

$$\varphi(\theta, \phi) = A \sin^2 \theta \cos 2\phi$$

What values of L^2 and L_z will measurement find? To answer this question, in principle we should first evaluate the coefficients a_{lm} given by the operation (9.81). However, for the case at hand, reference to Table 9.1 reveals that φ is the simple superposition

$$\varphi = A'(Y_2^2 + Y_2^{-2})$$

where A and A' are constants. So the only coefficients that enter the expansion (9.80) are a_{22} and a_{2-2}. We may conclude that measurement will find the value of $L^2 = 6\hbar^2$ with probability 1 and the values $L_z = \pm 2\hbar$ with equal probabilities of $\frac{1}{2}$. No other values of L_z and L^2 would be found for a rotator in the state given above.

Second Construction of the Spherical Harmonics

Let us now turn to the second procedure for finding the Y_l^m eigenfunctions, initiated by (9.52). Consider that we have already solved for the eigenfunction of $\hat{L}_z$, so that φ_{lm} is known to be in the form given by (9.70). Equation (9.52) then becomes

(9.84)
$$\hat{L}_+ e^{il\phi} \Theta_l^l(\theta) = \hbar e^{i\phi}\left(\frac{\partial}{\partial\theta} + i \cot\theta \frac{\partial}{\partial\phi}\right) e^{il\phi} \Theta_l^l(\theta) = 0$$

Bringing the $\exp(il\phi)$ factor through the differential operator gives (deleting the l-scripts for the moment)

(9.85)
$$\frac{\partial}{\partial\theta}\Theta = l \cot\theta\, \Theta$$

Substituting the relation

(9.86)
$$l \cot\theta = \frac{\partial}{\partial\theta} \ln \sin^l \theta$$

and then dividing through by Θ gives

(9.87)
$$\frac{1}{\Theta}\frac{\partial}{\partial\theta}\Theta = \frac{\partial}{\partial\theta}\ln\Theta = \frac{\partial}{\partial\theta}\ln\sin^l\theta$$

This is simply integrated to yield

(9.88)
$$\Theta_l^l = A_{ll}\sin^l\theta$$

where A_{ll} is a normalization constant. It follows that Y_l^l is

(9.89)
$$Y_l^l = \frac{A_{ll}}{\sqrt{2\pi}}\sin^l\theta\,e^{il\phi}$$

which agrees with the values given in Table 9.1. The eigenfunction Y_l^{l-1} is obtained from Y_l^l through the operator $\hat{L}_-$.

(9.90)
$$\hat{L}_-Y_l^l = Y_l^{l-1}$$

In this manner we obtain

(9.91)
$$Y_l^{l-1} = A_{l,l-1}e^{i(l-1)\phi}\sin^{l-1}\theta\cos\theta$$

which is also in agreement with the values given in Table 9.1. The relations between $\hat{L}_+$, $\hat{L}_-$, and the Y_l^m functions with correct normalization factors are given in Table 9.4.

We conclude this section with the following example. Suppose that a rigid rotator is in the eigenstate of $\hat{L}^2$ and $\hat{L}_z$ corresponding to $l = 1$ and $m = 1$ (i.e., Y_1^1). What is the probability that measurement of L_x finds the respective values $m = 0$, ± 1? To answer this question we must expand Y_1^1 in the eigenfunctions of $\hat{L}_x$. These eigenfunctions are solutions to the equation

(9.92)
$$\hat{L}_x X(\theta, \phi) = \hbar\alpha X(\theta, \phi)$$

TABLE 9.4 Normalized relations between $\hat{L}_+$, $\hat{L}_-$, $\hat{L}_x$, $\hat{L}_y$ and the states $|lm\rangle$[a]

$\hat{L}_z|lm\rangle = m\hbar|lm\rangle$

$\hat{L}_+|lm\rangle = \hbar[(l-m)(l+m+1)]^{1/2}|l, m+1\rangle$

$\hat{L}_-|lm\rangle = \hbar[(1+m)(l-m+1)]^{1/2}|l, m-1\rangle$

$\hat{L}_x|lm\rangle = \dfrac{1}{2}\hbar[(l-m)(l+m+1)]^{1/2}|l, m+1\rangle + \dfrac{1}{2}\hbar[(l+m)(l-m+1)]^{1/2}|l, m-1\rangle$

$\hat{L}_y|lm\rangle = -\dfrac{1}{2}i\hbar[(l-m)(l+m+1)]^{1/2}|l, m+1\rangle + \dfrac{1}{2}i\hbar[(l+m)(l-m+1)]^{1/2}|l, m-1\rangle$

$\hat{L}_\pm|lm\rangle = \hbar[l(l+1) - m(m\pm1)]^{1/2}|l, m\pm1\rangle$

[a] These normalization relations also apply to the total angular momentum operators, $\hat{J}_\pm$, $\hat{J}_x$, $\hat{J}_y$, $\hat{J}_z$, and $\hat{J}^2$, where

$$\hat{J}^2|jm_j\rangle = \hbar^2 j(j+1)|jm_j\rangle$$
$$\hat{J}_z|jm_j\rangle = \hbar m_j|jm_j\rangle$$

The student may question why these functions are not simply the spherical harmonics Y_l^m. After all, there is no intrinsic difference between $\hat{L}_x$ and $\hat{L}_z$. The answer is that the eigenfunctions of $\hat{L}_x$ are the Y_l^m functions if we define the x axis as the polar axis, so that θ is angular displacement from the x axis. However, for the problem at hand the z axis is the polar axis and the X functions are a bit more complicated.

Writing $\hat{L}_x$ as

$$(9.93) \qquad \hat{L}_x = \frac{1}{2}(\hat{L}_+ + \hat{L}_-)$$

it is clear that $\hat{L}_x Y_l^m$ gives a combination of spherical harmonics with the same l value. Also, since all Y_l^m functions with $|m| \le l$ are eigenfunctions of $\hat{L}^2$ with eigenvalue $\hbar^2 l(l+1)$, any combination of such functions is an eigenfunction of $\hat{L}^2$ with eigenvalue $\hbar^2 l(l+1)$.

With these properties in mind we seek a solution to (9.92) in the form

$$(9.94) \qquad X = aY_1^1 + bY_1^0 + cY_1^{-1}$$

The problem of finding the eigenfunctions of $\hat{L}_x$ (corresponding to $l = 1$) is then reduced to finding the coefficients a, b, and c in the expression above.

From the properties of $\hat{L}_+$ and $\hat{L}_-$ listed in Table 9.4, we have

$$(9.95) \qquad \begin{aligned} \hat{L}_+ Y_1^0 &= \sqrt{2}\,\hbar Y_1^1 \\ \hat{L}_+ Y_1^{-1} &= \sqrt{2}\,\hbar Y_1^0 \\ \hat{L}_- Y_1^0 &= \sqrt{2}\,\hbar Y_1^{-1} \\ \hat{L}_- Y_1^1 &= \sqrt{2}\,\hbar Y_1^0 \end{aligned}$$

Substituting the expansion (9.94) into the eigenvalue equation (9.92) and using the relations above gives the equation

$$(9.96) \qquad (aY_1^0 + bY_1^1 + bY_1^{-1} + cY_1^0) = \sqrt{2}\,\alpha\,(aY_1^1 + bY_1^0 + cY_1^{-1})$$

Since the Y_l^m functions form a linearly independent sequence, it follows that the only way to guarantee equality for all values of θ and ϕ in the equation above is to set the coefficients of individual Y_l^m functions equal to zero. This gives the following set of three homogeneous algebraic equations:

$$(9.97) \qquad \begin{pmatrix} -\sqrt{2}\,\alpha & 1 & 0 \\ 1 & -\sqrt{2}\,\alpha & 1 \\ 0 & 1 & -\sqrt{2}\,\alpha \end{pmatrix} \begin{pmatrix} a \\ b \\ c \end{pmatrix} = 0$$

A nontrivial solution to these equations occurs only for values of α that make the determinant of the coefficient matrix vanish. Setting the determinant equal to zero, one obtains

(9.98)
$$\alpha(\alpha^2 - 1) = 0$$

which gives the eigenvalues

(9.99)
$$\alpha = 0, \qquad \alpha = 1, \qquad \alpha = -1$$

Substituting these values back into (9.97) gives the (normalized) eigenvectors

$$X_0 = \frac{1}{\sqrt{2}}(Y_1^1 - Y_1^{-1}), \qquad\qquad \alpha = 0$$

(9.100)
$$X_+ = \frac{1}{2}(Y_1^1 + \sqrt{2}\, Y_1^0 + Y_1^{-1}), \qquad \alpha = +1$$

$$X_- = \frac{1}{2}(Y_1^1 - \sqrt{2}\, Y_1^0 + Y_1^{-1}), \qquad \alpha = -1$$

With these eigenfunctions of $\hat{L}_x$ at hand it becomes a matter of inspection to construct the linear combination that gives Y_1^1. It is given by

(9.101)
$$Y_1^1 = \frac{1}{2}(X_+ + \sqrt{2}\, X_0 + X_-)$$

It follows that if the rotator is in the eigenstate of $\hat{L}^2$ and $\hat{L}_z$ corresponding to $l = 1$, $m = 1$, then the probability that measurement of L_x finds the value $+\hbar$ is $\frac{1}{4}$, the probability of finding $-\hbar$ is $\frac{1}{4}$, and the probability of finding 0 is $\frac{2}{4}$ (Fig. 9.11).

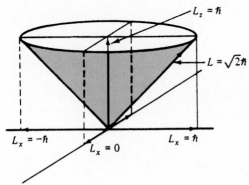

FIGURE 9.11 Given the state $L^2 = 2\hbar^2$, $L_z = \hbar$, what is the probability that measurement of L_x finds the values $\pm\hbar$, 0? Geometrical construction shows that two projections of L give $L_x = 0$, while one projection gives $L_x = +\hbar$ and one projection gives $L_x = -\hbar$.

PROBLEMS

9.14 Use transformation equations (9.55) to obtain the expression

$$\hat{L}_z = -i\hbar\frac{\partial}{\partial\phi}$$

Answer

From (9.55) we obtain the following useful relations.

$$r^2 = x^2 + y^2 + z^2, \qquad \cos\theta = \frac{z}{r}, \qquad \tan\phi = \frac{y}{x}$$

$$\frac{\partial\theta}{\partial x} = \frac{\cos\phi\cos\theta}{r} \qquad\qquad \frac{\partial\phi}{\partial x} = -\frac{y}{x^2}\cos^2\phi$$

$$\frac{\partial\theta}{\partial y} = \frac{\sin\phi\cos\theta}{r} \qquad\qquad \frac{\partial\phi}{\partial y} = \frac{\cos^2\phi}{x}$$

$$\frac{\partial\theta}{\partial z} = -\frac{\sin\theta}{r} \qquad\qquad \frac{\partial\phi}{\partial z} = 0$$

$$\frac{\partial r}{\partial x} = \frac{x}{r}, \qquad \frac{\partial r}{\partial y} = \frac{y}{r}, \qquad \frac{\partial r}{\partial z} = \frac{z}{r}$$

For example, from $\cos\theta = z/r$, one obtains

$$-\sin\theta\,\frac{\partial\theta}{\partial x} = z\,\frac{\partial}{\partial x}\frac{1}{r} = -\frac{zx}{r^3} = -\frac{(z/r)(x/r)}{r} = -\frac{\cos\theta\sin\theta\cos\phi}{r}$$

Substituting these expressions in the expansion

$$\hat{L}_z = -i\hbar\left(x\frac{\partial}{\partial y} - y\frac{\partial}{\partial x}\right)$$

$$= -i\hbar\left[x\left(\frac{\partial\theta}{\partial y}\frac{\partial}{\partial\theta} + \frac{\partial\phi}{\partial y}\frac{\partial}{\partial\phi} + \frac{\partial r}{\partial y}\frac{\partial}{\partial r}\right) - y\left(\frac{\partial\theta}{\partial x}\frac{\partial}{\partial\theta} + \frac{\partial\phi}{\partial x}\frac{\partial}{\partial\phi} + \frac{\partial r}{\partial x}\frac{\partial}{\partial r}\right)\right]$$

gives the desired result.

9.15 (a) What is $[\hat{\phi}, \hat{L}_z]$?

(b) Calculate the root-mean-square deviation $\Delta\phi$ for a particle in the uniform state $\varphi = 1/\sqrt{2\pi}$. (*Hint:* Perform your integrals over the interval $-\pi, \pi$.)

(c) Write down an uncertainty relation seemingly implied by your answer to part (a) and argue the physical inconsistency of this relation in view of your answer to part (b).

Answers

(a) $[\hat{\phi}, \hat{L}_z] = i\hbar$

(b) $\Delta\phi|_{max} = \pi/\sqrt{3}$

(c) One is tempted to write $\Delta\phi\,\Delta L_z \geq \hbar/2$; however, by virtue of the result in part (b), uncertainty in ϕ greater than $\pi/\sqrt{3}$ has little physical meaning. In the extreme that the system is in an eigenstate (e.g., Y_l^m) of $\hat{L}_z$, $\Delta L_z = 0$ and the uncertainty relation gives $\Delta\phi = \infty$. Thus we may conclude that the assumed uncertainty relation is erroneous. [*Note:* Consider the space of functions $\mathfrak{H}_\phi$ whose elements have finite norm on the finite interval $(0, 2\pi)$ (i.e., $\int_0^{2\pi} \varphi^*\varphi\, d\phi < \infty$). It has been pointed out by D. Judge[1] that $\hat{L}_z$ is not Hermitian on this space. As a consequence, the derivation of the uncertainty relation between ϕ and $\hat{L}_z$ from their commutator relation fails. The non-Hermiticity of $\hat{L}_z$ on $\mathfrak{H}_\phi$ may be seen as follows. It is evident that the Hermiticity condition $\langle \hat{L}_z\,\varphi_1 \mid \varphi_2 \rangle = \langle \varphi_1 \mid \hat{L}_z\,\varphi_2 \rangle$ is valid only on the subspace $\mathfrak{H}_\phi' \subset \mathfrak{H}_\phi$ whose elements are periodic: $\varphi(0) = \varphi(2\pi)$. Hence $\hat{L}_z$ is non-Hermitian on $\mathfrak{H}_\phi$. Specifically, note that even though $\varphi(\phi)$ is periodic, the product $\phi\varphi(\phi)$ is not periodic and one may not invoke Hermiticity of $\hat{L}_z$ with respect to functions of this type. This is the crux of the breakdown in the proof of the uncertainty relation. See Problem 5.42.]

9.16   In regard to inconsistencies presented by the azimuthal angle ϕ, as discussed in Problem 9.15, it has been pointed out by W. Louisell[2] that more consistent angle variables are $\sin\phi$ and $\cos\phi$.

(a)   Show that

$$[\sin\phi, \hat{L}_z] = i\hbar\cos\phi$$

$$[\cos\phi, \hat{L}_z] = -i\hbar\sin\phi$$

(b)   Use these commutator formulas to obtain uncertainty relations between $\sin\phi$, L_z and $\cos\phi$, L_z.

Answer (partial)

$$\Delta L_z\, \Delta \sin\phi \geq \frac{\hbar\langle\cos\phi\rangle}{2}$$

(b)

$$\Delta L_z\, \Delta \cos\phi \geq \frac{\hbar\langle\sin\phi\rangle}{2}$$

9.17   (a)   Show that the operator

$$\hat{R}_{\Delta\phi} \equiv \exp\left(\frac{i\,\Delta\phi\hat{L}_z}{\hbar}\right)$$

when acting on the function $f(\phi)$ changes f by a rotation of coordinates about the z axis so that the radius through ϕ is rotated to the radius through $\phi + \Delta\phi$. That is, show that

$$\hat{R}_{\Delta\phi} f(\phi) = f(\phi + \Delta\phi)$$

(b)   Show that the operator

$$\hat{R}_{\Delta\phi} = \exp\left(\frac{i\,\Delta\boldsymbol{\phi} \cdot \hat{\mathbf{L}}}{\hbar}\right)$$

[1] D. Judge, *Nuovo Cimento* **31**, 332 (1964). For further discussion and reference, see P. Carruthers and N. Nieto, *Rev. Mod. Phys.* **40**, 411 (1968).

[2] W. Louisell, *Phys. Lett.* **7**, 60 (1963).

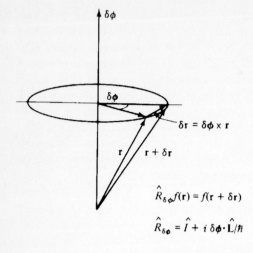

$$\hat{R}_{\delta\phi} f(\mathbf{r}) = f(\mathbf{r} + \delta\mathbf{r})$$

$$\hat{R}_{\delta\phi} = \hat{I} + i\,\delta\boldsymbol{\phi} \cdot \hat{\mathbf{L}}/\hbar$$

FIGURE 9.12   The rotation operator $\hat{R}_{\delta\phi}$ changes $f(\mathbf{r})$ by rotating r through the azimuthal increment $\delta\phi$. (See Problem 9.17.)

when acting on $f(\mathbf{r})$ changes f by rotating $\mathbf{r}$ to a new value on the surface of the sphere of radius r, but rotated away from $\mathbf{r}$ through the azimuth $\Delta\phi$, so that $\mathbf{r}(\theta, \phi) \to \mathbf{r}' = \mathbf{r}(\theta, \phi + \Delta\phi)$. For infinitesimal displacements $\delta\phi$, we may write

$$\hat{R}_{\delta\phi} f(\mathbf{r}) = f(\mathbf{r} + \delta\mathbf{r})$$

$$\delta\mathbf{r} = \delta\boldsymbol{\phi} \times \mathbf{r}$$

See Fig. 9.12.

Answers

(a) $\hat{R}_{\Delta\phi} f = \left[\exp\left(\Delta\phi \frac{\partial}{\partial\phi} \right) \right] f$

$$= f(\phi) + \Delta\phi \frac{\partial f}{\partial\phi} + \frac{(\Delta\phi)^2}{2} \frac{\partial^2 f}{\partial\phi^2} + \cdots = f(\phi + \Delta\phi)$$

(b)   Let $\delta\phi$ be an infinitesimal angle so that $\Delta\phi = n\delta\phi$ in the limit that $n \gg 1$. For the infinitesimal rotation

$$\mathbf{r}' = \mathbf{r} + \delta\mathbf{r} = \mathbf{r} + \delta\boldsymbol{\phi} \times \mathbf{r}$$

so that

$$f(\mathbf{r} + \delta\mathbf{r}) = f(\mathbf{r}) + \delta\boldsymbol{\phi} \times \mathbf{r} \cdot \boldsymbol{\nabla} f(\mathbf{r})$$

$$= f(\mathbf{r}) + \delta\boldsymbol{\phi} \cdot \mathbf{r} \times \boldsymbol{\nabla} f(\mathbf{r})$$

$$= f(\mathbf{r}) + \frac{i}{\hbar} \delta\boldsymbol{\phi} \cdot \mathbf{r} \times \hat{\mathbf{p}} f(\mathbf{r})$$

$$= f(\mathbf{r}) + \frac{i}{\hbar} \delta\boldsymbol{\phi} \cdot \hat{\mathbf{L}} f(\mathbf{r})$$

In the Taylor series expansion of $f(\mathbf{r} + \delta\mathbf{r})$ above we have only kept terms of $O(\delta\phi)$. [The expression $\delta\mathbf{r} = \delta\boldsymbol{\phi} \times \mathbf{r}$ is valid only to terms of $O(\delta\phi)$.] In this manner we obtain

$$f(\mathbf{r} + \delta\mathbf{r}) = \left(\hat{I} + \frac{i}{\hbar}\, \delta\boldsymbol{\phi} \cdot \hat{\mathbf{L}} \right) f(\mathbf{r}) = \hat{R}_{\delta\phi} f(\mathbf{r})$$

For a finite rotational displacement through the angle $\boldsymbol{\Delta\phi} = n\delta\phi$, we apply the operator $\hat{R}_{\delta\phi}$, n times:

$$\hat{R}_{n\delta\phi} = (\hat{R}_{\delta\phi})^n = \left(\hat{I} + \frac{i}{\hbar}\, \delta\boldsymbol{\phi} \cdot \hat{\mathbf{L}} \right)^n$$

and pass to the limit $n \to \infty$ or, equivalently, $\Delta\phi/\delta\phi \to \infty$.

$$\hat{R}_{\Delta\phi} = \lim_{\Delta\phi/\delta\phi \to \infty} \left(\hat{I} + \frac{i}{\hbar}\, \delta\boldsymbol{\phi} \cdot \hat{\mathbf{L}} \right)^{\Delta\phi/\delta\phi} = e^{i\Delta\boldsymbol{\phi} \cdot \hat{\mathbf{L}}/\hbar}$$

(*Note:* The operator $\hat{R}_{\delta\phi}$ rotates $\mathbf{r}$ to $\mathbf{r} + \delta\boldsymbol{\phi} \times \mathbf{r}$ with respect to a fixed coordinate frame. If, on the other hand, the coordinate frame is rotated through $\delta\boldsymbol{\phi}$ with $\mathbf{r}$ fixed in space, then in the new coordinate frame this vector has the value $\mathbf{r} - \delta\boldsymbol{\phi} \times \mathbf{r}$. Thus, rotation of coordinates through $\delta\boldsymbol{\phi}$ is generated by the operator $\hat{R}_{-\delta\phi}$.)

9.18   Show that $\hat{L}^2$ may be written as

$$\hat{L}^2 = -\hbar^2\left(\frac{\partial^2}{\partial\theta^2} + \cot\theta\, \frac{\partial}{\partial\theta} + \frac{1}{\sin^2\theta}\, \frac{\partial^2}{\partial\phi^2} \right)$$

9.19   Show by direct operation that

$$\hat{L}^2 Y_2^2 = 6\hbar^2 Y_2^2$$

$$\hat{L}_z Y_2^2 = 2\hbar Y_2^2$$

9.20   First calculate $P_2(\mu)$ using the generating function $(1 - 2\mu s + s^2)^{-1/2}$. Then obtain $P_2^1(\mu)$ using the relation between P_l and P_l^m given in Table 9.3. Having found P_2^1, form Θ_2^1 and then Y_2^1. Check your answers with the values given in Table 9.1.

9.21   Using the explicit form of Y_l^m, show that

$$\langle Y_l^m | Y_{l'}^{m'} \rangle = \langle lm | l'm' \rangle = 0 \qquad m \neq m'$$

9.22   Operate on Y_l^{l-1} with $\hat{L}_-$ to obtain the angular dependent factor of Y_l^{l-2}.

9.23   Assume that a particle has an orbital angular momentum with z component $\hbar m$ and square magnitude $\hbar^2 l(l + 1)$.

   (a)   Show that in this state

$$\langle L_x \rangle = \langle L_y \rangle = 0$$

   (b)   Show that

$$\langle L_x^2 \rangle = \langle L_y^2 \rangle = \frac{\hbar^2 l(l + 1) - m^2\hbar^2}{2}$$

[*Hints:* For part (a), use $\hat{L}_+$ and $\hat{L}_-$. For part (b), use $\hat{L}^2 = \hat{L}_x^2 + \hat{L}_y^2 + \hat{L}_z^2$.]

9.24   The same conditions hold as in Problem 9.23. What is the expectation of the operator $\frac{1}{2}(\hat{L}_x\hat{L}_y + \hat{L}_y\hat{L}_x)$ in the Y_l^m state?

9.25   A D_2 molecule at 30 K, at $t = 0$, is known to be in the state

$$\psi(\theta,\,\phi,\,0) = \frac{3Y_1^1 + 4Y_7^3 + Y_7^1}{\sqrt{26}}$$

(a)   What values of L and L_z will measurement find and with what probabilities will these values occur?

(b)   What is $\psi(\theta,\,\phi,\,t)$?

(c)   What is $\langle E \rangle$ for the molecule (in eV) at $t > 0$?

(*Note:* For the purely rotational states of D_2, assume that $\hbar/4\pi Ic = 30.4\text{ cm}^{-1}$.)

9.26   At a given instant of time, a rigid rotator is in the state

$$\varphi(\theta,\,\phi) = \sqrt{\frac{3}{4\pi}}\,\sin\phi\,\sin\theta$$

(a)   What possible values of L_z will measurement find and with what probability will these values occur?

(b)   What is $\langle \hat{L}_x \rangle$ for this state?

(c)   What is $\langle \hat{L}^2 \rangle$ for this state?

9.27   Suppose that a rotator is in the state Y_1^{-1}. What values will measurement of $\hat{L}_x$ find and with what probability will these values occur? (*Hint:* Most of the analysis in the text [(9.92) et seq.] involving the expansion of the state Y_1^1 may be used here.)

9.28   A one-particle system is in the angular state Y_l^m. Measurement is made of the component of **L** along the z' axis. The z' axis makes an angle λ with the z axis. What is the expectation of this component? What is the expectation of the square of this component? (See Fig. 9.13.)

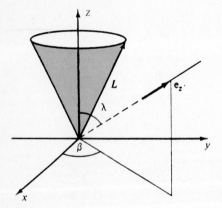

FIGURE 9.13   Configuration relevant to Problem 9.28.

Answer

For the first problem we must calculate $\langle \mathbf{e}_{z'} \cdot \hat{\mathbf{L}} \rangle$, where $\mathbf{e}_{z'}$ is the unit vector in the direction of the z' axis. For the second problem, we must calculate $\langle (\mathbf{e}_{z'} \cdot \hat{\mathbf{L}})^2 \rangle$. The components of $\mathbf{e}_{z'}$ are

$$\mathbf{e}_{z'} = (\sin \lambda \cos \beta, \sin \lambda \sin \beta, \cos \lambda)$$

where β is the azimuthal coordinate of $\mathbf{e}_{z'}$ with respect to the original axes.

$$\langle \mathbf{e}_{z'} \cdot \hat{\mathbf{L}} \rangle = \sin \lambda \cos \beta \langle L_x \rangle + \sin \lambda \sin \beta \langle L_y \rangle + \cos \lambda \langle L_z \rangle = \hbar m \cos \lambda$$

$$\langle (\mathbf{e}_{z'} \cdot \hat{\mathbf{L}})^2 \rangle = \sin^2 \lambda \langle L_x^2 \rangle + \cos^2 \lambda \langle L_z^2 \rangle$$

9.29 With $\Theta_l(\mu)$ replaced by $P_l(\mu)$ in (9.73), show that the single differentiation of this equation gives (9.72) with $\Theta(\mu) = P_l^1(\mu)$ and $m = 1$.

9.4 ADDITION OF ANGULAR MOMENTUM

Two Electrons

In this section we examine the relation between the angular momentum of a total system and that of its constituents. This problem is of practical importance in atomic and nuclear physics where one encounters systems of many particles (e.g., electrons, neutrons, protons, etc.). In many cases one is chiefly concerned with the resultant angular momentum of the atom or nucleus.

Consider two systems that are rotating about a common origin. They could be two rotators or two electrons in an atom (Fig. 9.14). We will speak in terms of an atom. If the angular momentum (neglecting spin) of the first electron is $\mathbf{L}_1$ and that of the second electron is $\mathbf{L}_2$, the magnitude and z component of the total angular momentum of the composite system of the two electrons is

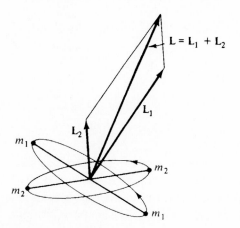

FIGURE 9.14 Classical addition of angular momentum. The two angular momentum vectors $\mathbf{L}_1$ and $\mathbf{L}_2$ add to give the resultant L.

$$\hat{L}^2 = (\hat{\mathbf{L}}_1 + \hat{\mathbf{L}}_2)^2 = \hat{L}_1^{\,2} + \hat{L}_2^{\,2} + 2\hat{\mathbf{L}}_1 \cdot \hat{\mathbf{L}}_2$$

(9.102)

$$\hat{L}_z = \hat{L}_{1z} + \hat{L}_{2z}$$

Suppose that the total system is in a state with definite values of L_{1z}, L_{2z} (e.g., $|m_1 m_2\rangle$). How much further may this state be resolved? Since there are only two good quantum numbers associated with each electron (i.e., $m_1 l_1$ and $m_2 l_2$), one suspects that the composite system will have no more than four good quantum numbers. As it turns out, the eigenstate $|m_1 m_2\rangle$ may further be resolved to the state $|l_1 l_2 m_1 m_2\rangle$. This state cannot be further resolved. For instance, one might wish to measure L^2. If the atom is in the state $|l_1 l_2 m_1 m_2\rangle$ before measurement, we are not assured that it will be in that state after measurement of L^2. That this is so follows from the fact that $\hat{L}^2$ does not commute with, say, $\hat{L}_{1z}$.

(9.103)
$$[\hat{L}_{1z}, \hat{L}^2] = [\hat{L}_{1z}, \hat{L}_1^{\,2} + \hat{L}_2^{\,2} + 2\hat{\mathbf{L}}_1 \cdot \hat{\mathbf{L}}_2]$$
$$= 2[\hat{L}_{1z}, \hat{\mathbf{L}}_1 \cdot \hat{\mathbf{L}}_2] = 2i\hbar(\hat{L}_{1y}\hat{L}_{2x} - \hat{L}_{1x}\hat{L}_{2y})$$

In order to establish that the set of eigenvalues (l_1, l_2, m_1, m_2) are *good quantum numbers* (i.e., that these values may be simultaneously specified in an eigenstate $|l_1 l_2 m_1 m_2\rangle$), one must show that the set of four operators $(\hat{L}_{1z}, \hat{L}_{2z}, \hat{L}_1^{\,2}, \hat{L}_2^{\,2})$ are a set of mutually commuting operators. The fact that no other commuting operators (restricting the discussion to the angular momentum properties of the system) can be attached to this set indicates that $(\hat{L}_{1z}, \hat{L}_{2z}, \hat{L}_1^{\,2}, \hat{L}_2^{\,2})$ is a *complete* set of commuting operators.

We wish to show that

(9.104)
$$[\hat{L}_{1z}, \hat{L}_{2z}] = [\hat{L}_{1z}, \hat{L}_1^{\,2}] = [\hat{L}_{1z}, \hat{L}_2^{\,2}] = [\hat{L}_{2z}, \hat{L}_1^{\,2}]$$
$$= [\hat{L}_{2z}, \hat{L}_2^{\,2}] = [\hat{L}_1^{\,2}, \hat{L}_2^{\,2}] = 0$$

The fact that $[\hat{L}_1^{\,2}, \hat{L}_{1z}] = 0$ was shown in Section 9.1. The commutators $[\hat{L}_{1z}, \hat{L}_{2z}]$ vanish because the coordinates of system 1 are independent of the coordinates of system 2, so that, for example,

$$\left[z_1, \frac{\partial}{\partial z_2} \right] = 0$$

All other terms in (9.104) vanish for similar reasons.

Suppose that we measure L^2 and L_z and establish the state $|lm\rangle$. Can this state be further resolved? Yes. One may subsequently measure $L_1^{\,2}$ and $L_2^{\,2}$ and not destroy the eigenvalues of L^2 and L_z already established. After measurement, the system is left in the state $|lml_1 l_2\rangle$. To show that l, m, l_1, and l_2 are good quantum numbers, we must establish that the set $(\hat{L}_1^2, \hat{L}_2^2, \hat{L}^2, \hat{L}_z)$ is a set of commuting operators. The only questionable pairs are of the form $[\hat{L}_1^2, \hat{L}^2]$ and $[\hat{L}_1^2, \hat{L}_z]$. Expanding these, we obtain

$$[\hat{L}_1^2, \hat{L}_1^2 + \hat{L}_2^2 + 2\hat{\mathbf{L}}_1 \cdot \hat{\mathbf{L}}_2] = 2[\hat{L}_1^2, \hat{\mathbf{L}}_1 \cdot \hat{\mathbf{L}}_2] = 2[\hat{L}_1^2, \hat{\mathbf{L}}_1] \cdot \hat{\mathbf{L}}_2 = 0$$

(9.105)

$$[\hat{L}_1^2, \hat{L}_z] = [\hat{L}_1^2, \hat{L}_{1z} + \hat{L}_{2z}] = [\hat{L}_1^2, \hat{L}_{1z}] = 0$$

Coupled and Uncoupled Representations

Thus we find, in quantum mechanics, that the angular momentum states for a composite system consisting of two subsystems are characterized by either of two sets of good quantum numbers. These correspond, respectively, to the eigenstates $|l_1 l_2 m_1 m_2\rangle$ and $|lml_1 l_2\rangle$. The latter states pertain to problems where the total angular momentum of the composite system is important. We will call this representation where L^2 and L_z (together with L_1^2 and L_2^2) are specified the *coupled representation*. The representation where the z component and magnitude of angular momentum are specified for all subcomponents (i.e., L_1^2, L_{1z}, L_2^2, L_{2z}) will be called the *uncoupled representation* (Fig. 9.15).

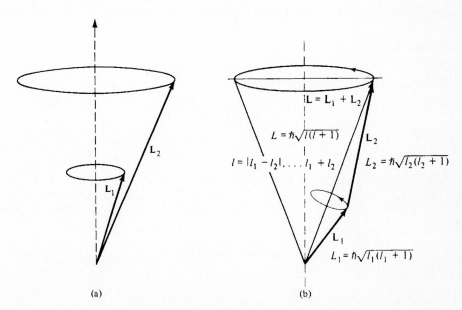

FIGURE 9.15   (a) In the uncoupled representation $\mathbf{L}_1$ and $\mathbf{L}_2$ move independently. Good quantum numbers are l_1, m_1, l_2, m_2. (b) In the coupled representation, $\mathbf{L}_1$ and $\mathbf{L}_2$ couple to give $\mathbf{L}$, which then exhibits discrete orientations along any prescribed axis. In this vector-model sketch of the state $|lml_1 l_2\rangle$, the z components of $\mathbf{L}_1$ and $\mathbf{L}_2$ are not conserved. This corresponds to the fact that most generally, $|lml_1 l_2\rangle$ is a superposition state involving all m_1, m_2 values with fixed $m_1 + m_2 = m$.

The eigenstates in either representation are constructed from products of the eigenstates $|m_1 l_1\rangle$ and $|m_2 l_2\rangle$. In the uncoupled representation the simultaneous eigenstates of $(\hat{L}_1{}^2, \hat{L}_{1z}, \hat{L}_2{}^2, \hat{L}_{2z})$ are given by the products

(9.106) $$|l_1 l_2 m_1 m_2\rangle = |l_1 m_1\rangle |l_2 m_2\rangle$$

or, equivalently,

$$Y_{l_1}{}^{m_1}(\theta_1, \phi_1)Y_{l_2}{}^{m_2}(\theta_2, \phi_2)$$

The spherical coordinates of electron 1 are θ_1, ϕ_1 while θ_2, ϕ_2 are the coordinates of electron 2 (Fig. 9.16). For given values of l_1 and l_2 there are $(2l_1 + 1)(2l_2 + 1)$ linearly independent eigenstates of the composite system each of the form (9.106) and each with specified values of (l_1, l_2, m_1, m_2).

Eigenstates $|lml_1 l_2\rangle$ of the coupled representation are simultaneous eigenstates of the commuting operators

(9.107)
$$\hat{L}^2 = \hat{L}_1{}^2 + \hat{L}_2{}^2 + 2\hat{\mathbf{L}}_1 \cdot \hat{\mathbf{L}}_2$$
$$\hat{L}_z = \hat{L}_{1z} + \hat{L}_{2z}$$
$$\hat{L}_1{}^2, \hat{L}_2{}^2 .$$

Any such state may be written as a superposition of the eigenstates of the uncoupled representation (9.106). In both representation l_1 and l_2 are good quantum numbers. It follows that in the expansion

(9.108) $$|lml_1 l_2\rangle = \sum \sum_{m_1+m_2=m} |l_1 l_2 m_1 m_2\rangle\langle l_1 l_2 m_1 m_2 | lml_1 l_2\rangle$$

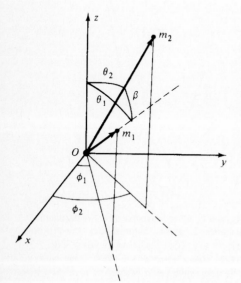

FIGURE 9.16 **Angular coordinates for particle** m_1 **and particle** m_2. **Two important addition theorems involving the angle** β **between** $0m_1$ **and** $0m_2$ **are**

(a) $\cos \beta = \cos \theta_1 \cos \theta_2$
$+ \sin \theta_1 \sin \theta_2 \cos(\phi_1 - \phi_2)$

(b) $P_l(\cos \beta) = \sum_{m=-l}^{l} \dfrac{(l - m)!}{(l + m)!}$
$\times P_l^m(\cos \theta_1)P_l^m(\cos \theta_2)e^{im(\phi_1 - \phi_2)}$
$= \dfrac{4\pi}{2l + 1}$
$\times \sum_{m=-l}^{l} [Y_l^m(\theta_1, \phi_1)]^* Y_l^m(\theta_2, \phi_2)$

summation can only run over the quantum numbers m_1 and m_2. The constraint $m_1 + m_2 = m$ stems from the middle equation (9.107) and the orthogonality of the states $|l_1 l_2 m_1 m_2\rangle$. Equation (9.108) may be rewritten

(9.109)
$$|lml_1 l_2\rangle = \sum_{m_1 + m_2 = m} \sum C_{m_1 m_2} |l_1 l_2 m_1 m_2\rangle$$

$$C_{m_1 m_2} \equiv \langle l_1 l_2 m_1 m_2 | lml_1 l_2\rangle$$

The expansion coefficients $C_{m_1 m_2}$ are called *Clebsch–Gordan* coefficients and their significance is as follows. Let the composite system be two electrons. In the state $|lml_1 l_2\rangle$ it is known that these electrons have respective angular momentum quantum numbers l_1 and l_2, and total angular momentum and z component quantum numbers l and m. The question then arises as to what measurement of L_{1z} and L_{2z} will find in the state $|lml_1 l_2\rangle$. The answer to this question is that

(9.110) $|C_{m_1 m_2}|^2 =$ probability that at fixed L and L_z
measurement finds one electron with $L_{1z} = m_1 \hbar$
and the other electron with $L_{2z} = m_2 \hbar$

As an elementary illustration of the technique employed to construct these coefficients, let us consider expansion of the state

$$|lml_1 l_2\rangle = |1, -1, 1, 1\rangle$$

With $m_1 + m_2 = -1$, the expansion (9.108) becomes

$$|1, -1, 1, 1\rangle = C_{0-1} |1, 0\rangle_1 |1, -1\rangle_2 + C_{-10} |1, -1\rangle_1 |1, 0\rangle_2$$

The coefficients C_{0-1} and C_{-10} are determined by normalization and propitious application of the $\hat{L}_+$ and $\hat{L}_-$ operators. For the case at hand, we operate on the last equation with

$$\hat{L}_- = \hat{L}_{1-} + \hat{L}_{2-}$$

There results (recall Table 9.4)

$\hat{L}_- |1, -1, 1, 1\rangle = 0$
$$= (\hat{L}_{1-} + \hat{L}_{2-})(C_{0-1} |1, 0\rangle_1 |1, -1\rangle_2 + C_{-10} |1, -1\rangle_1 |1, 0\rangle_2)$$
$$= \sqrt{2}(C_{0-1} + C_{-10}) |1, -1\rangle_1 |1, -1\rangle_2$$

We may conclude that

$$C_{0-1} = -C_{-10}$$

Normalization of the state $|1, -1, 1, 1\rangle$ gives $C_{0-1} = 1/\sqrt{2}$. So it is equally probable that measurement finds $(m_1, m_2) = (0, -1)$ or $(-1, 0)$.

Coordinate Representation

We have been writing $|lm\rangle$ for the eigenvectors of $\hat{L}^2$, $\hat{L}_z$. The coordinate representation of these states is given by the projection

$$\langle \theta, \phi | lm \rangle = Y_l^m(\theta, \phi)$$

Likewise, the coordinate representation of the composite state $|l_1 l_2 m_1 m_2\rangle$ is given by the projection

$$\langle \theta_1 \phi_1 \theta_2 \phi_2 | l_1 l_2 m_1 m_2 \rangle = Y_{l_1}^{m_1}(\theta_1 \phi_1) Y_{l_2}^{m_2}(\theta_2, \phi_2)$$

In this manner, the expansion (9.109) gives the coordinate representation

$$\langle \theta_1 \phi_1 \theta_2 \phi_2 | lm l_1 l_2 \rangle = \sum_{m_1 + m_2 = m} \sum C_{m_1 m_2} \langle \theta_1 \phi_1 \theta_2 \phi_2 | l_1 l_2 m_1 m_2 \rangle$$

$$= \sum \sum C_{m_1 m_2} Y_{l_1}^{m_1}(\theta_1, \phi_1) Y_{l_2}^{m_2}(\theta_2, \phi_2)$$

Thus we see that the Clebsch–Gordan expansion affords a means of obtaining the coordinate representation of the composite state $|lm l_1 l_2\rangle$.

The theory of representations is discussed further in Section 11.1 and Appendix A.

Values of l for Two Electrons

Next we consider the problem of finding the allowed values of (l, m), given (l_1, l_2). This problem is directly related to "two-electron atoms," such as He and Ca, whose energy levels are l-dependent. Suppose that one electron is a p electron (i.e., it is in a P state) and the other electron is a d electron. What values can result for l and m (still neglecting spin)?

Let us consider the general case where the two electrons have respective l values l_1 and l_2. Since

$$L_z = L_{1z} + L_{2z}$$

it follows that the maximum value m can have is

$$m_{\text{max}} = m_{1_{\text{max}}} + m_{2_{\text{max}}}$$

or, equivalently,

(9.111)
$$m_{\text{max}} = l_1 + l_2$$

It is clear that of the various values the total angular momentum quantum number l may assume, the maximum value is equal to m_{max}. With (9.111) we then obtain

(9.112)
$$l_{\text{max}} = l_1 + l_2$$

In that l is an angular momentum quantum number, successive values of l differ from l_{max} in unit steps down to some minimum value. What is this minimum value? To obtain it, we note the following.

As noted previously, in the uncoupled representation there are $(2l_1 + 1)(2l_2 + 1)$ independent, common eigenstates of $\hat{L}_1^2, \hat{L}_2^2, \hat{L}_{1z}$, and $\hat{L}_{2z}$ relevant to the two-electron system. These states span a $(2l_1 + 1)(2l_2 + 1)$-dimensional space. A change in representation[1] from this basis to the common eigenstates of $\hat{L}^2$, $\hat{L}_z, \hat{L}_1^2$, and $\hat{L}_2^2$ maintains the dimensionality of this space. This observation affords a method of obtaining l_{min}. That is, keep decreasing l_{max} in unit steps until the total number of independent states equals $(2l_1 + 1)(2l_2 + 1)$.

Now the number of independent eigenstates with a given l number is $2l + 1$. Then the value of l_{min} we seek satisfies the equation

$$\sum_{l=l_{min}}^{l_1+l_2} (2l + 1) = (2l_1 + 1)(2l_2 + 1)$$

This relation is satisfied if we set

(9.113)
$$l_{min} = |l_1 - l_2|$$

In this manner we find that the values of l corresponding to a system comprised of two electrons with respective l values l_1 and l_2 are

(9.114)
$$l = |l_1 - l_2|, \ldots, l_1 + l_2$$

For the problem cited above with one p electron ($l_1 = 1$) and one d electron ($l_2 = 2$), the total angular momentum may be any of the values

$$l = 1, 2, 3$$
$$L = \hbar\sqrt{2}, \qquad L = \hbar\sqrt{6}, \qquad L = \hbar\sqrt{12}$$

There are a totality of

$$N = (2 \times 1 + 1) + (2 \times 2 + 1) + (2 \times 3 + 1) = (2 \times 1 + 1)(2 \times 2 + 1)$$
$$= 15$$

states, corresponding to these three values of l. For the case of two p electrons ($l_1 = l_2 = 1$) there are nine eigenstates. These are listed in Table 9.5.

---

[1] The notion of changes in representation was discussed in Section 7.4 and will be developed further in Chapter 11.

TABLE 9.5 **Nine common eigenstates** $|lml_1l_2\rangle$ **of the operators** $\hat{L}^2, \hat{L}_z, \hat{L}_1^2, \hat{L}_2^2$ **for two** p **electrons**

Diagrammatic Representation[a]	$\|lml_1l_2\rangle$ $= \sum C_{m_1m_2}\|l_1m_1\rangle_1\|l_2m_2\rangle_2$
$\uparrow\uparrow$	$\|22\ 11\rangle\ = \|11\rangle_1\|11\rangle_2$
$\uparrow\cdot + \cdot\uparrow$	$\|21\ 11\rangle\ = \sqrt{\frac{1}{2}}\|11\rangle_1\|10\rangle_2 + \sqrt{\frac{1}{2}}\|10\rangle_1\|11\rangle_2$
$\uparrow\downarrow + \cdot\cdot + \downarrow\uparrow$	$\|20\ 11\rangle\ = \sqrt{\frac{1}{6}}\|11\rangle_1\|1,-1\rangle_2 + \sqrt{\frac{2}{3}}\|10\rangle_1\|10\rangle_2 + \sqrt{\frac{1}{6}}\|1,-1\rangle_1\|11\rangle_2$
$\cdot\downarrow + \downarrow\cdot$	$\|2,-1\ 11\rangle = \sqrt{\frac{1}{2}}\|10\rangle_1\|1,-1\rangle_2 + \sqrt{\frac{1}{2}}\|1,-1\rangle_1\|10\rangle_2$
$\downarrow\downarrow$	$\|2,-2\ 11\rangle = \|1,-1\rangle_1\|1,-1\rangle_2$
$\uparrow\cdot - \cdot\uparrow$	$\|11\ 11\rangle\ = \sqrt{\frac{1}{2}}\|11\rangle_1\|10\rangle_2 - \sqrt{\frac{1}{2}}\|10\rangle_1\|11\rangle_2$
$\uparrow\downarrow - \downarrow\uparrow$	$\|10\ 11\rangle\ = \sqrt{\frac{1}{2}}\|11\rangle_1\|1,-1\rangle_2 - \sqrt{\frac{1}{2}}\|1,-1\rangle_2\|11\rangle_2$
$\cdot\downarrow - \downarrow\cdot$	$\|1,-1\ 11\rangle\ = \sqrt{\frac{1}{2}}\|10\rangle_1\|1,-1\rangle_2 - \sqrt{\frac{1}{2}}\|1,-1\rangle_1\|10\rangle_2$
$\uparrow\downarrow - \cdot\cdot + \downarrow\uparrow$	$\|00\ 11\rangle\ = \sqrt{\frac{1}{3}}\|11\rangle_1\|1,-1\rangle_2 - \sqrt{\frac{1}{3}}\|10\rangle_1\|10\rangle_2 + \sqrt{\frac{1}{3}}\|1,-1\rangle_1\|11\rangle_2$

[a] The diagrammatic representation of states is such that an up-arrow, a down-arrow, and a dot represent, respectively, $m = 1, -1$, and 0 of individual electrons.

The distinction between the coupled and uncoupled representations is brought out in the following two sets of eigenstate equations.

Coupled Representation

$$\begin{pmatrix} \hat{L}^2 \\ \hat{L}_z \\ \hat{L}_1^2 \\ \hat{L}_2^2 \end{pmatrix} |lml_1l_2\rangle = \hbar^2 \begin{pmatrix} l(l+1) \\ m/\hbar \\ l_1(l_1+1) \\ l_2(l_2+1) \end{pmatrix} |lml_1l_2\rangle$$

Uncoupled Representation

$$\begin{pmatrix} \hat{L}_{1z} \\ \hat{L}_{2z} \\ \hat{L}_1^2 \\ \hat{L}_2^2 \end{pmatrix} |l_1l_2m_1m_2\rangle = \hbar^2 \begin{pmatrix} m_1/\hbar \\ m_2/\hbar \\ l_1(l_1+1) \\ l_2(l_2+1) \end{pmatrix} |l_1l_2m_1m_2\rangle$$

These equations are relevant, for example, to the case of two electrons, given that one is an l_1 electron and the other, an l_2 electron.

For three electrons, in the uncoupled representation the six operators

$$(\hat{L}_1{}^2, \hat{L}_{1z}, \hat{L}_2{}^2, \hat{L}_{2z}, \hat{L}_3{}^2, \hat{L}_{3z})$$

form a complete commuting set. Good quantum numbers associated with these states are $(l_1, m_1, l_2, m_2, l_3, m_3)$. In the more relevant coupled representation, the six operators

$$(\hat{L}^2, \hat{L}_z, \hat{L}_1{}^2, \hat{L}_2{}^2, \hat{L}_3{}^2, \hat{A}_1{}^2)$$

form a complete commuting set. The operator $\hat{A}_1{}^2$ is given by

(9.115) $$\hat{A}_1{}^2 \equiv a_{12}(\hat{\mathbf{L}}_1 + \hat{\mathbf{L}}_2)^2 + a_{13}(\hat{\mathbf{L}}_1 + \hat{\mathbf{L}}_3)^2 + a_{23}(\hat{\mathbf{L}}_2 + \hat{\mathbf{L}}_3)^2$$

where the a coefficients are arbitrary. If l' is the eigen-l-value related to $\hat{A}_1{}^2$, six good quantum numbers for the case at hand are $(l, m, l_1, l_2, l_3, l')$.

PROBLEMS

9.30 What are the eigenvalues of the set of operators $(\hat{L}_1{}^2, \hat{L}_{1z}, \hat{L}_2{}^2, \hat{L}_{2z})$ corresponding to the product eigenstate $|m_1 l_1\rangle |m_2 l_2\rangle$?

9.31 Let $\hat{\mathbf{J}}_1$ and $\hat{\mathbf{J}}_2$ be the respective angular momenta of the individual components of a two-component system. The total system has angular momentum $\hat{\mathbf{J}} = \hat{\mathbf{J}}_1 + \hat{\mathbf{J}}_2$. Show that

(a) $\hat{\mathbf{J}}_1 \cdot \hat{\mathbf{J}}_2 = \frac{1}{2}(\hat{J}_{1+}\hat{J}_{2-} + \hat{J}_{1-}\hat{J}_{2+}) + \hat{J}_{1z}\hat{J}_{2z}$

(b) $\hat{J}^2 = \hat{J}_1^2 + \hat{J}_2^2 + 2\hat{J}_{1z}\hat{J}_{2z} + (\hat{J}_{1+}\hat{J}_{2-} + \hat{J}_{1-}\hat{J}_{2+})$

9.32 (a) Using the expansions developed in Problem 9.31, operate on the coupled angular momentum eigenstates for two p electrons as listed in Table 9.5 with $\hat{L}^2$ and $\hat{L}_z$, respectively, to verify the lm entries in each of the nine $|lml_1 l_2\rangle$ eigenstates.

(b) What are the Clebsch–Gordan coefficients involved in the expansion of the state $|0011\rangle$?

(c) What is the inner product $\langle 2011|0011\rangle$?

9.33 (a) With respect to the diagrammatic representation of states depicted in Table 9.5, what are the states corresponding to the diagrams

$$\psi_0 = \cdot\cdot, \qquad \psi_1 = \uparrow\downarrow + \downarrow\uparrow, \qquad \psi_3 = \uparrow\downarrow?$$

(b) Expand each of these functions in terms of the nine diagrams listed in Table 9.5.

(c) Are any of these three states eigenstates of $\hat{L}^2$? [*Hint:* Use the expansions obtained in part (b).]

(d) Two electrons are known to be in the coupled state ψ_1. What values of total angular momentum L will measurement find and with what probabilities will these values occur?

9.34 Two p electrons are in the state $|lml_1 l_2\rangle = |1, -111\rangle$. If measurement is made of L_{1z} in this state, what values may be found and with what probability will these values occur?

9.35 Two p electrons are in the coupled angular momentum state $|lml_1 l_2\rangle = |2, -2, 11\rangle$. What is the joint probability of finding the two electrons with $L_{1z} = L_{2z} = -\hbar$?

9.36 How many independent eigenstates are there in the coupled representation for a two-component system, given that $l_1 = 5$ and $l_2 = 1$? Make a table listing the ml values for all these states.

9.37 Show that $\hat{A}_1{}^2$ as given by (9.115) commutes with $\hat{L}^2$ and $\hat{L}_z$.

9.38 The eigenstate corresponding to maximum l for the three-electron case is

$$|l, m, l_1, l_2, l_3, l'\rangle = |l_1, l_1\rangle|l_2, l_2\rangle|l_3, l_3\rangle$$

    (a) What are the eigenvalues of $\hat{L}^2$, $\hat{L}_z$ corresponding to this state?

    (b) What is the eigenvalue of the operator $\hat{A}_1{}^2$ given by (9.115) corresponding to this state?

9.5 TOTAL ANGULAR MOMENTUM FOR TWO OR MORE ELECTRONS

We are now concerned with the possible values the total angular momentum l numbers may assume for a system of N electrons with respective l_i values $l_1, l_2, \ldots, l_N$, in the coupled representation. A totality of $(2l_1 + 1)(2l_2 + 1) \ldots (2l_N + 1)$ product states may be formed which are simultaneous eigenstates of the set of operators

$$(\hat{L}^2, \hat{L}_z{}^2, \hat{L}_1{}^2, \hat{L}_2{}^2, \ldots, \hat{L}_N{}^2)$$

We must make sure that our procedure for calculating these l values preserves this number of states. This affords a check that we have found all l values.

    The possible values that l can assume may be obtained by one of two techniques. The first technique follows from the rule (9.114) for the addition of the angular momenta of two electrons with respective l values l_1 and l_2. In this case the combined angular momentum

$$\hat{L}^2 = (\hat{\mathbf{L}}_1 + \hat{\mathbf{L}}_2)^2$$

has eigenvalues, $\hbar^2 l(l + 1)$, where

$$l = |l_1 + l_2|, \ldots, |l_1 - l_2|$$

Consider the case of three electrons. Their total angular momentum is given by

$$\hat{L}^2 = (\hat{\mathbf{L}}_1 + \hat{\mathbf{L}}_2 + \hat{\mathbf{L}}_3)^2$$

This may be rewritten in the form

$$\hat{L}^2 = (\hat{\mathbf{L}}' + \hat{\mathbf{L}}_3)^2$$

$$\hat{\mathbf{L}}' = \hat{\mathbf{L}}_1 + \hat{\mathbf{L}}_2$$

Suppose that one of the l values corresponding to $\hat{L}'^2$ is l'. Then the l values corresponding to the total angular momentum $\hat{L}^2$ are

$$(9.116) \qquad l = |l' + l_3|, \ldots, |l' - l_3|$$

This again follows the rule of (9.114). For example, consider the case of three p electrons ($l_1 = l_2 = l_3 = 1$). Then for the first two electrons we have

$$l' = |l_1 + l_2|, \ldots, |l_1 - l_2| = 0, 1, 2$$

Adding the third electron gives [using (9.116)] the l values

$$l' = 0, 1, 2 \quad l = |l' + l_3|, \ldots, |l' - l_3|$$

$$l' = 0 \longrightarrow l = 1$$

$$l' = 1 \longrightarrow l = 0, 1, 2$$

$$l' = 2 \longrightarrow l = 1, 2, 3$$

Thus we obtain the result

$$\boxed{l = 0, 1, 2, 3 \qquad \text{for three } p \text{ electrons}}$$

There is a distinct eigenstate for each distinct manner in which l may be formed. This gives a total number of

$$(2 \times 0 + 1) + 3(2 \times 1 + 1) + 2(2 \times 2 + 1) + (2 \times 3 + 1)$$

$$= 1 + 9 + 10 + 7 = 27$$

states, which agrees with the product

$$(2l_1 + 1)(2l_2 + 1)(2l_3 + 1) = 3 \times 3 \times 3 = 27$$

For the case of N electrons with respective l values $l_1, l_2, \ldots, l_N$, we follow a similar procedure. First, we add the angular momenta of the first two electrons. This gives

$$l' = |l_1 + l_2|, \ldots, |l_1 - l_2|$$

To these values we add the angular momentum of the third electron to obtain

$$l'' = |l' + l_3|, \ldots, |l' - l_3|$$

There is a separate sequence of l'' values for each value of l'. Adding the angular momentum of the fourth electron gives

$$l''' = |l'' + l_4|, \ldots, |l'' - l_4|$$

We continue in this manner until all individual angular momentum l values are accounted for. The final sequence gives all possible values of l. For three electrons the sequence of l'' gives all the values of l. For four electrons the sequence for l''' gives the values of l.

Addition Rules

The values of total l obtained by sequential addition as described above may more simply be arrived at by the following rule. Consider N electrons with respective angular momentum values $l_1, l_2, \ldots, l_N$. These values may always be ordered so that

$$l_1 \leq l_2 \leq \ldots \leq l_N$$

Let

$$\Lambda = \sum_{i=1}^{N-1} l_i$$

Then we have the following:

    (a)   If $l_N - \Lambda > 0$,

(9.117) $$l^{\min} = l_N - \Lambda$$

    (b)   If $l_N - \Lambda \leq 0$,

(9.118) $$l^{\min} = 0$$

    (c)   In all cases

(9.119) $$l^{\max} = \sum_{i=1}^{N} l_i$$

    (d)   The possible values of l that give the values of total L,

$$L^2 = (\mathbf{L}_1 + \mathbf{L}_2 + \cdots + \mathbf{L}_N)^2 = \hbar^2 l(l + 1)$$

are given by

(9.120) $$l = |l^{\max}|, |l^{\max} - 1|, \ldots, |l^{\min}|$$

As a simple example of this technique, consider the case of two p electrons and one f electron ($l_1 = l_2 = 1$, $l_3 = 3$). Then

$$\Lambda = 1 + 1 = 2$$
$$l_3 - \Lambda = 3 - 2 = 1 = l^{\min}$$
$$l^{\max} = 1 + 1 + 3 = 5$$

Therefore,

$$l = 1, 2, 3, 4, 5 \qquad \text{for two } p \text{ electrons and}$$
$$\text{one } f \text{ electron}$$

The electron orbital angular momentum notation $s, p, d, f, \ldots$ stems from atomic physics. The correspondence between these letters and l values of individual electrons follows the scheme

Symbol	s	p	d	f	g	h	$\cdots$
l value	0	1	2	3	4	5	$\cdots$

This notation will be used again in Chapter 12.

In Chapter 10 we will see how $\hat{L}^2$ enters the Hamiltonian for one- and two-particle systems. The Y_l^m functions will take on further significance. They will emerge as the angular dependent factors of the energy eigenfunctions for these systems.

The topic of the addition of angular momentum is returned to in Chapter 11, where the rules developed above are applied to the addition of spin angular momentum. In Chapter 12 these rules are again applied to the addition of orbital and spin angular momentum as related to one- and two-electron atoms. In general, the rules developed above for the addition of angular momentum are valid for orbital, **L**, spin, **S**, and total angular momentum, **J**.

PROBLEMS

9.39 What are the possible values of l for
(a) Four p electrons?
(b) Three p and one $f (l_4 = 3)$ electrons?

9.40 What is the wavefunction (in Dirac notation) for three p electrons in the state with $l = m = 3$?

9.41 Show that the two schemes for obtaining the total l value for three electrons with respective l values l_1, l_2, and l_3, as described in the text, are equivalent.

9.42 (a) Show that the technique of sequential addition for obtaining total l values in the coupled representation gives

$$(2l_1 + 1)(2l_2 + 1) \cdots (2l_N + 1)$$

eigenstates. (*Hint:* Assume that $l_1 < l_2 < \cdots < l_N$.)
(b) How many eigenstates are there for three f electrons?

9.43 In the uncoupled representation, N electrons are described by the simultaneous eigenstates of the $2N$ operators

$$(\hat{L}_1^2, \hat{L}_{1z}, \hat{L}_2^2, \hat{L}_{2z}, \ldots, \hat{L}_N^2, \hat{L}_{Nz})$$

In the coupled representation, the $N + 2$ commuting operators

$$(\hat{L}^2, \hat{L}_z, \hat{L}_1^2, \hat{L}_2^2, \ldots, \hat{L}_N^2)$$

are relevant, and there are $N + 2$ good quantum numbers corresponding to these operators. One suspects that $2N - (N + 2) = N - 2$ operators may be added to this sequence, yielding a set of $2N$ commuting operators.

(a)   Construct such a set of $N - 2$ operators, $\{\hat{A}_i^2\}$.

(b)   Show explicitly that the terms in the sum $\hat{A}_2^2$ commute with the sequence of $N + 2$ operators given above.

Answer (partial)

(a)   The first operator is

$$\hat{A}_1^2 = a_{12}(\hat{\mathbf{L}}_1 + \hat{\mathbf{L}}_2)^2 + a_{13}(\hat{\mathbf{L}}_1 + \hat{\mathbf{L}}_3)^2 + \cdots = \sum_{i_1 \neq i_2}^{N} a_{i_1 i_2}(\hat{\mathbf{L}}_{i_1} + \hat{\mathbf{L}}_{i_2})^2$$

The second operator is

$$\hat{A}_2^2 = \sum_{i_1 \neq i_2 \neq i_3} a_{i_1 i_2 i_3}(\hat{\mathbf{L}}_{i_1} + \hat{\mathbf{L}}_{i_2} + \hat{\mathbf{L}}_{i_3})^2$$

The $(N - 2)$nd operator is

$$\hat{A}_{N-2}^2 = \sum_{i_1 \neq \cdots \neq i_{N-1}}^{N} a_{i_1 \cdots i_{N-1}}(\hat{\mathbf{L}}_{i_1} + \cdots \hat{\mathbf{L}}_{i_{N-1}})^2$$

9.44   The spherical harmonics $Y_l^m(\theta, \phi)$ are simultaneous eigenstates of $\hat{L}_z$ and $\hat{L}^2$. How must the Cartesian x, y, z axes be aligned with the spherical r, θ, ϕ frame in order for this to be true, or is the validity of this statement independent of the relative orientation of these two frames?

9.45   Suppose that L^2 is measured for a free particle and the value $6\hbar^2$ is found. If L_y is then measured, what possible values can result?

9.46   The parity operator, $\hat{\mathbb{P}}$, in three dimensions is defined by the equation $\hat{\mathbb{P}}f(r, \theta, \phi) = f(r, \pi - \theta, \pi + \phi)$. Show that $\hat{\mathbb{P}}Y_l^m = (-)^l Y_l^m$. That is, the parity of Y_l^m (odd or even) is the same as that of l. (Compare with Problem 6.23.)

9.47   Establish the following equalities.

$$x = -r\sqrt{\frac{2\pi}{3}}(Y_1^1 - Y_1^{-1}), \qquad y = -\frac{r}{i}\sqrt{\frac{2\pi}{3}}(Y_1^1 + Y_1^{-1}), \qquad z = r\sqrt{\frac{4\pi}{3}}Y_1^0$$

$$xy = \frac{r^2}{i}\sqrt{\frac{2\pi}{15}}(Y_2^2 - Y_2^{-2}), \qquad yz = -\frac{r^2}{i}\sqrt{\frac{2\pi}{15}}(Y_2^1 + Y_2^{-1}), \qquad zx = -r^2\sqrt{\frac{2\pi}{15}}(Y_2^1 - Y_2^{-1})$$

$$x^2 - y^2 = r^2\sqrt{\frac{8\pi}{15}}(Y_2^2 + Y_2^{-2}), \qquad 2z^2 - x^2 - y^2 = r^2\sqrt{\frac{16\pi}{5}}Y_2^0,$$

$$y^2 - z^2 = -r^2\sqrt{\frac{2\pi}{15}}(Y_2^2 + \sqrt{6}\,Y_2^0 + Y_2^{-2})$$

9.48 Using the preceding relations, argue that x/r is an eigenfunction of $\hat{L}_x$ and y/r is an eigenfunction of $\hat{L}_y$. What are the respective eigenvalues of these functions? (*Note:* Such functions are widely employed in quantum chemistry and are commonly called *orbitals*. These topics are returned to in our discussion of *hybridization* in the following chapter.)

9.49 The Clebsch–Gordan expansion (9.109) affords a coordinate representation of angular momentum states in the coupled scheme. What is the explicit θ_1, ϕ_1, θ_2, ϕ_2 representation of the coupled state $|lml_1l_2\rangle = |1, -1, 1, 1\rangle$, relevant to two p electrons?

9.50 (a) Show that the energy eigenvalues relevant to a rigid rotator (9.49) follow from the Bohr–Sommerfeld quantization rule (7.192) in the limit of large quantum numbers.

(b) What conclusion may be inferred from your answer to part (a) concerning the domain of relevance of the Bohr–Sommerfeld rules?

Answers

(a) We find

$$\oint L\, d\theta = \left(l + \frac{1}{2}\right)h$$

$$L = \left(l + \frac{1}{2}\right)\hbar$$

$$E = \left\langle \frac{L^2}{2I} \right\rangle = \frac{(2l + 1)^2\hbar^2}{8I}$$

Thus, for $l \gg 1$ (but not neglecting l compared with l^2),

$$E \sim \frac{l(l + 1)\hbar^2}{2I}$$

(b) If we associate large quantum numbers with the classical domain, the preceding example indicates that the Bohr–Sommerfeld rules are relevant to this same region. Note also that the first-order solution which enters the near-classical WKB analysis (7.172) is the *action integral*, $\int p\, dx$.

9.51 (a) Show that in the state $|lm\rangle$

$$\langle L_x^2 \rangle = \langle L_y^2 \rangle = \hbar^2[l(l + 1) - m^2] - \frac{1}{4}\langle [\hat{L}_+, \hat{L}_-]_+ \rangle$$

Here $[\,,\,]_+$ represents the *anticommutator*

$$[\hat{A}, \hat{B}]_+ \equiv \hat{A}\hat{B} + \hat{B}\hat{A}$$

(b) Evaluate $\langle [\hat{L}_+, \hat{L}_-]_+ \rangle$ in terms of a function of l and m.

9.52 Establish the following relations relevant to a system in a state with definite angular momentum $\hbar\sqrt{l(l+1)}$. The degeneracy factor is $g_l = 2l + 1$.

(a) $\hbar^{-2}\langle L^2 \rangle = \displaystyle\sum_{m=-l}^{l} |m|$

(b) $\hbar^{-4}\langle L^2 \rangle^2 = 2 \displaystyle\sum_{m=-l}^{l} |m|^3$

(c) $\hbar^{-2}g_l\langle L^2 \rangle = 3 \displaystyle\sum_{m=-l}^{l} m^2$

In addition, show the following:

(d) $\displaystyle\sum_{l=0}^{n-1} g_l = n^2$

(e) $\displaystyle\sum_{l=|l_1-l_2|}^{l_1+l_2} g_l = g_{l_1} g_{l_2}$

9.53 A student argues that the rotational kinetic energy of a classical rigid sphere, spinning about a fixed origin with angular frequency ω, may also be associated with quantum mechanical spin. Is the student correct? Explain your answer.

Answer

The student is incorrect. The kinetic energy of the sphere obeys the relation

$$E = \frac{\omega S}{2}$$

where S is the "spin" angular momentum of the sphere. When evaluating S, one integrates differential elements of orbital angular momentum over the volume of the sphere. So, as with all classical angular momentum, classical "spin" is orbital angular momentum. With the preceding relation, the kinetic energy of the sphere, E, is likewise associated with orbital angular momentum. (The concept of spin is fully developed in Chapter 11.)

9.54 Assuming the form

$$P_{|m|}^{m}(\cos \theta) = (-1)^m C_m \sin^{|m|} \theta$$

and the integral relation[1]

$$\int_0^{\pi} \sin^{2m+1} \theta \, d\theta = \frac{2^{m+1}m!}{(2m+1)!!}, \qquad m > 0$$

and the normalized expression for $Y_l^m(\theta, \phi)$ given in Table 9.1, obtain an expression for C_m.

---

[1] I. S. Gradshteyn and I. M. Ryzhik, *Tables of Integrals Series and Products*, Academic Press, New York, 1965, equation [3.63(5)].

Answer

$$C_m^2 = \frac{(2|m| - 1)!!(|m| + m)!}{2^{|m|}|m|!(|m| - m)!}$$

Thus, $C_1 = 1$, $C_2 = 3$, in agreement with values in Table 9.1.

9.55 The energy of a rigid molecule, free to rotate about its center of mass, is given by

$$E = \frac{L_x^2 + L_y^2}{2I_1} + \frac{L_z^2}{2I_3}$$

where moments of inertia (I_1, I_2, I_3) are evaluated in principal axis[1] with $I_1 = I_2$ and the origin at the center of mass.

(a) Is the preceding expression a valid Hamiltonian form?

(b) Write down the proper Hamiltonian of the system together with eigenfunctions and eigenenergies.

(c) What are the energy degeneracies of this system? In what manner is your answer to this question related to symmetries of the molecule?

(d) The molecule obeys the following relation.

$$\frac{2}{I_1} = \frac{I_3 - I_1}{I_1 I_3} = \frac{1}{8ma_0^2}$$

If the molecule is in its ground state, at what energy (in eV) will an incident photon raise the molecule to the second excited state (above the ground state)?

Answers (partial)

(a) As L_x, L_y, L_z are not canonical momenta, the given form is not a valid Hamiltonian. (See Problem 1.25.)

(b) The appropriate Hamiltonian is given by (dropping hats on operators)

$$H = \frac{L^2 - L_z^2}{2I_1} + \frac{L_z^2}{2I_3}$$

9.56 Show that the angular momentum commutator relations (9.8), as well as any one of the equations (9.13), are satisfied by corresponding Poisson-bracket relations [i.e., in (9.8), set $i\hbar = 1$]. Poisson brackets are defined in Problem 1.15.

9.57 For a rotating system with angular momentum L, show that the classical domain is described by the criterion, $L \gg \hbar$.

[1] H. Goldstein, *Classical Mechanics*, 2d ed., Addison-Wesley, Reading, Mass., 1980.

10

PROBLEMS IN THREE DIMENSIONS

In this chapter we discuss the structure of the Schrödinger equation for a particle moving in three dimensions. General properties are developed through examination of the free-particle problem in Cartesian and spherical coordinates. Separation of variables in spherical coordinates yields product solutions for the free-particle problem comprised of spherical harmonics and spherical Bessel functions. Solution to the corresponding radial wave equation for the hydrogen atom gives Laguerre polynomials. Application is also directed toward the motion of a charged particle in a magnetic field. An elementary description of the theory of radiation from atoms and the formulation of selection rules are given. The chapter concludes with a description of the Thomas–Fermi model important to atomic physics.

10.1 THE FREE PARTICLE IN CARTESIAN COORDINATES

We again recall that the linear momentum operator $\hat{\mathbf{p}}$ is given by

$$(10.1) \qquad \hat{\mathbf{p}} = -i\hbar\nabla$$

Inserting this form into the Hamiltonian for a free particle of mass m moving in three dimensions gives

(10.2)
$$\hat{H} = \frac{\hat{p}^2}{2m} = -\frac{\hbar^2}{2m} \nabla^2$$

It follows that the time-independent Schrödinger equation for this same particle appears as

(10.3)
$$\hat{H}\varphi = -\frac{\hbar^2}{2m} \left(\frac{\partial^2}{\partial x^2} + \frac{\partial^2}{\partial y^2} + \frac{\partial^2}{\partial z^2} \right) \varphi = E\varphi$$

or, alternatively,

(10.4)
$$\nabla^2 \varphi = -k^2 \varphi$$
$$E = \frac{\hbar^2 k^2}{2m}$$

Separating variables

(10.5)
$$\varphi \equiv X(x)Y(y)Z(z)$$

permits (10.4) to be rewritten as

(10.6)
$$\frac{X_{xx}}{X} + \frac{Y_{yy}}{Y} + \frac{Z_{zz}}{Z} = -k^2$$
$$-\frac{X_{xx}}{X} = k^2 + \left(\frac{Y_{yy}}{Y} + \frac{Z_{zz}}{Z} \right) \equiv k_x^2$$

In the last equation, the left-hand side is a function only of x, while the middle term is a function only of y and z. The only way for the equality to hold for all (x, y, z) is for both terms to be equal to the same constant. Labeling this constant k_x^2 gives the equation

(10.7)
$$X_{xx} + k_x^2 X = 0$$

which has a solution[1]

(10.8)
$$X = A' e^{ik_x x}$$

In similar manner we obtain

(10.9)
$$Y = B' e^{ik_y y}, \qquad Z = C' e^{ik_z z}$$

where

(10.10)
$$k^2 = k_x^2 + k_y^2 + k_z^2$$

---

[1] Here we consider only the forward propagating wave.

Combining all three factors X, Y, and Z gives the solution

(10.11) $\varphi = A'B'C' \exp[i(k_x x + k_y y + k_z z)] = \varphi_{\mathbf{k}} = Ae^{i\mathbf{k}\cdot\mathbf{r}}$

The wave vector $\mathbf{k}$ and position vector $\mathbf{r}$ have components

$$\mathbf{k} = (k_x, k_y, k_z)$$
(10.12)
$$\mathbf{r} = (x, y, z)$$

The function $\varphi_{\mathbf{k}}$ so obtained is an eigenfunction of $\hat{H}$ (10.2) with the eigenvalue

(10.13) $$E_k = \frac{\hbar^2 k^2}{2m}$$

Plane Waves

The corresponding solution to the time-dependent Schrödinger equation (3.52) appears as

(10.14)
$$\psi_{\mathbf{k}}(\mathbf{r}, t) = Ae^{i(\mathbf{k}\cdot\mathbf{r} - \omega t)}$$
$$\hbar\omega = E_k$$

This solution represents a propagating plane wave. At any instant of time, $\psi_{\mathbf{k}}(r, t)$ is constant on the surfaces $\mathbf{k}\cdot\mathbf{r} = $ constant. These are surfaces normal to $\mathbf{k}$. Consider one such surface. The projection of $\mathbf{r}$ onto $\mathbf{k}$

(10.15) $$r_{\parallel} = \frac{\mathbf{k}\cdot\mathbf{r}}{k}$$

from any point on this surface is constant. This is the normal displacement between the origin and the surface. See Fig. 10.1. Rewriting (10.14) in the form

(10.16) $$\psi_{\mathbf{k}}(\mathbf{r}, t) = Ae^{ik[r_{\parallel} - (\omega/k)t]}$$

reveals that the rate of increase of $r_{\parallel}$ with respect to a surface of constant $\psi_{\mathbf{k}}$ is the *wave speed*

(10.17) $$v = \frac{\omega}{k}$$

The normalization constant A may be chosen so that

(10.18) $\langle\psi_{\mathbf{k}}|\psi_{\mathbf{k}'}\rangle = \langle\varphi_{\mathbf{k}}|\varphi_{\mathbf{k}'}\rangle = \langle\mathbf{k}|\mathbf{k}'\rangle = \iiint \varphi_{\mathbf{k}}^*\varphi_{\mathbf{k}'}\, dx\, dy\, dz = \delta(\mathbf{k} - \mathbf{k}')$

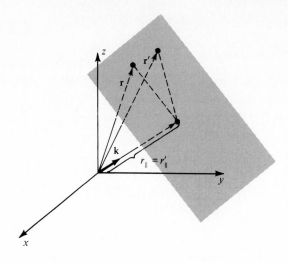

FIGURE 10.1 At any instant of time, the plane wave

$$\psi_{\mathbf{k}}(\mathbf{r}, t) = A \exp[i(\mathbf{k} \cdot \mathbf{r} - \omega t)]$$

is constant on the surface $\mathbf{k} \cdot \mathbf{r} =$ constant. These are surfaces normal to $\mathbf{k}$. At every point $\mathbf{r}$, on such a surface, the projection $r_\parallel = \mathbf{k} \cdot \mathbf{r}/k$ is constant.

The three-dimensional delta function is defined as the product

(10.19) $$\delta(\mathbf{r} - \mathbf{r}') = \delta(x - x')\delta(y - y')\delta(z - z')$$

and has the representation

$$\delta(\mathbf{r} - \mathbf{r}') = \frac{1}{(2\pi)^3} \iiint e^{i\mathbf{k} \cdot (\mathbf{r} - \mathbf{r}')} \, d\mathbf{k}$$

(10.20)
$$d\mathbf{k} = dk_x \, dk_y \, dk_z$$

Comparison of this representation with (10.18) yields the normalized wavefunction

(10.21) $$\varphi_{\mathbf{k}} = \frac{1}{(2\pi)^{3/2}} e^{i\mathbf{k} \cdot \mathbf{r}}$$

Superposition of Free-Particle States

A free-particle wave packet may be represented by the superposition

(10.22) $$\psi(\mathbf{r}, t) = \frac{1}{(2\pi)^{3/2}} \iiint b(\mathbf{k}, t) e^{i(\mathbf{k} \cdot \mathbf{r} - \omega t)} \, d\mathbf{k}$$

with corresponding inverse

(10.23) $$b(\mathbf{k}, t) = \frac{1}{(2\pi)^{3/2}} \iiint \psi(\mathbf{r}, t) e^{-i(\mathbf{k} \cdot \mathbf{r} - \omega t)} \, d\mathbf{r}$$

$$d\mathbf{r} = dx \, dy \, dz$$

As in the one-dimensional case, the coefficient b gives the probability

$$(10.24) \qquad P(\mathbf{k}) \, d\mathbf{k} = |b(\mathbf{k}, t)|^2 \, d\mathbf{k}$$

that measurement at the instant t finds the particle with momentum in the volume element $\hbar^3 \, d\mathbf{k}$ about the value $\hbar k$ (Fig. 10.2).

If the probability amplitude $b(\mathbf{k}, t)$ is peaked about a value of $\mathbf{k}$, say $\mathbf{k}_0$, then the three-dimensional wave packet (10.22) propagates with the *group velocity*

$$(10.25) \qquad \mathbf{v}_g = \boldsymbol{\nabla}_{\mathbf{k}} \, \omega(k) \big|_{\mathbf{k}=\mathbf{k}_0}$$

where $\boldsymbol{\nabla}_{\mathbf{k}}$ is written for the gradient with respect to $\mathbf{k}$. Inasmuch as (10.22) depicts the state of a free particle, for each $\mathbf{k}$-wave component one has

$$(10.26) \qquad E_k = \hbar\omega = \frac{\hbar^2 k^2}{2m}$$

This gives $\omega(k)$, which with (10.25) yields

$$(10.27) \qquad \mathbf{v}_g = \frac{\hbar \mathbf{k}_0}{m} = \mathbf{v}_{\text{CL}}$$

This is the classical velocity of a particle of mass m, moving with momentum $\hbar k_0$.

For a free particle

$$(10.28) \qquad [\hat{\mathbf{p}}, \hat{H}] = \frac{1}{2m} [\hat{\mathbf{p}}, \hat{p}^2] = 0$$

so that $\hat{\mathbf{p}}$ and $\hat{H}$ have simultaneous eigenstates. These are the functions $\varphi_{\mathbf{k}}(\mathbf{r})$. In the eigenstate $\varphi_{\mathbf{k}}$, the linear momentum $\hbar\mathbf{k}$ and energy $\hbar^2 k^2/2m$ are specified. The state

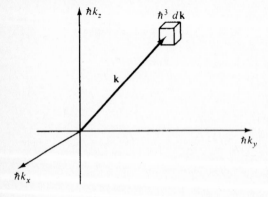

FIGURE 10.2 In the plane-wave decomposition of the free wave packet $\psi(\mathbf{r}, t)$, as given by (10.22), the Fourier amplitude $b(\mathbf{k}, t)$ is such that

$$|b(\mathbf{k}, t)|^2 \, d\mathbf{k}$$

is the probability that measurement finds the particle with momentum in the volume element $\hbar^3 \, d\mathbf{k}$ about the value $\hbar k$.

cannot be further resolved. For instance, suppose that we measure the z component of the angular momentum of the particle L_z. This measurement destroys the information in the state before measurement, relating to the linear momentum $\mathbf{p}$. The components of $\mathbf{p}$ and $\mathbf{L}$, in general, do not commute.

What, then, are the states for the free particle, which include specification of L^2 and L_z? To find these states it proves most convenient to express $\hat{H}$ in spherical coordinates. This is discussed in the next section.

PROBLEMS

10.1   If $\psi(\mathbf{r}, t)$ is a free-particle state and $b(\mathbf{k}, t)$ the momentum probability amplitude for this same state, show that

$$\iiint \psi^*\psi \, d\mathbf{r} = \iiint b^*b \, d\mathbf{k}$$

10.2   At time $t = 0$, a free particle is in the superposition state

$$\psi(\mathbf{r}, 0) = \frac{\pi^{-3/2}}{2} \sin 3x \, \exp[i(5y + z)]$$

(a)   If the energy of the particle is measured at $t = 0$, what value is found?

(b)   What possible values of momentum (p_x, p_y, p_z) will measurement find at $t = 0$, and with what probability will these values occur?

(c)   Given the above state $\psi(\mathbf{r}, 0)$, what is $\psi(\mathbf{r}, t)$?

(d)   If $\mathbf{p}$ is measured at $t = 0$ and the value $\mathbf{p} = \hbar(3\mathbf{e}_x + 5\mathbf{e}_y + \mathbf{e}_z)$ is found, what is $\psi(\mathbf{r}, t)$?

10.3   (a)   What is the Hamiltonian for N free, noninteracting particles of mass m?

(b)   What is the eigenstate of this Hamiltonian corresponding to the eigenvalue

$$E = \frac{\hbar^2}{2m} \sum_{j=1}^{N} k_j^2$$

(c)   Show that the eigenstate found in part (b) is also an eigenstate of the momentum of the center of mass. What is the velocity of the center of mass in this state?

Answers (partial)

(a) $\hat{H} = -\sum_{j=1}^{N} \frac{\hbar^2}{2m} \boldsymbol{\nabla}_j^2, \qquad \boldsymbol{\nabla}_j = \left(\frac{\partial}{\partial x_j}, \frac{\partial}{\partial y_j}, \frac{\partial}{\partial z_j} \right)$

(b) $\psi_{\mathbf{k}_1, \mathbf{k}_2, \dots, \mathbf{k}_N} = A \exp\left(i \sum_{j=1}^{N} \mathbf{k}_j \cdot \hat{\mathbf{r}}_j \right)$

(c) $\mathbf{v}_{CM} = \sum \hbar \mathbf{k}_j / M$

10.2   THE FREE PARTICLE IN SPHERICAL COORDINATES

Hamiltonian

We wish to express the free-particle Hamiltonian (10.2) in spherical coordinates, (r, θ, ϕ) (see Fig. 1.6). We have already found the classical expression for H in spherical coordinates in Chapter 1 [see (1.20)]. Let us again construct this classical form. However, in the present instance we wish H to include the angular momentum term

(10.29)
$$L^2 = (\mathbf{r} \times \mathbf{p})^2 = r^2 p^2 - (\mathbf{r} \cdot \mathbf{p})^2$$

The linear momentum is written $\mathbf{p}$. It follows that

(10.30)
$$H = \frac{p^2}{2m} = \frac{p_r^2}{2m} + \frac{L^2}{2mr^2}$$

where p_r is written for the radial component of the particle's momentum.

(10.31)
$$p_r = \frac{1}{r}(\mathbf{r} \cdot \mathbf{p})$$

If we wish to carry (10.29) over to quantum mechanics, we must make sure that all terms in $\hat{H}$ are Hermitian. The two operators in (10.30) are $r^{-2}\hat{L}^2$ and $\hat{p}_r^2$. To examine the Hermiticity of the first operator, we note the following.

Rotation and Angular Momentum

In Section 9.3 we found that the effect of the rotation operator $\hat{R}_{\delta\phi}$ when operating on a function $f(\mathbf{r})$ is to change f by rotating $\mathbf{r}$ to $\mathbf{r} + \delta\phi \times \mathbf{r}$. Suppose that a function f is isotropic[1] in $\mathbf{r}$; that is, f is independent of the direction of $\mathbf{r}$. It depends only on the magnitude of $\mathbf{r}$. Any function of the form $f(r^2)$ is isotropic in $\mathbf{r}$. For example, $f = ar^2 + br^4$, where a and b are constants, is isotropic in $\mathbf{r}$. What is the value of $f(r^2)$ on the surface of a sphere of radius r_0? The answer is, the constant $f(r_0^2)$. An isotropic function is constant on the surface of any sphere about the origin. Now suppose that we operate on an isotropic function with $\hat{R}_{\delta\phi}$. This causes f to change by rotating $\mathbf{r}$ to the value $\mathbf{r} + \delta\phi \times \mathbf{r}$. The new vector lies on the same sphere on which $\mathbf{r}$ lies.

$$(\mathbf{r} + \delta\phi \times \mathbf{r})^2 = r^2 + 2\mathbf{r} \cdot \delta\phi \times \mathbf{r} + O(\delta\phi^2) = r^2$$

Terms of $O(\delta\phi^2)$ are neglected while the middle term vanishes because $\mathbf{r}$ is normal to $\delta\phi \times \mathbf{r}$. It follows that the operator $\hat{R}_{\delta\phi}$ has no effect on $f(r^2)$.

(10.32)
$$\hat{R}_{\delta\phi} f(r^2) = f(r^2)$$

---

[1] One may also say that f is spherically symmetric.

Since $\hat{R}_{\delta\phi} = (1 + i\boldsymbol{\delta\phi} \cdot \mathbf{L}/\hbar)$, we may conclude that

$$(10.33) \qquad \frac{i\boldsymbol{\delta\phi} \cdot \hat{\mathbf{L}}}{\hbar} f(r^2) = 0$$

In that this statement is true for all axes of rotation about the origin, or equivalently, for all directions of the vector $\boldsymbol{\delta\phi}$, it follows that any isotropic function is a *null eigenstate* of the three components of angular momentum as well as of $\hat{L}^2$.

$$(10.34) \qquad \hat{L}_x f(r^2) = \hat{L}_y f(r^2) = \hat{L}_z f(r^2) = \hat{L}^2 f(r^2) = 0$$

As noted previously in Section 9.2, these spherically symmetric states are called S states.

If $g(\mathbf{r})$ is any function of $\mathbf{r}$ (for example, $g = x/r$) and $f(r^2)$ is any isotropic function, then owing to the conclusion immediately above,

$$\hat{L}^2 f(r^2)g(\mathbf{r}) = f(r^2)\hat{L}^2 g(\mathbf{r})$$

$$(\hat{L}^2 f(r^2) - f(r^2)\hat{L}^2)g(\mathbf{r}) = [\hat{L}^2, f(r^2)]g(\mathbf{r}) = 0$$

Since this latter equality holds for all differentiable functions g, we obtain

$$[\hat{L}^2, f(r^2)] = 0$$

Similarly,

$$(10.35) \qquad [\hat{L}_x, f(r^2)] = [\hat{L}_y, f(r^2)] = [\hat{L}_z, f(r^2)] = 0$$

We are now prepared to investigate the Hermiticity of the term $r^{-2}\hat{L}^2$ in the Hamiltonian (10.30). With $\hat{r}^{-2}$ denoting multiplication by r^{-2}, we write

$$(10.36) \qquad (\hat{r}^{-2}\hat{L}^2)^\dagger = \hat{L}^{2\dagger}\hat{r}^{-2\dagger} = \hat{L}^2 \hat{r}^{-2} = \hat{r}^{-2}\hat{L}^2$$

so that $\hat{r}^{-2}\hat{L}^2$ is Hermitian. In the last equality we used the fact that $\hat{L}^2$ commutes with the isotropic function r^{-2}.

Radial Momentum

Next we consider the operator

$$(10.37) \qquad \hat{p}_r = r^{-1}(\mathbf{r} \cdot \hat{\mathbf{p}}) = r^{-1}(x\hat{p}_x + y\hat{p}_y + z\hat{p}_z)$$

Forming the Hermitian adjoint of $\hat{p}_r$ gives

$$(10.38) \qquad (\hat{p}_r)^\dagger = [r^{-1}(\hat{x}\hat{p}_x + \hat{y}\hat{p}_y + \hat{z}\hat{p}_z)]^\dagger$$

$$= (\hat{x}\hat{p}_x)^\dagger(\hat{r}^{-1})^\dagger + (\hat{y}\hat{p}_y)^\dagger(\hat{r}^{-1})^\dagger + (\hat{z}\hat{p}_z)^\dagger(\hat{r}^{-1})^\dagger$$

$$= \hat{p}_x\hat{x}\hat{r}^{-1} + \cdots$$

$$\neq \hat{r}^{-1}\hat{x}\hat{p}_x + \cdots$$

The operators $\hat{r}^{-1}$ and $\hat{x}$ cannot be brought through $\hat{p}_x$. Similarly for the other two terms. We conclude that $\hat{p}_r$ is not Hermitian.

The more appropriate operator corresponding to radial momentum is given by the symmetric form (see Problems 10.5 and 10.6)

$$(10.39) \qquad \hat{p}_r = \frac{1}{2}(\hat{p}_r + \hat{p}_r{}^\dagger)$$

or, equivalently,

$$(10.40) \qquad \hat{p}_r = \frac{1}{2}\left(\frac{1}{r}\mathbf{r}\cdot\hat{\mathbf{p}} + \hat{\mathbf{p}}\cdot\mathbf{r}\frac{1}{r}\right)$$

The component of $\hat{\mathbf{p}}$ in the direction of $\mathbf{r}$ is given by

$$(10.41) \qquad \frac{1}{r}\mathbf{r}\cdot\hat{\mathbf{p}} = -i\hbar\frac{1}{r}\mathbf{r}\cdot\boldsymbol{\nabla} = -i\hbar\frac{\partial}{\partial r}$$

while the second term in $\hat{p}_r$ is given by

$$(10.42) \qquad \hat{\mathbf{p}}\cdot\mathbf{r}\frac{1}{r} = -i\hbar\boldsymbol{\nabla}\cdot\mathbf{e}_r$$

where $\mathbf{e}_r$ is written for the unit radius vector, $\mathbf{r}/r$. Let $f(\mathbf{r})$ be a differentiable function of the radius vector $\mathbf{r}$. Consider the operation

$$(10.43) \qquad \hat{p}_r f(r) = \frac{-i\hbar}{2}\left(\frac{\partial}{\partial r} + \boldsymbol{\nabla}\cdot\mathbf{e}_r\right)f$$

$$= \frac{-i\hbar}{2}\left(\frac{\partial f}{\partial r} + \mathbf{e}_r\cdot\boldsymbol{\nabla}f + f\boldsymbol{\nabla}\cdot\mathbf{e}_r\right)$$

$$= \frac{-i\hbar}{2}\left(\frac{\partial f}{\partial r} + \frac{\partial f}{\partial r} + \frac{2f}{r}\right) = -i\hbar\left(\frac{\partial f}{\partial r} + \frac{f}{r}\right)$$

Equivalently, we may write

$$\hat{p}_r f = -i\hbar\frac{1}{r}\frac{\partial}{\partial r}rf$$

$$(10.44)$$

$$\hat{p}_r = -i\hbar\frac{1}{r}\frac{\partial}{\partial r}r$$

With the above definition of $\hat{p}_r$, we may write the following for the Hamiltonian operator:

$$(10.45) \qquad \hat{H} = \frac{\hat{p}_r{}^2}{2m} + \frac{\hat{L}^2}{2mr^2}$$

The student may well ask the following question at this point. We know that the Hamiltonian $\hat{H}$ has the correct representation

(10.46)
$$\hat{H} = \frac{\hat{p}^2}{2m} = -\frac{\hbar^2 \nabla^2}{2m}$$

How are we assured that $\hat{H}$, as given by (10.45), with the definition of $\hat{p}_r$ obtained by symmetrization of $\hat{p}_r$, is equivalent to this correct form (10.46)? This question is answered by demonstration. The representation of the Laplacian operator ∇^2, in spherical coordinates, is

(10.47)
$$\nabla^2 = \frac{1}{r}\frac{\partial^2}{\partial r^2} r + \frac{1}{r^2}\left(\frac{1}{\sin\theta}\frac{\partial}{\partial\theta}\sin\theta\frac{\partial}{\partial\theta} + \frac{1}{\sin^2\theta}\frac{\partial^2}{\partial\phi^2}\right)$$

Noting the equality

(10.48)
$$\left(\frac{1}{r}\frac{\partial}{\partial r}r\right)^2 = \frac{1}{r}\frac{\partial}{\partial r}r\left(\frac{1}{r}\frac{\partial}{\partial r}r\right) = \frac{1}{r}\frac{\partial^2}{\partial r^2}r$$

and recalling the expression for $\hat{L}^2$, as given by (9.58), permits the equation

(10.49)
$$\hat{H} = -\frac{\hbar^2\nabla^2}{2m} = \frac{\hat{p}_r{}^2}{2m} + \frac{\hat{L}^2}{2mr^2}$$

This is the correct form of $\hat{H}$, in spherical coordinates. In the next section we will examine its eigenfunctions and eigenvalues.

PROBLEMS

10.4   What is the time-independent wavefunction in spherical coordinates of a free particle of mass m, zero angular momentum, and energy E which satisfies the property $|r\varphi| = 0$ at $r = 0$? (*Hint:* Introduce the function $u \equiv r\varphi$.)

10.5   (a)   Show that
$$[\hat{r}, \hat{p}_r] = i\hbar$$

(b)   What properties of φ and ψ insure that
$$\langle\varphi|\hat{p}_r\psi\rangle = \langle\hat{p}_r\varphi|\psi\rangle$$

[*Note:* $\langle\varphi|\psi\rangle = \int d\Omega \int dr r^2\varphi^*\psi$.]

10.6   The current vector $\mathbf{J}$ associated with a wavefunction $\psi(\mathbf{r}, t)$ is given by (7.107)
$$\mathbf{J} = \frac{\hbar}{2mi}(\psi^*\nabla\psi - \psi\nabla\psi^*)$$

The wavefunction $\psi(\mathbf{r}, t)$ may be termed *source-free*, if $\nabla \cdot \mathbf{J} = 0$ for all values of $\mathbf{r}$.

(a)   What is the eigenfunction of $\hat{p}_r$, corresponding to the eigenvalue $\hbar k$?

(b)   Calculate $\nabla \cdot \mathbf{J}$ for this eigenfunction of $\hat{p}_r$.

Answers

(a) Integration of the eigenvalue equation

$$-i\hbar \frac{1}{r} \frac{\partial}{\partial r} r\tilde{\varphi}_k = \hbar k \tilde{\varphi}_k$$

gives

$$\tilde{\varphi}_k = A \frac{e^{ikr}}{r}$$

The corresponding time-dependent solution is

$$\tilde{\varphi}_k = \frac{A}{r} e^{i(kr - \omega t)}$$

This "outgoing wave" is a solution to the time-dependent Schrödinger equation for a free particle with no angular momentum. It is important in the construction of scattering states, which will be discussed in Chapter 12.

(b) The current vector corresponding to $\tilde{\varphi}_k$ only has an r component.

$$\mathbb{J}_r = \frac{|A|^2 (\hbar k/m)}{r^2} \equiv \frac{\Gamma(0)}{4\pi r^2}$$

The divergence of this current is

$$\nabla \cdot \mathbf{J} = \frac{1}{r^2} \frac{\partial}{\partial r} r^2 \mathbb{J}_r = 0 \qquad (\text{for } r \neq 0)$$

Since $\mathbf{J}$ is radial and a function only of r, we may write

$$\int_V \nabla \cdot \mathbf{J} \, d\mathbf{r} = \int_{r=R} \mathbf{J} \cdot d\mathbf{S} = \int_{4\pi} \frac{\Gamma(0)}{4\pi r^2} r^2 \, d\Omega = \Gamma(0)$$

The spherical volume V has radius R and is centered at the origin, while $d\Omega$ is an element of solid angle about this same origin. Given these two properties of $\nabla \cdot \mathbf{J}$, it follows that

$$\nabla \cdot \mathbf{J} = \Gamma(0)\delta(\mathbf{r})$$

The three-dimensional Dirac delta function is $\delta(\mathbf{r})$ [see (10.19)].

[*Note:* Thus we see that the eigenstates of $\hat{p}_r$ have the unreasonable property of implying that a constant flux of particles, $\Gamma(0)$, emanates from the origin. We may infer from this that the operator $\hat{p}_r$, in spite of its symmetric form (10.39) and proper commutation property with $\hat{r}$ (Problem 10.5), is not a good observable equivalent. Nevertheless, for problems involving a central potential, in quantum mechanics the operator $\hat{p}_r$ proves to be a valuable tool. It is interesting to note that inconsistencies that accompany p_r are also found in classical mechanics. If a free point particle crosses the origin, p_r changes sign instantaneously. This jump in p_r stems from a choice of coordinate frame. It is in no way associated with a force (the particle is free).[1,2]]

10.7   Show that the kinetic-energy operator

$$\hat{T} = -\frac{\hbar^2}{2m}\nabla^2$$

is Hermitian for functions in $\mathfrak{H}_2$ (the space of square integrable functions—see Section 4.4). [*Hint:* Use *Green's theorem*

$$\int_V (f\nabla^2 g - g\nabla^2 f)\, d\mathbf{r} = \int_S (f\nabla g - g\nabla f)\cdot d\mathbf{S}$$

The volume V is enclosed by the surface S.]

10.8   Show that

$$\boldsymbol{\nabla}\cdot\mathbf{J}(\psi) = 0$$

for the superposition state

$$\psi = \psi_1 + \psi_2$$

provided that

$$\boldsymbol{\nabla}\cdot\mathbf{J}(\psi_1) = \boldsymbol{\nabla}\cdot\mathbf{J}(\psi_2) = 0$$

and

$$\mathrm{Im}(\psi_1{}^*\nabla^2\psi_2 + \psi_2{}^*\nabla^2\psi_1) = 0$$

10.3   THE FREE-PARTICLE RADIAL WAVEFUNCTION

The time-independent Schrödinger equation for a free particle in spherical coordinates appears as

(10.50) $$\frac{1}{2m}\left(\hat{p}_r{}^2 + \frac{\hat{L}^2}{r^2}\right)\varphi_{klm} = E_{klm}\,\varphi_{klm}$$

The quantum number k is defined below. The radial kinetic-energy operator $\hat{p}_r{}^2/2m$ is inferred from (10.48), while the angular momentum operator $\hat{L}^2$ is given by (9.58). Insofar as $\hat{p}_r{}^2$ is a function only of r, and $\hat{L}^2$ is a function only of the angle variables

---

[1] For further discussion of this problem, see R. L. Liboff, I. Nebenzahl, and H. A. Fleishmann, *Am. J. Phys.* **41**, 976 (1973).

[2] Related ambiguities of the radial momentum operator in cylindrical coordinates are discussed in Problems 10.73 and 11.90.

(θ, ϕ), one may seek solution to (10.50) by separation of variables. Substituting the product form

(10.51) $$\varphi_{klm}(r, \theta, \phi) = R_{kl}(r)Y_l^m(\theta, \phi)$$

into (10.50) gives

(10.52) $$\left[-\left(\frac{1}{r}\frac{d^2}{dr^2}r\right) + \frac{l(l+1)}{r^2}\right]R_{kl}(r) = \frac{2mE}{\hbar^2}R_{kl}(r)$$

In obtaining (10.52) we have recalled the eigenvalue equation for $\hat{L}^2$ (9.51). With the substitution

(10.53)
$$E \equiv \frac{\hbar^2 k^2}{2m}$$
$$x \equiv kr$$

(10.52) becomes the "spherical Bessel differential equation"

(10.54) $$\frac{d^2}{dx^2}R(x) + \frac{2}{x}\frac{dR(x)}{dx} + \left[1 - \frac{l(l+1)}{x^2}\right]R(x) = 0$$

Spherical Bessel Functions

This ordinary linear equation for the radial function R has two linearly independent solutions.[1] They are called spherical Bessel and Neumann functions and are denoted conventionally by the symbols $j_l(x)$ and $n_l(x)$, respectively. The first few values of these functions are

(10.55)

$$j_0(x) = \frac{\sin x}{x}$$ $$n_0(x) = -\frac{\cos x}{x}$$

$$j_1(x) = \frac{\sin x}{x^2} - \frac{\cos x}{x}$$ $$n_1(x) = -\frac{\cos x}{x^2} - \frac{\sin x}{x}$$

$$j_2(x) = \left(\frac{3}{x^3} - \frac{1}{x}\right)\sin x - \frac{3}{x^2}\cos x$$ $$n_2(x) = -\left(\frac{3}{x_3} - \frac{1}{x}\right)\cos x - \frac{3}{x^2}\sin x$$

These functions are sketched in Fig. 10.3, from which it is evident that of the two classes of functions, only the spherical Bessel functions $\{j_l\}$ are regular at the origin. These are the solutions appropriate to the Schrödinger equation (10.50) inasmuch as they are not singular anywhere. Some additional properties of these spherical Bessel and Neumann functions are listed in Table 10.1.

[1] One obtains these solutions by the method of series substitution. Details may be found in most books on mathematical physics, e.g., G. Goertzel and N. Tralli, *Some Mathematical Methods in Physics*, McGraw-Hill, New York, 1960.

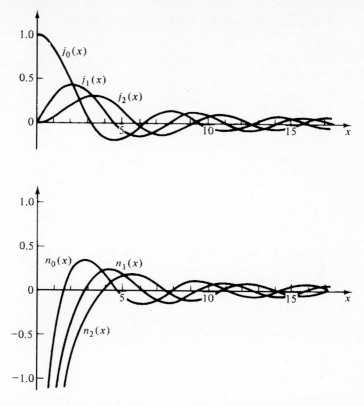

FIGURE 10.3   Spherical Bessel functions $j_l(x)$ and spherical Neumann functions $n_l(x)$ for $l = 0, 1, 2.$
Note that only $j_l(x)$ are regular at the origin.

In this manner we find that the eigenstates and eigenenergies of the free-particle
Hamiltonian in spherical coordinates are

(10.56)

$$\varphi_{klm}(r, \theta, \phi) = j_l(kr)Y_l^m(\theta, \phi)$$

$$E_k = \frac{\hbar^2 k^2}{2m}$$

The orthonormality of this sequence $\{\varphi_{klm}\}$ is given by the relation

(10.57) $\langle lmk \, | \, l'm'k' \rangle = \displaystyle\int_{4\pi} d\Omega [Y_l^m(\theta, \phi)]^* Y_{l'}^{m'}(\theta, \phi) \int_0^\infty j_l(kr) j_{l'}(k'r) r^2 \, dr$

$$= \delta_{ll'} \delta_{mm'} \frac{\pi}{2k^2} \delta(k - k')$$

TABLE 10.1 Properties of the spherical Bessel and Neumann functions

Spherical Bessel Functions

$$j_l(kr) = \left(-\frac{r}{k}\right)^l \left(\frac{1}{r}\frac{d}{dr}\right)^l j_0(kr)$$

$$j_0(kr) = \frac{\sin kr}{kr}$$

Spherical Neumann Functions

$$n_l(kr) = \left(-\frac{r}{k}\right)^l \left(\frac{1}{r}\frac{d}{dr}\right)^l n_0(kr)$$

$$n_0(kr) = -\frac{\cos kr}{kr}$$

Equation

$$f'' + \frac{2}{x}f + \left[1 - \frac{l(l+1)}{x^2}\right]f = 0$$

Asymptotic Values

$x \to 0$

$$j_l(x) \sim \frac{x^l}{1 \cdot 3 \cdot 5 \cdots (2l+1)}$$

$$n_l(x) \sim -\frac{1 \cdot 3 \cdot 5 \cdots (2l-1)}{x^{l+1}}$$

$x \to \infty$

$$j_l(x) \sim \frac{1}{x}\cos\left[x - \frac{\pi}{2}(l+1)\right]$$

$$n_l(x) \sim \frac{1}{x}\sin\left[x - \frac{\pi}{2}(l+1)\right]$$

Recurrence Relations (f is written for j or n)

$$f_{l-1}(x) + f_{l+1}(x) = (2l+1)x^{-1}f_l(x)$$

$$\frac{d}{dx}[x^{l+1}j_l(x)] = x^{l+1}j_{l-1}(x)$$

$$lf_{l-1}(x) - (l+1)f_{l+1}(x) = (2l+1)\frac{d}{dx}f_l(x)$$

$$\frac{d}{dx}[x^{-l}j_l(x)] = -x^{-l}j_{l+1}(x)$$

Generating Functions

$$\frac{1}{x}\cos\sqrt{x^2 - 2xs} = \sum_0^\infty \frac{s^l}{l!}j_{l-1}(x)$$

$$\frac{1}{x}\sin\sqrt{x^2 + 2xs} = \sum_0^\infty \frac{(-s)^l}{l!}n_{l-1}(x)$$

Orthogonality

$$\int_0^\infty j_l(kr)j_l(k'r)r^2\,dr = \frac{\pi}{2k^2}\delta(k - k')$$

$$\int j_1(x)\,dx = -j_0(x), \quad \int j_0(x)x^2\,dx = x^2 j_1(x)$$

$$\int f_l^2(x)x^2\,dx = \frac{x^3}{2}[f_l^2(x) - f_{l-1}(x)f_{l+1}(x)]$$

Connection to Bessel and Neumann Functions of Integral Order, J_l and N_l

$$j_l(kr) = \sqrt{\frac{\pi}{2kr}}J_{l+1/2}(kr)$$

$$n_l(kr) = \sqrt{\frac{\pi}{2kr}}N_{l+1/2}(kr)$$

The vector **r** has the spherical coordinates (r, θ, ϕ). This orthonormality condition is similar to that corresponding to the free-particle states expressed in Cartesian coordinates (10.18), as well as that corresponding to free-particle motion in one dimension (4.41). In all these cases, the allowed values of momentum, $\hbar k$, comprise a continuum.

Once again we note that the projection

$$\langle r\theta\phi \,|\, lmk \rangle = j_l(kr)Y_l^m(\theta, \phi)$$

gives the coordinate representation of the ket vector $|lmk\rangle$. In similar manner, the coordinate representation of the free-particle ket vector $|\mathbf{k}\rangle$ is given by the projection

$$\langle \mathbf{r} \,|\, \mathbf{k} \rangle = \frac{1}{(2\pi)^{3/2}} \, e^{i\mathbf{k}\cdot\mathbf{r}}$$

(See Problem 7.31.)

Measurements on a Free Particle

Given that a particle is in the eigenstate φ_{klm}, measurement of

(10.58)
$$
\begin{aligned}
E \quad &\text{gives} \quad \hbar^2 k^2/2m \\
L^2 \quad &\text{gives} \quad \hbar^2 l(l+1) \\
L_z \quad &\text{gives} \quad \hbar m
\end{aligned}
$$

How do we know that these values may be measured simultaneously? The answer is that φ_{klm} is a simultaneous eigenfunction of $\hat{H}$, $\hat{L}^2$, and $\hat{L}_z$. The existence of such common eigenfunctions follows from the fact that $\hat{H}$, $\hat{L}^2$, and $\hat{L}_z$ are a commuting set of operators. We have already discussed the commutability of $\hat{L}^2$ and $\hat{L}_z$ in Chapter 9. The fact that these operators commute with $\hat{H}$ follows if they commute with $\hat{p}_r^2$. But $\hat{p}_r^2$ is an isotropic operator; $\hat{p}_r^2 f(r)$ is constant on the surface of any given sphere. It follows that $\hat{p}_r^2$ is unaffected by rotations about the origin; hence

(10.59)
$$[\hat{p}_r^2, \hat{L}_z] = [\hat{p}_r^2, \hat{L}^2] = 0$$

and

(10.60)
$$[\hat{H}, \hat{L}_z] = [\hat{H}, \hat{L}^2] = 0$$

The solution $\varphi_{klm}(r, \theta, \phi)$ should be compared to the eigenstate of the free-particle Hamiltonian in Cartesian coordinates (10.11),

(10.61)
$$\varphi_{\mathbf{k}}(\mathbf{r}) = A e^{i\mathbf{k}\cdot\mathbf{r}}$$

Given that a particle is in this state, measurement of

(10.62)
$$
\begin{aligned}
E \quad &\text{gives} \quad \frac{\hbar^2 k^2}{2m} \\
p_x \quad &\text{gives} \quad \hbar k_x \\
p_y \quad &\text{gives} \quad \hbar k_y \\
p_z \quad &\text{gives} \quad \hbar k_z
\end{aligned}
$$

In the spherical representation, (L_z, L^2, E) are specified. In the Cartesian representation, $(\mathbf{p}, E)$ are specified. In the latter representation, E is redundant $(E = p^2/2m)$, but in the former representation it is not. It is not determined by L^2 and L_z. Thus we find that in either Cartesian or spherical representations, there are three good quantum numbers [recall (1.41)].

Free-Particle S States

The special case $L = 0$ is of interest. For this case the Schrödinger equation (10.50) becomes

$$(10.63) \qquad \left(\frac{\hat{p}_r{}^2}{2m}\right)\varphi_k = E_k\varphi_k$$

The radial kinetic-energy operator $\hat{p}_r{}^2/2m$ commutes with the radial momentum operator $\hat{p}_r$ and they have common eigenstates. Due to degeneracy, however (the eigenstates E_k are doubly degenerate), eigenfunctions of $\hat{p}_r{}^2$ are not necessarily eigenfunctions of $\hat{p}_r$ (see Fig. 10.4). Owing to the inadmissibility of the eigenfunctions of $\hat{p}_r$, it is the eigenstates of $\hat{p}_r{}^2$ alone which are physically relevant. Namely, these functions are

$$(10.64) \qquad \varphi_k = j_0(kr) = \frac{\sin kr}{kr}$$

Rewriting φ_k in the form

$$(10.65) \qquad \varphi_k = \tilde{\varphi}_{+k} + \tilde{\varphi}_{-k} = \frac{1}{2i}\left(\frac{e^{ikr}}{kr} - \frac{e^{-ikr}}{kr}\right)$$

reveals that it is a superposition of the *outgoing* wave, $\tilde{\varphi}_{+k}$, and the *ingoing* wave, $\tilde{\varphi}_{-k}$, which gives zero flux at the origin.

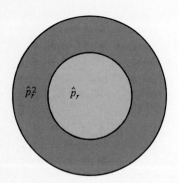

FIGURE 10.4 Central domain represents the eigenstates common to $\hat{p}_r{}^2$ and $\hat{p}_r$. Peripheral domain represents only eigenstates of $\hat{p}_r{}^2$, which alone are the physically relevant ones.

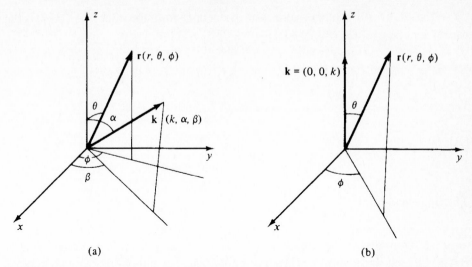

FIGURE 10.5   Coordinates relevant to the expansion of a plane wave in the eigenstates of $\hat{L}^2$ and $\hat{L}_z$.
(a) Direction of **k** is arbitrary. (b) **k** in the direction of the polar axis.

Measurement of L_z and L^2 for a Plane Wave

Next, we consider the following important problem. Suppose that a particle of mass
m is "prepared" so that it has momentum $\hbar\mathbf{k}$. Then we know that it is in the plane-
wave state (10.11).

$$\varphi_\mathbf{k} = Ae^{i\mathbf{k}\cdot\mathbf{r}} \qquad (10.66)$$

Measurement of E is certain to find $\hbar^2 k^2/2m$. Measurement of momentum is certain
to find $\hbar\mathbf{k}$. What will measurement of L_z or L^2 find? And in what states do such measurements leave the particle? To answer this question we must expand the given plane
wave in the simultaneous eigenstates of $\hat{H}$, $\hat{L}^2$, and $\hat{L}_z$, that is, φ_{klm}, as given by
(10.56). This expansion appears as[1]

$$e^{i\mathbf{k}\cdot\mathbf{r}} = \sum_{l=0}^{\infty} \sum_{m=-l}^{l} a_{lm}(\mathbf{k})\varphi_{klm} \qquad (10.67)$$

where the coefficients of expansion, a_{lm}, are

$$a_{lm} = 4\pi i^l [Y_l^m(\alpha, \beta)]^* \qquad (10.68)$$

and (k, α, β) are the spherical coordinates of **k** (Fig. 10.5).

---

[1] See Geortzel and Tralli, *Some Mathematical Methods in Physics*. This expression is also discussed in Problem
10.11.

The relative probability that measurement of L^2 finds the value $\hbar^2 l(l + 1)$ is the partial sum [see (9.82) and (9.83)]

$$(10.69) \qquad P[\hbar^2 l(l + 1)] = \sum_{m=-l}^{l} |a_{lm}|^2 = (4\pi)^2 \sum_{m=-l}^{l} |Y_l^m(\alpha, \beta)|^2$$

The relative probability that measurement of L_z finds the value $\hbar m$ is

$$(10.70) \qquad P[\hbar m] = (4\pi)^2 \sum_{l=|m|}^{\infty} |Y_l^m(\alpha, \beta)|^2$$

The Spherical Well

We consider a particle of mass M confined to the interior of a spherical well with impenetrable walls. In the domain $r \geq a$, the wavefunction vanishes. In the domain $r < a$, the time-independent Schrödinger equation is given by (10.50) with general solutions given by (10.56). To impose the boundary condition $\varphi(r = a) = 0$ we set

$$(10.71a) \qquad j_l(ka) = 0$$

With Fig. 10.3 we see that (10.71a) has an infinite number of solutions. To delineate these values we return to the notation $x \equiv kr$ in terms of which (10.71a) becomes

$$(10.71b) \qquad j_l(x_{ln}) = 0$$

where x_{ln} is the nth zero of $j_l(x)$.

Eigenfunctions and eigenenergies for the spherical well are then given by

$$(10.72a) \qquad \boxed{\varphi_{nlm}(r, \theta, \phi) = j_l\left(\frac{x_{ln}r}{a}\right) Y_l^m(\theta, \phi)}$$

$$(10.72b) \qquad \boxed{E_{nl} = \frac{\hbar^2 x_{ln}^2}{2Ma^2} = \epsilon \mathbb{R} \left(\frac{a_0}{a}\right)^2 x_{ln}^2}$$

where $\epsilon \equiv m_e/M$. Orthogonality of spherical Bessel functions is given by

$$(10.72c) \qquad \int_0^a dr\, r^2 j_l\left(\frac{x_{ln}r}{a}\right) j_l\left(\frac{x_{ln'}r}{a}\right) = \frac{a^3}{2}[j_{l+1}(x_{ln})]^2\, \delta_{nn'}$$

which indicates the nature of normalization of these functions. Note that the continuous spectrum of k values for the free particle in spherical coordinates translates to the discrete spectrum of the quantum number n for the finite spherical well problem.

The following is a table of the first number of zeros of $j_l(x)$.

(l, n)	$(0, 1)$	$(1, 1)$	$(2, 1)$	$(0, 2)$	$(3, 1)$	$(1, 2)$	$(4, 1)$
x_{ln}	π	4.49	5.76	2π	6.99	7.73	8.18

The ground-state wavefunction and eigenenergy for the spherical well are given by

(10.73a)
$$\varphi_G(r, \theta, \phi) = \frac{1}{\sqrt{4\pi}} j_0\left(\frac{\pi r}{a}\right)$$

(10.73b)
$$E_G = \hbar^2 \pi^2 / 2Ma^2$$

Note that this ground state is nondegenerate.

The Cylindrical Well

We next consider the case of a particle of mass M confined to a cylindrical box of radius a and length L. Expressed in cylindrical coordinates (Fig. 1.7), the potential of this configuration is given by (with r written for ρ as depicted in Fig. 1.7)

(10.74a) $V(r, z, \phi) = 0, \qquad r < a, \qquad 0 < z < L$

(10.74b) $\qquad\qquad = \infty \quad$ elsewhere

Employing the Hamiltonian given in Table 10.2 and the expression for the Laplacian in cylindrical coordinates given in Appendix D, the Schrödinger equation for the confined particle is given by

(10.75a)
$$\frac{\partial^2 \varphi}{\partial r^2} + \frac{1}{r}\frac{\partial \varphi}{\partial r} + \frac{1}{r^2}\frac{\partial^2 \varphi}{\partial \phi^2} + \frac{\partial^2 \varphi}{\partial z^2} + k^2 \varphi = 0$$

(10.75b)
$$\hbar^2 k^2 / 2M = E$$

With the separation of coordinates

(10.76a)
$$\varphi(r, z, \phi) = R(r) Z(z) \Phi(\phi)$$

(10.75) becomes

(10.76b)
$$\frac{1}{R}\left(\frac{\partial^2 R}{\partial r^2} + \frac{1}{r}\frac{\partial R}{\partial r}\right) + \frac{1}{r^2 \Phi(\phi)}\frac{\partial^2 \Phi}{\partial \phi^2} + \frac{1}{Z}\frac{\partial^2 Z}{\partial z^2} + k^2 = 0$$

It follows that

(10.77a)
$$\frac{1}{Z}\frac{d^2Z}{dz^2} = \text{constant} \equiv -k_z^2$$

(10.77b)
$$\frac{1}{\Phi}\frac{d^2\Phi}{d\phi^2} = \text{constant} \equiv -m^2$$

(10.77c)
$$\frac{1}{R}(r^2R'' + rR') + r^2(k^2 - k_z^2) = m^2$$

where a prime denotes differentiation with respect to r. With conditions (10.74) we find

(10.78a) $\quad Z(z) = A\sin k_z z, \qquad k_z L = n_z \pi, \qquad n_z = 1, 2, \ldots$

Furthermore, as $\Phi(\phi) = \Phi(\phi + 2\pi)$, we obtain

(10.78b) $\qquad\qquad \Phi(\phi) = Be^{im\phi}, \qquad m = 0, \pm 1, \pm 2, \ldots$

where A and B are constants. Returning to (10.77c) and labeling

(10.78c)
$$k^2 - k_z^2 \equiv K^2$$
$$\rho \equiv Kr$$

there results

(10.79) $\qquad\qquad \rho^2 R'' + \rho R' + (\rho^2 - m^2)R = 0$

which is known as *Bessel's equation*. General solutions to this equation are given by

(10.80) $\qquad\qquad R(\rho) = C_1 J_m(\rho) + C_2 N_m(\rho)$

where C_1 and C_2 are constants. The functions $J_m(\rho)$ and $N_m(\rho)$ are called *Bessel* and *Neumann functions of the first kind,* respectively. As $N_m(0) = -\infty$, it is omitted in (10.80). The functions $J_m(\rho)$ oscillate.[1] Graphs of these functions are similar to those for the spherical Bessel functions shown in Fig. 10.3. The remaining boundary conditions [first of (10.74a)] gives

(10.81a) $\qquad\qquad R(r = a) = 0 = C_1 J_m(aK)$

Let us call the sth finite zero of $J_m(\rho)$, x_{ms} so that

(10.81b) $\qquad\qquad J_m(aK_{ms}) \equiv J_m(x_{ms}) = 0$

---

[1] For integer m, $J_{-m}(\rho) = (-1)^m J_m(\rho)$, so that with no loss in generality, we consider only $m \geq 0$.

The following is a table of the first few zeros of $J_n(\rho)$.

	$s = 1$	$s = 2$	$s = 3$
x_{0s}	2.40	5.52	8.65
x_{1s}	3.83	7.02	10.17
x_{2s}	5.14	8.42	11.62
x_{3s}	6.30	9.76	13.02

Eigenenergies are given by

$$E = \frac{\hbar^2 k^2}{2M} = \frac{\hbar^2}{2M}(K^2 + k_z^2) = \frac{\hbar^2}{2M}\left[K^2 + \left(\frac{m\pi}{L}\right)^2\right]$$

With $x_{ms} \equiv aK_{ms}$, the preceding becomes

(10.82a)
$$E_{nln_z} = \frac{\hbar^2}{2M}\left[\frac{x_{ms}^2}{a^2} + \left(\frac{n_z\pi}{L}\right)^2\right]$$

Corresponding eigenfunctions are given by

(10.82b)
$$\varphi_{msn_z}(r, z, \phi) = AJ_m\left(\frac{rx_{ms}}{a}\right)\sin\left(\frac{n_z\pi z}{L}\right)e^{im\phi}$$
$$m \geq 0, \qquad s > 0, \qquad n_z \geq 1$$

All three parameters are integers.

The ground state of this system is given by

(10.83a)
$$E_G = \frac{\hbar^2}{2M}\left[\frac{(2.40)^2}{a^2} + \frac{\pi^2}{L^2}\right]$$

(10.83b)
$$\varphi_G(r, z, \phi) = AJ_0\left(\frac{2.40r}{a}\right)\sin\left(\frac{\pi z}{L}\right)$$

and is positive within the cylindrical box. Orthogonality of the Bessel functions in (10.82b) parallels that of (10.72c) and is given by

(10.84)
$$\int_0^a dr\, r J_m\left(\frac{rx_{ms}}{a}\right)J_m\left(\frac{rx_{ms'}}{a}\right) = \frac{a^2}{2}[J_{m+1}(x_{ms})]^2\,\delta_{ss'}$$

which, again, indicates the nature of normalization of these functions. A compilation of properties of the rectangular, spherical, and cylindrical quantum wells is listed in Table 10.2.

Note that for spherical Bessel functions (10.72a), x_{ln} represents the nth zero of $j_l(x)$ and l is orbital quantum number. For Bessel functions of the first kind (10.82b), x_{ms} represents the sth zero of $J_m(x)$ and m is azimuthal quantum number [see (9.51)]. Thus, for both spherical and cylindrical quantum wells, the orders of respective Bessel functions that enter the analysis are angular-momentum quantum numbers. Note that for $\hat{H}_{cyl}(r, z, \phi)$, $[\hat{L}_z, \hat{H}_{cyl}] = 0$, whereas for $H_{sph}(r, \theta, \phi)$, $[\hat{L}^2, \hat{H}_{sph}] = [\hat{L}_z, \hat{H}_{sph}] = 0$. For the latter case, l and m are good quantum numbers, corresponding to invariance of the Hamiltonian to rotations that leave r^2 invariant. For the cylindrical case, only m is a good (angular momentum) quantum number, corresponding to invariance of the Hamiltonian to rotations about the z axis.

PROBLEMS

10.9 Calculate the divergence of particle current, $\nabla \cdot \mathbf{J}$, for a collection of particles that are all in the state

$$\psi(r, t) = j_2(kr)e^{-i\omega t}$$

$$\hbar\omega = \frac{\hbar^2 k^2}{2m}$$

10.10 A spherically propagating shell contains N neutrons, which are all in the state

$$\psi(\mathbf{r}, 0) = 4\pi i \left[\frac{\sin kr}{(kr)^2} - \frac{\cos kr}{kr}\right] \frac{3Y_1^0(\theta, \phi) + 5Y_1^{-1}(\theta, \phi)}{\sqrt{34}}$$

at $t = 0$

(a) What is $\psi(\mathbf{r}, t)$?

(b) What is the expectation of the energy for this "beam"?

(c) What possible values of L^2 and L_z will measurement find and how many neutrons will have these values?

(d) If at $t = 0$, measurement of L^2 finds the value $2\hbar^2$, what is $\psi(\mathbf{r}, t)$?

(e) If at $t = 0$, measurement of L_z finds the value $-\hbar$, what is $\psi(\mathbf{r}, t)$?

10.11 Use the expansion of a plane wave in spherical harmonics,

$$e^{i\mathbf{k}\cdot\mathbf{r}} = 4\pi \sum_{l=0}^{\infty} \sum_{m=-1}^{l} i^l j_l(kr)[Y_l^m(\alpha, \beta)]^* Y_l^m(\theta, \phi)$$

together with the spherical coordinate representation of $\delta(\mathbf{r} - \mathbf{r}')$,

$$\delta(\mathbf{r} - \mathbf{r}') = \frac{\delta(r - r')\delta(\theta - \theta')\delta(\phi - \phi')}{r^2 \sin\theta} = \left(\frac{1}{2\pi}\right)^3 \int_0^{2\pi}\int_0^{\pi}\int_0^{\infty} e^{i\mathbf{k}\cdot(\mathbf{r}-\mathbf{r}')}k^2\, dk \sin\alpha\, d\alpha\, d\beta$$

(the spherical coordinates of $\mathbf{k}$ are k, α, β; see Fig. 10.5a) to obtain the orthonormality condition

$$\delta(\mathbf{r} - \mathbf{r}') = \frac{2}{\pi} \sum_l \sum_m [Y_l^m(\theta', \phi')]^* Y_l^m(\theta, \phi) \int_0^\infty j_l(kr) j_l(kr') k^2 \, dk$$

The spherical coordinates of $\mathbf{r}$ are (r, θ, ϕ), and those of $\mathbf{r}'$ are (r', θ', ϕ'). [Compare with (C.14) in Appendix C.]

10.12 Use the addition theorem for spherical harmonics (see Fig. 9.16) to reduce the first equation of Problem 10.11 to the expansion

$$e^{ikz} = e^{ikr \cos \theta} = \sum_{l=0}^\infty (2l + 1) i^l j_l(kr) P_l (\cos \theta)$$

Note that in this description, $\mathbf{k}$ is aligned with the polar (z) axis, so that $\mathbf{k} \cdot \mathbf{r} = kz$ (see Fig. 10.5b). This expansion is important to the theory of partial wave scattering and will be called upon in Chapter 14.

10.13 The expansion in Problem 10.12 of the plane wave e^{ikz} indicates that the probability of measuring $L^2 = \hbar^2 l(l + 1)$ is

$$P[\hbar^2 l(l + 1)] \simeq (2l + 1)^2$$

Give a semiclassical heuristic argument in support of this conclusion (i.e., that $P \sim l^2$).

Answer

Consider a surface S, of constant phase of the plane wave, $\exp(ikz)$. (See Fig. 10.6.) All points in the annular region $dS = 2\pi r_\perp \, dr_\perp = \pi d(r_\perp^2)$ correspond to angular momentum $L = r_\perp p_z = r_\perp \hbar k$. It follows that

$$dS = \pi d\left(\frac{L^2}{\hbar^2 k^2}\right)$$

The probability of finding such "points" is proportional to the annular surface dS, so (k^2 is constant)

$$dP \sim dS \sim dL^2$$

In the classical (correspondence) limit, $L^2 \sim \hbar^2 l^2$ and $P \sim l^2$.

10.14 How many independent eigenstates are there corresponding to a free particle moving with energy $E_k = \hbar^2 k^2 / 2m$ in
    (a) The Cartesian coordinate representation?
    (b) The spherical coordinate representation?
Give a classical description of the different orbits corresponding to these degenerate states.

Answers (partial)

    (a) In the Cartesian representation, any state

$$\varphi_k = A e^{i\mathbf{k} \cdot \mathbf{r}}, \qquad k^2 = \frac{2mE}{\hbar^2}$$

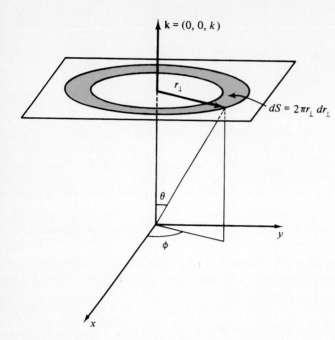

FIGURE 10.6 The probability of finding a particle in a plane-wave state with angular momentum $\sim\hbar l$ increases as l^2. (See Problem 10.13.)

is an eigenstate corresponding to the given value of E. These **k** vectors describe a sphere of radius $\sqrt{2mE/\hbar^2}$. This continuum of states corresponds to aiming the particle in different directions, while holding its speed, $\hbar k/m$, fixed.

(b) In the spherical representation, any state

$$\varphi_{klm} = j_l(kr)Y_l^m(\theta, \phi), \qquad k^2 = \frac{2mE}{\hbar^2}$$

is an eigenstate corresponding to the given value of E. Different states are obtained by choosing different values of l and m. This countable infinity of states corresponds to propitious choice of straight-line trajectories about the origin, all at constant speed, $\hbar k/m$.

10.15 An electron is confined to a spherical well of radius $a = 10$ Å. Calculate the emission frequency, f_1 (in Hz), of radiation for this system due to transition from the first excited state to the ground state. State the class of radiation to which this frequency belongs (IR, visible, UV, soft x rays, etc).

10.16 A "quantum dot," relevant to semiconductor devices, may be modeled as an electron in a spherical well with perfectly reflecting walls. Design a quantum dot whose characteristic frequency of emission is 10 GHz, where "characteristic frequency" corresponds to decay from the first excited state to the ground state. That is, obtain the radius, a, of the spherical cavity that has this property. Repeat this problem for the quantum dot that is a cube of edge length $2b$. In either case assume an effective mass of $m = 0.067m_e$. Give your answers in μm.

10.17 (a) What are the eigenfunctions and eigenenergies (in eV) of the ground and first excited states of an electron trapped in a cylindrical box of radius $a = 1.5$ Å and length $L = 3.0$ Å? What is the degeneracy of these two eigenenergies?

(b) An electron propagates in a "quantum wire" of circular cross-sectional radius 5.2 Å and z component of kinetic energy 1.03 eV. What is the lowest energy wavefunction and energy (in eV) corresponding to this condition?

(*Note:* A quantum wire is a rectilinear wirelike domain of narrow band-gap material immersed in a medium of wide band-gap material. In this problem we model the quantum wire as being of infinite length and assume that the propagating electron acts as a confined free particle.)

10.4   A CHARGED PARTICLE IN A MAGNETIC FIELD

A closely allied motion to that of a free particle is the motion of a charged particle (e.g., an electron) in a uniform, constant magnetic field $\mathcal{B}$. The Hamiltonian for the electron is given by

$$(10.85) \qquad H = \frac{1}{2m}\left(\mathbf{p} - \frac{e}{c}\mathbf{A}\right)^2$$

The magnetic field is related to the vector potential $\mathbf{A}$ through the relation

$$\mathcal{B} = \nabla \times \mathbf{A}$$

The Cartesian components of $\mathbf{A}$,

$$\mathbf{A} = (-y\mathcal{B}, 0, 0)$$

generate a uniform magnetic field which points in the z direction.

$$\mathcal{B} = (0, 0, \mathcal{B})$$

Substituting this value of $\mathbf{A}$ into the Hamiltonian above gives the time-independent Schrödinger equation

$$(10.86) \qquad \hat{H}\varphi = \left[\frac{1}{2m}\left(\hat{p}_x + \frac{ey\mathcal{B}}{c}\right)^2 + \frac{\hat{p}_y{}^2 2}{m} + \frac{\hat{p}_z{}^2}{2m}\right]\varphi = E\varphi$$

Since the coordinates x and z are missing from the Hamiltonian, it follows that

$$[\hat{p}_x, \hat{H}] = [\hat{p}_z, \hat{H}] = 0$$

and we may conclude that $\hat{p}_x$, p_z, and $\hat{H}$ have simultaneous eigenstates. The eigenstates of $\hat{p}_x$ and $\hat{p}_z$ appear as

$$\varphi_{k_x k_z} = e^{i(k_x x + k_z z)}$$

so that we may write the common eigenstates of $\hat{H}$, $\hat{p}_x$, and $\hat{p}_z$ in the form

$$(10.87) \qquad \varphi = e^{i(k_x x + k_z z)} f(y)$$

Substituting this product into (10.86) gives

(10.88)
$$\left[\frac{\hat{p}_y{}^2}{2m} + \frac{K}{2}(y - y_0)^2\right] f = \left(E - \frac{\hbar^2 k_z{}^2}{2m}\right) f$$

where we have set

$$y_0 \equiv -\frac{c\hbar k_x}{e\mathcal{B}}$$

$$\frac{K}{m} \equiv \left(\frac{e\mathcal{B}}{mc}\right)^2 \equiv \Omega^2$$

The frequency Ω is called the *cyclotron frequency*. This is the frequency of rotation corresponding to the classical motion of a charged particle in a uniform magnetic field (see Problem 10.18).

The Schrödinger equation (10.88) is the same as that for a simple harmonic oscillator constrained to move along the y axis, about the point y_0, with natural frequency Ω. From Section 7.2 we recall that the eigenenergies of this equation are

$$E_n - \frac{\hbar^2 k_z{}^2}{2m} = \hbar\Omega\left(n + \frac{1}{2}\right)$$

which gives the desired result

(10.89)
$$E_n = \hbar\Omega\left(n + \frac{1}{2}\right) + \frac{\hbar^2 k_z{}^2}{2m}$$

The kinetic-energy term $\hbar^2 k_z{}^2/2m$ corresponds to free, linear motion parallel to the z axis. Classically, such motion is unaffected by a magnetic field in the z direction. The first term in E_n corresponds to the rotational motion normal to the $\mathcal{B}$ field. In the corresponding classical motion the charged particle moves in a helix of constant radius, constant energy, constant rotational frequency, and constant z velocity. The projection of the motion onto the xy plane is a circle with a fixed center (Fig. 10.7). The energy levels (10.89) are commonly referred to as *Landau levels*.

The eigenfunction corresponding to the eigenenergy (10.89) is

$$f_n = A_n \mathcal{H}_n\left[\sqrt{\frac{m\Omega}{\hbar}}(y - y_0)\right] \exp\left[-\frac{1}{2}\frac{m\Omega}{\hbar}(y - y_0)^2\right]$$

[recall (7.59)]. The nth-order Hermite polynomial is written $\mathcal{H}_n$, while A_n is a normalization constant. Together with (10.87), this form for f_n gives the wavefunction

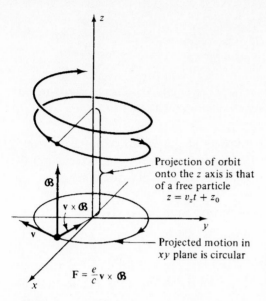

Projection of orbit
onto the z axis is that
of a free particle
$z = v_z t + z_0$

Projected motion in
xy plane is circular

FIGURE 10.7   **Helical motion of a positive charge in a uniform, constant magnetic field that points in the z direction.**

(10.90)

$$\varphi_n = A_n \mathcal{H}_n \left[\sqrt{\frac{m\Omega}{\hbar}} \, (y - y_0) \right] \exp \left[-\frac{1}{2} \frac{m\Omega}{\hbar} (y - y_0)^2 + i(k_x x + k_z z) \right]$$

for a charged particle moving in a uniform magnetic field which points in the z direction.

Degeneracy of Landau Levels

In this section we wish to discover the manner in which the continuous energy spectrum of a free electron changes when the electron is in the presence of a magnetic field. In the course of this discussion we will obtain an expression for the degeneracy of Landau levels.

To examine this problem, we consider the electron to be enclosed in a large cubical box of edge length L. Free-particle wavefunctions and energies are given by (10.14). To account for the finite enclosure, for L sufficiently large, say $L^2 \gg \hbar/m\Omega$, we impose *periodic boundary conditions*:

$$\varphi(x, y, z) = \varphi(x + L, y + L, z + L)$$

As the Landau levels given by (10.89) are evidently degenerate in k_x, we will focus on this wavenumber for the free-particle motion. Substituting the free-particle state (10.14) into the relation above gives

$$k_x = \frac{2\pi n_x}{L}$$

$$n_x = 0, \pm 1, \pm 2, \ldots$$

Thus we discover the following important fact. The continuous energy spectrum of a free, unconfined particle changes to a nearly continuous discrete spectrum when the particle is confined in a large enclosure.

In the presence of a magnetic field, the Schrödinger equation for the electron is given by (10.88). As noted in Problem 10.18, y_0 is associated with the center of the corresponding classical circular motion. Thus, for the present case we may take y_0 to lie between 0 and L.

It follows that the maximum value of this parameter $\bar{y}_0$ is given by

$$\bar{y}_0 = L = \frac{c\hbar \bar{k}_x}{|e|\mathcal{B}} = \frac{ch\bar{n}_x}{|e|\mathcal{B}L}$$

We may conclude that n_x values are positive numbers and have the maximum value[1]

(10.91)
$$\bar{n}_x \equiv \bar{g} = \left(\frac{|e|\mathcal{B}}{hc}\right)L^2$$

This is the desired expression for the degeneracy $\bar{g}$ of a Landau level. The presence of L^2 in (10.91) corresponds to the property that a given helical orbit can be displaced anywhere in the xy plane without changing the energy of the electron (see Fig. 10.7).

Thus, we find that the energy spectrum of a confined electron changes from a nearly continuous one for $\mathcal{B} = 0$ to a discrete spectrum for $\mathcal{B} > 0$, with degeneracy given by (10.91).

From (10.89) we see that the spacing between Landau levels, at fixed k_z, is the constant value

$$\Delta E = \hbar\Omega$$

The degeneracy $\bar{g}$ given by (10.91) gives the number of free-particle states that contribute to the increment ΔE. See Fig. 10.8. Note, in particular, the resemblance between the equally spaced Landau levels and the equally spaced levels of the harmonic oscillator shown in Fig. 7.8. This congruence of spectra stems from the previously described parallel structure of the two respective Hamiltonians. In either event, the density of states, $g(E)$ (Section 8.8), when plotted as a function of energy, is a series of equally spaced delta functions.

---

[1] We employ the barred variable $\bar{g}$ to distinguish it from the unbarred variable g used to denote density of states elsewhere in the text.

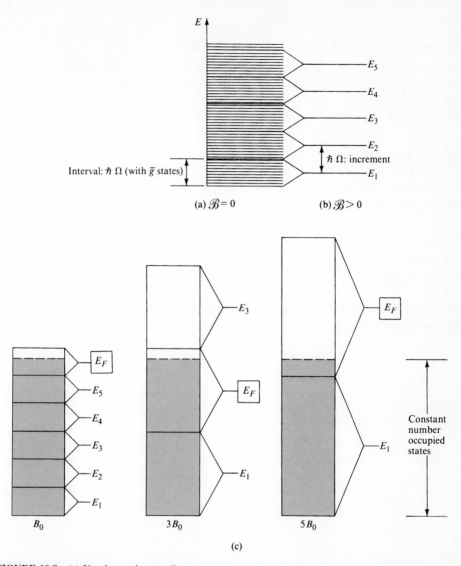

FIGURE 10.8 (a) **Nearly continuous discrete spectrum for a particle confined to a large box with** $\mathscr{B} = 0$. (b) **Equally spaced Landau levels corresponding to** $\mathscr{B} > 0$. **Each increment of energy,** $\hbar\Omega$, **corresponds to** $\bar{g}$ **free-particle states, which, in turn, is the degeneracy of each Landau level.** (c) **Variation of the Fermi energy with change in** $\mathscr{B}$.

Note further that in the classical limit $\hbar \rightarrow 0$, the degeneracy $\bar{g}$ grows infinite and the spacing between levels ΔE goes to zero.

Fermi Energy and Landau Levels

Application of the preceding results may be made to the conduction electrons in a two-dimensional metal. In this event, the degenerate states shown in Fig. 10.8a become filled with electrons. The highest energy state occupied (at 0 K) is that of the Fermi energy, E_F [Sections 2.3 (Fig. 2.3), 8.4, 12.9]. With a magnetic field $\mathcal{B}$ normal to the plane of the sample, we may identify E_F as the highest Landau level occupied. How does E_F change with change in $\mathcal{B}$? To answer this question, we first note that the maximum number of electrons at the Fermi level is given by the degeneracy factor $\bar{g}$(10.91). Let $n^{oc} \leq \bar{g}$ denote the number of occupied states in this level. The value of E_F remains constant as long as $n^{oc} > 0$. When $\mathcal{B}$ is increased, E_F changes to the value of the new partially occupied Landau level. See Fig. 10.8c. Such variation of E_F exhibits a periodicity with respect to $\mathcal{B}^{-1}$. This phenomenon is a component of the *de Haas–van Alphen effect.*[1]

PROBLEMS

10.18   The following is a problem in classical physics. The force on a charged particle in a uniform magnetic field $\mathcal{B}$ is

$$\mathbf{F} = \frac{d}{dt}(m\mathbf{v}) = \frac{e}{c}\mathbf{v} \times \mathcal{B}$$

with $\mathcal{B} = (0, 0, \mathcal{B})$.
   (a)   Show that

$$\frac{1}{2}mv^2 = \text{constant}$$

   (b)   Show that

$$p_z = mv_z = \text{constant}$$

   (c)   Show that the motion of the particle is that of a helix whose axis is parallel to $\mathcal{B}$ and whose projection onto the xy plane is circular with constant angular frequency Ω.
   (d)   Show that the center of this circle in the xy plane has coordinates

$$y_0 = \frac{-cp_x}{e\mathcal{B}} = \frac{-cmv_x}{e\mathcal{B}} + y$$

$$x_0 = \frac{cmv_y}{e\mathcal{B}} + x = \frac{cp_y}{e\mathcal{B}} + x$$

---

[1] For further discussion, see C. Kittel, *Introduction to Solid State Physics*, 6th ed., Wiley, New York, 1986, Chap. 9.

Note that p_x, canonical momentum, is not equal to mv_x for $\mathbf{A} = (A_x, 0, 0)$. The correct relation follows from (1.14) and (10.85).

10.19 Show that the operator

$$\hat{x}_0 \equiv \hat{x} + \frac{c\hat{p}_y}{e\mathcal{B}}$$

commutes with $\hat{H}$ as given in (10.86) but does not commute with

$$\hat{y}_0 = \frac{-c\hat{p}_x}{e\mathcal{B}}$$

These operators correspond to the coordinates of the center of the related projected classical motion in the xy plane. In quantum mechanics we see that although x_0 and E, or y_0 and E, may, respectively, be specified simultaneously, x_0 and y_0 may not be simultaneously specified.

10.20   (a) What is the vector potential $\mathbf{A}$ which gives the uniform $\mathcal{B}$ field $(0, 0, \mathcal{B})$ which includes $A_x = A_z = 0$?

(b)   What is the form of the wavefunctions φ_n corresponding to this choice of vector potential? How do they compare to the wavefunctions corresponding to $A_y = A_z = 0$ found in the text?

(c)   How do the eigenenergies compare to those found in the representation $A_y = A_z = 0$?

10.21   What is the nature of the frequency spectrum emitted by a charged particle moving in a uniform magnetic field? (Assume that the kinetic energy parallel to $\mathcal{B}$ does not change.) For an electron moving in a $\mathcal{B}$ field of 10^4 gauss, what type of radiation is this (x rays, micro-waves, etc.)?

10.5   THE TWO-PARTICLE PROBLEM

Coordinates Relative to the Center of Mass

When dealing with systems containing more than one particle (e.g., an atom), it is convenient to separate the motion into that of the center of mass of the system and motion relative to the center of mass. This separation is effected through a partitioning of the Hamiltonian into a part, H_{CM}, involving center of mass coordinates, and a part, H_{rel}, containing coordinates relative to the center of mass.

For example, consider the two-particle Hamiltonian

$$(10.92) \qquad H = \frac{p_1{}^2}{2m_1} + \frac{p_2{}^2}{2m_2} + V(|\mathbf{r}_1 - \mathbf{r}_2|)$$

The potential of interaction $V(|\mathbf{r}_1 - \mathbf{r}_2|)$ is a function only of the radial distance between the particles. For instance, for the hydrogen atom, the interaction V is the Coulomb potential

(10.92a)
$$V = -\frac{e^2}{r}$$

where we have written r for the distance between particles, $|\mathbf{r}_1 - \mathbf{r}_2|$. Such potentials, which are only a function of the scalar distance r, are called *central potentials*.

In the Hamiltonian above, $\mathbf{p}_1$ and $\mathbf{p}_2$ are the linear momenta of particle 1 and particle 2, respectively, while m_1 and m_2 are the respective masses of these particles.

A two-particle system has six degrees of freedom which are the number of coordinates in the state vector $(\mathbf{r}_1, \mathbf{p}_1; \mathbf{r}_2, \mathbf{p}_2)$. The partitioning of the Hamiltonian into $H_{CM} + H_{rel}$ is generated through the transformation of variables

(10.93)
$$(\mathbf{r}_1, \mathbf{p}_1; \mathbf{r}_2, \mathbf{p}_2) \rightarrow (\mathbf{r}, \mathbf{p}; \mathcal{R}, \mathcal{P})$$

where

$$\mathbf{r} = \mathbf{r}_2 - \mathbf{r}_1, \qquad \mathcal{P} = \mathbf{p}_1 + \mathbf{p}_2$$

(10.94)
$$\mathbf{p} = \frac{m_2 \mathbf{p}_1 - m_1 \mathbf{p}_2}{m_1 + m_2}, \qquad \mathcal{R} = \frac{m_1 \mathbf{r}_1 + m_2 \mathbf{r}_2}{m_1 + m_2}$$

Using these equations the Hamiltonian (10.92) is transformed to the sum

(10.95)
$$H = \frac{\mathcal{P}^2}{2M} + \left[\frac{p^2}{2\mu} + V(r) \right] \equiv H_{CM} + H_{rel}$$

where the reduced mass μ and the total mass M are

(10.96)
$$\mu = \frac{m_1 m_2}{m_1 + m_2}, \qquad M = m_1 + m_2$$

Equation (10.95) represents the desired separation of H into the Hamiltonian of the center of mass, H_{CM}, and the Hamiltonian of the coordinates relative to the center of mass, H_{rel} (Fig. 10.9). Since $\mathcal{R}$ is absent in H (i.e., $\mathcal{R}$ is a cyclic coordinate; see Section 1.2), the momentum of the center of mass, $\mathcal{P}$, is constant. The center of mass moves in straight rectilinear motion, characteristic of a free particle of mass M. The motion relative to the center of mass is that of a particle of mass μ moving in the central potential $V(r)$.

The Transformation of $\hat{H}$

For the quantum mechanical case, the transformation of $\hat{H}$ is again effected with the equations (10.94), which are now interpreted as operator relations. Cartesian components of "old" coordinate and momentum operators $(\hat{\mathbf{r}}_1, \hat{\mathbf{p}}_1; \hat{\mathbf{r}}_2, \hat{\mathbf{p}}_2)$ obey the commutation relations

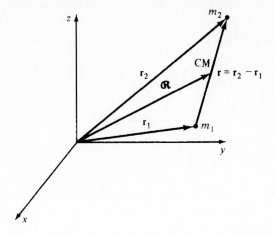

FIGURE 10.9   **The relative vector r and the center-of-mass vector $\mathfrak{R}$. In the classical motion, $\mathcal{P} = M\mathfrak{R} = \text{constant} = \mathcal{P}(0) = \text{the initial value of } \mathcal{P}(t)$. At any time t**

$$\mathfrak{R}(t) = \mathfrak{R}(0) + \frac{\mathcal{P}(0)t}{M}$$

Solving the dynamical equations (viz., Hamilton's equations) using H_{rel} gives $r(t)$, which when affixed to $\mathfrak{R}(t)$ gives the motion in the "lab frame."

(10.97)
$$[\hat{r}_{1j}, \hat{p}_{1j}] = i\hbar$$
$$[\hat{r}_{2j}, \hat{p}_{2j}] = i\hbar \qquad j = 1, 2, 3$$

These are the only nonvanishing commutators. With these relations and (10.94), one obtains that the only nonvanishing commutator relations for components of the "new" operators $(\hat{\mathbf{r}}, \hat{\mathbf{p}}; \hat{\boldsymbol{\mathfrak{R}}}, \hat{\boldsymbol{\mathcal{P}}})$ are

(10.98)
$$[\hat{r}_j, \hat{p}_j] = i\hbar$$
$$[\hat{\mathfrak{R}}_j, \hat{\mathcal{P}}_j] = i\hbar \qquad j = 1, 2, 3$$

Thus, in obtaining

(10.99)
$$\hat{H} = \hat{H}_{\text{CM}} + \hat{H}_{\text{rel}}$$
$$\hat{H}_{\text{CM}} = \frac{\hat{\mathcal{P}}^2}{2M}$$
$$\hat{H}_{\text{rel}} = \frac{\hat{p}^2}{2\mu} + V(r)$$

the Hamiltonian is separated into two parts involving components that are independent of one another. For such cases, the Schrödinger equation has product eigenfunctions

(10.100)
$$\overline{\varphi} = \varphi_{\text{CM}}(\boldsymbol{\mathfrak{R}})\varphi_{\text{rel}}(\mathbf{r})$$

and summational eigenvalues

(10.101)
$$\overline{E} = E_{\text{CM}} + E_{\text{rel}}$$

where

$$\hat{H}\,\overline{\varphi} = \overline{E}\,\overline{\varphi}$$

(10.102)
$$\hat{H}_{CM}\varphi_{CM} = E_{CM}\varphi_{CM}$$

$$\hat{H}_{rel}\varphi_{rel} = E_{rel}\varphi_{rel}$$

The Schrödinger equation for the center of mass appears explicitly as

(10.103)
$$\frac{\hat{\mathcal{P}}^2}{2M}\,\varphi_{CM} = E_{CM}\varphi_{CM}$$

This is the Schrödinger equation for a free particle of mass M. Its solution was obtained in the previous section. With the linear momentum $\mathcal{P}$ specified, the states are

(10.104)
$$\varphi_{CM} = Ae^{i\mathbf{K}\cdot\mathfrak{R}}$$

$$\mathcal{P} = \hbar\mathbf{K}, \qquad E_{CM} = \frac{\hbar^2 K^2}{2M}$$

In the representation where L_{CM}^2 and L_{CM_z} are specified, the eigenstates are

(10.105)
$$\varphi_{CM} = j_{l_C}(K\mathfrak{R})Y_{l_C}^{m_C}(\theta_C, \phi_C)$$

The spherical coordinates of $\mathfrak{R}$ are $(\mathfrak{R}, \theta_C, \phi_C)$ (see Fig. 10.10a).

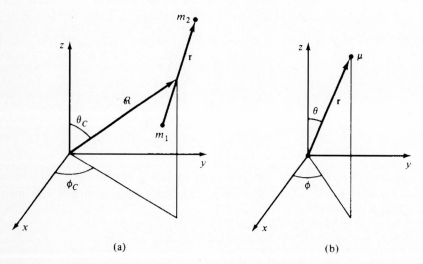

(a)   (b)

FIGURE 10.10 (a) Spherical angle variables for the center-of-mass radius vector $\mathfrak{R}$. (b) Coordinates relative to the center of mass.

Radial Equation for a Central Potential

The Schrödinger equation for φ_{rel} appears as (dropping the "rel" subscript)

$$(10.106) \qquad \left[\frac{\hat{p}^2}{2\mu} + V(r)\right]\varphi = E\varphi$$

For central potential functions $V(r)$, it proves most convenient to express the above Hamiltonian in spherical coordinates. The interparticle radius $\mathbf{r}$ has coordinates (r, θ, ϕ) with the polar axis depicted as lying in the z direction (see Fig. 10.10b).

In these coordinates the Schrödinger equation above becomes

$$(10.107) \qquad \hat{H}\varphi = \left[\frac{\hat{p}_r^2}{2\mu} + \frac{\hat{L}^2}{2\mu r^2} + V(r)\right]\varphi = E\varphi$$

First, we note that $\hat{L}^2$ and $\hat{L}_z$ both commute with $\hat{H}$. The remaining components in $\hat{H}$ are all isotropic forms and are therefore unaffected by angular momentum operators. It follows that $\hat{H}$, $\hat{L}^2$, and $\hat{L}_z$ have simultaneous eigenstates. These are given by the product form

$$(10.108) \qquad \varphi = R(r)Y_l^m(\theta, \phi)$$

Substituting this solution into the Schrödinger equation above gives the "radial" equation

$$(10.109) \qquad \left[\frac{\hat{p}_r^2}{2\mu} + \frac{\hbar^2 l(l+1)}{2\mu r^2} + V(r)\right]R(r) = ER(r)$$

This is an ordinary, second-order, linear differential equation for the radial dependent component of the wavefunction $R(r)$. Since only one variable is involved in (10.109), it is suggestive of one-dimensional motion with the effective potential

$$(10.110) \qquad V_{\text{eff}} = V(r) + \frac{\hbar^2 l(l+1)}{2\mu r^2}$$

The second term in this expression is called the "angular momentum barrier." It becomes infinitely high as $r \to 0$ and acts as a repulsive core, which for $l > 0$ prevents collapse of the system (see Fig. 10.11).

The normalization of the eigenstates (10.108) is given by the integral

$$(10.111) \qquad \langle RY_l^m | RY_l^m \rangle = \int_0^\infty dr\, r^2 \int_{4\pi} d\Omega\, |R(r)Y_l^m(\theta, \phi)|^2 = 1$$

$$= \int_0^\infty r^2 |R(r)|^2\, dr = 1$$

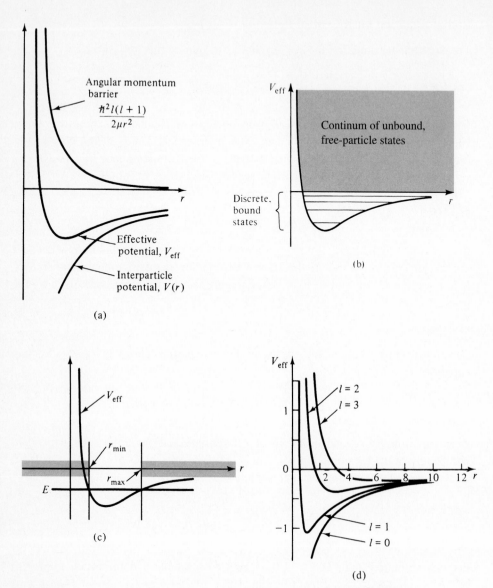

FIGURE 10.11 **(a) The effective potential in relation to the angular momentum barrier**

$$V_{\text{eff}} = V(r) + \frac{\hbar^2 l(l + 1)}{2\mu r^2}$$

(b) Nature of the quantum mechanical energy spectrum for central potential problems. (c) The classical motion corresponding to the energy E. Shaded regions define classically forbidden domains. (d) The effective potential energy V_{eff} for hydrogen for several values of the orbital quantum number l. Units of r are angstroms. V_{eff} is in units of 10^{-11} erg.

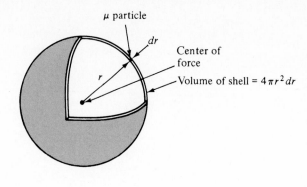

μ particle

dr

Center of force

Volume of shell = $4\pi r^2 dr$

r

FIGURE 10.12 Probability of finding the fictitious μ particle in a spherical shell between r and $r + dr$ is

$$P_r \, dr = |R(r)r|^2 \, dr$$

This is also the probability of finding m_2 in a spherical shell about m_1 in the configuration shown (or m_1 in a shell about m_2).

The radial displacement r separates the two particles m_1 and m_2. If we envision particle m_1 at the origin, then

$$(10.112) \qquad Pr^2 \, dr \, d\Omega = |R(r)Y_l^m(\theta, \phi)|^2 r^2 \, dr \, d\Omega$$

is the probability of finding m_2 in the *volume element* $r^2 \, dr \, d\Omega$ about m_1 (an equally valid statement is obtained with m_1 and m_2 reversed). What is the probability of finding m_2 in a *spherical shell* of radius between r and $r + dr$, about m_1? The answer is (Fig. 10.12)

$$(10.113) \qquad P_r \, dr = \left(\int_{4\pi} Pr^2 \, d\Omega \right) dr = |R(r)|^2 r^2 \, dr \equiv |u(r)|^2 \, dr$$

so that

$$\int_0^\infty |u(r)|^2 \, dr = 1$$

The classically forbidden domains (see Chapter 1) correspond to values of r for which $E < V$. The related property for a spherical quantum mechanical system is that the probability density $|u(r)|^2$ becomes small in these domains (see Figs. 10.11c and 10.13).

Having found the radial function $R(r)$, in a specific two-body problem, the wavefunction for the system relative to the laboratory frame (as opposed to the center-of-mass frame) is either of the forms

$$(10.114) \qquad \begin{aligned} \overline{\varphi} &= A e^{i\mathcal{P}\cdot\mathcal{R}/\hbar} R(r) Y_l^m(\theta, \phi) \\ \overline{\varphi} &= j_{l_C}(K\mathcal{R}) Y_{l_C}^{m_C}(\theta_C, \phi_C) R(r) Y_l^m(\theta, \phi) \end{aligned}$$

In the first representation, the six parameters $(\mathcal{P}; L^2, L_z, E)$ are specified. In the second representation, the six parameters $(E_{CM}, L_{CM}^2, L_{CM_z}; L^2, L_z, E)$ are specified.

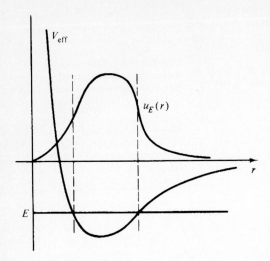

FIGURE 10.13 The radial probability amplitude u_E corresponding to the energy E decays to zero in the classically forbidden domain.

Continuity and Boundary Conditions

Some general properties of the radial wavefunction are as follows. With $R(r)$ everywhere bounded, we note first that $u(r) \equiv rR(r)$ must vanish at the origin.[1] For $r > 0$, with energy E and potential energy $V(r)$ bounded, the radial equation (10.109) indicates that

$$\hat{p}_r^2 R(r) = -\hbar^2 \frac{1}{r} \frac{\partial}{\partial r} \left[\frac{\partial}{\partial r} u(r) \right]$$

is likewise bounded. It follows that $\partial u / \partial r$ is continuous. The existence of this derivative implies that $u(r)$ is continuous. The latter two conditions infer continuity of the logarithmic derivative

$$\frac{1}{u} \frac{du}{dr} = \frac{d \ln u}{dr}$$

These conditions on the wavefunction in spherical coordinates are employed in obtaining the ground state of the deuteron (Problem 10.31) and in construction of the bound states of the hydrogen atom as described in the following section. They will also come into play in construction of the states for low-energy scattering from a spherical well (Section 14.2).

---

[1] Dirac obtains this boundary condition from the stipulation that solutions to the Schrödinger equation in spherical coordinates agree with those obtained in Cartesian coordinates. For further discussion, see P. A. M. Dirac, *The Principles of Quantum Mechanics*, 4th ed., Oxford University Press, New York, 1958.

PROBLEMS

10.22   Consider a two-particle system. The momenta of the particles are $\mathbf{p}_1$ and $\mathbf{p}_2$, respectively.

(a)   What is $[\hat{\mathbf{p}}_1, \hat{\mathbf{p}}_2]$? Explain your answer.

(b)   Use the answer to part (a) to show that $[\hat{\mathbf{p}}, \hat{\mathscr{P}}] = 0$, where $\mathbf{p}$ is the "relative" momentum [defined in (10.94)] and $\mathscr{P}$ is the momentum of the center of mass.

(c)   The particles interact under a potential that is a function of the distance between them. As shown in the text, transformation to coordinates relative to the center of mass effects a partitioning of the Hamiltonian, $\hat{H} = \hat{H}_{CM} + \hat{H}_{rel}$. What is $[\hat{H}_{CM}, \hat{H}_{rel}]$?

10.23   Prove that the following equations are compatible with the transformation equations (10.94).

(a) $\dfrac{p_1^2}{2m_1} + \dfrac{p_2^2}{2m_2} = \dfrac{p^2}{2\mu} + \dfrac{\mathscr{P}^2}{2M}$

(b) $m_1 r_1^2 + m_2 r_2^2 = \mu r^2 + M\mathscr{R}^2$

(c) $\mathbf{p}_1 \cdot \mathbf{r}_1 + \mathbf{p}_2 \cdot \mathbf{r}_2 = \mathbf{p} \cdot \mathbf{r} + \mathscr{P} \cdot \mathscr{R}$

(d) $\mathbf{L}_1 + \mathbf{L}_2 = \mathbf{L} + \mathbf{L}_{CM}$

In part (d),

$$\mathbf{L}_1 = \mathbf{r}_1 \times \mathbf{p}_1 \qquad\qquad \mathbf{L} = \mathbf{r} \times \mathbf{p}$$

$$\mathbf{L}_2 = \mathbf{r}_2 \times \mathbf{p}_2 \qquad\qquad \mathbf{L}_{CM} = \mathscr{R} \times \mathscr{P}$$

10.24   At a particular time, the wavefunction of a mass m moving in a three-dimensional potential field is

$$\varphi = A(x + y + z)e^{-k_0 r}$$

(a)   Calculate the normalization constant A.

(b)   What is the probability that measurement of L^2 and L_z finds $2\hbar^2$ and 0, respectively? (See Table 9.1.)

(c)   What is the probability of finding the particle in the sphere $r \leq k_0^{-1}$?

10.25   For a two-particle system (m_1, m_2), what is the fractional distance to the center of mass from m_1 and m_2, respectively? What are these numbers for hydrogen?

10.26   Let $\mathbf{e}$ be a unit vector in an arbitrary but fixed direction. Show that the commutators between the components of $\hat{\mathbf{r}}$ and $\hat{\mathbf{p}}$, respectively, with the component $\mathbf{e} \cdot \hat{\mathbf{L}}$ obey the relations

$$[\hat{\mathbf{p}}, \mathbf{e} \cdot \hat{\mathbf{L}}] = i\hbar \mathbf{e} \times \hat{\mathbf{p}}$$

$$[\hat{\mathbf{r}}, \mathbf{e} \cdot \hat{\mathbf{L}}] = i\hbar \mathbf{e} \times \hat{\mathbf{r}}$$

10.27   Use the results of Problem 10.26 to show that $\hat{p}^2$, $\hat{r}^2$, and $\hat{\mathbf{r}} \cdot \hat{\mathbf{p}}$ all commute with every component of $\hat{\mathbf{L}}$. Then show that every component of $\hat{\mathbf{L}}$ commutes with any isotropic function $f(r^2)$.

Answer (partial)

If the statement is true for arbitrary **e**, it is true for all components of $\hat{\mathbf{L}}$.

$$[\hat{p}^2, \mathbf{e} \cdot \hat{\mathbf{L}}] = \hat{\mathbf{p}}[\hat{\mathbf{p}}, \mathbf{e} \cdot \hat{\mathbf{L}}] + [\hat{\mathbf{p}}, \mathbf{e} \cdot \hat{\mathbf{L}}] \cdot \hat{\mathbf{p}}$$
$$= i\hbar(\hat{\mathbf{p}} \cdot \mathbf{e} \times \hat{\mathbf{p}} + \mathbf{e} \times \hat{\mathbf{p}} \cdot \hat{\mathbf{p}}) = 0$$

10.28 Two free particles of mass m_1 and m_2, respectively, move in 3-space. They do not interact. Write the eigenfunctions and eigenenergies of this system in as many representations as you can. Indicate the number of parameters specified in the eigenstates associated with these representations.

10.29 Write down the time-dependent wavefunction corresponding to eigenstates for a two-particle system in the two representations (10.114).

10.30 Consider two particles that attract each other through the potential

$$V(r) = -\frac{\hbar^2 K^2}{2\mu r^2}$$

The displacement between particles is r, K is a constant, and μ is reduced mass. In states of definite angular momentum, what are the values of the angular momentum quantum number l for which the effective force between particles is repulsive?

10.31 In Problem 8.1 the depth of the potential well appropriate to a deuteron was evaluated using a one-dimensional approximation. A more refined estimate may be obtained using a three-dimensional spherical well with characteristics

$$V(r) = -|V|, \qquad r < a \quad \text{region I}$$
$$V(r) = 0 \qquad r \geq a \quad \text{region II}$$

(a) Construct components of the ground-state wavefunction in regions I and II, respectively.

(b) Show that matching conditions at $r = a$ give the dispersion relation

$$\eta = -\xi \cot \xi$$
$$\rho^2 = \xi^2 + \eta^2$$

where

$$\rho^2 \equiv \frac{2m|V|a^2}{\hbar^2}, \qquad \xi \equiv ka, \qquad \eta \equiv \kappa a$$

$$|E| = \frac{\hbar^2 \kappa^2}{2m}, \qquad |V| - |E| = \frac{\hbar^2 k^2}{2m}$$

(c) To within the same approximation suggested in Problem 8.1, obtain a numerical value for the depth $|V|$ of the three-dimensional deuteron well. From the ratio $|E|/|V|$ for this bound state, would you say that the deuteron is a strongly or a weakly bound nucleus?

Answer (partial)

(a)   Component wavefunctions in the well domain, region I, are the spherical Bessel functions. The ground-state component is therefore

$$\varphi_I = \frac{\sin kr}{kr}$$

In region II the component ground-state wavefunction is exponentially damped.

$$\varphi_{II} = A\frac{e^{-\kappa r}}{\kappa r}$$

TABLE 10.2   Solutions to the three fundamental box problems in quantum mechanics

The Rectangular Box	The Cylindrical Box	The Spherical Box
Edge Lengths a_1, a_2, a_3	Radius a, Height b	Radius a

Hamiltonian

$\hat{H} = (\hat{p}_x^2 + \hat{p}_y^2 + \hat{p}_z^2)/2M$	$\hat{H} = (\hat{p}_\rho^2 + \hat{p}_z^2 + \hat{L}_z^2/\rho^2)/2M$	$\hat{H} = (\hat{p}_r^2 + \hat{L}^2/r^2)/2M$
$\hat{p}_x^2 = \left(-i\hbar\dfrac{\partial}{\partial x}\right)^2$	$\hat{p}_\rho^2 = -\hbar^2\dfrac{1}{\rho}\dfrac{\partial}{\partial \rho}\left(\rho\dfrac{\partial}{\partial \rho}\right)$	$\hat{p}_r^2 = -\hbar^2\left(\dfrac{1}{r}\dfrac{\partial}{\partial r}r\right)^2$

Eigenfunction

$\varphi_{qst} = A_{qst}\sin k_q x \sin k_s y \sin k_t z$	$\varphi_{qmn} = A_{qmn}J_m(K_{mn}\rho)\sin k_q z\, e^{im\phi}$	$\varphi_{nlm} = A_{nlm}j_l(k_{ln}r)Y_l^m(\theta, \phi)$
$(A_{qst})^2 = 8/a_1 a_2 a_3$	$(A_{qmn})^2 = 2/\pi b[aJ_m{}'(K_{mn}a)]^2$	$(A_{nlm})^2 = 2/a^3[j_l{}'(k_{ln}a)]^2$
$\sin k_q a_1 = \sin k_s a_2 = \sin k_t a_3 = 0$	$\sin k_q b = J_m(K_{mn}a) = 0$	$j_l(k_{ln}a) = 0$

Wave Equation   *Bessel's Equation*   *Spherical Bessel Equation*

$\left(\dfrac{d^2}{dx^2} + k^2\right)\sin kx = 0$	$\left[\dfrac{1}{x^2}\left(x\dfrac{d}{dx}\right)^2 + 1 - \dfrac{m^2}{x^2}\right]J_m(x) = 0$	$\left[\left(\dfrac{1}{x}\dfrac{d}{dx}x\right)^2 + 1 - \dfrac{l(l+1)}{x^2}\right]j_l(x) = 0$

Eigenenergy

$E_{qst} = \hbar^2(k_q^2 + k_s^2 + k_t^2)/2M$	$E_{qmn} = \hbar^2(K_{mn}^2 + k_q^2)/2M$	$E_{nl} = \hbar^2 k_{ln}^2/2M$

(b)   Here one must invoke continuity of $d \ln u/dr$.

(c)   You should obtain the answer $|E|/|V| = 0.08$. The binding energy of a deuteron is 2.2 MeV which, with the preceding estimate, indicates a relatively large well depth. A sketch of the normalized wavefunction further reveals that there is approximately only one chance in three that the nucleons are closer together than the well radius a.

10.32   Show by explicit calculation that the eigenfunctions and eigenenergies as given in Table 10.2 are correct for each of the three respective "box" configurations shown. (Primes denote differentiation.)

10.6   THE HYDROGEN ATOM

Hamiltonian and Eigenenergies

The (relative) Hamiltonian for the hydrogen atom (more accurately, for a "hydrogenic" atom[1] of atomic number Z) appears as

$$(10.115) \qquad \hat{H} = \frac{\hat{p}_r^2}{2\mu} + \frac{\hat{L}^2}{2\mu r^2} - \frac{Ze^2}{r}$$

The corresponding Schrödinger equation is

$$(10.116) \qquad \left(\frac{\hat{p}_r^2}{2\mu} + \frac{\hat{L}^2}{2\mu r^2} - \frac{Ze^2}{r} \right)\varphi = E\varphi = -|E|\varphi$$

We are seeking the *bound states* of hydrogen. These correspond to the negative eigenenergies, $E = -|E|$. Setting $\varphi = R(r)Y_l^m(\theta, \phi)$ in the latter equation gives the radial equation (10.109)

$$(10.117) \qquad \left[\frac{-\hbar^2}{2\mu} \left(\frac{1}{r}\frac{d^2}{dr^2}r \right) + \frac{\hbar^2 l(l+1)}{2\mu r^2} - \frac{Ze^2}{r} + |E| \right]R = 0$$

Changing the dependent variable to

$$u = rR$$

introduced previously in (10.113), gives

$$(10.118) \qquad \left(-\frac{d^2}{dr^2} + \frac{l(l+1)}{r^2} - \frac{2\mu}{\hbar^2}\frac{Ze^2}{r} + \frac{2\mu|E|}{\hbar^2} \right)u = 0$$

[1] Hydrogenic atoms are atoms that are ionized with all but one electron bound to the nucleus which carries the charge $+Ze$ (e.g., He^+, Li^{++}, etc.).

Introducing the notation

$$\rho \equiv 2\kappa r, \qquad \frac{\hbar^2 \kappa^2}{2\mu} = |E|$$

(10.119)
$$\lambda^2 = \left(\frac{Z}{\kappa a_0}\right)^2 = \frac{Z^2 \mathbb{R}}{|E|}$$

$$\mathbb{R} = \frac{\hbar^2}{2\mu a_0^2}, \qquad a_0 = \frac{\hbar^2}{\mu e^2}$$

where $\mathbb{R}$ is the Rydberg constant[1] (2.13) and a_0 is the Bohr radius (2.14), the radial equation may be further simplified to the form

(10.120)
$$\frac{d^2 u}{d\rho^2} - \frac{l(l + 1)}{\rho^2} u + \left(\frac{\lambda}{\rho} - \frac{1}{4}\right) u = 0$$

For large values of ρ this equation reduces to

$$\frac{d^2 u}{d\rho^2} - \frac{u}{4} = 0$$

so that

$$u \sim A e^{-\rho/2} + B e^{\rho/2}$$

In order that u vanish as $\rho \to \infty$, we set $B = 0$, so

$$u \sim e^{-\rho/2} \qquad (\rho \to \infty)$$

In the neighborhood of the origin, (10.120) reduces to

$$\frac{d^2 u}{d\rho^2} - \frac{l(l + 1)}{\rho^2} u = 0$$

Substitution of the trial solution $u = \rho^q$ gives

$$u \sim A\rho^{-l} + B\rho^{l+1}$$

In order for u to vanish at the origin, we must set $A = 0$. This gives

$$u \sim \rho^{l+1} \qquad (\rho \to 0)$$

---

[1] The Rydberg constant written with m in place of μ (i.e., assuming infinite proton mass) is sometimes written $\mathbb{R}_\infty$.

With these two asymptotic forms at hand, we are prepared to solve (10.120) through a polynomial expansion. Solution in the form

$$u(\rho) = e^{-\rho/2}\rho^{l+1}F(\rho)$$

(10.121)
$$F(\rho) = \sum_{i=0}^{\infty} C_i \rho^i$$

with F finite everywhere, gives the proper behavior at $\rho \sim 0$ and $\rho \sim \infty$. Substituting (10.121) for u into (10.120), we obtain

(10.122)
$$\left[\rho \frac{d^2}{d\rho^2} + (2l + 2 - \rho) \frac{d}{d\rho} - (l + 1 - \lambda) \right] F(\rho) = 0$$

Note that, for a given value of the orbital quantum number l, this is an eigenvalue equation with eigenvalue λ. The values of λ (or, equivalently, the eigenenergies, $|E|$) are those values which ensure that $F(\rho)$ is finite for all ρ. Substituting the series (10.121) into the latter equation and equating coefficients of equal powers in ρ gives the *recurrence relation*

(10.123)
$$C_{i+1} = \frac{(i + l + 1) - \lambda}{(i + 1)(i + 2l + 2)} C_i \equiv \Gamma_{il} C_i$$

In the limit that $i \rightarrow \infty$, this relation becomes

$$C_{i+1} \sim \frac{C_i}{i}$$

which is the same ratio of coefficients obtained in the expansion

$$e^\rho = \sum C_i \rho^i = \sum \frac{\rho^i}{i!}$$

$$\frac{C_{i+1}}{C_i} = \frac{i!}{(i + 1)!} = \frac{1}{i + 1} \sim \frac{1}{i}$$

It follows that the form of $u(\rho)$ generated by the series (10.121) behaves as

$$u(\rho) \sim e^{-\rho/2}\rho^{l+1}e^\rho = e^{\rho/2}\rho^{l+1}$$

which diverges for large ρ. To obtain a finite wavefunction, the expansion (10.121) for any given value of l must terminate at some finite value of i, which we will call i_{max}. At this value of i, $\Gamma_{il} = 0$. Since all parameters in (10.123) are positive, Γ_{il} can only vanish if

$$i_{max} + l + 1 = \lambda$$

The function u so generated is a polynomial and, due to the exponential term in the form (10.121), we see that, as demanded, the wavefunction is finite everywhere.

Since i and l are integers, it follows that λ is also an integer, which is called the *principal quantum number, n.*

$$n = i_{\text{max}} + l + 1$$

Thus the above cutoff condition on the series (10.121), which ensures that $u(\rho)$ is finite for all ρ, also serves to determine the eigenenergies λ.

$$\lambda_n^2 = n^2 = \frac{Z^2 \mathbb{R}}{|E_n|}$$

(10.124)
$$\boxed{E_n = -|E_n| = -\frac{Z^2 \mathbb{R}}{n^2}}$$

These are the same values found previously in the simpler Bohr model (Section 2.4).

Laguerre Polynomials

The hydrogen eigenfunction corresponding to the eigenvalue E_n is given by (10.121) with the series over i cut off at the value

(10.125)
$$i_{\text{max}} = n - l - 1$$

and the recurrence relation for the coefficients $\{C_i\}$ given by (10.123).

$$u_{nl}(\rho) = e^{-\rho/2}\rho^{l+1}F_{nl}(\rho) = A_{nl}e^{-\rho/2}\rho^{l+1} \sum_{i=0}^{n-l-1} C_i \rho^i$$

(10.126)
$$C_{i+1} = \Gamma_{il}C_i, \qquad \rho \equiv 2\kappa_n r, \qquad \kappa_n = \frac{Z}{a_0 n}$$

where A_{nl} is a normalization constant. The polynomials $F_{nl}(\rho)$ (of order $n - l - 1$) so obtained are better known as the *associated Laguerre polynomials*, L_{n-l-1}^{2l+1} (see Table 10.3). The reader should take note of the fact that the scale length of ρ changes with different values of n. This is because the radial displacement r is nondimensionalized through the wavenumber κ_n, which is dependent on n.

TABLE 10.3   Eigenfunctions of hydrogen in terms of associated Laguerre polynomials

The Normalized Eigenfunctions of Hydrogen $(Z = 1)$

$$\varphi_{nlm}(r, \theta, \phi) = (2\kappa)^{3/2} A_{nl} \rho^l e^{-\rho/2} F_{nl}(\rho) Y_l^m(\theta, \phi) = R_{nl}(r) Y_l^m(\theta, \phi)$$

$$\rho = 2\kappa r = \frac{2Z}{a_0 n} r \qquad \int_0^\infty |R_{nl}(r)|^2 r^2 \, dr = 1$$

$$A_{nl} = \sqrt{\frac{(n - l - 1)!}{2n[(n + 1)!]^3}} \qquad \boxed{\varphi_{100} = \frac{1}{\sqrt{8\pi}} \left(\frac{2Z}{a_0}\right)^{3/2} e^{-(Z/a_0)r}}$$

$$F_{nl}(\rho) = L_{n-l-1}^{2l+1}(\rho) = L_{i_{max}}^{2l+1}(\rho) = \sum_{i=0}^{n-l-1} \frac{(-1)^i [(n + l)!]^2 \rho^i}{i! \, (n - l - 1 - i)! \, (2l + 1 + i)!}$$

Associated Laguerre Polynomials $L_p^q(\rho)$ *and Laguerre Polynomials* $L_p(\rho)$

Differential equation

$$\left[\rho \frac{d^2}{d\rho^2} + (q + 1 - \rho) \frac{d}{d\rho} + p\right] L_p^q(\rho) = 0$$

Generating function

$$\frac{e^{-\rho s/(1-s)}}{(1 - s)^{q+1}} = \sum_{p=0}^{\infty} \frac{s^p}{(p + q)!} L_p^q(\rho), \qquad L_0^p(0) = p!$$

Orthonormality

$$\int_0^\infty e^{-\rho} \rho^q L_p^q L_{p'}^q \, d\rho = \frac{[(p + q)!]^3}{p!} \delta_{pp'}$$

Rodrigues's formula

$$L_p(\rho) \equiv L_p^0(\rho) = e^\rho \frac{d^p}{d\rho^p} (\rho^p e^{-\rho}), \qquad L_1(\rho) = 1 - \rho, \qquad L_2(\rho) = 2! \left(1 - 2\rho + \frac{\rho^2}{2}\right)$$

$$L_p^q(\rho) \equiv (-1)^q \frac{d^q}{d\rho^q} [L_{q+p}(\rho)]$$

Recurrence relations

$$\rho L_p^q(\rho) = (2p + q + 1) L_p^q(\rho) - [(p + 1)/(p + q + 1)] L_{p+1}^q(\rho) - (p + q)^2 L_{p-1}^q(\rho)$$

$$\left(\rho \frac{d}{d\rho} + q - \rho\right) L_p^q(\rho) = (p + 1) L_{p+1}^{q-1}(\rho)$$

$$\frac{d}{d\rho} L_p^q(\rho) = -L_{p-1}^{q+1}(\rho)$$

Relation to Other Notations

Alternative notations for the polynomial L_p^q may be found in other texts. The relation between this notation (B) and our own (A) is given by the following table.[a]

TABLE 10.3 *(Continued)*

Notation A (e.g., found in Merzbacher, Messiah, and here)	Notation B (e.g., found in Pauling and Wilson, Schiff, and Tomonaga)
$(-)^q L_p^q$	L_{p+q}^q
$\rho(L_p^q)'' + (q + 1 - \rho)(L_p^q)' + pL_p^q = 0$	$\rho(L_{p+q}^q)'' + (q + 1 - \rho)(L_{p+q}^q)' + pL_{p+q}^q = 0$
$R_{nl} = A(2\kappa)^{3/2}e^{-\rho/2}\rho^l L_{n-l-1}^{2l+1}(\rho)$	$R_{nl} = -A(2\kappa)^{3/2}e^{-\rho/2}\rho^l L_{n+l}^{2l+1}(\rho)$
L_p^q is a polynomial of order p	L_{p+q}^q is a polynomial of order p or, equivalently, L_b^a is a polynomial of order $(b - a)$

The first row in this table tells us that L_p^q is written L_{p+q}^q in notation B. The second row indicates that both L functions satisfy the same differential equation. The third row gives the forms of the radial solution R_{nl} in both notations. Still another notation appears in I. S. Gradshteyn and I. M. Ryzhik,[b] where

$$L_p^q(\rho)[\text{here}] = (p + q)! \, L_p^q(\rho)[\text{G and R}]$$

[a] E. Merzbacher, *Quantum Mechanics,* 2d ed., Wiley, New York, 1970.
 A. Messiah, *Quantum Mechanics,* Wiley, New York, 1966.
 L. Pauling and E. B. Wilson, *Introduction to Quantum Mechanics,* McGraw-Hill, New York, 1935.
 L. Schiff, *Quantum Mechanics,* 3d ed., McGraw-Hill, New York, 1968.
 S. Tomonaga, *Quantum Mechanics,* North-Holland, Amsterdam, 1966.

[b] I. S. Gradshteyn and I. M. Ryzhik, *Tables of Integrals, Series and Products,* Academic Press, New York, 1965.

Degeneracy

Since $i_{max} \geq 0$, with (10.125) we obtain

$$l \leq n - 1$$

So for a given value of the *principal quantum number n*, the *orbital quantum number l* cannot exceed the value

$$(10.127) \qquad l_{max} = n - 1$$

This corresponds to the values $l = 0, 1, 2, \ldots, (n - 1)$. Each of these l values corresponds to different values of i_{max} and therefore different wavefunctions. Inasmuch as the eigenenergy E_n depends only on the principal quantum number n, these n distinct orbital states are degenerate. For instance, there are three distinct radial functions that correspond to the eigenenergy E_3. These are $u_{3,0}$, $u_{3,1}$, and $u_{3,2}$.

The complete eigenstate of the Hamiltonian (10.115) contains the factor $Y_l^m(\theta, \phi)$ [see (10.108)]. For each value of l, there are $2l + 1$ values of m_l: $m_l = -l, \ldots, +l$, which correspond to distinct Y_l^m functions that give the same eigenvalues of $\hat{L}^2$ [i.e., $\hbar^2 l(l + 1)$]. All these $2l + 1$ functions when substituted into

TABLE 10.4 Allowed values of l and m_l for $n = 1, 2, 3$

n	1	2		3		
l	0	0	1	0	1	2
Spectroscopic notation of state	$1S$	$2S$	$2P$	$3S$	$3P$	$3D$
m_l	0	0	$-1, 0, +1$	0	$-1, 0, +1$	$-2, -1, 0, +1, +2$
Degeneracy of state (n^2)	1	4		9		

(10.116) give the same radial equation, (10.117), which contains only the orbital number l. It follows that for each solution u_{nl} of (10.120), there are $(2l + 1)$ solutions to the Schrödinger equation (10.116) corresponding to the same eigenenergy E_n (see Table 10.4). In this manner we obtain[1]

$$(10.128) \qquad \text{Degeneracy of } E_n = \sum_{l=0}^{n-1} (2l + 1) = n^2$$

To recapitulate, the allowed values of n, l, and m are (see Fig. 10.14)

$$n = 1, 2, 3, \ldots$$
$$(10.129) \qquad l = 0, 1, 2, \ldots, (n - 1)$$
$$m = -l, -l + 1, \ldots, 0, 1, 2, \ldots, +l$$

Additional Properties of the Eigenstates

The eigenfunctions and eigenenergies of the hydrogenic Hamiltonian (10.15) are

$$(10.130)$$

$$\varphi_{nlm}(r, \theta, \phi) = R_{nl}(r) \, Y_l^m(\theta, \phi)$$

$$R_{nl} = \frac{A_{nl} u_{nl}}{r}$$

$$E_n = -\frac{Z^2 \mathbb{R}}{n^2} = -\frac{\mu(Ze^2)^2}{2\hbar^2 n^2}$$

[1] Including spin, degeneracy of states is $2n^2$. This topic is more fully discussed in Section 12.4.

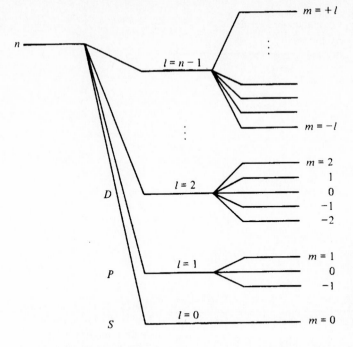

FIGURE 10.14 Term diagram for a hydrogenic atom illustrating all n^2 degenerate states corresponding to the principal quantum number n.

A term diagram of these energies is given in Fig. 10.15 (compare Fig. 2.8). The normalization constant A_{nl} (see Table 10.3) is determined by the condition

$$\langle \varphi_{nlm} | \varphi_{nlm} \rangle = \int_{4\pi} d\Omega \int_0^\infty r^2 \, dr \, \varphi_{nlm}{}^* \varphi_{nlm}$$

$$= |A_{nl}|^2 \int_0^\infty |u_{nl}|^2 \, dr = 1$$

Note also the orthogonality of these functions

$$\langle \varphi_{n'l'm'} | \varphi_{nlm} \rangle = \delta_{nn'} \, \delta_{ll'} \, \delta_{mm'}$$

The Ground State

To construct the *ground state* φ_{100} ($n = 1$, $l = 0$, $m = 0$) we must first find u_{10}. From (10.126), with $C_0 = 1$, and inserting the normalization constant[1] A_{10}, one obtains

$$u_{10} = A_{10} e^{-\rho/2} \rho$$

[1] Alternatively, we may take $C_0 = A_{10}$.

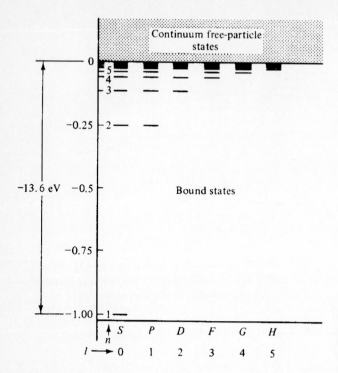

FIGURE 10.15 Energy-level diagram for hydrogen, including the $l \leq 5$ terms. Energy is measured in units of $\mathbb{R}$.

Normalization gives

$$A_{10}{}^2 \frac{1}{2\kappa_1} \int_0^\infty \rho^2 e^{-\rho}\, d\rho = 1$$

$$A_{10}{}^2 = \frac{1}{a_0}$$

This gives the normalized ground-state wavefunction (with $Z = 1$)

$$u_{10} = \frac{1}{\sqrt{a_0}} \rho e^{-\rho/2} = \frac{2r}{a_0{}^{3/2}} e^{-r/a_0}$$

$$R_{10} = \frac{u_{10}}{r} = \frac{2}{a_0{}^{3/2}} e^{-r/a_0}$$

$$\varphi_{100} = R_{10} Y_0{}^0 = \frac{2}{(4\pi)^{1/2} a_0{}^{3/2}} e^{-r/a_0}$$

Additional Properties

The first few normalized eigenstates of hydrogen, with corresponding eigenenergies obtained as outlined above, are listed in Table 10.5.

TABLE 10.5 Several normalized time-independent eigenstates of hydrogen

Spectroscopic Notation	Normalized Time-Independent Eigenstates
1S	$\varphi_{100} = \dfrac{2}{a_0^{3/2}} e^{-r/a_0} Y_0^0(\theta, \phi)$
2S	$\varphi_{200} = \dfrac{2}{(2a_0)^{3/2}} (1 - r/2a_0) e^{-r/2a_0} Y_0^0(\theta, \phi)$
2P	$\begin{pmatrix} \varphi_{211} \\ \varphi_{210} \\ \varphi_{21-1} \end{pmatrix} = \dfrac{1}{\sqrt{3}\,(2a_0)^{3/2}} \dfrac{r}{a_0} e^{-r/2a_0} \begin{pmatrix} Y_1^1(\theta, \phi) \\ Y_1^0(\theta, \phi) \\ Y_1^{-1}(\theta, \phi) \end{pmatrix}$
3S	$\varphi_{300} = \dfrac{2}{3(3a_0)^{3/2}} [3 - 2r/a_0 + 2(r/3a_0)^2] e^{-r/3a_0} Y_0^0(\theta, \phi)$
3P	$\begin{pmatrix} \varphi_{311} \\ \varphi_{310} \\ \varphi_{31-1} \end{pmatrix} = \dfrac{4\sqrt{2}}{9(3a_0)^{3/2}} \dfrac{r}{a_0} (1 - r/6a_0) e^{-r/3a_0} \begin{pmatrix} Y_1^1(\theta, \phi) \\ Y_1^0(\theta, \phi) \\ Y_1^{-1}(\theta, \phi) \end{pmatrix}$
3D	$\begin{pmatrix} \varphi_{322} \\ \varphi_{321} \\ \varphi_{320} \\ \varphi_{32-1} \\ \varphi_{32-2} \end{pmatrix} = \dfrac{2\sqrt{2}}{27\sqrt{5}\,(3a_0)^{3/2}} \left(\dfrac{r}{a_0}\right)^2 e^{-r/3a_0} \begin{pmatrix} Y_2^2(\theta, \phi) \\ Y_2^1(\theta, \phi) \\ Y_2^0(\theta, \phi) \\ Y_2^{-1}(\theta, \phi) \\ Y_2^{-2}(\theta, \phi) \end{pmatrix}$

In Fig. 10.16 nondimensionalized radial functions, $\bar{R}_{nl} = a_0^{3/2} R_{nl}$, are plotted together with the corresponding nondimensionalized probability density functions, $\bar{P}_{nl} = 4\pi a_0 u_{nl}^2 = 4\pi a_0 P_r$. These sketches reveal the shell structure of hydrogen found earlier in the Bohr theory.

The time development of the states of hydrogen follows from (3.70). Consider that

$$(10.131) \qquad \psi(\mathbf{r}, 0) = \varphi_{nlm} = R_{nl} Y_l^m$$

The state at time $t \geq 0$ is then

$$(10.132) \qquad \psi(\mathbf{r}, t) = e^{-i\hat{H}t/\hbar} \psi(\mathbf{r}, 0) = e^{-iE_n t/\hbar} \varphi_{nlm}$$

The charge density associated with this state is

$$(10.133) \qquad q(\mathbf{r}, t) = e|\psi_{nlm}(\mathbf{r}, t)|^2 = q(\mathbf{r}) = e|\varphi_{nlm}(\mathbf{r})|^2$$

which is independent of time. The electronic charge is e. Thus the atom suffers no radiation in these states. This topic will be returned to in the next section.

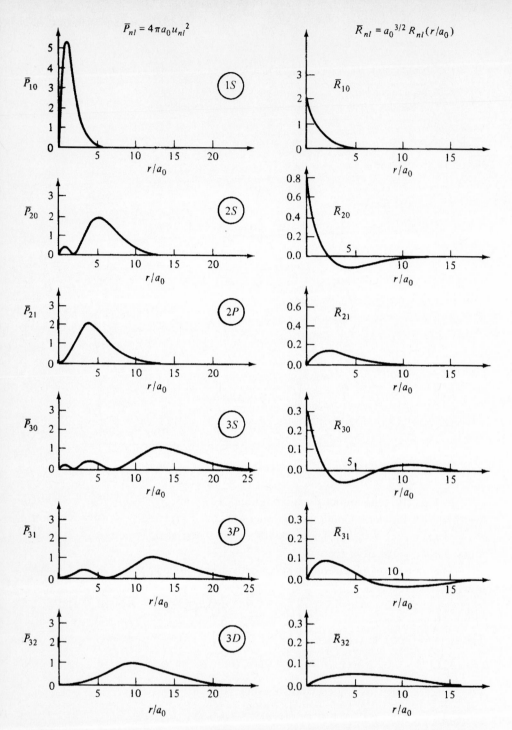

FIGURE 10.16 Nondimensional radial probability density $\bar{P}$ and nondimensional radial wavefunction $\bar{R}$, versus nondimensional radius r/a_0, for hydrogen. Note that the probability density $\bar{P}$ exhibits the shell structure of the atom.

The density configurations, $|\varphi_{nlm}|^2$, corresponding to some of the eigenstates of hydrogen are sketched in Fig. 10.17. Since the angular dependence of $|\varphi_{nlm}|^2$ is entirely contained in the factor $|Y_l^m|^2$, it follows that $|\varphi_{nlm}|^2$ is independent of the azimuthal angle ϕ [see (9.78)]. It is rotationally symmetric about the z axis. Thus we need only present a representation of $|\varphi|^2$ in any plane which includes the z axis, such as is depicted in Fig. 10.17. The value of $|\varphi|^2$ is proportional to the density of whiteness in each of the states depicted.

Hybridization

Having discovered the wavefunctions of hydrogen, a significant consequence of quantum mechanics emerges important to the formation of molecules. This property concerns the manner in which atomic wavefunctions exhibit geometric orientation in atom–atom binding.

Consider the $2P$ states of hydrogen listed in Table 10.5. Owing to their degeneracy, any linear combination of these states is an eigenstate corresponding to the same eigenenergy, $-\mathbb{R}/4$. Thus, for example, the following set of three orthogonal wavefunctions span the same subdimensional Hilbert space as the original three $2P$ states:

(10.134a)
$$\varphi_{2P_x} = -\frac{1}{\sqrt{2}}(\varphi_{211} - \varphi_{21-1}) = Axe^{-r/2a_0}$$

(10.134b)
$$\varphi_{2P_y} = \frac{1}{\sqrt{2}}(\varphi_{211} + \varphi_{21-1}) = -iAye^{-r/2a_0}$$

(10.134c)
$$\varphi_{2P_z} = \varphi_{210} = Aze^{-r/2a_0}$$

Angular plots (at fixed r) of the corresponding probability densities $|\varphi_{2P_x}|^2$, $|\varphi_{2P_y}|^2$, $|\varphi_{2P_z}|^2$, reveal that they are figure-eight surfaces of revolution about the x, y, and z axes, respectively (similar to those shown in Fig. 9.10).

Linear combinations of the wavefunctions (10.134) come into play in the formation of the methane molecule CH_4. The four outer-shell electrons of carbon are in the $2s^2 2p^2$ configuration (see Table 12.2). Wavefunctions appropriate to these four electrons are formed from linear combinations of the three $2P$ states (10.134) together with the $2S$ state of hydrogen and are given by

(10.135a)
$$\psi_1 = \frac{1}{\sqrt{4}}(\varphi_{2S} + \varphi_{2P_x} + \varphi_{2P_y} + \varphi_{2P_z})$$

(10.135b)
$$\psi_2 = \frac{1}{\sqrt{4}}(\varphi_{2S} + \varphi_{2P_x} - \varphi_{2P_y} + \varphi_{2P_z})$$

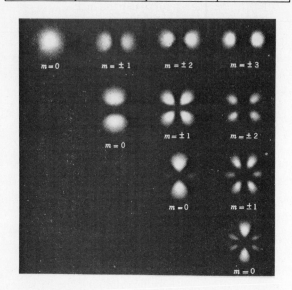

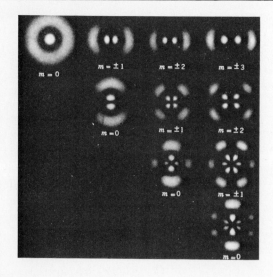

FIGURE 10.17 Probability density, $|\varphi_{nlm}|^2$, for various states of hydrogen. The plane of the paper contains the polar axis, which points from the bottom to the top of the figure. From *Principles of Modern Physics* by R. B. Leighton. Copyright 1959 by McGraw-Hill. Used with permission of the McGraw-Hill Book Company.

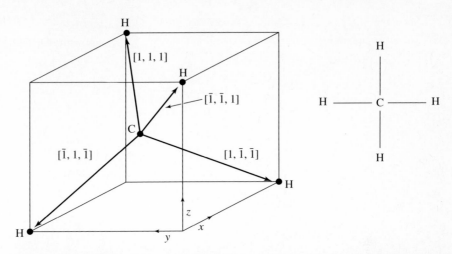

FIGURE 10.18 Orientation of C and H atoms in the CH_4 molecule. Here we have employed crystallographic notation for direction, where a bar over a direction number indicates negative direction. Shown also is the two-dimensional representation of this molecule. Generalizations of this diagram for $C_n H_{2n+2}$ hydrocarbons are evident. Shown also is the two-dimensional representation of this molecule. Generalizations of this diagram for $C_n H_{2n+2}$ hydrocarbons are evident. (Straight-chain alkanes.)

$$(10.135c) \qquad \psi_3 = \frac{1}{\sqrt{4}} \left(\varphi_{2S} + \varphi_{2P_x} - \varphi_{2P_y} - \varphi_{2P_z} \right)$$

$$(10.135d) \qquad \psi_4 = \frac{1}{\sqrt{4}} \left(\varphi_{2S} - \varphi_{2P_x} + \varphi_{2P_y} - \varphi_{2P_z} \right)$$

With the four outer electrons of carbon in these respective four ψ orbitals, the following picture emerges. Angular plots of the probability densities of these wavefunctions about a common origin reveal maxima along the $(1, 1, 1)$, $(-1, -1, 1)$, $(1, -1, -1)$, and $(-1, 1, -1)$ directions, respectively. With the carbon atom at the center of a cube, these orbitals reach out to four tetrahedral corners and covalently bond with hydrogen atoms at these sites to form the CH_4 molecule. See Fig. 10.18.

These orbitals come into play in the chaining of CH_3 molecules. Thus, for example, in the formation of C_2H_6, an uncoupled electronic ψ orbital of CH_3 bonds with a like orbital of an adjacent CH_3 molecule to form the ethane molecule. The concept of atomic binding in the formation of molecules is returned to in Chapter 12.

PROBLEMS

10.33 With $C_0 = 1$ in the recurrence relation (10.123), obtain C_1. Then use (10.126) to show that

$$u_{20} = A_{20} r e^{-r/2a_0} \left(1 - \frac{r}{2a_0} \right)$$

Calculate A_{20} and φ_{200}. Check your answer with the value given in Table 10.5.

10.34   Show that $u_{10}{}^2$ has its maximum at $r = a_0$, the Bohr radius.

10.35   Solve the equation $f_x + f = 0$ by expansion technique and check with the solution $f = Ae^{-x}$.

Answer

Assume that

$$f = \sum_0^\infty C_i x^i$$

to obtain

$$\sum_0^\infty C_i i x^{i-1} + \sum_0^\infty C_i x^i = 0$$

With $s = i - 1$ in the first series, we get

$$\sum_{s=-1}^\infty C_{s+1}(s + 1)x^s + \sum_{i=0}^\infty C_i x^i = 0$$

Since the first term in the first series is zero, we may write this equation in the form

$$\sum_{i=0}^\infty [C_{i+1}(i + 1) + C_i]x^i = 0$$

which is satisfied if and only if

$$C_{i+1} = -\frac{C_i}{i + 1}$$

10.36   The average energy of the hydrogen atom in an arbitrary bound state $\chi(\mathbf{r})$ is given by the integral

$$\langle E \rangle = \langle \chi(\mathbf{r}) | \hat{H} | \chi(\mathbf{r}) \rangle$$

Show that

$$\langle E \rangle \geq \langle 100 | \hat{H} | 100 \rangle = E_1$$

Answer

Since $\chi(\mathbf{r})$ lies in the Hilbert space spanned by the basis $\{\varphi_{nlm}\}$, we may expand

$$|\chi\rangle = \sum_{n=1}^\infty \sum_{l=0}^{n-1} \sum_{m=-l}^l b_{nlm} |nlm\rangle$$

so that

$$\langle \chi | \hat{H} | \chi \rangle = \sum_n \sum_l \sum_m |b_{nlm}|^2 E_n$$

$$\sum\sum\sum |b|^2 = 1$$

(The ket vector $|nlm\rangle$ represents the state φ_{nlm}.) Owing to the fact that all eigenenergies are negative, $E_n \leq 0$, the statement to be proven is equivalent to the inequality

$$|\langle E \rangle| \leq |E_1|$$

$$|\langle E \rangle| = \sum\sum\sum |b|^2 |E_n|$$

$$\leq |E|_{\max} \sum\sum\sum |b|^2 = |E|_{\max} = |E_1|$$

10.37   (a) What is the effective Bohr radius and ground-state energy for each of the following two-particle systems?

   (1) H^2, a deuteron and an electron (heavy hydrogen).

   (2) He^+, a singly ionized helium atom.

   (3) Positronium, a bound positron and electron.

   (4) Mesonium, a proton and negative μ meson. The μ meson has mass $207m_e$ and lasts $\sim 10^{-6}$ s.

   (5) Two neutrons bound together by their gravitational field.

   (b) Calculate the frequencies of the $(n = 2) \rightarrow (n = 1)$ transition for each of the systems above.

10.38   At time $t = 0$, a hydrogen atom is in the superposition state

$$\psi(\mathbf{r}, 0) = \frac{4}{(2a_0)^{3/2}} \left[\frac{e^{-r/a_0}}{\sqrt{4\pi}} + A \frac{r}{a_0} e^{-r/2a_0} (-iY_1^1 + Y_1^{-1} + \sqrt{7} Y_1^0) \right]$$

   (a)   Calculate the value of the normalization constant A.

   (b)   What is the probability that measurement of L^2 finds the value $\hbar^2 l(l + 1)$?

   (c)   What is the probability density $P_r(r)$ [see (10.113)] that the electron is found in the shell of thickness dr about the proton at the radius r?

   (d)   At what value of r is $P_r(r)$ maximum?

   (e)   Given the initial state $\psi(\mathbf{r}, 0)$, what is $\psi(\mathbf{r}, t)$?

   (f)   What is $\psi(\mathbf{r}, t)$ if at $t = 0$, measurement of L_z finds the value $\hbar$?

   (g)   What is $\psi(\mathbf{r}, t)$ if at $t = 0$, measurement of L_z find the value zero?

   (h)   What is the expectation of the "spherical energy operator," $\langle H_S \rangle$, where

$$\hat{H}_S \equiv \hat{H} - \frac{\hat{p}_r^2}{2\mu}$$

at $t = 0$?

   (i)   What is the lowest value of energy that measurement will find at $t = 0$? (Lowest means the negative value farthest removed from zero.)

10.39   Find the lowest energy and the smallest value for the classical turning radius of the H-atom electron in the state with $l = 6$ (see Fig. 10.11).

10.40   In what sense does the Bohr analysis of the hydrogen atom give erroneous results for the magnitude of angular momentum, L?

Answer

The Bohr analysis that yields the eigenenergies $-\mathbb{R}/n^2$ assumes circular orbits. Circular orbits do not exist in the Schrödinger theory. Quantization of the action, $\oint p_\theta \, d\theta$, in the Bohr theory gives $L = n\hbar$. In the Schrödinger theory, the maximum value of L is $\hbar\sqrt{n(n-1)}$, which is less than the value that L assumes $(n\hbar)$ in the Bohr theory.

10.41 What is the ionization energy of a hydrogen atom in the $3P$ state?

10.42 Show that $R_{nl}(r)$ has $(n - l - 1)$ zeros (not counting zeros at $r = 0$ and $r = \infty$).

10.43 (a) Show that the expectation of the interaction potential $V(r)$ for hydrogenic atoms is

$$\langle nlm | V(r) | nlm \rangle = -\left\langle \frac{Ze^2}{r} \right\rangle = -\frac{\mu Z^2 e^4}{\hbar^2 n^2} = -\frac{2Z^2 \mathbb{R}}{n^2}$$

(b) Calculate $\langle nlm | T | nlm \rangle$, where the kinetic-energy operator $\hat{T}$ is given by

$$\hat{T} = \frac{\hat{p}_r^2}{2\mu} + \frac{\hat{L}^2}{2\mu r^2}$$

What relation do $\langle T \rangle$ and $\langle V \rangle$ satisfy? (*the virial theorem*)

10.44 Obtain an explicit expression for the probability density $P_r(r)$ corresponding to the state whose energy is E_2, for a hydrogenic atom [see (10.113)].

Answer

There are four degenerate eigenstates corresponding to the energy E_2. Since no direction is preferred for a Hamiltonian whose only interaction term is the central potential $V(r)$, all these degenerate states carry the same "weight" (all lm states are equally probable). There results

$$4\pi P(r)r^2 \, dr = \int_{4\pi} \frac{1}{4}[\varphi_{200}{}^*\varphi_{200} + \varphi_{21-1}{}^*\varphi_{21-1} + \varphi_{210}{}^*\varphi_{210} + \varphi_{211}{}^*\varphi_{211}]r^2 \, dr \, d\Omega$$

$$= \int_{4\pi} \frac{1}{128\pi}\left(\frac{Z}{a_0}\right)^3 e^{-Zr/a}\left[\left(2 - \frac{Zr}{a^0}\right)^2 + \left(\frac{Zr}{a_0}\right)^2\left(\frac{1}{2}\sin^2\theta + \frac{1}{2}\sin^2\theta + \cos^2\theta\right)\right]$$

$$\times r^2 \, dr \, d\Omega$$

$$P(r) = \frac{1}{128\pi}\left(\frac{Z}{a_0}\right)^3 e^{-Zr/a_0}\left[\left(2 - \frac{Zr}{a_0}\right)^2 + \left(\frac{Zr}{a_0}\right)^2\right], \qquad P_r(r) = 4\pi r^2 P(r)$$

10.45 Give a physical argument in support of the conjecture that the sum

$$\Sigma_n \equiv \sum_{l=0}^{n-1} \sum_{m=-l}^{+l} [Y_l^m]^*[Y_l^m][R_{nl}]^2$$

is independent of θ or ϕ.

10.46   Show that for a hydrogen atom in the state corresponding to maximum orbital angular momentum ($l = n - 1$),

$$\langle n, n - 1 | r | n, n - 1 \rangle = a_0 n \left(n + \frac{1}{2} \right)$$

$$\langle n, n - 1 | r^2 | n, n - 1 \rangle = a_0^2 n^2 (n + 1) \left(n + \frac{1}{2} \right)$$

10.47   Use the result of Problem 10.46 to show that for large values of n and l,

$$\sqrt{\langle r^2 \rangle} \rightarrow a_0 n^2$$

$$\frac{\Delta r}{\langle r \rangle} \rightarrow 0$$

$$E_n \rightarrow -\frac{1}{2} \frac{e^2}{n^2 a_0}$$

That is, show that for large values of n, the electron is located near the surface of a sphere of radius $a_0 n^2$ and has energy which is the same as that of a classical electron in a circular orbit of the same radius. Recall that $(\Delta r)^2 = \langle r^2 \rangle - \langle r \rangle^2$.

10.48   Calculate $\langle \mathbf{r} \rangle$ in the state φ_{nlm} of hydrogen.

Answer

$$\langle \mathbf{r} \rangle = \mathbf{e}_x \langle x \rangle + \mathbf{e}_y \langle y \rangle + \mathbf{e}_z \langle z \rangle$$

$$\langle x \rangle = \int \int \int r \cos \phi \sin \theta \, | Y_l^m |^2 R_{nl}^2 r^2 \, dr \, d \cos \theta \, d\phi$$

$$= 0$$

since

$$\int_0^{2\pi} \cos \phi \, d\phi = 0$$

and $| Y_l^m |^2$ is independent of ϕ. Similarly, $\langle y \rangle = 0$. For $\langle z \rangle$ we must calculate

$$\langle z \rangle = \int_{-1}^1 | P_l^m |^2 \cos \theta \, d \cos \theta \int \ldots$$

Using the recurrence relations listed in Table 9.3, we find that $\langle z \rangle = 0$. It follows that $\langle \mathbf{r} \rangle = 0$.

10.49   Establish the following properties for hydrogen in the stationary state φ_{nlm}.

(a) $\dfrac{s + 1}{n^2} \langle r^s \rangle - (2s + 1) a_0 \langle r^{s-1} \rangle + \dfrac{s}{4} [(2l + 1)^2 - s^2] a_0^2 \langle r^{s-2} \rangle = 0, \; s > -2l - 1$

(b) $\langle r \rangle = n^2 \left[1 + \frac{1}{2} \left(1 - \frac{l(l+1)}{n^2} \right) \right] a_0$

(c) $\left\langle \frac{1}{r^2} \right\rangle = \frac{2}{(2l+1)n^3 a_0^2}$

(d) $\langle r^2 \rangle = \frac{1}{2} [5n^2 + 1 - 3l(l+1)] n^2 a_0^2$

(e) $\left\langle \frac{1}{r} \right\rangle = \frac{1}{n^2 a_0}$

(f) $\left\langle \frac{1}{r^3} \right\rangle = \frac{2}{a_0^3 n^3 l(l+1)(2l+1)}$

[*Hint:* Multiply (10.120) by $\{\rho^{s+1} u' + [(s+1)/2] \rho^s u\}$ and integrate by parts several times. Note that for hydrogenic atoms, a_0 is replaced by a_0/Z.]

10.50 Show that the most probable values of r for the $l = n - 1$ states of hydrogen are

$$\tilde{r} = n^2 a_0$$

These are values that satisfy the equation

$$\frac{d}{dr} (u_{nl})^2 = 0$$

10.7 ELEMENTARY THEORY OF RADIATION

In the last section we found that the hydrogen atom does not radiate in its eigen (stationary) states. The charge density (10.133) is fixed in space with configurations such as depicted in Fig. 10.17. In these states the hydrogen atom is stable against radiation. This is opposed to the classical description in which the electron loses kinetic energy to the radiation field and collapses to the nucleus (see Section 2.1).

The student may be perplexed about the absence of radiation from the state ψ_{nlm}. He/she may well ask: Doesn't the electron have a well-defined angular momentum in such a state, and doesn't this correspond to accelerated motion which gives rise to radiation? His/her friend answers: Maybe the orbit of the electron is so peculiar that, on the average, the radiation field washes out. After all, we know that if L^2 is specified, two of the three components of **L** remain uncertain.

The best way to see what the electron is doing in quantum mechanics is to calculate $\langle \mathbf{r} \rangle$. Specifically, we must calculate this expectation in the state ψ_{nlm}. Suppose that we find $\langle \mathbf{r} \rangle \sim \mathbf{e}_z \cos \omega t$. Then the electron is suffering linear, simple harmonic oscillation. Such oscillation gives rise to *dipole radiation*. But we have already calculated $\langle \mathbf{r} \rangle$ in Problem 10.48, where we found that $\langle \mathbf{r} \rangle = 0$. Not only is $\langle \mathbf{r} \rangle$ time-independent in the eigenstates of hydrogen, but it is also centered at the origin. Note that we may also reach this conclusion by the much simpler argument: Calculate

$$\langle \mathbf{r} \rangle = \langle \psi_{nlm} | \mathbf{r} | \psi_{nlm} \rangle = \iiint \mathbf{r} | \varphi_{nlm} |^2 \, d\mathbf{r}$$

The average of $\mathbf{r}$ is independent of time; hence it must also be zero since the Hamiltonian, (10.115), is isotropic. It contains no vectors. It in no way implies a "preferred" direction, so $\langle \mathbf{r} \rangle$ cannot be a finite constant vector.

Thus while the stability of the hydrogen atom to radiative collapse is totally inexplicable on classical grounds, our quantum mechanical model renders a denumerably infinite set of states $\{\psi_{nlm}\}$ in which the atom suffers no radiation.

How, then, does the atom radiate? In the Bohr theory of radiation, we recall that a photon is emitted when there is a transition from one eigenenergy state to a lower one. Such a decay might be induced by the collision of the atom with another atom in a gas. It might also be induced by collision with an electron in a discharge tube. It might also be induced by collision with a photon in the interior of a star.[1]

Suppose that at time $t = 0$, the atom is in an excited (stationary) state ψ_n (n denotes the sequence nlm). The atom is perturbed, emits radiation, and decays to the state $\psi_{n'}$ (Fig. 10.19). We may conclude that in the interim, the atom is in the superposition state

(10.136) $$\psi = a\psi_n + b\psi_{n'} \qquad |a|^2 + |b|^2 = 1$$

At any time t, in this interim (by the superposition principle) $|a|^2$ represents the probability that the atom is in the state ψ_n and $|b|^2$ represents the probability that it is in the state $\psi_{n'}$. These coefficients are therefore time-dependent. At $t = 0$, $(|a| = 1, |b| = 0)$. At $t = \infty$, $(|a| = 0, |b| = 1)$. Let us calculate the expected value of the position of the electron during this collapse.

$$\langle \mathbf{r} \rangle = \langle a\psi_n + b\psi_{n'} | \mathbf{r} | a\psi_n + b\psi_{n'} \rangle$$
$$= |a|^2 \langle \psi_n | \mathbf{r} | \psi_n \rangle + |b|^2 \langle \psi_{n'} | \mathbf{r} | \psi_{n'} \rangle$$
$$+ a^* b \langle \psi_n | \mathbf{r} | \psi_{n'} \rangle + b^* a \langle \psi_{n'} | \mathbf{r} | \psi_n \rangle$$

The first two terms are time-independent and do not contribute to radiation. The last two terms combine to yield

(10.137) $$\langle \mathbf{r}(t) \rangle = a^* b e^{i(E_n - E_{n'})t/\hbar} \langle \varphi_n | \mathbf{r} | \varphi_{n'} \rangle + b^* a e^{i(E_{n'} - E_n)t/\hbar} \langle \varphi_{n'} | \mathbf{r} | \varphi_n \rangle$$
$$= 2 \,\mathrm{Re}\, [a^* b \langle \varphi_n | \mathbf{r} | \varphi_{n'} \rangle e^{i(E_n - E_{n'})t/\hbar}]$$
$$= 2 |a^* b \langle \varphi_n | \mathbf{r} | \varphi_{n'} \rangle| \cos(\omega_{nn'} t + \delta) = 2 |\mathbf{r}_{nn'}| \cos(\omega_{nn'} t + \delta)$$

---

[1] Fundamentally, all these collision processes involve the exchange of photons. For further discussion, see E. G. Harris, *A Pedestrian Approach to Quantum Field Theory*, Wiley, New York, 1972.

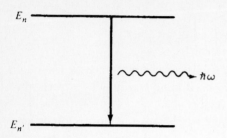

FIGURE 10.19 **Atom emits a photon in decaying from state ψ_n to $\psi_{n'}$.**

where $\omega_{nn'}$ is the Bohr frequency

$$\hbar\omega_{nn'} = E_n - E_{n'}$$

δ is a phase factor and $|a^*b|$ is assumed to be slowly varying and of order unity.

Atomic transitions typically occur in an interval of the order of 10^{-9} s. The frequency of emitted radiation, on the other hand, is typically of the order of 10^{15} s^{-1}, so the radiative oscillatory behavior of $\langle \mathbf{r}(t) \rangle$ is due almost exclusively to the cos term, with accompanying Bohr frequency $\omega_{nn'}$ (Fig. 10.20). When the atom is undergoing

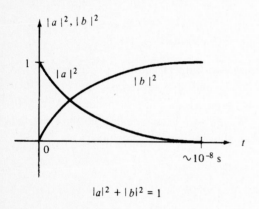

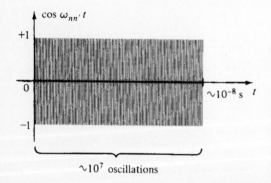

FIGURE 10.20 **Radiation decay from the ψ_n state to the $\psi_{n'}$ state involves the superposition state $\psi = a\psi_n + b\psi_{n'}$ with time-varying coefficients a and b. The "beat" frequency $\omega_{nn'}$ between the ψ_n and $\psi_{n'}$ states is much greater than the switchover frequency of ψ.**

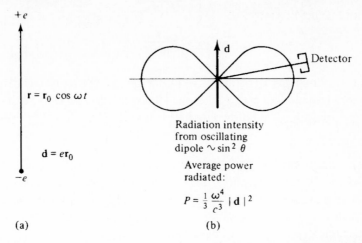

FIGURE 10.21 Energy characteristics of an oscillating dipole. (a) Dipole configuration. (b) Radiation profile of a dipole.

a transition between the states ψ_n and $\psi_{n'}$, the average position of the electron oscillates with the Bohr frequency corresponding to the energy difference between these states. At the beginning and conclusion of the transition, the atom is in stationary states in which it does not radiate. (These topics are returned to in Section 13.9.)

Selection Rules

Harmonic oscillation of an electron about a proton gives rise to what is commonly referred to as *dipole radiation*[1] (Fig. 10.21). The average radiated power from such an oscillating dipole is[2]

(10.138)
$$P = \frac{1}{3}\frac{\omega^4}{c^3}|\mathbf{d}|^2$$

where $\mathbf{d}$ is the dipole moment

(10.138a)
$$\mathbf{d} = e\mathbf{r}_0, \qquad \langle \mathbf{r} \rangle = \mathbf{r}_0 \cos \omega t$$

We may apply this formula to calculate the power radiated when the hydrogen atom decays from the nth state to the (n')th state. From (10.137) we obtain

(10.138b)
$$\mathbf{d} = 2e\mathbf{r}_{nn'}$$

---

[1] An atom also radiates in higher multipole channels (e.g., quadrupole). For the most part, dipole radiation is predominant.

[2] See J. D. Jackson, *Classical Electrodynamics,* 2d ed., Wiley, New York, 1975.

so that, with (10.138),

(10.139)
$$P = \frac{4}{3} \frac{(\omega_{nn'})^4 e^2}{c^3} |\mathbf{r}_{nn'}|^2$$

Calculation (see Problems 10.52 to 10.54) of the squared matrix element $|\mathbf{r}_{nn'}|^2$ with n standing for nlm and n' for $n'l'm'$ gives the following *selection rules*: The only conditions under which $|\mathbf{r}_{nn'}|^2$ (and therefore P) is not zero are

(10.140) $\Delta l = l' - l = \pm 1$ and $\Delta m_l = m' - m = 0, \pm 1$

For example, the transition $3S \to 1S (\Delta l = 0)$ is *forbidden,* as is the transition $3D \to 2S (\Delta l = 2)$. Such transitions are not accompanied by any (dipole) radiation and therefore are excluded by conservation of energy. The exclusion of the transitions between S states finds analogy with the classical theorem that spherically symmetric oscillatory charge distributions do not radiate.

The rule $\Delta l = \pm 1$, together with the law of conservation of angular momentum, indicates that for $\Delta l = -1$ the electromagnetic field (i.e., the photon) carries away angular momentum. As it turns out, photons have angular momentum quantum number equal to 1 and are therefore called *bosons.*

There are no restrictions on an atomic transition corresponding to change in the principal quantum number n. This is in agreement with the pre-Schrödinger spectral notation for emission from hydrogen: namely, the Lyman series corresponds to transitions from all n states to the ground state, the Balmer series corresponds to transitions to the $n = 2$ states, etc. (see Fig. 2.8).

PROBLEMS

10.51 What are the allowed transitions from the $5D$ states of hydrogen to lower states? Accompany your answer with a sketch representing these transitions.

10.52 With n representing the triplet nlm and n' the triplet $n'l'm'$, show that the matrix elements of $\mathbf{r}$ have the following complex representation:

$$|\mathbf{r}_{nn'}|^2 \equiv |\langle n|\mathbf{r}|n'\rangle|^2 = \frac{1}{2}\{|\langle n|x + iy|n'\rangle|^2 + |\langle n|x - iy|n'\rangle|^2\} + |\langle n|z|n'\rangle|^2$$

[*Hint:* Call $\langle n|x|n'\rangle \equiv x_{nn'}$, etc. Note also that $|\langle \mathbf{r}\rangle|$ denotes the magnitude of the vector $\langle \mathbf{r}\rangle$, while $|\langle x + iy\rangle|$ denotes the modulus of the complex variable $\langle x + iy\rangle$.]

10.53 Show that the matrix elements of $x \pm iy$ have the following integral representation:

(a) $\langle n|x \pm iy|n'\rangle = \displaystyle\int_{4\pi} [Y_l^m]^* Y_{l'}^{m'} \sin \theta \, e^{\pm i\phi} \, d\Omega \int_0^\infty R_{nl} R_{n'l'} r^3 \, dr$

(b) $\langle n|z|n'\rangle = \displaystyle\int_{4\pi} [Y_l^m]^* Y_{l'}^{m'} \cos \theta \, d\Omega \int_0^\infty R_{nl} R_{n'l'} r^3 \, dr$

10.54 Using the results of the last two problems and Tables 9.1 and 9.3, establish the selection rules for dipole radiation (10.140).

10.55 At the start of Section 10.7 it was noted that the energy eigenstates of an atom are stable against radiative decay, owing to their stationarity in time. However, these states do contain angular momentum, and one may argue that they therefore also contain rotating charge, which does radiate. Although the premise of this argument is correct, why is there still no radiation from the stationary states (from a classical point of view)?

Answer

An aggregate of N uniformly spaced point charges confined to move with fixed speed in a closed loop will radiate as a result of the acceleration of individual charges. However, in the limit that the charges approach a uniformly continuous distribution, $N \to \infty$, $\Delta \to 0$, $q \to 0$ (with total charge Nq and line charge density q/Δ constant), the radiation may be shown[1] to vanish. This limiting case closely resembles the state of affairs for the stationary states of an atom. Although there is rotating charge, such charge is continuously distributed and, in accord with the classical prescription, does not radiate.

10.56 The interaction potential of an electron moving in the far field of a dipole **d** is

$$V = -\frac{ed}{r^2} \cos \theta$$

The dipole is at the origin and points in the z direction. The spherical coordinates of the electron are (r, θ, ϕ).

    (a) Write down the time-independent Schrödinger equation for this system corresponding to zero total energy. What condition of the "atom" does this state describe?

    (b) Show that solutions to this equation are of the form $\varphi = r^s f(\theta, \phi)$. Obtain an equation for $f(\theta, \phi)$.

10.57 Two particles that are isolated from all other objects interact with each other through a central potential. As was established in Problem 10.23(d), the total angular momentum of the system may be written

$$\mathbf{L}_1 + \mathbf{L}_2 = \mathbf{L} + \mathbf{L}_{CM}$$

Show quantum mechanically that this total angular momentum is conserved.

10.58 (a) Prove that the *Runge–Lenz* vector

$$\hat{\mathbf{K}} = \frac{1}{2\mu e^2} [\mathbf{L} \times \hat{\mathbf{p}} - \hat{\mathbf{p}} \times \hat{\mathbf{L}}] + \frac{\mathbf{r}}{r}$$

commutes with the Hamiltonian of the hydrogen atom (10.115).

    (b) Show that the operator

$$\hat{\mathbf{A}} = \sqrt{-\mu e^4/2E}\, \hat{\mathbf{K}}$$

[1] See Jackson, *Classical Electrodynamics*, Chapter 14.

satisfies the commutation relations

$$[\hat{A}_x, \hat{A}_y] = i\hbar\hat{L}_z, \quad \text{etc.}$$

(c)   Use the result above to show that the operators

$$\hat{B}_+ = \frac{1}{2}(\hat{L} + \hat{A})$$

$$\hat{B}_- = \frac{1}{2}(\hat{L} - \hat{A})$$

obey the angular momentum commutation relations and the equality $\hat{B}_+{}^2 = \hat{B}_-{}^2$.

(d)   Derive the relation

$$\hat{B}_+{}^2 + \hat{B}_-{}^2 = -\frac{1}{2}\left(\hbar^2 + \frac{\mu e^4}{2E}\right)$$

and use it to obtain the Bohr formula for the energy levels of hydrogen.

10.59   The classical harmonic oscillator with spring constant K and mass m oscillates at the *single* frequency (independent of energy)

$$\omega_0 = \sqrt{K/m}$$

The quantum mechanical oscillator, on the other hand, gives frequencies at all integral multiples of ω_0, as follows directly from the eigenenergies

$$E_n = \hbar\omega_0\left(n + \frac{1}{2}\right) \qquad (n = 0, 1, 2, \ldots)$$

If one end of the oscillator is charged, dipole radiation is emitted. In the classical domain, this radiation has frequency ω_0. Show that selection rules that follow from calculation of the dipole matrix elements $x_{nn'}$ reduce the quantum mechanical spectrum to the classical one. (The concept of matrix elements of an observable is developed formally in Chapter 11.)

10.60   Consider a gas of noninteracting rigid dumbbell molecules with speeds small compared to the speed of light. The moment of inertia of each molecule is I.

(a)   What is the Hamiltonian of a molecule in the gas?

(b)   What are the eigenenergies of this Hamiltonian?

(c)   Let a molecule undergo spontaneous decay between two rotational states. Owing to the recoil of the center of mass, there is a change in momentum of the center of mass of the molecule as well. Show that the frequency of the photon emitted in this process is

$$\nu = \nu_l\left(1 - \frac{\hbar\mathbf{k} \cdot \mathbf{n}}{Mc}\right)$$

The initial momentum of the center of mass is $\hbar\mathbf{k}$, $\mathbf{n}$ is a unit vector in the direction of the momentum of the emitted photon, and $\{\nu_l\}$ is the rotational line spectrum.

(d)   What is the nature of the frequency spectrum emitted by the gas?

Answers

(a)   Let $\mathcal{P}$ and M denote the momentum and mass, respectively, of the center of mass. Then

$$\hat{H} = \frac{\mathcal{P}^2}{2M} + \frac{\hat{L}^2}{2I}$$

(b) $E_{k,l} = \dfrac{\hbar^2 k^2}{2M} + \dfrac{\hbar^2 l(l+1)}{2I}$

(c)   The frequency of photons emitted by a molecule is given by the change in energy

$$h\nu = \Delta E = \frac{\hbar^2 \mathbf{k} \cdot \Delta \mathbf{k}}{M} + h\nu_l$$

where $h\nu_l$ is written for the change in rotational energy. Since the molecule is a free particle, the momentum of the center of mass $\hbar \mathbf{k}$ can change only by virtue of the momentum carried away by the photon emitted in the transition. As a first approximation we will assume that the momentum carried away is $h\nu_l/c$. If $\hbar \mathbf{k}'$ is the momentum of the center of mass after emission, then by conservation of momentum we have (see Fig. 10.22)

$$\hbar \mathbf{k} = \hbar \mathbf{k}' + \frac{h\nu_l}{c} \mathbf{n}$$

so that

$$\hbar \, \Delta \mathbf{k} = \hbar(\mathbf{k}' - \mathbf{k}) = -\frac{h\nu_l}{c} \hat{\mathbf{n}}$$

Substituting this value in the expression above for $h\nu$ gives

$$\nu = \nu_l \left(1 - \frac{\hbar \mathbf{k} \cdot \mathbf{n}}{Mc} \right)$$

Since $\hbar k \ll Mc$, the assumption that momentum carried away by this radiation field is approximately $h\nu_l/c$ is justified.

(d)   The rotational spectrum $\{\nu_l\}$ remains a line spectrum with an infinitesimal broadening of lines.

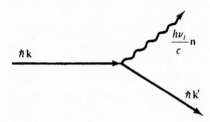

$\hbar \mathbf{k}$

$\dfrac{h\nu_l}{c}\mathbf{n}$

$\hbar \mathbf{k}'$

FIGURE 10.22   Change in linear momentum of rigid rotator due to recoil in the emission of a photon (see Problem 10.60).

10.61   Consider two identical rigid spheres of diameter a, which move in 3-space.
   (a)   What is the Hamiltonian of this system?
   (b)   Separate out the center-of-mass motion to obtain an equation for the wavefunction for relative motion, $\varphi(\mathbf{r})$.
   (c)   What boundary conditions must $\varphi(\mathbf{r})$ satisfy?
   (d)   What are the eigenstates and eigenenergies for this system?
   (e)   What are the parities of these eigenstates?
   (f)   What is the radial probability, $P(r)$, for this system?

Answers (partial)

   (a) $\hat{H}(1, 2) = \dfrac{\hat{p}_1^{\,2}}{2m} + \dfrac{\hat{p}_2^{\,2}}{2m} + V(|\mathbf{r}_1 - \mathbf{r}_2|)$

$$V(|\mathbf{r}_1 - \mathbf{r}_2|) \equiv V(r) = \infty, \qquad r \le a$$
$$V(r) = 0, \qquad r > a$$

   (b) $\left[\dfrac{\hat{p}_r^{\,2}}{2\mu} + \dfrac{\hat{L}^2}{2\mu r^2} + V(r) - E \right] \varphi(\mathbf{r}) = 0$

   (c) $\varphi_{klm}(\mathbf{r}) = 0, \qquad r \le a$

   (d) $\varphi_{klm}(\mathbf{r}) = \varphi_{lm}(kr)Y_l^m(\theta, \phi), \qquad r > a$
         $\varphi_{lm}(kr) = A[n_l(ka)j_l(kr) - j_l(ka)n_l(kr)]$

$$E_k = \dfrac{\hbar^2 k^2}{2\mu}$$

   (e)   Referring to Problem 9.46, we obtain

$$\hat{P}\varphi_{lm}(kr) = (-)^l \varphi_{lm}(kr)$$

10.8   THOMAS–FERMI MODEL

The Thomas–Fermi model is appropriate to atoms with sufficient number of electrons which permits the system to be treated in a statistical sense. The model views an atom as a spherically symmetric gas of Z electrons which surrounds a nucleus of charge Ze. It is further assumed that the potential in the medium, $\Phi(r)$, varies sufficiently slowly so that it may be taken as constant in small volumes of the electron gas. The description of the Fermi energy, such as given in Section 2.3, then applies. (See Fig. 10.23.). There, we recall the Fermi energy was equated to the maximum kinetic energy of an electron in an aggregate of electrons at 0 K. This description maintains in the present model since at normal temperature $k_B T \ll \Phi(r)$. (The Fermi energy is discussed further in Section 12.9.)

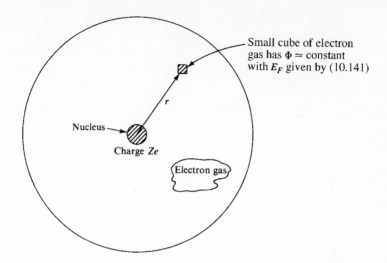

FIGURE 10.23 **The Thomas–Fermi statistical atom.**

The relation between E_F and electron density $n(r)$ is given in Problem 2.42, and with the preceding description we write

(10.141)
$$E_{\max}^{\text{Kin}} = E_F = \frac{\hbar^2}{2m} [3\pi^2 n(r)]^{2/3}$$

The energy of an electron in the medium at a distance r from the nucleus is

(10.142)
$$E(r) = \frac{p^2}{2m} - e\Phi(r)$$

where we have written $e = |e|$. The preceding equation indicates that for $p^2/2m > e\Phi(r)$, $E > 0$, which implies that the electron escapes the atom. As this does not occur in the present model, we may conclude that maximum kinetic energy may also be set equal to $e\Phi$. With (10.141) we may then write

$$e\Phi(r) = \frac{\hbar^2}{2m} [3\pi^2 n(r)]^{2/3}$$

An additional relation between the potential $\Phi(r)$ and density $n(r)$ is given by Poisson's equation (cgs),

$$\nabla^2 \Phi = 4\pi e n$$

Working in the given spherically symmetric geometry, and eliminating $n(r)$ from the latter two equations, gives

(10.143)
$$\nabla^2 \Phi = \frac{1}{r} \frac{d^2}{dr^2} (r\Phi) = \frac{4\pi e}{3\pi^2} \frac{(2me)^{3/2}}{\hbar^3} \Phi^{3/2}$$

This is the desired single equation for $\Phi(r)$.

The following boundary conditions apply to the potential. Near the origin $\Phi(r)$ is due predominantly to the nucleus, and we write

$$\Phi \simeq \frac{Ze}{r}$$

or, equivalently,

(10.144a)
$$\lim_{r \to 0} r\Phi(r) = Ze$$

At large r we obtain

(10.144b)
$$\lim_{r \to \infty} r\Phi(r) = 0$$

The resulting equation (10.143) may be written in a more concise form in terms of new dimensionless variables (x, χ) defined by

(10.145a)
$$r = ax$$

(10.145b)
$$a \equiv \frac{1}{2} \left(\frac{3\pi}{4} \right)^{2/3} \frac{\hbar^2/me^2}{Z^{1/3}} = \frac{0.885 a_0}{Z^{1/3}}$$

(10.145c)
$$\Phi(r) \equiv \frac{Ze}{r} \chi(x)$$

When written in terms of these new variables (see Problem 10.68), (10.143) reduces to

(10.146)
$$\boxed{x^{1/2} \frac{d^2\chi}{dx^2} = \chi^{3/2}}$$

This is the standard form of the (highly nonlinear) Thomas–Fermi equation. Boundary conditions (10.144), when written in terms of the new potential $\chi(x)$, appear as

(10.147)
$$\lim_{x \to 0} \chi = 1, \qquad \lim_{x \to \infty} \chi = 0$$

The semilogarithmic plot shown in Fig. 10.24 reveals the rapid decay of the potential χ with increase in x. Thus, for example, for $x > 4$, $\chi < 0.1$.[1]

---

[1] Numerical integration of the Thomas–Fermi equation (10.146) was originally performed by V. Bush and S. H. Caldwell, *Phys. Rev.* **38**, 1898 (1931).

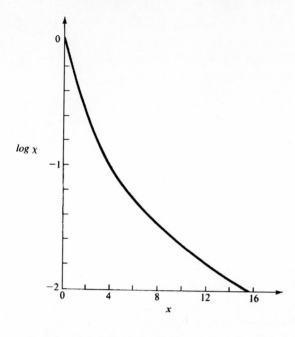

$\log \chi$

FIGURE 10.24   Semilogarithmic plot of solution to the Thomas–Fermi equation (10.146) subject to the boundary conditions (10.147).

Variation of Atomic Size with Z

An important conclusion of the preceding model is as follows: It is evident from (10.145b) that the parameter a may be viewed as an effective radius of the atom (i.e., the radius of a sphere which contains a fixed fraction of atomic electrons). Thus there is an implied decrease in effective atomic size with increase in Z. However, it should be borne in mind that the shell structure of atoms is lost in the Thomas–Fermi model. Thus one might expect the implied decrease in atomic size to be roughly valid within a given atomic shell of fixed principal quantum number. Observations find this property to be approximately obeyed. (See Fig. 12.11b.) The closely allied "Hartree–Fock" approximation is described in Section 13.10.

PROBLEMS

10.62   A hydrogen atom suffers a decay from the $3D$ to the $2P$ state and emits a photon.

(a)   What is the frequency of this photon (in Hz)?

(b)   Given that the decay time of this transition is $\simeq 10^{-10}$ s, what is the power (in W) emitted in the transition?

10.63   Show that the three φ_{2P} states (10.134) comprise an orthonormal set. What is the value of the constant A? Leave your answer in terms of square roots and a_0.

10.64   Of what angular momentum operators are the respective three φ_{2P} states (10.134) eigenfunctions?

Answer

As described in Chapter 9, when operating on a wavefunction $f(\mathbf{r})$, the operators $\hat{L}_x$, $\hat{L}_y$, $\hat{L}_z$ effect a rotation of $\mathbf{r}$ about the x, y, z axes, respectively. Thus φ_{2P_x} is an eigenstate of $\hat{L}_x$; φ_{2P_y} is an eigenstate of $\hat{L}_y$; and φ_{2P_z} is an eigenstate of $\hat{L}_z$. All three functions are eigenstates of $\hat{L}^2$.

10.65   Show that $\langle \psi_1 | \psi_2 \rangle = 0$ for the orbitals (10.135).

10.66   (a)   Show that the function φ_{2P_x} as given by (10.134a) is maximum along the x axis.

(b)   With this property at hand, argue that ψ_1 as given by (10.135a) exhibits maxima along the (1, 1, 1) directions.

Answers (partial)

(a)   We may write

$$\varphi_{2P_x} = A \sin \theta \cos \phi \; re^{-r/2a_0}$$

A polar plot of $\sin \theta$ reveals a circle tangent to the z axis at the origin. Thus $\sin \theta$ generates a circular torus about the z axis. A polar plot of $\cos \phi$ reveals a circle in the xy plane tangent to the y axis at the origin and with maximum at $\phi = 0$ (positive x axis). The intersection of these two factors, $\sin \theta \cos \phi$, is evidently maximum along the positive x axis. The remaining r-dependent component merely modulates this result.

10.67   This problem concerns annihilation and creation operators for spherical Bessel functions.[1] Introducing the operators (for $l > 0$)

$$b_l^{\pm} = -\frac{i}{x}\frac{d}{dx}x \pm i\frac{l}{x}$$

(a)   Show that

$$\hat{b}_l^{+}\hat{b}_l^{-} = \hat{b}_{l+1}^{-}\hat{b}_{l+1}^{+} = -\frac{1}{x}\frac{d^2}{dx^2} + \frac{l(l+1)}{x^2}$$

(b)   Show that Bessel's equation (10.54) may be written

$$(\hat{b}_l^{+}\hat{b}_l^{-} - 1)R_l(x) = 0$$

(c)   Show that

$$\hat{b}_{l+1}^{+}R_l = R_{l+1}$$

(d)   Show that

$$\hat{b}_l^{-}R_l = R_{l-1}$$

(e)   Apart from normalization, employ the preceding formalism to obtain:
(1) $j_1(x)$ from $j_0(x)$.
(2) $n_1(x)$ from $n_0(x)$.

[1] L. Infeld, *Phys. Rev.* **59**, 737 (1941).

10.68   The radial wavefunction satisfies the second-order differential equation (10.109). The student may recall that any such equation has two independent solutions. Show that for bound states there is only one linearly independent solution for given l and E values.

Answer

The equation may be written

$$\left[\hat{D}^2 - \frac{l(l+1)}{r^2} + \frac{2\mu}{\hbar^2} (E - V) \right] u = 0$$

$$\hat{D} \equiv \frac{d}{dr}$$

$$u \equiv rR$$

Let solutions to the preceding equation be written u_1 and u_2. Multiplying the equation for u_2 by u_1 and vice versa and subtracting the resulting equations gives

$$u_1 \hat{D}^2 u_2 - u_2 \hat{D}^2 u_1 = 0$$

or, equivalently,

$$\hat{D}[u_1 \hat{D} u_2 - u_2 \hat{D} u_1] = 0$$

Integrating, we find

$$u_1 \hat{D} u_2 - u_2 D u_1 = C$$

where C is a constant. For bound systems one may assume that either $u_1, u_2 \to 0$ or $\hat{D} u_1, \hat{D} u_2 \to 0$ as $r \to \infty$, which, in either case, gives $C = 0$. There results

$$\frac{d \ln u_2}{dr} = \frac{d \ln u_1}{dr}$$

which gives

$$u_2 = a u_1$$

where a is a constant. Thus, u_2 and u_1 are not independent.

10.69   Employing the transformation equations (10.145), show that (10.143) reduces to the Thomas–Fermi equation (10.146).

10.70   This problem addresses motion of an electron in a steady magnetic field as described in Section 10.4.

   (a)   Find the mean square radius a^2 of an electron moving in a plane normal to a steady magnetic field.

   (b)   Obtain your answer to part (a) once again, but now through use of the correspondence principle.

Answers

(a)   Referring to (10.86) et seq. indicates that

$$a^2 = \langle (y - y_0)^2 \rangle$$

Due to the harmonic oscillator structure of (10.88), we may write

$$\langle (y - y_0)^2 \rangle = \frac{2}{K} \langle V \rangle$$

where V is potential energy. With Problem 7.10 we may then write

$$a^2 = \frac{2}{K} \times \frac{\langle E \rangle}{2} = \frac{1}{K} \langle E \rangle$$

$$a^2 = \frac{\hbar}{m\Omega} \left(n + \frac{1}{2} \right)$$

(b)   The classical kinetic energy of the electron is

$$\frac{1}{2} mv^2 = \frac{1}{2} ma^2 \Omega^2 = \frac{1}{2} E$$

Thus

$$a^2 = \frac{E}{m\Omega^2}$$

Inserting the quantum value for E gives the desired result.

10.71   You are asked to design an absorbing sink for rotationally polarized radiation of frequency 2.5×10^{-10} Hz. You propose to use a gas of dipole molecules. What is the moment of inertia, I (cgs), of molecules that will perform this task in $l = 0 \to l = 1$ transitions? What is the effective radius, r_0 (in Å), of these molecules if their reduced mass is 30 amu? (See Problem 1.23.)

10.72   (a)   Show that $\varphi_{lm}(r, \theta, \phi) = r^l Y_l^m(\theta, \phi)$ are eigenfunctions of the free-particle Hamiltonian. What are the corresponding eigenenergies of these states?

(b)   Show that this finding is consistent with the free-particle eigenproperties (10.56).

(c)   What is the order of degeneracy of these eigenenergies? [*Hint:* For part (b), consult Table 10.1.]

10.73   (In the following, hats over operators are deleted.) The formulation for evaluating the momentum, p_n, conjugate to a given coordinate, q_n, for a system described by N curvilinear coordinates is based on the following three properties[1]: (i) the differential volume element: $dV = D \, dq_1 \ldots dq_N$; (ii) the commutator: $[q_n, p_n] = i\hbar$; (iii) Hermiticity: $\langle \varphi | p_n \psi \rangle = \langle p_n \varphi | \psi \rangle$. The momentum operator conjugate to q_n is then given by

$$p_n = -i\hbar \frac{1}{\sqrt{D}} \frac{d\sqrt{D}}{dq_n}$$

---

[1] W. Pauli, "Die Allgemeinen Prinzipien der Wellenmechanik," *Hand. d. Physik* **XXIV**, 83–272 (1933).

(a)   Obtain an expression for the radial momentum operator, p_ρ, appropriate to cylindrical coordinates (Fig. 1.7).

(b)   Show that for the operator you have found, $[\rho, p_\rho] = i\hbar$.

(c)   Show that the operator p_ρ is Hermitian (with respect to square integrable functions in a plane of constant z).

(d)   Show that this operator obeys the following relation. (An expression for $\nabla_\rho{}^2$ is given in Appendix D.)

$$p_\rho{}^2 = -\hbar^2\left[\nabla_\rho{}^2 - \frac{1}{4\rho^2}\right]$$

(e)   With your answer to part (d), write down a "corrected" Hamiltonian in cylindrical coordinates for a particle moving in the presence of a potential field, $V(\rho, \phi)$.

(f)   Show that your answer to part (e) gives the correct Schrödinger equation for a free particle in cylindrical coordinates (10.75a).

(g)   Show that your answer to part (e) reduces to the classical Hamiltonian in the classical limit.

Answers (partial)

(a) $p_\rho = -i\hbar\dfrac{1}{\rho^{1/2}}\dfrac{d\rho^{1/2}}{d\rho} = -i\hbar\left(\dfrac{1}{2\rho} + \dfrac{d}{d\rho}\right)$

(e) $H = \dfrac{1}{2m}\left[p_\rho{}^2 + \dfrac{p_\phi{}^2 - \hbar^2(1/2)^2}{\rho^2}\right] + V(\rho, \phi)$

(g)   Setting $\hbar = 0$ returns the classical Hamiltonian.

10.74   What are the eigenfunctions, good quantum numbers, and eigenenergy degeneracies for the following configurations?

(a)   A free particle moving in three dimensions in (i) Cartesian, (ii) spherical, and (iii) cylindrical coordinates.

(b)   A particle confined to a (i) cubical box, (ii) spherical box, and (iii) cylindrical box.

(c)   In each case, state which type of degeneracy is described. (See Section 8.5.) For infinite degeneracies discuss the nature of these infinities (i.e., continuous or discrete).

10.75   An electron is free to move in the domain exterior to a rigid, impenetrable, smooth-walled, charge-neutral sphere of radius a.

(a)   What are the eigenfunctions and eigenenergies of this system?

(b)   Write down the normalization condition for these wavefunctions. (*Hint:* See Problem 10.60.)

10.76   An electron is confined to the interior domain of a cylindrical cavity of radius a and length L. Photons at 3.0 eV are incident on the cavity and raise the electron from the ground state to the first excited state. What is the radius a (in Å) of the cylinder, given that $L = 4.7$ Å?

11

ELEMENTS OF MATRIX MECHANICS. SPIN WAVEFUNCTIONS

In this chapter some mathematical formalism is developed which is necessary for a more complete description of spin angular momentum. This formalism involves the theory of representations described briefly in Chapter 7 and matrix mechanics. Spin angular momentum operators are cast in the form of the Pauli spin matrices. The spinor wavefunction of a propagating spinning electron is constructed and used in problems involving the Stern–Gerlach apparatus. Examples involving the precession of an electron in a magnetic field and magnetic resonance are included as well as a prescription for adding spins. The coupled spin states so obtained are used extensively in the following chapter in conjunction with the Pauli principle in some basic atomic and molecular physics problems. The density matrix relevant to mixed states is introduced. Descriptions are included of the Heisenberg and interaction pictures in quantum mechanics. The chapter continues with a review of polarization states and a description of Bell's theorem in relation to present-day experiments concerning the notion of hidden variables. The chapter concludes with analysis of the transfer matrix method important to the study of particle transport through a periodic potential.

11.1 BASIS AND REPRESENTATIONS

Matrix Mechanics

At very nearly the same time that Schrödinger introduced his wavefunction development of quantum mechanics, an alternative but equivalent description of the same physics was formulated. It is known as *matrix mechanics* and is due to W. Heisenberg, M. Born, and P. Jordan.

We have already encountered the concept of representations in quantum mechanics in Section 7.4. There we noted that in the "A representation," states are referred to a basis comprised of the eigenfunctions of $\hat{A}$. What we will now find is that within any such representation it is always possible to express operators and wavefunctions as matrices. Operator equations become matrix equations. For example, an equation of the form $\psi = \hat{F}\psi'$ may be rewritten as a matrix equation with the wavefunctions ψ and ψ' written as column vectors and the operator $\hat{F}$ written as a square matrix.

Basis

Previously we have found that wavefunctions related to a given quantum mechanical problem must satisfy certain criteria. Examples include

Configuration	Wavefunction Criteria
(a) Particle in a box	$\psi(0) = \psi(a) = 0, \int \|\psi\|^2 \, dx < \infty$
(b) One-dimensional harmonic oscillator	
$V(x) = \dfrac{K}{2}x^2$	$\int \|\psi\|^2 \, dx < \infty, \ \psi \to 0 \text{ as } \|x\| \to \infty$
(c) Particle in a central potential	
$V = V(r)$	$\iint \|\psi\|^2 r^2 \, dr \, d\Omega < \infty, \ \|\psi\|^2 r^2 \to 0 \text{ as } r \to 0$

For a given problem each such set of conditions implies a related space of functions. Consider a specific problem. Let the space of functions relevant to that configuration be called $\mathfrak{H}$. Let the set of functions

$$(11.1) \qquad \qquad \mathfrak{B} = \{\varphi_1, \varphi_2, \ldots\}$$

be a basis of $\mathfrak{H}$. For instance, for a particle in a one-dimensional box, these functions are

$$\mathfrak{B} = \left\{ \sqrt{\frac{2}{a}} \sin\left(\frac{n\pi x}{a}\right) \right\}$$

For the hydrogen atom, they are

$$\mathfrak{B} = \{R_{nl}(r)Y_l^m(\theta, \phi)\}$$

whereas for a free particle (in spherical coordinates), they are

$$\mathfrak{B} = \{j_l(kr)Y_l^m(\theta, \phi)\}$$

Inasmuch as $\mathfrak{B}$ is a basis of $\mathfrak{H}$, any function ψ in $\mathfrak{H}$ may be expanded in terms of the basis functions φ_n:

(11.2)
$$\psi = \sum_n \varphi_n a_n$$

or, equivalently,

(11.3)
$$|\psi\rangle = \sum_n |\varphi_n\rangle\langle\varphi_n|\psi\rangle$$

The coefficients of expansion, a_n, represent ψ in the representation where $\mathfrak{B}$ is the basis. These coefficients are projections of ψ onto the basis vectors (see Fig. 4.6). The equivalence of $\{a_n\}$ to the state function ψ is akin to the equivalence between a three-dimensional vector $\mathbf{A}$ and its components (A_x, A_y, A_z).

If $\{a_n\}$ is equivalent to ψ, one should be able to rewrite equations involving ψ as equations involving only $\{a_n\}$. Consider the typical quantum mechanical equation, where $\hat{F}$ is an arbitrary operator:

(11.4)
$$\psi = \hat{F}\psi'$$
$$|\psi\rangle = \hat{F}|\psi'\rangle$$

Expanding the right-hand side of the latter equation in accordance with (11.3) and multiplying from the left with $\langle\varphi_q|$ gives

(11.5)
$$\langle\varphi_q|\psi\rangle = \sum_n \langle\varphi_q|\hat{F}|\varphi_n\rangle\langle\varphi_n|\psi'\rangle$$

or, equivalently,

(11.6)
$$a_q = \sum_n F_{qn}a_n{}'$$

where

(11.7)
$$F_{qn} \equiv \langle \varphi_q | \hat{F} | \varphi_n \rangle \equiv \int \varphi_q {}^* \hat{F} \varphi_n \, d\mathbf{r}$$

is *the matrix representation of the operator* $\hat{F}$ *in the basis* $\mathfrak{B}$. The term F_{qn} is also called the *matrix element connecting* φ_q *to* φ_n. Equation (11.6), involving only the expansion coefficients $\{a_q\}$, $\{a_n'\}$, and the matrix elements $\{F_{qn}\}$, is equivalent to (11.4) involving the wavefunctions ψ, ψ' and the operator $\hat{F}$. Equation (11.6) is called a *matrix equation.* It may be written in the form

(11.8)
$$\begin{pmatrix} a_1 \\ a_2 \\ \vdots \end{pmatrix} = \begin{pmatrix} F_{11} & F_{12} & \cdots \\ F_{21} & F_{22} & \cdots \\ \vdots & \vdots \end{pmatrix} \begin{pmatrix} a_1' \\ a_2' \\ \vdots \end{pmatrix}$$

In this equation the wavefunction ψ is represented by the column vector on the left, and ψ' is represented by the column vector on the right.[1]

(11.9)
$$\psi \rightarrow \begin{pmatrix} a_1 \\ a_2 \\ a_3 \\ \vdots \end{pmatrix} \qquad \psi' \rightarrow \begin{pmatrix} a_1' \\ a_2' \\ a_3' \\ \vdots \end{pmatrix}$$

The operator $\hat{F}$ is represented by the matrix F_{qn}.

(11.10)
$$\hat{F} = \begin{pmatrix} F_{11} & F_{12} & F_{13} & \cdots \\ F_{21} & F_{22} & F_{23} & \cdots \\ F_{31} & F_{32} & F_{33} & \cdots \\ \vdots & \vdots \end{pmatrix}$$

The infinite dimensionality of these matrix equations is a consequence of the infinite dimensionality of Hilbert space. Finite matrix equations are relevant to vector spaces of finite dimension.

Diagonalization of an Operator

Let the orthogonal basis $\mathfrak{B}$ be comprised of the eigenfunctions of a Hermitian operator $\hat{G}$:

(11.11)
$$\hat{G} \varphi_n = g_n \varphi_n$$

---

[1] The arrow in these identifications denotes "is represented by."

The matrix elements of $\hat{G}$ are

$$\langle \varphi_q | \hat{G} | \varphi_n \rangle = g_n \langle \varphi_q | \varphi_n \rangle = g_n \delta_{qn}$$

$$G_{qn} = g_n \delta_{qn}$$

(11.12)
$$\hat{G} = \begin{pmatrix} g_1 & 0 & 0 & \cdots \\ 0 & g_2 & 0 & \cdots \\ 0 & 0 & g_3 & \cdots \\ \vdots & \vdots & \vdots & \end{pmatrix}$$

Thus the matrix of an operator in a basis of the eigenfunctions of that operator is diagonal. The column vector representations of the eigenfunctions φ_n are the coefficients $\{a_q^{(n)}\}$ in the expansion

$$| \varphi_n \rangle = \sum_q a_q^{(n)} | \varphi_q \rangle$$

Multiplying from the left by $\langle \varphi_p |$ gives

$$\delta_{pn} = \sum_q a_q^{(n)} \delta_{pq} = a_p^{(n)}$$

(11.13)
$$a_p^{(n)} = \delta_{pn}$$

Thus the matrix representation of the eigenvector φ_n is a column vector with a single nonzero unit entry in the nth slot

$$\varphi_1 \rightarrow \begin{pmatrix} a_1^{(1)} \\ a_2^{(1)} \\ \vdots \end{pmatrix} = \begin{pmatrix} 1 \\ 0 \\ 0 \\ 0 \\ 0 \\ \vdots \end{pmatrix}, \qquad \varphi_2 \rightarrow \begin{pmatrix} a_1^{(2)} \\ a_2^{(2)} \\ \vdots \end{pmatrix} = \begin{pmatrix} 0 \\ 1 \\ 0 \\ 0 \\ 0 \\ \vdots \end{pmatrix},$$

(11.14)
$$\varphi_3 \rightarrow \begin{pmatrix} 0 \\ 0 \\ 1 \\ 0 \\ 0 \\ 0 \\ \vdots \end{pmatrix}, \qquad \varphi_4 \rightarrow \begin{pmatrix} 0 \\ 0 \\ 0 \\ 1 \\ 0 \\ 0 \\ \vdots \end{pmatrix}$$

The eigenvalue equation (11.11) can be written

$$\sum_q \langle \varphi_p | \hat{G} | \varphi_q \rangle \langle \varphi_q | \varphi_n \rangle = g_n \langle \varphi_p | \varphi_n \rangle$$

$$\sum_q G_{pq} a_q^{(n)} = g_n a_p^{(n)}$$

For $a^{(3)}$ it appears as

(11.15)
$$\begin{pmatrix} g_1 & 0 & 0 & 0 & \cdots \\ 0 & g_2 & 0 & 0 & \cdots \\ 0 & 0 & g_3 & 0 & \cdots \\ & & & & \\ \vdots & \vdots & \vdots & \vdots & \end{pmatrix} \begin{pmatrix} 0 \\ 0 \\ 1 \\ 0 \\ 0 \\ \vdots \end{pmatrix} = g_3 \begin{pmatrix} 0 \\ 0 \\ 1 \\ 0 \\ 0 \\ \vdots \end{pmatrix}$$

The "length" (squared) of a vector ψ is given by [recall (4.30)]

(11.16)
$$\|\psi\|^2 = \langle \psi | \psi \rangle = \sum_q \langle \psi | \varphi_q \rangle \langle \varphi_q | \psi \rangle$$

$$= \sum_q |a_q|^2$$

The lengths of the orthonormal basis vectors $\{\varphi_n\}$ are

(11.17)
$$\|\varphi_n\|^2 = \sum_q |a_q^{(n)}|^2 = 1$$

In matrix representation,

$$\|\varphi_4\|^2 = \langle \varphi_4 | \varphi_4 \rangle \to \overbrace{000100 \cdots} \begin{pmatrix} 0 \\ 0 \\ 0 \\ 1 \\ 0 \\ 0 \\ \vdots \end{pmatrix} = 1$$

Suppose that $\hat{G}$ is known to be diagonal in the basis $\{\varphi_n\}$. Then

(11.18)
$$G_{qn} = g_n \delta_{qn}$$

or, equivalently,

$$\langle \varphi_q | \hat{G} | \varphi_n \rangle = g_n \langle \varphi_q | \varphi_n \rangle$$

Multiplying from the left with the sum $\Sigma_q \, | \varphi_q \rangle$ gives

(11.19)
$$\sum_q | \varphi_q \rangle \langle \varphi_q | \hat{G} | \varphi_n \rangle = g_n \sum_q | \varphi_q \rangle \langle \varphi_q | \varphi_n \rangle$$

Recognizing the sum over φ_q products to be the unity operator $\hat{I}$ (Problem 11.1) allows this latter equation to be rewritten as

$$\hat{I}(\hat{G} | \varphi_n \rangle - g_n | \varphi_n \rangle) = 0$$

which in turn implies that

(11.20)
$$\hat{G} | \varphi_n \rangle = g_n | \varphi_n \rangle$$

Thus we find that if $\hat{G}$ is diagonal in a basis $\mathcal{B}$, then $\mathcal{B}$ is comprised of the eigenvectors of $\hat{G}$. One then notes the following important observation. The problem of finding the eigenvalues of an operator is equivalent to finding a basis which diagonalizes that operator.

Complete Sets of Commuting Operators

Suppose that $\hat{A}$, $\hat{B}$, and $\hat{C}$ are a "complete" set of three commuting operators. Let $\mathcal{B} = \{\varphi_1, \varphi_2, \ldots\}$ be a set of simultaneous eigenstates of these three operators. Then with respect to this basis, $\hat{A}$, $\hat{B}$, and $\hat{C}$ are all diagonal and one speaks of "working in a representation in which $\hat{A}$, $\hat{B}$, and $\hat{C}$ are diagonal." For example, for a free particle moving in 3-space, the representation in which $\hat{H}$, $\hat{L}^2$, and $\hat{L}_z$ are diagonal contains the basis (10.56), while the representation in which $\hat{p}_x$, $\hat{p}_y$, and $\hat{p}_z$ are diagonal contains the basis (10.11).

The Continuous Case

In some cases the indices of a matrix range over a continuum of values. Such, for example, is the Hamiltonian matrix for a free particle in the basis (10.11). In one dimension this basis is comprised of the states (3.24), and the Hamiltonian matrix assumes the continuous form

$$\langle k | \hat{H} | k' \rangle = \left\langle k \left| \frac{\hat{p}_x^2}{2m} \right| k' \right\rangle = \frac{\hbar^2 k'^2}{2m} \delta(k - k')$$

Summing over the index of a continuous matrix is equivalent to integration. For example,

$$\sum_{k'} \langle k | \hat{H} | k' \rangle \rightarrow \int_{-\infty}^{\infty} dk' \frac{\hbar^2 k'^2}{2m} \delta(k - k') = \frac{\hbar^2 k^2}{2m}$$

The matrix representation of a quantum mechanical equation involving a wavefunction ψ is the corresponding equation for the projection coefficients of ψ into the basis $\mathfrak{B}$. If this set of coefficients forms a discrete set, then equations are of the form (11.6), involving summations over a discrete index. For the continuous case, these sums become integrals. In Section 7.4 we considered the case of the simple harmonic oscillator in "momentum space." Since the eigenstates of the momentum operator form a continuous set, the Schrödinger equation for the projection coefficients $\{b(k)\}$ becomes the integral equation (7.81). Owing to the simple form of the harmonic oscillator potential, this in turn is reducible to the differential form

$$(7.84) \qquad \left(\frac{\hbar^2 k^2}{2m} - \frac{K}{2} \frac{\partial^2}{\partial k^2} \right) b(k) = Eb(k)$$

Thus the "matrix" form of the Schrödinger equation in the momentum representation remains a simple differential equation. Its argument is the single component $b(k)$ of the "column vector" $\{b(k)\}$.[1]

PROBLEMS

11.1   Let $\{\varphi_n\}$ be a *complete* orthonormal basis of a Hilbert space, $\mathfrak{H}$. Show that the identity operation $\hat{I}$ has the representation

$$\hat{I} = \sum_n | \varphi_n \rangle \langle \varphi_n |$$

in $\mathfrak{H}$. (This is sometimes called the *spectral resolution of unity*.)

Answer

Forming the matrix elements of $\hat{I}$, as given above, gives

$$I_{pq} = \langle \varphi_p | \hat{I} | \varphi_q \rangle = \sum_n \langle \varphi_p | \varphi_n \rangle \langle \varphi_n | \varphi_q \rangle$$

$$= \sum_n \delta_{pn} \delta_{nq} = \delta_{pq}$$

[1] For further remarks on the $\hat{x}$ and $\hat{p}$ representations, see Appendix A.

These are the matrix elements of the identity operator. This is a square matrix with unit entries along the diagonal.

$$\hat{I} = \begin{pmatrix} 1 & & & 0 \\ & 1 & & \\ & & 1 & \\ 0 & & & \ddots \end{pmatrix}$$

(*Note:* As described in Chapter 4, in order that an operator be a valid quantum mechanical representation of an observable, it must be Hermitian. To ensure further consistency of the theory, one also demands that the eigenstates of the operator comprise a complete set.[1])

11.2   Show that if $\psi = 0$, then $a_n = 0$, for all n, where $\psi = \Sigma_n a_n \varphi_n$ and $\{\varphi_n\}$ is an orthogonal sequence.

11.3   What is the matrix representation of the operator $\hat{p}_x$ in the momentum representation?

11.4   Show that the diagonal elements of $\hat{D} \equiv \partial/\partial x$, in $\mathfrak{H}_1$ (4.30), in any basis are purely imaginary.

11.5   Determine the wavefunctions $b(k)$ in the momentum representation for a particle of mass m in a homogeneous force field $\mathbf{F} = (F_0, 0, 0)$. (Compare with Problem 7.65.)

Answer

The Hamiltonian is

$$\hat{H} = \frac{p^2}{2m} - iF_0 \frac{\partial}{\partial k}$$

and the time-independent Schrödinger equation appears as

$$-iF_0 \frac{\partial b}{\partial k} + \left(\frac{\hbar^2 k^2}{2m} - E \right) b = 0$$

which has the solutions

$$b_E(k) = \frac{1}{\sqrt{2\pi F_0}} \exp\left[\frac{ik}{F_0} \left(E - \frac{\hbar^2 k^2}{6m} \right) \right]$$

These solutions obey the normalization

$$\int_{-\infty}^{\infty} b_E{}^*(k) b_{E'}(k)\, dk = \delta(E' - E)$$

---

[1] While completeness of eigenvectors is ensured for Hermitian operators in finite-dimensional spaces, this association is not guaranteed in infinite-dimensional spaces. For further discussion, see P. T. Matthews, *Introduction to Quantum Mechanics*, 2d ed., McGraw-Hill, New York, 1968.

11.2   ELEMENTARY MATRIX PROPERTIES

The following are a series of definitions and properties of matrices and operators relevant to the theory of matrix mechanics.

The Product of Two Matrices

(11.21)
$$(\hat{A}\hat{B})_{nq} = \sum_p A_{np} B_{pq}$$

As an example of matrix multiplication, consider the product of the two (finite) 2×2 matrices:

$$\begin{pmatrix} A_{11} & A_{12} \\ A_{21} & A_{22} \end{pmatrix} \begin{pmatrix} B_{11} & B_{12} \\ B_{21} & B_{22} \end{pmatrix} = \begin{pmatrix} (A_{11}B_{11} + A_{12}B_{21}) & (A_{11}B_{12} + A_{12}B_{22}) \\ (A_{21}B_{11} + A_{22}B_{21}) & (A_{21}B_{12} + A_{22}B_{22}) \end{pmatrix}$$

The Product of Two Wavefunctions

(11.22)
$$\langle \psi | \psi' \rangle = \sum_n \langle \psi | \varphi_n \rangle \langle \varphi_n | \psi' \rangle = \sum_n a_n{}^* a_n'$$
$$= \begin{pmatrix} a_1{}^* & a_2{}^* & a_3{}^* & \cdots \end{pmatrix} \begin{pmatrix} a_1' \\ a_2' \\ a_3' \\ \vdots \end{pmatrix} = a_1{}^* a_1' + a_2{}^* a_2' + \cdots$$

The Inverse of $\hat{A}$ The inverse of $\hat{A}$ is labeled $\hat{A}^{-1}$. It has the property that

(11.23)
$$\hat{A}^{-1}\hat{A} = \hat{A}\hat{A}^{-1} = \hat{I}$$

If $\hat{A}_c$ is a matrix composed of the cofactors of the elements of $\hat{A}$, then

(11.23a)
$$\hat{A}^{-1} = \frac{1}{D(A)} \tilde{\hat{A}}_c$$

where $D(A)$ is the determinant of $\hat{A}$. For example,

$$\begin{pmatrix} 1 & i \\ 2 & 4 \end{pmatrix}^{-1} = \frac{1}{4 - 2i} \begin{pmatrix} 4 & -i \\ -2 & 1 \end{pmatrix}$$

The Transpose of $\hat{A}$ The transpose of $\hat{A}$ is written $\tilde{\hat{A}}$. The matrix elements of $\tilde{\hat{A}}$ are obtained by "reflecting" the elements A_{nq} through the major diagonal of the matrix of $\hat{A}$.

(11.24)
$$(\tilde{A})_{nq} = A_{qn}$$

$\hat{A}$ *Is Symmetric or Antisymmetric*   If $\hat{A}$ is symmetric, then

(11.25)
$$\tilde{\hat{A}} = \hat{A}$$

If $\hat{A}$ is antisymmetric, then

(11.26)
$$\tilde{\hat{A}} = -\hat{A}$$

The Trace of $\hat{A}$ The trace of $\hat{A}$ is the sum over its diagonal elements. It is written

(11.27)
$$\operatorname{Tr}\hat{A} \equiv \sum_{q} A_{qq}$$

The Hermitian Adjoint of $\hat{A}$ This operator is written $\hat{A}^{\dagger}$. To construct $\hat{A}^{\dagger}$, one first forms the complex conjugate of $\hat{A}$ and then transposes.

(11.28)
$$\hat{A}^{\dagger} = \tilde{\hat{A}}*$$

Matrix elements of $\hat{A}^{\dagger}$ are given by

(11.29)
$$(A^{\dagger})_{nq} = (A_{qn})*$$

or, more explicitly,

$$\langle \varphi_{n} | \hat{A}^{\dagger} \varphi_{q} \rangle = \langle \varphi_{q} | \hat{A} \varphi_{n} \rangle * = \langle \hat{A} \varphi_{n} | \varphi_{q} \rangle$$

$\hat{A}$ *Is Hermitian*   If $\hat{A}^{\dagger} = \hat{A}$, then $\hat{A}$ is Hermitian or, equivalently, if

$$(A^{\dagger})_{nq} = A_{nq}$$

With (11.29), this definition becomes

(11.30)
$$(A_{qn})* = A_{nq}$$

or

$$\tilde{\hat{A}}* = \hat{A}$$

$\hat{U}$ *Is Unitary*   If the Hermitian adjoint $\hat{U}^{\dagger}$ of an operator $\hat{U}$ is equal to $\hat{U}^{-1}$, the inverse of $\hat{U}$, i.e.,

(11.31)
$$\hat{U}^{\dagger} = \hat{U}^{-1}$$

then $\hat{U}$ is said to be unitary. The matrix elements of $\hat{U}$ satisfy the relations

$$(U^{\dagger})_{nq} = (U^{-1})_{nq}$$

(11.32)
$$(U_{qn})* = (U^{-1})_{nq}$$
$$\tilde{\hat{U}}* = \hat{U}^{-1}$$

TABLE 11.1   Matrix properties

Matrix	Definition	Matrix Elements
Symmetric	$A = \tilde{A}$	$A_{pq} = A_{qp}$
Antisymmetric	$A = -\tilde{A}$	$A_{pp} = 0; A_{pq} = -A_{qp}$
Orthogonal	$A = \tilde{A}^{-1}$	$(\tilde{A}A)_{pq} = \delta_{pq}$
Real	$A = A^*$	$A_{pq} = A_{pq}{}^*$
Pure imaginary	$A = -A^*$	$A_{pq} = iB_{pq}; B_{pq}$ real
Hermitian	$A = A^\dagger$	$A_{pq} = A_{qp}{}^*$
Anti-Hermitian	$A = -A^\dagger$	$A_{pq} = -A_{qp}{}^*$
Unitary	$A = (A^\dagger)^{-1}$	$(A^\dagger A)_{pq} = \delta_{pq}$
Singular	$\det A = 0$	

If $\hat{U}$ is unitary, then

$$\hat{U}\hat{U}^\dagger = \hat{I}$$

(11.33)
$$(\hat{U}\hat{U}^\dagger)_{nq} = \delta_{nq}$$

$$\sum_p U_{np}(U_{qp})^* = \delta_{nq}$$

These matrix properties are summarized in Table 11.1.

PROBLEMS

11.6   Rotation of the xy axes about a fixed z axis through the angle ϕ_1 changes the components (x, y) of the vector $\mathbf{r}$ to (x', y') (see Fig. 11.1). These new components are related to the original components through the rotation matrix $\hat{R}(\phi_1)$ in the following way:

$$\mathbf{r}' = \begin{pmatrix} x' \\ y' \end{pmatrix} = \begin{pmatrix} \cos\phi_1 & \sin\phi_1 \\ -\sin\phi_1 & \cos\phi_1 \end{pmatrix} \begin{pmatrix} x \\ y \end{pmatrix} \equiv \hat{R}(\phi_1)\mathbf{r}$$

A second rotation through ϕ_2 gives

$$\mathbf{r}'' = \hat{R}(\phi_2)\mathbf{r}' = \hat{R}(\phi_2)\hat{R}(\phi_1)\mathbf{r}$$

(a)   Show that the rotation matrix $\hat{R}$ has the "group property"

$$\hat{R}(\phi_1)\hat{R}(\phi_2) = \hat{R}(\phi_1 + \phi_2)$$

(b)   Show that

$$[\hat{R}(\phi_1), \hat{R}(\phi_2)] = 0$$

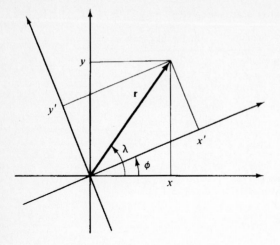

FIGURE 11.1 The rotation operator

$$\hat{R}(\phi) = \begin{pmatrix} \cos\phi & \sin\phi \\ -\sin\phi & \cos\phi \end{pmatrix}$$

is unitary,

$$\hat{R}^\dagger \hat{R} = I$$

and obeys the group property

$$\hat{R}(\phi_1)\hat{R}(\phi_2) = \hat{R}(\phi_1 + \phi_2)$$

(See Problem 11.6.)

(c) Show that $\hat{R}$ is an *orthogonal* matrix (see Table 11.1).

(d) In 3-space, rotation about the z axis is effected through the matrix

$$\hat{R} = \begin{pmatrix} \cos\phi & \sin\phi & 0 \\ -\sin\phi & \cos\phi & 0 \\ 0 & 0 & 1 \end{pmatrix}$$

Show that the eigenvalues of $\hat{R}$ have unit magnitude. (The angle λ in Fig. 11.1 remains fixed under rotation.)

11.7 Fill in the missing components of the matrix $\hat{C}$ which make $\hat{C}$ Hermitian.

$$\hat{C} = \begin{pmatrix} 1 & 3i & 4 & 2 \\ - & 2 & 7i & 3 \\ - & - & 4 & 9 \\ - & - & - & 6 \end{pmatrix}$$

11.8 Construct a 2×2 unitary matrix which has at least two imaginary elements.

11.9 What is the inverse of the matrix

$$\hat{A} = \begin{pmatrix} 2 & 3i \\ 4 & 6i \end{pmatrix}?$$

Answer

Calculation shows that $\hat{A}$ has no inverse. Under such circumstances $\hat{A}$ is said to be *singular*.

11.10 Show that if $\hat{U}$ is unitary, then the eigenvalues a_n of $\hat{U}$ are of unit magnitude.

Answer

$$\hat{U}\varphi_n = a_n\varphi_n$$

$$\langle\hat{U}\varphi_n|\hat{U}\varphi_n\rangle = \langle\varphi_n|\hat{I}\varphi_n\rangle = \langle\varphi_n|\varphi_n\rangle = a_n^*a_n\langle\varphi_n|\varphi_n\rangle$$

$$a_n^*a_n = |a_n|^2 = 1$$

11.3   UNITARY AND SIMILARITY TRANSFORMATIONS IN QUANTUM MECHANICS

The significance of unitary operators in quantum mechanics is due to the following. We are already aware of the fact that a given Hilbert space has many bases. This is similar to the fact that 3-space is spanned by one of a continuum of triad basis vectors. One can obtain a new orthogonal triad basis in 3-space through a rotation of axes about the origin. In the new basis a vector $\mathbf{V}$ has components V_x', V_y', and V_z'. The length of $\mathbf{V}$ remains the same ($\mathbf{V} \cdot \mathbf{V} = \mathbf{V}' \cdot \mathbf{V}'$). Furthermore, the angle between any two vectors remains fixed ($\mathbf{V} \cdot \mathbf{F} = \mathbf{V}' \cdot \mathbf{F}'$). The final orientation of the new Cartesian frame with respect to the old may be obtained by a single rotation about a fixed axis (through the origin). The eigenvectors of the rotation matrix lie along this axis. Any vector along this axis remains fixed during the rotation. It follows that the eigenvalues of the rotation matrix are all of unit magnitude.

The related transformation from one basis to another basis in Hilbert space is a unitary transformation. It has all the properties listed above that a rigid rotation in 3-space has. These properties are as follows (for the most part, proofs are left to the problems).

Transformation of Basis

Let the sequence $\{f_n\}$ denote a new basis. These are related to the old basis $\{\varphi_n\}$ through the unitary transformation $\hat{U}$,

(11.34) $$|f_n\rangle = \sum_p |\varphi_p\rangle\langle\varphi_p|f_n\rangle = \sum_p (U_{np})^*|\varphi_p\rangle$$

(11.35) $$(U_{np})^* = \langle\varphi_p|f_n\rangle, \qquad U_{np} = \langle f_n|\varphi_p\rangle$$

These matrix elements are the projections of old basis vectors into the new ones. (The fact that $\hat{U}$ is unitary is established in Problem 11.11.)

Transformation of the State Vector

Let us consider how the components of an arbitrary state vector ψ transform to components in the basis $\{f_n\}$. In the basis $\{f_n\}$, these components (elements of a column vector) are

$$(11.36) \qquad \psi_n' = \langle f_n | \psi \rangle$$

Taking the complex conjugate of (11.34) gives

$$(11.37) \qquad \langle f_n | = \sum_p U_{np} \langle \varphi_p |$$

Substituting into (11.36) we obtain

$$(11.38) \qquad \psi_n' = \sum_p U_{np} \langle \varphi_p | \psi \rangle = \sum_p U_{np} \psi_p$$

This is the matrix representation of the equation

$$(11.39) \qquad | \psi' \rangle = \hat{U} | \psi \rangle$$

This equation tells us how an arbitrary state vector transforms under a change of basis.

If $| \varphi \rangle$ and $| \psi \rangle$ are two arbitrary vectors in Hilbert space, then under a transformation of basis ($\hat{U}$), these vectors transform to $| \varphi' \rangle$ and $| \psi' \rangle$ according to (11.39).

$$| \varphi' \rangle = \hat{U} | \varphi \rangle$$
$$| \psi' \rangle = \hat{U} | \psi \rangle$$

Under such transformation the inner product $\langle \psi | \varphi \rangle$ is preserved.

$$\langle \psi' | \varphi' \rangle = \langle \psi | \varphi \rangle$$

Setting $| \psi \rangle = | \varphi \rangle$ gives $\langle \varphi' | \varphi' \rangle = \langle \varphi | \varphi \rangle$. Thus a unitary transformation preserves the length of vectors and the angle between vectors.

The Unitary-Similarity Transformation

Next we consider the manner in which operators transform under a change of basis. A typical quantum mechanical equation appears as

$$\hat{F} | \varphi \rangle = | \psi \rangle$$

In the new basis, the two state vectors transform according to (11.39). Multiplying these equations from the left with $\hat{U}^{-1}$ gives

$$|\varphi\rangle = \hat{U}^{-1}|\varphi'\rangle$$
$$|\psi\rangle = \hat{U}^{-1}|\psi'\rangle$$

Substituting these forms into our typical equation above gives

$$\hat{F}\hat{U}^{-1}|\varphi'\rangle = U^{-1}|\psi'\rangle$$

Multiplying from the left with $\hat{U}$, we obtain the result

$$\hat{F}'|\varphi'\rangle = |\psi'\rangle$$

where

(11.40) $$\hat{F}' = \hat{U}\hat{F}\hat{U}^{-1}$$

This transformation preserves the form of our typical equation. As a special case ($\varphi = \psi$) we see that the eigenvalue equation for $\hat{F}$ is preserved under such a transformation. Equation (11.40), which describes how an operator transforms under a change of basis, is called a *unitary-similarity transformation*. The more general class of transformations, $\hat{A} \rightarrow \hat{A}' = \hat{S}\hat{A}\hat{S}^{-1}$, where $\hat{S}$ is not necessarily unitary, are called *similarity transformations*. However, of these, the unitary-similarity transformations are more relevant to quantum mechanics.

Invariance of Eigenvalues

Since the eigenvalues of an operator corresponding to an observable are physically measurable quantities, one does not expect these values to be affected by a transformation of basis in Hilbert space. The eigenenergies of a harmonic oscillator are $\hbar\omega_0(n + \frac{1}{2})$ in all representations. In a similar vein, the eigenvalues of such Hermitian operators must be real. It follows that (1) the eigenvalues of a Hermitian operator are preserved under a unitary-similarity transformation, and (2) the Hermiticity of an operator is preserved under a unitary-similarity transformation.

PROBLEMS

11.11 Show that $\hat{U}$, with matrix elements

$$U_{np} = \langle f_n | \varphi_p \rangle$$

is unitary. The sequences $\{f_n\}$ and $\{\varphi_p\}$ are complete and orthonormal.

Answer

We must establish the property (11.32) for $\hat{U}$.

$$(\hat{U}^{-1})_{np} = (\hat{U}^{\dagger})_{np} = (U_{pn})^{*}$$

Equivalently, we must show that

$$I_{qp} = \delta_{qp} = \sum_{n} U_{qn}(U^{-1})_{np} = \sum_{n} U_{qn}(U_{pn})^{*}$$

$$= \sum_{n} \langle f_{q} | \varphi_{n} \rangle \langle \varphi_{n} | f_{p} \rangle$$

$$= \langle f_{q} | \left(\sum_{n} | \varphi_{n} \rangle \langle \varphi_{n} | \right) | f_{p} \rangle = \langle f_{q} | f_{p} \rangle = \delta_{qp}$$

11.12   Show that the inner product, $\langle \psi | \varphi \rangle$, is preserved under a unitary transformation.

Answer

$$\langle \psi' | \varphi' \rangle = \langle \hat{U}\psi | \hat{U}\varphi \rangle = \langle \psi | \hat{U}^{\dagger}\hat{U}\varphi \rangle$$
$$= \langle \psi | \hat{U}^{-1}\hat{U}\varphi \rangle = \langle \psi | \varphi \rangle$$

11.13   What is the matrix representation of the equation

$$| \varphi' \rangle = \hat{U} | \varphi \rangle$$

in the basis $\{f_{n}\}$? Write this equation explicitly, depicting elements of column and square matrices.

11.14   The matrix elements of $\hat{F}$ in the basis $\{\varphi_{n}\}$ are

$$F_{nq} = \langle \varphi_{n} | \hat{F} | \varphi_{q} \rangle$$

In the basis $\{f_{n}\}$ they are

$$F_{nq}' = \langle f_{n} | \hat{F} | f_{q} \rangle$$

Show that

$$F_{nq}' = (\hat{U}\hat{F}\hat{U}^{-1})_{nq} = \langle \varphi_{n} | \hat{U}\hat{F}\hat{U}^{-1} | \varphi_{q} \rangle$$

Answer

$$F_{nq}' = \langle f_{n} | \hat{F} | f_{q} \rangle = \sum_{r}\sum_{p} \langle f_{n} | \varphi_{r} \rangle \langle \varphi_{r} | \hat{F} | \varphi_{p} \rangle \langle \varphi_{p} | f_{q} \rangle$$

$$= \sum_{r}\sum_{p} U_{nr} F_{rp}(U_{qp})^{*} = \sum_{r}\sum_{p} U_{nr} F_{rp}(U^{\dagger})_{pq}$$

$$= \sum_{r}\sum_{p} U_{nr} F_{rp}(U^{-1})_{pq} = (\hat{U}\hat{F}\hat{U}^{-1})_{nq}$$

11.15   Let $\hat{A}' = \hat{U}\hat{A}\hat{U}^{-1}$, where $\hat{U}$ is unitary. Show that this transformation may be rewritten $\hat{A}' = \hat{T}^{-1}\hat{A}\hat{T}$, where $\hat{T}$ is unitary. (*Note:* It follows that both $\hat{U}\hat{A}\hat{U}^{-1}$ and $\hat{U}^{-1}\hat{A}\hat{U}$ represent unitary-similarity transformations. For demonstrating certain properties of the unitary-similarity transformation, it may prove more convenient to work with the form $\hat{U}^{-1}\hat{A}\hat{U}$.)

11.16   (a) Show that $\hat{A}$ and $\hat{U}^{-1}\hat{A}\hat{U}$ have the same eigenvalues. Must $\hat{U}$ be unitary for this to be true?

   (b) If the eigenvectors of $\hat{A}$ are $\{\varphi_n\}$, what are the eigenvectors of $\hat{U}^{-1}\hat{A}\hat{U}$?

11.17   If $\hat{U}$ is unitary and $\hat{A}$ is Hermitian, then show that $\hat{U}\hat{A}\hat{U}^{-1}$ is also Hermitian. That is, show that the Hermitian quality of an operator is preserved under a unitary-similarity transformation.

11.18   Show that the form of the operator equation

$$\hat{G} = \hat{A}\hat{B}$$

is preserved under a similarity transformation.

11.19   Consider the following decomposition of an arbitrary unitary operator $\hat{U}$:

$$\hat{U} = \frac{\hat{U} + \hat{U}^{\dagger}}{2} + i\frac{\hat{U} - \hat{U}^{\dagger}}{2i} \equiv \hat{A} + i\hat{B}$$

   (a)   Show that $\hat{A}$ and $\hat{B}$ are Hermitian.
   (b)   Show that $[\hat{A}, \hat{B}] = [\hat{A}, \hat{U}] = [\hat{B}, \hat{U}] = 0$.
   (c)   From part (b) we may conclude that $\hat{A}$, $\hat{B}$, and $\hat{U}$ have common eigenfunctions. Call them $|ab\rangle$. Use these eigenstates to show that the eigenvalues of $\hat{U}$ have unit magnitude.

11.20   (a) Show that diagonal matrices commute.
   (b) Let $A_{ik} = a_i\delta_{ik}$, $B_{jl} = b_l\delta_{jl}$, and $C_{nm} = c_n\delta_{nm}$ be three matrices. What are the components of ABC?
   (c)   Again consider the diagonal matrix $A_{ik} = a_i\delta_{ik}$. What is the matrix representation of $\exp \hat{A}$? What is the matrix representation of $\sin \hat{A}$?

11.21   If $\hat{A}$, $\hat{B}$, and $\hat{C}$ are three $n \times n$ square matrices, show that

$$\mathrm{Tr}\,(\hat{A}\hat{B}\hat{C}) = \mathrm{Tr}\,(\hat{C}\hat{A}\hat{B}) = \mathrm{Tr}\,(\hat{B}\hat{C}\hat{A})$$

11.22   Let $\hat{A}$ and $\hat{B}$ be two $n \times n$ square matrices. Employ the following property of determinants,[1]

$$\det \hat{A}\hat{B} = \det \hat{A} \det \hat{B}$$

to show that

$$\det \hat{A}\hat{B} = \det \hat{B}\hat{A}$$

---

[1] It is assumed that the student is familiar with the concept of a determinant. However, a definition of determinants may be found in Section 12.5. For further discussion, see G. Birkoff and S. MacLane, *A Survey of Modern Algebra*, Macmillan, New York, 1953.

11.23 A property of a matrix $\hat{A}$ which remains the same under a unitary transformation $\hat{A} \rightarrow \hat{U}\hat{A}\hat{U}^{-1}$ is called an *invariant*. Show that the trace of $\hat{A}$ is an invariant. That is, show that

$$\mathrm{Tr}\,\hat{A} = \mathrm{Tr}\,\hat{U}\hat{A}\hat{U}^{-1}$$

In your proof, establish that

$$\mathrm{Tr}\,\hat{A} = \sum_n a_n$$

where a_n are the eigenvalues of $\hat{A}$.

Answer (partial)

Let $\hat{U}$ diagonalize $\hat{A}$ so that the diagonal matrix

$$\hat{U}\hat{A}\hat{U}^{-1} = \hat{A}'$$

is comprised of the eigenvalues of $\hat{A}$. With Problem 11.21, we have

$$\sum_n a_n = \mathrm{Tr}\,\hat{A}' = \mathrm{Tr}\,\hat{U}\hat{A}\hat{U}^{-1} = \mathrm{Tr}\,\hat{A}\hat{U}^{-1}\hat{U} = \mathrm{Tr}\,\hat{A}$$

11.24 Let $\hat{A}$ be an $n \times n$ square matrix with eigenvalues $a_1, a_2, \ldots a_n$. Show that

$$\det \hat{A} = a_1 a_2 \cdots a_n$$

(*Hint:* Let $\hat{U}$ diagonalize $\hat{A}$ and refer to Problem 11.22.)

Note: This problem establishes that $\det \hat{A}$ is another invariant property of $\hat{A}$. Together with the trace, these two fundamental properties appear as

$$\mathrm{Tr}\,\hat{A} = \sum_i a_i$$

$$\det \hat{A} = \prod_i a_i$$

In general, an $n \times n$ matrix has n invariants, two of which are the trace and determinant. These n invariants are the coefficients of the characteristic equation for the eigenvalues of $\hat{A}$ [i.e., $\det (\hat{A} - \hat{I}a) = 0$], which itself is invariant. Namely, $\det (\hat{A} - \hat{I}a) = \det [\hat{U}(\hat{A} - \hat{I}a)\hat{U}^{-1}] = \det (\hat{U}\hat{A}\hat{U}^{-1} - \hat{I}a)$.

11.25 Let $\hat{A}$ be a Hermitian $n \times n$ matrix. Let the column vectors of the $n \times n$ matrix $\hat{S}$ be comprised of the orthonormalized eigenvectors of $\hat{A}$.

    (a)   Show that $\hat{S}$ is unitary.

    (b)   Show that $\hat{S}^{-1}\hat{A}\hat{S}$ is a diagonal matrix comprised of the eigenvalues of $\hat{A}$.

(*Note:* This establishes that a Hermitian matrix is always diagonalizable by a unitary-similarity transformation.)

11.26 Again, let $\hat{A}$ be a Hermitian $n \times n$ matrix. However, now let the column vectors of the $n \times n$ matrix $\hat{T}$ be comprised of the unnormalized, but still orthogonal eigenvectors of $\hat{A}$.

    (a)   Is $\hat{T}$ unitary?

    (b)   Is $\hat{T}^{-1}\hat{A}\hat{T} = \hat{A}'$ diagonal? If so, what are the elements of $\hat{A}'$?

    (c)   Is the inner product between two n-dimensional column vectors preserved under this transformation?

Answers

(a)   We note that although $\hat{T}^\dagger \hat{T}$ is diagonal, it is not the unit operator, so that $\hat{T}$ is not unitary.

(b)   Let the eigenvector equation for $\hat{A}$ be written

$$\hat{A}|n\rangle = a_n|n\rangle$$

It follows that the column vectors of $\hat{A}\hat{T}$ are $a_1|1\rangle, a_2|2\rangle, \ldots, a_n|n\rangle$. This matrix may be rewritten

$$\hat{A}\hat{T} = \hat{T}\hat{A}'$$

where the diagonal matrix $\hat{A}'$ is comprised of the eigenvalues of $\hat{A}$. We then have

$$\hat{T}^{-1}\hat{A}\hat{T} = \hat{A}'$$

(*Note:* Although the similarity transformation described in this problem diagonalizes $\hat{A}$ and yields a diagonal matrix comprised of the eigenvalues of $\hat{A}$, it is not a unitary-similarity transformation and therefore is not relevant to quantum mechanics. Changes in representations in quantum mechanics must preserve the inner product between state vectors, which in turn ensures preservation of the Hermiticity of operators. These invariances are maintained in a unitary-similarity transformation.)

11.27   In the Schrödinger description of quantum mechanics, the wavefunction evolves in time according to the equation (3.70)

$$\psi(\mathbf{r}, t) = e^{-i\hat{H}t/\hbar}\psi(\mathbf{r}, 0)$$

(a)   Show that the operator

$$\hat{U} = \exp\left(\frac{-i\hat{H}t}{\hbar}\right)$$

is unitary. (*Hint:* Use the property $\hat{H}^\dagger = \hat{H}$.)

(b)   Having shown that $|\psi(t)\rangle = \hat{U}|\psi(0)\rangle$, show that the normalization of ψ, $\langle\psi(t)|\psi(t)\rangle$, is constant.

[*Note:* In this description the state of the system is represented by a vector $|\psi(\mathbf{r}, t)\rangle$, which migrates in Hilbert space according to the unitary transformation above. This behavior is opposed to that of eigenvectors corresponding to observables (e.g., $\hat{L}^2$, $\hat{H}$, $\hat{p}_x$, etc.). These are fixed in the Hilbert space.]

11.28   Show that if $\hat{A}$ is Hermitian, then

$$\hat{U} = (\hat{A} + i\hat{I})(\hat{A} + i\hat{I})^{-1}$$

is unitary. [*Hint:* Multiply from the right with $(\hat{A} + i\hat{I})$.]

11.29   Show that if the unitary operator $\hat{U}$ does not have the eigenvalue 1, then

$$\hat{A} \equiv i(\hat{I} + \hat{U})(\hat{I} - \hat{U})^{-1}$$

is Hermitian.

11.30   Consider two Hermitian operators $\hat{A}$ and $\hat{B}$, which satisfy the commutation relation, $[\hat{A}, \hat{B}] = i\hbar$. Suppose a system is in an eigenstate $|a\rangle$ of $\hat{A}$. What can be said of the probability distribution relating to $\hat{B}$ (i.e., $|\langle a|b\rangle|^2$)? Does your argument apply to the observables ϕ and L_z? (Recall Problem 9.15.)

11.4   THE ENERGY REPRESENTATION

One-Dimensional Box

In the energy representation, the Hamiltonian is diagonal. This representation includes a basis comprised of the eigenfunctions of the Hamiltonian. For a particle in a one-dimensional box, the basis in which $\hat{H}$ is diagonal is

$$(11.41) \qquad \mathfrak{B} = \sqrt{\frac{2}{a}}\left\{\sin\frac{\pi x}{a}, \; \sin\frac{2\pi x}{a}, \; \sin\frac{3\pi x}{a}, \; \ldots\right\}$$

The Hamiltonian matrix in this representation is

$$(11.42) \qquad \hat{H} = E_1 \begin{pmatrix} 1 & & & & & & \\ & 4 & & & & 0 & \\ & & 9 & & & & \\ & & & 16 & & & \\ & & & & \cdot & & \\ 0 & & & & & n^2 & \\ & & & & & & \cdot \end{pmatrix}$$

Simple Harmonic Oscillator

For the one-dimensional harmonic oscillator, the basis that diagonalizes $\hat{H}$ is [recall (7.59)]

$$(11.43) \qquad \begin{aligned} \mathfrak{B} &= e^{-\xi^2/2}\{A_1 \mathcal{H}_1(\xi), \; A_2 \mathcal{H}_2(\xi), \; \ldots\} \\ &\equiv \{|0\rangle, \; |1\rangle, \; |2\rangle, \; \ldots\} \end{aligned}$$

$$\xi^2 \equiv \beta^2 x^2, \qquad \beta^2 \equiv \frac{m\omega_0}{\hbar}$$

The nth-order Hermite polynomial is written $\mathcal{H}_n(\xi)$. The Hamiltonian matrix in this representation is

$$(11.44) \qquad \hat{H} = \hbar\omega_0 \begin{pmatrix} 1/2 & & & & \\ & 3/2 & & & \mathbf{0} \\ & & \ddots & & \\ & & & \ddots & \\ \mathbf{0} & & & & (2n+1)/2 \\ & & & & & \ddots \end{pmatrix}$$

Position and Momentum Operators

Let us calculate the matrix representation of the position operator $\hat{x}$ for the harmonic oscillator in the energy representation. Recalling (7.61) and with k and n representing nonnegative integers, we have

$$(11.45) \qquad \langle n|\hat{x}|k\rangle = \frac{1}{\sqrt{2}\,\beta}\langle n|\hat{a}+\hat{a}^\dagger|k\rangle$$

$$= \frac{1}{\sqrt{2}\,\beta}[k^{1/2}\delta_{n,k-1} + (k+1)^{1/2}\delta_{n,k+1}]$$

This gives the matrix

$$(11.46) \qquad \hat{x} = \frac{1}{\sqrt{2}\,\beta} \begin{pmatrix} 0 & \sqrt{1} & 0 & 0 & 0 \\ \sqrt{1} & 0 & \sqrt{2} & 0 & 0 \\ 0 & \sqrt{2} & 0 & \sqrt{3} & 0 & \cdots \\ 0 & 0 & \sqrt{3} & 0 & \sqrt{4} \\ 0 & 0 & 0 & \sqrt{4} & 0 \\ & & & & & \ddots \end{pmatrix}$$

For the momentum operator $\hat{p}$, we find that

$$(11.47) \qquad \langle n|\hat{p}|k\rangle = \frac{m\omega_0}{\sqrt{2}\,i\beta}[k^{1/2}\delta_{n,k-1} - (k+1)^{1/2}\delta_{n,k+1}]$$

which gives the matrix

$$(11.48) \qquad \hat{p} = \frac{m\omega_0}{\sqrt{2}\,i\beta} \begin{pmatrix} 0 & \sqrt{1} & 0 & 0 & 0 \\ -\sqrt{1} & 0 & \sqrt{2} & 0 & 0 \\ 0 & -\sqrt{2} & 0 & \sqrt{3} & 0 & \cdots \\ 0 & 0 & -\sqrt{3} & 0 & \sqrt{4} \\ & & & & & \ddots \end{pmatrix}$$

Creation and Annihilation Operators

For the creation and annihilation operators we have [recall (7.61)]

(11.49)
$$a_{nk} = \langle n|\hat{a}|k\rangle = k^{1/2}\langle n|k - 1\rangle = k^{1/2}\delta_{n,k-1}$$
$$a_{nk}^{\dagger} = \langle n|\hat{a}^{\dagger}|k\rangle = (k + 1)^{1/2}\langle n|k + 1\rangle = (k + 1)^{1/2}\delta_{n,k+1}$$

which gives the matrices

$$\hat{a} = \begin{pmatrix} 0 & \sqrt{1} & 0 & 0 & 0 \\ 0 & 0 & \sqrt{2} & 0 & 0 & \cdots \\ 0 & 0 & 0 & \sqrt{3} & 0 \\ & & & \vdots \end{pmatrix}$$

(11.50)

$$\hat{a}^{\dagger} = \begin{pmatrix} 0 & 0 & 0 & 0 & 0 \\ \sqrt{1} & 0 & 0 & 0 & 0 \\ 0 & \sqrt{2} & 0 & 0 & 0 & \cdots \\ 0 & 0 & \sqrt{3} & 0 & 0 \\ & & & \vdots \end{pmatrix}$$

Let us check that these matrix operators promote and demote according to (7.61). The eigenfunctions $\{|n\rangle\}$ for the harmonic oscillator Hamiltonian are column vectors with the only nonzero entry in the $(n + 1)$st slot.

(11.51)
$$|0\rangle = \begin{pmatrix} 1 \\ 0 \\ 0 \\ 0 \\ 0 \\ 0 \\ 0 \\ \vdots \end{pmatrix}, \quad |1\rangle = \begin{pmatrix} 0 \\ 1 \\ 0 \\ 0 \\ 0 \\ 0 \\ 0 \\ \vdots \end{pmatrix}, \quad |2\rangle = \begin{pmatrix} 0 \\ 0 \\ 1 \\ 0 \\ 0 \\ 0 \\ 0 \\ \vdots \end{pmatrix}, \cdots$$

The time-dependent eigenstates of $\hat{H}$ appear as

$$\psi_0(x, t) = e^{-i\hat{H}t/\hbar}|0\rangle = e^{-i\omega_0 t/2}\langle 0|0\rangle = e^{-i\omega_0 t/2}\begin{pmatrix} 1 \\ 0 \\ 0 \\ 0 \\ 0 \\ 0 \\ \vdots \end{pmatrix}$$

$$(11.52) \quad \psi_1(x, t) = e^{-i3\omega_0 t/2} \begin{pmatrix} 0 \\ 1 \\ 0 \\ 0 \\ 0 \\ 0 \\ \vdots \end{pmatrix}, \qquad \psi_2(x, t) = e^{-i5\omega_0 t/2} \begin{pmatrix} 0 \\ 0 \\ 1 \\ 0 \\ 0 \\ 0 \\ \vdots \end{pmatrix}$$

Consider the operations $\hat{a}|2\rangle$ and $\hat{a}^\dagger|2\rangle$.

$$\hat{a}|2\rangle = \begin{pmatrix} 0 & \sqrt{1} & 0 & 0 & 0 & 0 \\ 0 & 0 & \sqrt{2} & 0 & 0 & 0 \\ 0 & 0 & 0 & \sqrt{3} & 0 & 0 \\ 0 & 0 & 0 & 0 & \sqrt{4} & 0 \\ & & & \vdots \end{pmatrix} \cdots \begin{pmatrix} 0 \\ 0 \\ 1 \\ 0 \\ 0 \\ \vdots \end{pmatrix} = \sqrt{2} \begin{pmatrix} 0 \\ 1 \\ 0 \\ 0 \\ 0 \\ \vdots \end{pmatrix} = \sqrt{2}|1\rangle$$

(11.53)

$$\hat{a}^\dagger|2\rangle = \begin{pmatrix} 0 & 0 & 0 & 0 & 0 \\ \sqrt{1} & 0 & 0 & 0 & 0 \\ 0 & \sqrt{2} & 0 & 0 & 0 \\ 0 & 0 & \sqrt{3} & 0 & 0 \\ 0 & 0 & 0 & \sqrt{4} & 0 \\ & & \vdots \end{pmatrix} \cdots \begin{pmatrix} 0 \\ 0 \\ 1 \\ 0 \\ 0 \\ \vdots \end{pmatrix} = \sqrt{3} \begin{pmatrix} 0 \\ 0 \\ 0 \\ 1 \\ 0 \\ \vdots \end{pmatrix} = \sqrt{3}|3\rangle$$

These equations very simply illustrate the promotion and demotion properties of the $\hat{a}^\dagger$ and $\hat{a}$ operators.

The Number Operator

In addition to the Hamiltonian (7.27)

$$\hat{H} = \hbar\omega_0 \left(\hat{a}^\dagger \hat{a} + \frac{1}{2} \right)$$

the number operator (7.28)

$$\hat{N} = \hat{a}^\dagger \hat{a}$$

is also diagonal in the energy representation.

$$\hat{N} = \begin{pmatrix} 0 & 0 & 0 & 0 \\ \sqrt{1} & 0 & 0 & 0 \\ 0 & \sqrt{2} & 0 & 0 \\ 0 & 0 & \sqrt{3} & 0 \\ & & & \vdots \end{pmatrix} \begin{pmatrix} 0 & \sqrt{1} & 0 & 0 & 0 \\ 0 & 0 & \sqrt{2} & 0 & 0 \\ 0 & 0 & 0 & \sqrt{3} & 0 \\ 0 & 0 & 0 & 0 & \sqrt{4} \\ & & \vdots & & \end{pmatrix} \cdots$$

(11.54)

$$= \begin{pmatrix} 0 & & & & & \\ & 1 & & & \mathbf{0} & \\ & & 2 & & & \\ & & & 3 & & \\ & & & & \ddots & \\ \mathbf{0} & & & & & n \\ & & & & & & \ddots \end{pmatrix}$$

The reader may readily check that the column vectors $\{|n\rangle\}$, as given by (11.51), are eigenvectors of both $\hat{H}$ as given by (11.44) and $\hat{N}$ as given by (11.54), with respective eigenvalues $\{\hbar\omega_0(n + \frac{1}{2})\}$ and $\{n\}$.

PROBLEMS

11.31 In Section 5.5 the importance of complete sets of commuting observables was discussed. The number of such variables (*good quantum numbers*, see Section 1.3) are analogous to the number of *canonical variables* relevant to the description of a classical system. It is important in classical physics that the number of such variables be preserved under a *canonical transformation*. In quantum mechanics it is equally significant that the number of operators comprising complete sets of commuting observables be preserved under a unitary transformation.

(a) Let such a set of compatible operators be $\hat{A}$, $\hat{B}$, $\hat{C}$, and $\hat{D}$. Show that this compatibility is preserved under a unitary transformation.

(b) Let $\hat{F}$ not commute with any element in the set $\hat{A}$, $\hat{B}$, $\hat{C}$, $\hat{D}$. Is this property preserved under a unitary transformation?

11.32 Show that

$$\det (\hat{I} + \epsilon\hat{A}) = 1 + \epsilon \operatorname{Tr} \hat{A} + O(\epsilon^2)$$

where $\hat{A}$ is an $n \times n$ matrix and $\hat{I}$ is the identity matrix in n dimensions.

11.33 Show that

$$\det (\exp \hat{A}) = \exp (\operatorname{Tr} \hat{A})$$

where $\hat{A}$ is an $n \times n$ matrix.

Answer

This equality may be established in two independent ways. In the first method we let $D(\lambda) \equiv \det [\exp (\lambda \hat{A})]$. Then with Problem 11.22 we obtain

$$\frac{dD}{d\lambda} = D \lim_{\epsilon \to 0} \left\{ \frac{\det [\exp (\epsilon \hat{A})] - 1}{\epsilon} \right\}$$

In the limit that ϵ goes to zero,

$$\det [\exp (\epsilon \hat{A})] = \det (1 + \epsilon \hat{A}) = 1 + \epsilon \operatorname{Tr} \hat{A}$$

where we have used the results of Problem 11.32 in establishing the second equality. It follows that

$$\frac{dD}{d\lambda} = D \operatorname{tr} \hat{A}$$

which has the solution

$$D(\lambda) = D(0) \exp (\lambda \operatorname{Tr} \hat{A})$$

But $D(0) = 1$, hence

$$D(1) = \det (\exp \hat{A}) = \exp (\operatorname{Tr} \hat{A})$$

In the second way we first note that if $\hat{U}$ diagonalizes $\hat{A}$, it also diagonalizes $e^{\hat{A}}$. Furthermore, $\det [\hat{U}(\exp \hat{A})\hat{U}^{-1}] = \det \exp \hat{A}$ (see Problem 11.23).

Now the diagonal matrix $U(\exp \hat{A})U^{-1}$ has the explicit form

$$\hat{U}(\exp \hat{A})\hat{U}^{-1} = \begin{pmatrix} e^{a_1} & 0 & 0 & \\ 0 & e^{a_2} & 0 & \\ 0 & 0 & e^{a_3} & \cdots \\ & & & \vdots \end{pmatrix}$$

where $\{a_i\}$ are the eigenvalues of A. The determinant of this matrix is

$$\det [U(\exp \hat{A})\hat{U}^{-1}] = e^{a_1}e^{a_2} \cdots = \exp \left(\sum_i a_i \right)$$

This is the value (in all representations) of the left-hand side of the equality to be established. For the right-hand side we recall that the trace is independent of representations (Problem 11.23) so that

$$\exp (\operatorname{Tr} \hat{A}) = \exp (\operatorname{Tr} \hat{A}') = \exp \left(\sum_i a_i \right) \qquad \text{Q.E.D.}$$

The matrix $\hat{A}'$ is the diagonal representation of $\hat{A}$.

11.34   Use the matrix representation (11.46) and (11.48) for $\hat{x}$ and $\hat{p}$ to obtain the matrix representation for the commutator $[\hat{x}, \hat{p}]$ for the harmonic oscillator in the energy representation.

11.35   Calculate the matrix representations for $\hat{x}^2$ and $\hat{p}^2$ for the harmonic oscillator in the energy representation.

11.36   Using the fact that any Hermitian matrix can be diagonalized by a unitary matrix, show that two Hermitian matrices, $\hat{A}$ and $\hat{B}$, can be diagonalized by the same unitary transformation $\hat{U}$ if and only if $[\hat{A}, \hat{B}] = 0$.

11.37   Consider the following equations:

$$\hat{A}^2 = 0, \qquad \hat{A}\hat{A}^\dagger + \hat{A}^\dagger\hat{A} = \hat{I}, \qquad \hat{B} = \hat{A}^\dagger\hat{A}$$

(a)   Show that $\hat{B}^2 = \hat{B}$.
(b)   Obtain explicit (2×2) matrices for $\hat{A}$ and $\hat{B}$.

Answer (partial)

(b)
$$\hat{A} = \frac{1}{2}\begin{pmatrix} 1 & i \\ i & -1 \end{pmatrix}$$

11.5   ANGULAR MOMENTUM MATRICES

The $\hat{L}$ Matrices

It was shown above that the matrix representation of an operator $\hat{A}$ in the basis consisting of the eigenvectors of $\hat{A}$, is diagonal. In Chapter 9 we found that the eigenfunctions of the angular momentum operators $\hat{L}^2$ and $\hat{L}_z$ are the spherical harmonics $Y_l^m(\theta, \phi)$. It follows that in the basis $\mathfrak{B} = \{Y_l^m\}$, the matrices $L^2_{lm, l'm'}$ and the matrices $(L_z)_{lm, l'm'}$ are diagonal. That is,

$$(11.55) \qquad L^2_{lm, l'm'} = \langle lm | \hat{L}^2 | l'm' \rangle = \int_{-1}^{1}\int_{0}^{2\pi} d\cos\theta \, d\phi (Y_l^m)^* \hat{L}^2 Y_{l'}^{m'}$$

$$= \hbar^2 l(l+1)\delta_{ll'}\delta_{mm'}$$

$$(11.56) \qquad (L_z)_{lm, l'm'} = \langle lm | \hat{L}_z | l'm' \rangle = \hbar m \delta_{ll'}\delta_{mm'}$$

The manner in which these elements are displayed is as follows. The rows and columns of a given matrix are ordered so that for every value of l, m_l runs from $(-l, \ldots, +l)$. For each of these m_l values, l is fixed. The diagonal matrix for $\hat{L}^2$ appears as

(11.57)

	$l\downarrow$	$m\downarrow$									
$l' \rightarrow$			0	1	1	1	2	2	2	2	2
$m' \rightarrow$			0	1	0	−1	2	1	0	−1	−2
	0	0	0		0				0		
	1	1		2	0	0					
	1	0	0	0	2	0			0		
	1	−1		0	0	2					
$L^2 = \hbar^2$	2	2					6	0	0	0	0
	2	1					0	6	0	0	0
	2	0	0		0		0	0	6	0	0
	2	−1					0	0	0	6	0
	2	−2					0	0	0	0	6

In this same scheme, we obtain the following diagonal matrix for $\hat{L}_z$:

$$(11.58) \quad L_z = \hbar \begin{pmatrix} 0 & & & & & & & & \\ & 1 & & & & & & & \\ & & 0 & & & & & & \\ & & & -1 & & & & & \\ & & & & 2 & & & & \\ & & & & & 1 & & & \\ & & & & & & -1 & & \\ & & & & & & & -2 & \\ & & & & & & & & \ddots \end{pmatrix}$$

To obtain the matrices for $\hat{L}_x$ and $\hat{L}_y$ in the representation in which $\hat{L}^2$ and $\hat{L}_z$ are diagonal, we first construct the matrices for the ladder operators $\hat{L}_+$ and $\hat{L}_-$ (see Section 9.2). Since $\hat{L}_- = \hat{L}_+^{\dagger}$, one merely needs to calculate the $\hat{L}_+$ matrix and then, from its Hermitian adjoint, find $\hat{L}_-$. Once these matrices are known, $\hat{L}_x$ and $\hat{L}_y$ are obtained from

$$\hat{L}_x = \tfrac{1}{2}(\hat{L}_+ + \hat{L}_-)$$

(11.59)

$$\hat{L}_y = -\frac{i}{2}(\hat{L}_+ - \hat{L}_-)$$

Using the relation (see Table 9.4)

(11.60) $$\hat{L}_- Y_l^m = [(l + m)(l - m + 1)]^{1/2}\hbar Y_l^{m-1}$$

one obtains

(11.61) $$(L_\pm)_{lm, l'm'} = [(l' \mp m')(l' \pm m' + 1)]^{1/2}\hbar\delta_{ll'}\delta_{m, m'\pm 1}$$

and the matrices [exhibiting only the $l \leq 2$ terms)

$$L_- = \hbar$$

0		0				0			
	0	0	0	0					
0	0	$\sqrt{2}$	0	0		0			
	0	0	$\sqrt{2}$	0					
					0	0	0	0	0
					2	0	0	0	0
0		0			0	$\sqrt{6}$	0	0	0
					0	0	$\sqrt{6}$	0	0
					0	0	0	2	0

(11.62)

$$L_+ = \hbar$$

0		0				0			
	0	$\sqrt{2}$	0						
0	0	0	$\sqrt{2}$			0			
	0	0	0						
					0	2	0	0	0
					0	0	$\sqrt{6}$	0	0
0		0			0	0	0	$\sqrt{6}$	0
					0	0	0	0	2
					0	0	0	0	0

Adding and subtracting these two matrices according to (11.59) gives (again exhibiting only the $l \leq 2$ terms)

$$L_x = \frac{\hbar}{2}
\begin{array}{|c|ccc|ccccc|}
\hline
0 & & 0 & & & & 0 & & \\
\hline
  & 0 & \sqrt{2} & 0 & & & & & \\
0 & \sqrt{2} & 0 & \sqrt{2} & & & 0 & & \\
  & 0 & \sqrt{2} & 0 & & & & & \\
\hline
  & & & & 0 & 2 & 0 & 0 & 0 \\
  & & & & 2 & 0 & \sqrt{6} & 0 & 0 \\
0 & & 0 & & 0 & \sqrt{6} & 0 & \sqrt{6} & 0 \\
  & & & & 0 & 0 & \sqrt{6} & 0 & 2 \\
  & & & & 0 & 0 & 0 & 2 & 0 \\
\hline
\end{array}$$

(11.63)

$$L_y = \frac{\hbar}{2i}
\begin{array}{|c|ccc|ccccc|}
\hline
0 & & 0 & & & & 0 & & \\
\hline
  & 0 & \sqrt{2} & 0 & & & & & \\
0 & -\sqrt{2} & 0 & \sqrt{2} & & & 0 & & \\
  & 0 & -\sqrt{2} & 0 & & & & & \\
\hline
  & & & & 0 & 2 & 0 & 0 & 0 \\
  & & & & -2 & 0 & \sqrt{6} & 0 & 0 \\
0 & & 0 & & 0 & -\sqrt{6} & 0 & \sqrt{6} & 0 \\
  & & & & 0 & 0 & -\sqrt{6} & 0 & 2 \\
  & & & & 0 & 0 & 0 & -2 & 0 \\
\hline
\end{array}$$

Next we consider the matrix representation of the eigenvectors of $\hat{L}^2$ and $\hat{L}_z$. These are column vectors whose elements are the coefficients of expansion of Y_l^m in the basis $\{Y_{l'}^{m'}\}$.

$$Y_l^m = \sum_{l'} \sum_{m'=-l'}^{l'} a_{lm,l'm'} Y_{l'}^{m'}$$

(11.64)

$$a_{lm,l'm'} = \delta_{ll'}\delta_{mm'}$$

The elements of these column vectors have zero entries for all values of l, m except at $l = l'$, $m = m'$, where the entry is unity. For example, the representations of the $l = 1$ eigenstates are [entries in these vectors follow the scheme of (11.57)]

$$(11.65) \quad Y_1^1 \rightarrow \begin{pmatrix} 0 \\ 1 \\ 0 \\ 0 \\ 0 \\ \vdots \end{pmatrix}, \qquad Y_1^0 \rightarrow \begin{pmatrix} 0 \\ 0 \\ 1 \\ 0 \\ 0 \\ \vdots \end{pmatrix}, \qquad Y_1^{-1} \rightarrow \begin{pmatrix} 0 \\ 0 \\ 0 \\ 1 \\ 0 \\ \vdots \end{pmatrix}$$

When the matrix for $\hat{L}^2$ operates on any of these three column vectors it gives $2\hbar^2$ times the vector. When $\hat{L}_z$ operates on them, it gives the respective values $(\hbar, 0, -\hbar)$.

Sub L Matrices

One often speaks of the submatrices of $\hat{L}^2$, $\hat{L}_z$, ... corresponding to a given value of l. For example, the $\hat{L}^2$ matrix for $l = 1$ is

$$\hat{L}^2 = \hbar^2 \begin{pmatrix} 2 & 0 & 0 \\ 0 & 2 & 0 \\ 0 & 0 & 2 \end{pmatrix}$$

while the $\hat{L}_x$, $\hat{L}_y$, and L_z matrices corresponding to $l = 1$ are

(11.66)

$$\hat{L}_x = \frac{\hbar}{\sqrt{2}} \begin{pmatrix} 0 & 1 & 0 \\ 1 & 0 & 1 \\ 0 & 1 & 0 \end{pmatrix}, \quad \hat{L}_y = \frac{\hbar}{\sqrt{2}} \begin{pmatrix} 0 & -i & 0 \\ i & 0 & -i \\ 0 & i & 0 \end{pmatrix}, \quad \hat{L}_z = \hbar \begin{pmatrix} 1 & 0 & 0 \\ 0 & 0 & 0 \\ 0 & 0 & -1 \end{pmatrix}$$

The $\hat{L}_x$, $\hat{L}_y$, and $\hat{L}_z$ matrices corresponding to $l = 2$ are

$$\hat{L}_x = \frac{\hbar}{2} \begin{pmatrix} 0 & 2 & 0 & 0 & 0 \\ 2 & 0 & \sqrt{6} & 0 & 0 \\ 0 & \sqrt{6} & 0 & \sqrt{6} & 0 \\ 0 & 0 & \sqrt{6} & 0 & 2 \\ 0 & 0 & 0 & 2 & 0 \end{pmatrix}, \quad \hat{L}_y = \frac{\hbar}{2} \begin{pmatrix} 0 & -i2 & 0 & 0 & 0 \\ i2 & 0 & -i\sqrt{6} & 0 & 0 \\ 0 & i\sqrt{6} & 0 & -i\sqrt{6} & 0 \\ 0 & 0 & i\sqrt{6} & 0 & -i2 \\ 0 & 0 & 0 & i2 & 0 \end{pmatrix},$$

(11.67)

$$\hat{L}_z = \hbar \begin{pmatrix} 2 & 0 & 0 & 0 & 0 \\ 0 & 1 & 0 & 0 & 0 \\ 0 & 0 & 0 & 0 & 0 \\ 0 & 0 & 0 & -1 & 0 \\ 0 & 0 & 0 & 0 & -2 \end{pmatrix}$$

We may consider the eigenvectors corresponding to these matrices. These are also subcomponents of the infinitely dimensional column vectors (11.64). For example, the eigenvectors of $\hat{L}_x$ for the case $l = 1$ appear as

$$\xi_x^0 = \frac{1}{\sqrt{2}} \begin{pmatrix} 1 \\ 0 \\ -1 \end{pmatrix}, \qquad \hat{L}_x \xi_x^0 = 0\hbar\xi_x^0$$

(11.68)

$$\xi_x^{-1} = \frac{1}{2} \begin{pmatrix} 1 \\ -\sqrt{2} \\ 1 \end{pmatrix}, \qquad \hat{L}_x \xi_x^{-1} = -\hbar\xi_x^{-1}$$

$$\xi_x^1 = \frac{1}{2} \begin{pmatrix} 1 \\ \sqrt{2} \\ 1 \end{pmatrix}, \qquad \hat{L}_x \xi_x^1 = +\hbar\xi_x^1$$

In the representation where $\hat{L}_x$ is the differential operator [recall (9.56)]

$$L_x = i\hbar\left(\sin\phi\,\frac{\partial}{\partial\theta} + \cot\theta\cos\phi\frac{\partial}{\partial\phi}\right)$$

the eigenvector ξ_x^1 corresponds to the linear combination of spherical harmonics

$$\xi_x^1 = \frac{1}{2}(Y_1^1 + \sqrt{2}\,Y_1^0 + Y_1^{-1})$$

which was previously labeled X_+ in (9.100).

The $\hat{\mathbf{J}}$ Matrices

In the preceding construction of the matrices for the angular momentum operators $\hat{\mathbf{L}}$ and $\hat{L}^2$, it proved convenient to work from the $Y_l^m(\theta, \phi)$ eigenstates relevant to the coordinate representation. As we recall from Chapter 9, the more inclusive angular momentum, $\hat{\mathbf{J}}$ (which may represent $\hat{\mathbf{L}}$, $\hat{\mathbf{S}}$, or $\mathbf{L} + \hat{\mathbf{S}}$), is defined in terms of the commutation relations (9.16) and the rule of length appropriate to vectors (9.17).

The procedure for obtaining the matrices for $\hat{\mathbf{J}}$ and $\hat{J}^2$ (in a representation where $\hat{J}^2$ and $\hat{J}_z$ are diagonal) parallels the construction above. In place of (11.60), one writes the operationally identical equations (see Table 9.4)

$$\hat{J}_+ | jm\rangle = \hbar[(j - m)(j + m + 1)]^{1/2} | j, m + 1\rangle$$
$$\hat{J}_- | jm\rangle = \hbar[(j + m)(j - m + 1)]^{1/2} | j, m - 1\rangle$$

Thus the matrices found above for $\hat{\mathbf{L}}$ and $\hat{L}^2$ are also valid for $\hat{\mathbf{J}}$ and $\hat{J}^2$. Such matrices have the correct commutation properties and obey the Pythagorean length rule (see

Problem 11.41). However, while $\hat{\mathbf{L}}$ matrices are restricted to integral l values and are therefore of odd $(2l + 1)$ dimension, $\hat{\mathbf{J}}$ matrices also incorporate j values that are half-odd integral. Such matices are of even dimension.

Since there are $2j + 1$ values of J_z for each value of j, the matrix of $\hat{J}_z$ has $2j + 1$ diagonal elements. For a given j value, the operators $\hat{\mathbf{J}}$ and $\hat{J}^2$ are $(2j + 1) \times (2j + 1)$ square matrices and operate on column vectors $2j + 1$ elements long. That is, $\hat{\mathbf{J}}$ and $\hat{J}^2$ operate on a $(2j + 1)$-dimensional space.

The diagonal matrices for $\hat{J}^2$ and $\hat{J}_z$ (for a given value of j) are simple to construct. The first four $(j = \frac{1}{2}, 1, \frac{3}{2}, 2)$ such pairs appear as

$$j = \frac{1}{2}: \quad \hat{J}^2 = \frac{3\hbar^2}{4}\begin{pmatrix} 1 & 0 \\ 0 & 1 \end{pmatrix}, \qquad \hat{J}_z = \frac{\hbar}{2}\begin{pmatrix} 1 & 0 \\ 0 & -1 \end{pmatrix}$$

$$j = 1: \quad \hat{J}^2 = 2\hbar^2\begin{pmatrix} 1 & 0 & 0 \\ 0 & 1 & 0 \\ 0 & 0 & 1 \end{pmatrix}, \qquad \hat{J}_z = \hbar\begin{pmatrix} 1 & 0 & 0 \\ 0 & 0 & 0 \\ 0 & 0 & -1 \end{pmatrix}$$

$$j = \frac{3}{2}: \quad \hat{J}^2 = \frac{15\hbar^2}{4}\begin{pmatrix} 1 & 0 & 0 & 0 \\ 0 & 1 & 0 & 0 \\ 0 & 0 & 1 & 0 \\ 0 & 0 & 0 & 1 \end{pmatrix}, \qquad \hat{J}_z = \frac{\hbar}{2}\begin{pmatrix} 3 & 0 & 0 & 0 \\ 0 & 1 & 0 & 0 \\ 0 & 0 & -1 & 0 \\ 0 & 0 & 0 & -3 \end{pmatrix}$$

$$j = 2: \quad \hat{J}^2 = 6\hbar^2\begin{pmatrix} 1 & 0 & 0 & 0 & 0 \\ 0 & 1 & 0 & 0 & 0 \\ 0 & 0 & 1 & 0 & 0 \\ 0 & 0 & 0 & 1 & 0 \\ 0 & 0 & 0 & 0 & 1 \end{pmatrix}, \quad \hat{J}_z = \hbar\begin{pmatrix} 2 & 0 & 0 & 0 & 0 \\ 0 & 1 & 0 & 0 & 0 \\ 0 & 0 & 0 & 0 & 0 \\ 0 & 0 & 0 & -1 & 0 \\ 0 & 0 & 0 & 0 & -2 \end{pmatrix}$$

When j is an integer and $\mathbf{J}$ represents orbital angular momentum, $\mathbf{L}$, one may transform from the $|jm\rangle$ column vector representation to the coordinate $Y_l^m(\theta, \phi)$ representation. In this representation the ladder operators $\hat{J}_{\pm}$ appear as [see (9.57)]

$$\hat{J}_{\pm} = \hbar e^{\pm i\phi}\left(i \cot\theta \frac{\partial}{\partial\phi} \pm \frac{\partial}{\partial\theta}\right)$$

and eigenstates of J^2 and J_z are the spherical harmonics. When $\hat{J}$ represents spin, $\hat{S}$, spherical harmonic eigenstates become inappropriate.

The Rotation Operator

A distinction between angular momenta corresponding to j integral or half-odd integral is found in the rotation operator,[1] described in Section 10.2.

$$\hat{R}_{\boldsymbol{\phi}} = \exp\left(\frac{i\boldsymbol{\phi} \cdot \hat{\mathbf{J}}}{\hbar}\right)$$

When $\hat{R}$ operates on a function $f(\mathbf{r})$, it rotates $\mathbf{r}$ through the angle $\boldsymbol{\phi}$. If rotation is solely about the z axis, $\boldsymbol{\phi} = \mathbf{e}_z\phi$, then $\hat{R}$ becomes

$$\hat{R}_{\phi} = \exp\left(\frac{i\phi\hat{J}_z}{\hbar}\right)$$

Let $|jm\rangle$ denote a common eigenstate of $\hat{J}^2$ and $\hat{J}_z$. Then, in particular,

$$\hat{J}_z|jm\rangle = \hbar m|jm\rangle$$

and

$$\hat{R}_{\phi}|jm\rangle = e^{i\phi m}|jm\rangle$$

For the case that j, and therefore m, is half-odd integral, $e^{i2\pi m} = -1$, and one obtains the somewhat surprising result

$$(2j + 1 = \text{even no.}) \qquad \hat{R}_{2\pi}|jm\rangle = -1|jm\rangle$$

That is, the eigenstates of $\hat{J}^2$ and $\hat{J}_z$ corresponding to half-odd integral j values change sign under complete rotation of axes. If, on the other hand, j is integral, one obtains

$$(2j + 1 = \text{odd no.}) \qquad \hat{R}_{2\pi}|jm\rangle = +1|jm\rangle$$

In the coordinate representation, eigenstates for this case become spherical harmonics. These functions return to their original values under complete rotation.

The smallest finite value j may assume is $j = \frac{1}{2}$. This spin quantum value is a profoundly important case and is developed in detail in the next section. Eigenstates corresponding to $j = \frac{1}{2}$ are called *spinors*. We will find (Problem 11.76), in accord with the discussion above, that spinors change sign under complete rotation of axes.

---

[1] A more fundamental distinction involves the theory of group representations. For a concise, self-contained discussion of this topic, see L. I. Schiff, *Quantum Mechanics*, 3d ed., McGraw-Hill, New York, 1968, Chapter 7.

PROBLEMS

11.38 The state column vectors ξ, corresponding to the case $l = 1$, exist in a three-dimensional vector space. Any element ξ of this space is a set of three numbers of the form

$$\xi = \begin{pmatrix} a \\ b \\ c \end{pmatrix}$$

Write the vector

$$\xi = \frac{1}{\sqrt{14}} \begin{pmatrix} 1 \\ 2 \\ 3 \end{pmatrix}$$

as a linear combination of the vectors $(\xi_x^0, \xi_x^{-1}, \xi_x^1)$, as given by (11.68).

Answer

$$\xi = -\frac{1}{\sqrt{7}}\xi_x^0 + \frac{1}{\sqrt{7}}(\sqrt{2} - 1)\xi_x^{-1} + \frac{1}{\sqrt{7}}(\sqrt{2} + 1)\xi_x^1$$

11.39 Use the results of Problem 11.38 to answer the following question. A rigid rotator with moment of inertia I is in the state

$$\psi(t) = \frac{1}{\sqrt{14}} \begin{pmatrix} 1 \\ 2 \\ 3 \end{pmatrix} e^{-iEt/\hbar}, \qquad E = \frac{\hbar^2}{I}$$

(a) What is the probability that measurement of L_x finds the value $-\hbar$?

(b) What is the column vector representation of the time-dependent state of the rotator after measurement finds the value $L_x = -\hbar$?

11.40 What is the column vector representation of the angular momentum state

$$\psi = \frac{1}{\sqrt{24}}(Y_2^2 + 3Y_2^1 + 2Y_2^0 + 3Y_2^{-1} + Y_2^{-2})$$

in the representation in which $\hat{L}_z$ is diagonal?

11.41 Show that the $l = 2$ angular momentum matrices satisfy the relation

$$\hat{L}^2 = \hat{L}_x^2 + \hat{L}_y^2 + \hat{L}_z^2 = 6\hbar^2 \begin{pmatrix} 1 & 0 & 0 & 0 & 0 \\ 0 & 1 & 0 & 0 & 0 \\ 0 & 0 & 1 & 0 & 0 \\ 0 & 0 & 0 & 1 & 0 \\ 0 & 0 & 0 & 0 & 1 \end{pmatrix}$$

11.42 Show directly, by matrix multiplication, that $\langle \hat{L}_x^2 \rangle$ in the state ξ_x^{-1} is $\hbar^2$.

11.43   What are the matrix representations of $\hat{L}^2$, $\hat{L}_z$, $\hat{L}_+$, and $\hat{L}_-$ for the case $l = 3$?

11.44   What are the column eigenvectors of $\hat{L}_z$ corresponding to $l = 3$? To what combinations of Y_l^m functions do these correspond?

11.6   THE PAULI SPIN MATRICES

Spin Operators

In Section 9.2 it was concluded that there are two classes of angular momentum. These, we recall, stem from the fact that orbital angular momentum l values cover only a subset of the spectrum of j values appropriate to $\hat{J}^2$ and $\hat{J}_z$. The second class includes angular momentum called *spin*. Spin, as described in Section 9.1, is not related to the spatial coordinates of a particle as is orbital angular momentum. A coordinate representation of spin wavefunctions does not exist. Spin is an intrinsic or internal property. Other intrinsic properties of a particle are charge, mass, dipole moment, and so forth. Values of such parameters comprise *internal degrees of freedom* for a particle.

Spin angular momentum is denoted by the symbol $\hat{\mathbf{S}}$. The Cartesian components of $\hat{S}$, being angular momentum components, obey the commutation rules

$$(11.69) \quad [\hat{S}_x, \hat{S}_y] = i\hbar\hat{S}_z, \qquad [\hat{S}_y, \hat{S}_z] = i\hbar\hat{S}_x, \qquad [\hat{S}_z, \hat{S}_x] = i\hbar\hat{S}_y$$

These are the fundamental relations from which all other properties of spin follow. Similar to the introduction of $\hat{J}_\pm$ and $\hat{L}_\pm$ in Chapter 9, one may also introduce ladder operators for spin.

$$(11.70) \qquad \hat{S}_\pm = \hat{S}_x \pm i\hat{S}_y$$

Furthermore, since the eigenvalue equations (9.43) for $\hat{J}^2$ and $\hat{J}_z$ were derived from the fundamental angular momentum commutator relations (9.16), we may conclude that a similar structure exists for the eigenvalues of $\hat{S}^2$ and $\hat{S}_z$.

$$(11.71) \quad \hat{S}^2 |sm_s\rangle = \hbar^2 s(s + 1)|sm_s\rangle, \qquad \hat{S}_z |sm_s\rangle = \hbar m_s |sm_s\rangle$$

For a given value of s, the azimuthal quantum number m_s runs in integral steps from $-s$ to $+s$. The lowest value s can have is zero. Mesons are particles that have zero spin.[1] Photons have unit spin. For $s = 1$ there are three values of m_s: $-1, 0, 1$.

---

[1] More precisely, *vector mesons* have spin one; *pseudoscalar* mesons have spin zero.

Spin Eigenstates

Electrons, protons, and neutrons have a spin of one-half. There are two values of m_s for $s = \frac{1}{2}$. These are $m_s = +\frac{1}{2}, -\frac{1}{2}$. Let us call the eigenstate corresponding to $\left(s = \frac{1}{2}, m_s = \frac{1}{2}\right) \alpha$ (also α_z) and the eigenstate corresponding to $\left(s = \frac{1}{2}, m_s = -\frac{1}{2}\right) \beta$ (also β_z). These eigenstates obey the eigenvalue equations

$$\hat{S}^2 \alpha = \frac{3}{4}\hbar^2 \alpha, \qquad \hat{S}_z \alpha = \frac{\hbar}{2}\alpha$$

(11.72)

$$\hat{S}^2 \beta = \frac{3}{4}\hbar^2 \beta, \qquad \hat{S}_z \beta = -\frac{\hbar}{2}\beta$$

The raising and lowering operators have the property that[1]

(11.73)
$$\hat{S}_+ |s, m_s\rangle = \hbar\sqrt{s(s+1) - m_s(m_s+1)}\,|s, m_s + 1\rangle$$
$$\hat{S}_- |s, m_s\rangle = \hbar\sqrt{s(s+1) - m_s(m_s-1)}\,|s, m_s - 1\rangle$$

It follows that

(11.74)
$$\hat{S}_+ \alpha = 0 \qquad \hat{S}_+ \beta = \hbar\alpha$$
$$\hat{S}_- \alpha = \hbar\beta \qquad \hat{S}_- \beta = 0$$

Matrix Representation

We now wish to construct the matrix elements of $\hat{S}_x$, $\hat{S}_y$, and $\hat{S}_z$ for $s = \frac{1}{2}$ in the α, β basis. In this basis $\hat{S}^2$ and $\hat{S}_z$ are diagonal. The first two equations on the left in (11.74) appear explicitly as

$$(\hat{S}_x + i\hat{S}_y)\alpha = 0$$
$$(\hat{S}_x - i\hat{S}_y)\alpha = \hbar\beta$$

Adding and subtracting these two equations, respectively, gives

$$\hat{S}_x \alpha = \frac{1}{2}\hbar\beta$$

(11.75)

$$\hat{S}_y \alpha = \frac{i}{2}\hbar\beta$$

---

[1] See Table 9.4.

In similar manner, addition and subtraction of the remaining two equations in (11.74) gives

(11.76)

$$\hat{S}_x \beta = \frac{\hbar}{2} \alpha$$

$$\hat{S}_y \beta = -\frac{i}{2} \hbar \alpha$$

Combining these with the following equations from (11.72),

(11.77)

$$\hat{S}_z \alpha = \frac{\hbar}{2} \alpha$$

$$\hat{S}_z \beta = -\frac{\hbar}{2} \beta$$

establishes all the six equations needed to calculate the matrix elements of $\hat{S}_x$, $\hat{S}_y$, and $\hat{S}_z$. For example, the element $\langle \alpha | \hat{S}_x | \alpha \rangle$ is [using the first equation in (11.75)]

$$\langle \alpha | \hat{S}_x | \alpha \rangle = \frac{1}{2} \hbar \langle \alpha | \beta \rangle = 0$$

The vectors α and β are eigenvectors of a Hermitian operator (i.e., $\hat{S}_z$) and are therefore orthogonal. We also take them individually to be normalized. Continuing in this way we find that

(11.78)

$$\hat{S}_x = \begin{pmatrix} \langle \alpha | \hat{S}_x | \alpha \rangle & \langle \alpha | \hat{S}_x | \beta \rangle \\ \langle \beta | \hat{S}_x | \alpha \rangle & \langle \beta | \hat{S}_x | \beta \rangle \end{pmatrix} = \frac{\hbar}{2} \begin{pmatrix} 0 & 1 \\ 1 & 0 \end{pmatrix}$$

$$\hat{S}_y = \frac{i\hbar}{2} \begin{pmatrix} 0 & -1 \\ 1 & 0 \end{pmatrix}, \qquad \hat{S}_z = \frac{\hbar}{2} \begin{pmatrix} 1 & 0 \\ 0 & -1 \end{pmatrix}$$

The matrix representations of the eigenvectors α and β are the two-dimensional column vectors

(11.79)

$$\alpha = \begin{pmatrix} 1 \\ 0 \end{pmatrix}, \qquad \beta = \begin{pmatrix} 0 \\ 1 \end{pmatrix}$$

In this representation the orthonormal relations between α and β appear as

(11.80)

$$\langle \alpha | \beta \rangle = \overbrace{1 \quad 0} \binom{0}{1} = 0 + 0 = 0$$

$$\langle \beta | \alpha \rangle = \overbrace{0 \quad 1} \binom{1}{0} = 0 + 0 = 0$$

$$\langle \alpha | \alpha \rangle = \overbrace{1 \quad 0} \binom{1}{0} = 1 + 0 = 1$$

$$\langle \beta | \beta \rangle = \overbrace{0 \quad 1} \binom{0}{1} = 0 + 1 = 1$$

The operator

(11.81)
$$\hat{\sigma} \equiv \frac{2}{\hbar} \hat{S}$$

is called the *Pauli spin operator*. The matrix representation of its Cartesian components (in the basis that diagonalizes $\hat{S}^2$ and $\hat{S}_z$) is

(11.82) $\quad \hat{\sigma}_x = \begin{pmatrix} 0 & 1 \\ 1 & 0 \end{pmatrix}, \qquad \hat{\sigma}_y = i \begin{pmatrix} 0 & -1 \\ 1 & 0 \end{pmatrix}, \qquad \hat{\sigma}_z = \begin{pmatrix} 1 & 0 \\ 0 & -1 \end{pmatrix}$

These are called the *Pauli spin matrices*. They will be brought into play shortly when we consider the quantum mechanical motion of a spinning electron in a magnetic field.

PROBLEMS

11.45 (a) For spin corresponding to $s = \frac{1}{2}$, show that the eigenvectors of $\hat{S}_x$ and $\hat{S}_y$ are

$$\alpha_x = \frac{1}{\sqrt{2}} \binom{1}{1}, \qquad \beta_x = \frac{1}{\sqrt{2}} \binom{1}{-1}$$

$$\alpha_y = \frac{1}{\sqrt{2}} \binom{1}{i}, \qquad \beta_y = \frac{1}{\sqrt{2}} \binom{1}{-i}$$

(b) What are the eigenvalues corresponding to these eigenvectors?

(c) Show that these eigenvectors comprise two sets of orthonormal vectors.

11.46 Spin-$\frac{1}{2}$ state vectors $\binom{a}{b}$ are called *spinors*. Spinors exist in a two-dimensional, complex space. Any element ξ of this space is a set of two complex numbers. Any two linearly independent spinors span this space. In particular, show that this space is spanned by any of the three pairs of eigenstates $(\alpha_x, \beta_x; \alpha_y, \beta_y; \alpha_z, \beta_z)$. That is, show that any spinor $\binom{a}{b}$ may be expressed as a linear combination of any one of these three pairs of eigenstates.

Answer

$$\begin{pmatrix} a \\ b \end{pmatrix} = a\alpha_z + b\beta_z$$

$$\begin{pmatrix} a \\ b \end{pmatrix} = \frac{1}{\sqrt{2}}[(a + b)\alpha_x + (a - b)\beta_x]$$

$$\begin{pmatrix} a \\ b \end{pmatrix} = \frac{1}{\sqrt{2}}[(a - ib)\alpha_y + (a + ib)\beta_y]$$

11.47 Measurement of the z component of the spin of a neutron finds the value $S_z = \hbar/2$.
    (a) What spin state is the particle in after measurement?
    (b) Show that in this state

$$\langle S_x^2 \rangle = \langle S_y^2 \rangle = \frac{1}{2}\langle S^2 - S_z^2 \rangle = \frac{\hbar^2}{4} = \langle S_z^2 \rangle$$

by direct calculation.

11.48 An electron is known to be in the spin state α_z. Show that in this state

$$\langle S_x \rangle = \langle S_y \rangle = 0$$

Explain this result geometrically (Fig. 11.2).

11.49 Show that it is impossible for a spin-$\frac{1}{2}$ particle to be in a state $\xi = \begin{pmatrix} a \\ b \end{pmatrix}$ such that

$$\langle S_x \rangle = \langle S_y \rangle = \langle S_z \rangle = 0$$

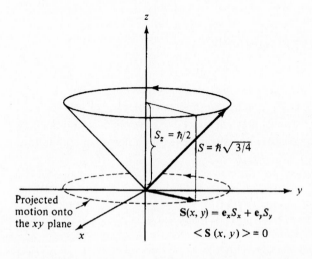

FIGURE 11.2 Dynamical conception of the spin-$\frac{1}{2}$ state. The angular momentum vector precesses maintaining a constant component about an axis. The projection of S onto a plane normal to the precession axis averages to zero. (See Problem 11.48.)

Answer

Since $\hat{\boldsymbol{\sigma}}$ and $\hat{\mathbf{S}}$ are related through a constant (11.81), we may work with $\hat{\boldsymbol{\sigma}}$ instead of $\hat{\mathbf{S}}$. First we find the relation between a and b which gives $\langle \hat{\sigma}_z \rangle = 0$.

$$\langle \xi | \hat{\sigma}_z | \xi \rangle = \begin{matrix} (a^* & b^*) \end{matrix} \begin{pmatrix} 1 & 0 \\ 0 & -1 \end{pmatrix} \begin{pmatrix} a \\ b \end{pmatrix} = |a|^2 - |b|^2 = 0$$

Setting $|a| = 1$, with no loss of generality, this condition implies that ξ is of the form

$$\xi = \begin{pmatrix} e^{i\phi} \\ e^{i\psi} \end{pmatrix}$$

where ϕ and ψ are the phases of a and b, respectively. Substituting this vector into $\langle \hat{\sigma}_x \rangle = 0$ gives

$$\langle \xi | \hat{\sigma}_x | \xi \rangle = \begin{matrix} (e^{-i\phi} & e^{-i\psi}) \end{matrix} \begin{pmatrix} 0 & 1 \\ 1 & 0 \end{pmatrix} \begin{pmatrix} e^{i\phi} \\ e^{i\psi} \end{pmatrix} = 2 \cos \alpha = 0$$

$$\alpha \equiv \psi - \phi$$

while $\langle \hat{\sigma}_y \rangle = 0$ gives

$$\langle \xi | \hat{\sigma}_y | \xi \rangle = 2 \sin \alpha = 0$$

Since there is no value of α for which $\sin \alpha = \cos \alpha = 0$, we conclude that the stated hypothesis is correct.

11.50   (a) Show that the components of $\hat{\boldsymbol{\sigma}}$ anticommute. For example, show that

$$\hat{\sigma}_x \hat{\sigma}_y + \hat{\sigma}_y \hat{\sigma}_x = 0$$

(b)   Show that $\hat{\sigma}_y$ is Hermitian.

11.7   FREE-PARTICLE WAVEFUNCTIONS, INCLUDING SPIN

The coordinates of a particle include the spin variables (s, m_s) and the position variables (x, y, z). The operators corresponding to these variables $(\hat{S}^2, \hat{S}_z, \hat{x}, \hat{y}, \hat{z})$ are assumed to commute. Their eigenvalues may therefore be prescribed simultaneously (one may locate a particle without destroying its spin state). Another set of commuting operators for a free particle is $(\hat{S}^2, \hat{S}_z, \hat{p}_x, \hat{p}_y, \hat{p}_z)$.

The Hamiltonian of a free particle is

$$\hat{H} = \frac{\hat{p}^2}{2m}$$

The reason that $\hat{S}$ does not enter in the Hamiltonian of a free particle is that the spin manifests itself only in the presence of an electromagnetic field.[1] It follows that $\hat{H}$ also commutes with $\hat{S}^2$ and $\hat{S}_z$. These operators have simultaneous eigenstates. Let $\varphi_\mathbf{k}$ be an eigenstate of $\hat{H}$ corresponding to the eigenvalue $\hbar^2 k^2/2m$ and (α, β) be the eigenstates of $\hat{S}^2$, $\hat{S}_z$ corresponding to the respective eigenvalues $3\hbar^2/4$ and $\pm\hbar/2$. Then

$$(11.83) \qquad \varphi_+ \equiv \varphi_\mathbf{k}(\mathbf{r})\alpha = A e^{i\mathbf{k}\cdot\mathbf{r}} \begin{pmatrix} 1 \\ 0 \end{pmatrix}$$

gives

$$\hat{H}\varphi_+ = \frac{\hbar^2 k^2}{2m}\varphi_+, \qquad \hat{S}^2\varphi_+ = \frac{3\hbar^2}{4}\varphi_+, \qquad \hat{S}_z\varphi_+ = \frac{\hbar}{2}\varphi_+$$

The eigenstate

$$(11.84) \qquad \varphi_- = \varphi_\mathbf{k}(\mathbf{r})\beta = A e^{i\mathbf{k}\cdot\mathbf{r}} \begin{pmatrix} 0 \\ 1 \end{pmatrix}$$

gives

$$\hat{H}\varphi_- = \frac{\hbar^2 k^2}{2m}\varphi_-, \qquad \hat{S}^2\varphi_- = \frac{3\hbar^2}{4}\varphi_-, \qquad \hat{S}_z\varphi_- = -\frac{\hbar}{2}\varphi_-$$

Consider that $\mathbf{k}$ is unidirectional. Then the state φ_+ gives the same energy $\hbar^2 k^2/2m$ for the two vectors $\pm\mathbf{k}$. The same is true for φ_-. Thus we find that eigenenergies of the spinning free particle propagating in one dimension are fourfold degenerate. This is illustrated below.

$$E_k = \frac{\hbar^2 k^2}{2m} \quad \begin{cases} \varphi_+(+\mathbf{k}) \\ \varphi_+(-\mathbf{k}) \\ \varphi_-(+\mathbf{k}) \\ \varphi_-(-\mathbf{k}) \end{cases}$$

The time-dependent wavefunction for an electron with momentum $\hbar\mathbf{k}$ and with z component of spin $-\hbar/2$ is the column vector

$$\psi_\mathbf{k}(\mathbf{r}, t) = A e^{i(\mathbf{k}\cdot\mathbf{r}-\omega t)} \begin{pmatrix} 0 \\ 1 \end{pmatrix}$$

$$\hbar\omega = \frac{\hbar^2 k^2}{2m}$$

---

[1] For relativistic electrons this is not the case. In this event, the free-particle Hamiltonian includes a spin-dependent term. See Section 15.3.

With $A = (2\pi)^{-3/2}$, one obtains

$$\langle \mathbf{k}' | \mathbf{k} \rangle = \iiint \psi_{\mathbf{k}'}{}^* \psi_{\mathbf{k}} \, d\mathbf{r} = \delta(\mathbf{k}' - \mathbf{k})$$

thereby regaining the normalization (10.18) relevant to free-particle motion.

PROBLEMS

11.51 A beam propagating in the x direction is comprised of 1.5-keV electrons, 20% of which have spin polarized in the $+z$ direction and 80% of which have spin polarized in the $-z$ direction.
  (a) What is the wavefunction of an electron in the beam?
  (b) What are the values of the wavenumber k and frequency ω of these electrons?

11.52 A free electron is known to have the following properties:
  1. Its orbital angular momentum about a prescribed origin is
$$L = \hbar \sqrt{l(l + 1)}$$

  2. Its z component of orbital angular momentum is $\hbar m_l$.
  3. Its z component of spin is $-\hbar/2$.
  4. It has kinetic energy
$$E = \frac{\hbar^2 k^2}{2m}$$

  (a) What is the time-dependent wavefunction, $|\psi\rangle = |k, l, m_l, s, m_s, t\rangle$, for this electron?
  (b) If in an ideal measurement, the x component of linear momentum is measured and the value
$$p_x = \hbar k_x$$

is found, what time-dependent state is the electron left in?
  (c) What is the degeneracy of the eigenenergy corresponding to part (b)?

11.8 THE MAGNETIC MOMENT OF AN ELECTRON

Bohr Magneton

The student will recall that a circular loop of wire carrying a current I, which is of cross-sectional area A, produces a magnetic moment[1]

(11.85) $$\mu = IA$$

The magnetic field due to this current loop is sketched in Fig. 11.3. In the limit that $A \to 0$ and $I \to \infty$ such that the product $IA = \mu$ remains finite, the magnetic field

---

[1] In cgs, A is in cm^2 and I is in emu/s. The units of μ are erg/gauss. Also recall that 1 emu = 1 esu/c.

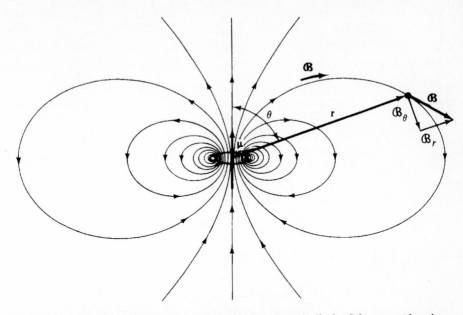

FIGURE 11.3   Magnetic field lines of a current loop. The magnetic dipole of the current loop is μ. The magnitude of μ is IA, where A is the area of the loop. At distances far removed from the origin, the $\mathcal{B}$ field is that due to a magnetic dipole at the origin. In spherical coordinates, with μ parallel to the polar axis, the components of the dipole magnetic field are

$$\mathcal{B}_r = \frac{2\mu \cos \theta}{r^3}, \qquad \mathcal{B}_\theta = \frac{\mu \sin \theta}{r^3}$$

These values are computed from the Biot–Savart law

$$\mathcal{B}(\mathbf{r}) = \frac{1}{c} \int \mathbf{J}(\mathbf{r}') \times \frac{(\mathbf{r} - \mathbf{r}')}{|\mathbf{r} - \mathbf{r}'|^3} \, d\mathbf{r}'$$

which gives the magnetic field at r, due to the current density J distributed over the source points, r′.

generated by the loop becomes a dipole field, whose components are also given in Fig. 11.3.

For an electron, one finds that the magnetic moment is directly proportional to its spin angular momentum. It is given by[1]

(11.86)
$$\boldsymbol{\mu} = \frac{e}{mc} \mathbf{S} = \frac{e\hbar}{2mc} \boldsymbol{\sigma} = -\mu_b \boldsymbol{\sigma}$$

---

[1] A more detailed analysis, including radiative corrections, finds a slightly larger value of electron magnetic moment and (11.86) is more accurately written $\boldsymbol{\mu} = (1.001\mu_b)\boldsymbol{\sigma}$. Thus to within 0.1% accuracy, the electron magnetic moment has the value of 1 Bohr magneton. See also related discussion in Chapter 15.

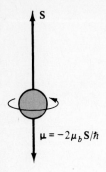

FIGURE 11.4   **For an electron, the spin and magnetic moment are antialigned.**

The quantity $e\hbar/2mc$ is called a *Bohr magneton*. It has the value

(11.87)
$$\mu_b = \frac{|e|\hbar}{2mc} = 0.927 \times 10^{-20} \text{ erg/gauss}$$

Since the charge of the electron is negative, (11.86) may be written

$$\boldsymbol{\mu} = -\mu_b\boldsymbol{\sigma} = -2\left(\frac{\mu_b}{\hbar}\right)\mathbf{S}$$

That is, the spin and magnetic moment of an electron are antialigned (Fig. 11.4).

If a magnetic moment is placed in a uniform, constant magnetic field, a torque is exerted on it about its origin, given by

(11.88)
$$\mathbf{N} = \boldsymbol{\mu} \times \boldsymbol{\mathcal{B}}$$

(Fig. 11.5). It follows that a magnetic moment tends to align itself with a magnetic field in which it is immersed. We may use (11.88) to calculate the work done in rotating the moment from the parallel orientation ($\theta = 0$) to the inclination $\theta > 0$ (Fig. 11.6):

$$V = \int_0^\theta N\, d\theta = -\mu\mathcal{B}\cos\theta + \text{constant}$$

(11.89)
$$V = -\boldsymbol{\mu} \cdot \boldsymbol{\mathcal{B}}$$

If we plot this potential versus θ (Fig. 11.7), it is evident that $\theta = 0$ is the stable orientation of the dipole. While the torque vanishes at $\theta = \pi$ ($\boldsymbol{\mu}$ antiparallel to $\boldsymbol{\mathcal{B}}$), any fluctuation about this orientation will cause the moment to "flip" to the stable position[1] $\theta = 0$. Although there is a torque on a magnetic moment in a uniform $\boldsymbol{\mathcal{B}}$

---

[1] This is so if we neglect the angular momentum of the dipole (due to the rotating charge); if the angular momentum of the moment is brought into play, precession results.

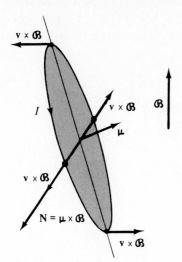

FIGURE 11.5   Torque on a magnetic dipole in a uniform $\mathcal{B}$ field. The force on a point charge moving in a $\mathcal{B}$ field, with velocity v, is

$$\mathbf{F} = \frac{e}{c}\mathbf{v} \times \mathcal{B}$$

Four components of this force along the ring current are shown. These forces tend to align μ with $\mathcal{B}$ so that μ and $\mathcal{B}$ are in the same direction.

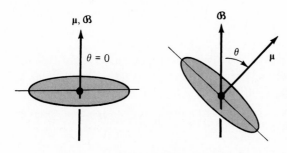

FIGURE 11.6   The energy of interaction between a magnetic dipole μ and a magnetic field $\mathcal{B}$ is a function of the inclination angle θ.

$$V = -\mu \cdot \mathcal{B}$$

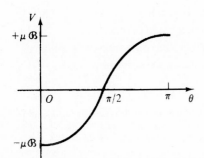

FIGURE 11.7 V versus θ for a magnetic dipole in a uniform, constant $\mathcal{B}$ field.

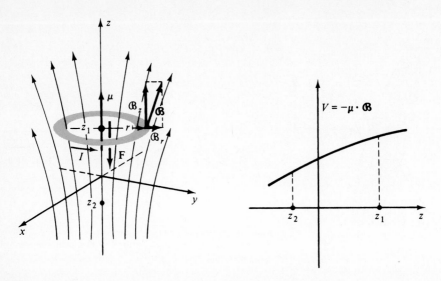

FIGURE 11.8 Force on a magnetic dipole in an inhomogeneous magnetic field. For the configuration shown, it is the r component of magnetic field, $\mathscr{B}_r$, which causes the downward force on the ring. This may be seen by calculating the force $(e/c)\mathbf{v} \times \mathscr{B}$ on the ring, due to the r and z components of $\mathscr{B}$, respectively. In terms of interaction energy, $V = -\boldsymbol{\mu} \cdot \mathscr{B}$, V is larger at z_1 than at z_2. This causes the dipole to "fall" from z_1 to z_2.

field, there is no net force on the dipole. However, the expression (11.89) for the interaction energy[1] between $\boldsymbol{\mu}$ and $\mathscr{B}$ suggests that there is a net force on a dipole in a nonuniform $\mathscr{B}$ field. This force causes the dipole, at a given orientation with $\mathscr{B}$, to "fall" to a neighboring point in space where the interaction energy, $-\boldsymbol{\mu} \cdot \mathscr{B}$, is smaller (Fig. 11.8). The net force on the dipole in a nonuniform $\mathscr{B}$ field is given by

$$(11.90) \qquad \mathbf{F} = -\boldsymbol{\nabla}V = -\boldsymbol{\nabla}(-\boldsymbol{\mu} \cdot \mathscr{B}) = \boldsymbol{\nabla}(\boldsymbol{\mu} \cdot \mathscr{B})$$

Stern–Gerlach Experiment

Equation (11.90) reveals the nature of the force which occurs in the Stern–Gerlach (S-G) experiment originally performed in 1922 using a beam of silver atoms. The spin of a silver atom is due to its outer $5s$ electron.[2] (Thus in the following we simply refer to electron spins in the beam.) A prototype of the experiment is shown in Fig. 11.9. The predominant component of $\mathscr{B}$ is $\mathscr{B}_z$. Furthermore, $\mathscr{B}_z$ varies most strongly with

[1] This expression for the energy does not take into account the energy supplied by the source that maintains the current in the dipole. It gives correct forces, however, if current and $\mathscr{B}$ field are constant in time. For further discussion on this topic, see R. Feynman, *Lectures on Physics*, Vol. II, Addison-Wesley, Reading, Mass., 1964.

[2] See Problem 11.91.

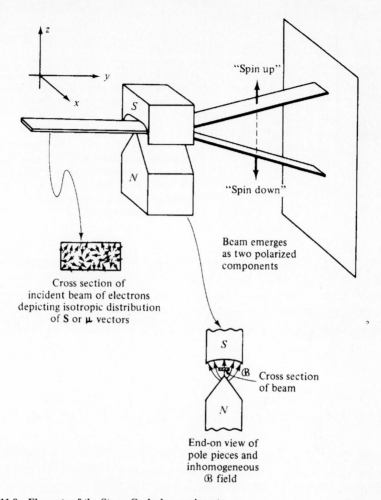

FIGURE 11.9 **Elements of the Stern–Gerlach experiment.**

changes in z so that $\nabla\mathscr{B}_z \simeq \mathbf{e}_z \, \partial\mathscr{B}_z/\partial z$. It follows that the force on electrons as they pass through the pole pieces is

$$(11.91) \qquad \mathbf{F} = \nabla\boldsymbol{\mu} \cdot \mathscr{B} \simeq \mathbf{e}_z \mu_z \frac{\partial\mathscr{B}_z}{\partial z} = \mathbf{e}_z F_z$$

The predominant force on the electrons is in the z direction. In addition, the sign of this force is solely determined by the sign of μ_z.

This z component of force causes electrons in the beam to be deflected. They strike the detection screen "off-axis." If we know where an electron strikes the screen,

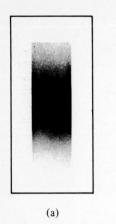

FIGURE 11.10 Traces of electron beam on detection screen in the Stern–Gerlach experiment. (a) The trace predicted from classical physics,

$$F_z = -\left(\mu_b \frac{\partial \mathcal{B}}{\partial z}\right) \cos \theta$$

Continuous distribution of $\cos \theta$ values implies a continuous trace. (b) The trace observed experimentally. If $s = \frac{1}{2}$ for an electron, then quantum mechanics predicts that two discrete beams emerge from the Stern–Gerlach apparatus.

(a)   (b)

we can[1] calculate F_z. Since $\partial \mathcal{B}_z / \partial z$ is known (it is a property of the apparatus) (11.91) then allows one to determine the z component of magnetic moment. Thus the S-G apparatus is a device that measures the z component of magnetic moment. More significantly, since $\boldsymbol{\mu}$ is directly proportional to the spin $\mathbf{S}$, the S-G apparatus becomes an instrument that measures the z component of spin. Written in terms of spin of the electron, the force (11.90) becomes

$$(11.92) \qquad \mathbf{F} = -\frac{2\mu_b}{\hbar} \boldsymbol{\nabla} \mathbf{S} \cdot \boldsymbol{\mathcal{B}} = -\mathbf{e}_z \frac{2\mu_b}{\hbar} S_z \frac{\partial \mathcal{B}_z}{\partial z}$$

At the start of the description of this experiment, we remarked that the incident beam carries an isotropic distribution of directions of magnetic moments or, equivalently, spin vectors (see Fig. 11.9). This means that at any point in the beam it is equally likely to find μ_z with any value from $-\mu$ to $+\mu$. Thus according to (11.91) one expects a uniform distribution of deflections. The pattern that the deflected particles in the beam make on the screen should be a continuous one. However, experiment finds that the beam divides into two discrete components (Fig. 11.10). Thus experiment indicates that μ_z or, equivalently, S_z has only two components. But this is precisely what one expects if $s = \frac{1}{2}$, for in this case S_z has only the two values $\pm \hbar/2$. So according to (11.92), an electron going through the Stern–Gerlach $\mathcal{B}$ field is acted on by only one of two possible values of force:

$$F_z = \pm \frac{\mu_b \partial \mathcal{B}_z}{\partial z}$$

These two oppositely directed components of force divide the beam into two separate components.

[1] The kinetic energy of the incident electron is assumed known.

Superposition Spin State

Instead of the beam containing an isotropic distribution of spin orientations, let electrons in the incident beam all be polarized with spins in the $+x$ direction. That is, each electron is in the eigenstate α_x of $\hat{S}_x$ and has the wavefunction

(11.93)
$$\psi = Ae^{i(ky-\omega t)}\alpha_x = \frac{A}{\sqrt{2}}e^{i(ky-\omega t)}\begin{pmatrix}1\\1\end{pmatrix}$$

The beam propagates in the $+y$ direction. Let the beam be incident on a S-G apparatus whose $\mathscr{B}$ field is aligned with the z axis. What portion of electrons in the incident beam emerges with spins in the $+z$ and $-z$ directions, respectively? Equivalently, we may ask, what is the probability that measurement of the z component of spin of an electron in the beam finds the respective values of $+\hbar/2$ or $-\hbar/2$? To answer this question, we call on the superposition principle. Expanding the column vector α_x in terms of the eigenvectors of $\hat{S}_z$ gives

(11.94)
$$\alpha_x = \frac{1}{\sqrt{2}}\begin{pmatrix}1\\1\end{pmatrix} = \frac{1}{\sqrt{2}}\left[\begin{pmatrix}1\\0\end{pmatrix} + \begin{pmatrix}0\\1\end{pmatrix}\right] = \frac{1}{\sqrt{2}}(\alpha_z + \beta_z)$$

The two coefficients of expansion are equal. It follows that the probability of measuring $S_z = +\hbar/2$

$$P\left(+\frac{\hbar}{2}\right) = \frac{1}{2} = |\langle\alpha_x|\alpha_z\rangle|^2$$

is equal to that of measuring $S_z = -\hbar/2$

$$P\left(-\frac{\hbar}{2}\right) = \frac{1}{2} = |\langle\alpha_x|\beta_z\rangle|^2$$

Thus if a beam of polarized electrons all in the state α_x enters a S-G apparatus, two equally populated beams emerge. These beams are also polarized, with electrons in spin states α_z and β_z, respectively.

PROBLEMS

11.53 Show that the magnetic moment due to the orbital motion (as opposed to the spin) of the electron in the Bohr-model hydrogen atom is given by

$$\boldsymbol{\mu} = \frac{\mu_b \mathbf{L}}{\hbar}$$

The orbital angular momentum of the electron is $\mathbf{L}$. (*Hint:* If the electron moves in a circle with velocity $\mathbf{v}$ and radius r, the related current is

$$I = \frac{ev}{2\pi rc} \quad \text{(emu/s)}$$

11.54 Consider that a polarized beam containing electrons in the α_z state is sent through a S-G analyzer which measures S_x. What values will be found, and with what probabilities will these values occur?

Answer

We write α_z as a linear combination of α_x and β_x.

$$\alpha_z = \begin{pmatrix} 1 \\ 0 \end{pmatrix} = \frac{1}{\sqrt{2}}(\alpha_x + \beta_x) = \frac{1}{2}\left\{ \begin{pmatrix} 1 \\ 1 \end{pmatrix} + \begin{pmatrix} 1 \\ -1 \end{pmatrix} \right\}$$

The coefficients of expansion are equal so that it is equally likely to find the values $S_x = +\hbar/2$ or $S_x = -\hbar/2$.

11.55 A proton is in the spin state α_y. What is the probability that measurement finds each of the following?
   (a) $S_y = -\hbar/2$
   (b) $S_x = +\hbar/2$
   (c) $S_x = -\hbar/2$
   (d) $S_z = +\hbar/2$
   (e) $S_z = -\hbar/2$

Answers
   (a) 0
   (b) $\frac{1}{2}$
   (c) $\frac{1}{2}$
   (d) $\frac{1}{2}$
   (e) $\frac{1}{2}$

11.56 A collection of electrons has an isotropic distribution of spin values. For an electron chosen at random, what is the probability of finding it with the following spin components?
   (a) $S_x = +\hbar/2, -\hbar/2$
   (b) $S_y = +\hbar/2, -\hbar/2$
   (c) $S_z = +\hbar/2, -\hbar/2$

Answers

If the x component of spin is measured, only two values can be found. If spin is isotropic, these two values are equally likely; hence there is a probability of $\frac{1}{2}$ that $S_x = \hbar/2$ and a probability of $\frac{1}{2}$ that $S_x = -\hbar/2$; similarly for S_y and S_z.

11.9   PRECESSION OF AN ELECTRON IN A MAGNETIC FIELD

In this section we consider the motion of a spinning but otherwise fixed electron which is in a constant uniform magnetic field that points in the z direction. Suppose that the electron is initially in the α_x state

(11.95)
$$\xi(0) = \alpha_x = \frac{1}{\sqrt{2}}\begin{pmatrix} 1 \\ 1 \end{pmatrix}$$

What is $\xi(t)$? To answer this question, we write down the time-dependent Schrödinger equation for the state $\xi(t)$.

(11.96)
$$i\hbar\,\frac{\partial}{\partial t}\,\xi = \hat{H}\xi$$

The Hamiltonian is the interaction energy

$$\hat{H} = -\hat{\boldsymbol{\mu}}\cdot\boldsymbol{\mathcal{B}} = +\mu_b\hat{\boldsymbol{\sigma}}\cdot\boldsymbol{\mathcal{B}} = \mu_b\mathcal{B}\hat{\sigma}_z$$

In the matrix representation with $\hat{S}_z$ diagonal, $\hat{H}$ becomes

(11.97)
$$\hat{H} = \mu_b\mathcal{B}\begin{pmatrix} 1 & 0 \\ 0 & -1 \end{pmatrix}$$

We seek the solution to (11.96) with this Hamiltonian, for the state vector

$$\xi(t) = \begin{pmatrix} a(t) \\ b(t) \end{pmatrix}$$

Substitution of this form into (11.96) with (11.97) substituted for $\hat{H}$ gives the column vector equation

$$\begin{pmatrix} \dot{a} \\ \dot{b} \end{pmatrix} = -\frac{i\Omega}{2}\begin{pmatrix} a \\ -b \end{pmatrix}$$

where $\Omega/2$ is the *Larmor frequency* and

(11.98)
$$\Omega = \frac{|e|\mathcal{B}}{mc}$$

is commonly referred to as the *cyclotron frequency* (previously encountered in Section 10.4). The column vector equation above is equivalent to the two independent equations

$$\dot{a} = -\frac{i\Omega}{2}a$$

$$\dot{b} = +\frac{i\Omega}{2}b$$

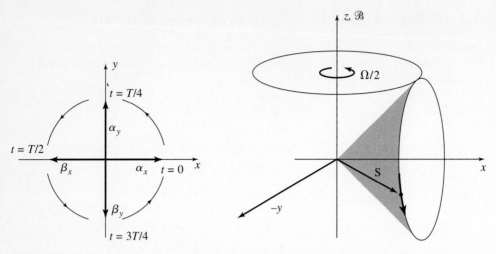

FIGURE 11.11 **Precession of a spinning but otherwise fixed electron in a constant, uniform magnetic field. Electron is initially in the α_x state.**

which has the solution

(11.99)
$$\xi(t) = \begin{pmatrix} a \\ b \end{pmatrix} = \frac{1}{\sqrt{2}} \begin{pmatrix} e^{-i(\Omega/2)t} \\ e^{+i(\Omega/2)t} \end{pmatrix}$$

This solution is compatible with the initial conditions (11.95). At later times (including $t = 0$) one obtains

$$\xi(0) = \alpha_x$$
$$\xi(T/4) = e^{-i\pi/4}\alpha_y$$
$$\xi(T/2) = e^{-i\pi/2}\beta_x$$
$$\xi(3T/4) = e^{-i3\pi/4}\beta_y$$

where we have written T for the period $2\pi/\Omega$. Apart from constant phase factors, we may conclude the following. At $t = 0$, the electron is in an eigenstate of S_x corresponding to the eigenvalue $+\hbar/2$. At $T/4$ it is in an eigenstate of S_y corresponding to the eigenvalue $+\hbar/2$. Proceeding in this manner one finds that the spin of the electron precesses about the z axis with angular frequency $\Omega/2$ (see Fig. 11.11).

Eigenenergies

We now consider the problem of calculating the eigenstates and eigenenergies of this same system, i.e., a spinning but otherwise fixed electron in a constant uniform magnetic field that points in the z direction. To solve this problem we use the time-independent Schrödinger equation. For the case at hand, it appears as

(11.100)
$$\hat{H}\xi = E\xi$$
$$\hat{H} = \mu_b \mathscr{B}\hat{\sigma}_z$$

Setting $\xi = \binom{a}{b}$ gives

$$\mu_b \mathscr{B}\begin{pmatrix} 1 & 0 \\ 0 & -1 \end{pmatrix}\begin{pmatrix} a \\ b \end{pmatrix} = E\begin{pmatrix} a \\ b \end{pmatrix}$$

or, equivalently,

$$(\mu_b \mathscr{B} - E)a = 0$$
$$(\mu_b \mathscr{B} + E)b = 0$$

If $a \neq 0$, $E = +\mu_b \mathscr{B}$ and $b = 0$. If $b \neq 0$, $E = -\mu_b \mathscr{B}$, and $a = 0$. Thus we obtain the (normalized) eigenstates, and eigenenergies

(11.101)
$$\alpha = \begin{pmatrix} 1 \\ 0 \end{pmatrix}, \qquad E = \mu_b \mathscr{B} = \frac{\hbar\Omega}{2}$$
$$\beta = \begin{pmatrix} 0 \\ 1 \end{pmatrix}, \qquad E = -\mu_b \mathscr{B} = -\frac{\hbar\Omega}{2}$$

In the state of higher energy, the spin of the electron is parallel to $\mathscr{B}$, so the magnetic moment is antiparallel to $\mathscr{B}$ and the interaction energy $-\boldsymbol{\mu} \cdot \mathscr{B}$ is maximum. In the state of lower energy, the spin is antiparallel to $\mathscr{B}$, so the magnetic moment is parallel to $\mathscr{B}$ and $-\boldsymbol{\mu} \cdot \mathscr{B}$ is minimum (Fig. 11.12).

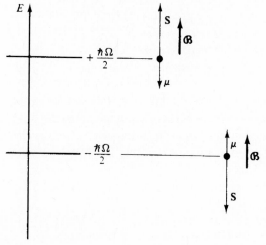

FIGURE 11.12 Energy eigenvalues of a spinning but otherwise fixed electron in a uniform, constant magnetic field. The orientation of $\boldsymbol{\mu}$ with respect to $\mathscr{B}$ is also shown for these two states.

Magnetic Resonance

As found above (11.86), the relation between the magnetic moment of an electron and its spin is given by

$$\boldsymbol{\mu} = -2\left(\frac{\mu_b}{\hbar}\right)\mathbf{S}$$

This expression may be written in terms of the *Landé g factor*,

$$\boldsymbol{\mu} = g\left(\frac{\mu_b}{\hbar}\right)\mathbf{S}$$

with $g = -2$. For nuclear particles such as a proton or a neutron, the relevant unit of magnetic moment is the *nuclear magneton*

$$\mu_N = \frac{e\hbar}{2M_p c} = 0.505 \times 10^{-23} \text{ erg/gauss}$$

where M_p is the mass of the proton. The magnetic moment of a nuclear particle is written

$$\boldsymbol{\mu} = g\left(\frac{\mu_N}{\hbar}\right)\mathbf{S}$$

For a proton $g = 2(2.79)$, while for a neutron $g = 2(-1.91)$. We wish now to describe a technique of measuring g or, equivalently, the magnetic moment.

As found in the first part of this section, if a magnetic moment is immersed in a steady $\mathscr{B}$ field in the z direction, the moment will precess about the z axis with the "Larmor" frequency $g\Omega/2$. Measurement of the magnetic moment is made through inducing a spin flip of the magnetic moment between its two energy states (for a spin-$\frac{1}{2}$ particle) parallel and antiparallel to $\mathscr{B}$ (Fig. 11.12). As in the corresponding classical configuration, in order to change the angle that $\boldsymbol{\mu}$ makes with the z axis, it is necessary to impose an additional transverse magnetic field normal to the plane through the z axis and $\boldsymbol{\mu}$ (Fig. 11.13). Since this plane rotates with the Larmor frequency, in the corresponding quantum mechanical motion one may expect a spin flip of the magnetic moment to be induced when the frequency of an imposed rotating transverse magnetic field is equal to the Larmor precessional frequency. Let us examine quantitatively the manner in which this resonance occurs for a spin-$\frac{1}{2}$ nuclear particle with magnetic moment $g(\mu_N/\hbar)\mathbf{S}$.

Since $\mathscr{B}$ has three components, the Schrödinger equation (11.96) takes the form

$$i\hbar \frac{\partial}{\partial t}\begin{pmatrix} a \\ b \end{pmatrix} = -\frac{g\mu_N}{2}\left[\begin{pmatrix} 0 & 1 \\ 1 & 0 \end{pmatrix}\mathscr{B}_x + i\begin{pmatrix} 0 & -1 \\ 1 & 0 \end{pmatrix}\mathscr{B}_y + \begin{pmatrix} 1 & 0 \\ 0 & -1 \end{pmatrix}\mathscr{B}_z \right]\begin{pmatrix} a \\ b \end{pmatrix}$$

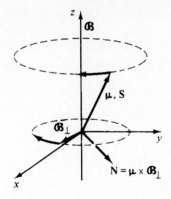

FIGURE 11.13   Transverse field $\mathscr{B}_\perp$ imposes a torque N that causes μ (or, equivalently, S) to change its orientation with respect to the z axis.

$$N = \mu \times \mathscr{B}_\perp$$

Let the transverse magnetic field $(\mathscr{B}_x, \mathscr{B}_y)$ rotate with frequency ω and let $\mathscr{B}_z$ maintain at the constant value $\mathscr{B}_\parallel$.

$$\mathscr{B}_x = \mathscr{B}_\perp \cos \omega t, \qquad \mathscr{B}_y = -\mathscr{B}_\perp \sin \omega t, \qquad \mathscr{B}_z = \mathscr{B}_\parallel$$

Substituting these values into the preceding equation gives the coupled equations

(11.102)
$$\frac{\partial a}{\partial t} = i(\Omega_\perp b e^{i\omega t} + \Omega_\parallel a)$$

$$\frac{\partial b}{\partial t} = i(\Omega_\perp a e^{-i\omega t} - \Omega_\parallel b)$$

The frequencies $\Omega_\perp$ and $\Omega_\parallel$ are defined through the relations

$$2\hbar\Omega_\perp = g\mu_N \mathscr{B}_\perp, \qquad 2\hbar\Omega_\parallel = g\mu_N \mathscr{B}_\parallel$$

We seek solution to equations (11.102) corresponding to the initial conditions

$$a = 1, \qquad b = 0 \qquad \text{at } t = 0$$

Let us look for solutions in the form

(11.103)
$$a = \bar{a}e^{i\omega_a t}$$

$$b = \bar{b}e^{i\omega_b t}$$

The coefficients $\bar{a}$ and $\bar{b}$ are assumed independent of time. Substituting these forms into (11.102) gives the homogeneous matrix equation

(11.104)
$$\begin{pmatrix} (\omega_a - \Omega_\parallel) & -\Omega_\perp e^{-i\phi t} \\ -\Omega_\perp e^{i\phi t} & (\omega_b + \Omega_\parallel) \end{pmatrix} \begin{pmatrix} \bar{a} \\ \bar{b} \end{pmatrix} = 0$$

$$\phi \equiv \omega_a - \omega_b - \omega$$

In order that $\bar{a}$ and $\bar{b}$ be independent of time, we must set $\phi = 0$, which gives

$$\omega_a = \omega_b + \omega$$

Setting the determinant of the coefficient matrix in (11.104) equal to zero gives the frequencies

(11.105)
$$\omega_b = -\frac{\omega}{2} \pm \bar{\omega}, \qquad \omega_a = \frac{\omega}{2} \pm \bar{\omega}$$

$$\bar{\omega}^2 \equiv \left(\Omega_\parallel - \frac{\omega}{2}\right)^2 + \Omega_\perp^2$$

It is only for these values of ω_a and ω_b that the proposed forms (11.103) are solutions to (11.102). The fact that two frequencies emerge for ω_a and ω_b corresponds to the property that (11.102) represents two independent, second-order differential equations in time. The general solution for, say, $b(t)$ is a linear combination of the two frequency components.

$$b(t) = b_1 \exp\left[-i\left(\frac{\omega}{2} - \bar{\omega}\right)t\right] + b_2 \exp\left[-i\left(\frac{\omega}{2} + \bar{\omega}\right)t\right]$$

To match this solution to the initial condition $b(0) = 0$, we choose $b_1 = -b_2 = C/2i$. There results

$$b(t) = C \sin \bar{\omega}t \, e^{-i(\omega/2)t}$$

To fix the coefficient C, we insert this solution into the second of equations (11.102) and set $t = 0$, $a(0) = 1$, to find

$$C = \frac{i\Omega_\perp}{\bar{\omega}}$$

The solution $a(t)$ corresponding to the specific form $b(t)$ obtained above is simply constructed from the second equation of (11.102).

Combining our results we obtain

(11.106)
$$|\xi\rangle = \begin{pmatrix} a(t) \\ b(t) \end{pmatrix} = \frac{\sin \bar{\omega}t}{\bar{\omega}} \begin{pmatrix} e^{i(\omega/2)t}[i(\Omega_\parallel - \omega/2) + \bar{\omega}\cot\bar{\omega}t] \\ ie^{-i(\omega/2)t}\Omega_\perp \end{pmatrix}$$

Since normalization of the spinor $|\xi\rangle$ is obeyed at $t = 0$,

$$\langle\xi(0)|\xi(0)\rangle = |a(0)|^2 + |b(0)|^2 = 1$$

we should find that it is maintained for all time. From (11.106) we obtain

$$|a|^2 + |b|^2 = \frac{\sin^2 \bar{\omega}t}{\bar{\omega}^2}\left[\left(\Omega_\parallel - \frac{\omega}{2}\right)^2 + \Omega_\perp^2\right] + \cos^2 \bar{\omega}t$$

Recalling the definition of $\bar{\omega}$ (11.105), we find that

$$\langle \xi(t) \,|\, \xi(t) \rangle = \sin^2 \bar{\omega}t + \cos^2 \bar{\omega}t = 1$$

so that normalization is maintained for all time.

Let us see how the solution (11.106) implies the resonant spin-flip behavior when the driving frequency ω of the transverse field $\mathscr{B}_\perp$ is equal to the frequency $2\Omega_\parallel$. From the form of $|b|^2$,

$$(11.107) \qquad |b(t)|^2 = \left[\frac{\Omega_\perp{}^2}{(\Omega_\parallel - \omega/2)^2 + \Omega_\perp{}^2} \right] \sin^2 \bar{\omega}t$$

we may infer the following. In an ensemble of spins that are all pointing in $+z$ direction at $t = 0$, the fraction $|b(t)|^2$ will be found pointing in the $-z$ direction at the time t. From the preceding expression for $|b|^2$, it is clear that the number of such spin flips is maximized at the resonant frequency

$$(11.108) \qquad \omega = 2\Omega_\parallel = \frac{g\mu_N \mathscr{B}_\parallel}{\hbar}$$

which, as expected, corresponds to the precessional frequency. The structure of the amplitude of $|b(t)|^2$ further indicates that this resonant phenomenon may be made more pronounced by choosing the transverse field small in comparison to the steady parallel field $\mathscr{B}_\parallel$.

Experimental Description

The molecules in a sample of water have zero magnetic moment save for that carried by the protons of the hydrogen nuclei. When placed in a steady magnetic field these protons align in one of two possible orientations. In thermal equilibrium there are slightly more protons in the lower-energy orientation with magnetic moment parallel to $\mathscr{B}$ (Fig. 11.12). One may utilize the presence of this net excess of magnetic moment to measure the g factor of protons. As described above, spin flips of the aligned protons may be induced by an additional rotating transverse magnetic field. This effect is maximized at the resonant frequency (11.108)

$$\omega = 2\Omega_\parallel = g \frac{e}{2M_p c} \mathscr{B}_\parallel$$

With the value of this frequency observed, the relation above may be solved for g, since all other quantities in the equation are known.

In inducing a transition from the lower to the upper energy state, energy in the amount

$$\hbar\Omega = \hbar g \frac{e}{2M_p c} \mathscr{B}_{\parallel}$$

is absorbed by the transition. This energy comes from the supporting transverse magnetic field coils. Spin flips to the lower parallel orientation expel energy $\hbar\Omega$. Since there are slightly more protons in the lower energy state than in the upper energy state in thermal equilibrium, there will be a net detectable absorption of energy from the transverse coils at the resonant frequency. Measurement of this frequency yields the value $g = 5.58$ for the proton.

In concluding this discussion we note the following convention regarding magnetic moment. Consider a particle with intrinsic spin $\mathbf{S}$. Its magnetic moment is given by

$$\boldsymbol{\mu} = g\left(\frac{\mu_N}{\hbar}\right)\mathbf{S}$$

The experimentally quoted value of magnetic moment refers to the expectation of the z component of $\boldsymbol{\mu}$ with $\mathbf{S}$ inclined maximally toward the z axis, that is, in the state $m_s = s$. With the wavefunction of the particle written $|s, m_s\rangle$ this value is given by

$$\mu = \langle s, s| g\left(\frac{\mu_N}{\hbar}\right)\hat{S}_z|s, s\rangle$$

$$= g\left(\frac{\mu_N}{\hbar}\right)\hbar s = g s \mu_N$$

For a proton

$$\boldsymbol{\mu}_p = 2(2.79)\left(\frac{\mu_N}{\hbar}\right)\mathbf{S}$$

which gives the value

$$\mu_p = 2.79\mu_N$$

For a neutron

$$\boldsymbol{\mu}_n = -2(1.91)\left(\frac{\mu_N}{\hbar}\right)\mathbf{S}$$

to which corresponds the value

$$\mu_n = -1.91\mu_N$$

Similarly, for an electron we find that

$$\mu_e = -\mu_b$$

PROBLEMS

11.57   What frequencies are emitted by an electron gas of low density ($\sim 10^8$ cm^{-3}) which is immersed in a uniform magnetic field of strength 10^4 gauss due to spin flips? What type of radiation is this (x ray, microwaves, etc.)? How does this emission frequency compare to the classical frequency emitted when an electron is executing Larmor rotation? (See Section 10.4.)

11.58   In a nuclear magnetic resonance (NMR) experiment with $\mathcal{B}_{\|}$ set at 5000 gauss, resonant energy absorption by a sample of water is detected when the frequency of the transverse components of magnetic field passes through the value 21.2 MHz. What value of g for a proton does this data imply?

11.10   THE ADDITION OF TWO SPINS

In Section 9.4 it was concluded that when adding the angular momenta of two components of a system, one may work in one of two representations: the *coupled representation* or the *uncoupled representation*. This also holds, of course, for the addition of spin angular momenta. As in the orbital case, the construction of wavefunctions in the uncoupled representation proves simpler. In this representation, wavefunctions for the two-electron system are simultaneous eigenstates of the four commuting operators $\hat{S}_1^2$, $\hat{S}_2^2$, $\hat{S}_{1z}$, and $\hat{S}_{2z}$. They may be written

(11.109) $$\xi = |s_1 s_2 m_{s_1} m_{s_2}\rangle = \left|\frac{1}{2}, \frac{1}{2}, \pm\frac{1}{2}, \pm\frac{1}{2}\right\rangle$$

where, for example,

$$\hat{S}_1^2 |s_1 s_2 m_{s_1} m_{s_2}\rangle = \hbar^2 s_1(s_1 + 1)|s_1 s_2 m_{s_1} m_{s_2}\rangle$$

$$= \frac{3}{4}\hbar^2 |s_1 s_2 m_{s_1} m_{s_2}\rangle$$

These eigenstates ($\xi_1, \ldots, \xi_4$) are simple products of the eigenstates $\alpha(1)$, $\beta(1)$ of $\hat{S}_1^2$, $\hat{S}_{1z}$ and $\alpha(2)$, $\beta(2)$ of $\hat{S}_2^2$, $\hat{S}_{2z}$. They are listed in Table 11.2. The column on the left of Table 11.2 contains diagrammatic representations of these states in which the relative orientation of the two spins is suggested.

TABLE 11.2 **Spin wavefunction for two electrons in the uncoupled representation**

Spin Alignment	Wavefunction $\xi = \lvert s_1 s_2 m_{s_1} m_{s_2}\rangle$	m_{s_1}	m_{s_2}	s_1	s_2
↑ ↑	$\xi_1 = \alpha(1)\alpha(2)$	$+\frac{1}{2}$	$+\frac{1}{2}$	$\frac{1}{2}$	$\frac{1}{2}$
↓ ↓	$\xi_2 = \beta(1)\beta(2)$	$-\frac{1}{2}$	$-\frac{1}{2}$	$\frac{1}{2}$	$\frac{1}{2}$
↑ ↓	$\xi_3 = \alpha(1)\beta(2)$	$+\frac{1}{2}$	$-\frac{1}{2}$	$\frac{1}{2}$	$\frac{1}{2}$
↓ ↑	$\xi_4 = \beta(1)\alpha(2)$	$-\frac{1}{2}$	$+\frac{1}{2}$	$\frac{1}{2}$	$\frac{1}{2}$

In the coupled representation, one constructs simultaneous eigenstates of the four commuting operators

$$\hat{S}^2 = (\hat{\mathbf{S}}_1 + \hat{\mathbf{S}}_2)^2 = \hat{S}_1^{\,2} + \hat{S}_2^{\,2} + 2\hat{\mathbf{S}}_1 \cdot \hat{\mathbf{S}}_2$$

$$\hat{S}_z = \hat{S}_{1z} + \hat{S}_{2z}$$

$$\hat{S}_1^{\,2}, \hat{S}_2^{\,2}$$

The simultaneous eigenstates of these four operators may be written $\lvert sm_s s_1 s_2\rangle$. Since $s = (0, 1)$, again there are four such independent eigenstates. In constructing these states we show first that two of the states appropriate to the uncoupled representation are also eigenstates of the coupled representation. Toward these ends note that all four uncoupled states are eigenstates of the total z component of spin.

(11.110)
$$\hat{S}_z \xi_1 = \hbar \xi_1$$
$$\hat{S}_z \xi_2 = -\hbar \xi_2$$
$$\hat{S}_z \xi_3 = \hat{S}_z \xi_4 = 0$$

The eigenstate ξ_1 is an eigenvector of $\hat{S}_z$, $\hat{S}_{1z}$, and $\hat{S}_{2z}$. If it is also an eigenstate of $\hat{S}^2$, it is one of the eigenstates appropriate to the coupled representation. To see if this is indeed the case, we employ the relation [see (9.22), Table 9.4, and Problem 9.31]

(11.111)
$$\hat{\mathbf{S}}_1 \cdot \hat{\mathbf{S}}_2 = (\hat{S}_{1x}\hat{S}_{2x} + \hat{S}_{1y}\hat{S}_{2y}) + \hat{S}_{1z}\hat{S}_{2z}$$

$$= \frac{1}{2}(\hat{S}_{1+}\hat{S}_{2-} + \hat{S}_{1-}\hat{S}_{2+}) + \hat{S}_{1z}\hat{S}_{2z}$$

in the cross term of $\hat{S}^2$ to obtain the expansion

(11.111a) $$\hat{S}^2 = \hat{S}_1^{\,2} + \hat{S}_2^{\,2} + 2\hat{S}_{1z}\hat{S}_{2z} + (\hat{S}_{1+}\hat{S}_{2-} + \hat{S}_{1-}\hat{S}_{2+})$$

The raising operator S_{1+} is defined by[1]

$$\hat{S}_{1+} = \hat{S}_{1x} + i\hat{S}_{1y}$$

with similar definitions carrying over to $\hat{S}_{2+}$, $\hat{S}_{1-}$, and $\hat{S}_{2-}$. Using the above representation of $\hat{\mathbf{S}}_1 \cdot \hat{\mathbf{S}}_2$, we find that

$$\hat{\mathbf{S}}_1 \cdot \hat{\mathbf{S}}_2 \xi_1 = \left[\frac{1}{2}(\hat{S}_{1+}\hat{S}_{2-} + \hat{S}_{1-}\hat{S}_{2+}) + \hat{S}_{1z}\hat{S}_{2z} \right] \alpha(1)\alpha(2)$$

$$= \left[\frac{1}{2}(0 + 0) + \frac{\hbar^2}{4} \right] \xi_1 = \frac{\hbar^2}{4} \xi_1$$

so that

$$\hat{S}^2 \xi_1 = (\hat{S}_1^2 + \hat{S}_2^2 + 2\hat{\mathbf{S}}_1 \cdot \hat{\mathbf{S}}_2)\xi_1 = \left(\frac{3}{4}\hbar^2 + \frac{3}{4}\hbar^2 + \frac{2}{4}\hbar^2 \right)\xi_1$$

$$= 2\hbar^2 \xi_1 = \hbar^2 1(1 + 1)\xi_1$$

We conclude that ξ_1 is also an eigenstate of $\hat{S}^2$, hence it is one of the eigenstates of $\hat{S}^2$ and $\hat{S}_z$ for two spin-$\frac{1}{2}$ particles in the coupled representation. We relabel this eigenstate $\xi_S^{(1)}$, for reasons that will become apparent immediately.

$$\xi_1 = \alpha(1)\alpha(2) \equiv \xi_S^{(1)}$$

Similarly, we find that ξ_2 is also a common eigenstate of $\hat{S}^2$ and S_z. This eigenstate is relabeled $\xi_S^{(-1)}$.

$$\xi_2 \equiv \beta(1)\beta(2) \equiv \xi_S^{(-1)}$$

However, ξ_3 and ξ_4 are not eigenstates of $\hat{S}^2$.

The Exchange Operator

To find the remaining two eigenstates (which will be called $\xi_S^{(0)}$ and ξ_A) of the coupled representation we introduce the *exchange operator*, $\hat{x}$. When $\hat{x}$ operates on a function of coordinates (spin or space) of two particles, it exchanges these coordinates. If $\varphi(1, 2)$ is a function of the coordinates of two particles (the spin coordinates of particle 1 are labeled "1," those of particle 2 are labeled "2"), then

(11.112) $$\hat{x}\varphi(1, 2) = \varphi(2, 1)$$

From the definition of $\hat{x}$, one obtains

$$\hat{x}^2 \varphi(1, 2) = \hat{x}\varphi(2, 1) = \varphi(1, 2)$$

---

[1] See Table 9.4.

so that the eigenvalues of $\hat{\mathscr{X}}$ are ± 1. One may construct the corresponding eigenfunctions of $\hat{\mathscr{X}}$ from any state $\varphi(1, 2)$ as follows:

$$\varphi_S = \varphi(1, 2) + \varphi(2, 1) \qquad (\text{``symmetric''})$$

clearly has eigenvalue $+1$, whereas

$$\varphi_A = \varphi(1, 2) - \varphi(2, 1) \qquad (\text{``antisymmetric''})$$

has eigenvalue -1.

Since $\hat{S}^2$ and $\hat{S}_z$ commute with $\hat{\mathscr{X}}$, it follows that $\hat{S}^2$, $\hat{S}_z$, and $\hat{\mathscr{X}}$ have common eigenstates. We already know two of them, $\xi_S^{(1)}$ and $\xi_S^{(-1)}$.

$$\hat{S}^2 \xi_S^{(1)} = 2\hbar^2 \xi_S^{(1)}, \qquad \hat{S}_z \xi_S^{(1)} = \hbar \xi_S^{(1)}, \qquad \hat{\mathscr{X}} \xi_S^{(1)} = +1 \xi_S^{(1)}$$
$$\hat{S}^2 \xi_S^{(-1)} = 2\hbar^2 \xi_S^{(-1)}, \qquad \hat{S}_z \xi_S^{(-1)} = -\hbar \xi_S^{(-1)}, \qquad \hat{\mathscr{X}} \xi_S^{(-1)} = +1 \xi_S^{(-1)}$$

These equations serve to explain our notation. The superscript on $\xi_S^{(1)}$ (i.e., $+1$) is the eigenvalue of $\hat{S}_z$, while the subscript S denotes "symmetric." The eigenstates $\xi_S^{(1)}$ and $\xi_S^{(-1)}$ are symmetric with respect to particle exchange.

Of the remaining two simultaneous eigenstates of $\hat{S}^2$, $\hat{S}_z$, and $\hat{\mathscr{X}}$, one is symmetric, $\xi_S^{(0)}$, and one is antisymmetric, ξ_A. Their construction follows simply from that of φ_S and φ_A given above, using the two degenerate eigenstates of $\hat{S}_z$ (i.e., ξ_3 and ξ_4):

(11.113)

$$\xi_S^{(0)} = \frac{1}{\sqrt{2}}(\xi_3 + \xi_4) = \frac{1}{\sqrt{2}}[\alpha(1)\beta(2) + \beta(1)\alpha(2)]$$

$$\xi_A = \frac{1}{\sqrt{2}}(\xi_3 - \xi_4) = \frac{1}{\sqrt{2}}[\alpha(1)\beta(2) - \beta(1)\alpha(2)]$$

Using the expansion (11.111) for $\hat{S}^2$, one finds that

$$\hat{S}^2 \xi_S^{(0)} = 2\hbar^2 \xi_S^{(0)}$$
$$\hat{S}^2 \xi_A = 0\hbar^2 \xi_A$$

while

$$\hat{S}_z \xi_S^{(0)} = 0\hbar \xi_S, \qquad \hat{\mathscr{X}} \xi_S^{(0)} = +1 \xi_S(0)$$
$$\hat{S}_z \xi_A = 0\hbar \xi_A, \qquad \hat{\mathscr{X}} \xi_A = -1 \xi_A$$

In this manner we obtain that the four independent eigenstates of $(\hat{S}_1^2, \hat{S}_2^2, \hat{S}^2, \hat{S}_z)$ are $(\xi_S^{(1)}, \xi_S^{(0)}, \xi_S^{(-1)}, \xi_A)$. Properties of these eigenstates are listed in Table 11.3. The

TABLE 11.3   **Spin wavefunctions for two electrons in the coupled representation**

Spin Alignment	Wavefunction $\xi = \lvert s_1 s_2 s m_s \rangle$	s	m_s	s_1	s_2
$\uparrow\ \ \uparrow$	$\xi_S^{(1)} = \alpha(1)\alpha(2)$	1	1	$\frac{1}{2}$	$\frac{1}{2}$
$\uparrow\ \downarrow + \downarrow\ \uparrow$	$\xi_S^{(0)} = \dfrac{1}{\sqrt{2}}[\alpha(1)\beta(2) + \beta(1)\alpha(2)]$	1	0	$\frac{1}{2}$	$\frac{1}{2}$
$\downarrow\ \ \downarrow$	$\xi_S^{(-1)} = \beta(1)\beta(2)$	1	-1	$\frac{1}{2}$	$\frac{1}{2}$
$\uparrow\ \downarrow - \downarrow\ \uparrow$	$\xi_A = \dfrac{1}{\sqrt{2}}[\alpha(1)\beta(2) - \beta(1)\alpha(2)]$	0	0	$\frac{1}{2}$	$\frac{1}{2}$

three ξ_S states correspond to $s = 1$, whereas the ξ_A state corresponds to $s = 0$. That is,

$$\hat{S}^2 \xi_S = 2\hbar^2 \xi_S$$
$$\hat{S}^2 \xi_A = 0$$

Spin Values for Two and Three Electrons

In the coupled representation one may speak of the *total* angular momentum of the two-electron system. In Section 9.4 we concluded that the total angular momentum quantum numbers for a two-component system vary in unit steps from $\lvert l_1 + l_2 \rvert$ to $\lvert l_1 - l_2 \rvert$ [see (9.114)]. In similar manner the total spin angular momentum of a two-electron system has spin quantum numbers varying from $\lvert \frac{1}{2} + \frac{1}{2} \rvert$ to $\lvert \frac{1}{2} - \frac{1}{2} \rvert$ in unit intervals. This gives the two values $s = 0$ and $s = 1$. For $s = 1$ there are three m_s values: $m_s = 1, 0,$ and -1. For $s = 0$ there is one m_s value, $m_s = 0$. This partitioning of states according to total spin number s is depicted in Table 11.3 in the (s, m_s) column.

If we use the angular momentum addition rules of Section 9.4 to calculate the spin quantum number corresponding to the possible total spin values for three electrons, one obtains the series

$$s = \left\lvert \frac{1}{2} + \frac{1}{2} + \frac{1}{2} \right\rvert, \ldots, \left\lvert \frac{1}{2} + \frac{1}{2} - \frac{1}{2} \right\rvert$$

which gives

$$s = \frac{3}{2}, \frac{1}{2}$$

There are four m_s values for $s = \frac{3}{2}$ and two for $s = \frac{1}{2}$.

These values of spin quantum number are appropriate to the coupled representation. The corresponding eigenstates are of the form

$$\xi = |s, m_s, s_1, s_2, s_3\rangle$$

which are simultaneous eigenstates of the five commuting operators

(11.114) $$\{\hat{S}^2, \hat{S}_z, \hat{S}_1^{\,2}, \hat{S}_2^{\,2}, \hat{S}_3^{\,2}\}$$

These concepts of spin angular momentum are very relevant to topics in atomic physics, as will be discussed further in Chapter 12.

PROBLEMS

11.59 Show that the coupled spin states $\xi_S^{(0)}$ and ξ_A given in (11.113) are eigenvectors of $\hat{S}^2$ with eigenvalues $2\hbar^2$ and 0, respectively.

11.60 Obtain the four spin states $|sm_s s_1 s_2\rangle$, listed in Table 11.3, relevant to two electrons in the coupled representation, through a Clebsch–Gordan expansion of the form

$$|sm_s s_1 s_2\rangle = \sum_{m_1 + m_2 = m_s} C_{m_1 m_2} |s_1 m_1\rangle |s_2 m_2\rangle$$

11.61 Consider a spin-1 particle. For integral angular momentum quantum number, the matrices developed in Section 11.5 for orbital angular momentum also apply to spin angular momentum. Using the three-component column eigenvectors of the Cartesian components of $\hat{\mathbf{S}}$ [e.g., (11.68)], determine if it is possible for a spin-1 particle to be in a state

$$\xi = \begin{pmatrix} a \\ b \\ c \end{pmatrix}$$

such that

$$\langle \hat{S}_x \rangle = \langle \hat{S}_y \rangle = \langle \hat{S}_z \rangle = 0$$

11.62   What is the form of a spin eigenstate, in Dirac notation, in the *uncoupled* representation for the three-electron case? How many "good" quantum numbers are there?

Answer

$$\xi = \left| s_1, s_2, s_3, m_{s_1}, m_{s_2}, m_{s_3} \right\rangle$$

There are six good quantum numbers.

11.63   Using the results of Problem 9.43 (and the discussion preceding), construct an operator, $\hat{A}_1{}^2$, which commutes with the five commuting operators (11.114) relevant to the addition of three electron spins in the coupled representation.

11.64   For the case of four electrons, in the coupled representation,

(a)   What are the s eigenvalues?

(b)   Write down the form of six commuting operators explicitly in terms of the vector operators $\hat{\mathbf{S}}_1, \hat{\mathbf{S}}_2, \hat{\mathbf{S}}_3$, their inner products, and their z components.

(c)   What is the form of an eigenstate in Dirac notation? How many independent states of this form are there?

Answer (*partial*)

(a) $s = \left| \frac{1}{2} + \frac{1}{2} + \frac{1}{2} + \frac{1}{2} \right|, \ldots, \left| \frac{1}{2} + \frac{1}{2} - \frac{1}{2} - \frac{1}{2} \right|$
$= 2, 1, 0$

11.65   The expression for the component of spin along an axis z', which makes respective angles (θ, β, γ) with the (x, y, z) axes, is

$$\hat{\sigma}_{z'} = \hat{\sigma}_x \cos \theta + \hat{\sigma}_y \cos \beta + \hat{\sigma}_z \cos \gamma$$

$$\cos^2 \theta + \cos^2 \beta + \cos^2 \gamma = 1$$

See Fig. 11.14. This relation follows from the vector quality of $\hat{\boldsymbol{\sigma}}$.

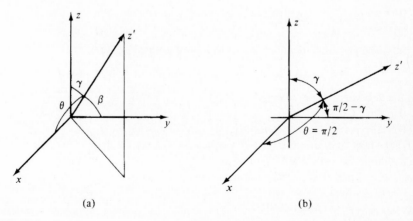

(a)   (b)

FIGURE 11.14   (a) Direction angles (θ, β, γ) locate the z' axis with respect to the original Cartesian frame. (b) The angles (θ, β, γ) for the case described in Problem 11.65. The z' axis lies in the yz plane.

(a)   Write the matrix for $\hat{\sigma}_z$ in terms of θ, β, γ and show that the eigenvalues of $\hat{\sigma}_{z'}$ are the same as those of $\hat{\sigma}_z$.

(b)   An electron is known to be in the state α_z. What is the probability that measurement of $S_{z'}$ finds the respective values $\pm\hbar/2$ for the angular displacements $\theta = \pi/2, \beta = \pi/2 - \gamma$? See Fig. 11.14b.

(c)   Describe a double S-G experiment in which such measurement may be effected.

Answer (partial)

(b)   To obtain the answer, one must find the eigenvectors of $\hat{\sigma}_{z'}$ corresponding to the eigenvalues ±1 and then expand α_z as a superposition of these two states. The squares of the coefficients of expansion give the respective probabilities

$$P\left(+\frac{\hbar}{2}\right) = \frac{1 + \cos\gamma}{2}$$

$$P\left(-\frac{\hbar}{2}\right) = \frac{1 - \cos\gamma}{2}$$

As $\gamma \to 0$, the z' and z axes merge and $P(+\hbar/2) \to 1, P(-\hbar/2) \to 0$. For $\gamma = \pi/2$, the z' axis is normal to the z axis and $P(+\hbar/2) = P(-\hbar/2) = \frac{1}{2}$.

11.66   Show, employing explicit matrix representations, that

$$\hat{S}_+\alpha = \hat{S}_-\beta = 0$$

$$\hat{S}_+\beta = \hbar\alpha, \qquad \hat{S}_-\alpha = \hbar\beta$$

11.67   Show that the Pauli spin operators obey the relation

$$\hat{\sigma}_x\hat{\sigma}_y\hat{\sigma}_z = i\hat{I}$$

11.68   Establish the following properties of the Pauli spin matrices:
(a) $\hat{\sigma}_x^2 = \hat{\sigma}_y^2 = \hat{\sigma}_z^2 = \hat{I}$ $(\hat{\sigma}^2 = 3\hat{I})$.
(b) $\{\hat{\sigma}_x, \hat{\sigma}_y\} = 0$, etc., where $\{,\}$ denotes the anticommutator.
(c) $\hat{\sigma}_x\hat{\sigma}_y = i\hat{\sigma}_z$, etc.
(d)   The preceding relations are included in the equality

$$(\hat{\sigma} \cdot a)(\hat{\sigma} \cdot b) = a \cdot b\hat{I} + i\hat{\sigma} \cdot (a \times b)$$

where **a** and **b** are arbitrary vectors.
(e) $[\hat{\sigma}_x, \hat{\sigma}_y] = i2\hat{\sigma}_z$, etc.
(f)   If **e** is a unit vector in an arbitrary direction, what is $(\hat{\sigma} \cdot e)^2$?
(g)   Show that an arbitrary but traceless 2×2 matrix may be written as a linear combination of the three Pauli spin matrices. [*Hint:* Write your test matrix with diagonal elements $(z, -z)$ and arbitrary off-diagonal elements and solve for the coefficients of combination of Pauli matrices.]

11.69   Measurement is made of the sum of the x and y components of the spin of an electron. What are the possible results of this experiment? After this measurement, the z component of spin is measured. What are the respective probabilities of obtaining the values $\pm\hbar/2$?

11.11   THE DENSITY MATRIX

Pure and Mixed States

In our description of quantum mechanics to this point, we have considered systems which by and large have satisfied the idealization of being isolated and totally uncoupled to any external environment. The free particle, the particle in a box, the harmonic oscillator, and the hydrogen atom are cases in point. Any such isolated system possesses a wavefunction of the coordinates only of the system. This wavefunction determines the state of the system.

Consider, on the other hand, a system that is coupled to an external environment. Such, for example, is the case of a gas of N particles maintained at a constant temperature through contact with a "temperature bath." Very simply, if x denotes coordinates of a system and y coordinates of its environment, then whereas the closed composite of system plus environment has a self-contained Hamiltonian and wavefunction $\psi(x, y)$, this wavefunction does not, in general, fall into a product $\psi_1(x)\psi_2(y)$. Under such circumstances, we may say that the system does not have a wavefunction. A system that does not have a wavefunction is said to be in a *mixed state*. A system that does have a wavefunction is said to be in a *pure state*.

It may also be the case that, owing to certain complexities of the system, less than complete knowledge of the state of the system is available. In quantum mechanics, such maximum information is contained in a wavefunction that simultaneously diagonalizes a complete set of commuting operators relevant to the system, such as that described in Section 5.5. Let a set of such operators be $\hat{A}$, $\hat{B}$, $\hat{C}$ with common wavefunction ψ_{abc}. Now suppose that the system is such that it is virtually impossible to measure A, B, and C in an appropriately small interval of time. Then the wavefunction ψ_{abc} cannot be determined and under such circumstances one also speaks of the system being in a mixed state. As in classical statistical mechanics, this situation arises for systems with a very large number of degrees of freedom, such as, for example, a mole of gas. The quantum state of such a system involves specification of $\sim 10^{23}$ momenta. The study of such complex systems is called *quantum statistical mechanics* and it is in this discipline that the density operator finds its greatest use.

The Density Operator

In dealing with situations where less than maximum information on the state of the system is available, one takes the point of view that a wavefunction for the system exists but that it is not completely determined. In place of the wavefunction, one introduces the density operator $\hat{\rho}$. If A is some property of the system, the density operator determines the expectation of A through the relation

$$\langle A \rangle = \mathrm{Tr}\, \hat{\rho}\hat{A} \tag{11.115}$$

and

(11.116) $$\text{Tr } \hat{\rho} = 1$$

The trace operation, written Tr, denotes summation over diagonal elements [see (11.27)]. From the density operator we may calculate expectation values of all relevant properties of the system.

Let us calculate the matrix elements of $\hat{\rho}$ for the case of a system whose wavefunction ψ is known. Then

$$\langle A \rangle = \langle \psi | \hat{A} \psi \rangle$$

Let the basis $\{ | n \rangle \}$ span the Hilbert space containing ψ. One may expand ψ in this basis to obtain

$$| \psi \rangle = \sum_n | n \rangle \langle n | \psi \rangle$$

Substituting this expansion into the preceding equation gives

$$\langle A \rangle = \sum_q \sum_n \langle \psi | q \rangle \langle q | \hat{A} | n \rangle \langle n | \psi \rangle$$

$$= \sum_q \sum_n \rho_{nq} A_{qn} = \text{Tr } \hat{\rho} \hat{A}$$

Here, we have made the identification

$$\rho_{nq} = \langle q | \psi \rangle^* \langle n | \psi \rangle = a_q{}^* a_n$$

The coefficient a_n represents the projection of the state ψ onto the basis vector $| n \rangle$. The nth diagonal element of $\hat{\rho}$ is

(11.117) $$\rho_{nn} = | \langle \psi | n \rangle |^2 = a_n{}^* a_n = P_n$$

which we recognize to be the probability P_n of finding the system in the state $| n \rangle$. Thus, *the diagonal elements of $\hat{\rho}$ are probabilities* and must sum to 1. This is the rationale for the property (11.116), $\text{Tr } \hat{\rho} = 1$.

The Mixed State

Now consider a system that is in a mixed state. Let $\hat{N}$ be a measurable property of the system such as its energy. Let $\{ | n \rangle \}$ be the eigenstates of $\hat{N}$. Since the system is in a mixed state, we may assume that the projections a_n are not determined quantities. In this case we define the elements ρ_{nq} to be the *ensemble averages* (see Section 5.1),

(11.118) $$\rho_{nq} = \overline{a_q{}^* a_n}$$

The diagonal element

(11.119) $$\rho_{nn} = \overline{a_n{}^* a_n}$$

represents the probability that a system chosen at random from the ensemble is found in the nth state.

Equation of Motion

Suppose again that a system is in a pure state and has the wavefunction ψ. Again, let $\hat{N}$ be of a measurable property of the system with eigenstates $\{|n\rangle\}$. Expanding ψ in terms of the projections a_n gives

$$|\psi\rangle = \sum_n a_n(t)|n\rangle$$

From the Schrödinger equation for ψ, we obtain

$$i\hbar \sum_n \frac{\partial a_n}{\partial t}|n\rangle = \sum_n a_n \hat{H}|n\rangle$$

Operating on this equation from the left with $\langle l|$ gives

(11.120) $$i\hbar \frac{\partial a_l}{\partial t} = \sum_n H_{ln} a_n, \qquad -i\hbar \frac{\partial a_l{}^*}{\partial t} = \sum_n H_{ln}{}^* a_n{}^*$$

We may use these relations to obtain an equation of motion for the matrix elements of $\hat{\rho}$.

$$i\hbar \frac{\partial \rho_{nl}}{\partial t} = i\hbar \frac{\partial a_l{}^* a_n}{\partial t} = i\hbar \left(\frac{\partial a_l{}^*}{\partial t} a_n + a_l{}^* \frac{\partial a_n}{\partial t} \right)$$

Substituting the expressions (11.120) for the time derivatives of the projections a_n and setting $H_{ik}{}^* = H_{kl}$ gives

(11.121) $$i\hbar \frac{\partial \rho_{nl}}{\partial t} = \sum_k (H_{nk} \rho_{kl} - \rho_{nk} H_{kl})$$

This may be written in operator form as

(11.122) $$i\hbar \frac{\partial \hat{\rho}}{\partial t} = [\hat{H}, \hat{\rho}]$$

Random Phases

Consider the matrix elements of $\hat{\rho}$ (11.118),

$$\rho_{nm} = \overline{a_m{}^* a_n}$$

The indeterminacy of the state of the system may be manifest in a corresponding indeterminacy of the phases $\{\phi_n\}$ of the projections $\{a_n\}$. These phases are defined by the relation

$$a_n = c_n e^{i\phi_n}$$

where c_n and ϕ_n are real. What is the consequence of assuming that the phases $\{\phi_n\}$ are random over the sample systems in the ensemble? Consider the matrix element

$$\rho_{nm} = \overline{c_m{}^* c_n \exp\left[i(\phi_n - \phi_m)\right]} = \overline{c_m{}^* c_n \left[\cos(\phi_n - \phi_m) + i \sin(\phi_n - \phi_m)\right]}$$

If phases are random, then in averaging over the ensemble, $\cos(\phi_n - \phi_m)$ will enter with positive value equally often as with negative value. The same is true for $\sin(\phi_n - \phi_m)$, so that

$$\overline{\cos(\phi_n - \phi_m)} = \overline{\sin(\phi_n - \phi_m)} = 0$$

except when $n = m$. In this case

$$\rho_{nn} = \overline{c_n{}^* c_n \cos(\phi_n - \phi_n)} = \overline{c_n{}^* c_n}$$

It follows that for the case of random phases, $\hat{\rho}$ is diagonal.

(11.123)
$$\rho_{nm} = \rho_{nn} \delta_{nm}$$

Evolution in Time of Diagonal Density Matrix

Suppose that $\hat{\rho}$ is diagonal. How does $\hat{\rho}$ then evolve in time? Specifically, does it remain diagonal? From (11.121) we conclude that the evolution in time of the diagonal elements of $\hat{\rho}$ depends on the off-diagonal elements of $\hat{\rho}$. If these off-diagonal elements begin to grow away from zero, $\{\rho_{ll}\}$ will change. The equation for the off-diagonal elements is obtained from (11.121). If at $t = 0$, $\hat{\rho}$ is diagonal, then

(11.124)
$$i\hbar \frac{\partial \rho_{nl}}{\partial t} = H_{nl}(\rho_{ll} - \rho_{nn}) \neq 0 \qquad (t = 0)$$

Thus the diagonal distribution is not, in general, constant in time. However, it is quite clear that the *uniform* distribution

$$\rho_{nl} = \rho_0 \delta_{nl}$$

(all states equally populated) is stationary in time.

Density Matrix for a Beam of Electrons

An electron beam generated at a cathode is known to have an isotropic distribution of spins, so that

$$\langle S_x \rangle = \langle S_y \rangle = \langle S_z \rangle = 0$$

Let us calculate a density matrix that gives this property in a representation in which $\hat{S}_z$ is diagonal. The matrices for $\hat{S}_x$, $\hat{S}_y$, and $\hat{S}_z$ are given by (11.78). Since the spins are isotropically oriented, the probabilities of finding S_z with values $\pm\hbar/2$ are both $1/2$. In the said representation, and with the property (11.117), we conclude that the diagonal elements of $\hat{\rho}$ are $\frac{1}{2}$ and $\frac{1}{2}$. It follows that the matrix

(11.125)
$$\hat{\rho} = \frac{1}{2}\begin{pmatrix} 1 & 0 \\ 0 & 1 \end{pmatrix}$$

gives

$$\langle S_z \rangle = \mathrm{Tr}\,\hat{\rho}\hat{S}_z = \mathrm{Tr}\,\frac{\hbar}{4}\begin{pmatrix} 1 & 0 \\ 0 & 1 \end{pmatrix}\begin{pmatrix} 1 & 0 \\ 0 & -1 \end{pmatrix} = 0$$

From (11.78) we quickly conclude that this choice of $\hat{\rho}$ also renders $\langle S_x \rangle = \langle S_y \rangle = 0$.

Projection Representation

As noted above, if the wavefunction of a system is indeterminate, one may describe properties of the system through the use of an ensemble of replica systems. Consider that states ψ of the ensemble systems are distributed with the probability P_ψ. An alternative form of the density operator is given by the projection sum over states of the ensemble.

(11.126)
$$\hat{\rho} = \sum_\psi |\psi\rangle P_\psi \langle\psi|$$

In the preceding example of a beam of isotropic spins, we found the density operator to be given by (11.125). This operator is written in a representation in which $\hat{S}_z$ is diagonal, for which the states corresponding to $S_z = \pm\hbar/2$ are $|\alpha\rangle$ and $|\beta\rangle$ given by (11.79). Since the ensemble of systems contains only two values of P_{S_z}, the summation (11.126) for this example may be written

(11.127)
$$\hat{\rho} = |\alpha\rangle\frac{1}{2}\langle\alpha| + |\beta\rangle\frac{1}{2}\langle\beta|$$

Let us use this form to calculate $\langle S_x \rangle$.

$$\langle S_x \rangle = \mathrm{Tr}\,\hat{\rho}\hat{S}_x = \langle\alpha|\,\hat{\rho}\hat{S}_x\,|\alpha\rangle + \langle\beta|\,\hat{\rho}\hat{S}_x\,|\beta\rangle = \frac{1}{2}\langle\alpha|\hat{S}_x|\alpha\rangle + \frac{1}{2}\langle\beta|\hat{S}_x|\beta\rangle$$

$$= \frac{\hbar}{4}\left[\begin{pmatrix} 1 & 0 \end{pmatrix}\begin{pmatrix} 0 & 1 \\ 1 & 0 \end{pmatrix}\begin{pmatrix} 1 \\ 0 \end{pmatrix} + \begin{pmatrix} 0 & 1 \end{pmatrix}\begin{pmatrix} 0 & 1 \\ 1 & 0 \end{pmatrix}\begin{pmatrix} 0 \\ 1 \end{pmatrix}\right] = 0$$

In like manner, we find that (11.127) gives $\langle\hat{S}_y\rangle = 0$.

Let a system be in a mixed state. Although the wavefunction is not determined, it is known that the probability that measurement of energy finds the value E_n is P_n. In this case (11.126) becomes

$$(11.128) \qquad \hat{\rho} = \sum_n |\psi_n\rangle P_n \langle \psi_n|$$

Since $\{|\psi_r\rangle\}$ is an orthonormal sequence, it follows that $\hat{\rho}$ is diagonal in this representation with diagonal elements equal to P_n. If the system is a gas of N particles, the number of particles with energy E_n is NP_n, which for nondegenerate states is the same as the number of particles in the state $|\psi_n\rangle$. Thus the diagonal elements of $\hat{\rho}$ in the case at hand give the occupation numbers for the states of a system.[1]

PROBLEMS

11.70   Is a system that is in a superposition state in a pure or a mixed state?

Answer

A system in a superposition state is in a pure state. The wavefunction of the system is known, and all properties of the system may be determined to the maximum degree that quantum mechanics allows. Let $\hat{A}$, $\hat{B}$ be a complete set of compatible operators for the system. Let the common eigenstates $\{\psi\}$ of $\hat{A}$, $\hat{B}$ span the Hilbert space $\mathfrak{H}$. Then ψ', which is a superposition state with respect to the observables A and B, does not lie along any of the basis vectors $\{\psi\}$. As described earlier in the chapter, ψ', which exists in $\mathfrak{H}$, is related to ψ through a unitary transformation, $\psi' = \hat{U}\psi$. In this manner, one may obtain a new set of states $\{\psi'\}$ which also span $\mathfrak{H}$. In this new basis, the operator $\hat{A}$ has the value $\hat{A}' = \hat{U}\hat{A}\hat{U}^{-1}$. So $\hat{A}'\psi' = \hat{U}\hat{A}\hat{U}^{-1}U\psi = \hat{U}a\psi = a\psi'$. Furthermore, $[A', B'] = 0$ if $[A, B] = 0$. Thus we may conclude that ψ' is a common eigenstate of A', B'.

11.71   What is the spin polarization of a beam of electrons described by the density operator

$$\hat{\rho} = \frac{1}{2}\begin{pmatrix} 1 & 1 \\ 1 & 1 \end{pmatrix}?$$

11.72   (a)   What is the density operator corresponding to an isotropic distribution of deuterons (spin 1) in the representation in which $\hat{S}_z$ is diagonal?
   (b)   What is the value of $\langle \hat{S}_y \rangle$?
(*Hint:* Your answer should appear as a 3×3 matrix.)

11.73   Consider a particle in a one-dimensional box with walls at $x = 0$ and $x = L$. Eigenenergies are $E_n = n^2 E_1$. It is known that the probability of finding the particle with energy E_1 is $\frac{1}{2}$ and that of finding it with energy E_5 is $\frac{1}{2}$.
   (a)   What is the density matrix for this system in the energy representation?

---

[1] For further discussion and problems on the density matrix, see R. H. Dicke and J. P. Wittke, *Introduction to Quantum Mechanics*, Addison-Wesley, Reading, Mass., 1960.

(b) Construct two normalized wavefunctions that give the same probabilities and, therefore, the same density matrix.

11.74 The *canonical form*[1] of the density operator is given by

$$\hat{\rho} = A \exp\left(\frac{-\hat{H}}{k_B T}\right)$$

where k_B is Boltzmann's constant and T denotes temperature. Consider that $\hat{H}$ is the Hamiltonian of a one-dimensional harmonic oscillator with fundamental frequency ω_0. Working in the energy representation:

   (a) Find the diagonal elements of $\hat{\rho}$.
   (b) Determine the normalization constant A.
   (c) Calculate the expectation $\langle E \rangle$ of the oscillator.
   (d) Construct the projection representation of $\hat{\rho}$ (11.126).

(*Hint:* For summation of series, see Problem 2.36.)

11.75 Show that

$$(\hat{\sigma}_z)^{2n} = \hat{I}, \qquad (\hat{\sigma}_z)^{2n+1} = \hat{\sigma}_z$$

where n is an integer. (This result also holds for the operator $\mathbf{e} \cdot \hat{\boldsymbol{\sigma}}$, where $\mathbf{e}$ is any fixed unit vector.)

11.76 (a) Using the results of Problem 11.75, show that

$$\hat{R}_\phi = \exp\left(\frac{i\phi\hat{S}_z}{\hbar}\right) = \exp\left(i\frac{\phi}{2}\hat{\sigma}_z\right)$$

$$= \left(\cos\frac{\phi}{2}\right)\hat{I} + i\left(\sin\frac{\phi}{2}\right)\hat{\sigma}_z$$

(Recall Problem 9.17.) This rotation operator tells us how spinors transform under rotation. The transformed spinor is given by $\xi' = \hat{R}_\phi \xi$.

   (b) What is the matrix form of $\hat{R}_\phi$?
   (c) Show that $\hat{R}_\phi$ preserves the length of ξ. That is, show that $\langle \xi' | \xi' \rangle = \langle \xi | \xi \rangle$.
   (d) Show that $\hat{R}_\phi$, at most, changes the phases of the eigenvectors of $\hat{\sigma}_x$, $\hat{\sigma}_y$, and $\hat{\sigma}_z$.
   (e) Show that under a complete rotation ($\phi = 2\pi$) about the z axis, $\xi \rightarrow \xi' = -\xi$.

11.77 Prove the general expansion

$$\exp\left(i\mathbf{e} \cdot \hat{\boldsymbol{\sigma}}\phi\right) = (\cos\phi)\hat{I} + i(\sin\phi)\mathbf{e} \cdot \hat{\boldsymbol{\sigma}}$$

where, again, $\mathbf{e}$ is an arbitrarily oriented unit vector.

11.78 (a) Show that the spin-exchange operator $\hat{x}$ for two electrons has the representation

$$\hbar^2\hat{x} = 2\hat{\mathbf{S}}_1 \cdot \hat{\mathbf{S}}_2 + \frac{\hbar^2}{2}$$

---

[1] This density matrix is relevant to a system with Hamiltonian $\hat{H}$, which is maintained in equilibrium at the temperature T through contact with a heat reservoir. For a further discussion, see K. Huang, *Statistical Mechanics*, Wiley, New York, 1963.

(b) Show that $\hat{\mathfrak{X}}$ may also be written

$$\hbar^2 \hat{\mathfrak{X}} = \hat{S}^2 - h^2$$

[*Hint:* For part (a), let $\hat{\mathfrak{X}}$ operate on $\alpha(1)\beta(2)$ and $\alpha(2)\beta(1)$, respectively.]

11.79 A beam of neutrons with isotropically distributed spins has the density matrix

$$\hat{\rho} = \frac{1}{2}\begin{pmatrix} 1 & a^* \\ a & 1 \end{pmatrix}$$

in a representation where $\hat{S}^2$ and $\hat{S}_z$ are diagonal. From the condition $\langle \mathbf{S} \rangle = 0$, show that $a = 0$.

11.80 (a) If $\boldsymbol{\mu}$ is the magnetic moment of the electron, in what state will $|\langle \mu_z \rangle| = \mu_b$?
(b) What is the value of $\langle \mu^2 \rangle$ in this state?

11.81 Consider a process in which an electron and a positron are emitted collinearly in the $+y$ and $-y$ directions, respectively. Spins are polarized to lie in the $\pm z$ directions. The pair is emitted with zero linear and spin-angular momentum and with total energy $\hbar\omega$. With the electron labeled 1 and the positron labeled 2:

(a) Write down a spin-coordinate, time-dependent product wavefunction for the electron positron pair which contains these properties.

(b) What is the probability that measurement finds the electron's z component of spin equal to $+\hbar/2$?

(c) Suppose that measurement of the positron's z component of spin finds the value $-\hbar/2$. What is the wavefunction for the pair immediately after this measurement?

(d) What will measurement of the electron's z component of spin now find?

Answers

(a) The appropriate zero spin factor of the wavefunction is found in Table 11.3.

$$\psi(1, 2) = \frac{1}{\sqrt{2}\,(2\pi)}(\alpha(1)\beta(2) - \alpha(2)\beta(1))\, \exp\left[i(ky_1 - ky_2 - \omega t)\right]$$

$$2\left(\frac{\hbar^2 k^2}{2m}\right) = \hbar\omega$$

(b) It is equally likely to find the values $\pm\hbar/2$ for the electron's z component of spin. The same is true of the positron's z component of spin.

(c) $$\psi_{\text{after}}(1, 2) = \frac{1}{2\pi}\alpha(1)\beta(2)\, \exp\left[i(k(y_1 - y_2) - \omega t)\right]$$

Here we are assuming that measurement preserves S_z and energy.

(d) $+\hbar/2$.

(*Note:* This problem contains a key tool of the Einstein–Podolsky–Rosen paradox.[1] Consider two observers O_1 and O_2 positioned along the y axis equipped, respectively, with detectors S-G$_1$

---

[1] A. Einstein, B. Podolsky, and N. Rosen, *Phys. Rev.* **47**, 777 (1935). For further discussion and references on this topic, see M. O. Scully, R. Shea, and J. D. McCullen, *Phys. Repts.* **43**, 501 (1978). See also B. d'Espagnat, "The Quantum Theory and Reality," *Scientific American* (Nov. 1979), p. 158. [*Note:* The configuration described in Problem 11.81 is due to D. Bohm (1952).] These topics are returned to in Section 11.13.

and S-G$_2$ oriented for measurement of S_z. Up until the time that O_1 makes his measurement, O_2 is equally likely to measure $\pm\hbar/2$. Once O_1 makes the measurement and measures, say, $+\hbar/2$, O_2 is certain to find the value $-\hbar/2$ upon measurement. This situation, presumably, maintains for O_1 and O_2 sufficiently far from each other with electron and positron beyond each other's range of interaction, thereby violating the *principle of locality*.)

11.82   Consider the antisymmetric spin zero state

$$\xi_A(x) = \frac{1}{\sqrt{2}}[\alpha_x(1)\beta_x(2) - \beta_x(1)\alpha_x(2)]$$

relevant to two spin-$\frac{1}{2}$ particles labeled 1 and 2, respectively. This state is an eigenstate of $\hat{S}_x$ and $\hat{S}^2$ with eigenvalue zero.

(a)   Show that $\xi_A(x)$ is also an eigenstate of $\hat{S}_z$, thereby establishing that it is a common eigenstate of the two noncommuting operators, $\hat{S}_x$ and $\hat{S}_z$.

(b)   What property renders the result in part (a) compatible with the commutator theorem (Section 5.2)?

Answers

(a)   From the forms of the spinors α_x, β_x as given in Problem 11.45, we obtain

$$\alpha_x = \frac{1}{\sqrt{2}}(\alpha_z + \beta_z)$$

$$\beta_x = \frac{1}{\sqrt{2}}(\alpha_z - \beta_z)$$

Substituting into $\xi_A(x)$ gives $\xi_A(x) = a\xi_A(z)$, where a is a constant.

(b)   As pointed out previously in Fig. 9.1 and Section 10.2, the null eigenstates of angular momentum are common eigenstates of the individual Cartesian components of angular momentum.

11.83   Employing the rotation operator obtained in Problem 11.76 relevant to spinors, show that the null eigenstate $\xi_A(x)$ of S_x and S^2, introduced in Problem 11.82, is invariant under rotation about the z axis.

Answer

The rotation operator is given by

$$\hat{R}_\phi = \cos\frac{\phi}{2}\hat{I} + i\sin\frac{\phi}{2}\hat{\sigma}_z$$

When applied to the product state $\alpha_x(1)\beta_x(2)$, we obtain

$$\hat{R}_\phi[\alpha_x(1)\beta_x(2)] = [\hat{R}_\phi\alpha_x(1)][\hat{R}_\phi\beta_x(2)]$$

$$= \left[\left(\cos\frac{\phi}{2}\right)\alpha_x(1) + i\left(\sin\frac{\phi}{2}\right)\beta_x(1)\right]\left[\left(\cos\frac{\phi}{2}\right)\beta_x(2) + i\left(\sin\frac{\phi}{2}\right)\alpha_x(2)\right]$$

Applying a similar rotation to the product $\alpha_x(2)\beta_x(1)$ and carrying out the multiplication gives

$$\hat{R}_\phi[\alpha_x(1)\beta_x(2) - \alpha_x(2)\beta_x(1)] = \left(\sin^2\frac{\phi}{2} + \cos^2\frac{\phi}{2}\right)[\alpha_x(1)\beta_x(2) - \alpha_x(2)\beta_x(1)]$$

$$\hat{R}_\phi\xi_A(x) = \xi_A(x)$$

[*Note:* This problem establishes that zero spin states are invariant under rotation of coordinates and thus transform as a scalar. A like quality is shared by the null orbital angular momentum states which, we recall, are given by any spherically symmetric function $f(r^2)$. See discussion preceding (10.34). For both null spin and null orbital angular momentum states, $\langle J_x \rangle = \langle J_y \rangle = \langle J_z \rangle = 0$. There is no preferred direction for a system in any of these states.]

11.84 One of the puzzles of the early theory of neutron decay, $n \to p + e$, was the fact that such a process could not conserve spin angular momentum. The neutron n, proton p, and electron e each have spin $\frac{1}{2}$. To answer this objection, Pauli[1] proposed that together with the proton and electron, a massless, chargeless, spin-$\frac{1}{2}$ particle was emitted, which he called a neutrino.

(a) Explain how the original process cannot conserve angular momentum.

(b) Explain how the corrected process, $n \to p + e + \bar{\nu}$, can conserve angular momentum.

11.12 OTHER "PICTURES" IN QUANTUM MECHANICS

Schrödinger, Heisenberg and Interaction Pictures

In addition to representations in quantum mechanics stemming from transformation of bases in Hilbert space, one also speaks of different "pictures" of the theory. These alternative formulations stem from the fact that wavefunctions and operators are not objects of measurement. As we have found previously, the superposition theorem specifies the possible outcome of a measurement in terms of certain projections in Hilbert space. Thus, for example, if $\hat{A}|a\rangle = a|a\rangle$, and the system is in the state ψ, the probability that measurement of A finds the value a is given by the absolute square, $|\langle a|\psi\rangle|^2$.

The so-called *Schrödinger picture* refers to the formulation which is based on the Schrödinger equation (3.45). With the preceding description one requires that any alternative picture satisfy two basic requirements: (1) Eigenvalues of operators corresponding to observables in the new picture must be the same as in the Schrödinger picture; (2) inner products of wavefunctions must maintain their values as well.

---

[1] W. Pauli, *Rapports du Septième Conseil de Physique, Solvay, Brussels, 1933*, Gauthier-Villars, Paris, 1934.

In Section 11.3 these properties were found to be obeyed by any unitary transformation. In particular, the so-called *Heisenberg picture* stems from the unitary operator (3.66). If initial time $t = 0$ is labeled $t = t_0$, then this operator becomes

(11.129)
$$\hat{U}(t, t_0) = \exp\left[-\frac{i}{\hbar}(t - t_0)\hat{H}\right]$$
$$\hat{U}^\dagger(t, t_0) = \exp\left[\frac{i}{\hbar}(t - t_0)\hat{H}\right] = U^{-1}(t, t_0)$$

where we have assumed that $\hat{H}$ is not an explicit function of time. With (3.70) we may write

(11.130)
$$\psi(t) = \hat{U}(t, t_0)\psi(t_0)$$
$$\psi(t_0) = \hat{U}^{-1}(t, t_0)\psi(t)$$

Let ψ' be a wavefunction in the Heisenberg picture. Then

(11.131)
$$\psi' = \hat{U}^{-1}\psi$$
$$\hat{U}\psi' = \psi$$

It follows that

(11.132)
$$\psi'(t) = \hat{U}^{-1}\psi(t) = \hat{U}^{-1}\hat{U}\psi(t_0)$$
$$\psi'(t) = \psi(t_0)$$

That is, in the Heisenberg picture the wavefunction remains constant. On the other hand, operators, which in the Schrödinger picture are constant, vary in time in the Heisenberg picture. This follows from the transformation (11.40).

(11.133)
$$\hat{A}'(t) = \hat{U}^{-1}(t)\hat{A}\hat{U}(t)$$

An important exception to this conclusion is the Hamiltonian. If $\hat{H}$ is constant in the Schrödinger picture, it remains constant in the Heisenberg picture.

An equation of motion for an operator in the Schrödinger picture is given by (6.68). To obtain an equation of motion for an operator in the Heisenberg picture, first with (11.133), we write

(11.134)
$$\frac{d\hat{A}'}{dt} = \frac{d\hat{U}^{-1}}{dt}\hat{A}\hat{U} + \hat{U}^{-1}\hat{A}\frac{d\hat{U}}{dt} + \hat{U}^{-1}\frac{\partial\hat{A}}{\partial t}\hat{U}$$

In obtaining the last term in this equation, we have noted the following. Consider, for example, that $\hat{A} = \hat{A}(\hat{q}, \hat{p}, t)$. Then in the Schrödinger picture, as $\hat{q}$ and $\hat{p}$ are stationary in time,

(11.134a)
$$\frac{d\hat{A}(\hat{q}, \hat{p}, t)}{dt} = \frac{\partial\hat{A}(\hat{q}, \hat{p}, t)}{\partial t}$$

To find a relation for the time derivative of $\hat{U}$, we consider the following sequence of events. Consider a particle which at the time t_0 is in the state $\psi(t_0)$. At time $t_1 > t_0$ it evolves to the state

$$\psi(t_1) = \hat{U}(t_1, t_0)\psi(t_0)$$

At time $t_2 > t_1$ it is in the state

$$\psi(t_2) = \hat{U}(t_2, t_1)\psi(t_1) = \hat{U}(t_2, t_1)\hat{U}(t_1, t_0)\psi(t_0)$$
$$= \hat{U}(t_2, t_0)\psi(t_0)$$

Thus we may write[1]

(11.135) $$\hat{U}(t_2, t_0) = \hat{U}(t_2, t_1)\hat{U}(t_1, t_0)$$

Recalling the definition of differentiation, we write

(11.136) $$\frac{d\hat{U}(t, t_0)}{dt} = \lim_{\varepsilon \to 0} \frac{U(t + \varepsilon, t_0) - U(t, t_0)}{\varepsilon}$$

which, with (11.134), gives

(11.137) $$\hat{U}(t + \varepsilon, t_0) = U(t + \varepsilon, t)U(t, t_0)$$

Substitution into (11.136) gives

$$\frac{d\hat{U}(t, t_0)}{dt} = \lim_{\varepsilon \to 0} \frac{[\hat{U}(t + \varepsilon, t) - 1]\hat{U}(t, t_0)}{\varepsilon}$$

For sufficiently small ε we may write

$$\hat{U}(t + \varepsilon, t) = \exp\left(\frac{-itH}{\hbar}\right) = 1 - \frac{i\varepsilon\hat{H}}{\hbar}$$

When substituted into (11.137), this relation gives

(11.138a) $$i\hbar\frac{d\hat{U}}{dt} = \hat{H}\hat{U}$$

Taking the Hermitian adjoint gives

(11.138b) $$-i\hbar\frac{d\hat{U}^\dagger}{dt} = \hat{U}^\dagger\hat{H}$$

---

[1] Steps leading to (11.135) remain valid for $\hat{H}$ an explicit function of time. See Problem 3.18.

TABLE 11.4   **Elements of Schrödinger and Heisenberg pictures**

	Wavefunction	Operator
Schrödinger picture	Varying	Constant
Heisenberg picture	Constant	Varying

Substitution of these latter two equations into (11.134) gives the desired result:

(11.139a)
$$i\hbar \frac{d\hat{A}'}{dt} = [\hat{A}', \hat{H}'] + i\hbar \frac{\partial \hat{A}'}{\partial t}$$

where we have set

(11.139b)
$$\frac{\partial \hat{A}'}{\partial t} \equiv \hat{U}^{-1} \frac{\partial \hat{A}}{\partial t} \hat{U}$$

Note in particular the similarity between (11.139a) and the equation of motion in the Schrödinger picture (6.68). Properties of wavefunctions and operators in these two representations are illustrated in Table 11.4. The right-hand column in this table refers to operators that are time-independent in the Schrödinger picture. It should be noted, however, that operators may be time-dependent in the Schrödinger picture. Such cases are discussed in Section 13.5 as well as in the description below of the interaction representation.

The Heisenberg equation of motion (11.39) often comes into play in discussions concerning the correspondence principle. If one makes the identification

(11.140)
$$\frac{1}{i\hbar}[\hat{A}, \hat{H}]_{QM} \rightarrow \{A, H\}_{CL}$$

then the equation of motion for an operator in the Heisenberg picture (11.139a) is seen to reduce to the corresponding classical equation of motion for a dynamical function as given in Problem 1.15.[1]

Interaction Picture

Finally, we turn to an approximation scheme important to time-dependent perturbation theory known as the *interaction picture*. Consider the Hamiltonian

$$\hat{H} = \hat{H}_0 + \lambda\hat{V}(t)$$

[1] For further discussion, see R. L. Liboff, *Found. Phys.* **17**, 981 (1987).

where the "unperturbed" Hamiltonian $\hat{H}_0$ is assumed to be time-independent and λ is a nondimensional parameter of smallness.

In analogy to the Heisenberg picture (11.131), wavefunctions in the new picture are given by

(11.141) $$\psi_I = \hat{U}_0^{-1}\psi$$

where

(11.141a) $$\hat{U}_0 \equiv \exp\left[-\frac{i}{\hbar}(t - t_0)\hat{H}_0\right]$$

and, in analogy with (11.133),

(11.142) $$\hat{A}_1 = \hat{U}_0^{-1}\hat{A}\hat{U}_0$$

Taking the time derivative of (11.141), and employing the Schrödinger equation (3.45) and noting that

$$[\hat{U}_0, \hat{H}_0] = 0$$

gives the desired equation of motion,

(11.143) $$\boxed{i\hbar\,\frac{\partial}{\partial t}\,\psi_I = \lambda\hat{V}_I\psi_I}$$

which is Schrödinger-like in form but only involves the interaction potential V_I. Integrating the preceding equation, we obtain

(11.144) $$\psi_I(t) = \psi_I(t_0) + \frac{\lambda}{i\hbar}\int_{t_0}^{t}\hat{V}_I(t')\psi_I(t')\,dt'$$

This equation leads naturally to a series solution. Namely, substituting $\psi_I(t)$ as given by the RHS of (11.144) for $\psi_I(t')$ in the integrand and continuing this iteration gives the series

(11.145) $$\psi_I(t) = \psi_I(t_0) + \frac{\lambda}{i\hbar}\int_{t_0}^{t}dt'\,\hat{V}_I(t')\psi_I(t_0)$$

$$+ \left(\frac{\lambda}{i\hbar}\right)^2\int_{t_0}^{t}dt'\int_{t_0}^{t'}dt''\,\hat{V}_I(t')\hat{V}_I(t'')\psi_I(t_0) + \cdots$$

Other techniques in time-dependent perturbation theory are discussed in Section 13.5 et seq. Application of the preceding formalism is given in Problem 11.102.[1]

---

[1] The interaction representation is encountered again in Chapter 13 (Problem 13.57) and in Section 14.6 in derivation of the Lippmann–Schwinger equation.

PROBLEMS

11.85 With regard to the interaction picture, derive (11.143) starting with (11.141).

11.86 Show that the matrix representation for $\hat{p}$, (11.48), is Hermitian.

11.87 (a) Integrate Heisenberg's equation of motion (11.139a) for a free particle to obtain $\hat{q}(t)$ and $\hat{p}(t)$ as functions of $\hat{q}(0)$ and $\hat{p}(0)$, respectively.

(b) Show that in this case

$$[\hat{q}(0), \hat{p}(0)] = -\frac{i\hbar}{m} t$$

(c) How does your answer to part (b) relate to a localized free-particle wave packet propagating in one dimension?

Answers (partial)

(a) We find

$$\hat{p} = \hat{p}(0), \qquad \hat{q}(t) = \hat{q}(0) + \frac{\hat{p}(0)t}{m}$$

(c) We obtain $\Delta q(0)\, \Delta q(t) \geq \hbar t/2m$, indicating that the wave packet spreads in time.

11.88. (In the following three problems, hats over operators have been deleted.) Show that

(a) $[x_i, F(\mathbf{p})] = i\hbar \dfrac{\partial F}{\partial p_i}$

(b) $[p_i, G(\mathbf{x})] = -i\hbar \dfrac{\partial G}{\partial x_i}$

where x_i and p_i denote Cartesian components of respective operators and the functions $F(\mathbf{p})$ and $G(\mathbf{x})$ have power series expansions.

(c) Evaluate the operator $[p_x, e^{-\kappa x}]$ in the Heisenberg picture.

11.89 Assume a free-particle Hamiltonian, with unprimed and primed variables denoting operators in the Schrödinger and Heisenberg pictures, respectively.

(a) How are the operators p and p' related?

(b) How are the operators x and x' related?

[*Hint:* You should find (a) $p = p'$ and (b) $x' = x + (pt/m)$, which, with Problem 11.87, indicates that $x(t) = x'(0)$. This result is consistent with the property that operators in the Schrödinger picture are constant in time. See Table 11.4.]

11.90 In Problem 10.73 it was found that the radial momentum operator in cylindrical coordinates is ambiguous.

(a) Working in cylindrical coordinates, what is the Hamiltonian, $H(\rho, z, \phi)$, of a particle of mass m moving in the potential field, $V(\rho, z, \phi)$?

(b) Employing this Hamiltonian, write down the Heisenberg equations of motion for momenta, p_ρ, p_z, p_ϕ, respectively. Do ambiguities arise in this representation?

[*Hint:* In part (b), for evaluation of $\dot{p}_\phi$, recall discussion of Problem 9.16(a). In the evaluation of $\dot{p}_\rho$ and $\dot{p}_z$, recall Problem 11.88(b).]

11.13 POLARIZATION STATES. EPR REVISITED

In this section we introduce eigenstates of photon polarization. This formalism is then applied to the Einstein–Padolsky–Rosen (EPR) paradox (see Problem 11.81) and closely allied notion of *hidden variables* briefly discussed in Chapter 2.

Polarization States

We recall that photons have zero rest mass and are spin-1 bosons with spin either aligned or antialigned with photon linear momentum.

The states of polarization of a photon may be written in terms of the two basis eigenstates $|H\rangle$ and $|V\rangle$, representing horizontal and vertical polarization, respectively. With x and y axes taken as horizontal and vertical directions, respectively, related electric fields at fixed z are given by

(11.146)
$$\mathcal{E}_H = \mathcal{E}_0[\cos \omega t, 0, 0] \Rightarrow |H\rangle$$
$$\mathcal{E}_V = \mathcal{E}_0[0, \cos \omega t, 0] \Rightarrow |V\rangle$$

Let us consider the field of the superposition state

(11.147)
$$|R\rangle = \frac{1}{\sqrt{2}} (|H\rangle + i|V\rangle)$$

Since $i|V\rangle = \exp(i\pi/2)|V\rangle$ corresponds to $|V\rangle$ shifted by $\pi/2$ radians, the field corresponding to $|R\rangle$ is

(11.148)
$$\mathcal{E}_R = \mathcal{E}_0[\cos \omega t, \sin \omega t, 0] \Rightarrow |R\rangle$$

which we recognize to be right-handed, circularly polarized radiation—viewing the photon head-on. The wave (11.148) propagates in the $+z$ direction. The closely allied field

(11.149)
$$\mathcal{E}_L = \mathcal{E}_0[\cos \omega t, -\sin \omega t, 0] \Rightarrow |L\rangle$$

represents left-handed, circularly polarized radiation corresponding to the superposition state

(11.150)
$$|L\rangle = \frac{1}{\sqrt{2}} (|H\rangle - i|V\rangle)$$

Note that these polarization states have the orthonormal properties

(11.151)
$$\langle H|H\rangle = \langle V|V\rangle = \langle R|R\rangle = \langle L|L\rangle = 1$$
$$\langle H|V\rangle = \langle R|L\rangle = 0$$

The preceding states of polarization are depicted graphically in Fig. 11.15.

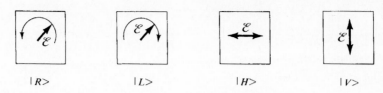

|R> |L> |H> |V>

FIGURE 11.15 **Diagrammatic representation of four photon states of polarization (viewing the photon head-on).**

Experimental Setup

With these preliminaries at hand, the question we examine relevant to the EPR paradox is whether quantum theory is consistent or whether it is necessary to bolster the theory with hidden variables. An operational means of answering this question is given by Bell's theorem.[1]

Two processes relevant to this theorem are considered. The first of these is described by the experimental setup depicted in Fig. 11.16 in which an atom undergoes a two-photon decay with no net change in angular momentum. The notation is such that $|R_1\rangle$ is the polarization state of photon 1, etc.

The zero net change in angular momentum of the atomic decay indicates that the photons must be in the symmetric state [analogous to the antisymmetric spin state of Problem 11.81(a)]:

$$(11.152) \qquad |P_{12}\rangle = \frac{1}{\sqrt{2}} [\,|R_1\rangle |R_2\rangle + |L_1\rangle |L_2\rangle]$$

In this expression $|R_1\rangle$ and $|R_2\rangle$ are antialigned states so that their total angular momentum is zero. The same is true of $|L_1\rangle$ and $|L_2\rangle$.

Now consider a polarizer which may be set to pass only $|R\rangle$, $|L\rangle$, $|H\rangle$, $|V\rangle$, or $|H_\theta\rangle$. Here we have written $|H_\theta\rangle$ for a linearly polarized state at an angle θ from the horizontal. Two such polarizers are placed in front of photon detectors as shown in

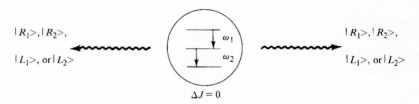

$|R_1\rangle, |R_2\rangle,$

$|L_1\rangle,$ or $|L_2\rangle$

ω_1

ω_2

$|R_1\rangle, |R_2\rangle,$

$|L_1\rangle,$ or $|L_2\rangle$

$\Delta J = 0$

FIGURE 11.16 **Schematic of two-photon atomic decay and possible circular polarization states of emitted photons.**

[1] J. S. Bell, *Physics* **1**, 195 (1964). This paper and other early works of Bell are reprinted in *Quantum Theory of Measurement*, J. A. Wheeler and W. H. Zurek, eds., Princeton University Press, Princeton, N.J., 1983.

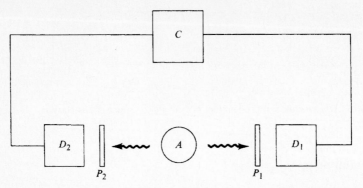

FIGURE 11.17 Photons from source A pass through the polarizers P to detectors D. Signals are then transmitted to coincidence counter C.

Fig. 11.17. Note in particular that the photons emanating from A in Fig. 11.17 are in the superposition state (11.152).

The polarizers are set so that P_1 only passes $|H_{\theta_1}\rangle$ and P_2 only passes $|H_{\bar{\theta}_2}\rangle$. (The motivation for the special notation $\bar{\theta}_2$ is given below.) The counter C responds positively when photons are detected in D_1 and D_2 simultaneously.

We wish to obtain the number of coincidence counts N as a function of the angles θ_1 and $\bar{\theta}_2$. To obtain this relation, first we recall the superpositions (11.147) and (11.150):

$$|R\rangle = \frac{1}{\sqrt{2}}(|H\rangle + i|V\rangle)$$

$$|L\rangle = \frac{1}{\sqrt{2}}(|H\rangle - i|V\rangle)$$

Substituting these transformations into (11.152) gives

(11.153) $$|P_{12}\rangle = \frac{1}{\sqrt{2}}[|H_1\rangle|H_2\rangle - |V_1\rangle|V_2\rangle]$$

The two-body polarization state so written indicates, for example, that if the polarization of photon 1 is measured to be H_1, then measurement of the polarization of photon 2 is certain to find H_2.

The function $N(\theta_1, \bar{\theta}_2)$ we seek is given by

(11.154) $$N(\theta_1, \bar{\theta}_2) \propto |\langle H_{\theta_1} H_{\theta_2}|P_{12}\rangle|^2$$

We note that photon polarization states rotate like a spin-1 vector.[1] Thus

(11.155) $$|H_\theta\rangle \pm i|V_\theta\rangle = e^{\pm i\theta}(|H\rangle + i|V\rangle)$$

---

[1] See Section 11.5 concerning the rotation operator.

where $|V_\theta\rangle$ represents a polarization state at θ radians from the vertical. There results

(11.156) $$|H_\theta\rangle = \cos\theta|H\rangle - \sin\theta|V\rangle$$

These expressions together with (11.153) give

(11.157) $$|\langle H_{\theta_1} H_{\bar\theta_2}|P_{12}\rangle|^2 = |(\cos\theta_1\langle H_1| - \sin\theta_1\langle V_1|)(\cos\bar\theta_2\langle H_2| - \sin\bar\theta_2\langle V_2|)$$
$$\times (|H_1\rangle|H_2\rangle - |V_1\rangle|V_2\rangle)|^2$$
$$= [\cos\theta_1\cos\bar\theta_2 - \sin\theta_1\sin\bar\theta_2]^2 = \cos^2(\theta_1 + \bar\theta_2)$$

Polarizer Angles

An important point is now made concerning the appropriate angle for polarizer 2 (in the configuration of Fig. 11.17). We have been working in the convention where the field description corresponding to the polarization state of a photon is written with respect to viewing the photon head-on. The Cartesian axes at detectors D_1 and D_2 are shown in Fig. 11.18.

With reference to this diagram we see that the angle polarizer 2 makes with the common axis defined by polarizer 1 is θ_2. From the diagram we also note

$$\bar\theta_2 = \pi - \theta_2$$

Thus (11.157) gives

(11.158) $$N(\theta_1, \theta_2) \propto \cos^2(\theta_1 - \theta_2 + \pi) = \cos^2(\theta_1 - \theta_2)$$

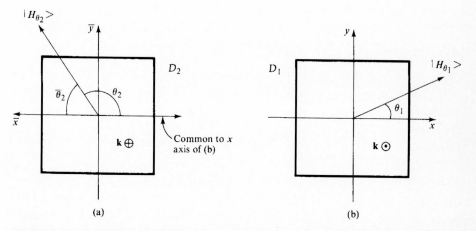

FIGURE 11.18 (a) Coordinate frame at D_2. The propagation vector **k** is into the page. (b) Coordinate frame at D_1. The vector **k** is out of the page. Note that looking at (a) from the back of the page gives a right-handed frame with **k** pointed toward the reader. It is with respect to this frame that $H_{\bar\theta_2}$ is defined in (11.157).

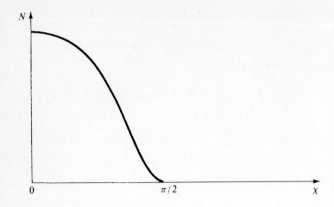

FIGURE 11.19   Coincidence counts N as a function of the angle $\chi \equiv \theta_1 - \theta_2$.

This is our desired relation. It is plotted in Fig. 11.19. The functional dependence of N on $(\theta_1 - \theta_2)$ reflects the fact that the experimental arrangement of Fig. 11.17 is invariant with respect to rotation about the axis $D_2 A D_1$.

For $\chi \equiv \theta_1 - \theta_2 = 0$, polarizers P_1 and P_2 are aligned and there is maximum coincidence. As is evident from (11.153), if measurement of photon 1 finds H_1, then measurement of photon 2 is certain to find H_2. At $\chi = \pi/2$ polarizers are set at right angles and there is no coincidence.

For further reference we set

$$(11.159) \qquad\qquad N(\chi) = \frac{1}{2}\cos^2\chi$$

See Fig. 11.19. Application of this result is made below.

Bell's Theorem

For discussion of Bell's theorem, first we return to the configuration of Problem 11.81 relevant to electron–positron decay. The appropriate antisymmetric spin-zero state is given by

$$(11.160) \qquad\qquad \xi_A(1, 2) = \frac{1}{\sqrt{2}}[\alpha(1)\beta(2) - \alpha(2)\beta(1)]$$

Consider that two Stern–Gerlach devices are positioned at equal but opposite distances from the source A, as illustrated in Fig. 11.20. The setup is such that S-G(1) detects the projection of particle spin on the unit vector $\mathbf{a}$. Similarly, S-B(2) detects the projection of particle spin on the unit vector $\mathbf{b}$. We consider the expectation

$$(11.161) \qquad\qquad P(\mathbf{a}, \mathbf{b}) = \langle \xi_A | (\hat{\boldsymbol{\sigma}}_1 \cdot \mathbf{a})(\hat{\boldsymbol{\sigma}}_2 \cdot \mathbf{b}) | \xi_A \rangle$$

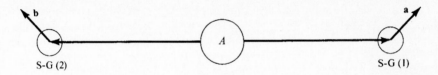

FIGURE 11.20   **Two S-G devices situated at equal but opposite distances from the decay process at** *S*.

where $\hat{\boldsymbol{\sigma}}$ is the Pauli spin matrix (11.82). Inserting the spin state (11.160) into the preceding gives (see Problem 11.103)

$$(11.162) \qquad P_{QM}(\mathbf{a}, \mathbf{b}) = -\mathbf{a} \cdot \mathbf{b}$$

Here we have inserted the subscript QM to denote that the expectation is quantum-mechanical.

   A formulation of this average stemming from hidden variables proceeds as follows. We introduce a hidden variable λ such that to each state of the two-particle system there corresponds a definite value of λ. Let $\rho(\lambda)$ denote the probability density of λ. Then with A and B denoting the results of measurements of the spin projections onto the directions $\mathbf{a}$ and $\mathbf{b}$, respectively, the expectation $P(\mathbf{a}, \mathbf{b})$ becomes

$$(11.163) \qquad P_h(\mathbf{a}, \mathbf{b}) = \int d\lambda \, \rho(\lambda) A(\mathbf{a}, \lambda) B(\mathbf{b}, \lambda)$$

where h denotes "hidden variables," and

$$(11.164a) \qquad \int \rho(\lambda) \, d\lambda = 1$$

Furthermore, with the above interpretation of A and B we set

$$(11.164b) \quad |A(\mathbf{a}', \lambda)| \leq 1, \qquad |B(\mathbf{b}', \lambda)| \leq 1, \qquad \text{for all } \mathbf{a}', \mathbf{b}'$$

Note that in writing (11.163), it is tacitly assumed that measurements of the two spin projections do not influence one another. With apparatuses for these measurements spatially removed from one another, such as in the present situation, the preceding assumption is an example of the *principle of locality*, which is a basic premise in physics.

   Stemming from (11.163) and the conditions (11.164b), we obtain *Bell's inequality* (see Problem 11.105).

$$(11.165) \quad S \equiv |P(\mathbf{a}, \mathbf{b}) - P(\mathbf{a}, \mathbf{b}')| + |P(\mathbf{a}', \mathbf{b}') + P(\mathbf{a}', \mathbf{b})| \leq 2$$

where $\mathbf{a}'$, $\mathbf{b}'$ are arbitrary unit vectors. Note in particular that this result is independent of the probability density $\rho(\lambda)$.

We may also form the ratio (11.165) employing $P_{QM}(\mathbf{a}, \mathbf{b})$ as given by (11.162). Thus, with $\cos^{-1}(\mathbf{a}' \cdot \mathbf{a}) = \phi$, $\cos^{-1}(\mathbf{a} \cdot \mathbf{b}) = \theta$, and $\cos^{-1}(\mathbf{b}, \mathbf{b}') = \gamma$, and assuming that these angles are coplanar, the difference S becomes

$$(11.166) \quad S_{QM} = |\cos\theta - \cos(\theta + \gamma)| + |\cos(\theta - \phi + \gamma) + \cos(\theta - \phi)| \leq 2$$

In what sense is the expression (11.166) quantum mechanical? The answer is that it stems from the quantum mechanical expectation (11.161), which in turn involves the superposition state ξ_A given by (11.160). This state, as described in Section 5.1, is purely quantum-mechanical. It describes each particle as being partly in the α ("spin up") state and the β ("spin down") state.

With specific choice of unit vectors: $\mathbf{b} = \mathbf{a}'$, $\mathbf{a} \cdot \mathbf{b} = \mathbf{b} \cdot \mathbf{b}' = \cos\theta$, the preceding relation becomes (see Problem 11.106)

$$(11.167) \qquad\qquad S_{QM} = 2\cos\theta - \cos 2\theta \leq 1$$

This inequality is violated over an interval of θ with maximum violation at $\theta = \pi/3$, where $S_{QM} = \frac{3}{2}$. We may conclude that quantum mechanics is in conflict with assumptions employed in the derivation of Bell's inequality (11.165) (e.g., the principle of locality). The proof of the existence of such conflicts with quantum mechanics is the essence of Bell's theorem.

We return now to the two-photon decay process, discussed at the start of this section. In applying Bell's analysis to this process, the unit vectors, $\mathbf{a}$, $\mathbf{b}$, introduced above, are associated with respective polarization axes of the two polarizers (see Fig. 11.17). In this representation relevant to Bell's analysis, the particle decay process depicted in Fig. 11.17 and the two-photon decay process depicted in Fig. 11.20 are effectively equivalent. Thus, one expects the two-photon decay process also to carry an inequality violation paralleling (11.167).

A generalization of Bell's inequality (11.165) was obtained by Clauser and Horn.[1] When applied to the two-photon decay process one finds

$$(11.168) \qquad\qquad 3N(\chi) - N(3\chi) \leq 1$$

where $N(\chi)$, as given by (11.159), is interpreted as probability of coincidence. Substituting this value for $N(\chi)$ into the preceding equation indicates that the inequality is violated over angular intervals at $\chi \geq 0$ and $\chi \leq \pi$. Again, we are confronted with conflicts between assumptions leading to the inequality (11.168) and quantum mechanics.

---

[1] J. F. Clauser and M. A. Horn, *Phys. Rev.* **D 10**, 526 (1974). See also, L. E. Ballentine, *Quantum Mechanics*, Prentice-Hall, Englewood Cliffs, N.J., 1990, Chapter 20.

However, it should be borne in mind that the preceding statements of inequality are idealized in the sense that they do not incorporate limitations of the apparatus of an actual experiment. Results to date, stemming from experiments performed in the laboratory, indicate that quantum mechanics is consistent.[1]

PROBLEMS

11.91   In actual S-G experiments, measurement of particle spin through direct observation of deflection of such particles runs into difficulty with the uncertainty relation if particle mass is too small.[2] Thus, in the original S-G experiment, electron spin was measured through observation of a beam of silver atoms. Such atoms contain an uncoupled electron in the $5s$ shell.

Consider a beam of particles of mass M and spin $\frac{1}{2}$, propagating in the $+x$ direction. The beam has cross section d. It interacts with an S-G apparatus whose field is in the z direction. Employing relevant uncertainty relations, show that the smallest uncertainty in normal displacements (Δz) grows large with decreasing mass M.

Answer

The uncertainty relation indicates that the minimum perpendicular component of momentum of a particle in a beam of cross section d is

$$p_z \simeq \frac{\hbar}{d}$$

Let the beam pass through the S-G apparatus in time t. Then the related spread in normal displacements of particles is given by

$$\Delta z \simeq \frac{p_z}{M} t = \frac{\hbar}{Md} t$$

We may view t as an interval over which energy states of the particles are measured. For the case at hand, particles have two energy states and $\Delta E = E_+ - E_-$. With Problem 5.28 we then write

$$t \gtrsim \frac{\hbar}{\Delta E}$$

---

[1] A compilation of experimental results in this context is given by B. d'Espagnat, *Scientific American* **241**, 158 (Nov. 1979). See also, J. F. Clauser and A. Shimony, *Repts. Prog. Phys.* **41**, 1881 (1978); M. Redhead, *Incompleteness, Nonlocality, and Realism*, Clarendon Press, Oxford, 1987, Chapter 4.

[2] I am indebted to Norman Ramsey for bringing this problem to my attention and for reference to Pauli's discussion on this topic; see *Hand. der Physik* **XXIV**, 1st part, 83–272 (1933).

Substituting in the preceding equation gives

$$\Delta z \simeq \frac{\hbar t}{Md} \gtrsim \frac{\hbar^2}{Md\,\Delta E}$$

Thus we find that the smallest transverse uncertainty spread in the beam grows large with decreasing mass.

11.92 (a) How many spin states are there in the uncoupled representation for three electrons?

(b) Calling individual particle states $\alpha(1)$, $\beta(1)$, $\alpha(2)$, etc., write down these uncoupled spin states.

11.93 With the states found in Problem 11.92 used as basis states, employ a Clebsch–Gordan expansion to find the normalized coupled spin states for three electrons. In listing these states, in addition to spin quantum numbers (s, m_s), designate also $s_{1,2}$, the quantum number corresponding to the subcomponent spin, $\mathbf{s}_{1,2} = \mathbf{s}_1 + \mathbf{s}_2$.

Answer

With the notation

$$(\alpha\beta\beta) \equiv \alpha(1)\beta(2)\beta(3)$$

the eight coupled spin states are given by

$$|s_{12}, s, m_s\rangle = \left|1, \frac{3}{2}, \frac{3}{2}\right\rangle = (\alpha\alpha\alpha)$$

$$\left|1, \frac{3}{2}, \frac{1}{2}\right\rangle = \frac{1}{\sqrt{3}}[(\alpha\alpha\beta) + (\beta\alpha\alpha) + (\alpha\beta\alpha)]$$

$$\left|1, \frac{3}{2}, -\frac{1}{2}\right\rangle = \frac{1}{\sqrt{3}}[(\beta\beta\alpha) + (\alpha\beta\beta) + (\beta\alpha\beta)]$$

$$\left|1, \frac{3}{2}, -\frac{3}{2}\right\rangle = (\beta\beta\beta)$$

$$\left|1, \frac{1}{2}, \frac{1}{2}\right\rangle = \sqrt{\frac{2}{3}}(\alpha\alpha\beta) - \sqrt{\frac{1}{6}}[(\beta\alpha\alpha) + (\alpha\beta\alpha)]$$

$$\left|1, \frac{1}{2}, -\frac{1}{2}\right\rangle = \sqrt{\frac{1}{6}}[(\alpha\beta\beta) - (\beta\alpha\beta)] - \sqrt{\frac{2}{3}}(\beta\beta\alpha)$$

$$\left|0, \frac{1}{2}, \frac{1}{2}\right\rangle = \frac{1}{\sqrt{2}}[(\alpha\beta\alpha) - (\beta\alpha\alpha)]$$

$$\left|0, \frac{1}{2}, -\frac{1}{2}\right\rangle = \frac{1}{\sqrt{2}}[(\alpha\beta\beta) - (\beta\alpha\beta)]$$

11.94   An important theorem in quantum mechanics addresses coupled spin states for three electrons. The theorem states that a *coupled* spin state which is antisymmetric with respect to all three particles vanishes identically.[1] Another theorem states that if $u(1, 2, 3)$ is antisymmetric with respect to interchange of particles 1 and 2, then

$$w(1, 2, 3) = u(1, 2, 3) + u(2, 3, 1) + u(3, 1, 2)$$

is totally antisymmetric.

Consider the spin state $|0, \frac{1}{2}, \frac{1}{2}\rangle$ given in Problem 11.93, which is antisymmetric with respect to particles 1 and 2 but not 3. Show that construction of the totally antisymmetric function $w(1, 2, 3)$ for this case gives $w = 0$.

11.95   (a)   Repeat Problem 11.92 for four electrons.

(b)   What are the possible total s values for this situation?

(c)   Working in the notation of Problem 11.93, in the coupled representation, write down an antisymmetric spin state for four electrons. For which particles is your state antisymmetric?

11.96   In relativistic formulations one works with observables called *four-vectors*. For example, an event which occurs in a frame S at the coordinates x, y, z at the time t is represented by the four-vector $X = (x, y, z, ict)$. Let the frame S' be in relative motion with respect to S with velocity v parallel to the z axis. If at $t = 0$ the two frames are coincident, then the event X as observed in S' has components $X' = (x', y', z', ict')$ which are given by the *Lorentz transformation* (where $\beta \equiv v/c$ and $\gamma^2 \equiv 1 - \beta^2$)

$$X' = \hat{L}X, \qquad \hat{L} = \begin{pmatrix} 1 & 0 & 0 & 0 \\ 0 & 1 & 0 & 0 \\ 0 & 0 & \gamma & i\beta\gamma \\ 0 & 0 & -i\beta\gamma & \gamma \end{pmatrix}$$

(The Lorentz transformation is discussed in detail in Section 15.1.)

(a)   Show that $\hat{L}$ is orthogonal and that $\hat{L}$ preserves the inner product (i.e., $X_1 \cdot X_2 = X'_1 \cdot X'_2$).

(b)   Show that the Lorentz transformation reduces to the *Galilian transformation*, $z' = z - vt$, $t' = t$, in the limit $\beta \ll 1$.

11.97   With reference to (11.140) it is generally the case that the commutator $[\hat{A}, \hat{B}]$ is equal to $i\hbar$ times the Poisson bracket considered as an operator.[2] For this property to be satisfied it is required that the Poisson bracket be symmetrized in its arguments. Show that

$$[x^2, p^2] = i\hbar[x^2, p^2]_{\mathrm{Cl}}$$

---

[1] For further discussion see, R. L. Liboff, *Am. J. Phys.* **52**, 561 (1984). This topic is returned to in Section 12.4.

[2] For further discussion, see D. Bohm, *Quantum Theory*, Prentice-Hall, New York, 1951.

Answer

First, we write

$$[x^2, p^2]_{\text{cl}} = \frac{1}{2}([x^2, p^2] - [p^2, x^2]) = 2(xp + px)$$

where the symmetrized term on the right is considered as an operator. For the commutator we find

$$[x^2, p^2] = 2i\hbar(xp + px)$$

which is seen to agree with the given rule.

11.98 Establish the following two vector-operator equalities:

(a)
$$[\hat{\mathbf{J}} \cdot \hat{\mathbf{A}}, \hat{\mathbf{J}}] = \hat{\mathbf{J}} \times \hat{\mathbf{A}}$$

where $\hat{\mathbf{A}}$ is a constant operator.

(b)
$$[\hat{J}_x, \hat{V}_y] = i\hbar\hat{V}_z$$

where $\hat{\mathbf{V}}$ is a vector operator.

11.99 Write down the Clebsch–Gordan expansion for two coupled electrons with $j = 2$, $m_j = 0$, $l = 1$, $s = 1$, in terms of the product states $|lm_l\rangle|sm_s\rangle$. Evaluate coefficients.

11.100 Show that the total angular momentum of two fermions is always an integer. (*Note:* This problem establishes that two coupled fermions constitute a *composite boson.*)

Answer

The total angular momentum operator of two fermions is

$$\hat{\mathbf{J}} = \hat{\mathbf{S}}_1 + \hat{\mathbf{S}}_2 + \hat{\mathbf{L}}$$

where **L** is the orbital angular momentum of the two-particle system. With $s_1 = n_1/2$ and $s_2 = n_2/2$, where n_1 and n_2 are odd integers, we find the total spin numbers

$$s = \left(\frac{n_1}{2} - \frac{n_2}{2}\right), \ldots, \left(\frac{n_1}{2} + \frac{n_2}{2}\right)$$

which are all integers. Total j is then given by

$$j = l - s, \ldots, l + s$$

which, for all s given above, are integers.

11.101 Employing the vector operator formula given in Problem 11.68, show that

$$\frac{\{\hat{\boldsymbol{\sigma}} \cdot [\hat{\mathbf{p}} - (e/c)\hat{\mathbf{A}}]\}^2}{2m} = \frac{[\hat{\mathbf{p}} - (e/c)\hat{\mathbf{A}}]^2}{2m} - \frac{e\hbar}{2mc}\hat{\boldsymbol{\sigma}} \cdot \hat{\mathbf{B}}$$

where $\hat{\mathbf{A}}$ is vector potential.

11.102 A one-dimensional harmonic oscillator is initially in the state $|\psi(0)\rangle = |1\rangle$ [in the notation of (7.61)]. The system is acted upon by an exponentially decaying potential in time. The total Hamiltonian is given by

$$\hat{H} = \hbar\omega_0\left(\hat{N} + \frac{1}{2}\right) + (\hat{a} + \hat{a}^\dagger)V_0 e^{-(t/\tau)}$$

where $V_0 \ll \hbar\omega_0$ and τ is a constant relaxation time.

(a) Working in the interaction picture, calculate $|\psi_I(t)\rangle$ to first order in $\lambda \equiv V_0/\hbar\omega_0$.

(b) From your answer to part (a), obtain the wavefunction in the Schrödinger picture, $|\psi(t)\rangle$.

(c) What is the probability $P_{1\to n}$ that a transition to the state $|n\rangle$ has occurred after time t due to the perturbation?

Answers

(a) First note that

$$|\psi_I(0)\rangle = |\psi(0)\rangle = |1\rangle$$

and

$$\begin{aligned}\hat{V}_I(t)|\psi_I(0)\rangle &= e^{it\hat{H}_0/\hbar}\hat{V}e^{-it\hat{H}_0/\hbar}|1\rangle \\ &= e^{it\hat{H}_0/\hbar}(\hat{a} + \hat{a}^\dagger)V_0 e^{-t/\tau}e^{-i3\omega_0 t/2}|1\rangle \\ &= e^{it\hat{H}_0/\hbar}(|0\rangle + \sqrt{2}|2\rangle)V_0 e^{-t/\tau}e^{-i3\omega_0 t/2} \\ &= V_0[e^{-(t/\tau)-i\omega_0 t}|0\rangle + \sqrt{2}e^{-(t/\tau)+i\omega_0 t}|2\rangle]\end{aligned}$$

Substituting this expression into (11.145) and integrating, we obtain

$$|\psi_I(t)\rangle = |1\rangle + \frac{iV_0}{\hbar\omega_0}[\Lambda(t)|0\rangle + \Lambda^*(t)\sqrt{2}|2\rangle]$$

where

$$\Lambda(t) \equiv \frac{e^{-t\phi} - 1}{\phi/\omega_0}$$

$$\phi \equiv \frac{1}{\tau} + i\omega_0$$

(b) This answer is obtained by substituting the preceding expression in the inverse of (11.141). There results

$$|\psi(t)\rangle = \Lambda e^{-it\omega_0/2}|0\rangle + \Lambda^* e^{-it5\omega_0/2}\sqrt{2}|2\rangle$$

(c) We may view the preceding result as an expansion of $|\psi(t)\rangle$ in the basis $\{|n\rangle\}$ and further consider the state $|\psi(t)\rangle$ as given in part (b) to be normalized to $O(\lambda)$, where $\lambda \equiv \hbar\omega_0/V_0$. There results

$$P_{1 \to 1} = 1 + O(\lambda^2)$$

$$P_{1 \to 0} = |\langle 0 | \psi(t) \rangle|^2 = \lambda^2 |\Lambda|^2 + O(\lambda^3)$$

$$P_{1 \to 2} = |\langle 2 | \psi(t) \rangle|^2 = 2\lambda^2 |\Lambda|^2 + O(\lambda^3)$$

where

$$|\Lambda|^2 = \frac{1 - e^{-2t/\tau} - 2e^{-t/\tau} \cos \omega_0 t}{1 + (1/\omega_0 \tau)^2}$$

Thus we find it is most probable for the oscillator to remain in the first excited state. The probability for a transition to the second excited state is twice as probable as that for a transition to the ground state.

11.103 Employing the spin-zero antisymmetric state (11.160), evaluate (11.161) to obtain (11.162). *Hint:* See Problem 11.68.

11.104 Show that $|P_{12}\rangle$ as given by (11.153) follows from (11.152) through the transformation (11.147–150).

11.105 Employing the conditions (11.164b), establish Bell's inequality (11.165).

Answer

Introduce the notation $A(\lambda, \mathbf{a}) = A(\mathbf{a})$. First we note

$$|P(\mathbf{a}, \mathbf{b}) - P(\mathbf{a}, \mathbf{b}')| = \left| \int A(\mathbf{a})B(\mathbf{b})\rho(\lambda) \, d\lambda - \int A(\mathbf{a})B(\mathbf{b}')\rho(\lambda) \, d\lambda \right.$$

$$\pm \left. \left[\int A(\mathbf{a})B(\mathbf{b})A(\mathbf{a}')B(\mathbf{b}')\rho(\lambda) \, d\lambda - \int A(\mathbf{a})B(\mathbf{b}')A(\mathbf{a}')B(\mathbf{b})\rho(\lambda) \, d\lambda \right] \right|$$

$$= \left| \int A(\mathbf{a})B(\mathbf{b})[1 \pm A(\mathbf{a}')B(\mathbf{b}')]\rho(\lambda) \, d\lambda - \int A(\mathbf{a})B(\mathbf{b}')[1 \pm A(\mathbf{a}')B(\mathbf{b})]\rho(\lambda) \, d\lambda \right|$$

$$\leq \int |A(\mathbf{a})B(\mathbf{b})| [1 \pm A(\mathbf{a}')B(\mathbf{b}')]\rho(\lambda) \, d\lambda + \int |A(\mathbf{a})B(\mathbf{b}')| [1 \pm A(\mathbf{a}')B(\mathbf{b})]\rho(\lambda) \, d\lambda$$

$$\leq 2 \pm [P(\mathbf{a}', \mathbf{b}') + P(\mathbf{a}', \mathbf{b})]$$

where we have used $|AB| \leq 1$. Thus

$$|P(\mathbf{a}, \mathbf{b}) - P(\mathbf{a}, \mathbf{b}')| \pm [P(\mathbf{a}', \mathbf{b}') + P(\mathbf{a}', \mathbf{b})] \leq 2$$

which implies (11.165).

11.106 Derive (11.167) from (11.166). *Hint:* Recall the rules:

   (i)   If $|A| + |B| \leq n$, then $|A + B| \leq n$.

   (ii) If $|A| \leq n$, then $-n \leq A$.

11.14 THE TRANSFER MATRIX

With matrix properties developed in this chapter, we return to Section 7.7 addressing the problem of transmission of particles through a potential barrier. Matrices are introduced which connect coefficients of incident and reflected waves. Employing reflection symmetry (Section 6.4), time reversibility (Problem 6.33), and conservation of current, properties of these matrices are developed which allow wavefunctions to be constructed. In the following section these rules are applied to the Kronig–Penny model (Section 8.2), and the general dispersion relation implying band structure is regained. This technique of solution is commonly referred to as the *transfer matrix method* and is important in the study of quantum-well microdevices[1] as well as in basic science.[2]

We recall (7.138), which are generalized to read

$$(11.169) \qquad \varphi(x) = \begin{cases} Ae^{ikx} + Be^{-ikx} & (x < -a) \\ Ce^{-\kappa x} + De^{\kappa x} & (-a < x < a) \\ Fe^{ikx} + Ge^{-ikx} & (x < a) \end{cases}$$

(with k_1 written as k). These relations pertain to a potential barrier of width $2a$ centered at the origin, $x = 0$. Furthermore, $(A \exp ikx)$ is an incident wave from the left, $(B \exp -ikx)$ is the reflected wave, and $(F \exp ikx)$ is the transmitted wave. In Section 7.7, the transmission coefficient was obtained by matching wavefunction components at $x = -a$ and $x = a$. The present analysis takes the point of view that F and G coefficients are given and that a matrix $\hat{S}$ exists that determines the coefficients A and B in terms of F and G. This relation is written

$$(11.170) \qquad \begin{pmatrix} A \\ B \end{pmatrix} = \begin{pmatrix} S_{11} & S_{12} \\ S_{21} & S_{22} \end{pmatrix} \begin{pmatrix} F \\ G \end{pmatrix} \equiv \hat{S} \begin{pmatrix} F \\ G \end{pmatrix}$$

which defines the "transfer matrix" $\hat{S}$. We wish to determine general properties of the $\hat{S}$ matrix from fundamental properties of the system.

Reflection Symmetry

We note that the potential barrier of Fig. 7.22 is an even function of x. It follows that if $\varphi(x)$ is a solution of the wave equation, then so is $\varphi(-x)$. Replacing x by $-x$ in (11.169) gives

$$(11.171) \qquad \overline{\varphi} = \begin{cases} Ae^{-ikx} + Be^{+ikx} & (x > a) \\ Ce^{\kappa x} + De^{-\kappa x} & (-a < x < a) \\ Fe^{-ikx} + Ge^{ikx} & (x < -a) \end{cases}$$

---

[1] R. C. Leavens and R. Taylor, eds., *Interfaces, Quantum Wells and Superlattices,* Plenum, New York, 1988.

[2] M. L. Mehta, *Random Matrices,* Academic Press, Boston, 1991.

In this relation $(G \exp ikx)$ is the incident wave from the left, $(F \exp -ikx)$ is the reflected wave, and $(B \exp ikx)$ is the transmitted wave. It follows that the relevant matrix equation connecting these coefficients may be obtained from (11.170) by the replacements $A \leftrightarrow G$, $B \leftrightarrow F$. There results

(11.172a)
$$\begin{pmatrix} G \\ F \end{pmatrix} = \begin{pmatrix} S_{11} & S_{12} \\ S_{21} & S_{22} \end{pmatrix}\begin{pmatrix} B \\ A \end{pmatrix}$$

or, equivalently

(11.172b)
$$\begin{pmatrix} F \\ G \end{pmatrix} = \begin{pmatrix} S_{22} & S_{21} \\ S_{12} & S_{11} \end{pmatrix}\begin{pmatrix} A \\ B \end{pmatrix}$$

Substituting this relation into (11.170) gives

(11.173a)
$$\begin{pmatrix} S_{11} & S_{12} \\ S_{21} & S_{22} \end{pmatrix}\begin{pmatrix} S_{22} & S_{21} \\ S_{12} & S_{11} \end{pmatrix} = \begin{pmatrix} 1 & 0 \\ 0 & 1 \end{pmatrix} \equiv \hat{I}$$

where $\hat{I}$ is the identity operator. It follows that

(11.173b)
$$\hat{S}^{-1} = \begin{pmatrix} S_{22} & S_{21} \\ S_{12} & S_{11} \end{pmatrix}$$

Time Reversibility

With Problem 7.82 we note that the time-reversed solution corresponding to (11.169) is given by

(11.174)
$$\varphi_T(x) = \begin{cases} A^*e^{-ikx} + B^*e^{ikx} & (x < -a) \\ C^*e^{-\kappa x} + D^*e^{\kappa x} & (-a < x < a) \\ F^*e^{-ikx} + G^*e^{ikx} & (a < x) \end{cases}$$

In this picture $(B^* \exp ikx)$ is the incident wave from the left, $(A^* \exp -ikx)$ is the reflected wave, and $(G^* \exp ikx)$ is the transmitted wave. With reference to (11.170) we write

(11.175a)
$$\begin{pmatrix} B^* \\ A^* \end{pmatrix} = \begin{pmatrix} S_{11} & S_{12} \\ S_{21} & S_{22} \end{pmatrix}\begin{pmatrix} G^* \\ F^* \end{pmatrix}$$

Taking the complex conjugate of this equation and rewriting the relation in normal form gives

(11.175b)
$$\begin{pmatrix} A \\ B \end{pmatrix} = \begin{pmatrix} S_{22}^* & S_{21}^* \\ S_{12}^* & S_{11}^* \end{pmatrix}\begin{pmatrix} F \\ G \end{pmatrix}$$

Comparison with (11.170) gives

(11.175c)
$$\begin{pmatrix} S_{22}{}^* & S_{21}{}^* \\ S_{12}{}^* & S_{22}{}^* \end{pmatrix} = \begin{pmatrix} S_{11} & S_{12} \\ S_{21} & S_{22} \end{pmatrix}$$

so that

(11.175d)
$$S_{11}{}^* = S_{22}, \qquad S_{12}{}^* = S_{21}$$

With (11.173) these latter relations give

(11.176a)
$$\begin{pmatrix} S_{11} & S_{12} \\ S_{12}{}^* & S_{11}{}^* \end{pmatrix}\begin{pmatrix} S_{11}{}^* & S_{12}{}^* \\ S_{12} & S_{11} \end{pmatrix} = \begin{pmatrix} 1 & 0 \\ 0 & 1 \end{pmatrix}$$

Multiplying matrices we find

(11.176b)
$$|S_{11}|^2 + |S_{12}|^2 = 1$$

(11.176c)
$$S_{11}(S_{12} + S_{12}{}^*) = 0$$

With these conditions on the components of the transfer matrix at hand, we turn to the third invariance property of the system.

Current Continuity

For time-independent configurations in one dimension, the continuity equation (7.97) gives constant current which, when applied to the wavefunction (11.171), gives

(11.177a)
$$|A|^2 - |B|^2 = |F|^2 - |G|^2$$

or, equivalently,

(11.177b)
$$(A \quad B)\begin{pmatrix} 1 & 0 \\ 0 & -1 \end{pmatrix}\begin{pmatrix} A^* \\ B^* \end{pmatrix} = (F \quad G)\begin{pmatrix} 1 & 0 \\ 0 & -1 \end{pmatrix}\begin{pmatrix} F^* \\ G^* \end{pmatrix}$$

Substituting the transpose of (11.170) and its complex conjugate into the left side of the preceding equation indicates that

(11.177c)
$$\hat{S}\begin{pmatrix} 1 & 0 \\ 0 & -1 \end{pmatrix}\hat{S}^* = \begin{pmatrix} 1 & 0 \\ 0 & -1 \end{pmatrix}$$

This relation gives the additional condition

(11.177d)
$$|S_{11}|^2 - |S_{12}|^2 = 1$$

This latter equation together with (11.175d) and (11.176c) indicates that S_{12} is purely imaginary.

Writing

(11.178)
$$\hat{S} \equiv \begin{pmatrix} \alpha_1 + i\beta_1 & \alpha_2 + i\beta_2 \\ \alpha_3 + i\beta_3 & \alpha_4 + i\beta_4 \end{pmatrix}$$

we may conclude that

(11.179a) $S_{12} = -S_{21} = i\beta_2,$ $S_{11} = S_{22}{}^* = \alpha_1 + i\beta_1$

When combined with (11.177d) the preceding relations give

(11.179b)
$$\alpha_1{}^2 + \beta_1{}^2 - \beta_2{}^2 = 1$$

We may then finally write

(11.180a)
$$\hat{S} = \begin{pmatrix} \alpha_1 + i\beta_1 & i\beta_2 \\ -i\beta_2 & \alpha_1 - i\beta_1 \end{pmatrix}$$

With (11.179b) we find

(11.180b)
$$\det \hat{S} = 1$$

and with (11.173b) we find

(11.180c)
$$\hat{S}^{-1} = \hat{S}*$$

The transfer matrix and related properties given by (11.180) are derived from fundamental principles and are independent of wavefunction boundary conditions. The transfer matrix is relevant to a wave packet incident on a barrier of arbitrary but smooth form, which vanishes at large displacement from the barrier center. Wave functions far removed from the barrier have the structure given by the first and third equations in (11.169). For general solution to a given problem one must express the parameters of $\hat{S}$ in terms of energy, for which boundary conditions come into play.

Let us apply the preceding formalism to construct expressions for the transmission and reflection coefficients T and R, respectively, in terms of parameters of α and β. With reference to (11.170) and matrix values (11.180a), and in the notation of Problem 7.44, we write

(11.181)
$$\sqrt{T}e^{i\phi_T} = \frac{F}{A} = \frac{1}{\alpha_1 + i\beta_1}$$

$$\sqrt{R}e^{i\phi_R} = \frac{B}{A} = \frac{-i\beta_2}{\alpha_1 + i\beta_1}$$

These expressions, with (11.179b), return the equality

(11.182)
$$T + R = \frac{1 + \beta_2{}^2}{\alpha_1{}^2 + \beta_1{}^2} = 1$$

[This equality follows also from (11.177a) with $G = 0$.]

Kronig–Penny Model Revisited

We wish to reformulate the Kronig–Penny model in terms of the transfer matrix formalism. Our aim is to rederive the basic dispersion relation (8.53) which, as previously noted, implies a band structure of allowed energies.

For consistency, we choose the periodic potential in accord with (11.169), namely, there is a potential barrier in the interval $(-a, a)$ and a potential valley in the interval $(a, a + 2b)$. The period of the periodic array is $d = 2(a + b)$. In the valley domains, $(a - d) < (x - nd) < -a$, the potential $V = 0$. Solutions in these domains are given by the free-particle solutions

$$(11.183) \qquad \varphi_n(x) = A_n e^{ik(x-nd)} + B_n e^{-ik(x-nd)}$$

With barrier centers at $x = nd$, successive values of coefficients may be obtained employing the transfer matrix S, as given by (11.180a). There results

$$(11.184) \qquad \begin{pmatrix} A_n \\ B_n \end{pmatrix} = \begin{pmatrix} \alpha_1 + i\beta_1 & i\beta_2 \\ -i\beta_2 & \alpha_1 - i\beta_1 \end{pmatrix} \begin{pmatrix} A_{n+1} e^{-ikd} \\ B_{n+1} e^{ikd} \end{pmatrix}$$

or, equivalently,

$$(11.185a) \qquad \begin{pmatrix} A_{n+1} \\ B_{n+1} \end{pmatrix} = \hat{Q} \begin{pmatrix} A_n \\ B_n \end{pmatrix}$$

where the reduced transfer matrix, $\hat{Q}$, is given by

$$(11.185b) \qquad \hat{Q} \equiv \begin{pmatrix} (\alpha_1 - i\beta_1)e^{ikd} & -i\beta_2 e^{ikd} \\ i\beta_2 e^{-ikd} & (\alpha_1 + i\beta_1)e^{-ikd} \end{pmatrix}$$

With the constraint (11.179b), we find

$$(11.185c) \qquad \det Q = 1$$

Note that consistent with the symmetry of the periodic array, $\hat{Q}$ is independent of the barrier number, n. Note further that $\hat{Q}$ may be rewritten,

$$\hat{Q} = \hat{D}\hat{S}$$

$$(11.185d)$$

$$\hat{D} \equiv \begin{pmatrix} e^{ikd} & 0 \\ 0 & e^{-ikd} \end{pmatrix}$$

In accord with Problem 11.22 we regain (11.185c). Furthermore, with reference to Problem 4.34, we may write

$$\hat{Q}^{-1} = \hat{S}^{-1}\hat{D}^{-1}$$

which gives

$$(11.185e) \qquad \hat{Q}^{-1} = \begin{pmatrix} (\alpha_1 + i\beta_1)e^{-ikd} & i\beta_2 e^{ikd} \\ -i\beta_2 e^{-ikd} & (\alpha_1 - i\beta_1)e^{ikd} \end{pmatrix}$$

Criteria for Bounded Wavefunction

Iterating (11.185a) we obtain

(11.186)
$$\begin{pmatrix} A_n \\ B_n \end{pmatrix} = \hat{Q}^n \begin{pmatrix} A_0 \\ B_0 \end{pmatrix}$$

for wavefunction amplitudes in the nth cell of the lattice.

In the present formulation it is important to ensure that wavefunctions remain finite in the limit of large n. This condition may be examined in terms of the eigenvalues of $\hat{Q}$, which in turn are roots of the characteristic equation

(11.187a)
$$q^2 - q \operatorname{Tr} \hat{Q} + \det \hat{Q} = 0$$

or, equivalently

(11.187b)
$$q^2 - 2q(\alpha_1 \cos kd + \beta_1 \sin kd) + 1 = 0$$

This quadratic expression has the roots

(11.187c)
$$q_{\pm} = \frac{1}{2}\left[\operatorname{Tr} \hat{Q} \pm \sqrt{(\operatorname{Tr} \hat{Q})^2 - 4} \right]$$

We may choose constants of the initial state (A_0, B_0) so that these states are eigenstates of $\hat{Q}$ corresponding to the eigenvalues, $q_{\pm}$. That is,

(11.188a)
$$\hat{Q}\begin{bmatrix} A_0^{(\pm)} \\ B_0^{(\pm)} \end{bmatrix} = q_{\pm}\begin{bmatrix} A_0^{(\pm)} \\ B_0^{(\pm)} \end{bmatrix}$$

For these solutions, (11.186) gives

(11.188b)
$$\begin{bmatrix} A_n^{(\pm)} \\ B_n^{(\pm)} \end{bmatrix} = (q_{\pm})^n \begin{bmatrix} A_0^{(\pm)} \\ B_0^{(\pm)} \end{bmatrix}$$

With (11.187c) we note that $q_{\pm}$ eigenvalues are real if $|\operatorname{Tr} Q| > 2$. In this case $|q_{\pm}|^n$ goes to infinity in either of the two limits, $n \to \pm\infty$, in which case wavefunctions grow unbounded. A finite solution is guaranteed, providing

(11.189a)
$$|\operatorname{Tr} \hat{Q}| \le 2$$

which is satisfied if, for real γ,

(11.189b)
$$\frac{1}{2} \operatorname{Tr} \hat{Q} = \cos \gamma$$

which gives

(11.189c)
$$q_+ = e^{i\gamma}, \qquad q_- = e^{-i\gamma}$$

Recalling the expression for Tr Q [see (11.187b)], the preceding identifications give the following condition for allowed energy eigenvalues.

(11.190)
$$\cos \gamma = \alpha_1 \cos k_1 d + \beta_1 \sin k_1 d$$
$$|\cos \gamma| \leq 1$$

To identify k with the valley wave vector, we have relabeled it k_1. To proceed further, boundary conditions are brought into play. With reference to (7.140) relevant to propagation through a square barrier of width $2a$, and (11.185c) relating β_i parameters, one may express α_i and β_i in terms of k-propagation parameters. Substituting these values into (11.190) and setting $\gamma = kd$, returns the dispersion relation (8.53).[1]

PROBLEMS

11.107   Show explicitly that
   (a)   det $\hat{S} = 1$
   (b) $\hat{S}^{-1} = \hat{S}*$
where $\hat{S}$ is the transfer matrix (11.170).

11.108   Show explicitly that the reduced transfer matrix $\hat{Q}$, (11.185b), has the following properties:
   (a)   Tr $\hat{Q} = 2[\alpha_1 \cos (kd) + \beta_1 \sin (kd)]$
   (b)   Det $\hat{Q} = 1$
   (c) $\hat{Q}^{-1}\hat{Q} = \hat{I}$ [where $\hat{Q}^{-1}$ is given by (11.185e)]

11.109   Employing (7.140) and (11.185c), express the coefficients α_1 and β_2 in terms of $\mathbf{k}_1$ and $\mathbf{k}_2$ wave vectors.

11.110   List three properties on which the derivation of the transfer matrix (11.180a) is based.

11.111   An electron propagates through a semi-infinite periodic potential: $V(x) = V(x + a)$ for $x \geq 0$, $V(x) = V_0$ for $x < 0$, where $V_m < V_0$ and V_m represents the amplitude of the periodic component. Does the transfer matrix (11.180a) apply to this situation? Explain your answer.

11.112   Show that the expectation of an operator in the Schrödinger picture is equal to expectation in the Heisenberg picture.

Answer
First note that if $\langle \psi' | = \langle \psi | \hat{U}$, then $| \psi' \rangle = \hat{U}^{-1} | \psi \rangle$. There follows
$$\langle \psi' | \hat{A}' | \psi' \rangle = \langle \psi | \hat{U} \hat{U}^{-1} \hat{A} \hat{U} \hat{U}^{-1} | \psi \rangle = \langle \psi | \hat{A} | \psi \rangle$$

---

[1] For further discussion, see E. Merzbacher, *Quantum Mechanics*, 2d ed., Wiley, New York, 1970. *Note*: Merzbacher uses M, P where we use $\hat{S}$, $\hat{Q}$.

11.113 Consider a homogeneous rigid spherical particle of finite radius that is free to rotate about a fixed center.
    (a) What is the Hamiltonian of this system?
    (b) What are the eigenfunctions of this system?
    (c) Discuss the correspondence between quantum mechanical and classical motion of this system.

Answers

    (a) $H = L^2/2I$ (This equation follows from the equality of the three components of moments of inertia of a sphere.)
    (b) Any wavefunction describing this system must be invariant under arbitrary rotations about the origin. This leaves only the spherically symmetric S state, corresponding to $l = 0$. It should be noted that the same Hamiltonian as that given above applies to any rigid body with three equal moments of inertia. However, it is only for the rigid sphere that the said angular invariance applies.
    (c) As a classical sphere may rotate at any angular momentum about a fixed center, there is no correspondence with the quantum situation in which the S state is the only permitted state.

11.114 A uniform non-interacting beam of electrons with an isotropic distribution of spins, propagates in the z direction in a medium which includes a uniform magnetic field,

$$\mathbf{B} = (B_0, 0, 0)$$

    (a) Write down the Hamiltonian of an electron in this beam.
    (b) What are the eigenvalues, $E(k, s)$, and time-dependent eigenstates $\psi(k, s, t)$ of this Hamiltonian?
    (c) If the translational motion of the beam is constant and $B_0 = 1.4$ kG, at what frequency would incident electromagnetic radiation on the beam be absorbed (in Hz)?

12

APPLICATION TO ATOMIC, MOLECULAR, SOLID-STATE, AND NUCLEAR PHYSICS. ELEMENTS OF QUANTUM STATISTICS

Having developed methods for addition of spin angular momentum in Chapter 11 and properties of the three-dimensional Hamiltonian in Chapter 10, these formalisms, together with the Pauli principle, are now applied to some practical problems. Symmetry requirements on the wavefunction of the helium atom imposed by the Pauli principle serve to couple electron spins. This coupling results in a separation of the spectra into singlet and triplet series for helium as well as other two-electron atoms. Symmetrization requirements stemming from the Pauli principle are also maintained in calculation of the binding of the hydrogen molecule. The relevance of Bose–Einstein condensation to superconductivity and superfluidity is described. Application of the Pauli principle is further noted in two examples. In the first of these, semiconductor theory, previously encountered in Chapter 8, is extended to extrinsic (impurity) conductivity, in which the Fermi–Dirac distribution comes into play. The second example addresses nuclear physics, and with concepts of the nucleon and isotopic spin, the totally antisymmetric ground state of the deuteron is constructed.

12.1 THE TOTAL ANGULAR MOMENTUM, J

In this section we consider the addition of spin and orbital angular momentum for atomic systems. As previously noted, the total angular momentum of a system (e.g., an atom or a single electron) which has both orbital angular momentum **L** and spin angular momentum **S** is called **J**.

$$(12.1) \qquad \hat{\mathbf{J}} = \hat{\mathbf{L}} + \hat{\mathbf{S}}$$

It has components

$$\hat{J}_x = \hat{L}_x + \hat{S}_x$$
$$\hat{J}_y = \hat{L}_y + \hat{S}_y$$
$$\hat{J}_z = \hat{L}_z + \hat{S}_z$$

Its square appears as

$$(12.2) \qquad \hat{J}^2 = \hat{L}^2 + \hat{S}^2 + 2\hat{\mathbf{L}} \cdot \hat{\mathbf{S}}$$

In obtaining this expression we have used the fact that **L** and **S** commute.

L-S Coupling

Individual electrons in an atom have both orbital and spin angular momentum. Among the lighter atoms, individual electrons' **L** vectors couple to give a resultant **L** and individual **S** vectors couple to give a resultant **S**. These two vectors then join to give a total angular momentum **J** (Fig. 12.1). This is called the *L-S* or *Russell–Saunders coupling scheme.*[1]

Eigenstates in this representation are simultaneous eigenstates of the four commuting operators

$$(12.3) \qquad \{\hat{J}^2, \hat{J}_z, \hat{L}^2, \hat{S}^2\}$$

There are six pairs of operators in this set which must be checked for commutability.

(i) $[\hat{L}^2 + \hat{S}^2 + 2\hat{\mathbf{L}} \cdot \hat{\mathbf{S}}, \hat{L}^2] = 0$ (iv) $[\hat{L}^2, \hat{J}_z] = [\hat{L}^2, \hat{L}_z + \hat{S}_z] = 0$

(ii) $[\hat{L}^2 + \hat{S}^2 + 2\hat{\mathbf{L}} \cdot \hat{\mathbf{S}}, \hat{S}^2] = 0$ (v) $[\hat{S}^2, \hat{J}_z] = [\hat{S}^2, \hat{L}_z + \hat{S}_z] = 0$

(iii) $[\hat{J}^2, \hat{J}_z] = 0$ (vi) $[\hat{L}^2, \hat{S}^2] = 0$

---

[1] This scheme is also relevant to "one- or two-electron atoms." More generally, in heavy elements with large Z, the spin-orbit coupling (Section 12.2) becomes large and serves to couple $\mathbf{L}_i$ and $\mathbf{S}_i$ vectors of individual electrons, giving resultant $\mathbf{J}_1, \mathbf{J}_2, \ldots$, values. These individual electron $\mathbf{J}_i$ values then combine to give a resultant **J**. This coupling scheme is known as *j-j* coupling.

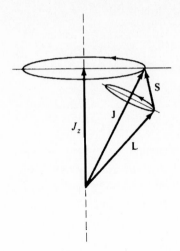

FIGURE 12.1 Schematic vector representation of the L-S scheme of angular momentum addition. J^2 and J_z are fixed, as are L^2 and S^2.

In (i), $\hat{L}^2$ commutes with all its components. In (ii), $\hat{S}^2$ commutes with all its components. The remaining relations are self-evident.

Eigenvalue equations related to the commuting operators (12.3) appear as

$$\hat{J}^2 |\, jm_j ls \rangle = \hbar^2 j(j+1)|\, jm_j ls \rangle$$

$$\hat{J}_z |\, jm_j ls \rangle = \hbar m_j |\, jm_j ls \rangle$$

$$\hat{L}^2 |\, jm_j ls \rangle = \hbar^2 l(l+1)|\, jm_j ls \rangle$$

$$\hat{S}^2 |\, jm_j ls \rangle = \hbar^2 s(s+1)|\, jm_j ls \rangle$$

For a given value of J, m_j runs in integral steps from $-j$ to $+j$.

A very important operator that commutes with all four operators (12.3) is $\hat{\mathbf{L}} \cdot \hat{\mathbf{S}}$.

$$\hat{\mathbf{L}} \cdot \hat{\mathbf{S}} |\, jm_j ls \rangle = \frac{1}{2}(\hat{J}^2 - \hat{L}^2 - \hat{S}^2)|\, jm_j ls \rangle$$

$$= (\hbar^2/2)[\, j(j+1) - l(l+1) - s(s+1)]|\, jm_j ls \rangle$$

Eigen-j-values and Term Notation

In the L-S representation, l and s are known. What are the possible j values corresponding to these values of l and s? Since **J** is the resultant of two angular momentum vectors, the rules of Section 9.4 apply. These rules indicate that j values run from a maximum of

$$j_{\max} = l + s$$

to a minimum of

$$j_{min} = |l - s|$$

in integral steps.

$$(l + s) \geq j \geq |l - s|$$
$$j = l + s, l + s - 1, l + s - 2, \ldots, |l - s| + 1, |l - s|$$

For $s < l$, there are a total of $(2s + 1) j$ values. The number $(2s + 1)$ is called the *multiplicity*. In the section to follow, these different j values are shown to correspond to distinct energy values for the atom. Thus for a one-electron atom $(s = \frac{1}{2})$, a state of given l splits into the doublet corresponding to the two values

$$j = l + \frac{1}{2}, \qquad j = l - \frac{1}{2}$$

(Fig. 12.2). The *term notation* for these states is given by the following symbol

$$^{2s+1}\mathscr{L}_j$$

where $\mathscr{L}$ denotes the letter corresponding to the orbital angular momentum l value according to the following scheme:

l	0	1	2	3	4	5	6	7	8	9	10	$\cdots$
Letter ($\mathscr{L}$)	S	P	D	F	G	H	I	K	L	M	N	$\cdots$

The doublet P states of one-electron atoms are denoted by the terms

$$^2P_{1/2}, \, ^2P_{3/2}$$

The doublet F states $(l = 3)$ are denoted by

$$^2F_{7/2}, \, ^2F_{5/2}$$

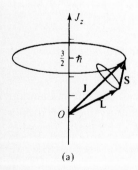

(a)

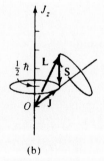

(b)

FIGURE 12.2   Diagrams depicting coupling of the L and S vectors of a single p electron, in the L-S scheme. The doublet contains two values of j.

In two-electron atoms, the resultant spin quantum number is either 0 or 1. These, we recall (Section 11.10), are the resultant s values corresponding to the addition of two $\frac{1}{2}$ spins. These two values of s give rise to two types of spectra (this is the case, for example, for He):

$$s = 0 \rightarrow \text{singlet series:} \qquad {}^1S, \, {}^1P, \, {}^1D, \ldots$$

$$s = 1 \rightarrow \text{triplet series:} \qquad {}^3S, \, {}^3P, \, {}^3D, \ldots$$

The j values of the ${}^3D$ states are $j = 1, 2, 3$. These correspond to the states

$$ {}^3D_1, \, {}^3D_2, \, {}^3D_3 $$

In general, any state with $l > 1$ becomes the triplet

$$ j = l + 1, l, l - 1 $$

(Fig. 12.3).

The multiplicity corresponding to the case $s > l$ is $2l + 1$. For example, if $s = \frac{3}{2}$ and $l = 1$, there are three j values: $j = \frac{5}{2}, \frac{3}{2}, \frac{1}{2}$. However, the notation for this state remains ${}^4P_{5/2, 3/2, 1/2}$, with 4 written for $2s + 1$, although in fact the multiplicity is $2l + 1$.

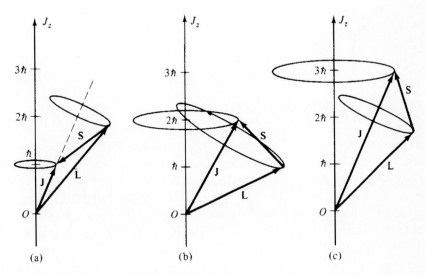

(a)   (b)   (c)

FIGURE 12.3   Diagrams depicting coupling of L and S vectors for two electrons in an orbital D state and a spin-1 state. The resultant triplet of j values is

$$ j = 1, 2, 3 $$

PROBLEMS

12.1   (a)   For given values of L and S, show that the four operators

$$\{\hat{L}^2, \hat{L}_z, \hat{S}^2, \hat{S}_z\}$$

form a commuting set of observables. This representation is akin to the "uncoupled" representation discussed in Sections 9.4 and 11.9, while the L-S representation (12.3) compares to the "coupled" representation.

(b)   Which of these operators are incompatible with those of (12.3)?

12.2   Show that
(a) $[\hat{J}^2, \hat{L}_z] = 2i\hbar e_z \cdot (\hat{\mathbf{L}} \times \hat{\mathbf{S}})$
(b) $[\hat{J}^2, \hat{S}_z] = 2i\hbar e_z \cdot (\hat{\mathbf{S}} \times \hat{\mathbf{L}})$

12.3   What are the multiplicities of the $G(l = 4)$ and $H(l = 5)$ states for the two spectral series related to a three-electron atom? What is the complete term notation for all these states?

12.4   What kind of terms can result from the following values of l and s?

(a) $l = 2, s = \dfrac{7}{2}$

(b) $l = 5, s = \dfrac{3}{2}$

(c) $l = 3, s = 3$

12.5   What are the l, s, j values and multiplicities of the following terms?
(a) $^3D_2$
(b) $^4P_{5/2}$
(c) $^2F_{7/2}$
(d) $^3G_3$

12.2   ONE-ELECTRON ATOMS

In this section we consider the manner in which the spin of the valence electron in one-electron atoms interacts with the shielded Coulomb field due to the nucleus and remaining electrons of the atom. One-electron atoms are better known as the alkali-metal atoms.[1] In such atoms all but one electron are in closed "shells" (to be discussed below). These "core" electrons, together with the nucleus, present a radial electric field to the outer valence electron (Fig. 12.4). Furthermore, the total orbital and spin angular momentum of a closed shell is zero, so that the angular momentum of the atom is determined by the valence electron.

---

[1] This analysis is also relevant to hydrogenic atoms (Section 10.6).

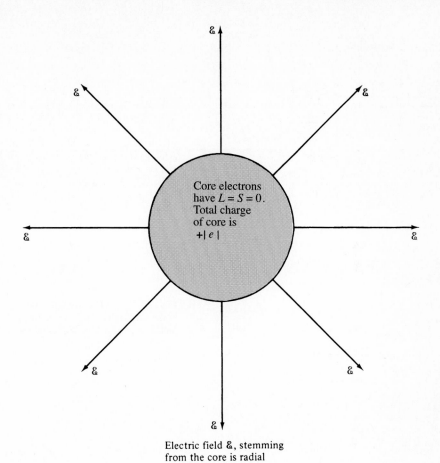

Electric field &, stemming
from the core is radial

FIGURE 12.4 Properties of "one-electron" atoms.

Spin-Orbit Coupling

The interaction between the spin of the valence electron and the shielded Coulomb field arises from the orbital motion of this electron through the Coulomb field. When an observer moves with velocity **v** across the lines of a static electric field $\mathscr{E}$, special relativity reveals that in the frame of the observer, a magnetic field

$$\mathscr{B} = -\gamma\boldsymbol{\beta} \times \mathscr{E}$$

$$\boldsymbol{\beta} \equiv \frac{\mathbf{v}}{c}, \qquad \gamma^{-2} \equiv 1 - \beta^2$$

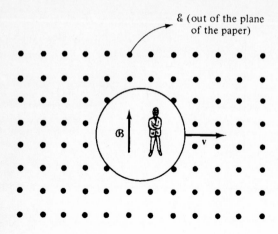

FIGURE 12.5 An observer in the (perfectly transparent) sphere which is moving with velocity **v** across the electric field $\mathscr{E}$ detects a magnetic field

$$\mathscr{B} = -\frac{\mathbf{v}}{c} \times \mathscr{E}$$

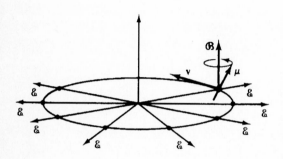

FIGURE 12.6 The magnetic moment $\boldsymbol{\mu}$ of the satellite electron sees a magnetic field due to its orbital motion across the radial, static Coulomb lines of force which emanate from the nucleus. The resulting torque on the moment produces a precession of the spin axis of the electron as shown.

is detected (Fig. 12.5). Keeping terms to first order in β gives

$$\mathscr{B} = -\frac{\mathbf{v}}{c} \times \mathscr{E}$$

It follows that (to this order), if an electron moves with momentum **p** across a field $\mathscr{E}$, the electron will feel a magnetic field[1]

(12.4)
$$\mathscr{B} = -\frac{\mathbf{p}}{mc} \times \mathscr{E}$$

This is the nature of the magnetic field with which the magnetic moment of the orbiting valence electron interacts (Fig. 12.6). The interaction energy between $\boldsymbol{\mu}$ and

---

[1] Relativistic momentum is $\mathbf{p} = \gamma m\mathbf{v}$, so that to terms of $O(\beta^2)$, $\mathbf{p} = m\mathbf{v}$. In these formulas m is the rest mass of an electron: $mc^2 = 0.511$ MeV. For further discussion see Chapter 15.

$\mathscr{B}$ is given by (11.89), modified by the Thomas factor, $\frac{1}{2}$. This correction factor represents an additional relativistic effect due to the acceleration of the electron.[1] Thus the interaction energy between the spin of the orbiting electron and the magnetic field (12.4) appears as

(12.5)
$$H' = -\frac{1}{2}\boldsymbol{\mu} \cdot \mathscr{B} = \frac{1}{2}\boldsymbol{\mu} \cdot \left(\frac{\mathbf{v}}{c} \times \mathscr{E}\right)$$

$$= -\frac{1}{2}\frac{\boldsymbol{\mu}}{mc} \cdot (\mathscr{E} \times \mathbf{p})$$

In spherical coordinates, $\mathscr{E}$ has only a radial component

$$\mathscr{E}_r = -\frac{d}{dr}\Phi(r)$$

where $\Phi(r)$ is the static Coulomb potential (ergs/esu) seen by the valence electron. Substituting this expression for $\mathscr{E} = (\mathscr{E}_r, 0, 0)$ into (12.5) gives

$$H' = \frac{1}{2}\frac{1}{mc}\left[\frac{1}{r}\frac{d\Phi(r)}{dr}\right](\mathbf{r} \times \mathbf{p}) \cdot \boldsymbol{\mu}$$

Recalling the linear relation (11.86) between $\boldsymbol{\mu}$ and $\mathbf{S}$,

$$\boldsymbol{\mu} = \left(\frac{e}{mc}\right)\mathbf{S}$$

and rewriting $\mathbf{L}$ for $\mathbf{r} \times \mathbf{p}$ permits H' to be rewritten in terms of its operational equivalent.

(12.6)
$$\hat{H}' = \frac{e}{2m^2c^2}\left[\frac{1}{r}\frac{d\Phi}{dr}\right]\hat{\mathbf{L}} \cdot \hat{\mathbf{S}} \equiv f(r)\hat{\mathbf{L}} \cdot \hat{\mathbf{S}}$$

which serves to define the scalar function $f(r)$.

Approximate Wavefunction

The Hamiltonian of a typical one-electron atom, neglecting the L-S coupling term just discovered, appears as [recall (10.107)]

$$\hat{H}_0 = \frac{\hat{p}^2}{2m} + V(r) = \frac{\hat{p}_r^2}{2m} + \frac{\hat{L}^2}{2mr^2} + V(r)$$

[1] L. H. Thomas, *Nature* **117**, 514 (1926).

where $V = e\Phi$, $\mathbf{L}$ is the orbital angular momentum of the valence electron, and p_r is its radial momentum. The eigenstates of $\hat{H}_0$ are hydrogenlike in structure. They are comprised of the eigenstates of $\hat{L}^2$ (spherical harmonics) and solutions to the radial equation [recall (10.109)]. Incorporating the spin-orbit interaction (12.6) gives the total Hamiltonian

$$\hat{H} = \hat{H}_0 + \hat{H}' = \frac{\hat{p}_r^2}{2m} + \frac{\hat{L}^2}{2mr^2} + V(r) + f(r)\hat{\mathbf{L}} \cdot \hat{\mathbf{S}}$$

Rewriting $\mathbf{L} \cdot \mathbf{S}$ in terms of J^2, $\hat{L}^2$, S^2 (12.2) permits this Hamiltonian to be rewritten

(12.7) $$\hat{H} = \hat{H}_0 + \frac{f(r)}{2}[\hat{J}^2 - \hat{L}^2 - \hat{S}^2]$$

In the preceding section we showed that $(\hat{J}^2, \hat{J}_z, \hat{L}^2, \hat{S}^2)$ comprise a set of commuting operators. Since these operators also commute with $\hat{H}_0$, approximate eigenstates of $\hat{H}$ may be taken to be of the form[1]

(12.8) $$|\varphi\rangle = |nl\rangle |jm_jls\rangle$$

where $|nl\rangle$ represents the *radial* component of the eigenstates of $\hat{H}_0$.

$$\hat{H}_0|nl\rangle = E_n|nl\rangle$$

For hydrogen, for example, $|nl\rangle$ are solutions to the radial equation (10.117), that is, weighted Laguerre polynomials (Table 10.3).

Substituting the product form (12.8) into the Schrödinger equation

$$\hat{H}|\varphi\rangle = E|\varphi\rangle$$

with $\hat{H}$ given by (12.7) gives

(12.9) $$\left\{ E_n + \frac{\hbar^2}{2}f(r)\left[j(j + 1) - l(l + 1) - \frac{3}{4} \right] \right\} |\varphi\rangle = E|\varphi\rangle$$

It follows that the product solutions (12.8) are *not* eigenstates of $\hat{H}$ (i.e., $\hat{H}|\varphi\rangle \neq$ constant $\times |\varphi\rangle$). But due to the fact that the spin-orbit correction to E_n is small compared to E_n, they do serve as approximate solutions. Approximate eigenvalues of $\hat{H}$ may then be obtained by constructing the expectation of $\hat{H}$ in these states.

(12.10) $$E_{nlj} = \langle \varphi | H | \varphi \rangle$$

$$= E_n + \frac{\hbar^2}{2}\left[j(j + 1) - l(l + 1) - \frac{3}{4} \right]\langle f(r)\rangle_{nl}$$

---

[1] The states $|jm_jls\rangle$ may be constructed from the product states $|lm_l\rangle|sm_s\rangle$ in a Clebsch–Gordan expansion (with $m_s + m_l = m_j$).

Since j can have two values $(l \pm \frac{1}{2})$ for a given value of l, it follows that an energy state of given l separates into a doublet when the spin-orbit interaction is "turned on." The two corresponding values of energy are

(12.11)

$$E_{nlj}^{(+)} \equiv E_{j=l+1/2} = E_n + \frac{\hbar^2}{2} l \langle f \rangle_{nl}$$

$$E_{nlj}^{(-)} \equiv E_{j=l-1/2} = E_n - \frac{\hbar^2}{2} (l + 1)\langle f \rangle_{nl}$$

An estimate of $\langle f \rangle_{nl}$ may be obtained using hydrogen wavefunctions and assuming the Coulomb potential for V,

$$V = -\frac{Ze^2}{r}$$

where Z is an effective atomic number. (See Problem 12.13.) Substituting this potential into $f(r)$ as given in (12.6) gives

$$f(r) = \frac{1}{2m^2c^2}\frac{1}{r}\frac{dV}{dr} = \frac{Ze^2}{2m^2c^2}\frac{1}{r^3}$$

(12.12)

$$\langle f \rangle_{nl} = \frac{Ze^2}{2m^2c^2}\int_0^\infty \frac{|R_{nl}(r)|^2}{r^3} r^2\, dr$$

$$\frac{\hbar^2}{2n}\langle f \rangle_{nl} = \frac{(me^4Z^2/2\hbar^2n^2)^2}{mc^2(l + \frac{1}{2})(l + 1)l}$$

(where we have used the results of Problem 10.49).

Fine Structure of Hydrogen

Recalling the energy eigenvalues of hydrogen (10.124)

$$|E_n| = \frac{m(Ze^2)^2}{2\hbar^2n^2}$$

permits the correction factor (12.12) to be written

$$\frac{\hbar^2\langle f \rangle_{nl}}{|E_n|} = \frac{2n}{l(l + \frac{1}{2})(l + 1)}\frac{|E_n|}{mc^2} = \frac{(Z\alpha)^2}{n}\frac{1}{l(l + \frac{1}{2})(l + 1)}$$

where α is the *fine-structure constant* (see Problems 2.20 and 2.29),

$$\alpha = \frac{e^2}{\hbar c} = \frac{1}{137.037}, \qquad \alpha^2 = 5.33 \times 10^{-5}$$

Substituting these forms into the doublet energies (12.11) gives the values

(12.13) E_n

"spin up"
$j = l + \frac{1}{2}$

$$E_{nlj}^{(+)} = -|E_n|\left[1 - \frac{1}{(2l + 1)(l + 1)} \frac{(Z\alpha)^2}{n}\right]$$

$$E_{nlj}^{(-)} = -|E_n|\left[1 + \frac{1}{l(2l + 1)} \frac{(Z\alpha)^2}{n}\right]$$

"spin down"
$j = l - \frac{1}{2}$

(where we have written $E_n = -|E_n|$). Thus we see that the spin-orbit corrections to the "unperturbed" energies E_n are about 1 part in 10^5. The fact that these corrections are indeed small lends consistency to our original assumption that the product eigenstates (12.8) closely approximate the eigenstates of the total Hamiltonian $\hat{H}$ (12.7).

The two energies (12.13) correspond to the two possible orientations of **S** with respect to **L** (see Fig. 12.2). When the spin is "down," the magnetic moment of the electron ($\boldsymbol{\mu} \sim -\mathbf{S}$) is aligned with the magnetic field $\mathscr{B}$ (12.4) of the relative electron-nucleus motion. This is the configuration of minimum energy. Thus the correction increment corresponding to the "spin down" case is smaller than that due to the "spin up" case, in agreement with the expressions (12.13).

The spin-orbit interaction serves to remove the l degeneracy of the eigenenergies of hydrogenic atoms. If the spin-orbit interaction is neglected, energies are dependent only on the principal quantum number n and are independent of l (and m_l). In the L-S representation, $nljm_j$ (and $s = \frac{1}{2}$) are good quantum numbers, and the l degeneracy is removed. Degeneracy with respect to m_j, however, remains. Eigenenergies are dependent only on (n, l, j), as indicated by expression (12.10). For a given principal quantum number n, the orbital quantum number is restricted to the values $l = 0, 1, \ldots, (n - 1)$ [recall (10.127)], while for a given l, the total angular momentum quantum number j can take the two values $j = l \pm \frac{1}{2}$.

The partial energy-level diagram[1] for potassium, depicting the transition from the doublet ^2P states to the ground state, is shown in Fig. 12.7. The corresponding radiation lies in the near infrared.

The selection rules for dipole radiation developed in Section 10.7 are generalized to the following, for one-electron atoms.[2]

---

[1] It was in explanation of such doublet spectra that G. E. Uhlenbeck and S. Goudsmit first postulated the existence of electron spin. *Naturwiss.* **13**, 953 (1925), and *Nature* **117**, 264 (1926).

[2] Derivation of these selection rules may be found in R. H. Dicke and J. P. Wittke, *Introduction to Quantum Mechanics*, Addison-Wesley, Reading, Mass., 1960.

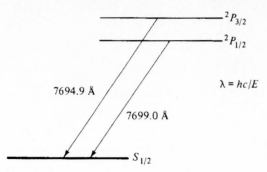

$\lambda = hc/E$

FIGURE 12.7 Wavelengths, in Å, corresponding to the transition from the lowest
^2P states to the ground state, for potassium.

$$\Delta l = \pm 1$$

$$\Delta j = \pm 1, 0 \qquad \text{(but } j = 0 \to 0 \text{ is forbidden)}$$

$$\Delta m_j = \pm 1, 0$$

$$\Delta n \text{ is unrestricted}$$

Photons are emitted only for transitions between states which obey these conditions.

Relativistic Corrections

The spin-orbit corrections to the energies of hydrogen are the same order of magnitude as the corrections due to the relativistic speed of the electron in its orbit. This small correction to the "unperturbed" energies E_n may also be obtained using the technique developed above to find the spin-orbit correction. The relativistic Hamiltonian for a particle of mass m moving in a potential field V is

(12.14)

$$H = (p^2c^2 + m^2c^4)^{1/2} - mc^2 + V$$

$$V = -\frac{Ze^2}{r}$$

If $p \ll mc$, then the radical may be expanded to obtain

(12.15)

$$H = \frac{p^2}{2m} - \frac{p^4}{8m^3c^2} + \cdots + V$$

$$\hat{H} = \left(\frac{\hat{p}^2}{2m} + V\right) - \frac{\hat{p}^4}{8m^3c^2} \equiv \hat{H}_0 + \hat{H}'$$

Again, as was done above, the correction to the eigenenergies E_n of $\hat{H}_0$ due to $\hat{H}'$ may be obtained by calculating $\langle H' \rangle$ using the hydrogen wavefunctions (10.126). There results

$$\langle H' \rangle = \langle nl | \hat{H}' | nl \rangle$$

$$= -\frac{1}{8m^3c^2} \langle \hat{p}^4 \rangle_{nl}$$

To evaluate this expectation value, we recall that the eigenstates $|nl\rangle$ satisfy the equation

$$\hat{H}_0 |nl\rangle = E_n |nl\rangle$$

$$\left(\frac{\hat{p}^2}{2m} + V \right) |nl\rangle = E_n |nl\rangle$$

so that (writing $|nl\rangle \equiv |\varphi_{nl}\rangle$)

$$\hat{p}^2 |\varphi_{nl}\rangle = 2m(E_n - V)|\varphi_{nl}\rangle$$

Owing to the Hermiticity of $\hat{p}$,

$$\langle \varphi_{nl} | \hat{p}^2\hat{p}^2 | \varphi_{nl} \rangle = \langle \hat{p}^2\varphi_{nl} | \hat{p}^2\varphi_{nl} \rangle$$

It follows that

$$\langle p^4 \rangle_{nl} = \int_0^{\infty} [2m(E_n - V)\varphi_{nl}^{\ast}][2m(E_n - V)\varphi_{nl}]r^2 \, dr$$

Hence

$$\langle H' \rangle = -\frac{1}{2mc^2} (E_n^2 - 2E_n\langle V \rangle_{nl} + \langle V^2 \rangle_{nl})$$

The integrals $\langle r^{-1} \rangle_{nl}$, $\langle r^{-2} \rangle_{nl}$ may be obtained using the results of Problem 10.49. There results

$$\langle H' \rangle = -|E_n|\frac{\alpha^2 Z^2}{4n^2}\left(\frac{8n}{2l + 1} - 3 \right)$$

As with the case of spin-orbit correction, we again find that $\langle H' \rangle$ is smaller than E_n by a factor of the order of α^2. Combining this correction with the expressions for the spin-orbit correction (12.13) gives the result

(12.16)

$$E_{nlj}(\text{spin-orbit} + \text{rel.}) = -|E_n|\left[1 + \left(\frac{Z\alpha}{2n} \right)^2 \left(\frac{4n}{j + \frac{1}{2}} - 3 \right) \right] \qquad \left(j = l \pm \frac{1}{2} \right)$$

This expression for the energies of hydrogen include the "fine-structure" corrections, due to spin-orbit and relativistic effects. The energies so found are in quite good agreement with observed hydrogen emission spectra.[1]

PROBLEMS

12.6   The magnetic field (12.4) due to the relative nucleus-electron motion may also be thought of as arising in the following way. If one "sits" on the electron, the nucleus is seen to move in orbital motion about this position. This nuclear orbit constitutes a current loop, which in turn generates a magnetic field. Calculate the value of this magnetic field for a given value of **L** and compare it to the value obtained from (12.4).

12.7   In quantum mechanics, when one says that the vector **J** is conserved, one means that for any state the system is in, the expectations of the three components of **J** are constant. This follows if these three operators all commute with the Hamiltonian. Show that for a one-electron atom with spin-orbit coupling, **L** and **S** are not conserved.

12.8   There is no spin-orbit interaction if an electron is in an S state ($l = 0$). Why?

12.9   What is the difference in energy between the two states of a doublet for a typical one-electron atom as a function of n and l?

12.10   What is the wavelength of a photon emitted by a typical one-electron atom when the valence electron undergoes a spin flip from the $2^2P_{3/2}$ to the $2^2P_{1/2}$ state? In this notation, 2 is the value of the principal quantum number n. Give your answer in terms of Z. According to the selection rules cited for dipole radiation, is such a transition allowed?

12.11   Make an estimate of the rotational kinetic energy, $L^2/2mr^2$, of an electron in a $2P$ state of hydrogen. What is the ratio of this energy to the rest-mass energy $mc^2 = 0.511$ MeV?

12.12   (a)   What are the respective frequencies (Hz) emitted in the transitions (1) $2P_{1/2} \to 1S_{1/2}$, (2) $2P_{3/2} \to 1S_{1/2}$, for lithium?
(b)   What is the percentage change between these frequencies?

12.13   In the theory developed in Section 12.2, the effect of the inner core electrons of an alkali-metal atom on the energy spectrum was described by an effective atomic number. Owing to penetration of the valence electron wavefunction into the core, such a simple model proves insufficient. A more quantitative model which includes the effects of this *quantum defect* is described in the following example.

One assumes that the potential seen by the outer valence electron is of the form

$$V = \frac{-Z'e^2}{r}\left(1 + \frac{b}{r}\right)$$

This modified potential has the effect of making the force of attraction between the valence electron and the nucleus (of charge Ze) grow with penetration of the valence electron into the core. The deeper this penetration, the larger is the net positive charge "seen" by the valence

---

[1] Calculation of higher-order effects may be found in L. I. Schiff, *Quantum Mechanics,* 3d ed., McGraw-Hill, New York, 1968.

electron. The effective nuclear charge $Z'e$ and displacement b may be chosen so as to give the best fit with observed spectral data.

(a)   Show that the method used to solve for the energy levels of the hydrogen atom can be applied to this problem with only slight modifications to give energy levels of the form

$$E_{nl} = \frac{-Ze^2}{2a_0[n + D(l)]^2}$$

Here a_0 is the hydrogen Bohr radius and

$$D(l) \equiv \sqrt{\left(l + \frac{1}{2}\right)^2 - \frac{2bZ'}{a_0}} - \left(l + \frac{1}{2}\right)$$

represents the l-dependent quantum defect.

(b)   How do these E_n energy states vary with increasing l? Give a physical explanation of this variation in terms of core penetration of the valence electron.

12.14   At sufficiently high temperatures, a diatomic dumbbell molecule may suffer vibrational modes of excitation above the normal rotational modes. Consider that the two atomic nuclei are bound through a central potential $V(r)$ which has a strong minimum at the separation $r = a$. At low temperatures the nuclei stay at this interparticle spacing and the effective Hamiltonian is

$$\hat{H} = \frac{\hat{L}^2}{2\mu a^2} + V(a)$$

where μ is reduced mass. At higher temperatures, the particles separate and the Hamiltonian becomes

$$\hat{H} = \frac{\hat{L}^2}{2\mu(a + \xi)^2} + V(a + \xi)$$

where ξ is the (small) radial deviation from the equilibrium separation, a. For the case that $\langle L^2/2\mu a^2 \rangle \ll (\xi/a)V(a)$:

(a)   Obtain a form of $\hat{H}$ that is valid to order of $(\xi/a)^2$.

(b)   By appropriate choice of product wavefunctions, obtain the eigenenergies E_{ln} of the Hamiltonian you have constructed.

(c)   Argue the consistency of omitting the radial kinetic energy $\hat{p}_r^2/2\mu$ in the Hamiltonians written above.

Hint [for part (a)]: Since $V(r)$ is minimum at $r = a$, it follows that $V'(a) = 0$.

12.15   In Chapter 11 [equation (11.86)] it was noted that the magnetic moment of an electron due to its spin is

$$\boldsymbol{\mu}_s = \frac{e}{mc}\mathbf{S}$$

Consider that the electron moves in a circle with angular momentum $\mathbf{L}$. Show classically that the magnetic moment due to such orbital motion is

$$\boldsymbol{\mu}_L = \frac{e}{2mc}\,\mathbf{L}$$

[*Note:* Thus we see that the total magnetic moment of an atomic electron

$$\boldsymbol{\mu} = \frac{e}{2mc}\,(\mathbf{J} + \mathbf{S}) = -\frac{\mu_b}{\hbar}(\mathbf{J} + \mathbf{S})$$

may not in general be assumed to be parallel to its total angular momentum **J**.]

12.16   Consider an atom whose electrons are L-S coupled so that good quantum numbers are $jlsm_j$ and eigenstates of the Hamiltonian $\hat{H}_0$ may be written $|\,jlsm_j\rangle$. In the presence of a uniform magnetic field $\mathscr{B}$, the Hamiltonian becomes

$$\hat{H} = \hat{H}_0 + \hat{H}'$$

$$\hat{H}' = -\hat{\boldsymbol{\mu}} \cdot \mathscr{B} = \frac{e}{2mc}(\hat{\mathbf{J}} + \hat{\mathbf{S}}) \cdot \mathscr{B}$$

where **J** and **S** are total and spin angular momenta, respectively, and e has been written for $|e|$. Before the magnetic field is turned on, **L** and **S** precess about **J** as depicted in Fig. 12.1. Consequently, $\boldsymbol{\mu} = -(\mu_b/\hbar)(\mathbf{L} + 2\mathbf{S})$ also precesses about **J**.

After the magnetic field is turned on, if it is sufficiently weak compared to the coupling between **L** and **S**, the ensuing precession of **J** about $\mathscr{B}$ is slow compared to that of $\boldsymbol{\mu}$ about **J**, as depicted in Fig. 12.8a.

(a)   In the same limit show that time averages obey the relation

$$\langle \boldsymbol{\mu} \cdot \mathscr{B} \rangle \simeq \left\langle \frac{(\boldsymbol{\mu} \cdot \mathbf{J})(\mathbf{J} \cdot \mathscr{B})}{J^2} \right\rangle$$

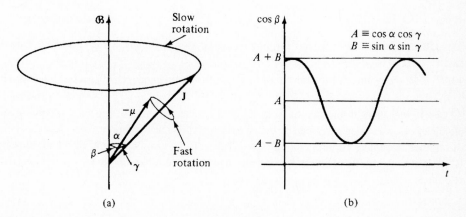

(a)   (b)

FIGURE 12.8   **(a) In the presence of a weak $\mathscr{B}$ field, the precession of J about $\mathscr{B}$ is slow compared to that of $\boldsymbol{\mu}$ about J. (See Problem 12.16.) (b) Variation of $\cos\beta$.**

(b) Assuming, as in the text, that eigenstates $| jlsm_j \rangle$ are still appropriate to the perturbed Hamiltonian $\hat{H}_0 + \hat{H}'$, show that an eigenenergy $E_{jls}{}^0$ of $\hat{H}_0$ splits into $2j + 1$ equally spaced levels

$$E_{jlsm_j} = E_{jls}{}^0 + \Delta E_{m_j}$$

$$\Delta E_{m_j} = \frac{\hbar\Omega}{2} g(jls)m_j \qquad (m_j = -j, \ldots, j)$$

where $\Omega/2$ is the Larmor frequency introduced in Section 11.9 and g is the *Landé g factor*

$$g(jls) = 1 + \frac{j(j + 1) + s(s + 1) - l(l + 1)}{2j(j + 1)}$$

also briefly discussed in Section 11.9. (*Note:* This splitting of lines due to the presence of the magnetic field is called the *Zeeman effect*. Note the inferred relation $\langle \mu_z \rangle = -g(\mu_b/\hbar)\langle J_z \rangle$.)

Answers (partial)

(a) From the orientation of vectors shown in Fig. 12.8a we see that the relation to be established is correct provided that time averages satisfy

$$\overline{\cos \beta} \simeq \overline{\cos \gamma \cos \alpha}$$

Again in reference to the figure one finds that

$$\beta_{\max} = \alpha + \gamma$$

$$\beta_{\min} = \alpha - \gamma$$

The variation of $\cos \beta$ between these extremum values is very nearly harmonic (Fig. 12.8b), and forming the time average of $\cos \beta$ gives the desired result.

(b) With the given approximation, one obtains (show this)

$$\hat{H}' = \left(\frac{\Omega}{2}\right) \frac{\hat{J}^2 + \hat{\mathbf{J}} \cdot \hat{\mathbf{S}}}{J^2} \hat{J}_z$$

Expanding the vector $(\mathbf{J} - \mathbf{S})^2$ and forming the expectation $\langle jlsm_j | \hat{H}' | jsm_j \rangle$ gives the desired result.

12.17 Use the results of Problem 12.16 to obtain the Zeeman pattern of spectral lines which stem from the transition ${}^4F_{3/2} - {}^2D_{5/2}$.

12.18 In the classical formulation of the Zeeman effect, one views the orbital motion of an atomic electron as being perturbed by the imposed magnetic field, thereby altering frequencies of rotation. Assuming circular motion of unperturbed frequency ω_0, show classically that in the presence of a sufficiently weak magnetic field ($\Omega \ll \omega_0$), this frequency divides into the two lines

$$\omega_\pm = \omega_0 \pm \frac{\Omega}{2}$$

while radiation polarized parallel to $\mathscr{B}$ remains with frequency ω_0.

Answer

The atomic component of force maintaining circular motion before the magnetic field is applied may be written $m\omega_0^2 r$. If the magnetic field is normal to the plane of the orbit, the magnetic force is also along the radius and, provided that the field is sufficiently weak, we may assume the perturbed motion to be slightly altered with small variation in frequency. Let the new frequency of rotation be ω. The total force may then be written

$$F = m\omega^2 r = m\omega_0^2 r + m\Omega\omega r$$

Solving for ω gives the roots

$$2\omega = \Omega \pm \sqrt{\Omega^2 + 4\omega_0^2}$$

The assumption of a weak magnetic field ($\Omega \ll 2\omega_0$) allows the radical to be expanded, giving the roots

$$\omega = \omega_0 \pm \frac{\Omega}{2}$$

Components of motion parallel to $\mathcal{B}$ are unaffected by $\mathcal{B}$ so that frequency ω_0 maintains for polarization parallel to $\mathcal{B}$.

12.19   In Problem 12.16 it was discovered that a magnetic field will split an eigenenergy corresponding to given *jls* values into $2j + 1$ levels

$$\Delta E_{m_j} = \left(\frac{\hbar\Omega}{2}\right)gm_j$$

Show that this proliferation of levels leads to the classical Zeeman splitting of a single frequency into three new lines (as demonstrated in Problem 12.18) in the event that the Landé g factors of both levels of the transition are the same. (*Hint:* Frequency displacements $\Delta\nu$ from the original unperturbed frequency are given by

$$h\,\Delta\nu = \Delta E_{m_{j'}} - \Delta E_{m_j}$$

This, together with the selection rules $\Delta m_j = 0, \pm 1$, gives the desired result.)

12.3   THE PAULI PRINCIPLE

Indistinguishable Particles

The concept of symmetric and antisymmetric wavefunctions was encountered in Section 8.5. These wavefunctions are appropriate to systems containing identical particles. There is a very fundamental distinction between the quantum and classical descriptions of such systems. At the quantum mechanical level of description, identical particles are also *indistinguishable*. In the classical description of a system of identical particles, one may conceptually label such particles and follow their respective motion. This is impossible at the quantum level. There is no experimental result that distinguishes between two states obtained by exchange (interchange) of identical particles.

Consider a system that consists of two identical particles (e.g., electrons) moving in one dimension (x). Let x_1 be the coordinate of the first particle and x_2 be the coordinate of the second particle. Then

$$P_{12}\, dx_1\, dx_2 = |\varphi(x_1, x_2)|^2\, dx_1\, dx_2$$

denotes the probability of finding particle 1 in the volume element dx_1 about the point x_1 and particle 2 in the volume element dx_2 about the point x_2 [recall (8.98)]. In this notation the first slot in the wavefunction $\varphi(\quad,\quad)$ is reserved for the position of particle 1, while the second slot is reserved for the position of particle 2. Now if these two particles are truly indistinguishable, then it is impossible to discern between the two states:

$$\begin{pmatrix} \text{No. 1 at } x_1 \\ \text{No. 2 at } x_2 \end{pmatrix} \quad \text{and} \quad \begin{pmatrix} \text{No. 2 at } x_1 \\ \text{No. 1 at } x_2 \end{pmatrix}$$

It follows that the probability of finding these two configurations is the same (Fig. 12.9).

(12.17) $$|\varphi(x_1, x_2)|^2 = |\varphi(x_2, x_1)|^2$$

Only wavefunctions with this exchange-symmetry property are valid wavefunctions for a system of identical particles. Of these, it turns out experimentally that wavefunc-

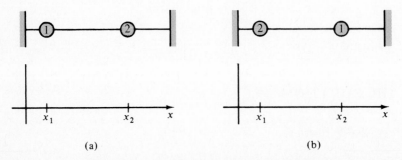

(a)   (b)

FIGURE 12.9 Two classically distinct configurations of two identical particles on a wire. The probability density $|\varphi(x_1, x_2)|^2$ pertains to configuration (a) and $|\varphi(x_2, x_1)|^2$ to configuration (b). In quantum mechanics, the identical particles 1 and 2 are also indistinguishable, so the probability densities associated with these configurations must be the same.

$$|\varphi(x_1, x_2)|^2 = |\varphi(x_2, x_1)|^2$$

tions relevant to quantum mechanics fall into two categories: symmetric (φ_S) and antisymmetric (φ_A). These functions have the respective properties

$$\varphi_S(x_1, x_2) = \varphi_S(x_2, x_1)$$
$$\varphi_A(x_1, x_2) = -\varphi_A(x_2, x_1)$$

which both obey (12.17). For particles free to move in three dimensions, we write

(12.18)
$$\varphi_S(\mathbf{r}_1, \mathbf{r}_2) = +\varphi_S(\mathbf{r}_2, \mathbf{r}_1)$$
$$\varphi_A(\mathbf{r}_1, \mathbf{r}_2) = -\varphi_A(\mathbf{r}_2, \mathbf{r}_1)$$

Let the Hamiltonian $\hat{H}$ describe a system that contains two identical particles: 1 and 2. If these two particles are truly indistinguishable, the Hamiltonian $\hat{H}$ must be symmetric with respect to the positions of these particles, that is,

$$\hat{H}(\mathbf{r}_1, \mathbf{r}_2) = \hat{H}(\mathbf{r}_2, \mathbf{r}_1)$$

Exchange Including Spin

In addition to space coordinates $\mathbf{r}$, a particle also has spin coordinates $\mathbf{S}$. The state of a free particle, for example, may be given in terms of the eigenvalues of these commuting observables (e.g., $\hat{S}^2$, $\hat{S}_z$, $\hat{x}$, $\hat{y}$, $\hat{z}$). Thus, more generally, the Hamiltonian $\hat{H}$ must be symmetric with respect to spin as well as position coordinates of particles. This symmetry property for $\hat{H}$ appears as

(12.19)
$$\hat{H}(\mathbf{r}_1, \mathbf{S}_1; \mathbf{r}_2, \mathbf{S}_2) = \hat{H}(\mathbf{r}_2, \mathbf{S}_2; \mathbf{r}_1, \mathbf{S}_1)$$

The properties (12.18) for φ_S and φ_A become

(12.20)
$$\varphi_S(\mathbf{r}_1, \mathbf{S}_1; \mathbf{r}_2, \mathbf{S}_2) = +\varphi_S(\mathbf{r}_2, \mathbf{S}_2; \mathbf{r}_1, \mathbf{S}_1)$$
$$\varphi_A(\mathbf{r}_1, \mathbf{S}_1; \mathbf{r}_2, \mathbf{S}_2) = -\varphi_A(\mathbf{r}_2, \mathbf{S}_2; \mathbf{r}_1, \mathbf{S}_1)$$

Again, as in the one-dimensional case (12.17), the probability densities associated with these wavefunctions are totally symmetric. Writing "1" for $(\mathbf{r}_1, \mathbf{S}_1)$, and "2" for $(\mathbf{r}_2, \mathbf{S}_2)$, this symmetry property appears as

$$|\varphi_S(1, 2)|^2 = |\varphi_S(2, 1)|^2$$
$$|\varphi_A(1, 2)|^2 = |\varphi_A(2, 1)|^2$$

These symmetry concepts are conveniently expressed in terms of the properties of the exchange operator $\hat{x}$ [recall (11.112) et seq.], which is defined by the equation

(12.21)
$$\hat{x}\varphi(1, 2) = \varphi(2, 1)$$

The operator $\hat{x}$ has two eigenvalues:

$$\hat{x}\varphi_S(1, 2) = \varphi_S(2, 1) = +\varphi_S(1, 2)$$
$$\hat{x}\varphi_A(1, 2) = \varphi_A(2, 1) = -\varphi_A(1, 2)$$

The eigenfunction φ_S corresponds to the eigenvalue $+1$. It is even under particle exchange. The state φ_A corresponds to the eigenvalue -1. It is odd under particle exchange.

Owing to the exchange symmetry of the Hamiltonian (12.19), $\hat{H}$ commutes with $\hat{x}$.

(12.22)
$$[\hat{x}, \hat{H}] = 0$$

It follows that $\hat{x}$ is a constant of the motion.

$$\frac{d}{dt}\langle\hat{x}\rangle = 0$$

If at time $t = 0$, $\varphi(0)$ is such that

$$\hat{x}\varphi(0) = +\varphi(0)$$

then the two-particle system remains with this property for all time. At time $t > 0$,

$$\hat{x}\varphi(t) = +\varphi(t)$$

Since $\hat{x}$ commutes with $\hat{H}$, it is possible to find simultaneous eigenstates of both these operators (Fig. 12.10). These common wavefunctions are the eigenstates appropriate to systems of identical particles. Furthermore, such eigenstates may be classified in terms of their symmetric or antisymmetric properties under particle exchange.

Bosons and Fermions

The constancy of $\langle\hat{x}\rangle$ is an immutable property of a system of identical particles. Owing to the permanence of this property, one may assume that it is a property of the

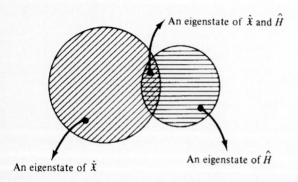

An eigenstate of $\hat{x}$ and $\hat{H}$

An eigenstate of $\hat{x}$

An eigenstate of $\hat{H}$

FIGURE 12.10   Venn diagram exhibiting the simultaneous eigenstates of the exchange operator $\hat{x}$ and the Hamiltonian $\hat{H}$.

particles themselves (as opposed to a property of the wavefunction). Particles characterized by the eigenvalue $+1$ of $\hat{x}$ are called *bosons*. The wavefunction for a system
of bosons is symmetric (φ_S). Particles characterized by the eigenvalue -1 of $\hat{x}$ are
called *fermions*. The wavefunction for fermions is antisymmetric (φ_A).

The characteristic of a particle that determines to which of these categories it
belongs is given by the *spin* of the particle.[1] Bosons have integral spin, while fermions
have half-integral spin. Electrons and neutrons are examples of fermions. Photons, π,
and K mesons are examples of bosons.[2]

Antisymmetric Wavefunctions

The Pauli principle is obeyed by fermions. It states, as described above, that the wavefunction for a system of identical fermions is antisymmetric. Consider a system of two
fermions. They are in the state $\varphi_A(1, 2)$. This state has the property that

$$\varphi_A(\mathbf{r}_1, \mathbf{S}_1; \mathbf{r}_2, \mathbf{S}_2) = -\varphi_A(\mathbf{r}_2, \mathbf{S}_2; \mathbf{r}_1, \mathbf{S}_1)$$

Consider that both particles have the coordinates $\mathbf{r}_1$ and $\mathbf{S}_1$. That is, the system is in
the state, $\varphi_A(1, 1)$. This state has the property

$$\varphi_A(\mathbf{r}_1, \mathbf{S}_1; \mathbf{r}_1, \mathbf{S}_1) = -\varphi_A(\mathbf{r}_1, \mathbf{S}_1; \mathbf{r}_1, \mathbf{S}_1)$$

The only value of φ_A which has this property is

$$\varphi_A(\mathbf{r}_1, \mathbf{S}_1; \mathbf{r}_1, \mathbf{S}_1) = 0$$

There is zero probability of finding the particles at the same point in space, with the
same value of spin. This is the essence of the Pauli principle: *two fermions cannot exist
in the same quantum state.*[3]

Let us consider a problem: What is the wavefunction of two *free* electrons
moving in 3-space? The Hamiltonian for this system is

$$(12.23) \qquad \hat{H}(1, 2) = \frac{\hat{p}_1^{\,2}}{2m} + \frac{\hat{p}_2^{\,2}}{2m}$$

The space component wavefunctions of this Hamiltonian are the product states

$$\varphi_{\mathbf{k}_1}(\mathbf{r}_1)\varphi_{\mathbf{k}_2}(\mathbf{r}_2) = \frac{1}{(2\pi)^3} e^{i\mathbf{k}_1 \cdot \mathbf{r}_1} e^{i\mathbf{k}_2 \cdot \mathbf{r}_2}$$

$$E_{\mathbf{k}_1\mathbf{k}_2} = \frac{\hbar^2}{2m}(k_1^{\,2} + k_2^{\,2})$$

---

[1] See Appendix B.

[2] Writing eigenvalues of $\hat{x}$ as $\exp(i\alpha\pi)$, for fermions $\alpha = 1$ and for bosons $\alpha = 2$. Particles, called "anyons," for
which α is fractional, are hypothesized to exist.

[3] This property of fermions is also called the *Pauli exclusion principle*.

This same value of energy characterizes the exchange state,

$$\varphi_{\mathbf{k}_1}(\mathbf{r}_2)\varphi_{\mathbf{k}_2}(\mathbf{r}_1) = \frac{1}{(2\pi)^3} e^{i\mathbf{k}_1 \cdot \mathbf{r}_2} e^{i\mathbf{k}_2 \cdot \mathbf{r}_1}$$

From these two (degenerate) states one may form the symmetric and antisymmetric eigenstates

(12.24)

$$\varphi_S(\mathbf{r}_1, \mathbf{r}_2) = \frac{1}{\sqrt{2}} [\varphi_{\mathbf{k}_1}(\mathbf{r}_1)\varphi_{\mathbf{k}_2}(\mathbf{r}_2) + \varphi_{\mathbf{k}_1}(\mathbf{r}_2)\varphi_{\mathbf{k}_2}(\mathbf{r}_1)]$$

$$\varphi_A(\mathbf{r}_1, \mathbf{r}_2) = \frac{1}{\sqrt{2}} [\varphi_{\mathbf{k}_1}(\mathbf{r}_1)\varphi_{\mathbf{k}_2}(\mathbf{r}_2) - \varphi_{\mathbf{k}_1}(\mathbf{r}_2)\varphi_{\mathbf{k}_2}(\mathbf{r}_1)]$$

Since the two particles in this problem are fermions, wavefunctions for the system must be antisymmetric with respect to exchange of particle spin and position. Inasmuch as the Hamiltonian (12.23) does not contain the spin, it commutes with all spin functions. Thus, if ξ denotes a spin state for the two-electron system, then

$$\xi\varphi_S(\mathbf{r}_1, \mathbf{r}_2) \quad \text{or} \quad \xi\varphi_A(\mathbf{r}_1, \mathbf{r}_2)$$

are possible eigenstates of $\hat{H}$. In Section 11.10 we found that in the coupled representation, two spin-$\frac{1}{2}$ particles combine to give three symmetric ($s = 1$) states, $\xi_S^{(1)}$, $\xi_S^{(0)}$, and $\xi_S^{(-1)}$, and one antisymmetric ($s = 0$) state, $\xi_A^{(0)}$. Combining these with the space state (12.24), one obtains the four antisymmetric states

(12.25)

$$^1\chi_A(\mathbf{r}_1, \mathbf{S}_1; \mathbf{r}_2, \mathbf{S}_2) = \varphi_S(\mathbf{r}_1, \mathbf{r}_2)\xi_A(1, 2)$$

$$^3\chi_A(\mathbf{r}_1, \mathbf{S}_1; \mathbf{r}_2, \mathbf{S}_2) = \varphi_A(\mathbf{r}_1, \mathbf{r}_2)\left\{\begin{array}{c} \xi_S^{(1)}(1, 2) \\ \xi_S^{(0)}(1, 2) \\ \xi_S^{(-1)}(1, 2) \end{array}\right\}$$

$$E_{\mathbf{k}_1\mathbf{k}_2} = \frac{\hbar^2}{2m}(k_1^2 + k_2^2)$$

Here we are using the simple rule that the product of a symmetric and antisymmetric state is antisymmetric.

This technique of incorporating spin states to ensure antisymmetry of a given state also applies to Hamiltonians that include interaction between particles, or interaction between particles and a central force field, but which are otherwise spin-independent. For such cases, the structure (12.25) of antisymmetric eigenstates is maintained. This concept finds direct application below in discussion of the helium atom. Symmetrization of the wavefunction for bosons will be applied in construction of the nuclear component eigenstates of the deuterium molecule.

PROBLEMS

12.20 Show that the two operators

$$\hat{x}_{\pm} = \frac{\hat{I} \pm \tilde{x}}{\sqrt{2}}$$

have the projection property

$$\hat{x}_{+}\varphi(1, 2) = \varphi_S(1, 2)$$
$$\hat{x}_{-}\varphi(1, 2) = \varphi_A(1, 2)$$

That is, $\hat{x}_{+}$ projects φ onto φ_S and $\hat{x}_{-}$ projects φ onto φ_A.

12.21 (a) Consider a two-particle system with relative radius vector $\mathbf{r} = \mathbf{r}_2 - \mathbf{r}_1$ (Figs. 10.8 and 10.9). Show that $\hat{x}\psi(\mathbf{r}) = \hat{P}\psi(\mathbf{r})$, where $\hat{P}$ is the parity operator introduced in Section 6.4 (see Problem 9.46).

(b) Is the parity of a two-particle system a "good" quantum number in the energy representation?

(c) Two particles interact under a central potential $V(r)$. It is known that the system is in the state $R_{43}(r)Y_3^0(\theta, \phi)$. What is the symmetry of the state?

Answer (partial)

(a)
$$\hat{x}\psi(\mathbf{r}) = \hat{x}\psi(\mathbf{r}_2 - \mathbf{r}_1) = \psi(\mathbf{r}_1 - \mathbf{r}_2)$$
$$\hat{P}\psi(\mathbf{r}_2 - \mathbf{r}_1) = \psi[-\mathbf{r}_2 - (-\mathbf{r}_1)] = \psi(\mathbf{r}_1 - \mathbf{r}_2)$$

12.4 THE PERIODIC TABLE

Central Field Approximation

In this model it is assumed that each electron "sees" only the electrostatic field due to the nucleus and remaining electrons and that this combined field is spherically symmetric. Owing to this spherical symmetry, the operators

$$\hat{H}, \hat{L}^2, \hat{L}_z, \hat{S}_z$$

relevant to a single electron form a set of commuting operators so that the state of each such electron is specified in terms of the eigenvalues

$$n, l, m_l, m_s$$

These quantum numbers correspond to a wavefunction for the ith electron of the form

(12.26)
$$\varphi_{nlm_lm_s}(\mathbf{r}_i, \mathbf{S}_i) = R_{nl}(r_i)Y_l^m(\theta_i, \phi_i)\xi_z^{\pm}(i)$$
$$\xi_z^{\pm} \equiv \alpha(i) \text{ or } \beta(i)$$

The Pauli principle precludes any two electrons being in the same state [i.e., having the same set of (n, l, m_l, m_s) values]. In our discussion of the hydrogen atom, we found that there are n^2 distinct states corresponding to a given value of n (10.128). When spin dependence is included in these states [e.g., (12.26)], this degeneracy is doubled. There are two values that m_s may assume for any set of values nlm_l. Thus, corresponding to any value of n, there are $2n^2$ functions of the form (12.26) which give the same energy for the ith electron.

This degeneracy rule, taken together with the Pauli principle, serves to explain the "shell structure" of the electronic configurations of the elements. As the atomic number Z increases, electrons fill the one-electron states of lowest energy first. For the lighter elements these are shells of lower n values. Within an n shell, for any given value of l there is an l subshell with $2(2l + 1)$ states, corresponding to two m_s values and $2l + 1$ values of m_l. Atoms with filled n shells have a total angular momentum and total spin of zero (see Table 12.1). Electrons exterior to these closed shells ("valence" electrons) determine the chemical properties of the atom. The "periodicity" of these properties owes to the fact that valence numbers repeat after shells become closed.

For $n = 1$, l can only be 0, but m_s may take on the two values, $\pm \frac{1}{2}$. Therefore, there can be, at most, only two electrons in the $n = 1$ "shell." (This is called the "K shell" in x ray notation.) In the ground state of helium ($Z = 2$), the $n = 1$ shell is filled. The electronic configuration for this state is described by the notation $1s^2$. This is read: there are two (the exponent) electrons in the state with $n = 1$ and $l = 0$ (denoted by the letter s). The term notation for this state is $^1S_0$.

The electronic configuration of the ground state of lithium ($Z = 3$) is

$$1s^2 2s^1$$

TABLE 12.1 **Diagrammatic enumeration of states available in the first three atomic shells[a]**

$m_l \rightarrow$	-2	-1	0	$+1$	$+2$	$l\downarrow$	Number of Available States in Each Shell
M shell, $n = 3$	↑↓	↑↓	↑↓	↑↓	↑↓	2	10 ⎫
		↑↓	↑↓	↑↓		1	6 ⎬ 18 = 2 × 3²
			↑↓			0	2 ⎭
L shell, $n = 2$		↑↓	↑↓	↑↓		1	6 ⎫ 8 = 2 × 2²
			↑↓			0	2 ⎭
K shell, $n = 1$			↑↓			0	2 = 2 × 1²

[a] Vertical arrows represent S_z values. The total orbital and spin angular momentum is zero for a closed shell.

There are two electrons with $n = 1$, $l = 0$ and one electron with $n = 2$, $l = 0$. The uncoupled spin gives a j value of $\frac{1}{2}$ with corresponding term notation, $^2S_{1/2}$. In beryllium ($Z = 4$) the ground-state configuration is $1s^2 2s^2$. All spins are paired and the ground state is given by the term $^1S_0$.

The electronic configurations and corresponding ground states for the first 36 elements are given in Table 12.2. Note that ground states follow the L-S coupling scheme. For example, for nitrogen ($Z = 7$), whose electronic configuration is

$$1s^2 2s^2 2p^3$$

the ground state is $^4S_{3/2}$. The three p electrons can have total spin values

$$s = \frac{1}{2}, \frac{3}{2}$$

and total l values (see Section 9.5)

$$l = 0, 1, 2, 3$$

The corresponding j values are given by the sequence

$$j = |l + s|, \ldots, |l - s|$$

for all pairs of l, s values. The doublet ($s = \frac{1}{2}$) states thus obtained are

$$^2S_{1/2}, \, ^2P_{1/2,3/2}, \, ^2D_{3/2,5/2}, \, ^2F_{5/2,7/2}$$

The quartet ($s = \frac{3}{2}$) states are

$$^4S_{3/2}, \, ^4P_{1/2,3/2,5/2}, \, ^4D_{1/2,3/2,5/2,7/2}, \, ^4F_{3/2,5/2,7/2,9/2}$$

Of these 19 possible states, the exclusion principle permits only the ^2D, ^2P, and ^4S states.

To understand this reduction in terms, consider the simpler case of the carbon atom which has two p electrons in its outer shell (with $n = 2$; see Table 12.2). Following the preceding construction, we find that the total l values, $l = 0, 1, 2$, and total s values, $s = 0, 1$, give the following terms:

$$^1S, \, ^1P, \, ^1D, \, ^3S, \, ^3P, \, ^3D$$

The allowed states for these two electrons must be antisymmetric under particle exchange. Wavefunctions have the product structure

(12.27) $$|n_1 n_2 jm_j ls\rangle = R_{[^S_A]}(1, 2)|jm_j ls\rangle$$

We assume for the moment that the two electrons are not "equivalent," i.e., that they have different principal quantum numbers, $n_1 \neq n_2$. The radial functions then have the form

TABLE 12.2 **Distribution of electrons in the atoms from $Z = 1$ to $Z = 36$**

X ray Notation	K	L		M			N				
Values of n, l	1,0	2,0	2,1	3,0	3,1	3,2	4,0	4,1	4,2	4,3	
Spectral Notation	$1s$	$2s$	$2p$	$3s$	$3p$	$3d$	$4s$	$4p$	$4d$	$4f$	
Element / Atomic number Z / Ionization potential (eV)[a]											
H 1 13.595	1										$^2S_{1/2}$
He 2 24.481	2										$^1S_0$
Li 3 5.39	2	1									$^2S_{1/2}$
Be 4 9.32	2	2									$^1S_0$
B 5 8.296	2	2	1								$^2P_{1/2}$
C 6 11.256	2	2	2								$^3P_0$
N 7 14.53	2	2	3								$^4S_{3/2}$
O 8 13.614	2	2	4								$^3P_2$
F 9 17.418	2	2	5								$^2P_{3/2}$
Ne 10 21.559	2	2	6								$^1S_0$
Na 11 5.138				1							$^2S_{1/2}$
Mg 12 7.644		Neon		2							$^1S_0$
Al 13 5.984		configuration		2	1						$^2P_{1/2}$
Si 14 8.149				2	2						$^3P_0$
P 15 10.484				2	3						$^4S_{3/2}$
S 16 10.357		10-electron		2	4						$^3P_2$
Cl 17 13.01		core		2	5						$^2P_{3/2}$
Ar 18 15.755				2	6						$^1S_0$
K 19 4.339							1				$^2S_{1/2}$
Ca 20 6.111							2				$^1S_0$
Sc 21 6.54						1	2				$^2D_{3/2}$
Ti 22 6.82						2	2				$^3F_2$
V 23 6.74						3	2				$^4F_{3/2}$
Cr 24 6.764						5	1				$^7S_3$
Mn 25 7.432		Argon				5	2				$^6S_{5/2}$
Fe 26 7.87		configuration				6	2				$^5D_4$
Co 27 7.86						7	2				$^4F_{9/2}$
Ni 28 7.633						8	2				$^3F_4$
Cu 29 7.724		18-electron				10	1				$^2S_{1/2}$
Zn 30 9.391		core				10	2				$^1S_0$
Ga 31 6.00						10	2	1			$^2P_{1/2}$
Ge 32 7.88						10	2	2			$^3P_0$
As 33 9.81						10	2	3			$^4S_{3/2}$
Se 34 9.75						10	2	4			$^3P_2$
Br 35 11.84						10	2	5			$^2P_{3/2}$
Kr 36 13.996						10	2	6			$^1S_0$

TABLE 12.2   *(Continued)*

Electronic configurations for the alkali metals

Shell	K	L	M	N	O	P
Li	$1s^2$	$2s$				
Na	$1s^2$	$2s^2 2p^6$	$3s$			
K	$1s^2$	$2s^2 2p^6$	$3s^2 3p^6$	$4s$		
Rb	$1s^2$	$2s^2 2p^6$	$3s^2 3p^6 3d^{10}$	$4s^2 4p^6$	$5s$	
Cs	$1s^2$	$2s^2 2p^6$	$3s^2 3p^6 3d^{10}$	$4s^2 4p^6 4d^{10}$	$5s^2 5p^6$	$6s$

[a]Data obtained from *Handbook of Chemistry and Physics,* 56th ed., CRC Press, Cleveland, Ohio, 1976.

$$(12.28) \qquad R_{\left[\begin{smallmatrix}S\\A\end{smallmatrix}\right]} = \frac{1}{\sqrt{2}} [R_1(n_1 l)R_2(n_2 l) \pm R_1(n_2 l)R_2(n_1 l)]$$

where the R functions are hydrogenlike wavefunctions. Angular momentum components $|jm_j ls\rangle$ are constructed as follows. In the Russell–Saunders coupling scheme total **L** adds to total **S** to give the atomic total angular momentum **J**. Wavefunctions are then given by the Clebsch–Gordan expansion,

$$(12.29) \qquad |jm_j ls\rangle = \sum_{m_j=m_l+m_s} C_{m_l m_s} |lm_l\rangle |sm_s\rangle$$

The $|lm_l\rangle$ and $|sm_s\rangle$ components as well are given by Clebsch–Gordan expansions, which for the case at hand are displayed in Tables 9.5 and 11.3, respectively. We see that orbital states of odd l are antisymmetric and those of even l are symmetric. All three $s = 1$ states are symmetric, whereas the singlet $s = 0$ state is antisymmetric. So we may construct the following list:

$$(12.30) \qquad \begin{aligned} {}^3D: &\quad l = 2 \text{(sym)}, &\quad s = 1 \text{(sym)} &\rightarrow |jm_j\rangle_S \rightarrow R_A \\ {}^1P: &\quad l = 1 \text{(anti)}, &\quad s = 0 \text{(anti)} &\rightarrow |jm_j\rangle_S \rightarrow R_A \\ {}^3S: &\quad l = 0 \text{(sym)}, &\quad s = 1 \text{(sym)} &\rightarrow |jm_j\rangle_S \rightarrow R_A \\ {}^1D: &\quad l = 2 \text{(sym)}, &\quad s = 0 \text{(anti)} &\rightarrow |jm_j\rangle_A \rightarrow R_S \\ {}^3P: &\quad l = 1 \text{(anti)}, &\quad s = 1 \text{(sym)} &\rightarrow |jm_j\rangle_A \rightarrow R_S \\ {}^1S: &\quad l = 0 \text{(sym)}, &\quad s = 0 \text{(anti)} &\rightarrow |jm_j\rangle_A \rightarrow R_S \end{aligned}$$

For the np^2 configuration under consideration, $n_1 = n_2 = n = 2$, so that $R_A = 0$. Only the terms which include the R_S factor survive. These are the ${}^1D$, ${}^3P$, and ${}^1S$ terms, which in all comprise 15 independent states corresponding to five energy levels. Of these, Hund's rules (Problem 12.25) determine the ground state. The first rule indicates that the ground state is among the $^3P_{0,1,2}$ states. Since there are only two electrons in the $2p$ shell of carbon, and this shell can accommodate six electrons, the third rule indicates that $^3P_0$ is the ground state of carbon. The same is true for the two-electron (np^2) atoms, Si and Ge. Remaining allowed atomic states for configurations up to the nd^{10} shell are shown in Table 12.2.

A note of caution is in order for three (or more)-electron atoms. For such cases spin states which are antisymmetric with respect to exchange of *any two* of the three particles do not exist (see Problems 11.92 et seq.). In this case, generalizations of the product form (12.27) comes into play.[1]

The symmetry of the states shown in Table 12.3 about the midvalue of occupation in a given l shell is due to the following. Consider, for example, the p shell which

TABLE 12.3   States allowed by the exclusion principle in the L-S coupling scheme

ns^0	^1S			
ns^1		^2S		
ns^2	^1S			
np^0	^1S			
np^1		^2P		
np^2	$^1S, {}^1D$		^3P	
np^3		$^2P, {}^2D$		^4S
np^4	$^1S, {}^1D$		^3P	
np^5		^2P		
np^6	^1S			
nd^0	^1S			
nd^1		^2D		
nd^2	$^1S, {}^1D, {}^1G$		$^3P, {}^3F$	
nd^3		$^2D, {}^2P, {}^2D, {}^2F, {}^2G, {}^2H$	$^4P, {}^4F$	
nd^4	$^1S, {}^1D, {}^1G, {}^1S, {}^1D, {}^1G, {}^1F, {}^1I$		$^3P, {}^3F, {}^3P, {}^3D, {}^3F, {}^3G, {}^3H$ ^5D	
nd^5		$^2D, {}^2P, {}^2D, {}^2F, {}^2G, {}^2H, {}^2S, {}^2D, {}^2F, {}^2G, {}^2I$	$^4P, {}^4F, {}^4D, {}^4G$ ^6S	
nd^6	$^1S, {}^1D, {}^1G, {}^1S, {}^1D, {}^1G, {}^1F, {}^1I$		$^3P, {}^3F, {}^3P, {}^3D, {}^3F, {}^3G, {}^3H$ ^5D	
nd^7		$^2D, {}^2P, {}^2D, {}^2F, {}^2G, {}^2H$	$^4P, {}^4F$	
nd^8	$^1S, {}^1D, {}^1G$		$^3P, {}^3F$	
nd^9		^2D		
nd^{10}	^1S			

[1] For further discussion, see M. Weissbluth, *Atoms and Molecules*, Academic Press, New York, 1978. Recall also Problem 11.88.

can be occupied by six electrons. In what manner does the p^2 configuration resemble the p^4 configuration? The p^4 configuration may be viewed as one in which two holes occupy the p shell. This atomic configuration yields the same allowed states as does the p^2 configuration (see Problem 12.49). The same holds true for the d^3 and d^7 configurations, and so on.

A complete description of an atomic state involves the total L, S, and J of the atom in addition to the quantum numbers of individual electrons. The first of these is given by the term notation of the state (e.g., $^5F_{7/2}$), whereas the second is given by the electronic configuration (e.g., $1s^2 2s^2 2p^3$).

The periodic chart is shown in Table 12.4. The ground states of elements and outer electron shell configurations[1,2] are listed, as well as atomic numbers.

As stated above, chemical properties of elements are determined by the electron configuration in the unfilled shell. An important related chemical reaction is that of the oxidation-reduction process. For example, in the *oxidation* of iron, two $4s$ electrons of an iron atom are transferred to an oxygen atom to complete its $2p$ shell, thereby forming the strong ionic bond of ferrous oxide. (The oxygen atom is *reduced* in this process.)

Atoms with similar valence electron configuration have nearly the same chemical properties. The properties of atoms in some of these groupings are described below.

The Alkali Metals. Group I

The alkali metals are the atoms with one valence electron: Li[3], Na[11], K[19], Rb[37], Cs[55], and Fr[87]. The ground state of these elements is $^2S_{1/2}$. Ionization energy is low. The spectra of these "one-electron" elements resemble that of hydrogen. Valence is $+1$.

The Alkaline Earths. Group II

All the alkaline earths have two s electrons outside a closed p subshell. They are Be[4], Mg[12], Ca[20], Sr[38], Ba[56], and Ra[88]. The ground state is $^1S_0$. Ionization remains relatively

---

[1] These data were obtained from G. Baym, *Lectures on Quantum Mechanics*, W. A. Benjamin, New York, 1969; S. Fraga, J. Karwowski, and K. Saxena, *Handbook of Atomic Data*, Elsevier, New York, 1976; and H. Gray, *Electrons and Chemical Bonding*, W. A. Benjamin, New York, 1964.

[2] A periodic chart with additional atomic and material data appears on the inside of the front cover.

TABLE 12.4 The periodic table

Period	Group I	II	Transition elements										III	IV	V	VI	VII	VIII
1	H^1 $1s^1$ $^2S_{1/2}$																	He^2 $1s^2$ $^1S_0$
2	Li^3 $1s^2 2s^1$ $^2S_{1/2}$	Be^4 $1s^2 2s^2$ $^1S_0$											B^5 $2s^2 2p^1$ $^2P_{1/2}$	C^6 $2s^2 2p^2$ $^3P_0$	N^7 $2p^3$ $^4S_{3/2}$	O^8 $2p^4$ $^3P_2$	F^9 $2p^5$ $^2P_{3/2}$	Ne^{10} $2p^6$ $^1S_0$
3	Na^{11} $3s^1$ $^2S_{1/2}$	Mg^{12} $3s^2$ $^1S_0$											Al^{13} $3s^2 3p^1$ $^2P_{1/2}$	Si^{14} $3s^2 3p^2$ $^3P_0$	P^{15} $3p^3$ $^4S_{3/2}$	S^{16} $3p^4$ $^3P_2$	Cl^{17} $3p^5$ $^2P_{3/2}$	Ar^{18} $3s^2 3p^6$ $^1S_0$
4	K^{19} $4s^1$ $^2S_{1/2}$	Ca^{20} $4s^2$ $^1S_0$	Sc^{21} $4s^2 3d^1$ $^2D_{3/2}$	Ti^{22} $4s^2 3d^2$ $^3F_2$	V^{23} $4s^2 3d^3$ $^4F_{3/2}$	Cr^{24} $4s^1 3d^5$ $^7S_3$	Mn^{25} $4s^2 3d^5$ $^6S_{5/2}$	Fe^{26} $4s^2 3d^6$ $^5D_4$	Co^{27} $4s^2 3d^7$ $^4F_{9/2}$	Ni^{28} $4s^2 3d^8$ $^3F_4$	Cu^{29} $4s^1 3d^{10}$ $^2S_{1/2}$	Zn^{30} $4s^2 3d^{10}$ $^1S_0$	Ga^{31} $4s^2 3d^{10} 4p^1$ $^2P_{1/2}$	Ge^{32} $3d^{10} 4p^2$ $^3P_0$	As^{33} $3d^{10} 4p^3$ $^4S_{3/2}$	Se^{34} $3d^{10} 4p^4$ $^3P_2$	Br^{35} $3d^{10} 4p^5$ $^2P_{3/2}$	Kr^{36} $4s^2 4p^6$ $^1S_0$
5	Rb^{37} $5s^1$ $^2S_{1/2}$	Sr^{38} $5s^2$ $^1S_0$	Y^{39} $5s^2 4d^1$ $^2D_{3/2}$	Zr^{40} $5s^2 4d^2$ $^3F_2$	Nb^{41} $5s^1 4d^4$ $^6D_{1/2}$	Mo^{42} $5s^1 4d^5$ $^7S_3$	Tc^{43} $5s^2 4d^5$ $^6S_{5/2}$	Ru^{44} $5s^1 4d^7$ $^5F_5$	Rh^{45} $5s^1 4d^8$ $^4F_{9/2}$	Pd^{46} $4d^{10}$ $^1S_0$	Ag^{47} $5s^1 4d^{10}$ $^2S_{1/2}$	Cd^{48} $5s^2 4d^{10}$ $^1S_0$	In^{49} $5s^2 4d^{10} 5p^1$ $^2P_{1/2}$	Sn^{50} $4d^{10} 5p^2$ $^3P_0$	Sb^{51} $4d^{10} 5p^3$ $^4S_{3/2}$	Te^{52} $4d^{10} 5p^4$ $^3P_2$	I^{53} $4d^{10} 5p^5$ $^2P_{3/2}$	Xe^{54} $5s^2 5p^6$ $^1S_0$
6	Cs^{55} $6s^1$ $^2S_{1/2}$	Ba^{56} $6s^2$ $^1S_0$	La^{57} $6s^2 5d^1$ $^2D_{3/2}$	Hf^{72} $6s^2 5d^2$ $^3F_2$	Ta^{73} $6s^2 5d^3$ $^4F_{3/2}$	W^{74} $6s^2 5d^4$ $^5D_0$	Re^{75} $6s^2 5d^5$ $^6S_{5/2}$	Os^{76} $6s^2 5d^6$ $^5D_4$	Ir^{77} $6s^2 5d^7$ $^4F_{9/2}$	Pt^{78} $6s^1 5d^9$ $^3D_3$	Au^{79} $6s^1 5d^{10}$ $^2S_{1/2}$	Hg^{80} $6s^2 5d^{10}$ $^1S_0$	Tl^{81} $6s^2 5d^{10} 6p^1$ $^2P_{1/2}$	Pb^{82} $6p^2$ $^3P_0$	Bi^{83} $6p^3$ $^4S_{3/2}$	Po^{84} $6p^4$ $^3P_2$	At^{85} $6p^5$ $^2P_{3/2}$	Rn^{86} $6p^6$ $^1S_0$
7	Fr^{87} $7s^1$ $^2S_{1/2}$	Ra^{88} $7s^2$ $^1S_0$	Ac^{89} $7s^2 6d^1$ $^2D_{3/2}$															

Rare earths^a

Ce^{58} $6s^2 5d^1 4f^1$ $^3H_5$	Pr^{59} $6s^2 4f^3$ $^4I_{9/2}$	Nd^{60} $6s^2 4f^4$ $^5I_4$	Pm^{61} $6s^2 4f^5$ $^6H_{5/2}$	Sm^{62} $6s^2 4f^6$ $^7F_0$	Eu^{63} $6s^2 4f^7$ $^8S_{7/2}$	Gd^{64} $6s^2 5d^1 4f^7$ $^9D_2$	Tb^{65} $6s^2 5d^1 4f^8$	Dy^{66} $6s^2 4f^{10}$	Ho^{67} $6s^2 4f^{11}$	Er^{68} $6s^2 4f^{12}$	Tm^{69} $6s^2 4f^{13}$ $^2F_{7/2}$	Yb^{70} $6s^2 4f^{14}$ $^1S_0$	Lu^{71} $6s^2 5d^1 4f^{14}$ $^2D_{3/2}$

Heavy elements^b

Th^{90} $7s^2 6d^2$	Pa^{91} $6d^3$	U^{92} $6d^1 5f^3$ $^5L_4$	Np^{93} $5f^5$	Pu^{94} $5f^6$	Am^{95} $5f^7$ $^8S_{7/2}$	Cm^{96} $6d^1 5f^7$	Bk^{97} $5f^9$	Cf^{98} $5f^{10}$	E^{99} $5f^{11}$	Fm^{100} $5f^{12}$	Md^{101} $5f^{13}$

^{a}With La^{57} included, this group is also called the *lanthanides*.
^{b}With Ac^{89} included, this group is also called the *actinides*.

small. Their valence is $+2$. When singly ionized, these atoms are known as "hydrogenic" ions; their spectra resemble that of hydrogen.

The Halogens. Group VII

These are the elements: F^9, Cl^{17}, Br^{35}, I^{53}, and At^{85}. They are all missing one electron in the outermost p subshell and therefore have a valence of -1. Halogens form stable molecules with the one-electron (alkali-metal) atoms (e.g., NaCl) through ionic bonding.

The Noble Elements. Group VIII

The noble elements are also called the "rare gases" or the "inert elements." They are He^2, Ne^{10}, Ar^{18}, Kr^{36}, Xe^{54}, and Rn^{86}. Except for He, all these atoms have a completed outermost p subshell. The ground state of these elements is $^1S_0$. Total spin and orbital angular momentum are zero, so the atom has no magnetic moment. Ionization energy is large (see Table 12.2); electrical conductivity is low. Noble elements are chemically inert and have low boiling points.

The Transition Group

In the transition-group elements the incomplete $3d$ subshell is filled while 2 (or 1) electrons remain in the outer $4s$ subshell. The incomplete $3d$ subshell gives rise to magnetic properties. For example, the ground state of Cr^{24} is $^7S_3$. The spin of the atom is $s = 3$, which implies that the five $3d$ electrons and one $4s$ electron all have their spins aligned. These parallel spins contribute to the large magnetic moment of Cr. The chemical properties of elements in the transition group are due primarily to the $4s$ electron(s).

The grouping of atoms according to their shell structure is further motivated by two very significant sets of observational and inferred data. These are, respectively, first ionization energies and radii of atoms. Thus, in Fig. 12.11a we see that binding in a given n shell grows roughly as the shell is completed, with one-electron atoms most weakly bound and noble elements most strongly bound. In Fig. 12.11b we note that the radii of atoms in a given n shell decrease roughly with increase in Z, with one-electron atoms having the largest radii and halogens the smallest radii. These properties become less valid at higher Z where outer electron shells of atoms in a given row of the periodic chart (Table 12.3) are not characterized by constant principal quantum number n.

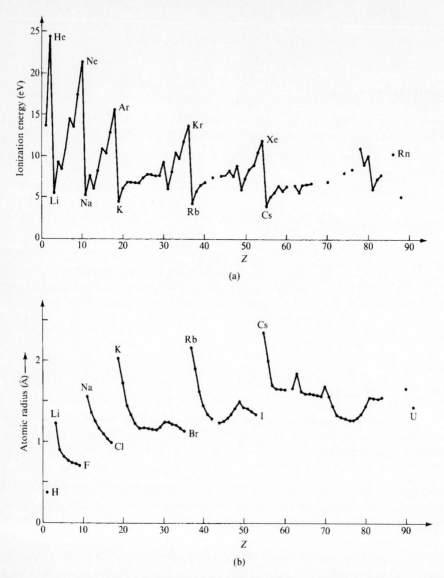

FIGURE 12.11 (a) First ionization energies versus atomic number Z. (b) Atomic radii versus atomic number. Recall conclusions of the Thomas–Fermi model concerning variation of atomic radius with increase in Z discussed in Section 10.8.

PROBLEMS

12.22   Show that the ground-state electronic configuration of Cr^{24} does not violate the Pauli principle. *Hint:* Make a list of the (n, l, m_l, m_s) values for the $3d^5 4s^1$ electrons.

12.23   What are the possible states for the ground configuration of O^{16} which includes four p electrons in its outermost shell? Check that the $^3P_2$ ground state is included in your list.

12.24   Describe the energy band structure of the metal lithium. Specifically, indicate which electrons fill the valence band and which electrons contribute to the conduction band. How full is the conduction band? (The band theory of conduction was discussed previously in Section 8.4.)

12.25   Show that the ground states for the first three elements in the "neon configuration" ($Z = 11$ to 18) are consistent with *Hund's rules:*

   (i)   The lowest energy state is the LS multiplet with largest s value.

   (ii)   When more than one value of l is associated with this maximum s value, the lowest energy state (among those satisfying the exclusion principle) is the one with largest l value.

   (iii)   For a given l subshell containing n_e electrons, in the lowest energy state the total angular momentum number j has the value $|l - s|$ for $n_e < N/2$ and $|l + s|$ for $n_e > N/2$, where N is the number of electrons in the completed subshell.

12.5   THE SLATER DETERMINANT

In the central field approximation, the Hamiltonian for an N-electron atom is written

$$(12.31) \qquad \hat{H}(1, 2, \ldots, N) = \hat{H}_1(1) + \hat{H}(2) + \cdots + \hat{H}_N(N)$$

In this notation the number 2 denotes the coordinates of the second electron. The eigenstates of the individual Hamiltonians are of the form (12.26). Calling the set of eigenvalues (n, l, m_l, m_s) of the ith electron ν_i, these eigenstates obey the equations

$$\hat{H}(1)\varphi_{\nu_1}(1) = E_{\nu_1}\varphi_{\nu_1}(1)$$
$$\hat{H}(2)\varphi_{\nu_2}(2) = E_{\nu_2}\varphi_{\nu_2}(2)$$
$$(12.32) \qquad\qquad\qquad \vdots$$
$$\hat{H}(N)V_{\nu_N}(N) = E_{\nu_N}\varphi_{\nu_N}(N)$$

In this notation the product eigenstates of $H(1, \ldots, N)$ appear as

$$(12.33) \qquad \varphi_{(\nu_1, \ldots, \nu_N)}(1, 2, \ldots, N) = \varphi_{\nu_1}(1)\varphi_{\nu_2}(2) \cdots \varphi_{\nu_N}(N)$$

However, this function is not properly antisymmetric. If $\hat{\mathfrak{x}}_{1,3}$ denotes the exchange operation of the coordinates of electrons 1 and 3, the correct antisymmetric wavefunctions of $\hat{H}(1, \ldots, N)$ have the property

$$(12.34) \quad \hat{\mathfrak{x}}_{1,3}\varphi(1, 2, 3, \ldots, N) = \varphi(3, 2, 1, \ldots, N) = -\varphi(1, 2, 3, \ldots, N)$$

The normalized wavefunction that obeys this rule (for all pairs of particles) and is an eigenstate of $\hat{H}(1, \ldots, N)$ is given by the *Slater determinant,*

$$(12.35) \quad \varphi_A(1, 2, \ldots, N) = \frac{1}{\sqrt{N!}} \begin{vmatrix} \varphi_{\nu_1}(1) & \varphi_{\nu_2}(1) & \cdots & \varphi_{\nu_N}(1) \\ \varphi_{\nu_1}(2) & \varphi_{\nu_2}(2) & \cdots & \varphi_{\nu_N}(2) \\ \vdots & \vdots & & \vdots \\ \varphi_{\nu_1}(N) & \varphi_{\nu_2}(N) & \cdots & \varphi_{\nu_N}(N) \end{vmatrix}$$

This determinant has four outstanding properties, which we discuss next.

Eigenvalues

It is an eigenstate of (12.31) with eigenvalue

$$(12.36) \quad E_{(\nu_1, \nu_2, \ldots, \nu_N)} = E_{\nu_1} + E_{\nu_2} + \cdots + E_{\nu_N}$$

The explicit form of φ_A appears as

$$(12.37) \quad \varphi_A = \frac{1}{\sqrt{N!}} \sum_{P(\nu_1, \nu_2, \ldots, \nu_N)} (-1)^{|P|} \varphi_{\nu_1}(1)\varphi_{\nu_2}(2) \cdots \varphi_{\nu_N}(N)$$

The sum is over all permutations P of the quantum indices $(\nu_1, \nu_2, \ldots, \nu_N)$. The symbol $|P|$ is 0 or 1. It is zero if the permutation $P(\nu_1, \ldots, \nu_N)$ can be obtained from $(\nu_1, \ldots, \nu_N)$ through an even number of exchanges of two indices. It is 1 if $P(\nu_1, \ldots, \nu_N)$ involves an odd number of exchanges. For example, the term

$$-\varphi_{\nu_4}(1)\varphi_{\nu_1}(2)\varphi_{\nu_2}(3)\varphi_{\nu_3}(4) \cdots \varphi_{\nu_N}(N)$$

corresponds to the permutation

$$P(\nu_1, \ldots, \nu_N) = \nu_4, \nu_1, \nu_2, \nu_3, \ldots, \nu_N$$

This sequence only involves permutation of the first four indices. To obtain $|P|$ we must rearrange the sequence $(4, 1, 2, 3)$ in the form of the original sequence $(1, 2, 3, 4)$

through exchanges of two integers only and count the minimum number of such exchanges which do the job.

$$(4, 1, 2, 3) \rightarrow (1, 4, 2, 3) \rightarrow (1, 2, 4, 3) \rightarrow (1, 2, 3, 4)$$

Three exchanges suffice so that $(-1)^{|P|} = -1$. We conclude that the preceding product wavefunction carries a minus sign, as written.

Each such N-particle product function is an eigenstate of $\hat{H}$ corresponding to the degenerate eigenenergy (12.36), whence the determinantal form (12.37), which is merely a linear combination of these product states, is also an eigenstate of $\hat{H}$ corresponding to the eigenenergy (12.36).

Orthonormality

The second property that the determinantal states (12.35) have is that they form an orthonormal sequence. That is,

$$
\begin{aligned}
\langle \varphi_A | \varphi_A \rangle &= \frac{1}{N!} \sum_P \langle \varphi_{\nu_1}(1) \cdots \varphi_{\nu_N}(N) | \varphi_{\nu_1}(1) \cdots \varphi_{\nu_N}(N) \rangle \\
&= 1
\end{aligned}
$$

since there are $N!$ terms of unit value in the sum. Furthermore, owing to the orthogonality of single-particle eigenstates,

$$
\langle \varphi_{A(\nu_1, \ldots, \nu_N)} | \varphi_{A(\nu_1', \ldots, \nu_N')} \rangle = 0
$$
$$
(\nu_1, \ldots, \nu_N) \neq (\nu_1', \ldots, \nu_N')
$$

This establishes the orthogonality of these states.

Antisymmetry

The third property concerns the symmetry of φ_A. If $\hat{x}_{ij}$ denotes the exchange of the ith-particle coordinates with those of the jth particle, then

$$\hat{x}_{ij} \varphi_A(i, j) = \varphi_A(j, i) = -\varphi_A(i, j)$$

Exchanging particle coordinate numbers in φ_A, as given by (12.35), is effected by an exchange of two rows of the determinant. But a determinant changes sign under interchange of two rows. It follows that $\varphi_A(i, j) = -\varphi_A(j, i)$. For example, for the exchange particles 1 and 2,

$$
\hat{x}_{1,2}
\begin{vmatrix}
\varphi_{\nu_1}(1) & \varphi_{\nu_2}(1) & \cdots & \varphi_{\nu_N}(1) \\
\varphi_{\nu_1}(2) & \varphi_{\nu_2}(2) & \cdots & \varphi_{\nu_N}(2) \\
\vdots & \vdots & & \vdots \\
\varphi_{\nu_1}(N) & \varphi_{\nu_2}(N) & \cdots & \varphi_{\nu_N}(N)
\end{vmatrix}
=
\begin{vmatrix}
\varphi_{\nu_1}(2) & \varphi_{\nu_2}(2) & \cdots & \varphi_{\nu_N}(2) \\
\varphi_{\nu_1}(1) & \varphi_{\nu_2}(1) & \cdots & \varphi_{\nu_N}(1) \\
\vdots & \vdots & & \vdots \\
\varphi_{\nu_1}(N) & \varphi_{\nu_2}(N) & \cdots & \varphi_{\nu_N}(N)
\end{vmatrix}
$$

$$
= -
\begin{vmatrix}
\varphi_{\nu_1}(1) & \varphi_{\nu_2}(1) & \cdots & \varphi_{\nu_N}(1) \\
\varphi_{\nu_1}(2) & \varphi_{\nu_2}(2) & \cdots & \varphi_{\nu_N}(2) \\
\vdots & \vdots & & \vdots \\
\varphi_{\nu_1}(N) & \varphi_{\nu_2}(N) & \cdots & \varphi_{\nu_N}(N)
\end{vmatrix}
$$

This property establishes the antisymmetry of the state φ_A.

Exclusion Principle

Finally, we note that if particle 2 has the same quantum numbers as particle 1, then $\varphi_A = 0$. This property also follows from the determinantal structure of φ_A: namely, if two particles are in the same eigenstate, then two columns of the determinant (12.31) are equal and φ_A vanishes. Thus φ_A written in the Slater determinant form is consistent with the Pauli exclusion principle. The Hartree–Fock approximation (Section 13.10) offers a refinement of the calculation of wavefunctions in the central field approximation.

PROBLEMS

12.26   Which purely determinantal properties related to the Slater determinant are involved in (a) the antisymmetry of φ_A? (b) the Pauli exclusion principle?

12.27   Two spin-$\frac{1}{2}$ neutrons move in a two-dimensional box of edge length L and impenetrable walls. If the neutrons do not interact with each other, construct the antisymmetric determinantal spin and coordinate-dependent energy eigenstates for the system.

12.6   APPLICATION OF SYMMETRIZATION RULES TO THE HELIUM ATOM

We have already found the Pauli principle to be an important rule in forming the periodic chart of the elements and in construction of properly antisymmetrized wavefunctions for atomic electrons in the central field approximation.

In this and the next section we will further demonstrate the important role played by symmetrization principles in analysis of two elemental systems in nature. The first of these is the helium atom, which has two outer electrons. Among other properties we will find how the Pauli principle influences the coupling between these electrons in the construction of properly antisymmetrized wavefunctions for the atom. The second example is the deuterium molecule, whose two deuteron nuclei are bosons. Consequently, the nuclear component of the molecular wavefunction must be symmetrized with respect to exchange of space and spin coordinates. Construction of such properly symmetrized states leads very simply to intensity rules for emission.

The Helium Atom

The Hamiltonian of helium, in a frame where the nucleus is at rest, is

$$(12.38) \qquad \hat{H} = \left(\frac{\hat{p}_1^2}{2m} - \frac{2e^2}{r_1} \right) + \left(\frac{\hat{p}_2^2}{2m} - \frac{2e^2}{r_2} \right) + \frac{e^2}{r_{12}} + \hat{H}_{SO}$$

$$= \hat{H}_0(1) + \hat{H}_0(2) + \hat{H}_{ES} + \hat{H}_{SO}$$

The last term, H_{SO}, is written for the spin-orbit interaction between the electrons and the nucleus, while H_{ES} is written for the electrostatic interaction, e^2/r_{12}, between the two electrons. The interelectron displacement is

$$r_{12} = |\mathbf{r}_2 - \mathbf{r}_1|$$

(Fig. 12.12). If the electrostatic as well as spin-orbit terms are neglected, $\hat{H}$ reduces to the sum of two hydrogenic Hamiltonians (each with $Z = 2$).

$$(12.39) \qquad \hat{H}_0(1, 2) = \hat{H}_0(1) + \hat{H}_0(2)$$

This Hamiltonian (as well as the total Hamiltonian) is symmetric with respect to the interchange of the two electrons

$$(12.40) \qquad [\hat{x}_{12}, \hat{H}_0(1, 2)] = 0$$

This merely reflects the indistinguishability of the two electrons. This property must be maintained in the eigenstates that we construct for $\hat{H}_0(1, 2)$. With the abbreviations

$$\nu_1 \equiv (n_1, l_1, m_{l_1}); \qquad \nu_2 \equiv (n_2, l_2, m_{l_2})$$

these symmetrized eigenstates of $H_0(1, 2)$ appear as[1]

$$(12.41) \qquad \varphi_{S, A}(\mathbf{r}_1, \mathbf{r}_2) = \frac{1}{\sqrt{2}} [\varphi_{\nu_1}(1)\varphi_{\nu_2}(2) \pm \varphi_{\nu_1}(2)\varphi_{\nu_2}(1)]$$

---

[1] In the event that $\nu_1 = \nu_2$, then $\varphi_S(\mathbf{r}_1, \mathbf{r}_2) = \varphi_{\nu_1}(1)\varphi_{\nu_1}(2)$.

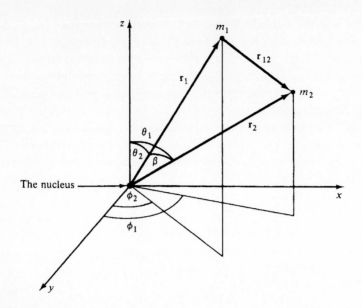

FIGURE 12.12   **The coordinates of the two electrons in helium. The six-dimensional volume element $(d\mathbf{r}_1\, d\mathbf{r}_2)$ is given by**

$$d\mathbf{r}_1\, d\mathbf{r}_2 = r_1^2\, d\Omega_1\, dr_1\, r_2^2\, d\Omega_2\, dr_2 = r_1^2 r_2^2\, dr_1\, dr_2\, d\cos\theta_1\, d\cos\theta_2\, d\phi_1\, d\phi_2$$

The potential of interaction between electrons is given by

$$V = \frac{e^2}{r_{12}} = \frac{e^2}{\sqrt{r_1^2 + r_2^2 - 2r_1 r_2 \cos\beta}}$$

(See Fig. 9.16 for addition formulas connecting β to θ_1, θ_2, ϕ_1, ϕ_2.)

The plus sign gives a symmetric state, φ_S, while the minus sign gives an antisymmetric state, φ_A. The energy eigenvalue corresponding to either of these states is

(12.42) $$E^{(0)}_{n_1, n_2} = -4\mathbb{R}\left(\frac{1}{n_1^2} + \frac{1}{n_2^2}\right) = -2mc^2\alpha^2\left(\frac{1}{n_1^2} + \frac{1}{n_2^2}\right)$$

Separation of Multiplets Due to Spin Symmetry

Although the spin-orbit correction to the Hamiltonian of helium is small $[\Delta E_{SO}/E \sim \alpha^2$; see (12.13)] and may well be neglected in a first approximation, the spin of the electrons still has an important influence on the properties of helium. This occurs through a combination of Pauli antisymmetrization requirements on wavefunctions with respect to exchange of space and spin coordinates and the relatively large electrostatic interaction between electrons (see Problem 12.28). In what follows, first we will construct the properly antisymmetrized space-spin dependent eigenstates of $\hat{H}_0(1, 2)$. This immediately implies a coupling between the electron spins. Three states emerge with $s = 1$ (the triplet series) and one state emerges with $s = 0$ (the singlet series). When the electrostatic interaction e^2/r_{12} is brought into play, it is found that the triplet states all lie lower in energy than the singlet states. In this manner we will find that symmetry requirements couple the electron spins and electrostatic interaction separates the resulting singlet and triplet states.

Insofar as $\hat{H}_0(1, 2)$ does not contain the spin, spin-dependent eigenstates of $\hat{H}_0(1, 2)$ are quite simple to construct. If $\varphi(\mathbf{r}_1, \mathbf{r}_2)$ is a space-dependent eigenstate (12.41) of $H_0(1, 2)$, then following the procedure described in Section 12.3 for two free electrons, we find that the properly antisymmetrized wavefunctions are given by

$$^1\chi = \varphi_S(\mathbf{r}_1, \mathbf{r}_2)\xi_A(1, 2) \qquad\qquad (s = 0)$$

(12.43)
$$^3\chi = \varphi_A(\mathbf{r}_1, \mathbf{r}_2)\left\{\begin{array}{l} \xi_S^{(1)}(1, 2) \\ \xi_S^{(0)}(1, 2) \\ \xi_S^{(-1)}(1, 2) \end{array}\right\} \qquad (s = 1)$$

The ξ-spin functions are listed in Table 11.3. We see how symmetrization requirements, together with the Pauli principle, effect a coupling between the spins of the two electrons in helium. In the triplet state ($s = 1$) the spins are aligned, whereas in the singlet state ($s = 0$) the spins are antialigned.

Electrostatic Interaction

To understand how the coupling augments the energies of helium, we recall the following property of symmetrized states (see Problem 8.32): namely, two particles in a symmetric state attract one another (in a statistical sense). It follows that the two electrons in the singlet $^1\chi$ state, which contains the symmetric φ_S state, are closer to each other than they are in the triplet $^3\chi$ state, which contains the antisymmetric φ_A state. Thus, owing to the positive repulsive energy of the electrostatic interaction; e^2/r_{12}, the triplet states lie lower in energy than do the singlet states (Fig. 12.13).

This is the mechanism behind Hund's first rule (see Problem 12.25)—that the total spin assumes the maximum value consistent with the Pauli principle. In this symmetric spin state, the space component of the wavefunction must be antisymmetric so that electrons are further removed from each other than in the corresponding symmetric space state.

Exchange and Coulomb Interaction Energies

Let us consider the space component eigenstates (12.41) of helium in more detail. As it turns out, the only states of helium that are of practical significance are those for

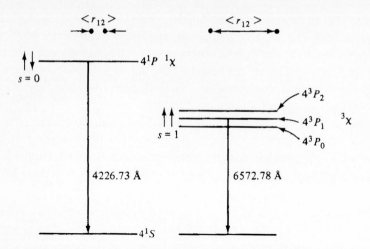

FIGURE 12.13 The 4P states of calcium, illustrating the fact that, in two-electron atoms, triplet states lie lower than the corresponding singlet state. This is due to the fact that the average interelectron distance is smaller in the symmetric space state, φ_S, than in the antisymmetric space state, φ_A, thereby increasing the singlet electrostatic contribution to the total energy compared to the corresponding triplet contribution.

$$\langle {}^1\chi | H_{ES} | {}^1\chi \rangle > \langle {}^3\chi | H_{ES} | {}^3\chi \rangle$$

$$H_{ES} = +\frac{e^2}{r_{12}}$$

which one of the two electrons is in its own ground state, with $n = 1$, $l = m_l = 0$. The reason for this is that it takes less energy to ionize a helium atom from the ground state than it does to raise both electrons to excited levels (Fig. 12.14). This means that one is more likely to find an He^+ ion (hydrogenic ion with $Z = 2$) than a helium atom with both electrons in excited states. It follows that the space-dependent states of helium atoms that exist under natural conditions are mostly of the form

$$(12.44) \qquad \varphi_{S,A}(\mathbf{r}_1, \mathbf{r}_2) = \frac{1}{\sqrt{2}} \left[\varphi_{100}(\mathbf{r}_1)\varphi_{nlm}(\mathbf{r}_2) \pm \varphi_{100}(\mathbf{r}_2)\varphi_{nlm}(\mathbf{r}_1) \right]$$

One may use these eigenstates to calculate the corrections to the eigenenergies of $\hat{H}_0$ due to the electrostatic interaction, e^2/r_{12}. As described above, we expect the triplet states to lie lower in energy than the singlet states.

One obtains

$$(12.45) \qquad \left\langle \frac{e^2}{r_{12}} \right\rangle_{singlet} = \langle {}^1\chi | \left(\frac{e^2}{r_{12}} \right) | {}^1\chi \rangle$$

$$= A + B$$

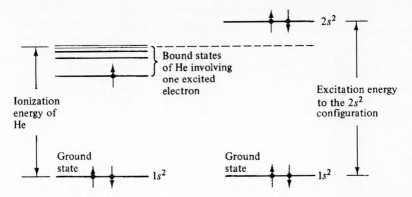

FIGURE 12.14 The ionization energy of He is smaller than the energy of the lowest energy state of He with both electrons excited. It is possible for He in this excited state to decay to He$^+$ and a free electron.

where

$$A = \langle \varphi_{100}(1)\varphi_{nlm_l}(2) \,|\, \frac{e^2}{r_{12}} \,|\, \varphi_{100}(1)\varphi_{nlm_l}(2) \rangle$$

$$= \int d\mathbf{r}_1 \, d\mathbf{r}_2 \,|\, \varphi_{100}(1)\,|^2 \,|\, \varphi_{nlm_l}(2)\,|^2 \, \frac{e^2}{r_{12}}$$

(12.46)

$$B = \langle \varphi_{100}(1)\varphi_{nlm_l}(2) \,|\, \frac{e^2}{r_{12}} \,|\, \varphi_{100}(2)\varphi_{nlm_l}(1) \rangle$$

$$= \int d\mathbf{r}_1 \, d\mathbf{r}_2 \, \varphi_{100}^*(1)\varphi_{nlm_l}(1)\varphi_{100}(2)\varphi_{nlm_l}^*(2) \, \frac{e^2}{r_{12}}$$

The energy A is called the *Coulomb interaction energy*. It is akin to the classical interaction potential of two electron clouds with respective charge densities $e\,|\,\varphi_{100}(1)\,|^2$ and $e\,|\,\varphi_{nlm}(2)\,|^2$. The second term, B, has no counterpart in classical physics. It is called the *exchange interaction energy*.

In the triplet state the Coulomb interaction energy becomes

(12.47)
$$\left\langle \frac{e^2}{r_{12}} \right\rangle_{triplet} = \langle\, {}^3\chi \,|\, \left(\frac{e^2}{r_{12}}\right) \,|\, {}^3\chi \rangle$$

$$= A - B$$

For A and B positive this correction energy is smaller than the corresponding singlet correction energy, $A + B$ (see Problem 12.31).[1]

Thus we find that the electrostatic interaction separates the singlet and triplet states of helium. Furthermore, when spin-orbit interaction is brought into play, differ-

---

[1] The positivity of these integrals is demonstrated in J. L. Slater, *Quantum Theory of Atomic Structure*, Vol. 1, McGraw-Hill, New York, 1960, Appendix 19.

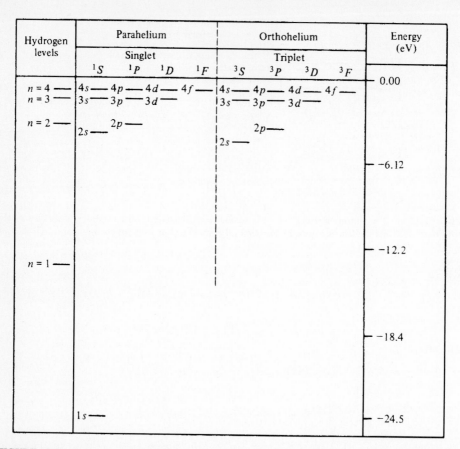

Hydrogen levels	Parahelium				Orthohelium				Energy (eV)
	Singlet				Triplet				
	^1S	^1P	^1D	^1F	^3S	^3P	^3D	^3F	

FIGURE 12.15 Energy levels of helium, illustrating the singlet and triplet series. The fine structure of the triplet levels is not shown. The energies of hydrogen appear at the left.

ent j values have slightly different energies in the triplet states. For a given value of l, the total angular momentum j number has the three values

$$j = l - 1, l, l + 1$$

which in turn give three distinct values for the spin-orbit coefficient $\langle \mathbf{L} \cdot \mathbf{S} \rangle$, thereby splitting l levels into triplets.

Helium in antisymmetric spin states (singlet series) is called *parahelium*. Helium in symmetric spin states is called *orthohelium*. The distinct spectra associated with these different atomic configurations are shown in Fig. 12.15.

Similar descriptions apply to the heavier two-electron atoms. Their spectra also are observed to separate into singlet and triplet series. The corresponding energy-level diagram for Ca^{20} is shown in Fig. 12.16.

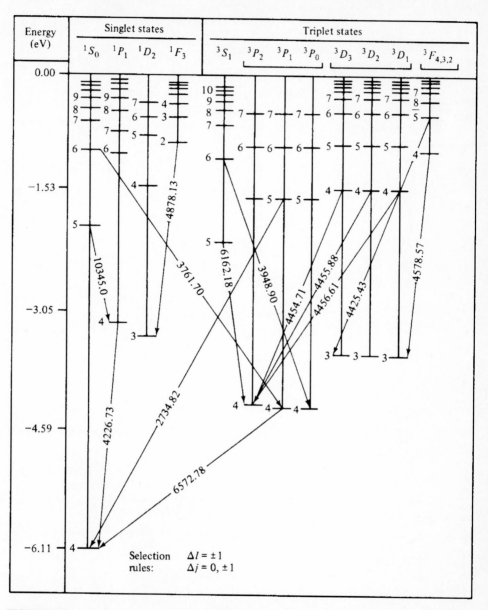

FIGURE 12.16 Energy levels of calcium, exhibiting the fine structure of the triplet states. A few typical transitions are also shown. Transition wavelengths are in Å. Principal quantum numbers appear at the left of levels.

PROBLEMS

12.28 (a) What is the ground-state wavefunction for helium in the approximation $\hat{H}_{ES} = \hat{H}_{SO} = 0$ in (12.38)?

(b) What is the ground-state energy in this approximation?

(c) What is the correction to this ground-state energy due to the electrostatic interaction, e^2/r_{12}? What is the total ground-state energy obtained in this manner?

(d) In view of your answer to part (c), is the ground state of $\hat{H}_0(1, 2)$ a good guess for the ground state of $\hat{H}_0(1, 2) + \hat{H}_{ES}$?

Answers

(a) The ground state is $\varphi_S(\mathbf{r}_1, \mathbf{r}_2)$, with $n_1 = n_2 = 1$ and $l_1 = l_2 = 0$. This gives $(Z = 2)$

$$\varphi_S = \frac{Z^3}{\pi a_0^3} \exp\left(-\frac{r_1 + r_2}{a_0/Z}\right)$$

(b) $E_{11}^{(0)} = -4\mathbb{R}(1 + 1) = -4mc^2\alpha^2 = -108.8$ eV

(c)
$$\Delta E = \langle \varphi_S | \frac{e^2}{r_{12}} | \varphi_S \rangle$$

$$= \iint |\varphi_S|^2 \frac{e^2}{r_{12}} d\mathbf{r}_1 d\mathbf{r}_2 = \frac{5e^2}{4a_0} = 34 \text{ eV}$$

$$d\mathbf{r}_1 = r_1^2 dr_1 d\Omega_1, \qquad d\mathbf{r}_2 = r_2^2 dr_2 d\Omega_2$$

$$r_{12}^2 = r_1^2 + r_2^2 - 2r_1 r_2 \cos\beta$$

(See Figs. 9.9 and 12.12, and recall the generating function for Legendre polynomials given in Table 9.2.) This gives the corrected ground-state energy

$$E = E_{11}^{(0)} + \Delta E = -74.8 \text{ eV}$$

(d) The fact that ΔE is not small compared to $E_{11}^{(0)}$ means that φ_S is not a good guess for the ground state of $\hat{H}_0 + \hat{H}_{ES}$. (*Note:* $E_{obs} = -78.98$ eV.)

12.29 Which of the following operators are diagonalized by the states $^1\chi$ and $^2\chi$ (12.43) relevant to helium?

$$\hat{L}^2, \hat{L}_z, \hat{J}^2, \hat{J}_z, \hat{S}^2, \hat{S}_z, \hat{H}_0, \hat{\mathscr{X}}_{12}$$

12.30 The spin-orbit interaction in two-electron atoms gives three distinct energies in the triplet series. What are the values of $\langle \mathbf{L} \cdot \mathbf{S} \rangle$ if one electron is an s electron and the other is a d electron.

12.31 Show that the integrals A and B in (12.45) are both positive. [*Hint:* Recall the Fourier transform of the Coulomb potential,

$$\frac{1}{r_{12}} = \frac{1}{2\pi^2} \int \exp(i\mathbf{k} \cdot \mathbf{r}_{12}) \frac{d\mathbf{k}}{k^2}.]$$

12.32 A positron is the antiparticle of an electron. When in the presence of an electron, it may bind to the electron, forming a *positronium atom* that is unstable to positron-electron annihilation. Prior to annihilation, the energies and wavefunctions of the atom may be approximated by those of hydrogen with the Bohr radius replaced by $2a_0 = 2\hbar^2/me^2$, owing to the change in reduced mass. What are the spin-dependent components of the wavefunctions of positronium? (*Note:* The annihilation time of the ^1S state of positronium for decay into two photons is $\approx 1.2 \times 10^{-10}$ s. The ^3S state decays into three photons and lasts $\approx 1.4 \times 10^{-7}$ s. The fact that positronium in the ^3S state must annihilate through the emission of three photons is due to the principle of charge conjugation. This principle states that electromagnetic interactions are invariant under change of all particles to their antiparticles.[1]) The positron is further discussed in Section 15.3.

12.7 THE HYDROGEN AND DEUTERIUM MOLECULES

Exchange Binding

Another important area in which symmetrization requirements imposed by the spin of constituent particles plays a significant role is that of the theory of diatomic molecules. The simplest of these is the hydrogen molecule, H_2. The fact that the proton nuclei are extremely more massive than the electrons permits analysis of the molecule to be divided into two parts.[2] The first of these concerns the chemical binding between the two atoms due to electron coupling. The second addresses the motion of the nuclei within the bound configuration.

The Hamiltonian of the molecule, neglecting all but electrostatic interaction, is (Fig. 12.17)

(12.48)
$$\hat{H} = \hat{H}_{\text{atom }a} + \hat{H}_{\text{atom }b} + V_{ee} + V_{pp} + V_{ep} + \hat{T}_{\text{nuc}}$$

$$\hat{H}_{\text{atom }a} = \frac{\hat{p}_1^2}{2m} - \frac{e^2}{r_1} \qquad V_{ee} = +\frac{e^2}{r_{12}}$$

$$\hat{H}_{\text{atom }b} = \frac{\hat{p}_2^2}{2m} - \frac{e^2}{r_2} \qquad V_{pp} = +\frac{e^2}{r_{ab}}$$

$$\hat{T}_{\text{nuc}} = \frac{1}{2M}(\hat{p}_a^2 + \hat{p}_b^2) \qquad V_{ep} = -e^2\left(\frac{1}{r_{1b}} + \frac{1}{r_{2a}}\right)$$

The mass of an electron is m, nuclear mass is M, and r_1 is the distance from electron 1 to nucleus a. The repulsion between the two electrons is given by the positive

[1] For additional discussion, see S. Gasiorowicz, *Quantum Physics*, Wiley, New York, 1974.
[2] This approximation is due to M. Born and J. Oppenheimer, *Ann. Physik* **84**, 457 (1927).

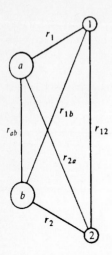

FIGURE 12.17 Radial distances appropriate to the hydrogen molecule.

potential e^2/r_{12}, while the repulsion between the nuclei is e^2/r_{ab}. The potential V_{ep} represents the cross attraction between electrons and protons. The kinetic energy of the two nuclei is written $\hat{T}_{\text{nuc}}$. Since $m/M \ll 1$, it is consistent to view the electrons as moving in the field of two fixed protons ($p_a{}^2 = p_b{}^2 = 0$, $r_{ab} = $ constant).

To uncover the binding between the two atoms, one constructs the wavefunctions of $\hat{H}$, further neglecting the interaction $V_{ee} + V_{ep}$. The coupling between atoms which follows is then due primarily to antisymmetrization requirements imposed on these wavefunctions in accord with the Pauli principle. The residual Hamiltonian appears as

$$(12.49) \qquad \hat{H} = \hat{H}_{\text{atom}\,a} + \hat{H}_{\text{atom}\,b}$$

with eigenstates

$$(12.50) \qquad \varphi_{S,A} = \frac{1}{\sqrt{2}}\left[\varphi_{\nu_a}(\mathbf{r}_1)\varphi_{\nu_b}(\mathbf{r}_2) \pm \varphi_{\nu_a}(\mathbf{r}_2)\varphi_{\nu_b}(\mathbf{r}_1)\right]$$

Following previous notation [e.g., (12.41)] atomic eigenvalues have been written $\nu_{a,b}$. Spin-dependent states parallel those constructed for the helium atom (12.43) in that both problems address the antisymmetric states for two electrons. There results

$$(12.51) \qquad \begin{aligned} {}^1\chi &= \varphi_S(\mathbf{r}_1, \mathbf{r}_2)\xi_A \\ {}^2\chi &= \varphi_A(\mathbf{r}_1, \mathbf{r}_2)\xi_S \end{aligned}$$

Using these state functions, it is possible to calculate the expectation of the total potential of the hydrogen molecule contained in $\hat{H}$ given by (12.48).

$$(12.52) \qquad \langle V \rangle = e^2 \left\langle \frac{1}{r_{12}} + \frac{1}{r_{ab}} - \frac{1}{r_1} - \frac{1}{r_2} - \frac{1}{r_{1b}} - \frac{1}{r_{2a}} \right\rangle$$

In calculating this average, r_1 and r_2 dependence is lost to integration, leaving only dependence on the internuclear distance r_{ab}. Thus we may write

$$\langle V \rangle = \bar{V}(r_{ab})$$

The resulting two curves for the triplet and singlet states are shown in Fig. 12.18. The potential of interaction is seen to have a minimum for the singlet state corresponding to antiparallel spins and symmetric space dependence as given by φ_S. Thus, binding of the atoms is possible in the singlet state. As discussed in the previous section, electrons in the state φ_S tend to occupy the same region of space. This common domain lies between the nuclei. At this location the electrons serve to attract each of the protons and bind the molecule. The same mechanism, we recall, is responsible for the triplet states of two-electron atoms lying lower in energy than the singlet states (see Fig. 12.13). However, in the present case the positive energy of repulsion between electrons is overbalanced by the negative energy of attraction of the protons toward the overlap domain. If r_{ab} is decreased beyond the minimum in $\bar{V}(r_{ab})$, the nuclear repulsion begins to overcome the binding afforded by the intermediary electrons and the atoms repel.

In the triplet state the antisymmetric wavefunction $\varphi_A(\mathbf{r}_1, \mathbf{r}_2)$ is appropriate, for which case $\varphi_A(\mathbf{r}_1, \mathbf{r}_1) = 0$, so that electrons do not tend to occupy the common domain between nuclei and there is no binding. This repulsion in the triplet state is evidenced by the monatonic increase of $\bar{V}(r_{ab})$ with decreasing internuclear distance, as shown in Fig. 12.18.

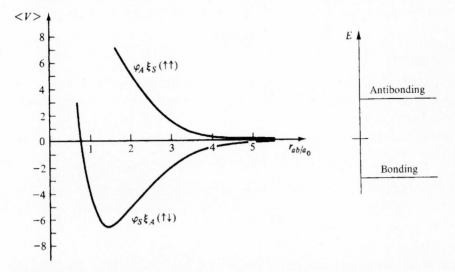

FIGURE 12.18   **Expectation of the hydrogen molecule potential (in units of 10^{-12} erg) versus internuclear distance r_{ab} (in units of Bohr radii). Shown also is a schematic of the corresponding "bonding" and "antibonding" levels of the molecule.**

In the symmetric electronic state, electrons may be said to be "shared" by the two hydrogen atoms, thereby allowing each atom a completed $(1s^2)$ K shell. The bond so effected is called a *covalent bond*. This bonding is to be differentiated from that which couples, say, the NaCl molecule. In this case the sodium atom gives its isolated $3s^1$ electron to the vacancy in the $(3s^2 3p^5)$ M shell of the chlorine atom. In the resulting configuration, the positively charged sodium ion is knitted to the negatively charged chlorine ion in what is termed an *ionic bond*.

Symmetric States for the Nuclear Motion of D_2

Having discovered the nature of the binding of the H_2 molecule, we turn next to a discussion of the nuclear motion within this bound configuration. However, in that we wish to address the construction of symmetric states relevant to bosons, we will consider the isotope of hydrogen, deuterium. The deuterium atom has at its nucleus a deuteron that has spin 1 and is therefore a boson. The Hamiltonian (12.48) carries over to D_2 with the change that M becomes the mass of a deuteron instead of a proton. The mass ratio then becomes $M/m \sim 34,000$ and approximations introduced above for H_2 are even more appropriate to D_2.

In the bound configuration the deuterons move within the effective potential field $\bar{V}(r_{ab})$, and the Hamiltonian for the deuteron motion may be written

$$(12.53) \qquad \hat{H}_{\text{nuc}} = \hat{T}_{\text{nuc}} + \bar{V}(r_{ab})$$

The kinetic energy of the two deuterons $\hat{T}_{\text{nuc}}$ may be rewritten in terms of center-of-mass motion and motion relative to the center of mass [see (10.95)]. There results

$$(12.54) \qquad \hat{H}_{\text{nuc}} - \hat{H}_{\text{CM}} = \frac{\hat{p}_r^2}{2\mu} + \frac{\hat{L}^2}{2\mu r^2} + \bar{V}(r)$$

The variable r is written for the interdeuteron distance r_{ab}, and $\mu = M/2$ is the reduced mass of the two deuterons. If $V(r)$ has its minimum at r_0, then near equilibrium one may write

$$(12.55) \qquad \bar{V}(r) = \bar{V}(r_0) + \frac{1}{2}(r - r_0)^2 \left(\frac{\partial \bar{V}}{\partial r} \right)_{r_0} + \cdots$$

This parabolic potential gives rise to vibrational motion. At moderately low temperature these vibrational modes are "frozen in." That is, they are not excited and the deuterons assume the shape of a rigid dumbbell which is free only to rotate.[1] The

---

[1] Characteristic rotational temperature for the D_2 molecule is $T = \hbar^2/2Ik_B = 44K$. Vibrational modes are excited at 4500 K and electron states are excited at temperatures several orders of magnitude larger. For further reference, see G. Herzberg, *Molecular Spectra and Structure*, Van Nostrand Reinhold, New York, 1950.

Hamiltonian in this temperature domain reduces to the simple rotational form

(12.56)
$$\hat{H}_{\text{nuc}} = \frac{\hat{L}^2}{2I}$$

where $I = \mu r_0^2$ is the moment of inertia of the dumbbell two-deuteron system. Eigenstates of this purely rotational Hamiltonian are the spherical harmonics $|lm_l\rangle$ with corresponding eigenenergies (9.49)

$$E_l = \frac{\hbar^2 l(l + 1)}{2I}$$

The frequencies of emission due to transitions between rotational states (see Problem 9.6) lie in the infrared and are clearly distinguished from frequencies due to transitions in the electron states which lie in the ultraviolet-visible portion of the spectrum.

Spin-dependent eigenstates of $\hat{H}_{\text{nuc}}$ (12.56) are simply constructed in the product form

$$\chi_{\text{nuc}} = |lm_l\rangle\xi$$

Since the deuteron has spin 1, it is a boson and χ_{nuc} must be properly symmetrized with respect to exchange of spin and space coordinates. The spin component ξ is composed of the nine states derived from the addition of two spin-1 particles. These states were previously constructed in Chapter 9 and are listed in Table 9.5. Of these nine states, six corresponding to $s = 0$ and $s = 2$ are symmetric with respect to exchange of spin coordinates, while the remaining three corresponding to $s = 1$ are antisymmetric. This separation of states is listed below.

Symmetric	Antisymmetric
$\hat{x}\lvert sm_s s_1 s_2\rangle = +\lvert sm_s s_1 s_2\rangle$	$\hat{x}\lvert sm_s s_1 s_2\rangle = -\lvert sm_s s_1 s_2\rangle$
$\begin{aligned}&\lvert 2211\rangle\\&\lvert 2111\rangle\\&\lvert 2011\rangle\\&\lvert 2-111\rangle\\&\lvert 2-211\rangle\end{aligned}\Big\}\ s=2$	$\begin{aligned}&\lvert 1111\rangle\\&\lvert 1011\rangle\\&\lvert 1-111\rangle\end{aligned}\Big\}\ s=1$
$\lvert 0011\rangle\quad \}\,s=0$	

In that the two deuterons form a dumbbell configuration, exchange of deuterons is equivalent to inversion through the origin. It follows that this operation is identical to the parity operation, so that the symmetry of a rotational state with angular momentum quantum number l is $(-1)^l$ (see Problem 12.21). Thus in order that the states χ_{nuc} be totally symmetric, they must be of the form

(12.57)
$$^{6}\chi_{\text{nuc}} = |l_{\text{even}}m_l\rangle\xi_S \qquad \text{(ortho)}$$
$$^{3}\chi_{\text{nuc}} = |l_{\text{odd}}m_l\rangle\xi_A \qquad \text{(para)}$$

For a given value of l, there are six χ_{nuc} states corresponding to l even and three states corresponding to l odd. As with the states relevant to helium, those containing an antisymmetric spin component are denoted as *para* states, while those containing a symmetric spin component are denoted as *ortho* states.[1]

Owing to the commutation property (12.22), the exchange operator $\hat{x}$ is a constant of the motion and one may expect transitions between states of different exchange symmetry to be forbidden. Assuming a uniform population of states, there are twice as many ortho states as there are para states. Thus in transitions between states of different angular momentum, radiation due to (even–even) decay is roughly twice as intense as that due to (odd–odd) decay.

A comparison of these properties of the nuclear wavefunctions for H_2 and D_2 is listed in Table 12.5. Since the proton nuclei of H_2 are fermions, the corresponding χ_{nuc} function must be antisymmetrized with respect to exchange of space and spin coordinates. This reversal of symmetry requirements on χ_{nuc} results in nearly a complete reversal of intensity rules obtained above for the D_2 molecule. (These properties of H_2 are further discussed in Problems 12.37 and 12.38.)

Finally, we note that the χ_{nuc} wavefunction of lowest angular momentum for the D_2 molecule is the ortho $^{1}S_0$ state. In this configuration both orbital and spin angular momentum vanish ($s = l = 0$). The relative orientation of the spins of the deuterons in this state may be described as antiparallel, although in fact this $|0011\rangle$ state is the superposition

$$\uparrow\downarrow + \cdot\cdot + \downarrow\uparrow = \sqrt{\tfrac{1}{3}}|11\rangle_1|1, -1\rangle_2 - \sqrt{\tfrac{1}{3}}|10\rangle_1|10\rangle_2 + \sqrt{\tfrac{1}{3}}|1, -1\rangle_1|11\rangle_2$$

(The diagrammatic representation on the left is explained in Table 9.5.)

In this section we have found how symmetry requirements imposed by the spin of constituent particles strongly influence the physical properties of diatomic molecules. Antisymmetrization of the electron wavefunctions was found to give rise to exchange binding of the hydrogen molecule. Symmetry requirements on the wavefunctions for the two-boson deuteron system were found to give rise to intensity rules for molecular radiation. Symmetrization requirements will enter again in the next

---

[1] States of greater statistical weight carry the prefix *ortho*, whereas those of smaller statistical weight carry the prefix *para*.

TABLE 12.5 Properties of the nuclear component wavefunctions for D_2 and H_2

D_2

$\chi_{\text{nuc}} = |m_l\rangle \xi$
χ_{nuc} is symmetric

s	Multiplicity $2s+1$	Symmetry of ξ_{spin}	Classification	l
2	5	Symmetric	ortho	even
1	3	Antisymmetric	para	odd
0	1	Symmetric	ortho	even

Intensity ratio for rotational transitions: odd-odd/even-even	1 : 2

State of lowest angular momentum: $^1S_0$(ortho)

H_2

$\chi_{\text{nuc}} = |m_l\rangle \xi$
χ_{nuc} is antisymmetric

s	Multiplicity $2s+1$	Symmetry of ξ_{spin}	Classification	l
1	3	Symmetric	ortho	odd
0	1	Antisymmetric	para	even

Intensity ratio for rotational transitions: odd-odd/even-even	3:1

State of lowest angular momentum: $^1S_0$(para)

section of this chapter, wherein the quantum mechanical basis of superconductivity and superfluidity will be described.

PROBLEMS

12.33 It is found experimentally that hydrogen atoms with parallel electron spins repel in scattering from each other. What is the reason for this repulsion?

12.34 Using moment of inertia values relevant to the H_2 molecule, show that the frequency $\hbar/I$ lies in the infrared.

12.35 (a) What is the spring constant K for the vibrational coupling between the deuterons in the D_2 molecule in terms of the effective potential $\overline{V}(r_{ab})$ and the equilibrium radius r_0?

(b) Is there a coupling between the spin of two deuterons and their vibrational motion (in one dimension)? Explain your answer.

12.36 (a) In what manner are the following two physical phenomena related? (1) The triplet states of He lie lower than the singlet states; (2) H_2 is bound in the singlet electronic state.

(b) What is the radial probability distribution for the nuclei of either H_2 or D_2 in the $^1S_0$ state? Where are the electrons with respect to this distribution? In what temperature domain is your description appropriate?

12.37 The nuclei of the ordinary H_2 molecule are protons that have spin $\frac{1}{2}$ and are therefore fermions.

(a) What exchange symmetry must ψ_{nuc} have for the H_2 molecule?

(b) How many antisymmetric and symmetric spin states are there for two spin-$\frac{1}{2}$ particles? (See Table 11.2.)

(c) What is the ratio of intensities of spectral lines due to transitions between even rotational states to that of lines due to transitions between odd rotational states? How does this ratio compare to that for the heavy hydrogen molecule D_2?

12.38 As discovered in Problem 12.37, the nuclear component of the wavefunction for the H_2 molecule must be antisymmetric. In that there are three symmetric, $s = 1$, triplet spin states and only one antisymmetric, $s = 0$, singlet spin state, the symmetric (or ortho) states are three times more prevalent than the antisymmetric (or para) states. At temperatures sufficiently high to populate the rotational levels, one expects to find molecules predominantly in odd rotational states. Describe qualitatively how this population of rotational and spin states changes with decrease in temperature. Specifically what rotational state should prevail near 0 K?

Answer

For $k_B T$ less than the energy between rotational levels, the molecules in an S rotational state cannot be excited out of that state. Owing to the Pauli principle, this symmetric rotational state must be accompanied by the antisymmetric (para) spin ($s = 0$, ^1S) state. Hydrogen at room temperature is a mixture of about 3:1 ortho to para molecules. However, near 20 K, the sample undergoes an ortho-para conversion. Beneath this temperature, molecules that are in the ^1S state become "frozen" in this state and the sample becomes comprised almost entirely of para molecules.

12.8 BRIEF DESCRIPTION OF QUANTUM MODELS FOR SUPERCONDUCTIVITY AND SUPERFLUIDITY

Bose–Einstein Condensation

The spin-statistics relation, which requires that fermions obey the Pauli exclusion principle and that bosons exist in totally symmetric states, has profound physical implications. We have seen that (Section 8.4) the mechanism of conduction in solids is intimately related to the fact that electrons, which have a spin of $\frac{1}{2}$, are fermions and therefore obey the Pauli exclusion principle.

Bosons (i.e., particles with integral spin values) have equally significant properties. Most interesting of these perhaps is the phenomenon of *Bose–Einstein condensation*. Since bosons do not obey an exclusion principle, a gas of such particles can conceivably be in a state in which all particles have the same momentum, same energy, and so on.

From kinetic theory we recall that the temperature T of a gas of particles is defined as[1]

$$(12.58) \qquad \frac{3}{2} k_B T = \frac{(\Delta p)^2}{2m} \equiv \frac{1}{2m} \langle (\mathbf{p} - \langle \mathbf{p} \rangle)^2 \rangle$$

The temperature is proportional to the mean-square deviation of the momentum of the particles in the gas. It follows that at low temperatures, there is a small spread of momentum values away from the mean. A gas of particles which all have the same momentum has zero temperature, even though this momentum value may be large. In a frame moving with the gas, however, zero temperature means that all momenta are zero[2] (Fig. 12.19).

Suppose that we look at a box of bosons. The box is fixed in space. Lower the temperature. At zero temperature, momentum values drop to a minimal value which is consistent with the uncertainty principle (see Problem 12.40). This collapse of an aggregate of bosons to a collective ground state in which they all have the same minimal ground-state momentum eigenvalues is called *Bose–Einstein condensation*. The interesting properties of a system in such a condensed state may be related to the uncertainty principle. As the momentum of a particle in the system falls to lower values, its uncertainty in position grows. Loosely speaking, it is in many places at the

---

[1] Recall also the thermodynamic identification, $T^{-1} = \partial S / \partial E$, where S is the entropy. This definition of temperature is more uniformly valid for low-temperature quantum systems than is (12.58). Thus, whereas the latter formula implies a finite temperature for a collection of fermions in the ground state, the thermodynamic relation gives the correct value, $T = 0$.

[2] For further discussion of the kinetic definition of temperature, see R. L. Liboff, *Kinetic Theory: Classical, Quantum and Relativistic Descriptions*, Prentice-Hall, Englewood Cliffs, N.J., 1990.

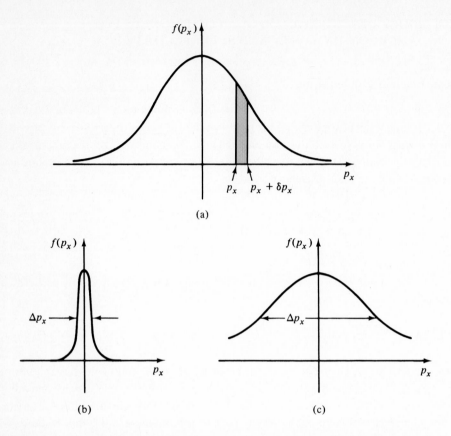

FIGURE 12.19 (a) The distribution function $f(p_x)$ relevant to a gas of particles constrained to move in one dimension. The function $f(p_x)$ is such that the number of particles with momentum in the interval p_x to $p_x + \delta p_x$ is $f(p_x)\,\delta p_x$ (the shaded area). The total number of particles in the gas is

$$N = \int_{-\infty}^{\infty} f(p_x)\,dp_x$$

The temperature of a gas is a measure of the mean-square deviation from the mean, $(\Delta p)^2$, of momentum values of the particles in the gas. Thus a one-dimensional gas of particles with the distribution (b) is colder than the same gas of particles in the distribution (c).

same time. There are two well-established phenomena in nature which are related to Bose–Einstein condensation: *superconductivity* and *superfluidity*.

At very low temperatures (near 0 K), certain metals (e.g., tin and lead) become superconductors.[1] If a current is established in a superconducting loop, it maintains itself with zero loss. No potential difference is needed to keep the current flowing. The resistance drops to zero below a certain critical temperature, T_C (Fig. 12.20).

---

[1] Superconductivity was discovered by K. H. Onnes in 1911.

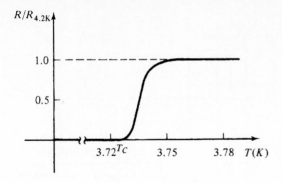

FIGURE 12.20   Resistance versus temperature for tin. The critical temperature for lead, $T = 7.18$ K, is higher than that of tin.

Also, magnetic fields become completely excluded from a superconducting sample for temperatures below T_C (the *Meissner effect*; see Problem 12.39). For tin, $T_C = 3.73$ K, and for mercury, $T_C = 4.17$ K.

Cooper Pairs

It has been established[1] that below the critical temperature, interaction between electrons and the vibrational modes (phonons) of the positive ion lattice of the metal results in a diminution of the Coulomb repulsion between electrons. Phonons are quantized lattice vibrations. When averaged over many such phonon emissions and absorptions, at sufficiently low temperature, the effects of these deformations overbalance the Coulomb repulsion and yield a net attraction between electrons.[2] This attraction allows pairs of electrons to couple with spins antialigned so that each pair carries zero net spin. In that their spin values are zero, these *Cooper pairs* act like bosons. Beneath the critical temperature, these pairs collapse to a collective ground state, χ_G. This state is described as follows. Let $\zeta_i \equiv (\mathbf{r}_i, \mathbf{s}_i)$, where $\mathbf{r}_i$ and $\mathbf{s}_i$, respectively, are the space and spin coordinates of the ith conduction electron in the sample. The ground state is comprised of the appropriately symmetrized sum of products

$$\chi_G = \sum_{P(1,2,\ldots,N)} (\pm)^P \varphi(\zeta_1, \zeta_2)\varphi(\zeta_3, \zeta_4) \cdots \varphi(\zeta_{N-1}, \zeta_N)$$

relevant to an N-particle fermion system [compare with (12.33)]. The spin components of the bound two-particle states $\varphi(\zeta_{i-1}, \zeta_i)$ is the singlet spin-zero state, ζ_A (see Table 11.3). In this sense the ground state may be thought of as being comprised of bosons. However, the space component of these two-particle states is too widely

[1] J. Bardeen, L. N. Cooper, and J. R. Schrieffer, *Phys. Rev.* **108**, 1175 (1957).

[2] This attractive mechanism was first suggested by H. Frohlich, *Phys. Rev.* **79**, 845 (1950).

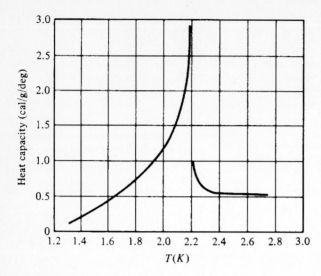

FIGURE 12.21 Heat capacity of liquid helium. The singularity near 2.19 K is evidence of a phase transition. The viscosity of the liquid above the transition temperature is similar to that of normal liquids, while below this temperature the viscosity is at least 10^6 times smaller than that above the transition.

spread to allow a consistent localized particle picture.[1] When Fourier-analyzed, it is found that electron pairs in the preceding ground state have equal and opposite **k** vectors corresponding to zero total momentum.

Excited states in the material are separated from the ground state by an energy gap which at temperature 0 K is centered about the Fermi energy. When Cooper pairs are excited above the gap, they break up into two electrons that exhibit normal conduction. However, in the superconducting state, current flowing in a current loop has been observed to persist indefinitely. Such current is attributed to a collective motion of Cooper pairs in the superconducting ground state.

Superfluidity

A second example of Bose–Einstein condensation is superfluidity. In 1932 Kapitza[2] discovered that the viscosity of liquid helium drops dramatically beneath the λ point (2.19 K). This absence of viscous effects allows the helium to flow freely through capillaries with diameters as small as 100 Å.

At pressures less than 25 atm, helium is a liquid at 0 K. If the heat capacity of liquid helium is measured, a singularity is observed at about 2.2 K, which suggests a phase transition (Fig. 12.21). The viscosity in the new phase (beneath T_λ) is essen-

---

[1] For further discussion, see N. W. Ashcroft and N. D. Merman, *Solid State Physics,* Holt, Rinehart and Winston, New York, 1976, Chapter 34.

[2] P. L. Kapitza, *Nature* **141**, 74 (1932).

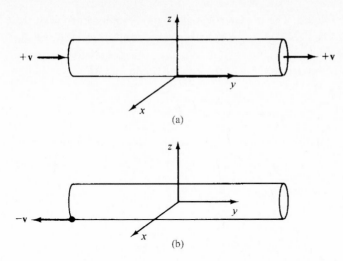

FIGURE 12.22 (a) In the lab frame the capillary is fixed. Fluid moves with velocity +v. (b) In the frame moving with the fluid, the fluid is at rest while the capillary moves with velocity −v.

tially zero, while the thermal conductivity is very high. In the new phase, helium is a *superfluid*. Below T_λ, helium is called *helium II*. Helium II is a mixture of superfluid and normal fluid.

The ground state of helium is $^1S_0$. The electron spins are antialigned with total spin zero. The orbital angular momentum is zero. The nucleus also has zero spin. The whole helium atom has zero angular momentum and is a boson. The superfluid component of helium II contains atoms condensed to the collective ground state.

Landau Theory

The first theoretical model related to superfluidity is due to L. Landau.[1] In this analysis the liquid interacts with the walls of the capillary through which it is flowing, via quantized vibrational modes of excitation generated in the liquid.

Consider that the superfluid is moving through the capillary with velocity **v**. In a frame moving with the fluid (in this frame the fluid is at rest) the capillary wall moves with velocity −**v** (Fig. 12.22). Owing to friction between the wall and the fluid, elementary excitations appear in the liquid. Let one such excitation be generated with momentum **p** and energy $\varepsilon(p)$. Transforming back to the frame where the tube is at rest (i.e., the lab frame), the energy of the liquid becomes

$$(12.59) \qquad E = \frac{1}{2}Mv^2 + [\varepsilon(p) + \mathbf{p} \cdot \mathbf{v}]$$

---

[1] L. Landau, *J. Phys.* **5**, 71 (1941); see also L. D. Landau and E. M. Lifshitz, *Statistical Physics,* Addison-Wesley, Reading, Mass., 1958.

where M is the mass of the liquid. The $\mathbf{p} \cdot \mathbf{v}$ term stems from Doppler shifting of the phonon frequency (see Problem 12.41). If no excitation is present, the energy of the fluid is $\frac{1}{2}Mv^2$. The presence of the phonon excitation causes this energy to change by the amount $[\varepsilon(p) + \mathbf{p} \cdot \mathbf{v}]$. Since the energy of the flowing liquid must decrease owing to such dissipative coupling,

$$\varepsilon(p) + \mathbf{p} \cdot \mathbf{v} < 0$$

This condition must be satisfied in order that an excitation appear in the liquid. Since

$$(\varepsilon + \mathbf{p} \cdot \mathbf{v}) \geq \varepsilon - pv$$

it follows that excitations have the property

(12.60)
$$v > \frac{\varepsilon(p)}{p}$$

We may conclude that an excitation of energy ε and momentum p cannot be created in a fluid moving past a wall with speed v unless the preceding inequality is satisfied. If ε/p has some minimum value greater than zero, then for small velocities of flow beneath this minimum, dissipative excitations will not appear in the liquid. That is, the liquid will exhibit superfluidity.

If the energy spectrum of excitations $\varepsilon(p)$ is plotted against p, then the minima of ε/p occur as those values of p where

(12.61)
$$\frac{d\varepsilon}{dp} = \frac{\varepsilon}{p}$$

that is, at points where a line drawn from the origin of the $p\varepsilon$ plane is tangent to the curve $\varepsilon(p)$ (see Problem 12.44).

The $\varepsilon(p)$ curve for liquid helium has been obtained by neutron scattering experiments.[1] It is sketched in Fig. 12.23. There are two values of p where ε/p is minimum, at $p = 0$ and $p = p_1$. The minimum at $p = 0$ is appropriate to temperatures near zero. At such temperatures superfluidity occurs for speeds

$$v < v_0 = \left.\frac{d\varepsilon}{dp}\right|_{p=0}$$

For slightly larger temperatures the minimum at p_1 comes into play. Superfluidity occurs for this branch of excitations at fluid speeds

$$v < v_1 = \left.\frac{d\varepsilon}{dp}\right|_{p=p_1} < v_0$$

[1] J. L. Yarnell et al., *Phys. Rev.* **113**, 1379, 1386 (1959).

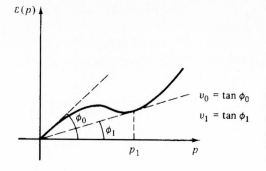

FIGURE 12.23 The minima of $\varepsilon(p)/p$ occur at values of p where a line drawn through the origin is tangent to $\varepsilon(p)$. For liquid helium this occurs at the origin and at p_1. Excitations at the origin are called phonons. Those at p_1 are called rotons.

The speed of phonons at $p = 0$ is that of sound in liquid helium at 0 K. The excitations at p_1 are called *rotons* corresponding, it is believed, to rotational motion of small clusters of helium atoms.[1]

The isotope of helium, He^3, has an unpaired neutron and is therefore a fermion. One would not expect this isotope to exhibit Bose–Einstein condensation. Recent experimental observation,[2] however, suggests the existence of a superfluid phase in this liquid as well. As with the case of superconductivity, such phenomena may be ascribed to a pairing process of fermions allowing for Bose–Einstein-like behavior.

PROBLEMS

12.39 Consider a sphere of tin immersed in a uniform magnetic field at $T > T_C$. The finite conductivity of the tin permits the $\mathcal{B}$ field to penetrate. Inside the tin the field has the value $\mathcal{B}'$. If the conductivity becomes infinite as $T \rightarrow T_C$, what happens to $\mathcal{B}'$?

Answer

From Faraday's law

$$\frac{\partial \mathcal{B}}{\partial t} = -\nabla \times \mathscr{E} = -\frac{1}{\sigma} \nabla \times \mathbf{J}$$

we obtain that

$$\frac{\partial \mathcal{B}}{\partial t} = 0 \qquad \text{for } \sigma = \infty$$

so that $\mathcal{B}$ is constant in time and is *trapped* inside an ordinary conductor. This is not the case for a superconductor. At $T = T_C$, a magnetic field is excluded from a superconductor, so that

---

[1] R. P. Feynman, *Phys. Rev.* **74**, 262 (1954); for further discussion of this topic, see D. L. Goodstein, *States of Matter*, Prentice-Hall, Englewood Cliffs, N.J., 1975.

[2] D. D. Osheroff, R. C. Richardson, and D. M. Lee, *Phys. Rev. Lett*, **28**, 885 (1972).

$\mathscr{B}' = 0$ for the configuration given. In this sense a superconductor is said to have perfect dia-magnetism. This effect is called the *Meissner effect*.

12.40 Consider a model of superconductivity where the critical temperature T_C describes the width of the Fermi sea (see Sections 2.3 and 8.4). Estimate the spatial spread (Δx) of the wave-functions for such electrons in tin.

Answer

$$\frac{1}{2m}(\Delta p)^2 \simeq k_B T_C$$

$$(\Delta p)^2 \simeq m v_F \, \Delta p$$

where

$$m v_F^{\,2} = 2E_F$$

Together with the uncertainty principle, this gives

$$\Delta x \geq \frac{\hbar}{\Delta p} \simeq \frac{\hbar v_F}{k_B T_C} \simeq 10^{-4} \text{ cm}$$

which is a macroscopic length.

12.41 An excitation has energy $\varepsilon = \hbar\omega$ and momentum $\mathbf{p} = \hbar\mathbf{k}$ in a frame S. Show that in a frame S' moving with velocity $\mathbf{v}$ with respect to S,

$$\varepsilon' = \varepsilon - \mathbf{p} \cdot \mathbf{v}$$

Answer

The energy of a photon of frequency ω in the S frame is

$$\varepsilon = \hbar\omega$$

In a frame moving with $\mathbf{v}$ with respect to S, the frequency is Doppler-shifted to the frequency

$$\omega' = \omega - \mathbf{k} \cdot \mathbf{v}$$

(If $\mathbf{k}$ is parallel to $\mathbf{v}$, ω' decreases. If $\mathbf{k}$ is antiparallel to $\mathbf{v}$, ω' increases.) Multiplying through by $\hbar$ gives the desired result.

12.42 (a) Use the uncertainty relation to estimate the lowest temperature that a collection of bosons confined to a box of edge length L can have.

(b) What temperature does this correspond to for helium in a box with edge length 1 cm?

12.43 A classical gas is said to be *degenerate* when thermodynamic properties of the system (equation of state, specific heat, conductivity, etc.) are governed by quantum statistics, as opposed to classical, *Boltzmann* statistics. A criterion that determines if a gas is degenerate may be obtained by comparing the mean interparticle distance $n^{-1/3}$ (n = particle number density) with the average de Broglie wavelength of particles.

(a) What is this criterion in terms of n, T, and m, where T is the temperature of the gas and m is the mass of a particle?

(b) Use this formula to estimate the temperature at which a *neutron star* of mass density 10^{14} g/cm^3 becomes degenerate.

12.44 Show that $\varepsilon(p)/p$ is minimum at those values of p for which

$$\frac{d\varepsilon}{dp} = \frac{\varepsilon}{p}$$

12.45 Show that superfluidity does not occur if excitations have the free-particle spectrum

$$\varepsilon = \frac{p^2}{2m}$$

[*Note:* If the liquid is a system of uncoupled bosons, one expects that excitations follow this spectrum. N. N. Bogoliubov[1] was the first to show that a gas of bosons with weak interactions has a spectrum of excitations $\varepsilon(p)$, which has a finite slope at $p = 0$ (such as sketched in Fig. 12.23).]

12.46 A certain Bose liquid has the excitation spectrum

$$\frac{\varepsilon(p)}{\varepsilon_0} = \tilde{p}\left[\frac{b^2}{3} + (\tilde{p} - b)^2\right]$$

$$\tilde{p} \equiv \frac{p}{p_0}$$

where ε_0, p_0, and b are constants.

(a) What are the maximum superfluid speeds for the phonon and roton branches of excitations, respectively?

(b) What is the energy gap for the roton branch of excitations?

12.47 Give a qualitative explanation of the fact that superconductors are poor normal conductors.

Answer

Superconductivity is due to electron-phonon interactions. Metals with strong electron-phonon interaction will show large resistance at room temperature and therefore be poor conductors. On the other hand, strong electron-phonon interaction will raise the critical temperature beneath which superconductivity becomes evident. Such, for example, is the case for lead, which is a poor conductor but has one of the highest critical temperatures. On the other hand, superconductivity in gold and silver, which are very good conductors at room temperature and must therefore be typical of a weak electron-phonon interaction, proves difficult to exhibit.

[1] N. N. Bogoliubov, *J. Phys. USSR* **11**, 23 (1947).

12.48   When He II is constrained to flow in a circular channel, the circulation maintains itself with no dissipation. Let particles in the ground state have the wavefunction $\varphi = A \exp[i\phi(\mathbf{r})]$.

(a)   The velocity field $\mathbf{u}$ of particles in this state is related to mass current[1] through the relation $\mathbf{J}_m = mA^2\mathbf{u}$. Show that $\mathbf{u} = (\hbar/m)\,\nabla\phi$.

(b)   What is the value of $\nabla \times \mathbf{u}$ for this flow?

(c)   Show that the values of the *circulation* are restricted to the discrete quantum values

$$\oint \mathbf{u} \cdot d\mathbf{l} = na \qquad (n = 0, \pm 1, \pm 2, \cdots)$$

$$a \equiv \frac{h}{m}$$

The constant $a \simeq 10^{-3}$ cm^2/s for He.

Answers

(a) $\mathbf{J} = \hbar A^2\,\nabla\phi$.

(b)   In that $\mathbf{u}$ is the gradient of some function, $\nabla \times \mathbf{u} = 0$.

(c)   Around the path of flow we have

$$\oint \mathbf{u} \cdot d\mathbf{l} = \frac{\hbar}{m}\oint \nabla\phi \cdot d\mathbf{l}$$

To ensure that the wavefunction is single-valued, change in ϕ about the closed loop is restricted to integral multiples of 2π.

12.49   The halogens, which comprise group VII of the periodic table, are characterized by the common property of missing one electron in the outermost p subshell. It is found that the ground states of these atoms are well described by the equivalent configuration of a single *hole* bound to an atom in an orbital p state. Using this model, obtain the possible ground states of a halogen atom. Check your answer with the ground states given in Table 12.3. The notion of a hole was introduced in Chapter 8 and is discussed further in the following section.

12.50   In the Heisenberg model for ferromagnetism, the Hamiltonian for an array of magnetic moments is given by

$$H = -\lambda \sum_i \sum_j \boldsymbol{\mu}_i \cdot \boldsymbol{\mu}_j - \sum_{i=1}^{N} \boldsymbol{\mu}_i \cdot \mathcal{B}$$

where λ is a positive constant. The first sum is over "nearest neighbors," and the remaining sum extends over the N moments in the sample. The form of the first sum in this Hamiltonian presupposes that aligned magnetic moments are lower in energy than antialigned moments.

(a)   Is this description consistent with classical physics?

(b)   Offer a quantum mechanical explanation for the Heisenberg model, citing the Pauli principle appropriate to electrons.

---

[1] Note the relation $\mathbf{J}_m = m\mathbf{J}$, where the particle current $\mathbf{J}$ is defined by (7.107).

12.51 (a) Referring to Fig. 12.15, give the "first" ionization energy of helium.

(b) Recalling that the remaining electron after ionization is left in the ground state of the $Z = 2$ atom, calculate the "second" ionization energy of helium. Compare your answer with the observed value, 54.4 eV.

12.52 The relativistic wave equation for bosons of rest mass m may be obtained by transforming the relation (see Problem 2.26)

$$E^2 = p^2 c^2 + m^2 c^4$$

through the identifications

$$E \to \hat{E} = i\hbar \frac{\partial}{\partial t}$$

$$\mathbf{p} \to \hat{\mathbf{p}} = -i\hbar \nabla$$

(a) Obtain the wave equation relevant to bosons of rest mass m. This equation is called the *Klein–Gordon* equation. (This equation is discussed in detail in Section 15.2.)

(b) What form does this equation assume for photons?

(c) Obtain a time-independent isotropic solution to the Klein–Gordon equation for bosons of finite rest mass. What characteristic decay length does this solution imply?

(d) What is this length for π mesons ($m_\pi \simeq 260 m_e$)?

Answers

(a) $\nabla^2 \psi - \dfrac{1}{c^2} \dfrac{\partial^2 \psi}{\partial t^2} = \dfrac{m^2 c^2}{\hbar^2} \psi$

(b) $\nabla^2 \psi - \dfrac{1}{c^2} \dfrac{\partial^2 \psi}{\partial t^2} = 0$

Equation (b) is appropriate to the propagation of electromagnetic fields in vacuum, for either the scalar or vector potential.

(c) Referring to (C.16), we find (for $r > 0$)

$$\psi(r) = \frac{A e^{-r/a}}{r}$$

$$a \equiv \frac{\hbar}{mc}$$

(d) $a_\pi = \lambda_C / 260 = 1.49 \times 10^{-13}$ cm $= 1.49$ fermi

12.53 A deuteron is comprised of a bound neutron and proton. Its angular momentum quantum number is $j = 1$. The spin-orbit component of the Hamiltonian is given by

$$\hat{H}' = -\frac{2\mu c^2}{\hbar^2} \hat{\mathbf{L}} \cdot \hat{\mathbf{S}}$$

where μ is the reduced mass of the two-particle system and c is the speed of light.

(a)   Neutron and proton both have spin $\frac{1}{2}$. What are the possible values for the total spin quantum number s of the deuteron?

(b)   Rewrite $\hat{H}'$ as a function of $\hat{\mathbf{J}}$, $\hat{\mathbf{L}}$, and $\hat{\mathbf{S}}$.

(c)   What are the possible eigenvalues of $\hat{H}'$? Leave your answer in terms of μc^2.

12.54   The outer shell of neon in the ground state is in the $2p^6$ configuration. If one of these electrons is excited, one of four lines are observed in subsequent decay. Employing a model similar to that discussed in Problem 12.49, argue how these four lines might emerge from a coupling between the excited electron and the $2p^5$ unexcited electrons.

12.55   Lithium is a three-electron atom. Working in the central field approximation and neglecting L-S coupling:

(a)   Write down an appropriately symmetrized ground-state wavefunction, $\psi_G(1, 2, 3)$, for lithium with spin taken into account. Use the notation $\psi_{nlm}(i)\alpha(i)$ and $\psi_{nlm}(i)\beta(i)$ for single-electron wavefunctions. The variable i is electron number and runs from 1 to 3.

(b)   What is the ground-state energy E_G of lithium corresponding to your answer in part (a).

(c)   Your answer to part (b) presumes that E_G is the eigenvalue of $\hat{H}$ corresponding to the eigenfunction ψ_G. That is,

$$\hat{H}\psi_G = E_G\psi_G$$

What is the explicit form of $\hat{H}(1, 2, 3)$, and what property of E_G allows it to be an eigenvalue of the extended wavefunction ψ_G?

(d)   What is the answer to part (b) if electrons have spin zero?

Answers (partial)

(a) $\psi_G(1, 2, 3) = \dfrac{1}{\sqrt{3!}} \begin{vmatrix} \psi_{100}(1)\alpha(1) & \psi_{100}(1)\beta(1) & \psi_{200}(1)\alpha(1) \\ \psi_{100}(2)\alpha(2) & \psi_{100}(2)\beta(2) & \psi_{200}(2)\alpha(2) \\ \psi_{100}(3)\alpha(3) & \psi_{100}(3)\beta(3) & \psi_{200}(3)\alpha(3) \end{vmatrix}$

(b) $E_G = -Z^2\mathbb{R}\left(\dfrac{2}{1^2} + \dfrac{1}{2^2}\right) = -\left(\dfrac{81}{4}\right)\mathbb{R}$

(c) E_G is degenerate, so that in the present instance, all six wavefunctions in $\psi_G(1, 2, 3)$ correspond to the same energy, $-9Z^2\mathbb{R}/4$.

12.9   IMPURITY SEMICONDUCTORS AND THE *p-n* JUNCTION

The concepts of valence and conduction bands were described in Section 8.4 in reference to electrical conductivity in solids. Having developed the theory of atomic structure in the early part of this chapter, we are now prepared to pursue the various aspects of impurity semiconductors. Semiconductors which conduct by virtue of impurities are said to be *extrinsic*. *Intrinsic* semiconductors, on the other hand, are not dependent on impurities for conduction.

Donor (n)-Type Impurities

Silicon, which is in the IVth group of the periodic chart, has four valence electrons $(3s^2 3p^2)$.

Antimony lies in the Vth group of the periodic chart and has five valence electrons $(5s^2 5p^3)$. Thus, if an impurity antimony atom replaces a silicon atom, four of the five valence electrons contribute to the bonding of the impurity atom to the silicon crystal. The fifth electron remains loosely bound to the parent antimony atom. A schematic two-dimensional illustration of this bonding is shown in Fig. 12.24. Under minor thermal agitation, the impurity atom is ionized and the loosely bound electron is excited to the conduction band. See Fig. 12.25. Once in the conduction band, the electron becomes a charge carrier and contributes to current. Since it is considerably more difficult to activate the valence electron of silicon to the conduction band, there will, in general, be a larger fraction of electrons in the conduction band from the impurity antimony component than from silicon atoms. This explains the experimentally observed large change in conduction for such crystals due to the presence of impurities.

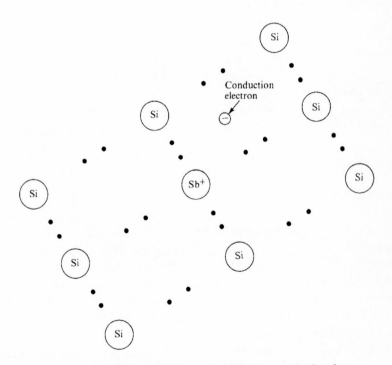

FIGURE 12.24   An antimony atom in a silicon crystal contributes a conduction electron.

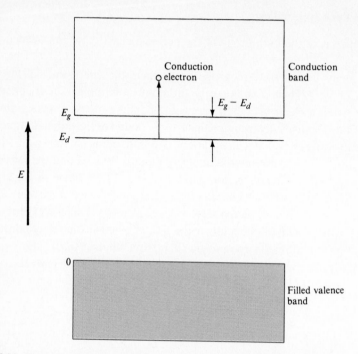

FIGURE 12.25 Energy level E_d of an impurity antimony atom in a silicon crystal lies close to the bottom of the conduction band. The energy gap between valence and conduction bands is E_g.

Since conduction electrons are supplied by the antimony atoms, this form of impurity is called a *donor-type impurity*. Extrinsic semiconductors which contain a donor-type impurity are called *n-type semiconductors*. (Think of n for negative.)

Acceptor (p)-Type Impurities

Consider now that a silicon atom is replaced with an atom from group III of the periodic chart, such as indium. Since In has a valence of three $(5s^25p^1)$, one of the four bonds to nearest neighbors will be missing an electron. See Fig. 12.26. This unsaturated indium bond effects an energy level slightly above the top of the valence band as shown in Fig. 12.27. Under thermal agitation an electron from the valence band of the host silicon crystal is excited to this level, leaving an unfilled level in the valence band. This unfilled level, which affects the presence of positive charge, is called a *hole*. If an electric field is applied, the hole will drift in the direction of the field and contribute to the current.

Since an electron is accepted by the impurity atom in the creation of a conducting hole, the impurity atom is called an *acceptor* and the semiconductor is said to be p type. (Think of p for positive.)

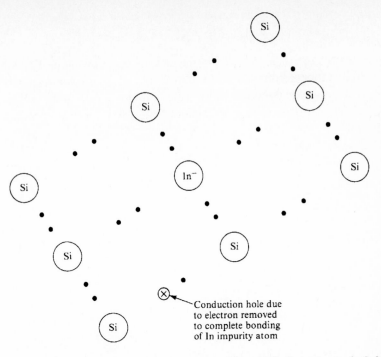

FIGURE 12.26 An indium impurity atom in a silicon crystal results in the presence of a hole.

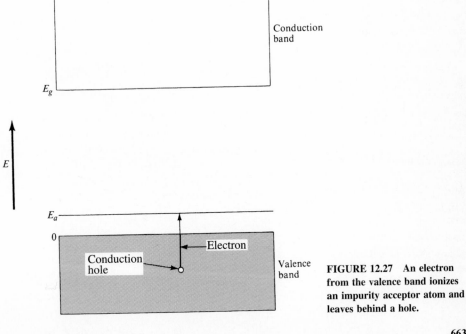

FIGURE 12.27 An electron from the valence band ionizes an impurity acceptor atom and leaves behind a hole.

Note that of the three atomic groups considered above, type IV atoms act as hosts to impurity atoms from type III (p-donor) and type V (n-donor) groups. Molecular semiconductors, on the other hand, are commonly comprised of type III and V atoms (e.g., GaAs, InP) and may be doped with impurity atoms from a number of other groups.[1]

For a semiconductor which contains both acceptor and donor impurities, one writes

$$\sigma = en\mu_n + ep\mu_p$$

for the conductivity, where μ denotes mobility and n, p denote electron and hole densities, respectively. (See Problem 8.27.) In germanium at room temperature (in MKS units),

$$\mu_n = 0.36(\text{m/s})/(\text{V/m})$$
$$\mu_p = 0.17(\text{m/s})/(\text{V/m})$$

General Results. Fermi–Dirac Distribution and Density of States

Consider a box of noninteracting electrons. The box has volume L^3 and is maintained at temperature T. Electrons may be found in any of a discrete infinity of states. Owing to the exclusion principle, not more than two electrons can be in any one energy state at any given time.

The *probability* that a given energy *level* at the value E is occupied is given by the *Fermi–Dirac* distribution relevant to fermions:[2,3]

(12.62)
$$f(E) = \frac{1}{e^{(E-E_F)/k_B T} + 1}$$

One may also interpret $f(E)$ as giving the fractional occupation of an electron in the energy level E. The Fermi energy E_F was discussed previously in Chapter 2.

At $T = 0$ K we see that $f(E) = 1$ for $E < E_F$ and $f(E) = 0$ for $E > E_F$. All levels up to E_F are occupied at zero temperature. See Figs. 2.5 and 12.28.

---

[1] This topic is discussed further by J. C. Phillips, *Bonds and Bands in Semiconductors*, Academic Press, New York, 1973.

[2] Recall Section 12.3 and see Appendix A.

[3] The parameter E_F in (12.62) is the Fermi energy at 0 K. For $T > 0$ K, E_F in (12.62) is more correctly termed the chemical potential. If two systems of a single species have equal chemical potentials, there is no particle flow between the systems. If chemical potentials are different, particle flow occurs in a direction to equalize the chemical potentials.

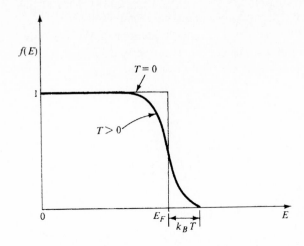

FIGURE 12.28   At 0 K all levels are occupied up to the Fermi level. At finite temperatures, levels above E_F are occupied.

To obtain the density (number/cm³) of electrons in the energy interval $(E, E + dE)$, one must multiply $f(E)$ by the density of states (number of states/energy − cm³), which we have previously labeled $g(E)$ (see Section 8.8). It has the value

$$(12.63) \qquad g(E) = \frac{1}{2\pi^2} \left(\frac{2m^*}{\hbar^2} \right)^{3/2} E^{1/2}$$

Here we have written m^* for the effective mass introduced previously in Section 8.2. Thus the density of electrons with energy in the interval $(E, E + dE)$ is given by $f(E)g(E)\, dE$.

General Results

a. Electron Density   First we turn our attention to the density of electrons, n, in the conduction band. This is obtained by integrating the product $f(E)g(E)\, dE$ over conduction band energies. There results

$$(12.64) \qquad n = \int_{\text{cond. band}} f(E)g(E)\, dE$$

If we consider electrons in the conduction band to be free, then $g(E)$ is given by (12.63). Since energies in (12.64) are measured from the bottom of the conduction band, we write

$$(12.65) \qquad g(E) = \frac{1}{2\pi^2} \left(\frac{2m_e^*}{\hbar^2} \right)^{3/2} (E - E_c)^{1/2}$$

where E_c denotes the energy at the bottom of the conduction band. Equation (12.65) reflects the property that there are no states available in the energy gap. With (12.62), we see that $f(E)$ decays rapidly with increasing E. This property permits the upper limit in (12.64) to be set equal to infinity. Combining results and setting

$$\beta \equiv \frac{1}{k_B T}$$

(12.64) becomes

(12.66)
$$n = \frac{1}{2\pi^2} \left(\frac{2m_e^*}{\hbar^2} \right)^{3/2} \int_{E_c}^{\infty} \frac{(E - E_c)^{1/2} \, dE}{e^{\beta(E - E_F)} + 1}$$

Here we have assumed that m_e^* is constant over the range of integration. Changing variables to

$$\eta = E - E_c$$

(12.66) becomes

(12.67)
$$n = \frac{1}{2\pi^2} \left(\frac{2m_e^*}{\hbar^2} \right)^{3/2} \int_0^{\infty} \frac{\eta^{1/2} \, d\eta}{e^{\beta\eta} e^{\beta(E_c - E_F)} + 1}$$

To simplify this integration, we assume that $E_c - E_F \gg k_B T$ or, equivalently, $\beta(E_c - E_F) \gg 1$. A material which satisfies this assumption is said to be *nondegenerate*.[1] If E_F lies in the conduction band, the material is said to be *degenerate*.

Given the preceding inequality, one may neglect the unity term in the denominator of (12.66) to obtain[2]

(12.68)
$$n = \frac{1}{2\pi^2} \left(\frac{2m_e^*}{\hbar^2} \right)^{3/2} e^{-\beta(E_c - E_F)} \int_0^{\infty} \eta^{1/2} e^{-\beta\eta} \, d\eta$$

which gives

(12.69)
$$n = N_c(T) e^{-(E_c - E_F)/k_B T}$$

where

(12.70)
$$N_c(T) = \frac{1}{4} \left(\frac{2m_e^* k_B T}{\pi \hbar^2} \right)^{3/2}$$

represents an *effective density of states* in the vicinity of $E \simeq E_c$, i.e., near the bottom of the conduction band.

---

[1] That is, classical statistics applies. See Problems 2.48 and 12.43. At room temperature, $k_B T \simeq 0.025$ eV. A measure of $E_c - E_F$ is roughly given by E_g. Here are some typical values: $E_g = 1.14$ eV for Si, 1.25 eV for InP, and 1.4 eV for GaAs, for which cases $k_B T \ll E_g$. Note that, in this limit, mean occupation numbers per state grow small, which is a basic property of the classical domain.

[2] Here we recall $\int_0^{\infty} \eta^{1/2} e^{-\beta\eta} \, d\eta = \sqrt{\pi}/2\beta^{3/2}$.

b. Hole Density   We turn next to a parallel calculation for density of holes in the valence band. A hole will be present at a given level providing an electron is missing from that level. So the Fermi distribution for holes becomes $1 - f(E)$. Thus the density of holes in the valence band is written

$$(12.71) \qquad p = \int_{\text{valence band}} g_p(E)[1 - f(E)] \, dE$$

When electrons in a partially filled band gain energy, they move upward in the band, leaving behind vacated states. Thus holes which gain energy move downward in their band. To incorporate this property and further stipulate that the zero of energy lies at the top of the valence band, the energy of a hole is written $E_v - E$, where E varies from E_v to $-\infty$. With this expression for energy substituted in the density of states (12.63), the preceding expression becomes

$$(12.72) \qquad p = \frac{1}{2\pi^2} \left(\frac{2m_h^*}{\hbar^2}\right)^{3/2} \int_{-\infty}^{E_v} (E_v - E)^{1/2} \left[1 - \frac{1}{e^{\beta(E - E_F)} + 1}\right] dE$$

Following steps leading to (12.65), we introduce the variable $\eta \equiv (E_v - E)$ and again assume nondegeneracy, i.e., $\beta(E_F - E_v) \gg 1$. There results

$$(12.73) \qquad p = N_v(T) e^{-(E_F - E_v)/k_B T}$$

where

$$(12.74) \qquad N_v(T) = \frac{1}{4} \left(\frac{2m_h^* k_B T}{\pi \hbar^2}\right)^{3/2}$$

is the effective density of holes near the top of the valence band. Multiplying (12.69) and (12.73), we find

$$(12.75) \qquad np = \frac{1}{2} \left(\frac{k_B T}{\pi \hbar^2}\right)^3 (m_e^* m_h^*)^{3/2} e^{-E_g/k_B T}$$

This gives the important result that the product of n and p for a given semiconductor is constant at a given temperature. The key assumption in obtaining the preceding result is that the semiconductor is nondegenerate. Thus, with the rule (12.75), we see that introducing an impurity which would, say increase n would diminish p, since the product np is constant. The generality of the result (12.75) follows from the observation that it is an application of the *law of mass action* to electron-hole dynamics.[1]

---

[1] That is, for the chemical reaction $A \rightleftharpoons B + C$, one may write $n_B n_C / n_A = f(T)$, where n_A denotes number density of A particles, etc., and $f(T)$ is an arbitrary function of temperature. Thus, with A denoting neutral atoms, B electrons, and C ionized atoms, one obtains the form of (12.75).

Expressions for the Fermi Energy

a. Intrinsic Case For intrinsic semiconductors, for every electron in the conduction band, there is a hole present in the valence band. Thus we may set

$$(12.76) \qquad n_i = p_i$$

where the subscript i denotes intrinsic. Equating (12.69) to (12.73) gives

$$(m_e^*)^{3/2} e^{-(E_c - E_F)/k_B T} = (m_h^*)^{3/2} e^{-(E_F - E_v)/k_B T}$$

Solving for E_F gives the *intrinsic* Fermi energy:

Intrinsic

$$(12.77) \qquad E_F = \frac{1}{2}(E_v + E_c) + \frac{3}{4} k_B T \ln\left(\frac{m_h^*}{m_e^*}\right)$$

If the zero of energy is taken to be at E_v, then E_c becomes the value of the energy gap E_g, and (12.77) may be rewritten in the more concise form

$$(12.78) \qquad E_F = \frac{1}{2}E_g + \frac{3}{4} k_B T \ln\left(\frac{m_h^*}{m_e^*}\right)$$

Note in particular that energy values in semiconductor theory are not absolute but are relative to a given reference level. Thus in (12.78), the $E = 0$ value is set at the top of the valence band and E_F (for $m_h^* = m_e^*$) has the value of $\frac{1}{2}E_g$, whereas in (12.77) E_F is equal to the average of E_v and E_c. In all reference schemes E_F (for equal m^*) lies halfway between E_v and E_c. However, in practical usage, this is not the case. For example, for Ge, $m_h^*/m_e^* = 0.67$ and for Si, $m_h^*/m_e^* = 0.54$. In both cases E_F falls slowly with increasing temperatures.

Returning to the density of charge carriers, for intrinsic semiconductors we may write

$$(12.79) \qquad n_i p_i = n_i^2 = p_i^2$$

With (12.69) and (12.73) there results

Intrinsic

$$(12.80a) \qquad \boxed{n_i^2 = N_c(T)N_v(T)e^{-E_g/k_B T}}$$

The effective density of states in the conduction band $N_c(T)$ is given by (12.70), and $N_v(T)$, corresponding to the valence band, is given by (12.74).

Returning to (12.75) we note that np at a given temperature is constant. Evaluating this constant in the intrinsic domain gives the relation

(12.80b)
$$np = n_i^2$$

This is a very useful relation as it is valid for both intrinsic and extrinsic semiconductors (in the nondegenerate limit and in the absence of recombination).

    b. n-Type Semiconductors. Low-, Intermediate-, and High-Temperature Parameters   Consider a crystal doped with donor-type impurities. Let N_d represent the density of impurity donor atoms. Then if density of unionized donor atoms is N_{d_0}, density of ionized donor atoms is $N_d - N_{d_0}$. Charge neutrality may then be written

(12.81)
$$n = p + (N_d - N_{d_0})$$

At relatively *low temperature* there are very few electron-hole pairs created, and the preceding equation becomes

(12.82)
$$n \simeq (N_d - N_{d_0})$$

Thus, in *n*-type semiconductors, electrons are *majority carriers* and holes are *minority carriers*.

    To calculate the Fermi energy for this case, first note that the density of neutral donors N_{d_0} is equal to the number of donor states per volume that are occupied. Thus we may write

(12.83)
$$N_{d_0} = N_d f(E_d)$$

Substitution of (12.82) gives

(12.84)
$$n = N_d \left[\frac{1}{1 + e^{\beta(E_F - E_d)}} \right]$$

Passing to the limit $k_B T \ll E_F - E_d$ gives

(12.85)
$$n = N_d e^{-(E_F - E_d)/k_B T}$$

At low temperatures, $n < N_d$ so that $E_F > E_d$, and we may conclude that for the present case E_F lies between E_d and E_c. See Fig. 12.29. Equating this latter expression to our previously derived expression for n (12.69) gives

(12.86)
$$N_d e^{-(E_F - E_d)/k_B T} = N_c e^{-(E_c - E_F)/k_B T}$$

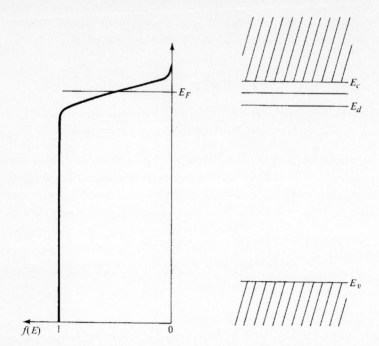

FIGURE 12.29 At low temperatures, E_F lies between E_d and E_c for an *n*-type semiconductor. A similar diagram applies to *p*-type semiconductors for which, under similar conditions, E_F lies between E_a and E_v. At such temperatures, electrons are said to be "frozen out" of the conduction band (donor levels remain occupied). Likewise, holes are frozen out of the valence band (acceptor levels remain occupied with holes).

from which we find

Low T

$$(12.87) \qquad E_F = \frac{1}{2}(E_c + E_d) + \frac{1}{2}k_B T \ln\left(\frac{N_d}{N_c}\right)$$

At $T = 0$, E_F is midway between E_c and E_d. As temperature rises, E_F falls since $N_c > N_d$.

At intermediate temperature nearly all donors are ionized. Assuming sufficiently large impurity density so that $N_d \gg n_i$, (12.82) gives

$$(12.88) \qquad n = N_d$$

Again employing (12.69), together with the preceding equation, we find

$$(12.89) \qquad N_d = N_c e^{-(E_c - E_F)/k_B T}$$

Solving for the Fermi energy gives

Int. T

(12.90)
$$E_F = E_c - k_B T \ln \frac{N_c}{N_d}$$

At sufficiently high temperature, electron-hole generation across the energy gap grows large compared with impurity contributions. In this limit the semiconductor behaves intrinsically and (12.77) et seq. apply.

c. *p-Type Semiconductors. Low-, Intermediate-, and High-Temperature Parameters*   For a semiconductor doped with acceptor-type impurities, charge neutrality is written

(12.91)
$$p = n + (N_a - N_{a_0})$$

Again note that at low temperature there are very few electron-hole pairs, and the preceding relation becomes

(12.92)
$$p = N_a - N_{a_0}$$

The right-hand side of this equation represents the density of ionized acceptor atoms. Each such ionized atom represents a filled acceptor state, and we may write

(12.93)
$$N_a - N_{a_0} = N_a f(E_a)$$

Again we assume $\beta(E_a - E_F) \gg 1$. Substituting the resulting expression into (12.92) and recalling (12.73) gives

Low T

(12.94)
$$p = N_a e^{-(E_a - E_F)/k_B T} = N_v e^{-(E_F - E_v)/k_B T}$$

(12.95)
$$E_F = \frac{1}{2}(E_v + E_a) + \frac{1}{2}k_B T \ln\left(\frac{N_v}{N_a}\right)$$

Again, the Fermi level at $T = 0$ is midway between the top of the valence band and the acceptor levels. Since $N_v > N_a$, the Fermi level rises with increasing temperature. At intermediate temperature, electrons in the valence band are thermally excited to the E_a level, ionizing all acceptor atoms. For sufficient doping density, charge neutrality becomes

$$p = N_a$$

or, equivalently,

$$N_a = N_v e^{-(E_F - E_c)/k_B T}$$

Taking the logarithm of both sides gives

Int T

(12.96) $$E_F = E_v + k_B T \ln (N_v/N_a)$$

At sufficiently high temperature, it is again the case that electron-hole generation across the gap grows large compared with impurity contributions, and the semiconductor behaves intrinsically and (12.77) et seq. apply.

These effects are depicted in Fig. 12.30 for the case $m_h^* \gtrsim m_e^*$. A general compilation of results is presented in Table 12.6.

Compensation

An intrinsic semiconductor section is often an important component of a semiconductor device.[1] However, it is frequently the case that semiconducting materials are not free of impurities, which render the material either n type or p type. Thus, for example, in high purity, both silicon and germanium tend to be p type. In this case, the method of compensation refers to the process of adding sufficient concentration of n-type impurities to counterbalance the p-type impurities and thereby produce an effectively intrinsic material. For the case at hand, in principle, alkali metals such as lithium, sodium, and potassium tend to form interstitial donors in silicon and germanium.

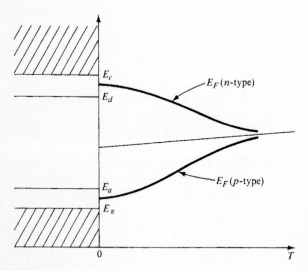

FIGURE 12.30   Change in E_F with temperature for n- and p-type semiconductors (for $m_h^* \gtrsim m_e^*$). Both curves are asymptotic to E_F (intrinsic).

[1] For example, this is the case for the p-i-n diode, important to infrared detection as well as particle detection. For further discussion, see S. M. Sze, *Physics of Semiconductor Devices*, 2d ed., Wiley, New York, 1981.

TABLE 12.6 Semiconductor relations

General Results (Nondegenerate Crystal)

Conduction Band

$$g_n(E) = \frac{1}{2\pi^2} \left(\frac{2m_e^*}{\hbar^2} \right)^{3/2} (E - E_c)^{1/2}$$

$$n(T) = N_c e^{-(E_c - E_F)/k_B T}$$

$$N_c = \frac{1}{4} \left(\frac{2m_e^* k_B T}{\pi \hbar^2} \right)^{3/2}$$

Valence Band

$$g_p(E) = \frac{1}{2\pi^2} \left(\frac{2m_h^*}{\hbar^2} \right)^{3/2} (E_v - E)^{1/2}$$

$$p(T) = N_v e^{-(E_F - E_v)/k_B T}$$

$$N_v = \frac{1}{4} \left(\frac{2m_h^* k_B T}{\pi \hbar^2} \right)^{3/2}$$

$$np = \frac{1}{2} \left(\frac{k_B T}{\pi \hbar^2} \right)^3 (m_e^* m_h^*)^{3/2} e^{-E_g/k_B T}$$

Intrinsic Semiconductor

$$E_F = \frac{1}{2}(E_v + E_c) + \frac{3}{4} k_B T \ln \left(\frac{m_h^*}{m_e^*} \right)$$

$$E_F = \frac{1}{2}E_g + \frac{3}{4} k_B T \ln \left(\frac{m_h^*}{m_e^*} \right)$$

Extrinsic, n type

T_{low}

$$n = N_d e^{-(E_F - E_d)/k_B T} = N_c e^{-(E_c - E_F)/k_B T}$$

$$E_F = \frac{1}{2}(E_c + E_d) + \frac{1}{2} k_B T \ln \left(\frac{N_d}{N_c} \right)$$

T_{int}[a]

$$n = N_d = N_c e^{-(E_c - E_F)/k_B T}$$

$$E_F = E_c - k_B T \ln \left(\frac{N_c}{N_d} \right)$$

Extrinsic, p type

T_{low}

$$p = N_a e^{-(E_a - E_F)/k_B T} = N_v e^{-(E_F - E_v)/k_B T}$$

$$E_F = \frac{1}{2}(E_v + E_a) + \frac{1}{2} k_B T \ln \left(\frac{N_v}{N_a} \right)$$

T_{int}[a]

$$p = N_a = N_v e^{-(E_F - E_v)/k_B T}$$

$$E_F = E_v + k_B T \ln \left(\frac{N_v}{N_a} \right)$$

[a]Presumes large doping density.

However, in practice only lithium can be introduced in these materials in sufficient concentration to produce practical compensation.[1]

The *p-n* Junction[2]

This section deals with a description of the *p-n* junction that plays a fundamental role in many solid-state devices and involves a number of properties of impurity semicon-

[1] G. F. Knoll, *Radiation Detection and Measurement,* 2d ed., Wiley, New York, 1989.

[2] For further discussion, see S. Wang, *Fundamentals of Semiconductor Theory and Device Physics,* Prentice-Hall, Englewood Cliffs, N.J., 1989.

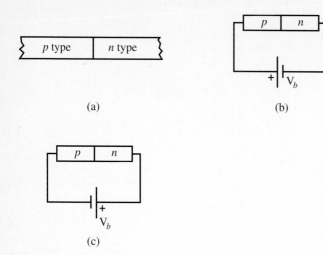

(a)

(b)

(c)

FIGURE 12.31 **(a) Unbiased** *p-n* **junction. (b) Forward-biased** *p-n* **junction in which electrons and holes are driven to the contact region where they recombine and current flows. (c) Reverse-biased** *p-n* **junction in which electrons and holes are driven away from the contact region and current drops dramatically.**

ductors developed above. The *p-n* junction is composed of a *p*-type semiconductor (uniformly doped with acceptors) joined across a plane interface with an *n*-type semiconductor (uniformly doped with donors). If no voltage is placed across the junction it is referred to as *unbiased*. A junction with voltage placed across it is called a *biased* junction (Fig. 12.31). In what follows, stemming from a basic property of the Fermi energy, we first describe the phenomenon of band bending. The carrier densities, n and p, are then obtained in terms of a potential function for the unbiased case, which leads to an expression for the "built-in potential," V_0, across the junction. The region of interaction of electrons and holes in the domain of the junction is called the *depletion zone*. Working in the *depletion-zone approximation,* expressions for the potential function $V(x)$ and n and p densities in the depletion zones are obtained as well as a relation between V_0 and the width of the depletion zone. The discussion continues with an estimate of the *I-V* characteristic for the biased *p-n* junction in the low-bias approximation. The resulting *I-V* relation is found to give good agreement with measurement in the small-bias potential domain.

Preliminary Relations

Here is a brief review of some physical relations relevant to the present topic. The *diffusion equation* relates to a fluid of particles in which a gradient of number density, $n(\mathbf{r})$, is present. It is given by

(12.97a)
$$n\mathbf{u} = -D\boldsymbol{\nabla}n$$

which indicates that a flux of particles comes into play to restore the fluid to uniformity. The fluid velocity is $\mathbf{u}(\mathbf{r})$ and D is the *diffusion coefficient*. Suppose the particles carry a charge, e. Then current density is given by

(12.97b)
$$\mathbf{J} = en\mathbf{u}$$

and (12.97a) becomes, for electron carriers (n) and hole carriers (p), respectively,

(12.97c)
$$\mathbf{J}_D{}^n = -D_n e \boldsymbol{\nabla} n = D_n |e| \boldsymbol{\nabla} n$$
$$\mathbf{J}_D{}^h = -D_h e \boldsymbol{\nabla} p = -D_h |e| \boldsymbol{\nabla} p$$

If an electric field, $\mathscr{E}$, is present in the device, then current is due to the applied field as well as diffusion current (12.97c). There results

(12.97d)
$$\mathbf{J}_n = en\mu_n \mathscr{E} + eD_n \boldsymbol{\nabla} n$$
$$\mathbf{J}_h = ep\mu_h \mathscr{E} - eD_h \boldsymbol{\nabla} p$$

where mobility, μ, is related to the diffusion coefficient through the *Einstein relations*

(12.97e)
$$D_n = \frac{\mu_n k_B T}{e}, \qquad D_h = \frac{\mu_h k_B T}{e}$$

Following (12.97c), e is written for $|e|$, so that electronic charge is $-e$, and holes have charge e. (With respect to this notation, mobilities in the preceding relations are positive.) Lastly we recall Poisson's equation (SI)

(12.97f)
$$\nabla^2 V = -\frac{\rho}{\varepsilon} = -\frac{e}{\varepsilon}(p + N_d - n - N_a)$$

where ρ represents charge density; N_d, N_a represent ionized donor-atom density and ionized acceptor-atom density, respectively; ε is the dielectric constant of the junction; and

(12.97g)
$$\mathscr{E} = -\boldsymbol{\nabla} V$$

We recall that (12.97f) stems from the preceding relation and the first of Maxwell's equations ($\boldsymbol{\nabla} \cdot \mathscr{E} = \rho/\varepsilon$).

Unbiased *p-n* Junction

When the n- and p-type materials are brought into contact, electrons diffuse to the p side and holes diffuse to the n side. Recall that holes on the p side stem from acceptors, and conduction electrons on the n side stem from donors. When an electron from the conduction band of the n side diffuses across the junction, it recombines with a hole

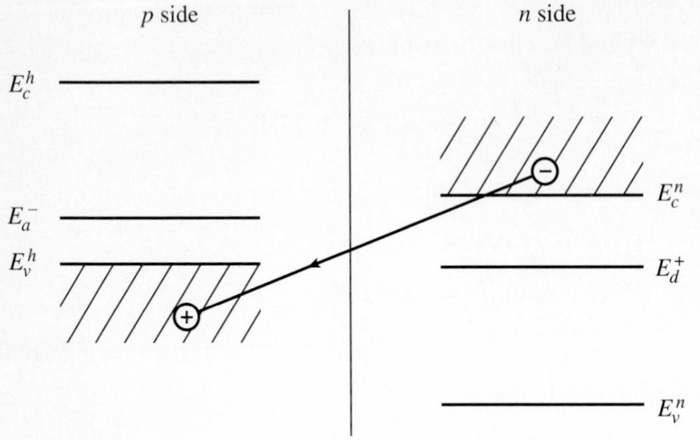

FIGURE 12.32 An electron from the n side of the depletion zone annihilates a hole from the p side, leaving an uncompensated, negatively charged acceptor on the p side and an uncompensated, positively charged donor on the n side.

in the valence band of the p side. This recombination leaves a negatively ionized acceptor on the p side and a positively ionized donor on the n side (Fig. 12.32). This charge buildup gives rise to an electric field that retards the diffusion. When current ceases, the domain of interaction about the junction is void of carriers and for this reason is called the *depletion zone*. In the state of no current flow, with (12.97d), and working in one dimension, we write

$$(12.98a) \qquad J_n = 0 = en\mu_n \mathscr{E} + eD_n \frac{dn}{dx}$$

$$(12.98b) \qquad J_h = 0 = ep\mu_h \mathscr{E} - eD_h \frac{dp}{dx}$$

Writing

$$(12.98c) \qquad \mathscr{E} = -\frac{dV}{dx}, \qquad \frac{1}{p}\frac{dp}{dx} = \frac{d \ln p}{dx}, \qquad \frac{1}{n}\frac{dn}{dx} = \frac{d \ln n}{dx}$$

together with the Einstein relation (12.97e), equations (12.98) give

$$(12.99a) \qquad \frac{d}{dx}\left(\frac{eV}{k_B T} + \ln p \right) = 0$$

$$(12.99b) \qquad \frac{d}{dx}\left(\frac{eV}{k_B T} - \ln n \right) = 0$$

Prior to integrating these equations, we note the following. At low temperature, the Fermi energy in the p-type material lies midway between the acceptor level and the top of the valence band. In the n-type material it lies midway between the donor

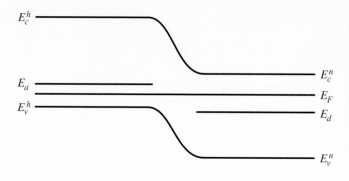

FIGURE 12.33 Band bending in the *p-n* junction maintaining constant E_F. At x_0, E_F is midway between E_v and E_c, which is identified as an intrinsic property. For equal doping densities, x_0 lies at the midpoint of the depletion zone.

level and the bottom of the conduction band (Fig. 12.30). When the materials are brought into contact, the relative positions of the conduction and valence bands adjust to maintain a constant Fermi energy[1] (Fig. 12.33). It follows from the continuity of the various levels in this diagram that there is value of x, which we label x_0, at which E_F lies at the midpoint of the energy gap. This condition was previously identified as an intrinsic property (12.78). It follows that at this point $n = p = n_i$. As we will find, it is consistent to set $V(x_0) = 0$. Integrating (12.99) from x to x_0, we obtain

(12.100a)
$$\frac{eV}{k_B T} - \ln n = 0 - \ln n_i, \qquad \frac{eV}{k_B T} = \ln \frac{n}{n_i}$$

$$\frac{eV}{k_B T} + \ln p = 0 + \ln n_i, \qquad \frac{eV}{k_B T} = -\ln \frac{p}{n_i}$$

or, equivalently,

(12.100b)
$$n(x) = n_i \exp \frac{eV(x)}{k_B T}$$

$$p(x) = n_i \exp -\frac{eV(x)}{k_B T}$$

Note that, as previously stated, at $x = x_0$, with $V = 0$, $n = p = n_i$. At $x = +\infty$, $n(x) \to N_d$ and at $x \to -\infty$, $p(x) \to N_a$, which, with (12.100a) give

(12.101a)
$$eV(\infty)/k_B T = \ln (N_d/n_i)$$

$$eV(-\infty)/k_B T = -\ln (N_a/n_i)$$

---

[1] At $T > 0$ K, the Fermi energy is more appropriately called the chemical potential. See footnote 3, page 664.

The built-in potential, V_0, is given by

(12.101b)
$$V_0 = V(\infty) - V(-\infty) = \frac{k_B T}{e} \ln \frac{N_a N_d}{n_i^2}$$

(In the present notation, N_d, N_a represent ionized donor and acceptor densities, respectively.)

Depletion-Zone Approximation

The relations (12.100b) are incomplete without knowledge of $V(x)$. A model that affords evaluation of the potential function is called the *depletion-zone approximation*. Profiles of various components of the *p-n* junction in this model are shown in Fig. 12.34. In this picture regions I and II denote the *p* and *n* domains of the

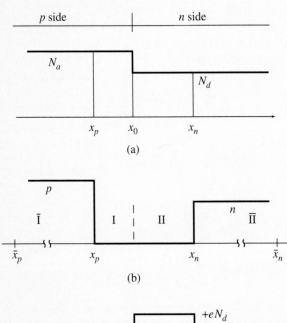

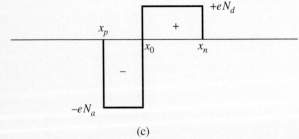

FIGURE 12.34 Densities in the depletion-zone approximation: (a) Doping densities. (b) Carrier densities. (c) Charge densities. The bulk material domains $\bar{\text{I}}$ and $\bar{\text{II}}$ are shown as well. Charge densities in these latter domains are zero as charges due to ionized impurity atoms are compensated by carrier charges.

depletion zone, respectively. As charge densities in the present approximation are constant in each p and n domain of the depletion zone, conservation of charge requires that

(12.102)
$$N_a(x_0 - x_p) = N_d(x_n - x_0)$$

As ρ is constant in regions I and II (see Fig. 12.34), the solution of Poisson's equation (12.97f) in each of these respective domains has the general form

$$V(x) = -\frac{\rho x^2}{2\varepsilon} + Bx + K$$

where B and K are arbitrary constants. Solutions of Poisson's equation over domains I and II of the depletion zone which include the built-in potential (12.101b) with respect to points x_p and x_n are then given by

(12.103a)
$$V_I(x) = \frac{eN_a}{2\varepsilon}(x - x_p)^2 - \frac{k_B T}{e}\ln\frac{N_a}{n_i}$$

(12.103b)
$$V_{II}(x) = \frac{eN_d}{2\varepsilon}(x - x_n)^2 + \frac{k_B T}{e}\ln\frac{N_d}{n_i}$$

Potentials are defined with respect to positive charge flow so that, as evident from Fig. 12.35, the preceding potentials retard positive charge flow from p to n regions. The potential "seen" by electrons is obtained by reflecting the curve of Fig. 12.35 through the x axis, from which it is evident that this potential equally retards negative charge flow from n to p regions. The constants in (12.103) return the correct built-in potential

$$V_{II}(x_n) - V_I(x_p) = V_0$$

where V_0 is given by (12.101b).

Electron and Hole Densities and the Depletion-Zone Width

Substituting the potentials (12.103) into the n and p expressions (12.100b) gives

(12.104a) $p(x) = N_a \exp\left[-\frac{(x - x_p)^2}{2\lambda_p^2}\right], \qquad x$ in region I $(x > x_p)$

(12.104b) $n(x) = N_d \exp\left[-\frac{(x - x_n)^2}{2\lambda_n^2}\right], \qquad x$ in region II $(x < x_n)$

where the λ lengths are given by

(12.104c)
$$\lambda_p^2 \equiv \frac{\varepsilon k_B T}{e^2 N_a}$$

(12.104d)
$$\lambda_n^2 \equiv \frac{\varepsilon k_B T}{e^2 N_d}$$

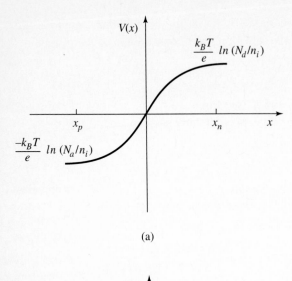

(a)

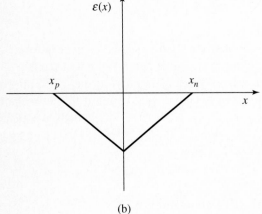

(b)

FIGURE 12.35 (a) Built-in potential in the depletion-zone approximation illustrating a barrier to positive charge flow from p to n domains or negative charge flow from n to p domains. (b) Built-in electric field corresponding to the potential in (a). This electric field drives holes to the left and electrons to the right, away from the contact region.

The relations (12.104a, b) indicate an exponential decay of charge carrier densities into the depletion zone. This decay increases with increase of the λ coefficients. This smooth decay is an improved estimate of carrier densities in the depletion zone compared with the sharp cutoff of starting values of these parameters depicted in Fig. 12.34.

We wish to apply these latter equations to obtain an expression for the depletion-zone width

$$W \equiv x_n - x_p$$

in terms of the built-in potential, V_0. This relation may be obtained from the constraint that V_I and V_{II} must match at x_0, which gives

(12.105a) $$\frac{eN_a}{2\varepsilon}(x_0 - x_p)^2 - \frac{k_B T}{e}\ln\frac{N_a}{n_i} = -\frac{eN_d}{2\varepsilon}(x_0 - x_n)^2 + \frac{k_B T}{e}\ln\frac{N_d}{n_i}$$

Referring to the charge-conservation equation (12.102) and the built-in potential relation (12.101b) we see that the latter equation implies the following two equations,

(12.105b) $$N_a^2(x_0 - x_p)^2 = N_d^2(x_n - x_0)^2 = \frac{2\varepsilon V_0 N_a N_d}{e(N_a + N_d)}$$

from which we find

(12.105c) $$V_0 = \frac{e}{2\varepsilon}\left(\frac{N_a N_d}{N_a + N_d}\right)W^2$$

This is the desired relation between the width of the depletion zone and the built-in potential. With (12.101b), the latter relation gives W in terms of known doping densities and intrinsic electron density.

The preceding results, (12.104) and (12.105c), represent equilibrium properties of an unbiased *p-n* junction at the temperature T.

One may associate a capacitance with a *p-n* junction. Thus, if A represents the cross-sectional area of the junction, the capacitance of an unbiased *p-n* junction is given by

(12.105d) $$C = \frac{\varepsilon A}{W}$$

where the width, W, of the depletion zone is given by the preceding relation, (12.105c).

Biased *p-n* Junction

A biased *p-n* junction is one with an external voltage, V_b, applied across the junction. Recall that the built-in voltage, V_0, opposes further current flow in the unbiased junction. The resulting voltage across the junction is

(12.106) $$\Delta V = V(x_n) - V(x_p) = V_0 - V_b$$

That is, the total potential drop occurs across the depletion zone. Again, potentials are defined with respect to positive charge flow. (Recall Fig. 12.35.) For forward bias, $V_b > 0$, and V_b diminishes V_0 and current flow is enhanced. For reversed bias, $V_b < 0$, and the built-in potential is enhanced, thereby inhibiting current flow. In forward bias, electrons and holes are driven together and they recombine. The external potential

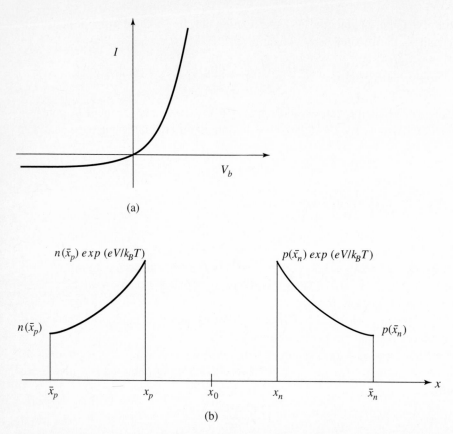

(a)

(b)

FIGURE 12.36 (a) *I-V* characteristic of a *p-n* junction. Typical currents are of the order of mA at 1 V. (b) Minority charge-carrier densities in the depletion-zone approximation from which it is seen that electrons and holes in their respective minority domains are driven to the depletion zone, thereby increasing the occurrence of recombination and current flow. For reverse bias, $(V \to -V)$, electrons and holes are repelled from the depletion zone, sharply diminishing the probability of recombination, with corresponding decrease of current flow.

resupplies electrons to the *n* domain and current flows. In reverse bias, electrons and holes are driven away from the junction domain in opposite directions and current falls dramatically. A typical *I-V* characteristic for a *p-n* junction is shown in Fig. 12.36. As evident from this figure, a *p-n* junction serves as a rectifier.

Low-Bias Approximation

We wish to obtain an estimate for the *I-V* characteristic of a *p-n* junction in the limit of small applied voltage. The estimate is based in large part in evaluating $n(x)$ and $p(x)$

in respective minority domains outside the depletion zone. The reason for this procedure is that, to this same order, carrier densities in their respective majority domains
are approximately constant and equal to related donor densities (Fig. 12.34b). In the
first step of this estimate, one assumes that (12.98) maintains, leading again to
(12.99). Integrating this expression between x_n and x_p gives

(12.107a)
$$p(x_n) = p(x_p) \exp\left\{ -\frac{e[V(x_n) - V(x_p)]}{k_B T} \right\}$$

(12.107b)
$$n(x_p) = n(x_n) \exp\left\{ -\frac{e[V(x_n) - V(x_p)]}{k_B T} \right\}$$

With (12.106) and recalling the expression (12.101b) for V_0, the preceding equations
are rewritten

(12.108a)
$$p(x_n) = p(x_p)\frac{n_i^2}{N_a N_d} \exp \frac{eV_b}{k_B T}$$

(12.108b)
$$n(x_p) = n(x_n)\frac{n_i^2}{N_a N_d} \exp \frac{eV_b}{k_B T}$$

With $p(x_p) \simeq N_a$, $n(x_n) \simeq N_d$, these latter equations become

(12.108c)
$$p(x_n) \simeq \frac{n_i^2}{N_d} \exp \frac{eV_b}{k_B T}$$

(12.108d)
$$n(x_p) \simeq \frac{n_i^2}{N_a} \exp \frac{eV_b}{k_B T}$$

Note that, as indicated above, $p(x_n)$ and $n(x_p)$ are minority carrier densities. We
introduce the two displacements:

$$\bar{x}_n \gg x_n, \qquad \bar{x}_p \ll x_p$$

These displacements lie far beyond the depletion zone (Fig. 12.34b). It is assumed that
gradients of minority carrier densities as well as drift terms in minority carrier
currents are negligibly small at respective displacements $\bar{x}_n$ and $\bar{x}_p$. As all doping
atoms are ionized, in region $\bar{\text{I}}$, $p = N_a$, and in region $\overline{\text{II}}$, $n = N_d$. With reference to
(12.80b), we then write

(12.108e)
$$p(\bar{x}_n) \simeq \frac{n_i^2}{N_d}, \qquad n(\bar{x}_p) \simeq \frac{n_i^2}{N_a}$$

The latter three relations are boundary conditions for $n(x)$ and $p(x)$ in their respective
minority domains.

Before proceeding further, we recall the continuity equation (7.97). In the present situation this equation is written for both n and p. However, for transport in a semiconductor, one must take into account the possibility of recombination of holes and electrons. Thus, for example, the continuity equation for holes is given by (in one dimension)

$$(12.109a) \qquad \frac{\partial p}{\partial t} + \frac{1}{e}\frac{\partial J_h}{\partial x} = -\frac{1}{\tau_h}[p - p(\bar{x}_n)]$$

This equation indicates that p in its minority bulk domain, $\overline{\mathrm{II}}$ (Fig. 12.34b), relaxes in the characteristic recombination time, τ_h, to its equilibrium value, $p(\bar{x}_n)$. [As we are employing (12.80b), in the following one assumes small recombination rates, $1/\tau$.] In steady state we set $\partial p/\partial t = 0$. Employing the drift current in (12.97c)

$$(12.109b) \qquad J_h = -eD_h\frac{dp}{dx}$$

together with the steady-state version of (12.109a) and (12.108e), we obtain

$$(12.109c) \qquad \frac{d^2p}{dx^2} - \kappa_h^2 p = -\frac{n_i^2}{N_d}\kappa_h^2$$

$$(12.109d) \qquad \kappa_h^2 \equiv \frac{1}{D_h\tau_h}$$

The solution to (12.109c), subject to the boundary conditions (12.108c, e), is given by

$$(12.110a) \qquad p(x) = \frac{n_i^2}{N_d}\left[1 + S(x)\left(\exp\frac{eV}{k_BT} - 1\right)\right]$$

$$(x \text{ in region } \overline{\mathrm{II}})$$

$$(12.110b) \qquad S(x) \equiv \frac{\sinh \kappa_h(\bar{x}_n - x)}{\sinh \kappa_h(\bar{x}_n - x_n)}$$

where we have set $V_b \equiv V$. At $x = \bar{x}_n$, the preceding solution returns (12.108e), whereas at $x = x_n$, it returns (12.108c). In this manner we find that for forward bias, $eV/k_BT > 0$, $p(x)$ grows from its value at $\bar{x}_n$ to its value at x_n.

With (12.109b) we obtain the minority current density on the n side of the junction (region $\overline{\mathrm{II}}$).

$$(12.110c) \qquad J_h = \frac{en_i^2D_h}{N_d}S'(x)\left(\exp\frac{eV}{k_BT} - 1\right)$$

where a prime denotes differentiation.

The equation for $n(x)$ in region $\bar{\mathrm{I}}$ is the same as (12.109c) with the parameter, κ_n^2, defined parallel to (12.109d). The solution for $n(x)$ on the p side of the junction (region $\bar{\mathrm{I}}$) is given by

(12.110d)
$$n(x) = \frac{n_i^2}{N_a}\left[1 + \tilde{S}(x)\left(\exp\frac{eV}{k_B T} - 1\right)\right]$$

(x in region $\bar{\text{I}}$)

(12.110e)
$$\tilde{S}(x) \equiv \frac{\sinh \kappa_n(x - \bar{x}_p)}{\sinh \kappa_n(x_p - \bar{x}_p)}$$

which returns the correct boundary conditions, (12.108d) and the second of (12.108e). We note that for forward bias, $eV/k_B T > 0$, $n(x)$ grows from its value at $\bar{x}_h$ to its value at x_h. In this manner we find that a forward bias forces both n and p minority carriers to the region of the depletion zone (Fig. 12.36a), in opposition to the built-in field (Fig. 12.35). For reverse bias ($V \rightarrow -V$), electrons and holes are driven away from the depletion zone, in accord with the built-in potential, diminishing the occurrence of recombination. That is, in reverse bias, $p(x)$ diminishes from its value at $\bar{x}_n$ to its value at x_n, and $n(x)$ diminishes from its value at $\bar{x}_h$ to its value at x_h.

The relations (12.110d, e) give the minority current density on the p side of the junction.

(12.110f)
$$J_n = \frac{en_i^2 D_n}{N_a}\tilde{S}'(x)\left(\exp\frac{eV}{k_B T} - 1\right)$$

If there is no generation or recombination of carriers in the depletion zone, $\kappa \propto \tau^{-1} \rightarrow 0$, and J_n and J_h are constant in these regions. To demonstrate this property consider the κ dependence of, say, J_n.

$$J_n \propto \frac{\cosh \kappa(x - \bar{x}_p)}{[\sinh \kappa(x_p - \bar{x}_p)]/\kappa}$$

In the limit, $\kappa \rightarrow 0$, this form goes to the constant, $1/(x_p - \bar{x}_p)$. It follows that

(12.111a)
$$J_n(x_p) = J_n(x_n), \qquad J_h(x_p) = J_h(x_n)$$

These relations permit preceding expressions for minority currents to be employed in calculation of total current, and we write

(12.111b)
$$J = J_n(x_p) + J_h(x_n)$$

so that with (12.110d) and (12.110f) there results

(12.111c)
$$J = J_s\left(\exp\frac{eV}{k_B T} - 1\right)$$

(12.111d)
$$J_s \equiv en_i^2\left[\frac{D_k \kappa_h}{N_d}S'(x_n) + \frac{D_n \kappa_n}{N_a}\tilde{S}'(x_p)\right]$$

With $I = JA$, where A is the cross-sectional area of the junction and I represents line current, we obtain

(12.112a)
$$I = I_s\left(\exp\frac{eV}{k_B T} - 1\right)$$

(12.112b)
$$I_s = AJ_s$$

This finding represents the first-order solution in the low-bias approximation above the zero-current starting value (12.98). With V replaced by V_b, the current (12.112) is illustrated by the small-bias portion of the I-V characteristic shown in Fig. 12.36 (where V is again written V_b). Replacing V with $-V$ in (12.112a) indicates that I_s is the saturation current in reverse bias.

We close this discussion with the generalization of our previous experience for the capacitance for an unbiased junction, (12.105b), to the case of the capacitance of a biased junction. With (12.106), we recall the potential across the depletion zone of a p-n junction changes in the presence of an applied potential: $V_0 \to V_0 - V$. The related generalized width of the depletion zone is given by

(12.113a)
$$W^2 = \frac{2\varepsilon(V_0 - V)}{e} \frac{(N_a + N_d)}{N_a N_d}$$

When inserted in (12.105b), the preceding relation gives the desired generalized expression for the capacitance of a biased p-n junction,

(12.113b)
$$C = A\left[\frac{e\varepsilon}{2(V_0 - V)} \frac{N_d N_a}{(N_d + N_a)}\right]^{1/2}$$

An important application of capacitance of a p-n junction is as follows. Consider a p^+-n junction (which denotes a p-n junction with a highly doped p side, i.e., $N_a \gg N_d$). In this limit the capacitance of the device becomes

(12.113c)
$$C = A\left[\frac{e\varepsilon}{2(V_0 - V)} N_d\right]^{1/2}$$

For large reverse bias, $V \to -V$, $|V| \gg V_0$, and V_0 may be neglected compared to V. For this case, the preceding relation affords a measurement of N_d in terms of the cross-sectional area of the device.

PROBLEMS

12.56 Consider an n-type semiconductor.

(a) Give two relations in terms of density of ionized donors, $N_d - N_{d_0}$, and conduction electrons n which are obeyed in the T_{low} domain.

(b) Give two relations involving these variables which are obeyed in the T_{int} domain.

12.57   A *p*-type semiconductor has a hole density $p = 1.7 \times 10^{13}$ cm^{-3} at $T = 388$ K. If $E_g = 1.8$ eV and $m_e^* m_h^* = 0.25\, m_e^2$ at this temperature, what is the electron carrier concentration n?

12.58   The effective density of states, $N_c(T)$, of a semiconductor at $T = 370$ K is 0.42×10^{20} cm^{-3}. What is the value of m_e^*/m_e for this material at this temperature?

12.59   An intrinsic semiconductor has an energy gap of 1.03 eV with effective masses $m_e^* = 1.8 m_h^*$.
    (a)   What are the values of E_F at 300 K and 500 K?
    (b)   What are the charge carrier densities n and p, respectively, at these temperatures?

12.60   In the text we discussed Sb and In as impurity atoms in an Si host crystal. With reference to the periodic chart and an appropriate handbook for required additional data, choose another host element from the same group that Si lies in and two other impurity elements which would serve as donors and acceptors, respectively, for your host element. The device is to operate in the temperature interval 300 K $\leq T \leq$ 500 K.

12.61   Where does the Fermi energy of a conductor lie at 0 K with respect to valence and conduction bands? Your answer should include reference to the Fermi–Dirac distribution.

12.62   A one-dimensional model of a certain intrinsic semiconductor with lattice constant a has the following E versus k relations for the conduction and valence bands, respectively.

$$E_c(k) = 2E_0 - bE_0 \cos^2 \frac{ka}{2}$$

$$E_v(k) = bE_0 \cos^2 \frac{ka}{2}$$

The dimensionless parameter b is less than 1. At 0 K the conduction band is empty.
    (a)   Sketch these curves on the same E versus k graph in the reduced zone.
    (b)   Obtain expressions for the effective electron and hole masses, m_e^* and m_h^*, respectively. (See appropriate expressions in Problem 8.57.)
    (c)   If $E_0 = 2.5$ eV, $b = \frac{1}{2}$, and $a = 1.5$ Å, obtain the value (in eV) of the Fermi energy for this semiconductor.
    (d)   At what frequency (Hz) would incident photons cause this semiconductor to conduct at 0 K?

12.63   A semiconductor is doped with donor impurity atoms of density N_d and acceptor atoms of density N_a. What is the equation of charge neutrality for this extrinsic semiconductor?

12.64   Electronegativity is a measure of the force required to remove the ionizing electron of negatively charged ions. Consider, for example, that an H atom in the ground state is negatively ionized, resulting in a $1s^2$ configuration.
    (a)   Repeating the calculation in the text relevant to helium, calculate the new ground-state energy of the H$^-$ ion.
    (b)   Is the H$^-$ a bound system?
    (c)   Repeat part (a) for a one-electron atom with nuclear charge number Z. How does the energy you obtain depend on Z?

Answer (partial)

(a) Referring to Problem 12.28 and Table 10.3, we find

$$E = 2\mathbb{R} + \Delta E = -\frac{2e^2}{2a} + \frac{5e^2}{8a} < 0$$

Thus, to within the stated approximation, we find the ion to be a bound system.

12.65 Employing rules of electrostatics, show that the Coulomb interaction energy A (12.45) may be written as a single integral over electric field.

Answer

This integral is in the form

$$A = \iint d\mathbf{r}\, d\mathbf{r}'\, \frac{\rho(\mathbf{r})\rho(\mathbf{r}')}{|\mathbf{r} - \mathbf{r}'|}$$

where $\rho = e|\varphi|^2$ is charge density. Introducing the field (cgs)

$$\nabla \cdot \mathscr{E} = 4\pi\rho$$

permits the preceding integral to be rewritten

$$A = 2 \int \frac{|\mathscr{E}(\mathbf{r})|^2}{8\pi}\, d\mathbf{r} > 0$$

12.66 Write down the condition of charge neutrality in a *p-n* junction in

(a) The p domain beyond the depletion zone, i.e., $x < x_p$.

(b) The n-domain beyond the depletion zone, i.e., $x > x_n$. (See Fig. 12.34.) In the depletion-zone approximation,

(c) What is the reason for evaluating only minority carrier densities in calculation of current across a *p-n* junction?

(d) What entities are depleted in the depletion zone of a *p-n* junction?

Answers (partial)

(a) $p(x < x_p) = N_a + n(x < x_p)$

(b) $n(x > x_n) = N_d + p(x > x_n)$

12.67 (a) Show that the potentials (12.103), relevant to the depletion-zone approximation, satisfy Poisson's equation (12.97f) (in one dimension) with correct charge densities.

(b) Again employing these potentials, obtain expressions for the electric field, $\mathscr{E}$, across the *p-n* junction.

12.68 Assuming doping densities, $N_a \simeq N_d \simeq 10^{17}$ cm^{-3} and $\varepsilon = 1.02$ at $T = 300$ K,

(a) Find values for the characteristic lengths, λ_p and λ_n (12.104).

(b) Make a rough sketch of $n(x)$ in region II and $p(x)$ in region I corresponding to a *p-n* junction with the preceding properties in the unbiased state. (See Fig. 12.34.)

12.69 (a) Obtain the relations (12.105b) from (12.105a).

(b) From these relations, obtain (12.105c) for V_0 in terms of W. A p-n junction operates at $T = 300$ K wth densities $n_i = 10^{18}$ cm^{-3}, $N_a = 10^{20}$ cm^{-3}, $N_d = 10^{19}$ cm^{-3}. In addition, $\varepsilon = 1.2$.

(c) What is the width of the depletion zone, W? Answer in μm.

(d) A reverse bias is applied across the junction so that $V_0 - V = 5$ V. The cross-sectional area of the junction is $A = 15$ μm^2. What is the capacitance of the device (in F)?

(e) If $N_d \gg N_a$, what is the value of N_d (in cm^{-3})?

12.10 ELEMENTS OF NUCLEAR PHYSICS. THE DEUTERON AND ISOSPIN

The Pauli principle was applied earlier in this chapter in constructing properly symmetrized atomic and molecular states. In the preceding section we saw how this principle, through use of the Fermi–Dirac distribution, comes into play in describing electron and hole properties in semiconductors. In this concluding section of the present chapter we will find how the Pauli principle, through the notion of isotopic spin, comes into play in nuclear physics. This example addresses construction of the ground state of the deuteron.

The first section of the present discussion addresses the angular and spin components of the ground state of the deuteron. Recall that an estimate of the radial component of the ground state was obtained in Problem 10.31. In the present discourse, the angular component is obtained by employing conservation of parity and magnetic moment data. In this construction, noncentral, spin-dependent forces are encountered.

In the second component of the discussion, the notion of isotopic spin is developed, which, as noted above, permits quantum statistics to be applied, and the appropriately symmetrized ground state of the deuteron is obtained.

Magnetic Moment of the Deuteron

The magnetic moment of the deuteron is related to its angular momentum. In general, the angular momentum of a nucleus may be written as a sum of orbital and spin angular momenta of nuclear constituents, and we may write[1]

(12.114) $$\mathbf{J} = \mathbf{L} + \mathbf{S}$$

For the deuteron, $\mathbf{L}$ is attributed to rotational motion of the neutron and proton about each other, while $\mathbf{S}$ is the resultant of their spins.

---

[1] Nuclear spin is often denoted by the symbol $\mathbf{I}$.

As noted in Section 12.7, the spin of the deuteron is unity (i.e., $j = 1$). The spin of the neutron and proton are both $\frac{1}{2}$ so that allowed coupled spin values are $s = 0, 1$. The values of orbital angular momenta l which combine with these spin values to give $j = 1$ are listed below.

$$
\begin{array}{llll}
s = 1 & l = 0: & j = \underline{1} & {}^3S_1 \\
& l = 1: & j = 0, \underline{1}, 2 & {}^3P_1 \\
(12.115) & l = 2: & j = \underline{1}, 2, 3 & {}^3D_1 \\
s = 0 & l = 0: & j = 0 & \\
& l = 1: & j = \underline{1} & {}^1P_1
\end{array}
$$

Thus, the experimentally observed nuclear spin of unity for the deuteron in its ground state implies that the angular component of the wavefunction is some superposition of ${}^3S_1$, ${}^3P_1$, ${}^3D_1$, and ${}^1P_1$ states. We will now find how parity conservation properties of the Hamiltonian of the deuteron further specify the composition of the ground-state wavefunction.

The Hamiltonian of the deuteron may be written $\hat{H}(\mathbf{r}_p, \mathbf{r}_n; \hat{\mathbf{S}}_p, \hat{\mathbf{S}}_n)$, where the subscripts n and p denote neutron and proton, respectively. The parity operator $\hat{\mathbb{P}}$ (see Section 6.4) which reflects $\mathbf{r}_p$ and $\mathbf{r}_n$ through the origin leaves $\hat{H}$ unchanged.[1]

$$
(12.116) \qquad
\begin{aligned}
\hat{\mathbb{P}}\hat{H}(\mathbf{r}_p, \mathbf{r}_n) &= \hat{H}(-\mathbf{r}_p, -\mathbf{r}_n) = \hat{H}(\mathbf{r}_p, \mathbf{r}_n) \\
[\hat{\mathbb{P}}, \hat{H}] &= 0
\end{aligned}
$$

It follows that the parity is a good quantum number and may be listed along with other quantum numbers in specifying the state of a deuteron. That is, states are of definite parity.

The combination of states ${}^3S_1$ and ${}^3D_1$ corresponds to even parity, while the combination ${}^3P_1$ and ${}^1P_1$ corresponds to odd parity (recall Problem 9.46). The specific combinations which comprises the angular component of the ground-state wavefunction is obtained by finding the combination which gives the closest value to that of the observed magnetic moment of the deuteron,

$$
\mu_d = 0.857\mu_N
$$

The *nuclear magneton* μ_N has the value

$$
\mu_N = \frac{\hbar e}{2M_p c} = 0.505 \times 10^{-23} \text{ erg gauss}^{-1}
$$

The proton rest mass is M_p.

---

[1] This statement is valid for the strong nuclear forces.

There are three distinct contributions to the deuteron's magnetic moment:

(12.117)
$$\boldsymbol{\mu}_d = \boldsymbol{\mu}_p + \boldsymbol{\mu}_n + \boldsymbol{\mu}_L$$

As with the electron, the proton and neutron magnetic moments are proportional to their respective spins. Experimentally, one obtains the values (noted previously in Section 11.9)

$$\boldsymbol{\mu}_p = g_p\left(\frac{\mu_N}{\hbar}\right)\mathbf{S}(p) \qquad g_p = 2(2.79)$$

$$\boldsymbol{\mu}_n = g_n\left(\frac{\mu_N}{\hbar}\right)\mathbf{S}(n) \qquad g_n = 2(-1.91)$$

The contribution to $\boldsymbol{\mu}_d$ due to orbital motion of the neutron and proton about each other is (see Problem 12.72)

$$\boldsymbol{\mu}_L = \frac{1}{2}\left(\frac{\mu_N}{\hbar}\right)\mathbf{L}$$

It follows that (12.117) may be rewritten

(12.118)
$$\boldsymbol{\mu}_d = \left(\frac{\mu_N}{\hbar}\right)[g_p\,\mathbf{S}(p) + g_n\,\mathbf{S}(n) + 0.50\mathbf{L}]$$

As stated in the previous chapter, it is conventional to measure the expectation of the z component of $\boldsymbol{\mu}_d$ in the state $m_j = j$ and to call this the magnetic moment of the deuteron μ_d.[1] We will calculate the contributions to this value from each of the four states listed in (12.115) and see which contribution of parity-conserving pairs of states ($^3P_1$ and $^1P_1$ or $^3S_1$ and $^3D_1$) gives a value closest to the experimentally observed value.

Introducing the spin operators

$$\mathbf{S} = \mathbf{S}(p) + \mathbf{S}(n)$$
$$\mathbf{S}^- = \mathbf{S}(p) - \mathbf{S}(n)$$

permits (12.118) to be rewritten

$$\boldsymbol{\mu}_d = \left(\frac{\mu_N}{\hbar}\right)\left[\left(\frac{g_p + g_n}{2}\right)\mathbf{S} + \left(\frac{g_p - g_n}{2}\right)\mathbf{S}^- + 0.50\mathbf{L}\right]$$

We are interested in the expectation of the z component of $\boldsymbol{\mu}_d$ in the three-triplet and one-singlet coupled spin states listed in Table 11.3. Calculation readily shows that $\langle S_z^- \rangle$ is zero in any of these four states (see Problem 12.70). Thus, for purposes of the said calculation, we are permitted to write

---

[1] See also H. Von Buttlar, *Nuclear Physics*, Academic Press, New York, 1968.

$$(12.119) \qquad \boldsymbol{\mu}_d = \left(\frac{\mu_N}{\hbar}\right)[0.88\mathbf{S} + 0.50\mathbf{L}]$$

$$= \left(\frac{\mu_N}{\hbar}\right)[0.38\mathbf{S} + 0.50\mathbf{J}]$$

Construction of the Ground State

Following the description given in Problem (12.16) relevant to the Zeeman effect, we again consider that rotation of $\mathbf{J}$ about the z axis (direction, say, of a measuring $\mathbf{B}$ field) is slow compared with that of $\boldsymbol{\mu}$ about $\mathbf{J}$. With (12.119) this gives the form

$$\langle \mu_{d_z} \rangle = \left\langle \frac{(\boldsymbol{\mu}_d \cdot \mathbf{J})J_z}{J^2} \right\rangle$$

$$= \mu_N \left\{ 0.50 + 0.38 \left[\frac{j(j+1) + s(s+1) - l(l+1)}{2j(j+1)} \right] \right\} m_j$$

Using this expression, one obtains the following contributions to $\langle \mu_d \rangle$ for the four states in question (setting $m_j = 1$ as indicated above):

$$\langle \mu_d(^3S_1) \rangle = \mu_N \left[0.50 + 0.38 \left(\frac{2+2}{2 \times 2} \right) \right] = 0.88\mu_N$$

$$\langle \mu_d(^1P_1) \rangle = \mu_N \left[0.50 + 0.38 \left(\frac{2-2}{2 \times 2} \right) \right] = 0.50\mu_N$$

$$(12.120)$$

$$\langle \mu_d(^3P_1) \rangle = \mu_N \left[0.50 + 0.38 \left(\frac{2+2-2}{2 \times 2} \right) \right] = 0.69\mu_N$$

$$\langle \mu_d(^3D_1) \rangle = \mu_N \left[0.50 + 0.38 \left(\frac{2+2-2 \times 3}{2 \times 2} \right) \right] = 0.31\mu_N$$

Since $\langle \mu_d(^3P_1) \rangle$ and $\langle \mu_d(^1P_1) \rangle$ are both less than $\langle \mu_d \rangle$, there is no combination of the form

$$\langle \mu_d \rangle = (a_{3_P})^2 \langle \mu_d(^3P_1) \rangle + (a_{1_P})^2 \langle \mu_d(^1P_1) \rangle$$

$$(a_{3_P})^2 + (a_{1_P})^2 = 1$$

which gives the observed value of $\langle \mu_d \rangle$. On the other hand, fitting $\langle \mu_d \rangle$ to the combination of even parity states,

$$\langle \mu_d \rangle = (a_S)^2 \langle \mu_d(^3S_1) \rangle + (a_D)^2 \langle \mu_d(^3D_1) \rangle = 0.86\mu_N$$

$$(a_S)^2 + (a_D)^2 = 1$$

does allow the solution

(12.121) $(a_S)^2 = 0.96, \qquad (a_D)^2 = 0.04$

One may conclude that the ground state is a mixture of $\sqrt{0.96}\,^3S_1$ and $\sqrt{0.04}\,^3D_1$ states. That is,

(12.122) $|\psi_G\rangle = a_S|^3S_1\rangle + a_D|^3D_1\rangle$

The coordinate representation of this state is given by

(12.122a) $\langle\theta\phi|\psi_G\rangle = a_S Y_0^0(\theta, \phi) + a_D Y_2^m(\theta, \phi)$

where the $Y_l^m(\theta, \phi)$ spherical harmonics are listed in Table 9.1.

As $a_D \ll a_S$, for purposes of calculating the magnetic moment of the deuteron, it suffices to take $|\psi_G\rangle$ to be entirely an S state. However, in evaluating the electric quadrupole moment, the D contribution to the ground state is necessary to obtain a finite result. Thus it is found that the superposition state (12.122) gives agreement with the experimentally observed quadrupole moment of the deuteron, $Q = 0.0027 \times 10^{-24}$ cm^2.[1] This moment is a measure of the deviation from spherical symmetry of the charge density of the deuteron. Furthermore, the sign $(+)$ of the quadrupole moment indicates that the charge distribution of the deuteron is prolate (resembling an egg) rather than oblate (resembling the earth).

Noncentral Forces

We note the following important fact about the neutron-proton interaction. Since the ground state of the deuteron is a superposition of states of different l, it is not an eigenstate of $\hat{L}^2$. If the interaction between these two particles were a central force,[2] then eigenstates of $\hat{H}$ would be common with eigenstates of $\hat{L}^2$, and $\hat{L}_z$ and would appear as in the form of (10.108). We conclude that the internucleon interaction is not a central-force field but rather is *spin-dependent*. That is, the potential of interaction between the particles is dependent on the relative orientation of the spins $\hat{\boldsymbol{\sigma}}_n$ and $\hat{\boldsymbol{\sigma}}_p$ of the neutron and proton, respectively, and the radius vector $\mathbf{r}$ separating them. An example of such a noncentral spin-dependent potential of interaction is given by the form[3]

(12.123) $\hat{H}_S = f_S(r)\left[\dfrac{(\hat{\boldsymbol{\sigma}}_n \cdot \mathbf{r})(\hat{\boldsymbol{\sigma}}_p \cdot \mathbf{r})}{r^2} - \dfrac{1}{3}\hat{\boldsymbol{\sigma}}_n \cdot \hat{\boldsymbol{\sigma}}_p\right]$

[1] This value takes electronic charge equal to 1.

[2] Central potential was previously discussed in Section 10.5.

[3] H. Feshbach and J. Schwinger, *Phys. Rev.* **84**, 194 (1951). An assortment of forms for two-body nuclear interactions is presented in L. Eisenbud and E. P. Wigner, *Nuclear Structure*, Princeton University Press, Princeton, N.J., 1958.

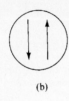

(a)   (b)

FIGURE 12.37 Relative possible orientations of the neutron and proton spins in the deuteron. The fact that the triplet spin state depicted by (a) is the ground state, whereas the singlet state depicted by (b) is not observed in nature, is evidence of the spin dependence of nuclear forces.

where $\hat{\boldsymbol{\sigma}}$ denotes the relevant Pauli spin matrices and $f_S(r)$ is a scalar function of r. Note that $\hat{H}_S$ preserves parity.

The fact that the deuteron has only one bound state (Problems 8.1 and 10.31) with parallel neutron-proton spin orientations (triplet state) is also evidence for the spin-dependent nature of the nuclear force, for if this were not the case, a bound state would also exist for the antiparallel case (singlet state). See Fig. 12.37.

Isotopic Spin

In nuclear physics, the neutron and proton are viewed as two distinct states of a single particled called a *nucleon*. This point of view (due to Heisenberg, 1932) stems from the assumption that the nuclear force is far larger than the Coulomb force within the nucleus, or, equivalently, that the internucleon force is *charge-independent*.

These features are incorporated in a property called *isotopic spin,* or, equivalently, *isospin*. The operator corresponding to this property is written $\hat{I}$ and has three orthogonal components which obey the relation

(12.124) $$\hat{I}^2 = \hat{I}_1^2 + \hat{I}_2^2 + \hat{I}_3^2$$

These operators exist in isotropic spin space. Eigenvalues of the third component $\hat{I}_3$, relevant to a particle of charge Q (measured in units of electronic charge), are given by

(12.125) $$I_3 = Q - \frac{1}{2}$$

It follows that $I_3 = +\frac{1}{2}$ for a proton and $I_3 = -\frac{1}{2}$ for a neutron.

Let us ascertain the isospin states for a two-nucleon system. For this purpose we introduce the single-particle isospin states (see Fig. 12.38)

(12.126)

$$\hat{I}^2|p\rangle = \frac{3}{4}|p\rangle \qquad \hat{I}^2|n\rangle = \frac{3}{4}|n\rangle$$

$$\hat{I}_3|p\rangle = \frac{1}{2}|p\rangle \qquad \hat{I}_3|n\rangle = -\frac{1}{2}|n\rangle$$

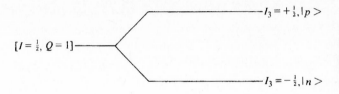

FIGURE 12.38   Nucleon with isospin $\frac{1}{2}$ and charge number 1 has two I_3 components corresponding to the eigenstates $|p\rangle$ and $|n\rangle$.

so that $|p\rangle$ and $|n\rangle$ correspond, respectively, to $I_3 = \frac{1}{2}$ and $I_3 = -\frac{1}{2}$. These states are written in analogy with the ordinary spin states α, β defined in Section 11.6. Thus, raising and lowering operators may be introduced, defined as

$$(12.127) \qquad\qquad \hat{I}_\pm = \hat{I}_1 \pm i\hat{I}_2$$

Applying these operators on the states $|p\rangle$ and $|n\rangle$ gives

$$(12.128) \qquad \begin{aligned} \hat{I}_+|p\rangle = \hat{I}_-|n\rangle = 0 \\ \hat{I}_+|n\rangle = |p\rangle, \qquad \hat{I}_-|p\rangle = |n\rangle \end{aligned}$$

   With these results at hand, paralleling the construction of the coupled states shown in Table 11.3 gives the coupled isospin states listed in Table 12.7.

   In the framework of isospin, as noted above, neutron and proton are viewed as separate states of a nucleon. As the nucleon we are considering has spin $\frac{1}{2}$, the coupled quantum states of a pair of such nucleons must be antisymmetric with respect to simultaneous interchange of particle coordinates, spins, and isospins.[1] Accordingly, the ground state of the deuteron is written

$$(12.129) \qquad\qquad |\overline{\psi}_G\rangle = |\psi_G\rangle|^3\chi\rangle|II_3\rangle$$

TABLE 12.7   Two-nucleon coupled isospin states

State	I	I_3
$\lvert pp\rangle_S = \lvert p\rangle_1 \lvert p\rangle_2$	1	1
$\lvert pn\rangle_S = \frac{1}{\sqrt{2}}[\lvert p\rangle_1\lvert n\rangle_2 + \lvert p\rangle_2\lvert n\rangle_1]$	1	0
$\lvert nn\rangle_S = \lvert n\rangle_1\lvert n\rangle_2$	1	−1
$\lvert pn\rangle_A = \frac{1}{\sqrt{2}}[\lvert p\rangle_1\lvert n\rangle_2 - \lvert p\rangle_2\lvert n\rangle_1]$	0	0

[1] I_3 is sometimes called the "charge variable" and is labeled I_ζ. For further discussion, see L. D. Landau, E. M. Lifshitz, and L. P. Pitaevskii, *Quantum Mechanics,* 3d ed., Pergamon Press, New York, 1974.

where the overbar on the left side indicates symmetrization. The component $|\psi_G\rangle$ represents the state (12.122) comprised of ^3S and ^3D states, with definite $j = 1$ and $s = 1$. The middle factor on the right side of (12.129) represents the triplet, symmetric, two-particle spin functions corresponding to spin 1 (see Table 11.3). Since these contributions are both symmetric states, the combined state is symmetric as well. It follows that $|II_3\rangle$ must be the antisymmetric state $|pn\rangle_A$.

Suppose this were not the case and that, instead, the proper isospin state were the symmetric $I = 1$ states. The Hamiltonian for our system is assumed to be rotationally invariant in isotopic spin space. Consequently, all three $I = 1$ states correspond to the same energy. This triplet includes the states $|pp\rangle_s$ or $|nn\rangle_s$ that correspond, respectively, to the diproton and dineutron which are not found in nature.

Thus, the appropriately symmetrized ground state of the deuteron (apart from the radial factor) is given by

$$(12.130) \qquad |\overline{\psi}_G\rangle = |\psi_G\rangle|^3\chi\rangle|pn\rangle_A$$

which is a common eigenstate of $\hat{J}^2, \hat{S}^2, \hat{S}_z, \hat{I}^2, \hat{I}_3$.

Summary

The preceding result (12.129) for the ground state of the deuteron was based on the following concepts and properties: (1) Parity is conserved for the strong nuclear forces; (2) the value 1 for the spin of the deuteron; (3) values of the magnetic dipole and electric quadrupole moments of the deuteron. These properties gave rise to the even-parity, unsymmetrized component $|\psi_G\rangle$, (12.121).

Notions of the nucleon, charge-independent nuclear forces, and isotopic spin were then introduced. With these concepts and the Pauli principle at hand, the isospin factor $|pn\rangle_A$ was obtained. When coupled with $|\psi_G\rangle$, this factor produced the appropriately symmetrized ground state (12.129).

It should be noted that until the notion of isotopic spin is introduced, there is no reason to incorporate the Pauli principle in the formalism, as neutron and proton are distinct particles. However, once isospin is introduced, neutron and proton become different states of the same particle, the nucleon. As this particle is a fermion, coupled states of such particles must be antisymmetric.

PROBLEMS

12.70 Show that $\langle S_z^-\rangle$ [defined following (12.118)] is zero for any of the four coupled spin states listed in Table 11.3.

12.71 Is the Pauli principle (neglecting isotopic spin) relevant to a proton-neutron system? Explain your answer.

12.72 Let the neutron and proton in a deuteron rotate about a common origin with combined orbital angular momentum **L**. Assuming circular motion, show classically that the magnetic moment due to this *orbital* motion is

$$\boldsymbol{\mu}_L = \left(\frac{e}{4M_p c}\right)\mathbf{L} = \frac{1}{2}\left(\frac{\mu_N}{\hbar}\right)\mathbf{L}$$

where μ_N is the nuclear magnetic moment. Note that the orbital neutron motion contributes zero current.

12.73 Consider an ideal experiment in which a gas of deuterons is immersed in a steady $\mathscr{B}$ field of 10^3 gauss. Radiation of frequency ν is transmitted through the sample. At what frequency might you expect to see a diminishment in the transmission?

12.74 (a) What is the value of $\hat{\boldsymbol{\sigma}}_p \cdot \hat{\boldsymbol{\sigma}}_n \xi$ for each of the four spin states ξ listed in Table 11.3?

(b) Let $f_C(r)$ and $f_S(r)$ represent central potentials. Is the interaction represented by the Hamiltonian

$$\hat{H} = f_C(r) + f_S(r)\hat{\boldsymbol{\sigma}}_p \cdot \hat{\boldsymbol{\sigma}}_n$$

a central potential? Explain your answer.

(c) What is $[\hat{H}, \hat{L}^2]$ for this Hamiltonian?

12.75 At which orientation of spins relative to interparticle radius does the interaction $\hat{H}_S$ given by (12.123) lose **r** dependence?

12.76 A two-nucleon system is in a state with zero orbital angular momentum.

(a) List the possible spin states for this system. (Each nucleon has spin $\frac{1}{2}$.)

(b) List the possible isotopic spin states for this system.

(c) Employing the preceding results, write down the appropriately symmetrized spin-isospin states for this system.

12.77 Derive an expression for the current vector, **J**, for a particle of charge q and mass m moving in a steady magnetic field with vector potential **A**. This derivation should follow (7.103) et seq., and the Hamiltonian (10.85) appropriate to a charge **q**. *Hint:* Your answer should agree with the form (7.107) with momentum replaced by $\mathbf{p} - (q\mathbf{A}/c)$.

12.78 Let the wavefunction for the particle described in Problem 12.77 be written $\psi(\mathbf{r}) = |\psi| \exp i\phi$ and assume further that the significant spatial variation of ψ is contained solely in the phase Φ. (Here we are outlining the Ginzburg–Landau[1] formalism in which ψ, called the *order parameter,* vanishes above T_c and whose magnitude below T_c measures superconducting order at **r**.) With these assumptions, show that your expression for **J**, obtained in Problem 12.77, reduces to

$$\mathbf{J} = -\left[\frac{q^2}{mc}\mathbf{A} + \frac{q\hbar}{m}\nabla\phi\right]|\psi|^2$$

---

[1] V. I. Ginzburg and L. D. Landau, *Zh. Exps. Teor. Fiz.* **20**, 1064 (1950). See also, N. W. Ashcroft and N. D. Mermin, *Solid State Physics,* Holt, Rinehart and Winston, New York, 1976, Chapter 34.

12.79   As appreciable currents in a superconductor can flow only near the surface, with the expression for **J** obtained in Problem 12.78 and the assumption that $|\psi|^2$ is essentially constant in the superconductor, we find

$$0 = \oint \mathbf{J} \cdot dl = \oint \left[\frac{q^2}{mc} \mathbf{A} + \frac{q\hbar}{m} \nabla\phi \right] \cdot dl$$

where the path of integration lies deep within the superconductor. Employing Stokes's theorem to reduce the **A** integral and the single-valuedness of ϕ to reduce the $\nabla\phi$ integral in the preceding relation, show that the magnetic flux enclosed by the ring, $\Phi = \iint \mathbf{B} \cdot d\mathbf{S}$, is quantized according to the rule $|\Phi| = n\Phi_0$, where n is an integer and $\Phi_0 = hc/q$. For Cooper pairs, one sets $q = 2e$. There results

$$\Phi_0 = hc/2e = 2.068 \times 10^{-7} \text{ gauss-cm}^2$$

which is known as the *fluxoid* or the *flux quantum*.

12.80   Consider a hydrogen atom confined to a spherical box of radius $a \gg a_0$, where a_0 is the Bohr radius. Separating the motion into center-of-mass and relative coordinates, obtain explicit expressions for wavefunctions and eigenenergies of the hydrogen atom. In what domain of the box is your solution valid?

Answer (partial)

With the boundary condition, $\varphi_{CM}(R = a) = 0$, this solution is valid in the domain $R < (a - a_0)$.

12.81   (a)   Show that all classical electrostatic equilibrium state are unstable (Earnshaw's theorem).

(b)   Does this theorem apply in quantum mechanics?

[*Hint:* For part (a), expand the potential in a Taylor series about the given equilibrium point. Then apply Laplace's equation and consider a propitious displacement away from equilibrium.]

12.82   (a)   What is the numerical value of the Fermi–Dirac function, $f(E)$, at $E = E_F$ at $T = 100$ K?

(b)   Given that the density of states of a given system is $g(E)$, write down an expression for the number of states, ΔN, this system has in the energy interval, $E = E_0$ to $10E_0$, where E_0 is the ground state of the system.

12.83   A schematic for the low-lying levels of barium ([Xe] $6s^2$) is shown in Fig. 12.39.

(a)   What transition out of the ground state of this atom is of minimum energy? Explain your answer.

(b)   If the atom is in one of the $^3P_1$ states, what related states may it fall to? If it is in the ^1D state, which states may it fall to? What transitions are possible out of the ^3D states? Explain your answers.

(c)   The decay time for the transition $^3P \rightarrow {}^1D$ is significantly shorter than that of the $^1D \rightarrow {}^3D$ transition. Incorporating this property and the preceding answers, suggest a three-level lasing scheme for ideally isolated barium atoms.

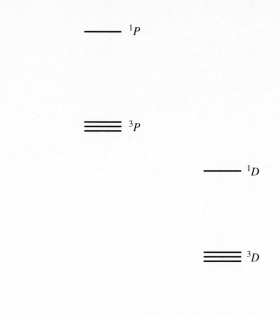

FIGURE 12.39   Schematic of the energy levels for the low-lying states of barium. (See Problem 12.83.)

12.84   The commutator relation for the operator[1] corresponding to the number of photons, N, in a cavity and that corresponding to the phase, ϕ, of these photons, is given by $[N, \phi] = -i$.

    (a)   What is the related uncertainty relation[2] for these variables?

    (b)   If $\Delta\phi = 10^{-4}$ rad, what is ΔN?

Note: With the answer to part (a) we note that ΔN may be made vanishingly small at the expense of very large $\Delta\phi$. The resulting field[3] is called "squeezed light." The field is "squeezed" in the sense that intensity fluctuations are vanishingly small.

12.85   An important test for the presence of atomic hydrogen in the universe in detection of the 21-cm radiative wavelength related to the electron-proton spin coupling in hydrogen. This radiation is emitted in the decay of the excited antiparallel electron-proton magnetic-moments state to the aligned electron-proton magnetic-moments ground state. The Hamiltonian for this interaction is given by

---

[1] The number operator for photons is discussed in Problem 13.38.

[2] The uncertainty relation derived here is relevant to a system of many quanta. For further discussion, see W. Heitler, *Quantum Theory of Radiation,* Oxford University Press, New York, 1944, page 68; W. H. Louisell, *Phys. Lett.* **7,** 60 (1963).

[3] For further discussion, see M. C. Teich and B. E. A. Saleh, *Physics Today,* June 1993, page 26.

$$H = -A\boldsymbol{\mu}_e \cdot \boldsymbol{\mu}_p$$

(Values of magnetic moments in the preceding expression are given in Section 11.9.)

(a) Given that the said transition emits 21-cm radiation, obtain a value of the constant A. State units of this constant. (The lifetime of this excited state is $\tau \simeq 10^{15}$ s $\simeq 10^8$ yr. The frequent detection[1] of this transition on earth indicates an abundant presence of atomic hydrogen in the universe.)

(b) What is the ratio, κ, of the energy of this 21-cm photon to that of a Lyman α photon? (The resulting small value of κ indicates that the energy of electron-nuclear transitions is far less than that of atomic transitions. Corresponding levels are labeled "hyperfine.")

12.86 A rare gas of lithium +2 ions ($M = 6.9$ amu; $Z = 3$; electronic configuration of $[1s^2 2s^1]$) is to be used to filter out radiation at a frequency corresponding to the transition from the ground to the first excited state of an ion.

(a) What is the ground-state energy of this ion? (Answer in eV.)

(b) What is the frequency corresponding to the above-stated transition? (Answer in Hz.) To what frequency domain does this radiation correspond?

[1] First detected by E. M. Purcell and H. I. Ewen, *Nature* **168**, 356 (1951).

13

PERTURBATION THEORY

In this chapter perturbation techniques are described which serve to generate approximate solutions to the Schrödinger equation. Such solutions appear in the form of an expansion away from known, unperturbed values. A special procedure is developed for systems with degenerate eigenenergies. Application is made to problems in atomic physics and the problem of an electron in a periodic potential, encountered previously in Chapter 8. Harmonic perturbation theory is applied in a rederivation of the Planck radiation formula and the theory of the laser. The chapter continues with a description of the interaction of a radiation field with an atom. Here we encounter the notion of oscillator strengths and the important Thomas–Reiche–Kuhn sum rule. The concluding section addresses the Hartree–Fock model for obtaining approximate atomic wavefunctions and eigenenergies.

13.1 TIME-INDEPENDENT, NONDEGENERATE PERTURBATION THEORY

Approximate methods of solution to an assortment of problems in quantum mechanics were described previously in Chapter 7 (WKB analysis), Chapter 8 (LCAO approximation), Chapter 10 (Thomas–Fermi model), and Chapter 11 (interaction picture).

Our present concern lies in refinement of the approximation method described in Chapter 12, used in calculating both the ground-state wavefunction of the helium atom (Problem 12.28) and the fine-structure spectrum of hydrogen (12.13). In both these problems the Hamiltonian encountered was of the form

$$\hat{H} = \hat{H}_0 + \hat{H}'$$

This breakup of a Hamiltonian into a part $\hat{H}_0$, whose eigenfunctions are known, and an addition term $\hat{H}'$, which is in some sense small compared to $\hat{H}_0$, is typical of many practical problems encountered in quantum mechanics. The theory that seeks approximate eigenstates of the *total Hamiltonian $\hat{H}$* is called *perturbation theory*. In the expression above, the Hamiltonian, $\hat{H}_0$, is called the *unperturbed Hamiltonian*, while $\hat{H}'$ is called the *perturbation Hamiltonian*. Some typical perturbation problems are listed in Table 13.1.

TABLE 13.1 Examples of perturbation Hamiltonians

Name	Description	Hamiltonian
L-S coupling	Coupling between orbital and spin angular momentum in a one-electron atom	$\hat{H} = \hat{H}_0 + f(r)\hat{\mathbf{L}} \cdot \hat{\mathbf{S}}$ $\hat{H}' = f(r)\hat{\mathbf{L}} \cdot \hat{\mathbf{S}}$ $\hat{H}_0 = \hat{p}^2/2m - e^2 Z/r$
Stark effect	One-electron atom in a constant, uniform electric field $\mathscr{E} = \mathbf{e}_z \mathscr{E}_0$	$\hat{H} = \hat{H}_0 + e\mathscr{E}_0 z$ $H' = e\mathscr{E}_0 z$ $\hat{H}_0 = (\hat{p}^2/2m) - e^2 Z/r$
Zeeman effect	One electron atom in a constant, uniform magnetic field $\mathscr{B}$	$\hat{H} = \hat{H}_0 + (e/2mc)\hat{\mathbf{J}} \cdot \mathscr{B}$ $\hat{H}' = (e/2mc)\hat{\mathbf{J}} \cdot \mathscr{B}$ $\hat{H}_0 = (\hat{p}^2/2m) - e^2 Z/r$
Anharmonic oscillator	Spring with nonlinear restoring force	$\hat{H} = \hat{H}_0 + K'x^4$ $\hat{H}' = K'x^4$ $\hat{H}_0 = (\hat{p}_x^2/2m) + \frac{1}{2}Kx^2$
Nearly free electron model	Electron in a periodic lattice	$\hat{H} = \hat{H}_0 + V(x)$ $V(x) = \sum_n V_n \exp\left[i(2\pi n x/a)\right]$ $\hat{H}_0 = \hat{p}_x^2/2m$

The perturbation analysis we will develop in this chapter divides into three categories: (1) time-independent, nondegenerate; (2) time-independent, degenerate; (3) time-dependent. In the last category one investigates the time development of a system in a given state due to a perturbation on the system which is turned on at a given instant of time.

Smallness of the Perturbation

Perturbation theory begins with the assumption that the perturbation Hamiltonian, $\hat{H}'$, is in some sense small compared to the unperturbed Hamiltonian, $\hat{H}_0$. The criterion that establishes the smallness of $\hat{H}'$ compared to $\hat{H}_0$ will emerge in the course of the analysis. Another underlying assumption in perturbation theory is that the eigenstates and eigenenergies of the total Hamiltonian $\hat{H}$, do not differ appreciably from those of the unperturbed Hamiltonian, $\hat{H}_0$. That is, suppose that $\{\varphi_n\}$ and $\{E_n\}$ are, respectively, the eigenstates and eigenenergies of the total Hamiltonian $\hat{H}$,

$$\hat{H}\varphi_n = (\hat{H}_0 + \hat{H}')\varphi_n = E_n\varphi_n$$

while $\{\varphi_n^{(0)}\}$ and $\{E_n^{(0)}\}$ are, respectively, the eigenstates and eigenenergies of the unperturbed Hamiltonian

$$\hat{H}_0\varphi_n^{(0)} = E_n^{(0)}\varphi_n^{(0)}$$

Then it is always possible to write

$$\varphi_n = \varphi_n^{(0)} + \Delta\varphi_n$$
$$E_n = E_n^{(0)} + \Delta E_n$$

where, owing to the smallness of $\hat{H}'$, $\Delta\varphi_n$ is a small correction to $\varphi_n^{(0)}$ and ΔE_n is a small correction to $E_n^{(0)}$.

To keep the smallness of $\hat{H}'$ in mind, we rewrite it as $\lambda\hat{H}'$, where λ is an infinitesimal parameter and is introduced for "bookkeeping" purposes only. The equation to which we seek a solution is of the form

(13.1)
$$(\hat{H}_0 + \lambda\hat{H}')\varphi_n = E_n\varphi_n$$

The Perturbation Expansion

The eigenstates and eigenenergies of $\hat{H}_0$ are assumed known. Since $\varphi_n \to \varphi_n^{(0)}$ as $\lambda \to 0$, it is consistent to seek solution to (13.1) in the form of a series with $\varphi_n^{(0)}$ entering as the leading term. In a similar manner, E_n is expanded, with $E_n^{(0)}$ entering as the leading term.

(13.2)
$$\varphi_n = \varphi_n^{(0)} + \lambda\varphi_n^{(1)} + \lambda^2\varphi_n^{(2)} + \cdots$$
$$E_n = E_n^{(0)} + \lambda E_n^{(1)} + \lambda^2 E_n^{(2)} + \cdots$$

Substituting these expansions into (13.1) and arranging terms according to powers in λ gives

$$(13.3) \quad [\hat{H}_0 \varphi_n^{(0)} - E_n^{(0)} \varphi_n^{0}] + \lambda[\hat{H}_0 \varphi_n^{(1)} + \hat{H}' \varphi_n^{(0)} - E_n^{(0)} \varphi_n^{(1)} - E_n^{(1)} \varphi_n^{(0)}]$$
$$+ \lambda^2[\hat{H}_0 \varphi_n^{(2)} + \hat{H}' \varphi_n^{(1)} - E_n^{(0)} \varphi_n^{(2)} - E_n^{(1)} \varphi_n^{(1)} - E_n^{(2)} \varphi_n^{(0)}]$$
$$+ \cdots = 0$$

This equation is of the form

$$F^{(0)} + \lambda F^{(1)} + \lambda^2 F^{(2)} + \lambda^3 F^{(3)} + \cdots = 0$$

If this equation is to be true for *arbitrarily* small values of λ, then

$$F^{(0)} = F^{(1)} = F^{(2)} = \cdots = 0$$

In this manner (13.3) gives the coupled set of equations

$$(13.4) \quad \begin{aligned} &\text{(a)} \ \hat{H}_0 \varphi_n^{(0)} = E_n^{(0)} \varphi_n^{(0)} \\ &\text{(b)} \ (\hat{H}_0 - E_n^{(0)})\varphi_n^{(1)} = (E_n^{(1)} - \hat{H}')\varphi_n^{(0)} \\ &\text{(c)} \ (\hat{H}_0 - E_n^{(0)})\varphi_n^{(2)} = (E_n^{(1)} - \hat{H}')\varphi_n^{(1)} + E_n^{(2)}\varphi_n^{(0)} \\ &\text{(d)} \ (\hat{H}_0 - E_n^{(0)})\varphi_n^{(3)} = (E_n^{(1)} - \hat{H}')\varphi_n^{(2)} + E_n^{(2)}\varphi_n^{(1)} + E_n^{(3)}\varphi_n^{(0)} \\ &\qquad\qquad\qquad\qquad \vdots \end{aligned}$$

In the lowest approximation, (13.4a) returns the information that $\{\varphi_n^{(0)}\}$ and $\{E_n^{(0)}\}$ are, respectively, the eigenstates and eigenenergies of $\hat{H}_0$. The second (as well as all of the higher-order equations) has the following interesting property. The left-hand side of this equation remains the same under the replacement

$$\varphi_n^{(1)} \rightarrow \varphi_n^{(1)} + a\varphi_n^{(0)}$$

where a is an arbitrary constant. Suppose that one solves (13.4b) for $\varphi_n^{(1)}$ and $E_n^{(1)}$. Then $\varphi_n^{(1)} + a\varphi_n^{(0)}$; $E_n^{(1)}$ is also a solution. An extra constraint is needed to remove this arbitrary quality of solution. This constraint may be taken as follows.[1] We assume that all corrections to $\varphi_n^{(0)}$ in (13.2) are normal to $\varphi_n^{(0)}$.

$$(13.5) \qquad \langle \varphi_n^{(s)} | \varphi_n^{(0)} \rangle = 0 \qquad \text{(for } s > 0 \text{ and all } n)$$

---

[1] Another popular constraint is the construct $\varphi_n^{(s)}$ so that it is normalized. Both choices of constraint yield the same corrections to the energy, $\{E_n^{(s)}\}$, while the wavefunctions that emerge differ by at most a phase factor (see Section 4.1).

In Hilbert space this relation indicates that $\Delta \varphi_n$ is normal to $\varphi_n^{(0)}$. This condition will aid us in the construction of $\varphi_n^{(s)}$.

Returning to (13.4b) we note that $\hat{H}_0$ operates on $\varphi_n^{(1)}$ in this equation, which suggests that the solution may be obtainable through expansion of $\varphi_n^{(1)}$ in a superposition of the eigenstates of $\hat{H}_0$.

(13.6)
$$|\varphi_n^{(1)}\rangle = \sum_i c_{ni} |\varphi_i^{(0)}\rangle$$

If this expansion is substituted into (13.4b), there results

$$(\hat{H}_0 - E_n^{(0)}) \sum_i c_{ni} |\varphi_n^{(0)}\rangle = (E_n^{(1)} - \hat{H}') |\varphi_n^{(0)}\rangle$$

Multiplying from the left with $\langle \varphi_j^{(0)}|$ gives

(13.7)
$$(E_j^{(0)} - E_n^{(0)})c_{nj} + H'_{jn} = E_n^{(1)}\delta_{jn}$$

where H'_{jn} are the matrix elements of $\hat{H}'$ in the $\{\varphi_n^{(0)}\}$ representation

$$H'_{jn} \equiv \langle \varphi_j^{(0)} | \hat{H}' | \varphi_n^{(0)} \rangle$$

First-Order Corrections

With $j \neq n$, (13.7) gives the coefficients, $\{c_{nj}\}$, which when substituted into (13.6) gives the first-order correction to φ_n.

(13.8)
$$c_{ni} = \frac{H'_{in}}{E_n^{(0)} - E_i^{(0)}}$$

$$\varphi_n^{(1)} = \sum_{i \neq n} \frac{H'_{in}}{E_n^{(0)} - E_i^{(0)}} \varphi_i^{(0)} + c_{nn}\varphi_n^{(0)}$$

Here one assumes that all corrections $\{\varphi_n^{(s)}\}$ lie in a Hilbert space that is spanned by the unperturbed wavefunctions, $\{\varphi_n^{(0)}\}$.

The coefficient c_{nn} is obtained from (13.5), which yields

$$c_{nn} = 0$$

With $j = n$, (13.7) gives the first-order corrections to the energy E_n.

(13.9)
$$E_n^{(1)} = H'_{nn}$$

These are the diagonal elements of $\hat{H}'$. Substituting (13.8) and (13.9) into (13.2) and setting $\lambda = 1$ gives

$$\varphi_n = \varphi_n^{(0)} + \sum_{i \neq n} \frac{H'_{in}}{E_n^{(0)} - E_i^{(0)}} \varphi_i^{(0)}$$

(13.10)

$$E_n = E_n^{(0)} + H'_{nn}$$

The first of these equations tells us that in order for the expansion (13.2) to make sense, the coefficients of expansion should be less than 1.

$$|H'_{in}| \ll |E_n^{(0)} - E_i^{(0)}|$$

The matrix elements of $\hat{H}'$ should be small compared to the difference between the corresponding unperturbed energy levels. In similar manner, the second equation in (13.10) reveals that

$$|H'_{nn}| \ll E_n^{(0)}$$

The diagonal elements of the perturbation Hamiltonian should be small compared to the corresponding unperturbed energy level.

Second-Order Corrections

To find the second-order correction to φ_n and E_n, we must solve (13.4c). Again we note that $\hat{H}_0$ operates on $\varphi_n^{(2)}$, and it is again advantageous to expand $\varphi_n^{(2)}$ in the eigenstates of $\hat{H}_0$.

(13.11)

$$\varphi_n^{(2)} = \sum_i d_{ni} \varphi_i^{(0)}$$

Substitution into (13.4c) gives

$$\sum_i E_i^{(0)} d_{ni} |\varphi_i^{(0)}\rangle + \hat{H}' |\varphi_n^{(1)}\rangle = E_n^{(0)} \sum_i d_{ni} |\varphi_i^{(0)}\rangle + E_n^{(1)} |\varphi_n^{(1)}\rangle$$
$$+ E_n^{(2)} |\varphi_n^{(0)}\rangle$$

Multiplying from the left with $\langle \varphi_j^{(0)} |$ gives

(13.12) $(E_j^{(0)} - E_n^{(0)})d_{nj} + \langle \varphi_j^{(0)} | H' | \varphi_n^{(1)} \rangle = E_n^{(2)}\delta_{jn} + E_n^{(1)}\langle \varphi_j^{(0)} | \varphi_n^{(1)} \rangle$

With $j = n$, this equation gives

$$E_n^{(2)} = \langle \varphi_n^{(0)} | \hat{H}' | \varphi_n^{(1)} \rangle$$

$$= \sum_{i \neq n} \langle \varphi_n^{(0)} | \frac{\hat{H}' H_{in}'}{E_n^{(0)} - E_i^{(0)}} | \varphi_i^{(0)} \rangle$$

$$= \sum_{i \neq n} \frac{H_{ni}' H_{in}'}{E_n^{(0)} - E_i^{(0)}}$$

Owing to the Hermiticity of $\hat{H}'$, this equation may be rewritten

(13.13) $$E_n^{(2)} = \sum_{i \neq n} \frac{|H_{ni}'|^2}{E_n^{(0)} - E_i^{(0)}}$$

Note that in obtaining this result we have used the result that $c_{nn} = 0$. Substituting this expression for $E_n^{(2)}$ into (13.2) together with the expression for $E_n^{(1)}$ given by (13.9) gives the following second-order expression for E_n:

(13.14) $$\boxed{E_n = E_n^{(0)} + H_{nn}' + \sum_{i \neq n} \frac{|H_{ni}'|^2}{E_n^{(0)} - E_i^{(0)}}}$$

To calculate the second-order corrections to the wavefunction φ_n, we must obtain the coefficients d_{ni} in (13.11). These are directly obtained from (13.12). With $n \neq j$ this equation gives

$$(E_n^{(0)} - E_j^{(0)})d_{nj} = \langle \varphi_j^{(0)} | \hat{H}' \sum_{k \neq n} \frac{H_{kn}'}{E_n^{(0)} - E_k^{(0)}} | \varphi_k^{(0)} \rangle$$

$$- H_{nn}' \langle \varphi_j^{(0)} | \sum_{k \neq n} \frac{H_{kn}'}{E_n^{(0)} - E_k^{(0)}} | \varphi_k^{(0)} \rangle$$

In the second sum, only the $k = j$ term survives the $\langle \varphi_j^{(0)} | \varphi_k^{(0)} \rangle$ inner product. All terms in the first term remain. There results

$$d_{nj} = \frac{1}{E_n^{(0)} - E_j^{(0)}} \left(\sum_{k \neq n} \frac{H_{jk}' H_{kn}'}{E_n^{(0)} - E_k^{(0)}} \right) - \frac{H_{nn}' H_{jn}'}{(E_n^{(0)} - E_j^{(0)})^2}$$

Again, with (13.5), one finds that

$$d_{nn} = 0$$

In this manner one obtains the following expression for φ_n, good to terms of second order in $\hat{H}'$.

(13.15)
$$\varphi_n = \varphi_n^{(0)} + \sum_{i \neq n} \left[\frac{H'_{in}}{E_n^{(0)} - E_i^{(0)}} - \frac{H'_{nn} H'_{in}}{(E_n^{(0)} - E_i^{(0)})^2} \right.$$
$$\left. + \sum_{k \neq n} \frac{H'_{ik} H'_{kn}}{(E_n^{(0)} - E_i^{(0)})(E_n^{(0)} - E_k^{(0)})} \right] \varphi_i^{(0)}$$

PROBLEMS

13.1 Calculate the first-order correction to $E_3^{(0)}$ for a particle in a one-dimensional box with walls at $x = 0$ and $x = a$ due to the following perturbations:
  (a) $H' = 10^{-3} E_1 x/a$
  (b) $H' = 10^{-3} E_1 (x/a)^2$
  (c) $H' = 10^{-3} E_1 \sin(x/a)$

13.2 What is the eigenfunction φ_n for the same configuration as in Problem 13.1, to terms of second order, for the constant perturbation

$$H' = 10^{-3} E_1?$$

13.3 Calculate the eigenenergies of the anharmonic oscillator whose Hamiltonian is listed in Table 13.1, to first order in $\hat{H}'$.

Answer

In terms of raising and lowering operators, $(a^\dagger, a)$, the perturbation Hamiltonian appears as

$$\hat{H}' = Kx^4 = \frac{K}{4\beta^4}(\hat{a} + \hat{a}^\dagger)^4$$

The corrections to $E_n^{(0)}$ which we seek are given by

$$E_n^{(1)} = H'_{nn} = \langle n | \hat{H}' | n \rangle$$

The only terms in the expansion of $(\hat{a} + \hat{a}^\dagger)^4$ which give nonzero contributions are those which maintain the eigenvector $|n\rangle$. All other terms vanish because of orthogonality with $\langle n |$. Of the 16 terms in the expansion of $(\hat{a} + \hat{a}^\dagger)^4$, only six survive this orthogonality condition. The energy $E_n^{(1)}$ may be determined by a graphical analysis, according to which a diagram is associated with each integral that contributes to $\langle n|\hat{H}'|n\rangle$. The eigenvector $|n\rangle$ is represented by a dot drawn at the right of the diagram. Another dot drawn at the left of the $|n\rangle$ dot, but on the same horizontal, represents the eigenbra $\langle n|$. The creation operator, $\hat{a}^\dagger$, is represented by a diagonal arrow from the right and inclined upward at $\pi/4$, while the annihilation operator, $\hat{a}$, is an arrow from the right at $-\pi/4$. Thus the diagram related to $\langle n|\hat{a}\hat{a}^\dagger|n\rangle$ is

$$\langle n|\hat{a}\hat{a}^\dagger|n\rangle$$

The diagram that represents the fourth-order term, $\langle n|\hat{a}^\dagger\hat{a}\hat{a}^\dagger\hat{a}|n\rangle$, is

$$\langle n|\hat{a}^\dagger\hat{a}\hat{a}^\dagger\hat{a}|n\rangle$$

while the second-order term, $\langle n|\hat{a}^2|n\rangle$, is represented by

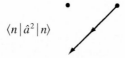

$$\langle n|\hat{a}^2|n\rangle$$

Any sequence of arrows that do not join the two horizontal dots represents a zero contribution. Continuing in this manner, we find that, in all, there are 16 fourth-order diagrams. Of these, only six gave nonzero contributions. These six diagrams are

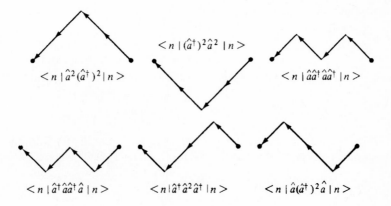

$$\langle n|\hat{a}^2(\hat{a}^\dagger)^2|n\rangle \qquad \langle n|(\hat{a}^\dagger)^2\hat{a}^2|n\rangle \qquad \langle n|\hat{a}\hat{a}^\dagger\hat{a}\hat{a}^\dagger|n\rangle$$

$$\langle n|\hat{a}^\dagger\hat{a}\hat{a}^\dagger\hat{a}|n\rangle \qquad \langle n|\hat{a}^\dagger\hat{a}^2\hat{a}^\dagger|n\rangle \qquad \langle n|\hat{a}(\hat{a}^\dagger)^2\hat{a}|n\rangle$$

Summing these contributions gives the desired result:

$$E_n^{(1)} = \frac{3K'}{4\beta^4}[2n(n+1)+1]$$

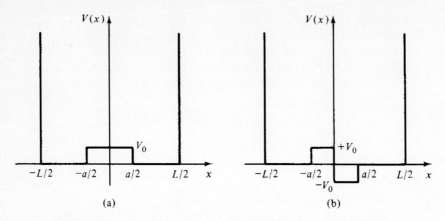

FIGURE 13.1 Potential configurations for Problems 13.4, 13.5, and 13.6.

13.4 Consider a particle of mass m in a potential well shown in Fig. 13.1a. Suppose that the bump at the bottom of the well can be considered a small perturbation.

(a) Calculate the corrected second eigenenergy and eigenfunction to first order in the perturbation. [Recall (6.100).]

(b) What dimensionless ratio must be small compared to 1 in order for your approximation to be valid?

(c) Is the parity of the unperturbed wavefunction maintained by this perturbation? Explain your answer.

13.5 (a) Consider the perturbation bump shown in Fig. 13.1b. What are the first-order corrected eigenenergies of a particle of mass m confined to this well?

(b) Calculate the eigenenergies of this configuration, in the domain $E \gg V_0$, using the WKB formula (7.191) and compare your results with those obtained in part (a).

13.6 Again consider the potential shown in Fig. 13.1a. What is the unperturbed ground state for:

(a) Two identical bosons of mass m confined to the box.

(a) Two identical fermions of mass m confined to the box.

(c) What are the unperturbed ground-state energies E_S and E_A for these two cases?

(d) Use first-order perturbation theory to obtain the new ground-state energies for these two cases.

Answers (partial)

(a) $\quad \varphi_S(x_1, x_2) = \dfrac{2}{L} \cos\left(\dfrac{\pi x_1}{L}\right) \cos\left(\dfrac{\pi x_2}{L}\right)$

(b) $\quad \varphi_A(x_1, x_2) = \dfrac{\sqrt{2}}{L} \left[\cos\left(\dfrac{\pi x_1}{L}\right) \sin\left(\dfrac{2\pi x_2}{L}\right) - \cos\left(\dfrac{\pi x_2}{L}\right) \sin\left(\dfrac{2\pi x_1}{L}\right) \right]$

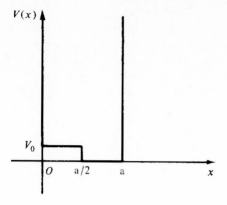

FIGURE 13.2 **Potential configuration for Problem 13.7.**

13.7 A particle of mass m is in an asymmetrical one-dimensional box, depicted in Fig. 13.2.
(a) Use first-order perturbation theory to calculate the eigenenergies of the particle.
(b) What are the first-order corrected wavefunctions?
(c) If the particle is an electron, how do the frequencies emitted by the perturbed systems compare with those of the unperturbed system?
(d) What smallness assumption is appropriate to V_0?

Answers

(a) $\quad E_n = E_n^{(0)} + E_n^{(1)} = n^2 E_1 + \dfrac{V_0}{2}$

(b) $\quad \varphi_n = \sqrt{\dfrac{2}{a}} \sin\left(\dfrac{n\pi x}{a}\right)\left[1 + \dfrac{V_0}{E_1} \sum_{l \neq n} \dfrac{l^2 n^2}{n^2 - l^2} \Lambda_{ln}\right]$

$\qquad \Lambda_{ln} \equiv \dfrac{1}{\pi}\left[\dfrac{\sin(n-l)\pi/2}{n-l} - \dfrac{\sin(n+l)\pi/2}{n+l}\right]$

(c) They are the same.
(d) $V_0 \ll E_1$

13.8 Show that the matrix elements of the perturbation Hamiltonian $\hat{H}'$ obey the equality

$$\sum_m |H'_{nm}|^2 = (H'^2)_{nn}$$

Answer

$$\sum_m |H'_{nm}|^2 = \sum_m \langle n|\hat{H}'|m\rangle\langle m|\hat{H}'|n\rangle$$

$$= \langle n|\hat{H}'\hat{I}\hat{H}'|n\rangle = \langle n|\hat{H}'^2|n\rangle = (H'^2)_{nn}$$

Here we have recalled the relation

$$\hat{I} = \sum_m |m\rangle\langle m|$$

13.9 A hydrogen atom in its ground state is in a constant, uniform electric field that points in the z direction. The electric field polarizes the atom. Show that there is no change in the ground-state energy of the atom to terms of first order in the electric field. The interaction energy is

$$H' = e\mathcal{E}z = e\mathcal{E}r\cos\theta$$

13.10 The conditions are those of Problem 13.9. In calculating second-order corrections to the ground state, one must know the values of the matrix elements $\langle 100|H'|nlm\rangle$, where $|nlm\rangle$ denotes a hydrogen eigenstate. Show that these matrix elements vanish if $l \neq 1$.

13.11 Consider again, as in Problem 13.9, that a hydrogen atom is in a constant, uniform electric field $\mathcal{E}$ that points in the z direction. If α is the *polarizability* of hydrogen, then the change in the ground-state energy is $\frac{1}{2}\alpha\mathcal{E}^2$, so we may write

$$\frac{1}{2}\alpha\mathcal{E}^2 = \sum_{n>1} \frac{|H'_{1n}|^2}{E_n^{(0)} - E_1^{(0)}}, \qquad H'_{1n} \equiv \langle 100|H'|n10\rangle$$

(a) Use the result of Problem 13.8 to show that

$$\frac{1}{2}\alpha\mathcal{E}^2 < \frac{(H'^2)_{11}}{E_2^{(0)} - E_1^{(0)}}$$

(b) What maximum value for the polarizability of hydrogen does this inequality imply? How does this value compare with the correct value of α?

Answers

(a) $$[E_n^{(0)} - E_1^{(0)}]_{min} = E_2^{(0)} - E_1^{(0)}$$

so

$$\frac{1}{2}\alpha\mathcal{E}^2 < \sum_{n>1} \frac{|H'_{1n}|^2}{E_2^{(0)} - E_1^{(0)}} = \frac{1}{E_2^{(0)} - E_1^{(0)}} \sum_{n>1} |H'_{1n}|^2$$

Using the results of Problem 13.8, we obtain

$$\sum_{n=1}^{\infty} |H'_{1n}|^2 = |H'_{11}|^2 + \sum_{n>1}^{\infty} |H'_{1n}|^2 = (H'^2)_{11}$$

so that

$$\sum_{n>1} |H'_{1n}|^2 = (H'^2)_{11} - |H'_{11}|^2 < (H'^2)_{11}$$

hence

$$\frac{1}{2}\alpha\mathscr{E}^2 < \frac{1}{E_2^{(0)} - E_1^{(0)}} \sum_{n>1} |H'_{1n}|^2 < \frac{(H'^2)_{11}}{E_2^{(0)} - E_1^{(0)}}$$

(b)

$$\frac{1}{2}\alpha\mathscr{E}^2_{max} = \frac{4}{3\mathbb{R}}(H'^2)_{11}$$

$$= \frac{4}{3\mathbb{R}}e^2\mathscr{E}^2 \iiint |\varphi^*_{100}|^2 r^4 \cos^2\theta \, d\cos\theta \, d\phi \, dr$$

$$= \frac{8}{3}a_0^3\mathscr{E}^2$$

which gives

$$\alpha_{max} = \frac{16}{3}a_0^3$$

The more correct value is $\alpha = 9a_0^3/2$.

13.2   TIME-INDEPENDENT, DEGENERATE PERTURBATION THEORY

Again we consider a system whose Hamiltonian has the form

$$\hat{H} = \hat{H}_0 + \hat{H}'$$

where $\hat{H}'$ is a small perturbation about the unperturbed Hamiltonian, $\hat{H}_0$. In the present case, however, $\hat{H}_0$ has degenerate eigenstates. We have found previously (see Section 8.5) that degeneracy in quantum mechanics stems from symmetries inherent to the system at hand. Any distortion of such symmetry should therefore tend to remove the related degeneracy.

Suppose, for example, that the ground state of $\hat{H}_0$ is q-fold-degenerate. If the symmetry producing this degeneracy is destroyed by the perturbation $\hat{H}'$, the ground state $E_n^{(0)}$ separates into q distinct levels (Fig. 13.3). The primary aim of degenerate perturbation theory is to calculate these new energies. Suppose that we proceed as in the nondegenerate case described in the previous section and expand the first-order wavefunctions of $\hat{H}$ in the eigenstates of $\hat{H}_0$ (13.6).

$$\varphi_n^{(1)} = \sum_i c_{ni}\varphi_i^{(0)}$$

The formula that emerges for the coefficients $\{c_{ni}\}$ is given by (13.8).

$$c_{ni} = \frac{H'_{in}}{E_n^{(0)} - E_i^{(0)}}$$

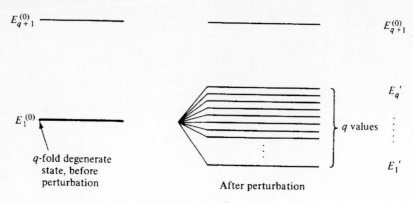

$E_{q+1}^{(0)}$

$E_1^{(0)}$

q-fold degenerate
state, before
perturbation

After perturbation

$E_{q+1}^{(0)}$

E_q'

q values

E_1'

FIGURE 13.3   Perturbation causes a removal of degeneracy.

If $E_1^{(0)}$ is q-fold-degenerate, then

$$E_1^{(0)} = E_2^{(0)} = \cdots = E_q^{(0)}$$

and c_{ni} is infinite for $n, i \leq q$. This situation is remedied by constructing a new set of basis functions from the set $\{\varphi_n^{(0)}\}$ which diagonalize the submatrix, H'_{in} (for $n, i \leq q$). With the off-diagonal elements of H'_{in} vanishing, the corresponding singular c_{in} coefficients also vanish and we may proceed as in the nondegenerate case.

Diagonalization of the Submatrix

Thus the primary aim in degenerate perturbation theory is to diagonalize this submatrix of H'_{in}. As it turns out, the diagonal elements so constructed are the incremental energies, which when added to $E_1^{(0)}$ separate the q energies contained in the ground state.

Let the q functions that diagonalize H'_{in} ($i, n \leq q$) be labeled $\bar{\varphi}_n$.

(13.16)
$$\bar{\varphi}_n = \sum_{i=1}^{q} a_{ni}\varphi_i^{(0)}$$

These linear combinations of the degenerate eigenstates $\{\varphi_i^{(0)}\}$ diagonalize H'_{in}, so

(13.17)
$$\langle \bar{\varphi}_n | \hat{H}' | \bar{\varphi}_p \rangle = H'_{np}\delta_{np} \qquad (n, p \leq q)$$

These functions, when joined with the complementary set of non-degenerate states, $\{\varphi_i^{(0)}, i > q\}$, give the basis[1]

(13.18)
$$\mathcal{B} = \{\bar{\varphi}_1, \bar{\varphi}_2, \ldots, \bar{\varphi}_q, \varphi_{q+1}^{(0)}, \varphi_{q+2}^{(0)}, \ldots\}$$

---

[1] The sequence (13.16) spans the same subspace of Hilbert space spanned by the degenerate states $\{\varphi_n^{(0)}\}$, $n \leq q$. Thus the basis (13.18) spans the same Hilbert space spanned by the basis $\{\varphi_i^{(0)}\}$, $i \geq 1$.

The matrix of $\hat{H}'$ calculated in this basis appears as

$$
(13.19) \qquad \hat{H}' = \begin{pmatrix} H'_{11} & & & \vdots & H'_{1,q+1} & \cdots \\ & H'_{22} & 0 & \vdots & & \\ 0 & & \ddots & \vdots & & \\ & & & H'_{qq} & \vdots & \\ \hline H'_{q+1,1} & & & & & \\ \vdots & & & & & \\ \vdots & & & & & \end{pmatrix}
$$

First-Order Energies

We will now show that the diagonal elements of the $q \times q$ submatrix of $\hat{H}'$ are the first-order energy corrections E_n' to $E_n^{(0)}$, $n \leq q$. That is,

$$
(13.20) \qquad E_n' = \langle \overline{\varphi}_n | \hat{H}' | \overline{\varphi}_n \rangle = H'_{nn} \qquad (n \leq q)
$$

If these diagonal elements are mutually distinct, then the q-fold degeneracy of $\hat{H}_0$ is removed by the perturbation $\hat{H}'$. To establish the equality (13.20), we proceed as follows.

The Schrödinger equation for the total Hamiltonian appears as

$$
\hat{H}\varphi_n = (\hat{H}_0 + \hat{H}')\varphi_n = E_n\varphi_n
$$

The ground state of $\hat{H}_0$ is a q-fold degenerate. Substituting

$$
(13.21) \qquad \left. \begin{array}{l} \varphi_n = \overline{\varphi}_n \\ E_n = E_n^{(0)} + E_n' \end{array} \right\} (n \leq q)
$$

into the Schrödinger equation gives

$$
(13.22) \qquad \hat{H}'\overline{\varphi}_n = E_n'\overline{\varphi}_n \qquad (n \leq q)
$$

Here we have recalled that $\overline{\varphi}_n$, being a linear combination of degenerate states, is itself a degenerate state (corresponding to the eigenvalue $E_1^{(0)}$). With the elements of $\{\overline{\varphi}_n\}$ (as well as those of $\{\varphi_n^{(0)}\}$) taken to comprise an orthogonal sequence,[1] one is able to identify (13.17) as being the matrix counterpart of the operator equation (13.22). This is so, provided that we set

$$
E_n' = H'_{nn} \qquad (n \leq q)
$$

which again is the relation (13.20). This equality establishes the fact that the diagonal elements of the submatrix H'_{np} are the first-order corrections to the total Hamiltonian $\hat{H}$ (for $n \leq q$).

---

[1] One may always construct an orthogonal set of functions from a given sequence of degenerate functions through the so-called Schmidt orthogonalization procedure. See Problem 13.53.

Let us now construct the new basis functions $\{\overline{\varphi}_n\}$ which diagonalize the said submatrix of $\hat{H}'$. These are given in terms of the a_{ni} coefficients in (13.16), which make $\{\overline{\varphi}_n\}$ obey the eigenvalue equation (13.22). Substituting the former into the latter gives

$$\hat{H}' \sum_{i=1}^{q} a_{ni} |\varphi_i^{(0)}\rangle = E_n' \sum_{i=1}^{q} a_{ni} |\varphi_i^{(0)}\rangle$$

Multiplying from the left with $\langle \varphi_p^{(0)} |$ gives

$$\sum_i a_{ni} H'_{pi} = E_n' \sum_i a_{ni} \delta_{pi} = E_n' a_{np} \qquad \text{(for fixed } n, p \leq q\text{)}$$

This equation may be rewritten as

(13.23)
$$\sum_{i=1}^{q} (H'_{pi} - E_n' \delta_{pi}) a_{ni} = 0 \qquad (n, p \leq q)$$

The coefficients $\{a_{ni}\}$ for a fixed value of n comprise the column vector representation of $\overline{\varphi}_n$ in the subbasis $\{\varphi_l^{(0)}, l \leq q\}$. Similarly,

$$H'_{pi} = \langle \varphi_p^{(0)} | \hat{H}' | \varphi_i^{(0)} \rangle$$

are the matrix elements of $\hat{H}'$ in this same basis.

The Secular Equation

For each value of n and p, (13.23) is one equation for E_n' and the q components $\{a_{ni}\}$. There are q such equations corresponding to the q values of p. For $n = 1$, for example, these equations appear as

(13.24)
$$\begin{pmatrix} H'_{11} - E_1' & H'_{12} & H'_{13} & \cdots & H'_{1q} \\ H'_{21} & H'_{22} - E_1' & H'_{23} & \cdots & H'_{2q} \\ \vdots & \vdots & \vdots & \vdots & \vdots \\ H'_{q1} & & & & \end{pmatrix} \begin{pmatrix} a_{11} \\ a_{12} \\ \vdots \\ a_{1q} \end{pmatrix} = 0$$

This is the matrix equivalent of (13.22) in the basis $\{\varphi_n^{(0)}, n \leq q\}$. Setting $n = 2$ in (13.23) generates (13.24), with the modifications that E_1' is replaced with E_2' and the column vector $\{a_{1i}\}$ is replaced by $\{a_{2i}\}$. As n runs from 1 to q, one obtains q such equations. The condition that there be a nontrivial solution $\{a_{ni}\}$ for any one of these q matrix equations is that the determinant of the coefficient matrix vanish, which gives the *secular equation*

(13.25)
$$\det |H'_{pi} - E_n' \delta_{pi}| = 0$$

This equation may be rewritten in a purely operational form,

$$\det |\hat{H}' - E_n'\hat{I}| = 0$$

The identity operator is $\hat{I}$ (or, equivalently, the $q \times q$ unit matrix). The q roots of the algebraic equation (13.25) are the eigenvalues of (13.22). They are the diagonal elements of the submatrix of $\hat{H}'$ depicted in (13.19). Substituting any value of E' so obtained, say E_1', back into (13.24) permits one to solve for the coefficients $\{a_{1i}\}$. In similar manner, E_2' permits calculation of $\{a_{2i}\}$, and so on. These coefficients, in turn, give the new basis functions $\{\overline{\varphi}_n\}$ in (13.18).

Using this new basis (13.18), the ambiguities due to the degeneracy of $\hat{H}_0$ are removed[1] and one may proceed with the analysis developed in the previous section for the nondegenerate case. For example, (13.2) appears as

$$\varphi_n = \overline{\varphi}_n \quad + \lambda\overline{\varphi}_n^{(1)} + \lambda^2\overline{\varphi}_n^{(2)} + \cdots \qquad n \le q$$

$$\varphi_n = \varphi_n^{(0)} + \lambda\varphi_n^{(1)} + \lambda^2\varphi_n^{(2)} + \cdots \qquad n > q$$

$$E_n = E_n^{(0)} + \lambda E_n'^{(1)} + \lambda^2 E_n'^{(2)} + \cdots \qquad n \le q \qquad (E_1^{(0)} = \cdots = E_q^{(0)})$$

$$E_n = E_n^{(0)} + \lambda E_n^{(1)} + \lambda^2 E_n^{(2)} + \cdots \qquad n > q$$

$$E_n' = \langle \overline{\varphi}_n | \hat{H}' | \overline{\varphi}_n \rangle \qquad n \le q$$

$$E_n^{(1)} = \langle \varphi_n^{(0)} | \hat{H}' | \varphi_n^{(0)} \rangle \qquad n > q$$

An outline of this analysis is shown in Fig. 13.4. An important feature of degenerate perturbation theory is that solution of the matrix equation (13.24) gives (1) first-order corrections to the energy; and (2) corrected wavefunctions which, together with the nondegenerate states, serve as a proper basis for higher-order calculations.

Two-Dimensional Harmonic Oscillator

In this section we are primarily concerned with the degeneracy-removing property of the perturbation $\hat{H}'$. As a simple example of these procedures, consider the case of the two-dimensional harmonic oscillator whose Hamiltonian is

$$\hat{H}_0 = \frac{\hat{p}_x^2 + \hat{p}_y^2}{2m} + \frac{K}{2}(x^2 + y^2)$$

or, equivalently,

$$\hat{H}_0 = \hbar\omega_0(\hat{a}^\dagger\hat{a} + \hat{b}^\dagger\hat{b} + 1)$$

---

[1] It may be that first-order calculation does not remove the degeneracies of H_0. For example, this occurs if the off-diagonal elements of $\hat{H}'$ are zero. For such cases it becomes necessary to include higher-order terms to remove the degeneracy. For a discussion and problems relating to such *second-order degenerate perturbation theory*, see L. I. Schiff, *Quantum Mechanics*, 3d ed., McGraw-Hill, New York, 1968.

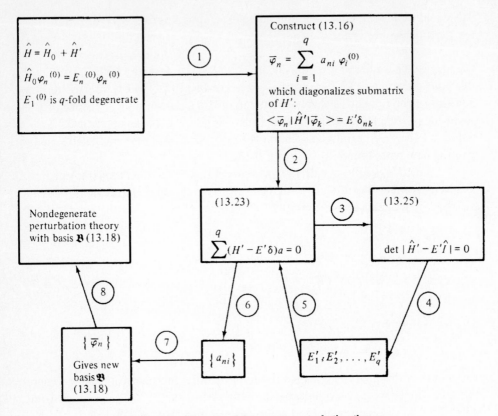

FIGURE 13.4   Elements of degenerate perturbation theory.

where

$$x = \frac{1}{\sqrt{2}\,\beta}(\hat{a} + \hat{a}^{\dagger}) \qquad y = \frac{1}{\sqrt{2}\,\beta}(\hat{b} + \hat{b}^{\dagger}) \qquad \beta^2 = \frac{m\omega_0}{\hbar}$$

The eigenstates of $\hat{H}_0$ are the product forms (8.111)

$$\varphi_{np} \equiv \varphi_n(x)\varphi_p(y)$$

which we will label $|np\rangle$. The corresponding eigenenergy

$$E_{np} = \hbar\omega_0(n + p + 1)$$

is $(n + p + 1)$-fold degenerate. It follows that the energy

$$E_{10} = E_{01} = 2\hbar\omega_0$$

is twofold degenerate, with corresponding eigenstates $|10\rangle$ and $|01\rangle$.

Let us apply the analysis developed above to determine how this energy separates due to the perturbing potential

$$H' = K'xy$$

Furthermore, we wish to find the two new wavefunctions that diagonalize $\hat{H}'$. These are given by the linear combinations (13.16)

$$\bar{\varphi}_1 = a\varphi_{10} + b\varphi_{01}$$

$$\bar{\varphi}_2 = a'\varphi_{10} + b'\varphi_{01}$$

The submatrix of $\hat{H}'$ in the basis $\{\varphi_{10}, \varphi_{01}\}$ appears as

$$\hat{H}' = K' \begin{pmatrix} \langle 10|xy|10 \rangle & \langle 10|xy|01 \rangle \\ \langle 01|xy|10 \rangle & \langle 01|xy|01 \rangle \end{pmatrix} = \mathbb{E} \begin{pmatrix} 0 & 1 \\ 1 & 0 \end{pmatrix}$$

$$\mathbb{E} = \frac{K'}{2\beta^2}$$

Consider, for example, the calculation of the $(1, 2)$ elements of H'.

$$\langle 10|\hat{x}\hat{y}|01 \rangle = \frac{1}{2\beta^2} \langle 10|\hat{a}\hat{b} + \hat{a}^\dagger\hat{b} + \hat{a}\hat{b}^\dagger + \hat{a}^\dagger\hat{b}^\dagger|01 \rangle$$

$$= \frac{1}{2\beta^2} \langle 10|\hat{a}^\dagger\hat{b}|01 \rangle = \frac{1}{2\beta^2}$$

With these values of the matrix elements of $\hat{H}'$, we are prepared to solve (13.25) for the incremental energies E'. This equation appears as

$$\begin{vmatrix} -E' & \mathbb{E} \\ \mathbb{E} & -E' \end{vmatrix} = 0$$

which has the solutions

$$E' = \pm\mathbb{E}$$

Thus we find that the perturbation separates the first excited state by the amount $2\mathbb{E}$.

(13.26)
$$E_{10} \left\{ \begin{array}{l} E_+ = E_{10} + \mathbb{E} \\ \\ E_- = E_{10} - \mathbb{E} \end{array} \right.$$

The corresponding new wavefunctions are obtained by substituting these values into the matrix equation (13.23), which for the present case takes the form

$$\begin{pmatrix} -E' & \mathbb{E} \\ \mathbb{E} & -E' \end{pmatrix} \begin{pmatrix} a \\ b \end{pmatrix} = 0$$

The two values of E' given above serve to determine two column vectors, (a, b) and (a', b'). The value $E' = \mathbb{E}$ gives $a = b$, while $E' = -\mathbb{E}$ gives $a' = -b'$. Thus we obtain for the new wavefunctions, $\overline{\varphi}_1$ and $\overline{\varphi}_2$, the values

$$E' = +\mathbb{E} \rightarrow \overline{\varphi}_1 = \frac{1}{\sqrt{2}}(\varphi_{10} + \varphi_{01})$$

(13.27)

$$E' = -\mathbb{E} \rightarrow \overline{\varphi}_2 = \frac{1}{\sqrt{2}}(\varphi_{10} - \varphi_{01})$$

These new wavefunctions serve to diagonalize the perturbation Hamiltonian H'.

PROBLEMS

13.12 How does the threefold-degenerate energy

$$E = 3\hbar\omega_0$$

of the two-dimensional harmonic oscillator separate due to the perturbation

$$H' = K'xy?$$

13.13 Consider a particle confined in a two-dimensional square well with faces at $x = 0, a$; $y = 0, a$ (see Section 8.5). The doubly degenerate eigenstates appear as

$$\varphi_{np}(x, y) = \frac{2}{a} \sin\left(\frac{n\pi x}{a}\right) \sin\left(\frac{p\pi y}{a}\right)$$

$$E_{np} = E_1(n^2 + p^2)$$

What do these energies become under the perturbation

$$H' = 10^{-3} E_1 \sin\left(\frac{\pi x}{a}\right)?$$

13.14 The eigenstates of a rotating dumbbell, with moment of inertia I,

$$E_l = \frac{\hbar^2 l(l + 1)}{2I}$$

are $(2l + 1)$-fold degenerate. In the event that the dumbbell is equally and oppositely charged at its ends, it becomes a dipole. The interaction energy between such a dipole and a constant, uniform electric field $\mathscr{E}$ is

$$\hat{H}' = -\mathbf{d} \cdot \mathscr{E} \qquad (\hat{H} = \hat{H}_0 - \mathbf{d} \cdot \mathscr{E})$$

The dipole moment of the dumbbell is $\mathbf{d}$. Show that to terms of first order, this perturbing potential *does not separate* the degenerate E_l eigenstates.

13.15 Consider again the dipole moment described in Problem 13.14. If both ends are equally charged, the rotating dipole constitutes a magnetic dipole. If the dipole has angular momentum **L**, the corresponding magnetic dipole moment is

$$\boldsymbol{\mu} = \frac{e}{2mc}\mathbf{L}$$

where e is the net charge of the dipole. The interaction energy between this magnetic dipole and a constant, uniform magnetic field $\mathcal{B}$ is

$$\hat{H}' = -\hat{\boldsymbol{\mu}} \cdot \mathcal{B} = -\frac{e}{2mc}\hat{\mathbf{L}} \cdot \mathcal{B} \qquad (\hat{H} = \hat{H}_0 - \hat{\boldsymbol{\mu}} \cdot \mathcal{B})$$

(a) If $\mathcal{B}$ points in the z direction, show that $\hat{H}'$ separates the $(2l + 1)$-fold degenerate E_l energies of the rotating dipole.

(b) Apply these results to one-electron atoms to find the splitting of the P states. (Neglect spin-orbit coupling.) (*Note:* This phenomenon is an example of the *Zeeman effect* discussed previously in Problems 12.15 et seq.)

13.3 THE STARK EFFECT

In Problem 13.9 we found that to within a first-order calculation, an electric field does not remove m_l degeneracy of states of definite orbital number l. However, as we shall now see, a similar calculation reveals that such a field will induce a partial separation of the n^2 degeneracy of eigenenergies related to one-electron atoms. This effect was first noticed in 1913 by Stark. He observed the splitting of the Balmer lines in a field of 100,000 V/cm. (The more readily observed Zeeman effect was first observed in 1897.)

The Hamiltonian of a one-electron atom in a constant, uniform electric field $\mathcal{E}$ which points in the z direction, neglecting spin, is

$$\hat{H} = \frac{\hat{p}_r^2}{2m} + \frac{\hat{L}^2}{2mr^2} - \frac{Ze^2}{r} - e\mathcal{E}z$$

$$= \hat{H}_0 + \hat{H}'$$

$$\hat{H}' = -e\mathcal{E}z = -e\mathcal{E}r\cos\theta$$

The eigenstates of the unperturbed Hamiltonian are n^2-fold degenerate. Let us consider how the perturbing electric field removes this degeneracy. Specifically, let us consider the fourfold degenerate $n = 2$ states. The related degenerate wavefunctions are, in the $|nlm\rangle$ notation,

$$|200\rangle, |211\rangle, |210\rangle, |21\text{-}1\rangle$$

To calculate the incremental changes in the energy E_2, we must solve the determinantal equation

(13.28)

$$0 = \begin{vmatrix} \langle 200|\hat{H}'|200\rangle - E' & \langle 200|\hat{H}'|211\rangle & \langle 200|\hat{H}'|210\rangle & \langle 200|\hat{H}'|21\text{-}1\rangle \\ \langle 211|\hat{H}'|200\rangle & \langle 211|\hat{H}'|211\rangle - E' & \langle 211|\hat{H}'|210\rangle & \langle 211|\hat{H}'|21\text{-}1\rangle \\ \langle 210|\hat{H}'|200\rangle & \langle 210|\hat{H}'|211\rangle & \langle 210|\hat{H}'|210\rangle - E' & \langle 210|\hat{H}'|21\text{-}1\rangle \\ \langle 21\text{-}1|\hat{H}'|200\rangle & \langle 21\text{-}1|\hat{H}'|211\rangle & \langle 21\text{-}1|\hat{H}'|210\rangle & \langle 21\text{-}1|\hat{H}'|21\text{-}1\rangle - E \end{vmatrix}$$

Only two elements survive integration. All elements with different m_l numbers vanish by orthogonality of the $|nlm_l\rangle$ states. Equivalently, one says, "$\hat{H}'$ does not *connect* states of different m_l." Integration gives[1]

$$\langle 210|\hat{H}'|200\rangle = \langle 200|\hat{H}'|210\rangle$$

$$= -\frac{ea_0\mathcal{E}}{32\pi} \int_0^\infty \rho^4(2 - \rho)e^{-\rho}\,d\rho \int_{-1}^1 d\cos\theta\,\cos^2\theta \int_0^{2\pi} d\phi$$

$$= \frac{e\mathcal{E}\hbar^2}{mZe} = -|e|3\mathcal{E}a_0 \equiv -\mathbb{E}$$

With these values inserted into the determinant above, (13.25) becomes

$$\begin{vmatrix} -E' & 0 & -\mathbb{E} & 0 \\ 0 & -E' & 0 & 0 \\ -\mathbb{E} & 0 & -E' & 0 \\ 0 & 0 & 0 & -E' \end{vmatrix} = 0$$

which has the four roots

(13.29)
$$E' = 0, 0, +\mathbb{E}, -\mathbb{E}$$
$$\mathbb{E} = 3|e|\mathcal{E}a_0$$

Thus we find that to terms of lowest order in the electric field $\mathcal{E}$, the degenerate $n = 2$ state separates into three states:

$$E_2 \left\{ \begin{array}{l} E_2^{(0)} + \mathbb{E} \\ E_2^{(0)} \\ E_2^{(0)} - \mathbb{E} \end{array} \right.$$

To calculate the new $n = 2$ wavefunctions

$$\varphi = a|200\rangle + b|211\rangle + c|210\rangle + d|21\text{-}1\rangle$$

[1] The nondimensional radius ρ is defined in Table 10.3. See also Tables 9.1 and 10.5.

we substitute the values (13.29) into the matrix equation

(13.30)
$$\begin{pmatrix} -E' & 0 & -\mathbb{E} & 0 \\ 0 & -E' & 0 & 0 \\ -\mathbb{E} & 0 & -E' & 0 \\ 0 & 0 & 0 & -E' \end{pmatrix} \begin{pmatrix} a \\ b \\ c \\ d \end{pmatrix} = 0$$

There results

$$E_2^+ = E_2^{(0)} + \mathbb{E} \rightarrow \varphi_+ = \frac{1}{\sqrt{2}}(|200\rangle - |210\rangle)$$

(13.31)
$$E_2^- = E_2^{(0)} - \mathbb{E} \rightarrow \varphi_- = \frac{1}{\sqrt{2}}(|200\rangle + |210\rangle)$$

$$E_2^0 = E_2^{(0)} \begin{cases} \varphi = |211\rangle \\ \varphi = |21\text{-}1\rangle \end{cases}$$

The perturbation mixes the $m = 0$ states, while the $m = 1, -1$ states are left degenerate. The values $\pm\mathbb{E}$ represent the average values of the interaction $\hat{H}'$ in the respective states, $\varphi_\pm$.

Symmetry Breaking and Removal of Degeneracy

Degeneracy, such as in the example of the two-dimensional harmonic oscillator considered above, is due to symmetry properties of $\hat{H}_0$. In this example the potential $x^2 + y^2$ is noted to be invariant under rotation about the origin in the plane. The perturbation xy destroys this symmetry, thereby removing the degeneracy. (Recall related discussions in Section 8.5.) It has been noted that an eigenenergy which is n-fold degenerate corresponds to n independent eigenfunctions. When the symmetry related to this degeneracy is broken, an n-fold degenerate eigenenergy may split into as many as n new eigenenergies. Another example addresses the case when a hydrogen atom is placed in an external field. This perturbation breaks rotational symmetry of the Hamiltonian about the nuclear center of the atom. In the problem of the Stark effect discussed immediately above, a uniform electric field breaks rotational symmetry and eigenenergies proliferate.[1] In addition, as noted in (13.31), there is a decrease in the number of "good quantum numbers" from (n, l, m) to (n). Consider the case in which the strength of the symmetry-breaking perturbation is represented by a parameter in the Hamiltonian, which we label g. If eigenenergies are plotted versus g,

---

[1] The order of degeneracy of a given system may exceed that implied by symmetry principles (i.e., group theory). Such, for example, is the case for the two-dimensional harmonic oscillator as well as for the hydrogen atom. For further discussion, see F. A. Cotten, *Chemical Applications of Group Theory*, 3d ed., Wiley, New York, 1993.

then at some value of g, eigenenergies bifurcate. Such behavior of eigenvalues is sometimes referred to as "level repelling." The properties of repelling of levels and loss of good quantum numbers are often associated with nonintegrable systems.

PROBLEMS

13.16   Show that the $n = 2$ matrix of the interaction Hamiltonian $\hat{H}'$, of the Stark effect, is diagonal in the basis (13.31).

13.17   What is the dipole moment of the hydrogen atom in the $\varphi_\pm$ states (13.31)?

Answer

The interaction energy between an electric dipole **d** and the electric field $\mathscr{E}$ is

$$H' = -\mathbf{d} \cdot \mathscr{E}$$

In the $\varphi_\pm$ states, the average value of H' is $E = \pm 3|e|a_0\mathscr{E}$. We may infer from this result that the magnitude of the dipole moment in the $\varphi_\pm$ states is $3|e|a_0$. The directions of these moments are parallel or antiparallel to the z axis (i.e., the direction of $\mathscr{E}$).

13.18   What is the charge density $q(r, \theta)$ of the hydrogen atom associated with the state φ_- (13.31)?

Answer

$$q(r, \theta) = e|\varphi_-|^2 = \frac{e}{16\pi a_0^3}\left[1 - \frac{r}{a_0}\sin^2\left(\frac{\theta}{2}\right)\right]^2 e^{-r/a_0}$$

13.19   Of the two states $\varphi_\pm$ in (13.31), φ_- is said to be more *stable* than φ_+. Why? Discuss your answers in light of the interaction energy, $-\mathbf{d} \cdot \mathscr{E}$.

13.4   THE NEARLY FREE ELECTRON MODEL

In this section we return to the problem of an electron in a periodic potential $V(x)$, discussed in Sections 8.2 and 12.9. Wavefunctions are in the Bloch form

$$\varphi(x) = u(x)e^{ikx}$$

where the periodic function $u(x)$ has the same period a as $V(x)$. Eigenenergies are functions of the crystal momentum wavenumber k. For a lattice of length L, the nearly continuous wavenumber k has the discrete values [see (8.42)]

$$k_j = \frac{j2\pi}{L}$$

We recall (see Problem 8.13) that in the high-energy domain $(E \gg V)$, the energy spectrum reduces to the free-particle values $\hbar^2 k^2 / 2m$, or, equivalently,

$$E_{k_j} = \frac{\hbar^2 k_j{}^2}{2m} = \frac{j^2 h^2}{2mL^2}$$

which is doubly degenerate: $E_{k_j} = E_{-k_j}$.

We now wish to obtain an expression for the energy gap δE_n at the nth band edge. A band edge, we recall, is a break in the energy spectrum which occurs at the k values $k_j a = n\pi$.

In the present analysis the periodic potential is considered a small perturbation to the free-particle Hamiltonian

$$\hat{H}_0 = \frac{\hat{p}^2}{2m}$$

The electron is "nearly free," which is the same as saying that $E \gg V$.

Unperturbed eigenstates are normalized to the sample interval L.

$$\varphi_{k_j}{}^{(0)} = \frac{1}{\sqrt{L}} \exp\,(ik_j x)$$

With $k_j = j(2\pi/L)$ these functions comprise an orthonormal sequence, as may be seen as follows:

$$(13.32) \quad \langle k_q | k_j \rangle = \frac{1}{L} \int_{-L/2}^{L/2} \exp\,[i(k_j - k_q)]\,dx = \frac{\sin\,[(k_j - k_q)L/2]}{(k_j - k_q)L/2}$$

$$= \frac{\sin\,(j - q)\pi}{(j - q)\pi} = \delta_{jq}$$

The Perturbation Potential

The perturbing potential is periodic and may be expanded in the Fourier series (see Problem 13.58).

$$(13.33) \quad H' = V(x) = \sum_{n=-\infty}^{\infty} V_n \exp\left[i2\pi n\left(\frac{x}{a}\right) \right]$$

The zero energy line in the present analysis is set at the average of $V(x)$. This ensures that the dc component V_0 of V vanishes (see Fig. 13.5).

$$V_0 = \int_{-L/2}^{L/2} V(x)\,dx = 0$$

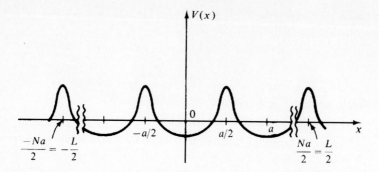

FIGURE 13.5 The zero in potential is chosen so as to eliminate V_0.

$$V_0 = \int_{-L/2}^{L/2} V(x)\, dx = 0$$

Application of first-order perturbation theory necessitates calculation of the matrix elements of H'.

(13.34)
$$H'_{qj} = \langle k_q | \sum n\, V_n \exp\left[i2\pi n\left(\frac{x}{a}\right)\right] | k_j \rangle$$

$$= \sum_n V_n \left\langle k_q \middle| k_j + \left(\frac{2\pi n}{a}\right) \right\rangle$$

$$= \sum_n V_n \delta_{k_q,\, k_j + (2\pi n/a)}$$

Substituting these matrix elements into the first equation of (13.10) gives the first-order corrected wavefunctions

(13.35)
$$\varphi_{k_j} = \varphi_{k_j}^{(0)} + \frac{1}{\sqrt{L}} \sum_{n=-\infty}^{\infty} \frac{V_n \exp\{ix[k_j + (2\pi n/a)]\}}{E_{k_j}^{(0)} - E_{k_j + (2\pi n/a)}^{(0)}}$$

Calculation of the first-order corrected eigenenergies as given by the second equation in (13.10) gives

(13.36)
$$E_{k_j} = E_{k_j}^{(0)} + \langle k_j | V | k_j \rangle = E_{k_j}^{(0)} + V_0$$

$$= E_{k_j}^{(0)} = \frac{\hbar^2 k_j^{\,2}}{2m}$$

To first order, the energy remains unperturbed. This free-particle spectrum was found in Chapter 8 to maintain in the domain $E \gg V$. However, in the present analysis there is explicit evidence which indicates that this result is invalid at the band edges of the energy spectrum. Namely, the summation in (13.35) for the first-order corrected wavefunctions becomes singular if the denominator of any term vanishes.

$$(13.37) \qquad E_{k_j}^{(0)} - E_{k_j+(2\pi n/a)}^{(0)} = -\frac{\hbar^2}{2m}\frac{4\pi n}{a^2}(k_j a + n\pi) = 0$$

This singularity arises from the zeros of (13.37), corresponding to the degeneracy at the band edges, $k_j a + n\pi = 0$. To obtain correct energies at these values of k_j, one must use degenerate perturbation theory. As described in Section 13.2, the first step in this procedure is to construct a new basis $\{\bar{\varphi}\}$ that diagonalizes the relevant 2×2 submatrix of H'. This new subbasis $\{\bar{\varphi}\}$ is constructed from linear combinations of the degenerate portion of the unperturbed basis and we may write

$$(13.38) \qquad \bar{\varphi} = \bar{a}e^{ik_j x} + \bar{b}e^{-ik_j x}$$

Diagonalization of H' in the basis (13.38) yields the matrix equation

$$(13.39) \qquad \begin{pmatrix} H'_{k_j k_j} - E_{k_j}' & H'_{k_j, -k_j} \\ H'_{-k_j, k_j} & H'_{-k_j, -k_j} - E_{k_j}' \end{pmatrix}\begin{pmatrix} \bar{a} \\ \bar{b} \end{pmatrix} = 0$$

$$k_j = -\frac{n\pi}{a}$$

The matrix elements of H' are given by (13.34), from which we find

$$H'_{k_j k_j} = H'_{-k_j, -k_j} = V_0 = 0$$

Only the off-diagonal elements survive. Choosing the origin so that $V(x)$ is an even function gives

$$H'_{k_j, -k_j} = V_n = V_{-n} = H'_{-k_j, k_j}$$

The determinant equation (13.25) then becomes

$$\begin{vmatrix} -E_{k_j}' & V_n \\ V_n & -E_{k_j}' \end{vmatrix} = 0$$

which gives the roots

$$E_{k_j}' = \pm V_n$$

Resubstituting into the matrix equation (13.39) gives two nondegenerate eigenstates, which written together with the first-order corrected energies appear as

$$E_{k_j}^+ = E_{k_j}^{(0)} + V_n \qquad \bar{\varphi}_+ = 2\bar{a}\sin k_j x$$
$$E_{k_j}^- = E_{k_j}^{(0)} - V_n \qquad \bar{\varphi}_- = 2\bar{a}\cos k_j x$$

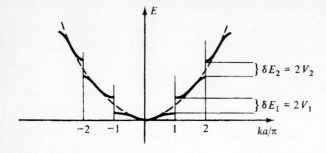

FIGURE 13.6   Energy gaps at the band edges in the nearly free electron model are given by twice the corresponding Fourier coefficient of the periodic potential.

These eigenstates, we recall, are the standing waves at the band edges previously obtained in Section 8.3 relevant to the Kronig–Penney potential. The present calculation affords an estimate (in the domain $E \gg V$) of the width of the energy gap at the band edges.

$$\delta E_n = E_{k_j}^{+} - E_{k_j}^{-} = 2|V_n| \qquad k_j a = \pm n\pi$$

Thus we find that the nth gap in the energy spectrum $E = E(k)$, at the values $k = \pm n\pi/a$, has width which is twice the nth Fourier coefficient of the potential $V(x)$ (Fig. 13.6).

In the nearly free electron model described above, conduction electrons are weakly bound to atoms of the lattice. This is opposite to the situation in the *tight-binding approximation* described in Problem 8.16 and in the LCAO analysis presented in Section 8.7.

PROBLEMS

13.20   Resketch the $E(k)$ curve shown in Fig. 13.6 for the case that the zeroth Fourier coefficient $V_0 > 0$.

13.21   A periodic potential has Fourier coefficients $V_n = V_1/n^2$. The width of the tenth energy gap is 0.031 eV. What is the value of V_1?

13.22   What are the energy gaps at the band edges for a particle in the periodic delta-function potential

$$V(x) = V_0 a \sum_{q=-\infty}^{\infty} \delta(x - qa)$$

defined over the entire x interval? [*Hint:* Consider the following delta-function representation

$$\delta(y) = \sum_{n=-\infty}^{\infty} \exp(i2\pi ny)$$

The right-hand side is periodic with period 1 and represents a delta function at $y = 0, \pm 1, \pm 2, \ldots$. The left-hand side is a delta function only at the origin. The domain of validity of the equation may be extended to the whole y axis if one writes

$$\sum_{-\infty}^{\infty} \delta(y - m) = \sum_{-\infty}^{\infty} \exp{(i2\pi n y)}. \Bigg]$$

13.23   (a)   Show that the first-order corrected wavefunction φ_k as given by (13.35) may be cast in the form of a Bloch wavefunction $\varphi_k = u(x)e^{ikx}$.

(b)   Show that the expression you obtain for $u(x)$ has period a.

13.24   Estimate the energy gaps at the band edges for the Kronig–Penney potential of potential height V_0, well width a, and barrier width b.

Answer

$$\delta E_n \simeq \frac{2V_0}{n\pi} \sin\left(\frac{n\pi a}{a + b}\right)$$

13.25   Estimate the band-gap widths for the potential

$$V(x) = 2V_0 \cos\left(\frac{2\pi x}{a}\right)$$

Answer

Rewriting the cos term in terms of exponentials,

$$V(x) = 2V_0 \cos\left(\frac{2\pi x}{a}\right) = V_0\left[\exp\left(\frac{i2\pi x}{a}\right) + \exp\left(\frac{-i2\pi x}{a}\right)\right]$$

reveals that this potential has only two nonvanishing Fourier coefficients. First-order perturbation theory implies the existence of the gaps at $k = \pm\pi/a$, of width $2V_0$. Higher-order perturbation theory would uncover energy gaps at subsequent band edges as well. This may be concluded directly by writing the Schrödinger equation for the given potential,[1]

$$\varphi_{xx} + \frac{2m}{\hbar^2}\left[E - 2V\cos\left(\frac{2\pi x}{a}\right)\right]\varphi = 0$$

This is a well-known equation in mathematical physics and is called the *Mathieu equation*. As with most such equations, it stems from writing the "wave equation," $(\nabla^2 + k^2)\varphi = 0$, in a particular orthogonal coordinate frame. For the Mathieu equation these are elliptic cylinder coordinates. Solutions of the equation are called *Mathieu functions* and have been

---

[1] This case has been studied in detail by P. M. Morse, *Phys. Rev.* **35**, 1310 (1930). Here it is also shown that the energy bands approach the line spectrum of an infinitely deep well as the amplitude V_0 of the periodic potential grows infinitely large.

studied in detail. These analyses reveal a sequence of intervals on the E axis in which solutions to the equation are unstable.[1]

13.26 At 0 K a certain semiconductor has its valence band filled and conduction band empty. The potential "seen" by electrons may be approximated by the function

$$V(x) = 2V_0 \cos\left(\frac{2\pi x}{a}\right)$$

$$V_0 = 0.1 \text{ eV}, \qquad a = 0.5 \text{ Å}$$

   (a) Assuming that the valence band is the band of lowest energy, estimate the width of the gap δE between the valence and conduction bands.

   (b) Setting $\delta E = k_B T$, estimate the temperature at which the sample will begin to conduct appreciably. ($k_B = 8.6 \times 10^{-5}$ eV/K.)

Answers

   (a) $\delta E \simeq 2V_0 = 0.20$ eV
   (b) $T \simeq 2300$ K

13.5 TIME-DEPENDENT PERTURBATION THEORY

Time-dependent perturbation theory addresses the following problem. Initially, the unperturbed system is in an eigenstate of $\hat{H}_0$. Then the perturbation, $\hat{H}'(t)$, is "turned on." What is the probability, after a time t, that transition to another state (of $\hat{H}_0$) occurs?

   The total Hamiltonian for these problems is of the form

(13.40) $$\hat{H}(\mathbf{r}, t) = \hat{H}_0(\mathbf{r}) + \lambda\hat{H}'(\mathbf{r}, t)$$

where λ is again a parameter of smallness.

   Let the time-dependent eigenstates of $\hat{H}_0$ be written

(13.41) $$\psi_n(\mathbf{r}, t) = \varphi_n(\mathbf{r})e^{-i\omega_n t}$$
$$\hat{H}_0\varphi_n = E_n^{(0)}\varphi_n \equiv \hbar\omega_n\varphi_n$$

   Suppose that at time $t > 0$, the system is in the state

(13.42) $$\psi(\mathbf{r}, t) = \sum_n c_n(t)\psi_n(\mathbf{r}, t)$$

---

[1] That is, series solutions in these domains do not converge. For further discussion of the properties of Mathieu functions, see P. M. Morse and H. Feshbach, *Methods of Theoretical Physics*, McGraw-Hill, New York, 1953, Chapter 5; also, E. T. Whittaker and G. N. Watson, *A Course of Modern Analysis*, Cambridge University Press, New York, 1952, Chapter 19.

Then, by the superposition principle, $|c_n|^2$ is the probability that measurement finds the system in the state ψ_n at the time t. Thus the primary aim of the present discussion is to calculate these coefficients. They are determined in the following manner. The wavefunction $\psi(\mathbf{r}, t)$ is a solution of

$$(13.43) \qquad i\hbar \frac{\partial \psi}{\partial t} = (\hat{H}_0 + \lambda \hat{H}')\psi$$

Substituting the expansion (13.42) into this equation and then operating from the left with $\int d\mathbf{r} \psi_k^*(\mathbf{r}, t)$ gives

$$(13.44) \qquad i\hbar \frac{dc_k}{dt} = \lambda \sum_n \langle k | H' | n \rangle c_n$$

This is an infinite sequence of coupled equations for the coefficients $\{c_k(t)\}$. In the limit that $\lambda \to 0$, the c_k coefficients are all constant. It is therefore consistent to seek solution in the form

$$(13.45) \qquad c_k(t) = c_k^{(0)} + \lambda c_k^{(1)}(t) + \lambda^2 c_k^{(2)}(t) + \cdots$$

Substituting this series into (13.44) and equating terms of equal powers in λ gives (with a dot denoting time differentiation and H'_{kn} written for the matrix elements of $\hat{H}'$)

$$i\hbar \dot{c}_k^{(0)} = 0$$

$$i\hbar \dot{c}_k^{(1)} = \sum_n H'_{kn} c_n^{(0)}$$

$$(13.46) \qquad i\hbar \dot{c}_k^{(2)} = \sum_n H'_{kn} c_n^{(1)}$$

$$\vdots$$

$$i\hbar \dot{c}_k^{(s+1)} = \sum_n H'_{kn} c_n^{(s)}$$

The lowest-order equations for $c_k^{(0)}$ indicate that these coefficients are all constant in time. They are the initial values of the coefficients $\{c_k(t)\}$.

We now specialize to the problem in which it is known that initially the system is in a definite eigenstate of $\hat{H}_0$, say $\psi_l(\mathbf{r}, t)$. With (13.42) this implies that as $t \to -\infty$,

$$\psi(\mathbf{r}, t) \sim \psi_l(\mathbf{r}, t) = \sum_n \delta_{nl} \psi_n(\mathbf{r}, t)$$

$$(13.47)$$

$$c_n^{(0)}(-\infty) = \delta_{nl}$$

Note that we have taken "initially" to denote the time $t = -\infty$. Substituting this value into the second equation in (13.46) gives (dropping the superscripts 0 and 1)

$$(13.48) \qquad i\hbar \dot{c}_k(t) = \sum_n H'_{kn} c_n(-\infty) = H'_{kl}$$

For $n \neq l$, $c_n(-\infty) = 0$, so the first-order solution for $c_k(t)$ is given by

$$(13.49) \qquad c_k(t) = \frac{1}{i\hbar} \int_{-\infty}^{t} H'_{kl}(\mathbf{r}, t')\, dt' \qquad (k \neq l)$$

If the time dependence of $\hat{H}'(\mathbf{r}, t)$ is factorable, then

$$\hat{H}'(\mathbf{r}, t) = \hat{\mathsf{H}}'(\mathbf{r}) f(t)$$

and the matrix elements of $\hat{H}$ become [with (13.41)]

$$H'_{kl}(t) \equiv \langle \psi_k | H'(\mathbf{r}, t) | \psi_l \rangle = \langle \varphi_k | \mathsf{H}'(\mathbf{r}) | \varphi_l \rangle e^{i\omega_{kl} t} f(t)$$

$$(13.50) \qquad = \mathsf{H}'_{kl} e^{i\omega_{kl} t} f(t)$$

$$\hbar\omega_{kl} \equiv \hbar(\omega_k - \omega_l) = E_k - E_l$$

(Note that we have deleted the zero superscripts of E_k and E_l.) This gives the more explicit form of $c_k(t)$,

$$(13.51) \qquad c_k(t) = \frac{\mathsf{H}'_{kl}}{i\hbar} \int_{-\infty}^{t} e^{i\omega_{kl} t'} f(t')\, dt'$$

These coefficients determine the effect of the perturbation on the initial state, ψ_l. The probability that the system has undergone a transition from this state to some other eigenstate of H_0, ψ_k, at the time t, is

$$(13.52) \qquad \boxed{ P_{l \to k} = |c_k|^2 = \left| \frac{\mathsf{H}'_{kl}}{\hbar} \right|^2 \left| \int_{-\infty}^{t} e^{i\omega_{kl} t'} f(t')\, dt' \right|^2 }$$

Application of these results follows. The transition probability $P_{l \to k}$ is hereafter written P_{lk}.

Matrix elements in time-dependent perturbation theory are conventionally written with the initial state on the right and the final state on the left, as illustrated in the following symbolic form:

$$\langle \text{final state} | \text{interaction} | \text{initial state} \rangle$$

This convention is employed in the remainder of the text.

PROBLEMS

13.27 A system with discrete eigenstates $\{\varphi_n\}$ and eigenenergies $\{E_n\}$ is exposed to the perturbation

$$\hat{H}' = \hat{H}'(\mathbf{r}) \frac{e^{-t^2/\tau^2}}{\tau\sqrt{\pi}}$$

The perturbation is turned on at $t = -\infty$, when the unperturbed system is in its ground state, ψ_0. What is the probability that at $t = +\infty$ the system suffers a transition to the state $\psi_k, k > 0$?

Answer

To obtain the answer to this problem, we must calculate the time integral in (13.52). Writing ω for ω_{k0}, we have

$$I = \frac{1}{\sqrt{\pi}} \int_{-\infty}^{\infty} e^{i\omega t} e^{-t^2/\tau^2} \, d(t/\tau)$$

$$= \frac{1}{\sqrt{\pi}} \int_{-\infty}^{\infty} e^{i\bar{\omega}\xi - \xi^2} \, d\xi \qquad (\xi \equiv t/\tau, \; \bar{\omega} \equiv \tau\omega)$$

$$= \frac{1}{\sqrt{\pi}} e^{-\bar{\omega}^2/4} \int_{-\infty}^{\infty} \exp\left[-\left(\xi - \frac{i\bar{\omega}}{2}\right)^2\right] d\xi$$

$$= e^{-\bar{\omega}^2/4} = e^{-(E_0 - E_k)^2\tau^2/4\hbar^2}$$

Substituting this into (13.52) gives the desired result

$$P_{0k} = \left|\frac{\mathsf{H}'_{k0}}{\hbar}\right|^2 e^{-(E_0 - E_k)^2\tau^2/2\hbar^2}$$

13.28 Consider that the unperturbed system in Problem 13.27 is a particle of mass m confined to a one-dimensional box of width a. The spatial factor in $H'(x, t)$ is

$$\hat{\mathsf{H}}'(x) = \frac{10^{-4}\hat{p}_x^2}{2m}$$

What state does the perturbation leave the system in at $t = +\infty$?

13.6 HARMONIC PERTURBATION

Stimulated Emission

As a first application of the analysis above, we consider a perturbation that is switched on at $t = 0$ and is subsequently monochromatically harmonic in time. The perturbation acts on a system whose Hamiltonian is $\hat{H}_0$. If it is definitely known that the unperturbed system was in one of its own stationary states before the perturbation was

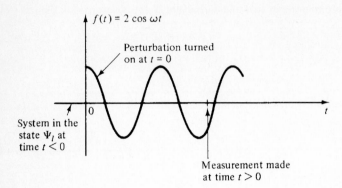

$$H' = \hat{H}'(\mathbf{r})f(t)$$

$f(t) = 2 \cos \omega t$

Perturbation turned on at $t = 0$

System in the state Ψ_l at time $t < 0$

Measurement made at time $t > 0$

FIGURE 13.7 Harmonic perturbation.

applied ($t < 0$), in what state will measurement find the system after the perturbation has been turned on for t seconds? This problem is appropriate, for example, to an atom that interacts with a (weak) electromagnetic field. The explicit form of the perturbation $\hat{H}'$ is (with $\omega > 0$)

$$(13.53) \qquad \hat{H}'(\mathbf{r}, t) = \begin{cases} 0 & t < 0 \\ 2\hat{H}'(\mathbf{r}) \cos \omega t & t \geq 0 \end{cases}$$

(see Fig. 13.7). Substituting this form into (13.51), with $f(t) = 2 \cos \omega t$, gives

$$c_k(t) = \frac{H'_{kl}}{i\hbar} \int_0^t e^{i\omega_{kl}t'}(e^{-i\omega t'} + e^{i\omega t'})\, dt'$$

$$= -\frac{H'_{kl}}{\hbar}\left[\frac{e^{i(\omega_{kl}-\omega)t} - 1}{\omega_{kl} - \omega} + \frac{e^{i(\omega_{kl}+\omega)t} - 1}{\omega_{kl} + \omega}\right]$$

Employing the relation (see Problem 1.21)

$$e^{i\theta} - 1 = 2ie^{i\theta/2} \sin (\theta/2)$$

permits the equation above to be rewritten

(13.54)

$$c_k(t) = -\frac{i2H'_{kl}}{\hbar}\left[\frac{e^{i(\omega_{kl}-\omega)t/2} \sin (\omega_{kl} - \omega)t/2}{\omega_{kl} - \omega} + \frac{e^{i(\omega_{kl}+\omega)t/2} \sin (\omega_{kl} + \omega)t/2}{\omega_{kl} + \omega}\right]$$

The dominant contributions to c_k come from the values $\omega \simeq \pm\omega_{kl}$. At these values, the moduli of the two bracketed terms in c_k, respectively, assume their maximum value, $t/2$. These *resonant* frequencies correspond to the energies

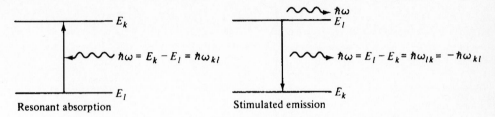

FIGURE 13.8 Dominant transition processes due to harmonic perturbation.

$$\omega \simeq +\omega_{kl} \quad \longrightarrow \quad E_k > E_l$$
$$\omega \simeq -\omega_{kl} \quad \longrightarrow \quad E_l > E_k$$

In the first case, the "final" energy E_k is larger than the "initial" energy E_l. The system absorbs energy and jumps to a higher energy level.

$$E_k = E_l + \hbar\omega$$

The energy absorbed, $\hbar\omega$, is that of a photon in the incident radiation field (Fig. 13.8).

For the second case the perturbation induces a decay in energy

$$E_k = E_l - \hbar\omega$$

A photon of energy $\hbar\omega$ is radiated away from the system. This decay process is *stimulated* by a photon of the same frequency in the perturbation field.

Let us consider the case that the incident radiation field excites only higher energies, so that $\omega_{kl} > 0$. Under such conditions, the first term in (13.564) dominates and the probability that the perturbing field causes a transition to the kth state becomes

$$(13.55) \qquad P_{lk} = |c_k|^2 = \frac{4|\mathbb{H}'_{kl}|^2}{\hbar^2(\omega_{kl} - \omega)^2} \sin^2\left[\frac{1}{2}(\omega_{kl} - \omega)t\right]$$

The frequency dependence of this function is plotted in Fig. 13.9.

Energy-Time Uncertainty

From this sketch it is evident that states falling in the interval

$$(13.56) \qquad |\hbar\omega_{kl} - \hbar\omega| = |E_k - (E_l + \hbar\omega)| \lesssim \frac{2\pi\hbar}{t} \simeq \Delta E$$

have the greatest probability of being excited, after the perturbing field has acted for t seconds. If this perturbation is applied many times to an ensemble of independent,

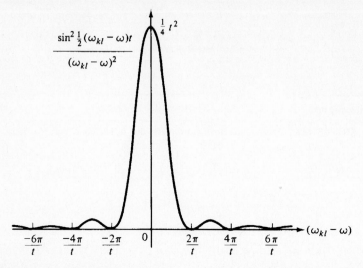

FIGURE 13.9 Frequency dependence of probability of transition from the lth to the kth state at the time t due to harmonic perturbation of frequency ω.

identical systems, all initially in the state ψ_l, then after a time t the energies excited among the members of the ensemble will lie primarily in the interval ΔE. Thus (13.56) gives the uncertainty in the values of energy observed. The final energies E_k are spread throughout the interval

$$\Delta E \simeq \frac{\hbar}{t}$$

(13.57)
$$E_k \simeq (E_l + \hbar\omega) \pm \Delta E$$

$$E_{\text{final}} \simeq E_{\text{initial}} \pm \Delta E$$

Note that E_{initial} refers to the energy of the incident photon plus that of the system in its inital state. Our perturbation analysis returns the principle of conservation of energy, properly modified by the uncertainty relation, $\Delta E \simeq \hbar/t$.

Long-Time Evolution

The transition probability formula for absorption (13.55) and its counterpart for stimulated emission may be written together as

(13.58)
$$P_{lk} = \frac{4|\mathsf{H}'_{kl}|^2}{\hbar^2(\omega_{kl} \mp \omega)^2} \sin^2\left[\frac{1}{2}(\omega_{kl} \mp \omega)t\right]$$

where the $\mp$ signs refer to absorption and stimulated emission, respectively (see Fig. 13.8). This expression takes a convenient form in the long-time or, equivalently, high-frequency limit. This form follows from the delta-function representation

$$(13.59) \qquad \delta(\omega) = \frac{2}{\pi} \lim_{t \to \infty} \frac{\sin^2 (\omega t/2)}{t\omega^2}$$

so that in the same limit,

$$(13.60) \qquad P_{lk} \to \frac{2\pi t}{\hbar^2} |H'_{kl}|^2 \delta(\omega_{kl} \mp \omega)$$

The corresponding transition probability *rate* appears as

$$(13.61) \qquad w_{lk} = \frac{2\pi}{\hbar^2} |H'_{kl}|^2 \delta(\omega_{kl} \mp \omega)$$

In this or the formula above, the delta function expresses the fact that in the long-time limit, the Fourier transform of a monochromatic perturbation becomes sharply peaked about the frequency of perturbation. The system "sees" only a single frequency. Since the uncertainty in energy $\hbar/t$ vanishes in this limit, the argument of the delta function is also an expression of conservation of energy.

Short-Time Approximation

If a harmonic perturbation is applied to a system for a short-time interval such that $(\omega_{kl} - \omega)t \ll 1$, (13.55) may be expanded to yield

$$(13.62) \qquad P_{lk} = \frac{t^2 |H'_{kl}|^2}{\hbar^2}$$

The related transition probability rate is

$$(13.63) \qquad w_{lk} = \frac{t |H'_{kl}|^2}{\hbar^2}$$

At early times, the rate at which transitions to the kth state occur grows linearly with time.

The Golden Rule

In many problems of practical interest, the final excited states lie in a band of energies (Fig. 13.10). Such is the case, for example, for ionization or free-particle scattering states. Such states comprise a continuum. If the density of final states is $g(E_k)$, then

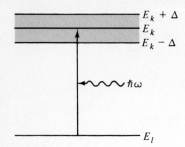

$E_k + \Delta$
E_k
$E_k - \Delta$

$\hbar\omega$

E_l

FIGURE 13.10 Resonant absorption to a band of energies. If the density of states about E_k is $g(E_k)$, then the probability rate of transition to the band is

$$\overline{w}_{lk} = \frac{2\pi}{\hbar} g(E_k) |\mathbb{H}'_{kl}|^2$$

$$dN = g(E_k) \, dE_k$$

is the number of energy states in the interval E_k to $E_k + dE_k$. The probability that a transition occurs to a state in a band of width 2Δ centered at E_k is

$$\overline{P}_{lk} = \int_{E_k - \Delta}^{E_k + \Delta} P_{lk} g(E'_k) \, dE'_k$$

Inserting (13.55) gives

$$\overline{P}_{lk} = \int_{E_k - \Delta}^{E_k + \Delta} dE'_k g(E'_k) \left| \frac{\mathbb{H}'_{kl}}{\hbar} \right|^2 \frac{\sin^2 \beta}{\beta^2/t^2}$$

$$2\hbar\beta \equiv \hbar(\omega_{kl} - \omega)t = (E'_k - E_l - \hbar\omega)t$$

For fixed E_l, t and ω

$$dE'_k = \frac{2\hbar \, d\beta}{t}$$

and

$$\overline{P}_{lk} = \frac{2t}{\hbar} \int_{-\delta}^{+\delta} g(E'_k) |\mathbb{H}'_{kl}|^2 \frac{\sin^2 \beta}{\beta^2} \, d\beta$$

where 2δ is the corresponding spread in β values. Owing to the rapid decay of the $\sin^2 \beta/\beta^2$ function (see Fig. 13.9), only a slight error is introduced in the expression above if we replace the interval $(-\delta, +\delta)$ by $(-\infty, +\infty)$. Furthermore, if we assume that g and $\mathbb{H}'_{kl}$ are slowly varying functions of E_k, they may be taken outside the integral. There results

$$\overline{P}_{lk} = \frac{2t}{\hbar} g(E_k) |\mathbb{H}'_{kl}|^2 \int_{-\infty}^{\infty} \frac{\sin^2 \beta}{\beta^2} \, d\beta$$

$$= t \left[\frac{2\pi}{\hbar} g(E_k) |\mathbb{H}'_{kl}|^2 \right]$$

The related transition probably rate is

$$(13.64) \qquad \overline{w}_{lk} = \frac{2\pi}{\hbar} g(E_k) |\mathbb{H}'_{kl}|^2$$

This formula was found to have such widespread application that Fermi[1] dubbed it "Golden Rule No. 2." It will be applied in Section 13.9 in study of atom-radiation interaction and in Chapter 14 in study of the Born approximation in the theory of scattering.

PROBLEMS

13.29 Show that if the perturbation field

$$\hat{H}' = 2\hat{\mathbb{H}}'(\mathbf{r}) \cos \omega t$$

acts on a system initially in the lth state for a very long time, and $\omega \simeq \omega_{kl}$, then the only state that will be excited is the kth state. Interpret this result in terms of the Fourier decomposition (in time) of $\hat{H}'$.

13.30 The expression in the text obtained for $\overline{P}_{lk}$, the probability of transition to a band of states, is seen to be independent of the frequency ω of the perturbing field. In what approximation is this result valid?

13.31 What does the transition probability P_{lk} (13.55) become if the perturbing field is precisely "on resonance"; that is, if $\omega = \omega_{kl}$?

13.32 A polarized beam of current, $\mathbf{J}$ cm^{-2}/s, contains electrons with spins aligned with a steady magnetic field of magnitude $\mathcal{B}$ which points in the z direction. The beam propagates in the x direction. The wavefunction for an electron in the beam is

$$\psi = \frac{1}{(2\pi)^{1/2}} e^{i(kx - \omega t)} \begin{pmatrix} 0 \\ 1 \end{pmatrix}$$

A monochromatic electromagnetic field of frequency

$$\hbar\omega = 2\mu_b \mathcal{B}$$

extends over a length of beam path, L cm long. A Stern–Gerlach analyzer is in the path of the beam at a point beyond the domain of the electromagnetic field. Its orientation is such that spins aligned with $\mathcal{B}$ are not deflected from the beam. If the beam moves with the speed v, what is the current of electrons scattered out of the beam by the S-G analyzer? Assume that the interaction between electrons in the beam and the electromagnetic wave is

$$\hat{H}' = -\boldsymbol{\mu} \cdot \mathcal{B}' \cos \omega t$$

where $\mathcal{B}'$ is the amplitude of magnetic field of the wave. The component of $\mathcal{B}'$ in the z direction is small compared to $\mathcal{B}$.

[1] E. Fermi, *Nuclear Physics,* University of Chicago Press, Chicago, 1950.

13.33 A one-dimensional harmonic oscillator of charge to mass ratio e/m and spring constant K is in its ground state. An oscillating uniform electric field

$$\mathscr{E}(t) = 2\mathscr{E} \cos \omega_0 t, \qquad \omega_0^2 = K/m$$

is applied for t seconds, parallel to the motion of the oscillator. What is the probability that the oscillator is excited to the nth state given that $(\omega_{n0} - \omega_0)t \ll 1$?

Answer

$$P_{0n} = \delta_{n,1} \left| \frac{e\mathscr{E}}{\sqrt{2}\hbar\beta} \right|^2 t^2, \qquad \beta^2 \equiv \frac{m\omega_0}{\hbar}$$

13.7 APPLICATION OF HARMONIC PERTURBATION THEORY

In this section we apply the transition probability formula (13.61) found above to Einstein's derivation of the Planck radiation formula (2.3) and to a brief qualitative description of the laser.

Einstein considered the equilibrium state between the walls of an enclosed cavity and the radiation field interior to the cavity. Atoms in the walls constantly exchange energy with the radiation field. The excited states of these atoms are very closely spaced and essentially comprise a continuum. Consider that the two energies E_l and E_k are representative of two states in the continuum (Fig. 13.11). Photons with energy $h\nu = E_k - E_l$ can be absorbed by atoms in the E_l state raising them to the E_k state. Atoms may decay from the E_k to the E_l state by *stimulated* or *spontaneous* emission. Stimulated emission was discussed in the preceding section. Spontaneous emission is related to the natural lifetime of the excited state and is more dependent on internal properties of the radiating system.

Einstein *A* and *B* Coefficients

The rate at which atoms in the E_k state decay by stimulated emission is proportional to the number of such atoms (N_k) and number of photons of frequency ν in the

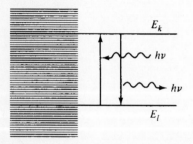

FIGURE 13.11 **Energy states of closely packed atoms in the walls of a radiation cavity.**

radiation field. This number of photons is proportional to photon energy density, $u(\nu)$. The rate at which the atoms in the E_k state decay by spontaneous emission, on the other hand, is proportional only to the number of atoms N_k in the E_k state. Thus the total transition rate of decay of atoms with energy E_k to energies E_l is

(13.65) $W_{kl} = [A_{kl} + B_{kl}u(\nu)]N_k = w_{kl}N_k$ $h\nu$

The transition probability rate per atom is written w_{kl}. The proportionality constants A and B are called *Einstein A and B coefficients*.

Atoms in the E_l state can be excited to the E_k state only by absorption of a photon of frequency $(E_k - E_l)/h$. Thus the rate of elevation of atoms in the E_l state to the E_k state is

(13.66) $W_{lk} = B_{lk}u(\nu)N_l = w_{lk}N_l$ $h\nu$

Planck Radiation Formula

In equilibrium, these rates must equal.

$$W_{lk} = W_{kl}$$

Equivalently, we may write

(13.67) $$\frac{N_l}{N_k} = \frac{A_{kl} + B_{kl}u(\nu)}{B_{lk}u(\nu)}$$

The ratio N_l/N_k may be obtained from elementary statistical mechanics. If the energy of an aggregate of atoms at the temperature T is partitioned such that N_1 of the atoms have energy E_1, N_2 have energy E_2, and so on, and the total energy of the aggregate is constant, then the most probable distribution of energies is given by the *Boltzmann distribution*.[1] In this distribution the number of atoms N_i with energy E_i is proportional

---

[1] This distribution was employed previously in Problem 2.36. For further discussion, see F. Reif, *Fundamentals of Statistical and Thermal Physics*, McGraw-Hill, New York, 1965.

to the Boltzmann factor, $\exp{(-E_i/k_B T)}$, where k_B is Boltzmann's constant. It follows that the ratio N_l/N_k has the value

$$\frac{N_l}{N_k} = e^{(E_k - E_l)/k_B T} = e^{h\nu/k_B T}$$

Substituting this into (13.67) gives

(13.68)
$$u(\nu) = \frac{A_{kl}}{B_{lk} e^{h\nu/k_B T} - B_{kl}}$$

The basic structure of the Planck radiation formula (2.3) follows if $B_{lk} = B_{kl}$. This equality may be obtained from the harmonic perturbation theory developed above. To these ends we consider the interaction between a wave mode in the radiation field,

$$\mathscr{E} = \mathscr{E}_0 \, 2 \cos \omega t$$

and an atom in the wall of the cavity. If $\mathbf{d}$ is the dipole moment of the atom, this interaction is

(13.69)
$$H' = -\mathbf{d} \cdot \mathscr{E} = -\mathscr{E}_0 \cdot \mathbf{d} \, 2 \cos \omega t$$

which is identical to the perturbation (13.53). Substituting into (13.61) gives the transition probability rate for radiative absorption,

$$w_{lk} = \frac{2\pi}{\hbar^2} |\langle k | \mathscr{E}_0 \cdot \mathbf{d} | l \rangle|^2 \delta(\omega_{kl} - \omega)$$

If the electric field is isotropic ($\mathscr{E}_0$ is randomly oriented), then

$$|\langle \mathbf{d} \cdot \mathscr{E}_0 \rangle|^2 = \frac{1}{3} \langle \mathscr{E}_0^2 \rangle |\mathbf{d}_{lk}|^2$$

Furthermore, the energy density associated with this mode is

$$U(\omega) = \frac{1}{2\pi} \langle \mathscr{E}_0^2 \rangle$$

It follows that

$$w_{lk} = \frac{4\pi^2}{3\hbar^2} U(\omega) |\mathbf{d}_{lk}|^2 \delta(\omega_{kl} - \omega)$$

With $u(\omega) = U(\omega)\delta(\omega_{kl} - \omega)$, (13.66) may be rewritten

$$w_{lk} = B_{lk} U(\omega) \delta(\omega_{kl} - \omega)$$

Equating this value to the preceding expression for w_{lk} gives the desired result,

(13.70)
$$B_{lk} = \frac{4\pi^2}{3\hbar^2} |\langle l | \mathbf{d} | k \rangle|^2$$

The square moduli of the matrix elements of $\mathbf{d}$ obey the relation

$$|\langle l | \mathbf{d} | k \rangle|^2 = \langle l | \mathbf{d} | k \rangle \cdot \langle k | \mathbf{d} | l \rangle$$
$$= \langle k | \mathbf{d} | l \rangle \cdot \langle l | \mathbf{d} | k \rangle = |\langle k | \mathbf{d} | l \rangle|^2$$

It follows that

$$B_{lk} = B_{kl}$$

and (13.68) reduces to the desired form,

(13.71)
$$u(\nu) = \frac{A/B}{e^{h\nu/k_B T} - 1}$$

To obtain the ratio A/B we will use the correspondence principle. This rule stipulates that (13.71) should reduce to the classical Rayleigh–Jeans law (discussed in Section 2.2) in the limit $h \to 0$.

$$u_{RJ} = \frac{8\pi\nu^2}{c^3} k_B T$$

Expanding the exponential in (13.71) in this limit gives

$$\frac{A/B}{h\nu/k_B T} = \frac{8\pi\nu^2}{c^3} k_B T$$

from which we obtain the desired result (see Problem 13.52)

$$\frac{A}{B} = \frac{8\pi h\nu^3}{c^3}$$

Substituting this value into (13.71) gives Planck formula, (2.3).

The Laser

The concepts of stimulated and spontaneous emission play an important role in the theory of the *laser*. A laser is a device for producing an intense beam of coherent, monochromatic light. A coherent beam may be defined as follows. Two or more collinear, unidirectional, monochromatic beams of electromagnetic radiation which propagate in the same region of space, and are in phase, form a coherent beam. In

1954, C. H. Townes[1] conceived of a process for the generation and amplification of such coherent radiation in the microwave domain. The device was called a *maser*. The term is an acronym for the words *m*icrowave *a*mplification by the *s*timulated *e*mission of *r*adiation. Shortly after, these concepts were extended to the optical region.[2] In this domain the corresponding device is called a *laser*.

Coherent Photons

The central principle in the realization of the laser is as follows. In constructing formula (13.65) for the transition rate from the E_k state to the E_l state, decay due to *stimulated* emission was taken to be proportional to the number of resonant photons present in the radiation field. Consider that a number of atoms in a gas are in the excited state E_k. Then when a photon of frequency

$$h\nu = E_k - E_l$$

falls on one of these atoms it stimulates the emission of another photon of the same frequency. These two photons, the emitted and incident, travel in phase in the same direction and combine coherently. A second atom, in the vicinity of the first and also excited to the E_k state, suddenly "sees" a duplication of resonant photons and is stimulated to emit another coherent photon, thereby adding to the intensity of the coherent radiation and amplifying it.

We found above that the matrix elements for radiative excitation, B_{lk}, and stimulated emission, B_{kl}, are equal [(13.70) et seq.]. To ensure that resonant photons stimulate decay to the E_l state more than they are absorbed in exciting the atom to the higher E_k state, the number of atoms, N_k, in the E_k state must outweigh the number of those in the lower E_l state, N_l. At any finite temperature, T, the ratio of these numbers of atoms, N_k/N_l, is given by the Boltzmann formula [preceding (13.68)].

$$\frac{N_k}{N_l} = e^{-h\nu/k_B T}$$

Population Inversion and Optical Pumping

The number of atoms in the higher E_k state decreases exponentially with energy difference, $h\nu$. To effect a *population inversion*, so that $N_k > N_l$, an outside source

---

[1] J. P. Gordon, H. J. Zeiger, and C. H. Townes, *Phys. Rev.* **99**, 1264 (1955).

[2] A. L. Schawlow and C. H. Townes, *Phys. Rev.* **112**, 1940 (1958). For further discussion and reference, see B. L. Lengyel, *Introduction to Laser Physics*, Wiley, New York, 1966; T. B. Melia, *An Introduction to Masers and Lasers*, Chapman & Hall, London, 1967.

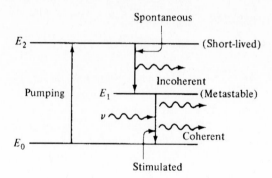

FIGURE 13.12 **Schematic for the three-level laser.**

must be brought into play.[1] In *optical pumping,* N_k is increased by irradiation with light of frequency $\nu \geq (E_k - E_0)/h$, where E_0 is the ground state of the atom. Another technique for effecting a population inversion is by bombardment of electrons with energy $E > E_k - E_0$, thereby exciting atoms in the ground state to higher energy states. In a third technique, inelastic collisions between atoms in the ground state with those of a foreign gas which are in an excited state $E > E_k - E_0$ serve to populate the higher-energy states.

Consider three atomic states: E_0, E_1, and E_2. The ground state is E_0, while E_2 is a short-lived state with a lifetime of the order of 10 to 100 ns. The more stable ("metastable") E_1 state has a lifetime of the order of μs to ms. The upper E_2 state replenishes the metastable state through *spontaneous decay* [the A coefficient in (13.65)]. These randomly emitted photons comprise an incoherent radiation field. In Fig. 13.12 a scheme is depicted where the E_2 state is populated through pumping with an external source. When an atom in the metastable E_1 state decays to the E_0 state, neighboring atoms in the populated E_1 state decay, through stimulated emission to the ground state, and a coherent beam is generated.

Such a device is realized in an optical cavity of cylindrical shape with end mirrors positioned so that the cavity is tuned to the frequency mode $(E_1 - E_0)/h$. Photons propagate parallel to the axis of the tube. The radiation field is coherently amplified by stimulated emission $(E_1 \rightarrow E_0)$ and may be tapped through a small aperture on the axis, in one of the mirrors. The spontaneously emitted photons $(E_2 \rightarrow E_1)$ propagate in random directions and are dissipated in collisions with the walls.

---

[1] Including the effects of degeneracy, the number of atoms in the nth state is given by $\bar{N}_n = g_n N_n$. The condition for growth of radiation then becomes $\bar{N}_k/g_k > \bar{N}_j/g_j$, which returns the inequality stated in the text. For further discussion, see A. Yariv, *Quantum Electronics,* Wiley, New York, 1968.

PROBLEMS

13.34   Obtain a relation between the spontaneous emission coefficient A_{kl} and the dipole matrix element $|d_{kl}|$, analogous to (13.70).

13.35   The electric dipole moment of the ammonia molecule, NH_3, has magnitude $d = 1.47 \times 10^{-18}$ esu-cm. A beam of these molecules with dipoles polarized in the $+z$ direction enters a domain of electric field of strength 1.62×10^4 V/cm, which points in the $-z$ direction. The resulting interaction between the molecules and the field gives rise to coherent radiation. What is the frequency of this radiation (Hz)? (*Note:* Units of voltage in cgs are statV; 1 statV = 300 V.)

Answer

$$\nu = 24 \text{ GHz}$$

13.36   In deriving Planck's formula (see Problem 2.36 et seq.) for the density of photons in frequency interval ν, $\nu + d\nu$, in a radiation field in equilibrium at the temperature T, in cubic volume $V = a^3$,

$$n(\nu)\, d\nu = \frac{8\pi\nu^2\, d\nu}{c^3} \frac{1}{e^{h\nu/k_B T} - 1}$$

one assumes the relation

$$n(\nu)\, d\nu = f_{BE}(\nu)g(\nu)\, d\nu$$

The term $g(\nu)\, d\nu$ is the density of available states in the said frequency interval, while f_{BE} is the *Bose–Einstein* factor, giving the average number of photons per state,

$$f_{BE}(\nu) = \frac{1}{e^{h\nu/k_B T} - 1}$$

Using the momentum-position uncertainty relation (appropriate to a three-dimensional box), derive the following expression for the *density of states g*, relevant to a radiation field:

$$g(\nu)\, d\nu = \frac{8\pi\nu^2\, d\nu}{c^3}$$

Answer

With $p = h\nu/c$, we look at Cartesian **p** space and note the volume of a spherical shell of radius p and thickness dp.

$$\text{Volume of shell} = 4\pi p^2\, dp$$

All momenta in this shell correspond to nearly the same energy—and therefore the same frequency. Owing to the uncertainty principle for a particle confined to a three-dimensional box (see Problem 5.42), the smallest volume (a cell in **p** space) that may be specified with certainty to contain a state is

$$(\Delta p)^3_{\min} = \frac{(h/2)^3}{a^3} = \frac{(h/2)^3}{V}$$

It follows that the number of states in the dp shell is

$$\frac{\text{Volume of shell}}{\text{Volume of cell}} = \frac{4\pi p^2 \, dp \, V}{(h/2)^3}$$

Finally, we note that one counts states only in the first quadrant of **p** space insofar as photons have only positive frequencies. Furthermore, to each such **p** state there are two photon states, corresponding to two possible polarizations. This gives

$$Vg(\nu) \, d\nu = \frac{1}{8} \times 2 \times \frac{4\pi V p^2 \, dp}{h^3/8} = \frac{8\pi V p^2 \, dp}{h^3}$$

The desired answer follows using the given relation between p and ν for a photon.

13.37 (a) What are the Hamiltonian $\hat{H}_0$, eigenstates and eigenvalues of a collection of harmonic oscillators with natural frequencies, $\omega_1, \omega_2, \ldots$, in the number operator representation?

(b) The Hamiltonian constructed in part (a) represents a radiation field in *second quantization*. In this representation the radiation field is viewed as a collection of harmonic oscillators. A perturbation $\hat{H}'$ is applied to the radiation field such that

$$[\hat{H}', \hat{N}_1] = \frac{i\hbar}{\tau} \hat{N}_1$$

$$\hat{N}_1 \equiv \hat{a}_1^\dagger \hat{a}_1$$

The subscript 1 denotes the frequency ω_1. Show that $[\hat{H}_0, \hat{N}_1] = 0$. Then show that the perturbation $\hat{H}'$ represents a sink that diminishes the number of photons of frequency ω_1 exponentially with an e-folding time τ.

Answers

(a)
$$\hat{H}_0 = \sum_i \hbar\omega_i \left(\hat{a}_i^\dagger \hat{a}_i + \frac{1}{2} \right)$$

Eigenstates take the form

$$|\psi_{n_1 n_2 \ldots}\rangle = |n_1, n_2, n_3, \ldots\rangle$$

The number of photons in the ω_i mode is n_i. These wavefunctions have the properties

$$\hat{a}_i | \ldots, n_i, \ldots \rangle = \sqrt{n_i} | \ldots, n_1 - 1, \ldots \rangle$$
$$\hat{a}_i^\dagger | \ldots, n_i, \ldots \rangle = \sqrt{n_i + 1} | \ldots, n_i + 1, \ldots \rangle$$

It follows that

$$\hat{H}_0 |n_1, n_2, \ldots\rangle = \sum_i \hbar\omega_i \left(n_i + \frac{1}{2} \right) |n_1, n_2, \ldots\rangle$$

(b) The average number of photons of frequency ω_1 is

$$\langle N_1 \rangle = \langle n_1, n_2, \ldots | \hat{N}_1 | n_1, n_2, \ldots \rangle = n_1$$

The equation of motion for $\langle N_1 \rangle$ is [recall (6.68)]

$$\frac{d\langle N_1 \rangle}{dt} = \frac{i}{\hbar} \langle [\hat{H}, \hat{N}_1] \rangle = \frac{i}{\hbar} \langle [\hat{H}', \hat{N}_1] \rangle = -\frac{1}{\tau} \langle N_1 \rangle$$

$$\frac{dn_1}{dt} = -\frac{1}{\tau} n_1 \qquad n_1 = n_1(0) e^{-t/\tau}$$

13.38   In Problem 13.37 we found that the Hamiltonian of an electromagnetic radiation field may be written[1]

$$\hat{H}_R = \sum_j \hbar\omega_j \hat{a}_j^\dagger \hat{a}_j$$

Recalling that the frequency of a photon of momentum $\hbar\mathbf{k}$ is $\omega = ck$ [see (2.28)], it follows that the operator corresponding to the total momentum of the radiation field is

$$\hat{\mathbf{P}} = \sum_j \hbar\mathbf{k}_j \hat{a}_j^\dagger \hat{a}_j$$

Here we are assuming that the field is contained in a large cubical box with perfectly reflecting walls. Boundary conditions then imply a discrete sequence of wavenumber vectors $\{\mathbf{k}_i\}$. (See Problem 2.37.) Consider that a charged particle bound to a point within the radiation field vibrates along the z axis. Its Hamiltonian is

$$\hat{H}_P = \hbar\omega_0 \left(\hat{a}^\dagger \hat{a} + \frac{1}{2} \right)$$

(a) If the oscillator and the field are uncoupled from each other, what are the Hamiltonian and eigenstates of the composite system?

(b) Consider now that a small coupling exists between the particle and the field whose interaction energy is proportional to the scalar product of the displacement of the particle and the total momentum of the field, with coupling constant $\alpha\hbar\omega_0^2/mc^2$, where α is the fine-structure constant. Assuming (13.61) to be appropriate, calculate the probability rate that as a result of this interaction, the oscillator emits a photon of energy $\hbar\omega_0$.

[1] Here we are omitting the infinite zero-point energy $\sum \frac{1}{2}\hbar\omega_j$.

Answers

(a)
$$\hat{H}_{PF} = \hat{H}_P + \hat{H}_F = \hbar\omega_0\left(\hat{a}^\dagger\hat{a} + \frac{1}{2}\right) + \sum_j \hbar\omega_j\hat{a}_j^\dagger\hat{a}_j$$

$$|\psi\rangle = |\psi_P\rangle|\psi_F\rangle = |n; n_1, n_2, \dots\rangle$$

$$\hat{H}_{PF}|\psi\rangle = \left[\hbar\omega_0\left(n + \frac{1}{2}\right) + \sum_j \hbar\omega_j n_j\right]|\psi\rangle$$

The notation is such that n_j denotes the number of photons with momentum $\hbar\mathbf{k}_j$.

(b) The interaction Hamiltonian is

$$\hat{H}' = \frac{\alpha\hbar\omega_0^2}{mc^2}\frac{1}{\sqrt{2}\beta}(\hat{a} + \hat{a}^\dagger)\sum_j \hbar k_{jz}\hat{a}_j^\dagger\hat{a}_j$$

where $\beta^2 \equiv m\omega_0/\hbar$ [recall (7.26)] and k_{jz} denotes the z component of $\mathbf{k}_j$. To apply (13.61) we must first calculate the matrix element of $\hat{H}'$ between the initial state

$$|\psi_i\rangle = |n; n_1, \dots, n_k, \dots\rangle, \qquad E_i = \hbar\omega_0\left(n + \frac{1}{2}\right) + \hbar\omega_0 n_k + \sum_{j\neq k} \hbar\omega_j n_j$$

and the final state

$$\langle\psi_f| = \langle n - 1; n_1, \dots, n_k + 1,$$

$$\dots|, \qquad E_f = \hbar\omega_0\left(n - \frac{1}{2}\right) + \hbar\omega_0(n_k + 1) + \sum_{j\neq k} \hbar\omega_j n_j$$

Here n_k represents the number of photons with wavenumber ω_0/c. This choice of $\langle\psi_f|$ guarantees conservation of energy in the transition, as prescribed by the delta-function factor $\delta(E_i - E_f)$ in (13.61). Since $|\psi_i\rangle$ (as well as $\langle\psi_f|$) is an eigenvector of $\sum \hat{a}_j^\dagger\hat{a}_j$, it may be brought through the field component of $\hat{H}'$. When completed with the ket vector $\langle\psi_f|$, the inner product of the field component factors of the wavefunctions vanishes and we conclude that for the given perturbation,

$$\langle\psi_f|H'|\psi_i\rangle = H'_{if} = 0$$

The hypothetical field-particle coupling does not induce a transition of the state of the harmonic oscillator.

13.8 SELECTIVE PERTURBATIONS IN TIME

The Adiabatic Theorem

Let us now consider the case that $\hat{H}'$ is *adiabatically* turned on, which means that $\hat{H}'$ changes slowly in time (Fig. 13.13). Consequently, at any instant of time, the Hamiltonian may be treated as constant and an approximate solution may be obtained

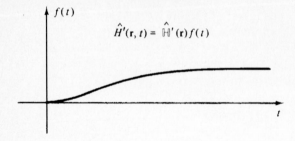

$$\hat{H}'(\mathbf{r}, t) = \hat{\mathbb{H}}'(\mathbf{r})f(t)$$

FIGURE 13.13   Adiabatic perturbation. The rise time, ω^{-1}, of the perturbation obeys the inequality

$$\omega \ll \omega_{kl}$$

where ω_{kl} are the natural frequencies of the unperturbed system.

by regarding the Schrödinger equation as time-independent. This will be shown below.

The slowly changing quality of $\hat{H}'$ may be incorporated into the analysis through a parts integration of (13.51).

$$(13.72) \quad c_k(t) = \frac{1}{i\hbar} \int_{-\infty}^{t} H'_{kl}(t')e^{i\omega_{kl}t'}dt' = -\frac{1}{\hbar\omega_{kl}} \int_{-\infty}^{t} H'_{kl}(t') \frac{\partial}{\partial t'} e^{i\omega_{kl}t'}dt'$$

$$= -\frac{1}{\hbar\omega_{kl}} \left\{ H'_{kl}(t)e^{i\omega_{kl}t} - \int_{-\infty}^{t} e^{i\omega_{kl}t'} \frac{\partial}{\partial t'} H'_{kl}(t') \, dt' \right\}$$

where we have set $H'_{kl}(t) \equiv \mathbb{H}'_{kl} f(t)$. If $\hat{H}'(t)$ is slowly varying, the second term is small compared to the first and $c_k(t)$ is well approximated by

$$(13.73) \qquad c_k(t) \simeq -\frac{1}{\hbar\omega_{kl}} H'_{kl}(t)e^{i\omega_{kl}t} = -\frac{\langle k|\hat{H}'|l\rangle e^{i\omega_{kl}t}}{E_k^{(0)} - E_l^{(0)}}$$

Let us recall that the analysis leading to (13.49) presumes that the system is in the stationary state ψ_l at $t < 0$. To terms of first order in the perturbation H', the series (13.42) then appears as

$$\psi(\mathbf{r}, t) = \psi_l + \sum_{k \neq l} c_k \psi_k$$

As the perturbation is slowly varying, we write $\psi(\mathbf{r}, t) \simeq \exp(-i\omega_l t)\varphi(\mathbf{r})$. Substituting this expression into the preceding and employing (13.73) gives

$$(13.74) \qquad \varphi(\mathbf{r}) = \varphi_l + \sum_{k \neq l} \frac{H'_{kl}\varphi_k}{E_l^{(0)} - E_k^{(0)}}$$

This is precisely the first-order result which stationary (time-independent) perturbation theory gives, (13.10). But the solution (13.10) represents an eigenfunction of the new Hamiltonian, which in the present case is $\hat{H} = \hat{H}_0 + \hat{H}'(t)$. That is, the wavefunction (13.74) represents the lth eigenstate of $\hat{H}_0 + \hat{H}'(t)$, to first order in H'. [Note that we are writing $\hat{H}'(t)$ for the operator $\hat{H}'$ evaluated at the specific time t.] Since the system was originally in the lth state of $\hat{H}_0$, we may conclude the following. A system originally in the lth state of an unperturbed Hamiltonian will at the end of an adiabatic perturbation be found in the lth state of the new Hamiltonian. One says that the system "remains in the lth state."

Having established that φ as given by (13.74) is an eigenstate of the new Hamiltonian, it follows that expectation of energy in this state is the same as the eigenenergy of the state. The expectation of energy to first order in the perturbation Hamiltonian is easily calculated with the explicit form of φ as given by (13.74). There results

(13.75)
$$E_l = E_l^{(0)} + H'_{ll}(t)$$

This is the lth eigenenergy of the Hamiltonian $\hat{H}_0 + \hat{H}'(t)$. So under an adiabatic perturbation, a system originally in the lth eigenstate of an unperturbed Hamiltonian will at time t be found in the lth eigenstate of the new Hamiltonian with the lth eigenenergy of the new Hamiltonian. With $\varphi_l(t)$ written for the time-independent wavefunction evaluated at the time t, we may write that under an adiabatic perturbation

(13.76a)
$$\varphi_l(0) \longrightarrow \varphi_l(t)$$
$$E_l(0) \longrightarrow E_l(t)$$

where

(13.76b)
$$[\hat{H}_0 + \hat{H}'(t)]\varphi_l(t) = E_l(t)\varphi_l(t)$$

This equation is *not* a time-dependent equation. One first evaluates $\hat{H}'(t)$ and then finds the solution $\varphi_l(t)$. These results constitute the *adiabatic theorem,* first proved by Born and Fock in 1928.[1]

As an example of the use of this theorem, consider that a particle of mass m is in a one-dimensional box of edge length a. The width of the box is slowly increased to αa over a long interval lasting t seconds, where $\alpha > 1$. Let us calculate the new wavefunction and new energy of the system given that the particle initially was in the ground state. The adiabatic theorem tells us that the new state is the ground state of the new Hamiltonian.

$$\varphi_1 \rightarrow \varphi_1(t) = \sqrt{\frac{2}{\alpha a}} \sin\left(\frac{\pi x}{\alpha a}\right)$$

[1] M. Born and V. Fock, Z. *Physik,* **51**, 165 (1928).

Here, as above, we are writing $\varphi(t)$ for the time-independent wavefunction evaluated at the time t. The new energy is the ground state of the new Hamiltonian.

$$E_1 \rightarrow E_1(t) = \frac{h^2}{8m(\alpha a)^2}$$

Note that in slowly expanding (as in the classical case), the particle loses energy to the receding walls in slowing down. The amount of work absorbed by the walls in the present example is

$$W = E_1 - E(t) = \frac{h^2}{8ma^2}\left(1 - \frac{1}{\alpha^2}\right)$$

Domain of Validity

Our conclusions regarding an adiabatic perturbation rest on neglecting the integral term in (13.72). Let us obtain a quantitative criterion which allows this term to be discarded. Let the perturbation be gradually turned on at $t = 0$. If $\hat{H}'$ changes slowly, then $\partial \hat{H}'/\partial t$ may be taken outside the integral term in (13.72) and we obtain

$$\left| \int_0^t e^{i\omega_{kl}t'} \frac{\partial H'_{kl}}{\partial t'} \, dt' \right| \simeq \left| \frac{\partial H'_{kl}}{\partial t} \right| \left| \int_0^t e^{i\omega_{kl}t'} \, dt' \right| \simeq \left| \frac{2}{\omega_{kl}} \frac{\partial H'_{kl}}{\partial t} \right| \left| \sin\left(\frac{\omega_{kl}t}{2}\right) \right|$$

It follows that the nonintegral term in (13.72) dominates provided that

(13.77)
$$|\omega_{kl} H'_{kl}| \gg 2 \left| \frac{\partial H'_{kl}}{\partial t} \right|$$

Thus a perturbation that undergoes a small fractional change in a typical period of the unperturbed system may be termed adiabatic.

Transition Probability

We have found that in the adiabatic limit, neglecting the integrated term in (13.72) leads to the system remaining in the initial state of the unperturbed Hamiltonian. What is the probability in this same limit that there is a transition out of the initial state? Suppose, for example, that an adiabatic perturbation is turned on at $t = 0$ and that the unperturbed system is originally in the lth state of the unperturbed Hamiltonian. The probability for a transition from the l to the k state is given by $|c_k|^2$. With (13.72) in the adiabatic limit, we obtain

$$|c_k|^2 = \frac{1}{\hbar^2(\omega_{kl})^2}\left[|H'_{kl}|^2 + \frac{4}{(\omega_{kl})^2}\left|\frac{\partial}{\partial t}H'_{kl}\right|^2 \sin^2\left(\frac{\omega_{kl}t}{2}\right)\right.$$
$$\left. - \frac{2}{\omega_{kl}}\frac{\partial(H'_{kl})^2}{\partial t}\sin\left(\frac{\omega_{kl}t}{2}\right)\cos\left(\frac{\omega_{kl}t}{2}\right)\right]$$

Over a long time interval, the last term averages to zero, whereas the $\sin^2$ term in the middle expression averages to $\frac{1}{2}$. As we have found above, the first term represents the probability that there is no transition out of the lth state. It follows that the probability that there is a transition out of the initial lth state to some kth state ($k \neq l$) in the adiabatic limit is

(13.78)
$$P_{lk} = \frac{2}{\hbar^2(\omega_{kl})^4}\left|\frac{\partial}{\partial t}H'_{kl}\right|^2$$

Once more we find that there is vanishingly small probability of transition out of the lth state for sufficiently slowly changing perturbation.

An instructive laboratory example of the adiabatic theorem is given by elastic collision of molecules in a gas. During the collision, electron states are acted upon by intermolecular forces. This collisional interaction varies roughly as the intermolecular velocity, which, typically, is far smaller than mean atomic electron speeds. Thus, collisional forces which act on atomic states may be viewed as being adiabatically switched on. Consistent with this observation, one finds that whereas final rotational and vibrational molecular states are altered in the collision, atomic states suffer minimal change.[1]

Sudden Perturbation

The next type of perturbation problem we wish to examine involves a sudden change in a parameter in $\hat{H}_0$. For example, suppose that the spring constant of a simple harmonic oscillator is suddenly doubled. If the oscillator is in its ground state before the perturbation, in what state is it after perturbation? For such problems it is presumed that the eigenstates of both Hamiltonians (i.e., before and after perturbation) are known. This, together with the assumption of an instantaneous change of $\hat{H}_0$, are basic to the *sudden approximation*.

---

[1] For further discussion, see R. D. Levine and R. B. Bernstein, *Molecular Reaction Dynamics and Chemical Reactivity*, Oxford, New York, 1987.

Let us call the Hamiltonian before the change in parameter, $\hat{H}$, and the Hamiltonian after the change in parameter, $\hat{H}'$, so that

$$\hat{H}\varphi_n = E_n\varphi_n \qquad t < 0$$
$$\hat{H}'\varphi_n' = E_n'\varphi_n' \qquad t \geq 0$$

(Note that $\hat{H}'$ is now a total Hamiltonian.) Initially, the system is in an eigenstate ψ_l of $\hat{H}$.

$$\psi'(\mathbf{r}, 0) = \varphi_l(\mathbf{r})$$

At later times the system is in a superposition state of $\hat{H}'$.

$$\psi'(\mathbf{r}, t) = \sum_n c_n\varphi_n'e^{-i\omega_n't}$$

Equating this function to its initial value at $t = 0$ gives

$$\varphi_l = \sum_n c_n\varphi_n'$$

so that

$$c_k = \langle\varphi_k'|\varphi_l\rangle$$

The probability that the sudden change from $\hat{H}$ to $\hat{H}'$ causes a transition from the lth state of $\hat{H}$ to the kth state of $\hat{H}'$ is

(13.79) $$P_{lk} = |c_k|^2 = |\langle\varphi_k'|\varphi_l\rangle|^2$$

A criterion for the validity of this approximation scheme may be drawn from the preceding relation. As described above, let the sudden change in Hamiltonian occur at $t = 0$. For the relation (13.79) to have meaning it is necessary that the initial wavefunction $\varphi_l(\mathbf{r})$: (1) maintain its form at $t = 0^+$ and (2) lie in the Hilbert space spanned by the eigenstates of the new Hamiltonian. Thus, for a particle confined to a rigid-walled box, we may conclude that the approximation is inappropriate to sudden compression of the box but is appropriate to sudden expansion of the box.

To further demonstrate this approximation, let us apply the formalism to the latter case. Thus we consider a particle of mass m confined to a one-dimensional box which suffers a sudden expansion. With no loss in generality we take the box to be of unit width ($a = 1$). The particle is in the energy eigenstate φ_l at $t < 0$. At $t = 0$ the box is suddenly expanded to the edge length $\alpha > 1$ (recall Problem 5.11). With the preceding formalism we may establish the somewhat surprising result that although no work is done on the particle in the expansion, energy is not necessarily conserved.

The new Hamiltonian $\hat{H}'$ (after expansion) is given by (4.2,3) with a replaced by α. New basis wavefunctions are the eigenfunctions

$$\varphi_k' = \sqrt{\frac{2}{\alpha}} \sin\left(\frac{k\pi x}{\alpha}\right)$$

The probability for transition is given by

$$(13.80) \qquad \sqrt{P_{lk}} = \langle \varphi_l | \varphi_k' \rangle = \frac{2}{\sqrt{\alpha}} \int_0^1 \sin l\pi x \sin\left(\frac{k\pi x}{\alpha}\right) dx$$

$$= \frac{1}{\sqrt{\alpha}} \left\{ \frac{\sin \pi[l - (k/\alpha)]}{\pi[l - (k/\alpha)]} - \frac{\sin \pi[l + (k/\alpha)]}{\pi[l + (k/\alpha)])} \right\}$$

Note that the integral in the preceding expression goes over $(0, 1)$ and not $(0, \alpha)$ because $\varphi_l(x)$ is zero over $(1, \alpha)$.

The transition which conserves energy occurs for the k value $\bar{k} = \alpha l$. With $A \equiv \hbar^2 \pi^2 / 2m$, we obtain

$$(13.81) \qquad E_k' \Big|_{k=\bar{k}} = A \frac{k^2}{\alpha^2} \Big|_{k=\bar{k}} = Al^2 = E_l$$

At this value of k, (13.80) gives

$$(13.81\text{a}) \qquad P_{l\bar{k}} = \frac{1}{\alpha} < 1$$

Thus, as stated, it is not certain that the transition conserves energy. However, it is simply shown that expectation of energy is conserved. To show this, we note that at $t = 0^+$ the wavefunction for the system at hand is given by

$$\bar{\varphi}_l = \varphi_l \qquad 0 \leq x \leq 1$$
$$\bar{\varphi}_l = 0 \qquad 1 < x \leq \alpha$$

It follows that

$$(13.82) \qquad E_l = \langle E \rangle = \langle \varphi_l | \hat{H} | \varphi_l \rangle = \langle \bar{\varphi}_l | \hat{H}' | \bar{\varphi}_l \rangle$$

Note that the third term is defined over $(0, 1)$ and is the value of $\langle E \rangle$ at $t < 0$. The last term is defined over $(0, \alpha)$ and is appropriate to $t = 0^+$. We conclude that $\langle E \rangle$ is conserved in the expansion. Further developing (13.82) gives

$$E_l = \sum_k \langle \overline{\varphi}_l | \hat{H}' | \varphi_k' \rangle \langle \varphi_k' | \overline{\varphi}_l \rangle$$

(13.83)

$$E_l = \sum_k P_{lk} E_k'$$

or, equivalently,

(13.84)

$$l^2 = \frac{1}{\alpha^2} \sum_{k=1}^{\infty} P_{lk} k^2$$

with P_{lk} given by (13.80). This result is valid for arbitrary $\alpha > 1$, i.e., for irrational as well as rational values. In particular, note that if $k/\alpha \neq l$ for all k, then no direct

---

TABLE 13.2   Transition probabilities for time-dependent perturbations

1. $\hat{H}'$ is separable. $\hat{H}'(\mathbf{r}, t) = \mathbb{H}'(\mathbf{r}) f(t)$

$$P_{ik} = \frac{|\mathbb{H}_{kl}'|^2}{\hbar^2} \left| \int_{-\infty}^{t} e^{i\omega_{kl} t'} f(t') \, dt' \right|^2$$

Harmonic perturbation, $f = 2 \cos \omega t$, turned on at $t = 0$:

$$P_{lk} = \frac{|\mathbb{H}_{kl}'|^2}{[\hbar(\omega_{kl} - \omega)/2]^2} \sin^2 [(\omega_{kl} - \omega)t/2]$$

DC perturbation ($\omega = 0$, $f = 1$) turned on at $t = 0$:

$$P_{lk} = \frac{|\mathbb{H}_{kl}'|^2 \sin^2 (\omega_{kl} t/2)}{(\hbar \omega_{kl}/2)^2}$$

Long-time or high-frequency limit:

$$P_{lk} = \frac{2\pi t |\mathbb{H}_{kl}'|^2}{\hbar^2} \delta(\omega_{kl} - \omega)$$

$$w_{kl} = \frac{2\pi |\mathbb{H}_{kl}'|^2}{\hbar^2} \delta(\omega_{kl} - \omega)$$

Short-time or low-frequency limit:

$$P_{lk} = \frac{t^2 |\mathbb{H}_{kl}'|^2}{\hbar^2}$$

$$w_{kl} = \frac{t |\mathbb{H}_{kl}'|^2}{\hbar^2}$$

Probability for transition to a band centered at E_k with $g |\mathbb{H}'|$ slowly varying in energy:

$$\overline{P}_{lk} = \frac{2\pi t}{\hbar} g(E_k) |\mathbb{H}_{kl}'|^2$$

$$\overline{w}_{kl} = \frac{2\pi}{\hbar} g(E_k) |\mathbb{H}_{kl}'|^2$$

2. $\hat{H}' = \hat{H}'(\mathbf{r}, t)$ is slowly changing
Probability for transition out of the initial l state:

$$P_{lk} \simeq \frac{2}{\hbar^2 (\omega_{kl})^4} \left| \frac{\partial}{\partial t} H_{kl}' \right|^2$$

Adiabatic theorem:

$$P_{kl} \simeq 0 \qquad (k \neq l)$$

3. $\hat{H}$ changes suddenly to $\hat{H}'$
Eigenstates of both $\hat{H}$ and $\hat{H}'$ are known.

$$P_{lk} = |\langle \varphi_k' | \varphi_l \rangle|^2$$

transition conserves energy. However, with (13.83), expectation of energy is still conserved.

Expressions obtained above for the transition probability in various limiting cases are listed in Table 13.2.

PROBLEMS

13.39   A neutron in a rigid spherical well of radius $a = 0.1$ Å is in the ground state. The radius of the well is slowly decreased to $0.9a$.

(a)   What is the energy and wavefunction of the neutron after the decrease in the well radius? (For normalization, see Table 10.2.)

(b)   How much work (in eV) is performed during the compression of the well?

13.40   A collection of $N_0 = 10^{13}$ independent electrons have spins polarized parallel to a uniform magnetic field that points in the z direction, of magnitude $\mathcal{B}_0$. A perturbation field is applied in the x direction of magnitude

$$\mathcal{B}'(t) = 10^{-3}\mathcal{B}_0(1 - e^{-t/\tau}) \qquad (t \geq 0)$$

(a)   Obtain a criterion involving τ which ensures that the perturbation is adiabatic.

(b)   Given that $\Omega\tau = 10^2$ and $\mathcal{B}_0 = 10^4$ gauss, estimate the number of electrons ΔN that are thrown out of the ground spin state at $t = 10\tau$.

Answers

(a) $\tau \gg \Omega^{-1} = mc/e\,\mathcal{B}_0$

(b)   A rough estimate is obtained from (13.78).

$$\Delta N \simeq 2N_0 \left| \frac{10^{-3}\mu_b\mathcal{B}_0}{\hbar\Omega} \right|^2 \left| \frac{e^{-t/\tau}}{\Omega\tau} \right|^2$$

13.41   A one-dimensional harmonic oscillator has its spring constant suddenly reduced by a factor of $\frac{1}{2}$. The oscillator is initially in its ground state. Show that the probability that the oscillator remains in the ground state is $P = 0.986$.

13.42   A particle of mass m in a one-dimensional box of width a is in the third excited state (above the ground state). The width of the box is suddenly doubled. What is the probability that the particle drops to the ground state?

13.43   A one-dimensional harmonic oscillator in the ground state is acted upon by a uniform electric field

$$\mathcal{E}(t) = \frac{\mathcal{E}_0}{\sqrt{\pi}} \exp\left[-\left(\frac{t}{\tau}\right)^2 \right]$$

switched on at $t = -\infty$. The field is parallel to the axis of the oscillator. What is the probability that the oscillator suffers a transition to its first excited state at $t = +\infty$ in the limits: (a) $\omega_0\tau \gg 1$, (b) $\omega_0\tau \approx 1$. For case (b), are any other transitions possible? [*Hint:* See Problem 13.27.]

13.44 Radioactive tritium, H^3, decays to light helium, (He^{3+}), with the emission of an electron. (This electron quickly leaves the atoms and may be ignored in the following calculation.) The effect of the β decay is to change the nuclear charge at $t = 0$ without effecting any change in the orbital electron. If the atom is initially in the ground state, what is the probability that the He^+ ion is left in the ground state after the decay?

13.45 A hydrogen atom in the ground state is placed in a uniform electric field in the z direction,

$$\mathcal{E} = \mathcal{E}_0 e^{-t/\tau}$$

which is turned on at $t = 0$. What is the probability that the atom is excited to the $2P$ state at $t \gg \tau$?

13.46 The perturbation

$$H' = \frac{A}{\tau\sqrt{\pi}} e^{i(\pi x/a)} e^{-t^2/\tau^2}$$

is applied to a particle of mass m in a one-dimensional box of width a at $t = -\infty$. At this time the particle is in the ground state. If $\hbar/\tau \ll E_1$, in what state is it most probable that the particle will be at $t = +\infty$?

13.47 An electron in a one-dimensional potential well

$$V = \frac{1}{2}Kx^2$$

is immersed in a constant, uniform electric field of magnitude $\mathcal{E}$ which points in the x direction. The corresponding perturbation to the Hamiltonian is

$$H' = e\mathcal{E}x$$

(a) Find the exact eigenenergies of the total Hamiltonian,

$$\hat{H} = \frac{\hat{p}^2}{2m} + \frac{1}{2}Kx^2 + e\mathcal{E}x$$

(see Problem 7.16). Discuss your findings with respect to the corresponding classical motion.

(b) Show that the first-order corrections to the energy vanish. Then calculate the second-order corrections. Show that these agree with your answer to part (a), so that the second-order corrections give the complete solution for this problem.

Answers

(a) Setting

$$\chi \equiv \frac{e\mathcal{E}}{K}$$

together with the transformation of variables

$$x' \equiv x + \chi$$

permits $\hat{H}$ to be rewritten

$$\hat{H} = \frac{p^2}{2m} + \frac{1}{2}K(x^2 + 2\chi x)$$

$$= \frac{p^2}{2m} + \frac{1}{2}Kx'^2 - \frac{1}{2}K\chi^2$$

$$= \hat{H}_0 - \mathbb{E}$$

Since

$$\mathbb{E} \equiv \frac{1}{2}K\chi^2 = \frac{e^2\mathscr{E}^2}{2K}$$

is constant, the eigenenergies of $\hat{H}$ are simply

$$E_n = E_n^0 - \mathbb{E} = \hbar\omega_0\left(n + \frac{1}{2}\right) - \frac{e^2\mathscr{E}^2}{2K}$$

All levels are equally depressed by the constant energy $\mathbb{E}$. The new wavefunctions are

$$\varphi_n = \varphi_n(x') = \varphi_n(x + \chi)$$

The center of symmetry of these wavefunctions is at $x = -\chi$.

In the corresponding classical problem, the potential of the electron in the presence of the uniform electric field is

$$V = \frac{1}{2}K(x + \chi)^2 - \mathbb{E}$$

This parabola is congruent to the original potential $Kx^2/2$. The new equilibrium at $x = -\chi$ occurs where the electric force is balanced by the spring force. This new potential minimum is lower than the original minimum by the amount $\mathbb{E}$. (Work must be done to move the electron, quasi-statically, from $x = -\chi$ to $x = 0$.) The classical analog of the quantum mechanical problem considered above involves an electric field that is very slowly (adiabatically) turned on. If the energy of the oscillator initially is $E^{(0)}$, what is it after the electric field is established? In classical mechanics, for such adiabatically changing harmonic motion, the ratio A/ω is constant (it is an *adiabatic invariant*). The amplitude of oscillation is A. Since the new potential well is congruent to the old well, ω is the same for both wells. This means that the amplitude of oscillation must also be preserved (during the adiabatic change). Owing again to congruency of the parabolas, this is ensured (only) if the distance between $E^{(0)}$ and the bottom of the new well is the same as the distance between $E^{(0)}$ and the bottom of the original well. That is, if

$$E^{(0)} = E + \mathbb{E}$$

It follows that the new energy of the oscillator

$$E = E^{(0)} - \mathbb{E}$$

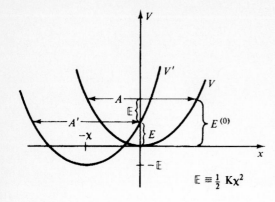

FIGURE 13.14 The classical adiabatic change from a parabolic potential, V, to a congruent ($\omega_0 = \omega_0'$) parabolic potential, V', with a new center of symmetry, preserves amplitude A. We see that the amplitude of oscillation is preserved, $A = A'$, provided that $E = E^{(0)} - \mathbb{E}$. This result for the new energy of oscillation, E, is identical to the quantum mechanical result. (See Problem 13.47.)

is depressed from the initial value, $E^{(0)}$, by the amount $\mathbb{E}$. This is identical to the quantum mechanical result (Fig. 13.14).

(b) The perturbation Hamiltonian may be rewritten

$$\hat{H} = \frac{e\mathscr{E}}{\sqrt{2}\beta}(\hat{a}^\dagger + \hat{a}) \equiv \mathscr{E}'(\hat{a}^\dagger + \hat{a})$$

$$\beta^2 \equiv \frac{m\omega_0}{\hbar}, \qquad \mathscr{E}' \equiv \frac{e\mathscr{E}}{\sqrt{2}\beta}$$

It follows immediately that

$$\langle n|\hat{H}'|n\rangle = 0$$

so there is no first-order correction to $E_n^{(0)}$. To calculate the second-order corrections, we must evaluate the off-diagonal matrix elements of $\hat{H}'$.

$$\langle n|\hat{H}|l\rangle = \mathscr{E}'\langle n|\hat{a}^\dagger + \hat{a}|l\rangle$$
$$= \mathscr{E}'(\sqrt{l+1}\,\delta_{n,l+1} + \sqrt{l}\,\delta_{n,l-1})$$

Substituting this expression into (13.14) gives the desired result.

13.48 A system with discrete energy states and Hamiltonian $\hat{H}_0$ has the density operator $\hat{\rho}_0$, which is diagonal. Furthermore, $[\hat{\rho}_0, \hat{H}_0] = 0$, so $\hat{\rho}_0$ is constant in time. Show that after a perturbation $\hat{H}'$ is applied, the diagonal elements of $\hat{\rho}$ change according to the *Pauli equation*

$$\frac{\partial \rho_{nn}}{\partial t} = \sum_k w_{nk}(\rho_{kk} - \rho_{nn})$$

The transition rates w_{nk} are given by (13.63).

$$w_{nk} = \lim_{\Delta t \to 0} \frac{\Delta t}{\hbar^2}|H_{nk}'|^2$$

The density operator was discussed in Section 11.11.

Answer

For a short-time interval after the perturbation $\hat{H}'$ is applied, we may expand $\hat{\rho}$ to obtain

$$\hat{\rho}(\Delta t) - \hat{\rho}(0) = \left(\frac{\partial \hat{\rho}}{\partial t}\right)_0 \Delta t + \left(\frac{\partial^2 \hat{\rho}}{\partial t^2}\right)_0 \frac{(\Delta t)^2}{2} + \cdots$$

Using the equation of motion (11.122) for $\hat{\rho}$ permits the last equation to be written, with $\hat{\rho}(0) = \hat{\rho}_0$,

$$\frac{\partial \hat{\rho}}{\partial t} = \lim_{\Delta t \to 0} \left[\frac{\hat{\rho}(\Delta t) - \hat{\rho}_0}{\Delta t} \right] = \lim_{\Delta t \to 0} \left\{ \frac{1}{i\hbar} [\hat{H}, \hat{\rho}]_0 - \frac{\Delta t}{2\hbar^2} [\hat{H}, [\hat{H}, \hat{\rho}]]_0 \right\}$$

The diagonal form of $\hat{\rho}_0$ leads to the following properties:

$$\langle n | [\hat{H}_0 + \hat{H}', \hat{\rho}_0] | n \rangle = 0$$
$$\langle n | [\hat{H}_0, [\hat{H}', \hat{\rho}_0]] | n \rangle = 0$$

Forming the diagonal elements of $\partial \hat{\rho} / \partial t$ then gives

$$\frac{\partial \rho_{nn}}{\partial t} = -\lim_{\Delta t \to 0} \frac{\Delta t}{2h^2} \langle n | [\hat{H}', [\hat{H}', \hat{\rho}_0]] | n \rangle$$

The diagonal element on the right-hand side reduces to

$$\langle n | [\hat{H}', [\hat{H}', \hat{\rho}_0]] | n \rangle = 2 \sum_k \{ |H'_{nk}|^2 \rho_{nn}(0) - |H'_{nk}|^2 \rho_{kk}(0) \}$$

which when substituted into the preceding equation gives the desired result.

13.9 ATOM-RADIATION INTERACTION

Our first description of the interaction between an atom and a radiation field was given in terms of Einstein's derivation of the Planck radiation law (Section 13.7). We now wish to present a Hamiltonian formulation of this problem. The Hamiltonian of an electron in an electromagnetic field with vector potential **A** (previously introduced in Section 10.4) and electric potential $V(\mathbf{r})$ is given by

(13.85)
$$\hat{H} = \frac{1}{2m}\left[\hat{\mathbf{p}} - \frac{e}{c}\mathbf{A}(\mathbf{r}, t) \right]^2 + V(\mathbf{r})$$

In the full quantum electrodynamic analysis of this problem, electrodynamic fields are quantized. The present analysis is termed a "semiclassical" description in that the vector potential **A** in (13.85) is taken as a classical field.[1]

---

[1] For a more complete quantum analysis, see E. G. Harris, *A Pedestrian Approach to Quantum Field Theory*, Wiley, New York, 1972; A. S. Davydov, *Quantum Mechanics*, 2d ed., Pergamon, Elmsford, N.Y., 1973.

Expanding the kinetic energy term in (13.85) gives

$$\hat{H} = \hat{H}^{(0)} + \hat{H}' + \hat{H}''$$

(13.86) $$\hat{H}^{(0)} \equiv \frac{\hat{p}^2}{2m} + V; \qquad \hat{H}' = -\frac{e}{2mc}[\hat{\mathbf{p}} \cdot \mathbf{A} + \mathbf{A} \cdot \hat{\mathbf{p}}]$$

$$H'' = \frac{e^2}{2mc^2} A^2$$

We neglect $\hat{H}''$ and consider $\hat{H}'$ a perturbation term to the unperturbed atomic Hamiltonian $\hat{H}^{(0)}$.

The matrix element of $\hat{H}'$ between initial $|n\rangle$ and final $|n'\rangle$ states of $\hat{H}^{(0)}$ is given by

(13.87) $$\langle n'|H'|n\rangle = -\frac{e}{2mc} \int \psi_{n'}^*(\hat{\mathbf{p}} \cdot \mathbf{A} + \mathbf{A} \cdot \hat{\mathbf{p}})\psi_n \, d\mathbf{r}$$

Note that the index n denotes a quantum state, not the principal quantum number. The first term in this integral may be written

(13.88) $$\hat{\mathbf{p}} \cdot \mathbf{A}\psi_n = \frac{\hbar}{i}\boldsymbol{\nabla} \cdot (\mathbf{A}\psi_n) = \frac{\hbar}{i}[\psi_n(\boldsymbol{\nabla} \cdot \mathbf{A}) + (\mathbf{A} \cdot \boldsymbol{\nabla})\psi_n]$$

The gradient and curl qualities of the potentials which enter electrodynamics permit certain gauge conditions to be imposed. In the so-called *Coulomb gauge* one sets

$$\boldsymbol{\nabla} \cdot \mathbf{A} = 0$$

so that (13.88) reduces to

$$\hat{\mathbf{p}} \cdot \mathbf{A}\psi_n = \frac{\hbar}{i}\mathbf{A} \cdot \boldsymbol{\nabla}\psi_n$$

Thus we may write (13.87) in the equivalent forms

(13.89a) $$\langle n'|H'|n\rangle = -\frac{e}{mc} \int \psi_{n'}^*\mathbf{A} \cdot \hat{\mathbf{p}}\psi_n \, d\mathbf{r}$$

(13.89b) $$\langle n'|H'|n\rangle = -\frac{e}{mc} \int \psi_{n'}^*\hat{\mathbf{p}} \cdot \mathbf{A}\psi_n \, d\mathbf{r}$$

A revealing insight into this analysis is gained if one associates the vector potential $\mathbf{A}$ with the wavefunction of a photon of energy $\hbar\omega$. This permits the preceding two matrix elements to be symbolically written

(13.90a) $$\langle n'|\hat{H}'|n\rangle \rightarrow -\frac{e}{mc}\langle n'; \hbar\omega|\hat{\mathbf{p}}|n\rangle \qquad (emission)$$

(13.90b) $$\langle n'|\hat{H}'|n\rangle \rightarrow -\frac{e}{mc}\langle n'|\hat{\mathbf{p}}|\hbar\omega; n\rangle \qquad (absorption)$$

Thus, (13.90a) corresponds to the case where a photon is emitted by the system, whereas (13.90b) corresponds to the case of the absorption of a photon. Since these two integrals are equal, we conclude that probabilities of emission and absorption, at the same frequency, are equal. This is an example of the principle of *microscopic reversibility*.

To obtain a more explicit representation of the matrix elements (13.89), we consider the atom in interaction with a plane electromagnetic wave whose vector potential is of the form

$$(13.91) \qquad \mathbf{A}(\mathbf{r}, t) = \mathbf{a} A_0 \cos (\mathbf{k} \cdot \mathbf{r} - \omega t)$$

where $\mathbf{a}$ is a unit polarization vector ($|\mathbf{a}|^2 = 1$) which is normal to the propagation vector $\mathbf{k}$. We wish to construct the amplitude A_0 so that the corresponding wave carries one photon per unit volume. The time-averaged energy density carried in a plane electromagnetic wave is

$$(13.92) \qquad \langle U \rangle = \frac{1}{4\pi} \langle \mathcal{B}^2 \rangle = \frac{1}{4\pi} \langle \mathcal{E}^2 \rangle$$

With

$$(13.93) \qquad \mathcal{B} = \boldsymbol{\nabla} \times \mathbf{A}$$

and given the form (13.91), we find that

$$(13.94) \qquad \langle U \rangle = \frac{1}{8\pi} k^2 A_0^2$$

For one photon per unit volume we set

$$(13.95) \qquad \frac{k^2 A_0^2}{8\pi} = \frac{\hbar \omega}{V}$$

which, with $\omega = ck$, gives

$$(13.96) \qquad A_0^2 = \frac{8\pi \hbar c^2}{\omega V}$$

Henceforth we set the volume $V = 1$.

In what follows, we will employ Fermi's golden rule (13.64). In formulating this relation it was found that for the time-dependent perturbation, $\cos \omega t$, the $\exp(-i\omega t)$ term was responsible for resonant absorption and the $\exp(+i\omega t)$ was responsible for decay. For consistent application of Fermi's rule, these observations must be incorporated into the present analysis. Thus, we first write (13.91) in the form

$$(13.97) \qquad \mathbf{A} = \frac{A_0}{2} \mathbf{a} [e^{i(\mathbf{k} \cdot \mathbf{r} - \omega t)} + e^{-i(\mathbf{k} \cdot \mathbf{r} - \omega t)}]$$

which permits the identification

(13.98)
$$\mathbf{A}_\pm = \frac{A_0}{2}\, \mathbf{a} e^{\pm i(\mathbf{k}\cdot\mathbf{r}-\omega t)}$$

where $\mathbf{A}_+$ corresponds to photon absorption and $\mathbf{A}_-$ to photon emission. So we may write

(13.99)
$$\mathbf{A}_\pm = c\left(\frac{2\pi\hbar}{\omega}\right)^{1/2} \mathbf{a} e^{\pm i(\mathbf{k}\cdot\mathbf{r}-\omega t)}$$

With

(13.100)
$$\hat{H}' = \hat{\mathsf{H}}_\pm(r) e^{\mp i\omega t}$$

and (13.89) we obtain

(13.101)
$$\langle n'|\,\mathsf{H}_\pm\,|n\rangle = -\frac{e}{m}\left(\frac{2\pi\hbar}{\omega}\right)^{1/2} \langle n'|\, \mathbf{a}\cdot\hat{\mathbf{p}} e^{\pm i\mathbf{k}\cdot\mathbf{r}}\,|n\rangle$$

The Dipole Approximation

For typical atomic transitions $\lambda \gg a_0$, where the Bohr radius a_0 is a length characteristic of atomic dimensions. Thus, over the domain of integration in the preceding matrix element, $\mathbf{k}\cdot\mathbf{r} \ll 1$, and we may write

$$e^{i\mathbf{k}\cdot\mathbf{r}} = 1 + i\mathbf{k}\cdot\mathbf{r} + \cdots$$

In the *dipole approximation*, one sets $e^{i\mathbf{k}\cdot\mathbf{r}} = 1$. There results

(13.102)
$$\langle n'|\,\mathsf{H}_\pm\,|n\rangle = -\frac{e}{m}\left(\frac{2\pi\hbar}{\omega}\right)^{1/2} \langle n'|\, \mathbf{a}\cdot\hat{\mathbf{p}}\,|n\rangle$$

With the aid of the commutator relation (5.62) this matrix element may be transformed to one only involving $\mathbf{r}$ (see Problem 13.60):

(13.103)
$$\hat{\mathbf{p}} = \frac{im}{\hbar}[\hat{H}^{(0)}, \mathbf{r}]$$

It follows that

(13.104)
$$\langle n'|\,\mathbf{p}\,|n\rangle = \frac{im}{\hbar}\langle n'|\, H^{(0)}\mathbf{r} - \mathbf{r}H^{(0)}\,|n\rangle$$

Recalling that $|n'\rangle$ and $|n\rangle$ are eigenfunctions of the unperturbed Hamiltonian $\hat{H}^{(0)}$, we obtain

(13.105)
$$\langle n'|\,\mathbf{p}\,|n\rangle = -\frac{im}{\hbar}(E_n^{(0)} - E_{n'}^{(0)})\langle n'|\,\mathbf{r}\,|n\rangle$$
$$\langle n'|\,\mathbf{p}\,|n\rangle = -im\omega\langle n'|\,\mathbf{r}\,|n\rangle$$

Note that we are writing ω for $\omega_{nn'} > 0$. With these relations at hand, (13.102) may be rewritten more explicitly as

(13.106) $\langle n' | \mathbb{H}_+ | n, \mathbf{k} \rangle = \langle n', \mathbf{k} | \mathbb{H}_- | n \rangle = ie(2\pi\hbar\omega)^{1/2}\langle n' | \mathbf{a} \cdot \mathbf{r} | n \rangle$

Here we have specified that in the absorption process, the initial state contains a photon of wavevector $\mathbf{k}$, whereas in the emission process, the final state contains a photon of wavevector $\mathbf{k}$. Again as in (13.90), we find that the related probabilities of these two processes are equal. Spontaneous decay may be associated with the matrix element $\langle n', \mathbf{k} | \mathbb{H}_- | n \rangle$.

Spontaneous Decay

As described previously, the Einstein A coefficient represents the probability rate for spontaneous decay. [See, for example, (13.65).] Ordinarily, one would suspect that such spontaneous decay occurs in the absence of any perturbation. However, in our earlier description of hydrogen (Section 10.6) we found atomic states to be stationary. So in reality, spontaneous decay must have a triggering mechanism. What is this mechanism?

The answer to this question stems from the observation that (as described in Problem 13.37) an electromagnetic field may be represented as a collection of harmonic oscillators. We have found previously that a harmonic oscillator always has a residual energy which is called its "zero-point energy." In like manner, no region of space is ever free of electromagnetic energy. It is such *vacuum fluctuations of electrodynamic fields* which are responsible for spontaneous decay.

Golden Rule Revisited

In describing spontaneous decay we concentrate on the emission matrix element $\langle n', \mathbf{k} | \mathbb{H}_- | n \rangle$, and consider that the emitted photon lies in the differential of solid angle $d\Omega$. The corresponding transition probability rate is given by Fermi's golden rule (13.64), which, in the present case, assumes the form

(13.107) $$dw_{nn'} = \sum_{\mathbf{a}_i} \frac{2\pi}{\hbar} |\mathbb{H}_{n'n}|^2 \overline{g}(E) \, d\Omega$$

Here $\mathbf{a}_i$ denotes possible photon polarization and $\overline{g}(E) \, d\Omega$ is the density of states of photons emitted in the solid angle $d\Omega$. This value of $\overline{g}(E)$ may be obtained from the value of $g(\nu)$ given in Problem 2.37 by setting

(13.108) $$Vg(\nu) \, d\nu = Vg(\omega) \, d\omega = 2\left[V \int \overline{g}(E) \, d\Omega \right] dE$$

The factor 2 in this equality accounts for photon polarizations. Setting the volume $V = 1$, we obtain

$$(13.109) \qquad \overline{g}(E) = \frac{E^2}{(2\pi\hbar c)^3} = \frac{\omega^2}{\hbar(2\pi c)^3}$$

Combining these results gives the following probability rate for spontaneous decay:

$$(13.110) \qquad dw_{nn'} = \sum_{\mathbf{a}_i} \frac{e^2\omega^3}{2\pi\hbar c^3} |\langle n', \mathbf{k} | \mathbf{a}_i \cdot \mathbf{r} | n \rangle|^2 \, d\Omega$$

The summation over polarizations in (13.110) is performed as follows. First, we note that

$$(13.111a) \qquad \langle n', \mathbf{k} | \mathbf{a} \cdot \mathbf{r} | n \rangle = \mathbf{a} \cdot \langle n', \mathbf{k} | \mathbf{r} | n \rangle$$

We recall that in the dipole approximation the presence of $\mathbf{k}$ in the preceding matrix element is cosmetic. It merely reminds us that the element is relevant to spontaneous decay and includes a photon of wavevector $\mathbf{k}$ in the final state. It follows that the only vector in the matrix element on the right side of (13.111a) is $\mathbf{r}$. Consequently, this matrix element is likewise in the direction of $\mathbf{r}$ and we may conclude that the entire matrix element on the left side of (13.111a) is proportional to $\mathbf{a} \cdot \mathbf{r}$. With this observation, the sum in (13.110) is evaluated as depicted in Fig. 13.15. The propagation vector of the emitted photon, $\mathbf{k}$, makes an angle θ with the radius vector $\mathbf{r}$, which is held fixed and taken as the polar axis. The unit polarization vectors $\mathbf{a}_1$, $\mathbf{a}_2$

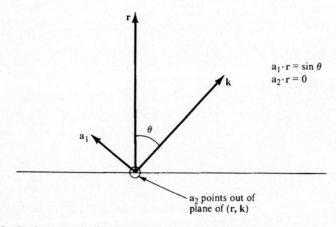

FIGURE 13.15 Configuration for summation over $\mathbf{a}_i$ in (13.110). The propagation vector k defines the direction of the differential of solid angle $d\Omega$.

are normal to each other as well as to the propagation vector $\mathbf{k}$. To facilitate calculation, the vectors $\mathbf{k}$, $\mathbf{a}_1$, and $\mathbf{r}$ are taken to lie in the same plane (see Fig. 13.15). It follows that the vector $\mathbf{a}_2$ points out of the plane of the paper, so that $\mathbf{a}_1 \cdot \hat{\mathbf{r}} = \sin\theta$, $\mathbf{a}_2 \cdot \hat{\mathbf{r}} = 0$ ($\hat{\mathbf{r}}$ is a unit vector). Thus we obtain

$$\sum_{\mathbf{a}_i} |\langle n', \mathbf{k} | \mathbf{a}_i \cdot \mathbf{r} | n\rangle|^2 = |\mathbf{a}_1 \cdot \langle n, \mathbf{k} | \mathbf{r} | n\rangle|^2 + |\mathbf{a}_2 \cdot \langle n', \mathbf{k} | \mathbf{r} | n\rangle|^2$$

$$= \sin^2\theta |\langle n', \mathbf{k} | \mathbf{r} | n\rangle|^2$$

Inserting this result into (13.110) gives the following probability rate for spontaneous decay:

$$(13.111b) \qquad dw_{nn'} = \frac{e^2\omega^3}{2\pi\hbar c^3} |\langle n', \mathbf{k} | \mathbf{r} | n\rangle|^2 \sin^2\theta \, d\Omega$$

So we reach the conclusion that the differential transition probability rate $dw_{nn'}$ depends only on the angle θ between $\mathbf{r}$ and $\mathbf{k}$. Consequently, the subsequent integration over $d\Omega$ is independent of the azimuthal angle, and with $\mathbf{r}$ still held fixed, we obtain

$$(13.112) \qquad A_{nn'} = \int dw_{nn'} = \frac{4}{3}\frac{e^2\omega^3}{\hbar c^3} |\langle n' | \mathbf{r} | n\rangle|^2$$

Note that we are writing

$$|\langle n' | \mathbf{r} | n\rangle|^2 = |\langle n' | x | n\rangle|^2 + |\langle n' | y | n\rangle|^2 + |\langle n' | z | n\rangle|^2$$

The coefficient $A_{nn'}$ represents the probable rate of transition for spontaneous decay and may be identified with the Einstein A coefficient in (13.65). To obtain the corresponding expression for radiated power, one multiplies $A_{nn'}$ by $\hbar\omega$. There results

$$(13.113) \qquad P_{nn'} = \frac{4}{3}\frac{e^2\omega^4}{c^3} |\langle n' | \mathbf{r} | n\rangle|^2$$

which agrees with our previous result (10.139).

The Einstein B coefficient may be found from our previously derived relation (see Problem 13.52):

$$\frac{A}{B} = \frac{\hbar\omega^3}{\pi^2 c^3}$$

which, with (13.112), returns our earlier finding (13.70).

It should be kept in mind that the preceding results are relevant to weak fields and the assumption that $\lambda \gg a_0$.

To apply the above formulas to an atom with Z electrons, one makes the replacement

$$(13.114) \qquad \mathbf{r}_{nn'} = \langle n' | \mathbf{r} | n \rangle \rightarrow \langle n' | \sum_{i=1}^{Z} \mathbf{r}_i | n \rangle$$

in the preceding expressions for A and P. The sum in (13.114) runs over all the electrons in the atom. For such cases $e\mathbf{r}_{nn'}$ represents the total dipole matrix element of the atom.

The total probable rate of spontaneous decay of an atom in the nth energy state is given by

$$(13.115) \qquad A_n = \sum_{E_{n'} < E_n} A_{nn'}$$

The summation runs over all states of lower energy than E_n. The corresponding mean lifetime of the nth excited state is then given by

$$(13.116) \qquad \tau_n = \frac{1}{A_n}$$

Oscillator Strengths

An important parameter in radiation analysis is the oscillator strength, defined as the dimensionless form

$$(13.117) \qquad f_{nn'} = \frac{2m\omega}{3\hbar} |\langle n' | \mathbf{r} | n \rangle|^2 = \frac{2m\omega}{3\hbar e^2} d_{nn'}^2$$

It follows that

$$(13.118) \qquad A_{nn'} = \frac{2e^2\omega^2}{mc^3} f_{nn'}$$

$$(13.119) \qquad P_{nn'} = \frac{2e^2\hbar\omega^3}{mc^3} f_{nn'}$$

Oscillator strengths obey the so-called Thomas–Reiche–Kuhn sum rule,

$$(13.120) \qquad \sum_{n'} f_{nn'} = 1$$

from which, together with (13.117), it may be concluded that $f_{nn'} < 1$.

The sum rule (13.120) is simply derived with the aid of the basic commutator relation (5.61), according to which, in one dimension, one obtains

$$(13.121) \qquad \langle n | [x, p_x] | n \rangle = i\hbar$$

Equivalently, we may write (see Problem 11.1)

(13.122) $$\sum_{n'} [\langle n|x|n'\rangle\langle n'|p_x|n\rangle - \langle n|p_x|n'\rangle\langle n'|x|n\rangle] = i\hbar$$

With (13.105) we obtain

$$\sum_{n'} [im\omega_{n'n}|\langle n'|x|n\rangle|^2 - im\omega_{nn'}|\langle n'|x|n\rangle|^2] = i\hbar$$

$$\sum_{n'} \frac{2m\omega_{n'n}}{\hbar}|\langle n'|x|n\rangle|^2 = 1$$

The same result follows with x replaced by y or z. There results

(13.123) $$\sum_{n'} \frac{2m\omega}{3\hbar}|\langle n'|\mathbf{r}|n\rangle|^2 = \sum_{n'} f_{nn'} = 1$$

For an atom with Z electrons, with (13.114) one finds

(13.124) $$\sum_{n'} f_{nn'} = Z$$

Values of the oscillator strengths of hydrogen corresponding to the first four $nP - 1S$ transitions are listed below.[1]

n	2	3	4	5
f_{nP-1S}	0.416	0.079	0.029	0.014

We conclude this section with an estimate of the lifetime of the excited states of one-electron atoms.[2] With $|\mathbf{r}_{nn'}| \simeq a_0$, we rewrite $A_{nn'}$ (13.112) as

(13.125) $$A_{nn'} \simeq \frac{4}{3}\frac{e^2\omega^3 a_0^2}{\hbar c^3}$$

Introducing the effective nuclear charge $\bar{Z}$,

(13.126) $$\hbar\omega \equiv \frac{\bar{Z}e^2}{a_0}$$

[1] R. Loudon, *The Quantum Theory of Light*, Clarendon Press, Oxford, 1973.

[2] More detailed discussions of these topics may be found in H. A. Bethe and E. E. Salpeter, *Quantum Mechanics of One- and Two-Electron Atoms*, Plenum, New York, 1977; F. Constantinescu and E. Magyari, *Problems in Quantum Mechanics*, Pergamon Press, Elmsford, N.Y., 1971; S. Flügge, *Practical Qauntum Mechanics*, Springer, New York, 1974.

permits the preceding formula to be written

(13.127) $$A_{nn'} \simeq \frac{4}{3}\alpha^3 \omega \bar{Z}^2$$

where $\alpha = e^2/\hbar c$ is the fine-structure constant. With $\bar{Z} \simeq 1$, and $\hbar\omega \simeq \mathbb{R} = \alpha^2 mc^2/2$, we find

(13.128) $$A_{nn'} \simeq \frac{\alpha^5}{\hbar}\frac{mc^2}{2} = \frac{\alpha^5 c}{2\lambda_C}$$

In this expression, $\lambda_C = \hbar/mc$ is the Compton wavelength. Inverting $A_{nn'}$ gives the lifetime[1]

(13.129) $$\tau \simeq \alpha^{-5}\left(\frac{2\lambda_C}{C}\right) \simeq 10^{-10} \text{ s} = 0.1 \text{ ns}$$

PROBLEMS

13.49   The radioactive isotope C^{11} decays through positron emission to B^{11-}. With the same assumptions holding as described in Problem 13.44, estimate the probability that B^{11-} is born in the ground state. (*Hint:* For your estimate, consult Problem 12.28 and pay attention to Z dependence of wavefunctions.)

Answer

As discussed in Problem 12.28, two-electron, ground-state wavefunctions (with $\hat{H}_{SO} = 0$) are given by

$$\varphi_Z(r_1, r_2) = \frac{1}{\pi a^3}\exp\left[\frac{-(r_1 + r_2)}{a}\right]$$

$$a \equiv \frac{a_0}{Z}$$

with normalization

$$\iint d\mathbf{r}_1\, d\mathbf{r}_2 |\varphi_Z|^2 = (4\pi)^2 \int_0^\infty dr_1 r_1^2 \int_0^\infty dr_2 r_2^2 |\varphi_Z|^2 = 1$$

The transition probability to the ground state of B^{11-} is given by

$$\sqrt{P} = \langle \varphi_C | \varphi_B \rangle = \frac{Z_C^3 Z_B^3}{\pi^2 a_0^6}\iint \exp\left[\frac{-Z(r_1 + r_2)}{a_0}\right] d\mathbf{r}_1\, d\mathbf{r}_2$$

---

[1] Nanosecond (10^{-9}), picosecond (10^{-12}), and femtosecond (10^{-15}) measurements are presently commonplace in many laboratories. A review of these topics is given by P. W. Smith and A. M. Weiner, *IEEE Circuits and Devices*, **4**, 3 (May 1988). Note the powers of 10 in physics: speed of light, 3×10^{10} cm/s; decay time of an atom, 10^{-10} s; age of the universe, 10^{10} yr.

where $Z \equiv Z_C + Z_B$. The preceding integral may be rewritten

$$\sqrt{P} = \frac{Z_C{}^3 Z_B{}^3}{\pi^2 a_0{}^6} I^2$$

$$I = a_0{}^3 \int_0^\infty dx\, x^2 e^{-Zx} = 2\left(\frac{a_0}{Z}\right)^3$$

There results

$$\sqrt{P} = \left(\frac{4Z_B Z_C}{Z^2}\right)^3$$

With $Z_B = 5$, $Z_C = 6$, $Z = 11$, we obtain $P = 0.951$.

13.50 (a) The density matrix (in a representation where $\hat{S}^2$ and $\hat{S}_z$ are diagonal) describing a beam of spinning electrons has the value

$$\hat{\rho} = \begin{pmatrix} 1 & \frac{1}{2} \\ \frac{1}{2} & 0 \end{pmatrix}$$

at $t < 0$. What are the values of $\langle S_z \rangle$, $\langle S_x \rangle$, and $\langle S_y \rangle$ for an electron in the beam?

(b) The beam interacts with a field which is turned on at $t = 0$. The corresponding interaction Hamiltonian has the matrix

$$\hat{H} = \hbar\omega_0 \begin{pmatrix} 0 & 1 \\ 1 & 0 \end{pmatrix}$$

where ω_0 is a characteristic frequency. Estimate the value of the matrix $\hat{\rho}$ at the time Δt, where $0 < \Delta t \ll \omega_0{}^{-1}$.

(c) What is the value of $\langle S_y \rangle$ at this value of time Δt? [*Hint:* Recall (11.122).]

Answers

(a) $\langle S_z \rangle = \langle S_x \rangle = \hbar/2$, $\langle S_y \rangle = 0$.

(b) Set

$$\hat{\rho}(\Delta t) = \hat{\rho}(0) + \left(\frac{\partial \hat{\rho}}{\partial t}\right)_0 \Delta t + \cdots$$

$$= \hat{\rho}(0) + \frac{1}{i\hbar}[\hat{H}, \hat{\rho}]_0 \Delta t + \cdots$$

Employing the given value of $\hat{\rho}(0)$ gives

$$\hat{\rho}(\Delta t) = \begin{pmatrix} 1 & 0.5 \\ 0.5 & 0 \end{pmatrix} - i\omega_0 \Delta t \begin{pmatrix} 0 & -1 \\ 1 & 0 \end{pmatrix} = \begin{pmatrix} 1 & a \\ a^* & 0 \end{pmatrix}$$

where

$$a \equiv 0.5 + i\omega_0 \Delta t$$

(c)   At $t = \Delta t$

$$\langle S_y \rangle = Tr \hat{\rho} \hat{S}_y = Tr\left[\hat{\rho} \frac{i\hbar}{2} \begin{pmatrix} 0 & -1 \\ 1 & 0 \end{pmatrix} \right]$$

$$= \frac{i\hbar}{2} Tr\left[\begin{pmatrix} 1 & a \\ a^* & 0 \end{pmatrix} \begin{pmatrix} 0 & -1 \\ 1 & 0 \end{pmatrix} \right]$$

$$= \frac{i\hbar}{2}(a - a^*) = -\hbar\omega_0 \Delta t$$

Thus, the perturbation causes a change in $\langle S_y \rangle$.

13.51   A one-dimensional harmonic oscillator of charge-to-mass ratio e/m, and spring constant K oscillates parallel to the x axis and is in its second excited state at $t < 0$, with energy

$$E_2 = \hbar\omega_0 \left(2 + \frac{1}{2} \right)$$

An oscillating, uniform electric field

$$\mathscr{E}(t) = 2\mathscr{E}_0 \cos \omega_0 t$$

is turned on at $t = 0$, parallel to the motion of the oscillator.

   (a)   What is the new Hamiltonian of the oscillator at $t > 0$?

   (b)   What are the matrix elements $\mathbb{H}'_{2n}$ for this system? You may leave your answer in terms of $\beta = \sqrt{m\omega_0/\hbar}$.

   (c)   What are the probabilities P_{2n} that the oscillator undergoes a transition to the nth state at the end of $t < \omega_0^{-1}$ seconds? [*Hint:* Use harmonic perturbation theory and look at the short-time limit.]

   (d)   Using your answer to part (c), offer a technique for 'pumping' a harmonic oscillator to higher states.

13.52   In our discussion of Planck's radiation law in Section 13.7, an expression for A/B was obtained as a function of frequency, ν. What is the corresponding expression for A/B in terms of angular frequency, ω? [*Hint:* See Problem 2.43.]

13.53   Suppose φ_1 and φ_2 are two normalized eigenfunctions of an operator $\hat{A}$ with the same eigenvalue. Construct two new normalized, orthogonal functions, ψ_1 and ψ_2, which are linear combinations of φ_1 and φ_2. Offer a geometrical description of construction in the appropriate Hilbert space.

Answer (partial)

Let

$$\psi_1 = \varphi_1, \qquad \psi_2 = \alpha\varphi_1 + \beta\varphi_2$$

and

$$\int \varphi_1^* \varphi_2 \, d\mathbf{r} = K \neq 0$$

Orthogonality of ψ_1 and ψ_2 gives

$$\alpha + \beta K = 0$$

whereas normalization of ψ_2 gives

$$|\beta|^2(1 - |K|^2) = 1$$

There results

$$\beta K = -\alpha = \frac{K}{\sqrt{1 - |K|^2}}$$

which determines ψ_2. This construction gives the essentials of the *Schmidt orthogonalization procedure.*

13.54 (a) Show that the time-dependent Schrödinger equation may be written [compare with (11.144)]

$$\psi(t) = \psi(0) + \frac{1}{i\hbar} \int_0^t \hat{H}(t')\psi(t') \, dt'$$

(b) Show that for small $\hat{H}$ the preceding equation gives the series *(Neumann–Liouville expansion)* with $t > t'$, etc.

$$\psi(t) = \left[1 + \frac{1}{i\hbar} \int_0^t dt' \, \hat{H}(t') + \frac{1}{(i\hbar)^2} \int_0^t dt' \int_0^{t'} dt'' \, \hat{H}(t')\hat{H}(t'') + \cdots \right]\psi(0)$$

(c) Show that the nth-order term in the preceding series may be written

$$\psi^{(n)}(t) = \frac{1}{n!} \frac{1}{(i\hbar)^n} \int_0^t \cdots \int_0^t dt_1 \cdots dt_n \, \hat{T}[\hat{H}(t_1) \cdots \hat{H}(t_n)]\psi(0)$$

where the time-ordering operator

$$\hat{T}f(t_a)g(t_b) = \begin{cases} f(t_a)g(t_b) & \text{for } t_a > t_b \\ g(t_b)f(t_a) & \text{for } t_b > t_a \end{cases}$$

13.55 A particle of mass m is confined to a one-dimensional partitioned box with walls at $(-L/2, 0, L/2)$. The Hamiltonian for this configuration is labeled $\hat{H}_0$. See Fig. 13.16a. One of two perturbations, $\hat{H}'_{a,b}$, is applied as shown in Fig. 13.16b, c. The total Hamiltonians for the particle are

$$\hat{H}_{a,b} = \hat{H}_0 + \hat{H}'_{a,b}$$

Let the perturbation potentials have magnitude $V_0 = 10^{-4} E_G$, where E_G is the ground state of $\hat{H}_0$, and let the normalized ground states of $\hat{H}_0$ be written $|r\rangle$ and $|l\rangle$ corresponding to the particle trapped in the right and left boxes, respectively.

(a) What are the coordinate representations of $|r\rangle$ and $|l\rangle$?

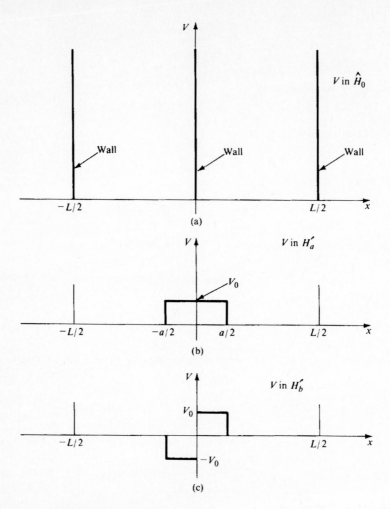

FIGURE 13.16 Potential configurations for Problem 13.55.

(b)   What is the ground-state eigenenergy E_G of $\hat{H}_0$? What is the order of the degeneracy of E_G?

(c)   What is $\langle l \,|\, H_0 \,|\, r \rangle$?

(d)   What are the values (zero or nonzero) of the commutators $[\hat{H}_0, \hat{\mathbb{P}}]$, $[\hat{H}_a, \hat{\mathbb{P}}]$, $[\hat{H}_b, \hat{\mathbb{P}}]$? Here we have written $\hat{\mathbb{P}}$ for the parity operator (6.90).

(e)   Construct two simultaneous eigenstates of $\hat{H}_0$ and $\hat{\mathbb{P}}$ in terms of $|l\rangle$ and $|r\rangle$. Label these states $|S\rangle$ (symmetric) and $|A\rangle$ (antisymmetric). What are $\hat{H}_0 |S\rangle$, $\hat{H}_0 |A\rangle$, $\mathbb{P} |S\rangle$, $\hat{\mathbb{P}} |A\rangle$?

(f)   Using degenerate perturbation theory, obtain the splitting of E_G due to $\hat{H}_a'$ and $\hat{H}_b'$, respectively. Do both $\hat{H}_a'$ and $\hat{H}_b'$ remove the degeneracy of E_G? Explain your answer. Call the new energies E_+, E_-. What effect does the nonsplitting perturbation have on the ground-state energy?

(g) The trapped particle is an electron and $2a = L = 10$ Å. What are E_+ and E_- (in eV)?

(h) Can you suggest an alternative way to evalute $E_\pm$ which does not employ degenerate perturbation theory?

Answers (partial)

(a) $\langle l | x \rangle = 0$ on $(0, L/2)$ and $\langle r | x \rangle = 0$ on $(-L/2, 0)$.

(e) $$|S\rangle = \sqrt{\frac{1}{2}} [|l\rangle + |r\rangle], \qquad |A\rangle = \sqrt{\frac{1}{2}} [|l\rangle - |r\rangle]$$

$$\hat{P}|S\rangle = |S\rangle, \qquad\qquad \hat{P}|A\rangle = -|A\rangle$$
$$\hat{H}_0|S\rangle = E_G|S\rangle, \qquad\qquad \hat{H}_0|A\rangle = E_G|A\rangle$$

(f) We must examine the secular equation

$$\begin{vmatrix} \langle l | H'_{a,b} | l \rangle - E' & \langle l | H'_{a,b} | r \rangle \\ \langle r | H'_{a,b} | l \rangle & \langle r | H'_{a,b} | r \rangle - E' \end{vmatrix} = 0$$

With

$$\mathscr{E}_a \equiv \langle l | H'_a | l \rangle = \langle r | H'_a | r \rangle$$
$$\mathscr{E}_b \equiv \langle l | H'_b | l \rangle = -\langle r | H'_b | r \rangle$$
$$\Delta\mathscr{E}_a \equiv \langle r | H'_a | l \rangle = \langle l | H'_a | r \rangle = 0$$
$$\Delta\mathscr{E}_b \equiv \langle r | H'_b | l \rangle = \langle r | H'_b | l \rangle = 0$$

We find that $\hat{H}'_b$ splits E_G but $\hat{H}'_a$ does not. The explanation of this result is that $\hat{H}'_a$ maintains the symmetry of $\hat{H}_0$. That is, for both $\hat{H}_0$ and $\hat{H}'_a$, there is no difference between right and left boxes. New energies (for $\hat{H}_b$) are given by

$$E_+ = E_G + \mathscr{E}_b$$
$$E_- = E_G - \mathscr{E}_b$$

13.56 An approximate form of the Hamiltonian of a positronium atom in the $1S$ state immersed in a weak magnetic field $\mathscr{B}$ is given by

$$\hat{H} = A\hat{\mathbf{S}}_1 \cdot \hat{\mathbf{S}}_2 + \mathbf{\Omega} \cdot (\hat{\mathbf{S}}_1 - \hat{\mathbf{S}}_2) \equiv H_0 + \lambda H'$$

$$\Omega = \frac{e\mathscr{B}}{\mu c}$$

where H' is the magnetic field term. The subscripts 1 and 2 denote electron and positron, respectively, A is a constant, and μ is reduced mass.

(a) Show that the coupled spin states ξ_S and ξ_A relevant to two spin-$\frac{1}{2}$ particles are eigenstates of $\hat{H}_0$. What is the ground-state energy of $\hat{H}_0$ and which eigenstate does this correspond to? [See Table 11.3 and recall (11.111).]

(b) Employing ξ_S and ξ_A as basis states, use first-order perturbation theory to obtain eigenvalues and eigenstates of $\hat{H}$. Assume that $\mathscr{B}$ is aligned with the z axis.

13.57   In Section 11.12 we encountered an approximation scheme centered about the interaction picture. Again consider the Hamiltonian given in Problem 11.102, which we now write in more symbolic form,

$$\hat{H} = \hat{H}_0 + \lambda \hat{V}(t)$$

Let the system be in the initial state

$$|\psi(0)\rangle = |n\rangle$$
$$\hat{H}_0|n\rangle = E_n|n\rangle$$

where, we recall $\hat{H}_0$ is appropriate to an harmonic oscillator with natural frequency ω_0.

(a)   Calculate the wavefunction $|\psi_1(t)\rangle$ to first order in λ.

(b)   For the problem at hand, show that

$$|\langle m|\psi(t)\rangle|^2 = |\langle m|\psi_I(t)\rangle|^2$$

(c)   Again, to $0(\lambda)$, obtain an integral expression for the time-dependent transition probabilities $P_{n\to m}$.

(d)   In the event that $\hat{V}$ is time-independent, show that your answer to part (c) reduces to (13.55) corresponding to the *DC* perturbation, $\omega = 0$.

Answers

(a)
$$|\psi_I(t)\rangle = |n\rangle + \sum_m \frac{\lambda}{i\hbar} \int_0^t e^{i\omega_{mn}t}|m\rangle\langle n|\hat{V}(t')|n\rangle \, dt'$$

$$\hbar\omega_{nn} = E_m - E_n = \hbar\omega_0(m - n)$$

(b)   First, note that

$$|\psi(t)\rangle = |e^{-i\hat{H}_0 t/\hbar}\psi_I(t)\rangle$$

The desired equality follows since $\hat{H}_0|m\rangle = E_m|m\rangle$.

(c)   With the result (b), we write

$$P_{n\to m} = |\langle m|\psi_I(t)\rangle|^2$$

where, for $m \neq n$,

$$\langle m|\psi_I(t)\rangle = \frac{\lambda}{i\hbar} \int_0^t e^{i\omega_{mn}t'}\langle m|\hat{V}(t')|n\rangle \, dt'$$

(d)   In the event that $\hat{V}$ is time-independent, integration of the preceding finding gives

$$\langle m|\psi_I(t)\rangle = \frac{V_{mn}}{\hbar\omega_{mn}}(1 - e^{i\omega_{mn}t})$$

$$V_{mn} \equiv \lambda\langle m|\hat{V}|n\rangle$$

which yields

$$P_{n\to m} = \left|\frac{2V_{mn}}{\hbar\omega_{mn}}\right|^2 \sin^2 \frac{\omega_{mn}t}{2}$$

This result is seen to agree with (13.55) for the *DC* perturbation, $\omega = 0$.

13.58   Establish the following relations for the coefficients of the Fourier expansion (13.33) of the real potential function $V(x)$.

(a) $V_n^* = V_{-n}$.

(b)   If $V(x)$ is even, then

$$V_n = V_{-n}$$

(c)   If $V(x)$ is odd, then

$$V_n = -V_{-n}$$

(d)   For $V(x)$ (even, odd) and with period $2a$, one writes

$$V(x)\binom{\text{even}}{\text{odd}} = \sum_{n=1}^{\infty} \begin{pmatrix} a_n \cos \dfrac{n\pi x}{a} \\ b_n \sin \dfrac{n\pi x}{a} \end{pmatrix}$$

Show that (with $a_0 = 0$)

$$a_n = 2V_n = 2V_{-n}$$
$$b_n = 2iV_n = -2iV_{-n}$$

Answers (partial)

(a)   As replacing n with $-n$ in the series (13.33) does not change the sum, we may set

$$\sum_{\forall n} V_n e^{i2\pi nx/a} = \sum_{\forall n} V_{-n} e^{-i2\pi nx/a}$$

For $V(x)$ real, $V(x) = V(x)^*$. With the preceding we then obtain

$$\sum_{\forall n} V_n^* e^{-i2\pi nx/a} = \sum_{\forall n} V_{-n} e^{-i2\pi nx/a}$$

whence $V_n^* = V_{-n}$.

(b)   If $V(x)$ is even, then

$$\sum_{\forall n} V_n e^{i2\pi nx/a} = \sum_{\forall n} V_n e^{-i2\pi nx/a} = \sum_{\forall n} V_{-n} e^{i2\pi nx/a}$$

whence $V_n = V_{-n}$.

(c)   If $V(x)$ is odd, then

$$\sum_{\forall n} V_n e^{i2\pi nx/a} = -\sum_{\forall n} V_n e^{-i2\pi nx/a} = -\sum_{\forall n} V_{-n} e^{i2\pi nx/a}$$

whence $V_n = -V_{-n}$.

13.59   (a)   Evaluate the average dipole moment of hydrogen for the $3P \rightarrow 1S$ transition from the corresponding value of oscillator strength given in the text. Work in cgs units and state the dimensions of your answer.

(b)   Compare your answer with the classical estimate, $a \simeq ea_0$.

13 PERTURBATION THEORY

13.60 Establish the commutator relation (13.103).

Answer

Since **r** commutes with $V(\mathbf{r})$, the relation reduces to

$$\hat{\mathbf{p}} = \frac{im}{\hbar}\left[\frac{\hat{p}^2}{2m}, \mathbf{r}\right]$$

With reference to Problem 5.45, the preceding is rewritten

$$\hat{\mathbf{p}} = \frac{i}{2\hbar}\{\hat{\mathbf{p}}[\hat{\mathbf{p}}, \hat{\mathbf{r}}] + [\hat{\mathbf{p}}, \hat{\mathbf{r}}]\hat{\mathbf{p}}\}$$

$$\hat{\mathbf{p}} = -\frac{i}{2\hbar}\{\hat{\mathbf{p}}i\hbar\hat{I} + i\hbar\hat{I}\hat{\mathbf{p}}\}$$

where $\hat{I}$ is the identity operator,

$$\hat{\mathbf{p}}\hat{I} = \hat{I}\hat{\mathbf{p}} = \hat{\mathbf{p}}$$

which establishes the said relation.

13.61 A particle of mass m is in a three-dimensional, rigid-walled cubical box of edge length a. Edges of the box are aligned with the Cartesian axes, with one corner of the box at the origin.
    (a) Write down the normalized ground state for this configuration.
    (b) The face of the cube at $x = a$ is suddenly displaced to $x = 2a$. Obtain an expression for the probability that the particle remains in the ground state.
    (c) Given that the particle is an electron and that $a = 2$ Å, what is the numerical value of this probability?

13.62 A particle of mass m is confined to the interior of a rigid-walled spherical cavity of radius a. The particle is in the ground state. At $t = 0$ the radius of the sphere begins to expand according to

$$a(t) = a_0 e^{-(t/\tau)^2}$$

where $\tau \gg ma_0^2/\hbar$. What are the wavefunction and energy of the particle at $t = \tau/2$?

13.63 At a given instant of time an harmonic oscillator undergoes a sudden change in spring constant from K to K'. Show that for energy to be conserved in the accompanying transition, $\sqrt{K/K'}$ must be the ratio of two odd numbers.

13.64 The same conditions as in Problem 13.26 apply. However, now the periodic potential is as shown in Fig. 13.17.
    If $V_0 = 1.5$ eV and $a = 2.3$ Å, at what energy (eV) will photons incident on the crystal cause it to conduct? [*Hint:* Before you start, decide on the period of $V(x)$.]

Answer

You should obtain

$$E_g = 2V_1 = \frac{4V_0}{\pi^2} = 0.61 \text{ eV}$$

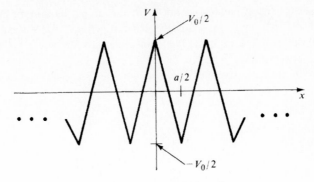

FIGURE 13.17   See Problem 13.64.

13.65   In our analysis of the *nearly free electron model* we set

(a) $H'_{k_j, k_j} = V_0 = 0$

(b) $H'_{k_j, -k_j} = V_n$

(c) $H'_{-k_j, k_j} = V_{-n}$

(d) $V_n = V_{-n}$

Establish the validity of these relations and state the relation between n and k_j.

13.66   The oscillator strength between the $4P$ and the $3S$ states of sodium is $f_{34} = 0.484.$[1]
Working in cgs units:

(a)   What are the dimensions of f_{34}?

(b)   What is the dipole moment, d_{34}, of the atom, corresponding to this transition?
State the units of d_{34} (in cgs).

(c)   Compare your values for d_{34} with the value of the dipole moment, $|e|a_0$, where
a_0 is the Bohr radius. [*Note:* Recall (10.138b).]

13.67   Derive the expression for time-averaged electromagnetic energy density, $\langle U \rangle$, as
given by (13.94), carried in the plane-wave vector potential (13.91).

13.68   Show that the basis (13.11) comprises an orthogonal sequence.

13.69   (a)   An electron with energy $E \gg V_0$ propagates through a one-dimensional crystal
with potential

$$V(x) = 2V_0 \left[\cos\left(\frac{2\pi x}{a}\right) + \frac{1}{2} \cos\left(\frac{4\pi x}{a}\right) \right]$$

What are the values of the nonzero band gaps, δE_n, for this material? How many energy bands
exist for this material?

---

[1] R. D. Cowan, *The Theory of Atomic Spectra and Structure*, University of California Press, Berkeley, Calif., 1981,
Section 14-14.

(b) The second energy band is identified with the valence band of the material and the next-higher band is identified with the conduction band, which at $T = 0$ K is empty. Given that $V_0 = 1.5$ eV and that $a = 2$ Å, estimate the temperature at which this material will start to conduct.

(c) What are the new wavefunctions related to the first energy gap of this system?

Answers (partial)

(b) $T_{\text{cond}} = 0.17 \times 10^5$ K

13.10 HARTREE–FOCK MODEL[1,2]

In Section 10.8 we encountered the Thomas–Fermi statistical model for determining an effective atomic potential. The Hartree–Fock self-consistent model more realistically takes into account the effects of atomic electrons. In addition to an effective atomic potential, the model also determines eigenenergies and wave-functions. Three primary elements of this model are: (1) Each electron moves in a central field equal to the nuclear potential and that due to the charge densities of the remaining atomic electrons. (2) The Schrödinger equation is solved for each electron in its own central field, and resulting wavefunctions are made self-consistent with the fields from which they are calculated. (3) The atomic wavefunction is a product of single-electron orthonormalized wavefunctions,

(13.130)
$$\psi(\mathbf{r}_1, \ldots, \mathbf{r}_Z) = \varphi_1(\mathbf{r}_1) \ldots \varphi_Z(\mathbf{r}_Z)$$

With statements (1) and (2) we note that each one-electron wavefunction satisfies the equation

(13.131a)
$$\hat{H}_k \varphi_k(\mathbf{r}_k) = E_k \varphi_k(\mathbf{r}_k)$$

(13.131b)
$$\hat{H}_k \equiv \left[-\frac{\hbar^2 \nabla_k^2}{2m} + W(r_k) \right]$$

(13.131c)
$$W(r_k) \equiv V_C(r_k) + V_k(r_k)$$

---

[1] This analysis is due to D. R. Hartree, *Proc. Cambridge Soc.* **24**, 111 (1929).

[2] Antisymmetrization of wavefunctions in this scheme was formulated by V. Fock, *Z. f. Phys.* **61**, 126 (1930).

(13.131d)
$$V_C(r_k) \equiv -\frac{Ze^2}{r_k}$$

(13.131e)
$$V_k(r_k) \equiv \int \sum_{j \neq k} |\varphi_j(\mathbf{r}_j)|^2 \frac{e^2}{r_{jk}} \, d\mathbf{r}_j$$

(13.131f)
$$r_{jk} \equiv |\mathbf{r}_j - \mathbf{r}_k|$$

The term V_C represents electron-electron Coulomb energy. Summation over k (or j) runs from 1 to Z relevant to an atom with Z electrons. Note that $W(r_k)$ is a central potential and that spin-orbit effects are neglected in the preceding equations. As there are Z electrons in the atom, (13.131) constitutes Z simultaneous nonlinear integrodifferential equations for the Z functions. An iterative scheme is used to solve these equations. This approximation involves the following four steps:

1.   An approximate central potential representing $W(r_k)$ is assumed, labeled $W^{(1)}(r_k)$. (Such a choice might be the Thomas–Fermi potential.)

   (13.132a)
   $$W(r_k) \rightarrow W^{(1)}(r_k)$$

2.   Electron wavefunctions are computed employing this approximate potential.

   (13.132b)
   $$\left[-\frac{\hbar^2 \nabla_k^2}{2m} + W^{(1)}(r_k) \right] \varphi_k^{(1)}(\mathbf{r}_k) = E_k^{(1)} \varphi_k^{(1)}(\mathbf{r}_k)$$

3.   With first-order wavefunctions, $\{\varphi_k^{(1)}(\mathbf{r}_k)\}$, at hand, charge densities, $e^2 |\varphi_k^{(1)}(\mathbf{r}_k)|^2$, are calculated.

4.   Employing the preceding results, the second-order atomic potential, $W^{(2)}(r_k)$, is calculated.

The iterative scheme (steps 1 through 4) is repeated until

(13.132c)
$$W^{(n+1)}(r_k) \simeq W^{(n)}(r_k)$$

that is, until $W^{(n)}(r_k)$ does not change appreciably with the next iterative cycle. At the conclusion of this iteration one has a self-consistent effective atomic potential together with eigenfunctions and eigenenergies relevant to an atom with atomic number Z, to within a given order of accuracy. A flowchart illustrating this scheme is shown in Fig. 13.18. In step 5 of this chart we have written $\|W\|$ for the norm of W (see Section 4.4). The value of the parameter ε determines the accuracy of the solution.

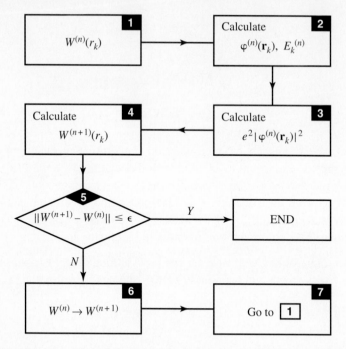

FIGURE 13.18 Flowchart illustrating the *n*th cycle of the iterative scheme for solution of the Hartree–Fock equation.

PROBLEMS

13.70 (a) Neglecting spin-orbit effects, write down the Hamiltonian, $\hat{H}_Z$, of an atom with atomic number Z.

(b) In what manner does $\hat{H}_Z$ differ from $\Sigma_k \hat{H}_k$ of the Hartree–Fock analysis?

Answers

(a)
$$\hat{H}_Z = \sum_k \left[-\frac{\hbar^2 \nabla_k^2}{2m} - \frac{Ze^2}{r_k} \right] + \sum_{j<k} \sum \frac{e^2}{r_{jk}}$$

The inequality $j < k$ restricts the double sum to distinct (j, k) pairs.

(b) The difference between $\hat{H}_Z$ and $\Sigma_k H_k$ is due to the substitution of the effective potential $\Sigma_k V_k (r_k)$ for the exact double-sum electron-electron Coulomb energy in the preceding expression.

13.71 With reference to the preceding problem, show that the expectation $\langle H_Z \rangle$, in the state given by the wavefunction (13.130), may be written as a sum over the single-particle expectations $\langle H_k \rangle$.

Answer

With orthonormality of $\varphi_k(\mathbf{r}_k)$ wavefunctions we obtain

$$\langle H \rangle = \int \cdots \int \psi^* \hat{H} \psi \, d\mathbf{r}_1 \cdots d\mathbf{r}_Z$$

$$= \sum_k \int\!\!\int \varphi_k^*(\mathbf{r}_k) \left(-\frac{\hbar^2 \nabla_k^2}{2m} - \frac{Ze^2}{r_k} \right) \varphi_k(\mathbf{r}_k) \, d\mathbf{r}_k$$

$$+ \sum_{j \neq k} \int\!\!\int \varphi_j^*(\mathbf{r}_j) \varphi_k^*(\mathbf{r}_k) \frac{e^2}{r_{jk}} \varphi_j(\mathbf{r}_j) \varphi_k(\mathbf{r}_k) \, d\mathbf{r}_k$$

With reference to (13.131b) et seq., and the preceding result, $\langle H \rangle$ may be written

$$\langle H \rangle = \sum_k \langle H_k \rangle = \sum_k \int \varphi_k^*(\mathbf{r}_k) \hat{H}_k \varphi_k(\mathbf{r}_k) \, d\mathbf{r}_k$$

13.72   (a)   Show that $\langle H \rangle$ may be made stationary through variation of individual one-electron atomic wavefunctions.

(b)   With reference to Problem 13.70, what can you conclude regarding the eigenstates of $\hat{H}_Z$?

Answer (partial)

(b)   In Problem 4.30 we found that wavefunctions that make $\langle H \rangle$ stationary are eigenstates of $\hat{H}$ and, conversely, eigenstates of $\hat{H}$ render $\langle H \rangle$ stationary. With part (a) of this problem, we may conclude that ψ as given by (13.130) is an eigenstate of $\hat{H}_Z$, providing $\varphi_k(\mathbf{r}_k)$ are eigenstates of $\hat{H}_k$.

13.73   Obtain an expression for the eigenenergies of $\hat{H}_Z$ in terms of the eigenenergies of $\hat{H}_k$ and electron-electron Coulomb energy.

Answer

Employing expressions obtained in Problem 13.71, we find

$$E_Z = \langle H_Z \rangle = \sum_k E_k + \sum_{j<k}\sum \int\!\!\int |\varphi_j(\mathbf{r}_j)|^2 \frac{e^2}{r_{jk}} |\varphi_k(\mathbf{r}_k)|^2 \, d\mathbf{r}_j \, d\mathbf{r}_k$$

13.74   A particle of mass m is confined to the interior of a spherical cavity of radius a.

(a)   What is the ground-state wavefunction, $\psi_G(r, \theta, \phi)$, and energy, E_G, of this system?

(b)   A potential sphere of radius $a/10$ and height V_0 is placed concentric and interior to the original sphere. Employing nondegenerate perturbation theory, evaluate the new first-order ground-state wavefunction, φ_G', and new ground-state energy, E_G'.

14

SCATTERING IN THREE DIMENSIONS

In this chapter an elementary description is offered of the quantum mechanical theory of scattering in three dimensions. Application of low-energy scattering is made to the Ramsauer effect, formerly encountered in Chapter 7, and scattering from a rigid sphere. The chapter continues with a discussion of the Born approximation. This important analysis permits certain scattering problems to be formulated in terms of harmonic perturbation theory developed previously in Chapter 13. The cross section of an atom interacting with a radiation field is obtained. For off-resonant incident phonons one encounters the line-shape factor. The chapter concludes with a description of the formal theory of scattering and derivation of the Lippmann–Schwinger equation in which the formalism of the interaction picture (Chapter 11) comes into play.

14.1 PARTIAL WAVES

The Rutherford Atom

One of the most fundamental tools of physics used for probing atomic and subatomic domains involves scattering of known particles from a sample of the element in question. Thus, for example, the description of an atom as being comprised of a positively charged central core of radius $\simeq 10^{-13}$ cm, with external satellite electrons, is due to scattering experiments performed by E. Rutherford in 1911. In these experiments α particles in an incident beam were deflected in passing through a thin metal foil. The prevalent model for an atom at the time was J. J. Thomson's "plum pudding" model, in which negative electrons floated in a ball of positive charge. The relatively large angle suffered by a small fraction of the α particles in the incident beam in Rutherford's experiments was found to be inconsistent with Thomson's model of the atom. For it is easily shown that α particles, after passing through hundreds of such spheres of distributed charge, are deflected at most only by a few degrees. On the other hand, the actual scattering data are consistent with an atomic model in which the positive charge is concentrated in a central core of small diameter. Large angle of scatter is then experienced by α particles which pass sufficiently close to the positive nucleus.

Scattering Cross Section

The typical configuration of a scattering experiment is shown in Fig. 14.1. A uniform monoenergetic beam of particles of known energy and current density $\mathbf{J}_{\text{inc}}$ (7.107) is incident on a target containing scattering centers. Such scattering centers might, for example, be the positive nuclei of atoms in a metal lattice. If the particles in the incident beam are, say, α particles, then when one such particle comes sufficiently close to one of the nuclei in the sample, it will be scattered. If the target sample is sufficiently thin, the probability of more than one such event for any particle in the incident beam is small and one may expect to obtain a valid description of the scattering data in terms of a single two-particle scattering event.

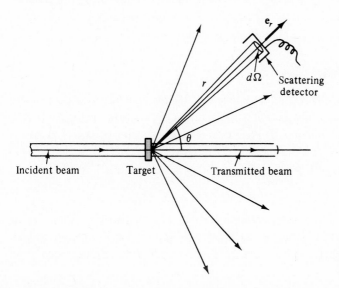

FIGURE 14.1 Scattering configuration.

Let the scattered current be $\mathbf{J}_{sc}$. Then the number of particles per unit time scattered through some surface element dS is $\mathbf{J}_{sc} \cdot d\mathbf{S}$. Let $d\mathbf{S}$ be at the radius $\mathbf{r}$ from the target. Then if $d\mathbf{\Omega}$ is the vector solid angle subtended by $d\mathbf{S}$ about the target origin, $d\mathbf{S} = r^2 \, d\mathbf{\Omega}$ (see Figs. 9.9 and 14.1). The vector solid angle $d\mathbf{\Omega}$ is in the direction of $\mathbf{e_r}$; that is, $d\mathbf{\Omega} = \mathbf{e_r} \, d\Omega$. It follows that

$$\left. \begin{array}{l} \text{Number of particles} \\ \text{passing through } d\mathbf{S} \\ \text{per second} \end{array} \right\} = dN = \mathbf{J}_{sc} \cdot d\mathbf{S} = r^2 \mathbf{J}_{sc} \cdot d\mathbf{\Omega}$$

Since the number of such scattered particles will grow with the incident current $\mathbf{J}_{inc}$, one may assume this number to be proportional to $\mathbf{J}_{inc}$ and can equate

(14.1)
$$dN = r^2 \mathbf{J}_{sc} \cdot d\mathbf{\Omega} = \mathbf{J}_{inc} \, d\sigma$$

The proportionality factor $d\sigma$ is called the *differential scattering cross section* and has dimensions of cm^2. It may be interpreted as an obstructional area which the scatterer presents to the incident beam. Particles taken out of the incident beam by this obstructional area are scattered into $d\mathbf{\Omega}$. The *total scattering cross section* σ represents the obstructional area of scattering in all directions.

(14.2)
$$\sigma = \int d\sigma = \int_{4\pi} \left(\frac{d\sigma}{d\Omega} \right) d\Omega$$

Scattering cross section has a classical counterpart. Classically, the total cross section seen by a uniform beam of point particles incident on a fixed rigid sphere of radius a is $\sigma = \pi a^2$. If the incident beam has current $\mathbb{J}_{inc}$, the number per second scattered out of the beam in all directions is $\pi a^2 \mathbb{J}_{inc}$.

The Scattering Amplitude

Returning to quantum mechanics, let the particles in the incident beam be independent of each other so that prior to interaction with the target a particle in the incident beam may be considered a *free particle*. If the z axis is taken to coincide with the axis of incidence, then a particle in the incident beam with momentum $\hbar \mathbf{k}$ and energy $\hbar^2 k^2 / 2m$ is in the plane-wave state,

(14.3)
$$\varphi_{inc} = e^{ikz}$$

When this wave interacts with a scattering center, an outgoing scattered wave φ_{sc} is initiated. If the scattering is *isotropic* so that scattering into all directions (all 4π steradians of solid angle) is equally probable, we can expect the scattered wave φ_{sc} to be a spherically symmetric outgoing wave. The specific form of an isotropic outgoing wave was described previously [(10.65) and Problem 10.6].

$$\varphi_{\text{sc, iso}} = \frac{e^{ikr}}{r}$$

More often, however, the scattered wave is anisotropic. Anisotropy of the scattering component wavefunction φ_{sc} may be described by a modulation factor $f(\theta)$, and in general we write

(14.4)
$$\varphi_{\text{sc}} = \frac{f(\theta)e^{ikr}}{r}$$

The modulation $f(\theta)$ is called the *scattering amplitude* and will be shown to determine the differential scattering cross section $d\sigma$.

The number of particles scattered into $d\Omega$, which is in the direction of $\mathbf{e}_r$, is obtained from the radial component of $\mathbf{J}_{\text{sc}}$ [recall (7.107)]:

(14.5)
$$\mathbb{J}_{\text{sc}, r} = \frac{\hbar}{2mi}\left(\varphi_{\text{sc}}^{*}\frac{\partial}{\partial r}\varphi_{\text{sc}} - \varphi_{\text{sc}}\frac{\partial}{\partial r}\varphi_{\text{sc}}^{*}\right)$$

$$= \frac{\hbar k}{mr^2}|f(\theta)|^2$$

Since the vector element of solid angle $d\Omega$ is in direction $\mathbf{e}_r$, it follows that

$$r^2\mathbf{J}_{\text{sc}} \cdot d\Omega = r^2\mathbb{J}_{\text{sc}, r}\, d\Omega = \mathbb{J}_{\text{inc}}\, d\sigma$$

In that the current vector of the incident beam only has a z component with magnitude $\hbar k/m$ [see Fig. 14.2], the preceding equation becomes

$$r^2\mathbb{J}_{\text{sc}, r}\, d\Omega = \frac{\hbar k}{m}\, d\sigma$$

Substituting (14.5) into this equation gives the desired relation,

(14.6)
$$\boxed{d\sigma = |f(\theta)|^2\, d\Omega}$$

Thus the problem of determining $d\sigma$ is equivalent to constructing the scattering amplitude $f(\theta)$.

Owing to the rotational symmetry of the scattering configuration about the axis of the incident beam and the assumed radial quality of the interaction potential between incident particle and scatterer, the scattering cross section depends only on

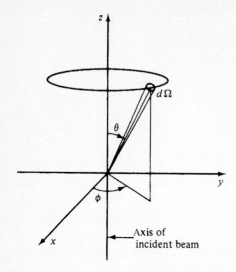

Axis of
incident beam

FIGURE 14.2 **The scattering cross section is independent of the azimuthal angle ϕ for central potentials of interaction $V(r)$.**

the scattering angle θ (and incident energy) and not on the azimuthal angle ϕ (see Fig. 14.2). It follows that, in integration (14.6) over all directions, the integration over $d\phi$ may be done separately to obtain 2π. There results

$$(14.7) \qquad \sigma = \int d\sigma = 2\pi \int_0^\pi |f(\theta)|^2 \sin\theta \, d\theta$$

The total cross section is a simple integral over the square modulus of the scattering amplitude. Referring again to Fig. 14.2, we see that the same symmetry implies that $f(\theta)$ is an even function of θ or, equivalently, $f(\theta) = f(\cos\theta)$.

Partial-Wave Phase Shift

The form of the wavefunction for the steady-state scattering configuration described above, at positions far removed from the scattering target, will contain a plane-wave incident component and an "outgoing" scattered component.

$$(14.8) \qquad \varphi(r, \theta) = e^{ikz} + \frac{f(\theta)e^{ikr}}{r} \qquad (r \to \infty)$$

(Fig. 14.3). The scattering amplitude is determined by matching (14.8) to the asymptotic form of the solution of the Schrödinger equation relevant to the configuration at hand. Such configuration includes a particle of mass m with known energy $\hbar^2 k^2/2m$,

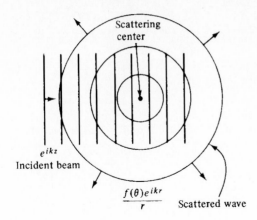

Scattering center

e^{ikz}
Incident beam

$\dfrac{f(\theta)e^{ikr}}{r}$ Scattered wave

FIGURE 14.3 Incident plane wave and scattered outgoing spherical wave.

interacting with a fixed scattering center through the central potential $V(r)$. The radial Schrödinger equation is given by (10.109).

$$(14.9) \qquad \left[\frac{1}{r}\frac{d^2}{dr^2}r - \frac{l(l+1)}{r^2} + k^2 - \frac{2mV}{\hbar^2}\right]R_{kl}(r) = 0$$

In the far field where $V(r)$ is rapidly approaching zero, one may expect the solution to this equation to be given approximately by the asymptotic form of the free-particle solution $j_l(kr)$ [see (10.55) and Table 10.1].

$$(14.10) \qquad R_{kl} \sim \frac{1}{kr}\sin\left(kr - \frac{l\pi}{2}\right)$$

Provided that $V(r)$ decreases faster than r^{-1}, this free-particle asymptotic form remains intact[1] save for a change in argument through a phase shift δ_l.

$$(14.11) \qquad R_{kl}{}^{\text{asm}} = \frac{1}{kr}\sin\left(kr - \frac{l\pi}{2} + \delta_l\right)$$

A superposition state comprised of these wavefunctions at fixed k has the form

$$(14.12) \qquad \varphi_k(r, \theta) = \sum_{l=0}^{\infty} C_l R_{kl}{}^{\text{asm}}P_l(\cos\theta)$$

The lth term in the sum is called the lth *partial wave* and δ_l is the phase shift that the partial wave incurs in scattering.

---

[1] For example, the analysis is not valid for the Coulomb potential $V(r) = r^{-1}$. Proof of the validity of the stated criterion may be found in L. Landau and E. Lifshitz, *Quantum Mechanics*, Addison-Wesley, Reading, Mass., 1958.

We must now match the asymptotic form of the general solution (14.12) to the form (14.8). With the expansion for exp (ikz) given in Problem 10.12, we obtain the asymptotic expression

$$e^{ikz} \sim \sum_{l=0}^{\infty} (2l + 1)i^l \frac{\sin (kr - l\pi/2)}{kr} P_l (\cos \theta)$$

The coefficients C_l and the scattering amplitude $f(\theta)$ are found from the matching equation

$$\sum_l C_l P_l (\cos \theta) \frac{\sin (kr - l\pi/2 + \delta_l)}{kr} = \sum_l (2l + 1)i^l P_l (\cos \theta) \frac{\sin (kr - l\pi/2)}{kr}$$
$$+ \frac{f(\theta)e^{ikr}}{r}$$

Expanding $f(\theta)$ in a series of Legendre polynomials, one obtains, after some trigonometric gymnastics,

(14.13)
$$C_l = i^l(2l + 1) \exp (i\delta_l)$$
$$f(\theta) = \frac{1}{k} \sum_{l=0}^{\infty} \frac{C_l}{i^l} \sin \delta_l P_l (\cos \theta)$$

The problem of calculating $d\sigma$ or, equivalently, $f(\theta)$ is reduced to one of constructing the phase shifts δ_l.

Two immediate results are evident: First, substituting the series (14.13) into (14.7) and taking advantage of the orthogonality of the $P_l (\cos \theta)$ polynomials, we obtain

(14.14)
$$\sigma = \frac{4\pi}{k^2} \sum_{l=0}^{\infty} (2l + 1) \sin^2 \delta_l$$

The second result follows from setting $\theta = 0$ in (14.13), which yields

$$f(0) = \frac{1}{k} \sum_l (2l + 1) \cos \delta_l \sin \delta_l + \frac{i}{k} \sum_l (2l + 1) \sin^2 \delta_l$$

Comparison with (14.14) reveals that

(14.15)
$$\sigma = \frac{4\pi}{k} \text{Im} [f(0)]$$

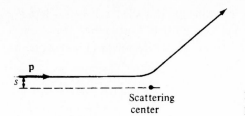

Scattering
center

FIGURE 14.4 **Classical trajectory and impact parameter** *s.*

This result is known as the *optical theorem*. It is a widely used relation connecting the forward scattering amplitude, $f(0)$, to the scattering in all directions, σ.[1]

Relative Magnitude of Phase Shifts

The problem of determining the partial wave phase shifts δ_l is often difficult. However, under certain conditions one may make simplifying assumptions which greatly facilitate calculation. In classical scattering one introduces the impact parameter. If L and p are the incident particle's angular momentum and linear momentum, respectively, then the impact parameter s is given by (see Fig. 14.4)

$$L = ps$$

Quite clearly, if the potential of interaction is appreciable only over the range r_0, then the interaction between incident particle and scatterer will be negligible for $s > r_0$. This criterion provides a useful rule of thumb applicable in quantum mechanics. With $L = \hbar\sqrt{l(l+1)} \simeq \hbar l$ and $p = \hbar k$, interaction will be negligible if

(14.16) $$l > r_0 k$$

The incident energy is $\hbar^2 k^2 / 2m$.

Each partial wave in the superposition (14.12) represents a state of definite angular momentum. From (14.16) we can expect that partial waves with l values in excess of $r_0 k$ will suffer little or no shift in phase. In the corresponding expansion of the scattering amplitude $f(\theta)$ as given by (14.13) it follows that only those δ_l values will contribute for which $l < r_0 k$. For low-energy scattering with $kr_0 \ll 1$, only the $l = 0$ phase shift will differ appreciably from zero. When such is the case (14.13) reduces to

(14.17) $$f(\theta) = \frac{1}{k} e^{i\delta_0} \sin \delta_0$$

---

[1] For inelastic scattering, (14.15) is still valid with σ replaced by the total cross section, $\sigma_T = \sigma_S + \sigma_A$, where σ_S is the elastic cross section and σ_A is the absorption cross section.

which is independent of θ. The scattering is isotropic and is called S-wave scattering. Only the S partial wave ($l = 0$) contributes to the scattering. In the opposite extreme of large incident energies, $kr_0 \gg 1$, we can expect all partial waves to suffer phase shifts and the cross section to be anisotropic.

PROBLEMS

14.1 From (14.1) we find that the number of particles scattered into the solid angle $d\Omega$ per second is

$$dN = \mathbb{J}\, d\sigma$$

or, equivalently,

$$\frac{dN}{\mathbb{J}} = \left(\frac{d\sigma}{d\Omega}\right) d\Omega$$

The Coulomb cross section for the scattering of a charged particle of energy E and charge q from a fixed charge Q is

$$\frac{d\sigma}{d\Omega} = \left(\frac{qQ}{4E}\right)^2 \frac{1}{\sin^4 (\theta/2)}$$

(a) What is the expression for the fraction of particles scattered into the differential cone $(\theta, \theta + d\theta)$ from a target comprised of Λ scattering centers per unit area?

(b) Employ the expression you have obtained to find the fraction of α particles with incident energy 5 MeV which are scattered into a differential cone $(\theta, \theta + d\theta)$ at $\theta = \pi/2$, in passing through a gold sheet 1 μm thick.

Answers

(a) If we assume that each particle in the incident beam sees only one scatterer and that there is a scattering event for each scatterer, then

$$\delta N = \Lambda \left(\frac{d\sigma}{d\Omega}\right) d\Omega$$

For the scattering into the cone $(\theta, \theta + d\theta)$

$$\delta N = \Lambda \int_0^{2\pi} d\phi \left(\frac{d\sigma}{d\Omega}\right) \sin\theta\, d\theta$$

$$= 2\pi\Lambda \left(\frac{d\sigma}{d\Omega}\right) \sin\theta\, d\theta$$

(b) For a sheet of mass density ρ, thickness l, comprised of atoms with atomic mass A,

$$\Lambda = \frac{\rho N_0 l}{A}$$

where N_0 is Avogadro's number (N_0 atoms have mass A grams). For a gold foil l cm thick with $\rho = 19.3$ g/cm^3 and $A = 197$, we obtain $\Lambda = 5.9 \times 10^{22}\,l$ atoms/cm^2. For α particles of energy 5 MeV scattered by the nuclei of gold atoms, $qQ/E = e^2 \cdot 2 \times 79/E = 4.6 \times 10^{-12}$ cm. Thus we obtain

$$\delta N(\pi/2) = \frac{\pi}{2}\left(\frac{qQ}{E}\right)^2 \frac{\rho N_0 l}{A}\,d\theta \simeq 2 \times 10^{-4}\,d\theta$$

14.2   S-WAVE SCATTERING

Let us consider the configuration of a low-energy beam of point particles of mass m scattering from a finite spherical attractive well of depth V_0 and radius a.

(14.18)
$$V(r) = \begin{cases} -V_0 & \text{for } r < a \\ 0 & \text{for } r > a \end{cases}$$

If we assume that energies are sufficiently small that $ka \ll 1$, we need only look at the S-wave scattering. The corresponding Schrödinger equation is obtained from (14.9). Setting $l = 0$ and $u \equiv rR$ there results, for $r < a$,

$$\frac{d^2 u_1}{dr^2} + k_1 u_1 = 0$$

$$\frac{\hbar^2 k_1^2}{2m} = E + V_0$$

The solution to this equation which corresponds to $R(r)$ remaining finite at $r = 0$ is

$$u_1 = A \sin k_1 r \qquad (r < a)$$

For $r > a$, $V = 0$ and we obtain the general solution

$$u_2 = B \sin (kr + \delta_0) \qquad (r > a)$$

$$\frac{\hbar^2 k^2}{2m} = E$$

Boundary conditions require continuity of $d \ln u/dr$ at $r = a$, which gives

(14.19)
$$k_1 \cot k_1 a = k \cot (ka + \delta_0)$$

In that k_1 is finite, in the limit that k goes to zero,

$$\cot (ka + \delta_0) = \frac{k_1 \cot k_1 a}{k}$$

grows large so that $\sin (ka + \delta_0)$ grows small and we may set

$$\sin (ka + \delta_0) \simeq ka + \delta_0$$

Since $ka \ll 1$, this equation implies that $\delta_0 \ll 1$ as well. Under these conditions (14.19) reduces to

$$k_1 \cot k_1 a \simeq \frac{k}{ka + \delta_0}$$

or equivalently

$$\delta_0 = ka \left(\frac{\tan k_1 a}{k_1 a} - 1 \right)$$

In that δ_0 is small, we may also set

$$\delta_0 \simeq \sin \delta_0 = ka \left(\frac{\tan k_1 a}{k_1 a} - 1 \right)$$

We may now construct the scattering amplitude (14.17) and cross section (14.7).

(14.20) $$\sigma = 4\pi a^2 \left(\frac{\tan k_1 a}{k_1 a} - 1 \right)^2$$

Two significant observations relevant to this study of attractive well scattering are discussed next.

S-Wave Resonances and Ramsauer Effect

First we note that when $k_1 a$ is an odd multiple of $(\pi/2)$, $\tan k_1 a$ is infinite and the cross section as given by (14.20) becomes singular. In that δ_0 is also infinite at these values of $k_1 a$, assumptions leading to (14.19) are violated and we must seek an alternative procedure to construct the cross section. Consider the relation (14.19), which assumes only that $ka \ll 1$. Let $k_1 a = n(\pi/2)$, where n is an odd number. At these values, (14.19) gives $\sin(\delta_0 + ka) = 1$, which with the condition $ka \ll 1$ yields $\sin \delta_0 \simeq 1$. Thus the maximum cross section at these *S-wave resonances* is

(14.21) $$\sigma_{max} = \frac{4\pi}{k^2}, \qquad k_1 a = n\left(\frac{\pi}{2} \right)$$

A more careful analysis pursued to higher angular momentum states, appropriate to larger incident energies, reveals corresponding resonances at $l = 1$, termed *P-wave resonances,* and so forth.

Whereas (14.20) suggests resonant scattering at odd multiples of $\pi/2$, it also indicates that the attractive scattering will become transparent to the incident beam at values of $k_1 a$ which satisfy the transcendental relation

$$\tan k_1 a = k_1 a$$

As noted in Section 7.8, such resonant transparency of an attractive well is experimentally corroborated in the scattering of low-energy electrons (~ 0.7 eV) by rare gas atoms and is termed the *Ramsauer effect.*

The Repulsive Sphere

The second observation related to our study of low-energy scattering by a scattering well is that merely changing the sign in the defining equations (14.18) produces the potential for a repulsive sphere of radius a. Solution for the corresponding scattering problem is effected by simply replacing k_1 by $i\kappa$, in the relations following (14.18). For the interior wavefunction we obtain

$$u_1 = A \sinh \kappa r \qquad (r < a)$$

$$\frac{\hbar^2 \kappa^2}{2m} = V_0 - E > 0$$

The exterior wavefunction u_2 maintains its sinusoidal dependence for $r > a$, as given in the equation preceding (14.19). Imposing boundary conditions at $r = a$ and, again assuming low-energy incident particles, we obtain the total scattering cross section,

$$(14.22) \qquad \sigma = 4\pi a^2 \left(\frac{\tanh \kappa a}{\kappa a} - 1 \right)^2$$

In the limit that $V_0 \to \infty$, the sphere becomes impenetrable and the total cross section reduces to

$$(14.23) \qquad \sigma = 4\pi a^2$$

Since this formula does not contain $\hbar$, our suspicion is that it is also appropriate to the classical domain. However, the obstructional area imposed by a rigid sphere of radius a to an incident beam of classical particles has the value πa^2, so the quantum cross section is larger than the classical one by a factor of 4. Although the cross section (14.23) does not contain Planck's constant, nevertheless one might still object to considering it a classical result in that it is relevant to the strictly nonclassical domain of large de Broglie wavelength. If a classical result is to be obtained, it should emerge in the limit of large incident energy, $ka \gg 1$. Such analysis, which includes the phase shifts of all waves,[1] again yields a cross section independent of $\hbar$, namely

$$\sigma = 2\pi a^2, \qquad ka \gg 1$$

which is still larger than the classical result. Thus the classical cross section does not emerge in the limit of large incident energy. This discrepancy may be ascribed to the sharp edge of the spherical potential barrier for the configuration at hand. Across the sharp potential step, dV/dx is infinite and it is impossible for the classical criterion (7.166) to be satisfied.

PROBLEMS

14.2 The scattering amplitude for a certain interaction is given by

$$f(\theta) = \frac{1}{k}(e^{ika} \sin ka + 3ie^{i2ka} \cos \theta)$$

where a is a characteristic length of the interaction potential and k is the wavenumber of incident particles.

(a) What is the S-wave differential cross section for this interaction?

(b) Suppose that the above scattering amplitude is appropriate to neutrons incident on a species of nuclear target. Let a beam of 1.3-eV neutrons with current 10^{14} cm^{-2} s^{-1} be incident on this target. What number of neutrons per second are scattered out of the beam into $4\pi \times 10^{-3}$ steradian about the forward direction?

14.3 Analysis of the scattering of particles of mass m and energy E from a fixed scattering center with characteristic length a finds the phase shifts

$$\delta_l = \sin^{-1}\left[\frac{(iak)^l}{\sqrt{(2l+1)l!}}\right]$$

(a) Derive a closed expression for the total cross section as a function of incident energy E.

(b) At what values of E does S-wave scattering give a good estimate of σ?

Answer (partial)

(a) $$\sigma = \frac{4\pi\hbar^2}{2mE} \exp\left(\frac{-2mEa^2}{\hbar^2}\right)$$

[1] The calculation may be found in L. I. Schiff, *Quantum Mechanics*, 3d ed., McGraw-Hill, New York, 1968.

14.3 CENTER-OF-MASS FRAME

In all of the preceding analysis, it has been assumed that the target particle remains fixed during the scattering process. This is the case if the mass of the target particle far exceeds that of the incident particle. More generally, however, the recoil motion of the target particle must be taken into account in any scattering analysis. Thus the general formulation of a scattering event involves two particles, of mass m_1 and m_2.

As described in Section 10.5, the motion of such two-particle systems may be described in terms of the motion of the center of mass and motion relative to the center of mass. The Hamiltonian of the relative motion (10.99) describes a single effective particle with reduced mass $\mu = m_1 m_2/(m_1 + m_2)$ at the radius $\mathbf{r} = \mathbf{r}_1 - \mathbf{r}_2$. This is the motion observed in a frame moving with the center of mass. So, in fact, in this center-of-mass frame, the scattering event may be described by a single particle of mass μ interacting with a potential $V(r)$ centered at a fixed origin. It follows that the preceding formulation of the cross section $\sigma(\theta)$ describing scattering from a fixed scattering center is appropriate to scattering in the center-of-mass frame. The only change is that the mass m of the incident particle is set equal to the reduced mass μ. In addition, we must note that the angle of deflection θ is measured in the center-of-mass frame. For example, in the expression (14.13) for the scattering amplitude, θ is the angle of scatter in the center-of-mass frame, which will henceforth be called θ_C. To obtain a relation between the scattering cross section $\sigma_L(\theta_L)$ in the frame of the experiment, or what is commonly called the *lab frame* and the cross section $\sigma_C(\theta_C)$ as measured in the center-of-mass frame, we note the following. The number of particles scattered into an element of solid angle in the lab frame $\mathbb{J}_{\text{inc}}\,(d\sigma_L/d\Omega_L)\,d\Omega_L$ is equal to the number scattered into the corresponding solid angle in the center-of-mass frame, $\mathbb{J}_{\text{inc}}\,(d\sigma_C/d\Omega_C)\,d\Omega_C$. This gives the equality

(14.24)
$$\frac{d\sigma_L}{d\cos\theta_L} = \frac{d\sigma_C}{d\cos\theta_C}\frac{d\cos\theta_C}{d\cos\theta_L}$$

The relation between $\cos\theta_C$ and $\cos\theta_L$ is obtained by examining the scattering in both frames. In transforming from one frame to the other, it is convenient to speak in terms of velocities. Such velocities are related to linear momentum through the prescription $\mathbf{v} = \hbar\mathbf{k}/m$. When one describes an "orbit" in this description, one has in mind a picture inferred by the direction of momentum $\mathbf{k}$ vectors. Thus, before collision, m_2 is at rest and the incident particle has velocity $\mathbf{v}_1 = \hbar\mathbf{k}/m_1$. After collision, m_1 is scattered through the angle θ_L.

The center-of-mass frame is characterized by the property that *total momentum in that frame is zero before and after collision* (Fig. 14.5). Letting barred variables denote values in the center-of-mass frame, and $\mathbf{v}$ the *relative velocity*,

$$\mathbf{v} \equiv \mathbf{v}_1 - \mathbf{v}_2$$

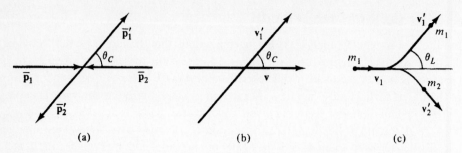

FIGURE 14.5 (a) In the center-of-mass frame, the total momentum is zero. (b) The relative velocity vector **v** rotates through the angle θ_C. (c) In the lab frame, m_2 is assumed to be at rest before collision.

one obtains, for before the collision,

$$\bar{\mathbf{p}}_1 = m_1(\mathbf{v}_1 - \mathbf{v}_{CM}) = \mu\mathbf{v}$$

We may immediately conclude that

$$\bar{\mathbf{p}}_2 = -\mu\mathbf{v}$$

In a similar manner, after collision we write

$$\bar{\mathbf{p}}_1{}' = \mu\mathbf{v}'$$

$$\bar{\mathbf{p}}_2{}' = -\mu\mathbf{v}'$$

or, equivalently,

$$\bar{\mathbf{v}}_1{}' = \frac{\mu}{m_1}\mathbf{v}'$$

$$\bar{\mathbf{v}}_2{}' = -\frac{\mu}{m_2}\mathbf{v}'$$

The corresponding relations in the lab frame are obtained by adding $\mathbf{v}_{CM}$ to the right-hand sides of these equations. Multiplying the resulting equations by m_1 and m_2, respectively, gives

$$\mathbf{p}_1{}' = \mu\mathbf{v}' + \frac{\mu}{m_2}\mathscr{P}$$

(14.25)
$$\mathbf{p}_2{}' = -\mu\mathbf{v}' + \frac{\mu}{m_1}\mathscr{P}$$

$$\mathscr{P} = \mathbf{p}_1 + \mathbf{p}_2$$

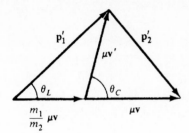

FIGURE 14.6   Orientation of momentum and relative velocity for $m_1/m_2 < 1$.

Since m_2 is at rest before scattering, $\mathbf{p}_1 = m_1\mathbf{v} = \mathscr{P}$. It follows that (14.25) may be rewritten

(14.26)
$$\mathbf{p}_1' = \mu\mathbf{v}' + \frac{m_1}{m_2}\mu\mathbf{v}$$

$$\mathbf{p}_2' = -\mu\mathbf{v}' + \mu\mathbf{v}$$

These vector equations imply the vector diagrams shown in Fig. 14.6. The desired relation between θ_L and θ_C is obtained by constructing $\tan\theta_L$ from the partial diagram shown in Fig. 14.7:

(14.27)
$$\tan\theta_L = \frac{\mu v' \sin\theta_C}{(m_1/m_2)\mu v + \mu v' \cos\theta_C}$$

Now H_{rel} is a conserved quantity throughout the scattering. Prior to, and after collision, H_{rel} is purely kinetic and has the respective values $\mu v^2/2$, $\mu v'^2/2$. It follows that the magnitude of the relative velocity is maintained in scattering

$$v = v'$$

Substituting this equality into (14.27) gives the desired relation,

(14.28)
$$\tan\theta_L = \frac{\sin\theta_C}{\epsilon + \cos\theta_C} \qquad \left(\epsilon \equiv \frac{m_1}{m_2}\right)$$

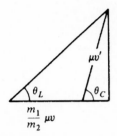

FIGURE 14.7   Triangle used to obtain relation between θ_L and θ_C.

This relation permits completion of (14.24):

$$(14.29) \qquad \frac{d\sigma_L}{d\cos\theta_L} = \frac{d\sigma_C}{d\cos\theta_C} \frac{(1 + \epsilon^2 + 2\epsilon\cos\theta_C)^{3/2}}{1 + \epsilon\cos\theta_C}$$

If the mass of the scatterer is very much larger than that of the incident particle, we may set $\epsilon = 0$ and the cross sections in both frames are equal. From (14.28) in this same extreme we obtain $\theta_L = \theta_C$.

In general, as (14.29) implies, scattering that is isotropic in the center-of-mass frame is not isotropic in the lab frame. For example, the isotropic cross section obtained for S-wave scattering (14.17),

$$\left(\frac{d\sigma}{d\Omega}\right)_C = |f(\theta)|^2 = \frac{\sin^2\delta_0}{k^2}$$

when substituted in (14.29) yields [with (14.28)] an anisotropic cross section in the lab frame.

$$(14.30) \qquad \left(\frac{d\sigma}{d\Omega}\right)_L = \frac{\sin^2\delta_0}{k^2} \frac{(1 + \epsilon^2 + 2\epsilon\cos\theta_C)^{3/2}}{1 + \epsilon\cos\theta_C}$$

Applications of results developed in this section appear in problems to follow. Whereas our primary example in the preceding analysis is relevant to low-energy scattering, where the potential of interaction plays a dominant role, the analysis to be developed in Section 14.4 addresses the case where the potential of interaction acts as a small perturbation on the incident plane-wave state. This analysis, known as the Born approximation, has many applications.

PROBLEMS

14.4 Assume that the differential cross section for a given interaction potential $d\sigma/d\Omega$ is isotropic in the center-of-mass frame. For mass ratio $\epsilon \ll 1$, what is the ratio of the differential cross section in the forward direction to that in the $\theta = \pi/2$ direction in the lab frame?

Answer

$$\frac{d\sigma(0)}{d\sigma(\pi/2)} = 1 + 2\epsilon$$

14.5 At what value of θ_C will the cross section vanish in the lab frame for S-wave scattering of two particles with mass ratio ϵ?

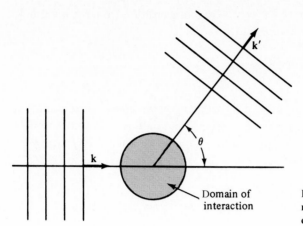

FIGURE 14.8 **In the Born approximation, incident and scattered particles are in plane-wave states.**

14.4 THE BORN APPROXIMATION

Harmonic perturbation theory, developed in Chapter 13, includes as a special case the example of a constant potential that has been turned on for t seconds. The perturbation Hamiltonian[1] is then given by (13.53) with $\omega = 0$. As was shown in Section 13.6, the theory of harmonic perturbation leads naturally to Fermi's formula (13.64) for cases where final states comprise a continuum. Such, of course, is the situation for scattering problems.

For these problems the perturbation Hamiltonian is the interaction potential, which is viewed as being "turned on" during the time that the incident particle is in the range of the potential. The incident particle enters the range of interaction with momentum $\hbar\mathbf{k}$ and leaves the range of interaction with momentum $\hbar\mathbf{k}'$. Such states of definite $\mathbf{k}$ before and after interaction correspond to plane-wave states (Fig. 14.8). Let us suppose that the scattering experiment is performed in a large cubical box of volume L^3. Normalized plane-wave states corresponding to $\mathbf{k}$ and $\mathbf{k}'$ are then given by

(14.31)

$$|\mathbf{k}\rangle = \frac{e^{i\mathbf{k}\cdot\mathbf{r}}}{L^{3/2}}$$

$$|\mathbf{k}'\rangle = \frac{e^{i\mathbf{k}'\cdot\mathbf{r}}}{L^{3/2}}$$

---

[1] For the case $\omega = 0$, the factor 2 is deleted in (13.53).

To apply Fermi's formula (13.64) for the rate of transition from the $\mathbf{k}$ to the $\mathbf{k}'$ state, caused by the perturbing potential $V(r)$,

(14.32)
$$\overline{w}_{\mathbf{kk}'} = \frac{2\pi}{\hbar} g(E_{k'}) |\langle \mathbf{k}' | V | \mathbf{k} \rangle|^2$$

we must know the density of final states $g(E_{k'})$. Having prescribed that the scattering experiment is performed in a large box of volume L^3, we may employ the expression for $g(E)$ as given by (8.141). Written in terms of final momentum $\hbar k'$, this expression becomes for nonspinning particles[1]

(14.33)
$$g(E_{k'}) = \frac{mL^3 k'}{2\pi^2 \hbar^2}$$

Now we wish to use the rate formula (14.32) to obtain an expression for the differential scattering cross section $d\sigma$. This parameter was defined by (14.1) according to which the number of particles scattered into $d\Omega$ per second is $\mathbb{J}_{\text{inc}} \, d\sigma$. To relate the transition rate $\overline{w}_{\mathbf{kk}'}$ to $d\sigma$, we note that the incident plane wave $|\mathbf{k}\rangle$ given in (14.31) corresponds to an incident current

(14.34)
$$\mathbb{J}_{\text{inc}} = \frac{\hbar \mathbf{k}}{mL^3}$$

In that $g(E_k)$ as given by (14.33) is isotropic in $\mathbf{k}'$, it represents the density of final $\mathbf{k}'$ states in all 4π solid angle. To select those scattered states that lie in the direction $d\Omega$ about $\mathbf{k}'$, we multiply g by the ratio $d\Omega/4\pi$. With $g(E_{k'})$ so augmented, $\overline{w}_{\mathbf{kk}'}$ then represents the rate at which particles of the incident flux (14.34) are scattered into $d\Omega$ in the direction of $\mathbf{k}'$. This rate is by definition the product $\mathbb{J}_{\text{inc}} \, d\sigma$. Thus we obtain the desired relation

$$\mathbb{J}_{\text{inc}} \, d\sigma = \frac{d\Omega}{4\pi} \overline{w}_{\mathbf{kk}'}$$

Inserting previous expressions, we obtain

(14.35)
$$\frac{d\sigma}{d\Omega} = \left(\frac{mL^3}{2\pi\hbar^2} \right)^2 \frac{k'}{k} |\langle \mathbf{k}' | V | \mathbf{k} \rangle|^2$$

Since particles suffer no loss in energy in the scattering process, we may equate

$$k = k'$$

---

[1] The g factor in Problem 2.42 represents density of states per unit volume.

The Scattering Amplitude

Recalling (14.6), which relates $d\sigma$ to the scattering amplitude $f(\theta)$, and inserting the explicit forms (14.31) for incident and scattered states into (14.35), allows the identification (with a conventional minus sign)

$$(14.36) \qquad f(\theta) = -\frac{m}{2\pi\hbar^2} \int V(r) e^{i\mathbf{r}\cdot(\mathbf{k}-\mathbf{k}')} \, d\mathbf{r}$$

This formula for the scattering amplitude may be further simplified through the substitution

$$\mathbf{K} = \mathbf{k} - \mathbf{k}'$$

As is evident in Fig. 14.9a, owing to the equal magnitudes of $\mathbf{k}$ and $\mathbf{k}'$, we may set

$$K = 2k \sin\left(\frac{\theta}{2}\right)$$

where θ is the angle of scatter. With the differential volume of integration $d\mathbf{r}$ in (14.36) written in spherical coordinates and the polar axis taken to be coincident with $\mathbf{K}$ (Fig. 14.9b), we obtain

$$f(\theta) = -\frac{m}{2\pi\hbar^2} \int_0^{2\pi} d\bar{\phi} \int_0^{\pi} d\bar{\theta} \, \sin\bar{\theta} \int_0^\infty dr \, r^2 V(r) e^{iKr\cos\bar{\theta}}$$

$$= -\frac{m}{\hbar^2} \int_0^\infty dr \, r^2 V(r) \int_{-1}^1 d\eta \, e^{iKr\eta}$$

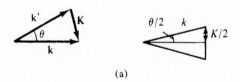

(a)

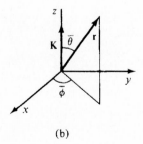

(b)

FIGURE 14.9   (a) Transformation $\mathbf{K} = \mathbf{k} - \mathbf{k}'$. (b) Spherical coordinate frame with $\mathbf{K}$ aligned with the polar axis.

Integration over $\eta \equiv \cos \bar{\theta}$ gives

(14.37)
$$f(\theta) = -\frac{2m}{\hbar^2 K} \int_0^\infty dr\, rV(r) \sin Kr$$

This expression for the scattering amplitude is called the *standard form of the Born approximation.*

In applying this formula for the scattering amplitude, one should keep in mind that it is derived on the basis of perturbation theory according to which the scattering potential should be small compared to the free-particle (unperturbed) Hamiltonian. This will be the case for sufficiently large incident energies or sufficiently weak strengths of potential.

The Shielded Coulomb Potential

Let us apply (14.37) to calculate the cross section for the shielded Coulomb potential,[1]

$$V(r) = -\frac{Ze^2 \exp(-r/a)}{r}$$

The exponential factor for $r > a$ acts to shield the bare Coulomb potential Ze^2/r between two particles with respective charges Ze and e. Thus beyond the range a, the potential is exponentially small. Within the range, $r < a$, the potential is essentially Coulombic. Substituting the shielded Coulomb potential into (14.37) gives

$$f(\theta) = \frac{2mZe^2}{\hbar^2 K} \int_0^\infty e^{-r/a} \sin Kr\, dr$$

$$= \frac{2mZe^2}{\hbar^2} \frac{1}{K^2 + (1/a)^2}, \qquad K = 2k \sin\left(\frac{\theta}{2}\right)$$

The corresponding scattering cross section is obtained from (14.6).[2]

(14.38)
$$\frac{d\sigma}{d\Omega} = \frac{(2mZe^2/\hbar^2)^2}{[K^2 + (1/a)^2]^2}$$

In the limit of large incident energies $K^2 \gg a^{-2}$, the predominant contribution to $d\sigma$ is due to the bare Coulomb potential. The resulting cross section, employed previously in Problem 14.1, appears as

---

[1] This potential enters in three independent areas of physics, where it carries the following names: in plasma physics, the *Debye potential;* in high-energy physics, the *Yukawa potential;* in solid-state physics, the *Thomas–Fermi potential.*

[2] With m replaced by the reduced mass μ, (14.38) represents the cross section in the center-of-mass frame.

(14.39)
$$\frac{d\sigma}{d\Omega} = \left[\frac{Ze^2}{4(\hbar^2k^2/2m)\sin^2(\theta/2)}\right]^2$$

$$= \left(\frac{Ze^2}{4E}\right)^2 \frac{1}{\sin^4(\theta/2)}, \qquad E = \frac{\hbar^2k^2}{2m}$$

This is the precise expression for the Rutherford cross section for the scattering of a charged particle with charge e and mass m from a fixed charge Ze, which is seen to be dominated by forward scattering ($\theta \approx 0$). Furthermore, the classical evaluation of the Rutherford cross section also gives (14.39), with $E = p^2/2m$.

PROBLEMS

14.6   Using the Born approximation, evaluate the differential scattering cross section for scattering of particles of mass m and incident energy E by the repulsive spherical well with potential

$$V(r) = \begin{cases} V_0, & 0 < r < a \\ 0, & r > a \end{cases}$$

Exhibit explicit E and θ dependence.

Answer

$$\frac{d\sigma}{d\Omega} = \left(\frac{2mV_0}{\hbar^2K^3}\right)^2 (\sin Ka - Ka\cos Ka)^2$$

$$\hbar K = 2\sqrt{2mE}\sin(\theta/2)$$

14.7   Using the Born approximation, obtain an integral expression for the total cross section for scattering of particles of mass m from the attractive Gaussian potential

$$V(r) = -V_0 \exp\left[-\left(\frac{r}{a}\right)^2\right]$$

14.8   An important parameter in scattering theory is the scattering length a. This length is defined as the negative of the limiting value of the scattering amplitude as the energy of the incident particle goes to zero.

$$a = -\lim_{k\to 0} f(\theta)$$

(a)   For low-energy scattering and relatively small phase shift, show that

$$a = -\lim_{k\to 0} \frac{\delta_0}{k}$$

(b) For the same conditions as in part (a), show that

$$\sigma = 4\pi a^2$$

(c) What is the scattering length for point particles scattering from a rigid sphere of arbitrary radius $\bar{a}$?

14.5 ATOMIC-RADIATIVE ABSORPTION CROSS SECTION

Returning to our analysis of Section 13.9, again we consider a flux of photons incident on an atom, carrying one photon per unit volume. Since the photon moves with speed c, this gives an incident photon current

(14.40)
$$J_{\text{inc}} = c \times \frac{1 \text{ photon}}{\text{cm}^3} = c\left(\frac{\text{photons}}{\text{cm}^2 \text{ s}}\right)$$

There is a probability rate for each photon in the incident current to be absorbed by the atom. The principle of microscopic reversibility allows us to equate this probability rate for absorption to the corresponding probability rate of atomic decay.

With Fermi's golden rule (13.64) we may then write the probability rate for atomic absorption as

(14.41)
$$w_{nn'} = \frac{2\pi}{\hbar} |\langle n', \mathbf{k} | \mathbb{H}_- | n \rangle|^2 \frac{g(\omega)}{\hbar}$$

where we have made the replacement

$$g(E) = \frac{g(\omega)}{\hbar}$$

[See (13.111b) and recall that $g(E) = 8\pi\bar{g}(E)$ and that $V = 1$.] With (13.106), our preceding equation (14.41) becomes

(14.42)
$$w_{nn'} = \frac{2\pi}{\hbar} e^2 (2\pi\hbar\omega) \frac{g(\omega)}{\hbar} |\langle n' | \mathbf{a} \cdot \mathbf{r} | n \rangle|^2$$

If the polarization unit vector $\mathbf{a}$ of the incident field is randomly oriented, then as was previously demonstrated in our discussion of the Einstein B coefficient (Section 13.7), we may set

(14.43)
$$|\langle n' | \mathbf{a} \cdot \mathbf{r} | n \rangle|^2 = \frac{1}{3} r_{nn'}^2 = \frac{1}{3} \frac{d_{nn'}^2}{e^2}$$

Here we have reintroduced the dipole moment $\mathbf{d}$ (13.69).

With (14.43) placed into (14.42) we find

(14.44)
$$w_{nn'} = \frac{4}{3} \frac{\pi^2 \omega d_{nn'}{}^2}{\hbar} g(\omega)$$

This is the probability rate for an atom to absorb a photon from an incident current carrying one photon per unit volume. Since the incident current in the present configuration is so normalized, $w_{nn'}$ represents the rate at which photons are absorbed by the atom from the incident beam. With our previous definition of cross section we may then write

(14.45)
$$J_{\text{inc}} \sigma_{nn'} = w_{nn'}$$

Note the dimensions:

$$J_{\text{inc}} \left(\frac{\text{photons}}{\text{cm}^2 \, \text{s}} \right) \times \sigma \, (\text{cm}^2) = w_{nn'} \left(\frac{1}{\text{s}} \right)$$

With (14.40) and (14.45) we obtain the total cross section,

(14.46)
$$\sigma_{nn'} = \frac{4}{3} \frac{\pi^2 \omega d_{nn'}{}^2}{\hbar c} g(\omega)$$

for resonant absorption at the incident frequency

$$\hbar \omega = E_n - E_{n'}$$

A more accurate description of this process includes the possibility of absorption of incident photons which are off-resonance. For such cases, with $\bar{\omega} \equiv \omega_{nn'}$, (14.46) is generalized to the form

(14.47)
$$\sigma(\bar{\omega}, \omega) = \frac{4}{3} \frac{\pi^2 d^2}{\hbar c} \bar{\omega} g(\bar{\omega}, \omega)$$

where we have set

$$g(\omega) \rightarrow g(\bar{\omega}, \omega)$$

The function $g(\bar{\omega}, \omega)$ is the so-called *line-shape factor*.

A realistic expression for $g(\bar{\omega}, \omega)$ which is appropriate to many line-broadening processes is the *Lorentzian line-shape factor*,

(14.48)
$$g_L(\bar{\omega}, \omega) = \frac{1}{\pi} \frac{\Delta \omega_L / 2}{(\bar{\omega} - \omega)^2 + (\Delta \omega_L / 2)^2}$$

See Fig. 14.10.

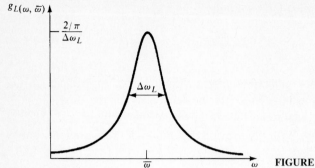

FIGURE 14.10 Lorentzian line shape.

The spreading of an absorption line is attributed to relaxation processes—such as, for example, the relaxation of excited atomic states incurred in atomic collisions. If τ represents the decay time for such processes, then one sets

$$\frac{\Delta\omega_L}{2} = \frac{1}{\tau}$$

In the idealized limit that these states last indefinitely, $\Delta\omega_L \to 0$, and $g_L(\overline{\omega}, \omega)$ becomes sharply peaked about $\omega = \overline{\omega}$. In this limit (14.47) becomes

(14.49)
$$\sigma(\overline{\omega}, \omega) = \frac{4}{3}\frac{\pi^2 d^2}{\hbar c}\overline{\omega}\delta(\omega - \overline{\omega})$$

Here we have employed the delta-function representation (C.9).

14.6 ELEMENTS OF FORMAL SCATTERING THEORY. THE LIPPMANN–SCHWINGER EQUATION

In this concluding section we present a brief introduction to the formal theory of scattering, central to which is the Lippmann–Schwinger equation.[1] An elementary derivation of this equation is presented based on the interaction picture described previously in Section 11.12. The Lippmann–Schwinger equation so derived appears in a form independent of representation. Writing this equation in coordinate representation is found to give an integral equation for scattered states, which in turn gives a general expression for the scattering amplitude. In the Born approximation this relation returns our previous expression for $f(\theta)$ given by (14.36).

---

[1] For further discussion, see E. Merzbacher, *Quantum Mechanics,* 2d ed., Wiley, New York, 1970, Chapter 19.

We consider the Hamiltonian

(14.50)
$$\hat{H} = \hat{H}_0 + \hat{V}e^{-\varepsilon|t|/\hbar}$$

where $\hat{H}_0$ is the free-particle Hamiltonian and the infinitesimal parameter ε has dimensions of energy. The interaction $\hat{V}$ is assumed to be independent of time. For small ε the exponential factor has the effect of "turning on" the interaction $\hat{V}$ in the interval about $t = 0$. We will also find that the presence of this factor insures convergence of integration in the derivation to follow (in both limits $t \rightarrow \pm\infty$).

Recall that the Schrödinger equation in the interaction picture has the integral form (11.144),[1]

(14.51)
$$|\psi_I(t)\rangle = |\psi_I(t_0)\rangle + \frac{1}{i\hbar} \int_{t_0}^{t} \hat{V}_I(t') |\psi_I(t')\rangle \, dt'$$

where, we recall,

(14.52a)
$$|\psi_I(t)\rangle = e^{it\hat{H}_0/\hbar} |\psi(t)\rangle$$

(14.52b)
$$\hat{V}_I(t) = e^{it\hat{H}_0/\hbar} \hat{V} e^{-itH_0/\hbar}$$

As we wish to apply (14.51) to scattering theory, we stipulate that $\hat{V}_I(t) \rightarrow 0$ in the limits $t \rightarrow \pm\infty$. At these asymptotic values, with the interaction vanishingly small, $|\psi_I(t)\rangle$ loses its time dependence, and we define

(14.53)
$$|\varphi\rangle \equiv |\psi_I(\pm\infty)\rangle$$

which may be identified as free-particle states.

Rewriting (14.51) over the interval $t_0 = \pm\infty$, $t = 0$, and identifying the scattering states

(14.54)
$$|\psi^{(\pm)}\rangle \equiv |\psi(0)\rangle$$

gives the equation

(14.55)
$$|\psi^{(\pm)}\rangle = |\varphi\rangle + \frac{1}{i\hbar} \int_{\mp\infty}^{0} \hat{V}_I(t) |\psi_I(t)\rangle \, dt$$

Note that with this choice of interval, $|\psi^{(\pm)}\rangle$, given by (14.54), represents scattered states in the domain of interaction. Furthermore, as $|\psi^{(+)}\rangle$ is relevant to the time interval $(-\infty \leq t \leq 0)$, we may identify it with incoming incident waves, commonly called the "in" solution. As $|\psi^{(-)}\rangle$ relates to the interval $(\infty, 0)$, it is the time-reversed

---

[1] Ket notation is employed to obtain a relation independent of representations.

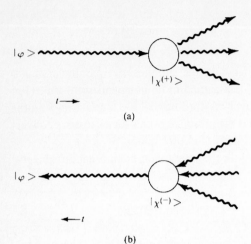

(a)

(b)

FIGURE 14.11   (a) "In" solution. (b) "Out" solution.

state of $|\psi^{(+)}\rangle$ and is commonly called the "out" solution. See Fig. 14.11. In the limit $\varepsilon \to 0$ we take $|\psi^{(\pm)}\rangle$ to be an eigenstate of the total Hamiltonian and write

$$(14.56) \qquad \hat{H}|\psi^{(\pm)}\rangle = E|\psi^{(\pm)}\rangle$$

Reduction of Interaction Integral

Consider the integrand in (14.55). With (14.52a, b) we write

$$(14.57) \qquad \hat{V}_I(t)|\psi_I(t)\rangle = e^{i\hat{H}_0 t/\hbar}\hat{V}e^{-\varepsilon|t|/\hbar}e^{-i\hat{H}_0 t/\hbar}e^{i\hat{H}_0 t/\hbar}|\psi(t)\rangle$$

$$= e^{i\hat{H}_0 t/\hbar}\hat{V}e^{-\varepsilon|t|/\hbar}e^{-i\hat{H}t/\hbar}|\psi(0)\rangle$$

and note

$$|\psi(0)\rangle = |\psi_I(0)\rangle = |\psi^{(\pm)}\rangle$$

Substituting this identification into (14.57) followed by replacement of (14.57) into (14.55) gives

$$(14.58) \qquad |\psi^{(\pm)}\rangle = |\varphi\rangle + \frac{1}{i\hbar}\int_{\mp\infty}^{0} e^{i\hat{H}_0 t/\hbar}\hat{V}e^{-\varepsilon|t|/\hbar}e^{-i\hat{H}t/\hbar}|\psi^{(\pm)}\rangle\, dt$$

With (14.56) the preceding becomes

$$(14.59) \qquad |\psi^{(\pm)}\rangle = |\varphi\rangle + \frac{1}{i\hbar}\int_{\mp\infty}^{0} e^{i\hat{H}_0 t/\hbar}\hat{V}e^{-\varepsilon|t|/\hbar}e^{-iEt/\hbar}|\psi^{(\pm)}\rangle\, dt$$

For "in" solutions we encounter the integral

$$\hat{G}^{(+)} = \frac{1}{i\hbar} \int_{-\infty}^{0} dt \, \exp\left[\frac{-t(i\hat{H}_0 - iE + \varepsilon)}{\hbar}\right]$$

$$= \frac{1}{E - \hat{H}_0 + i\varepsilon}$$

Similarly,

$$\hat{G}^{(-)} = \frac{1}{E - \hat{H}_0 - i\varepsilon}$$

Note that without the presence of ε the integrals $\hat{G}^{(\pm)}$ do not converge. Substituting these expressions for $\hat{G}^{(\pm)}$ into (14.59) gives the *Lippmann–Schwinger equation*,

(14.60)
$$\boxed{\; |\psi^{(\pm)}\rangle = |\varphi\rangle + \frac{1}{E - \hat{H}_0 \pm i\varepsilon} |\psi^{(\pm)}\rangle \;}$$

which, as previously noted, is independent of specific representation.

Scattering Amplitude Revisited

In Problem 14.11 you are asked to show that the coordinate representation of the "in" solution to (14.60) assumes the form

(14.61) $\psi_{\mathbf{k}}^{(+)}(\mathbf{r}) = \varphi_{\mathbf{k}}(r) - \dfrac{m}{2\pi\hbar^2} \displaystyle\int \dfrac{\exp\,(ik\,|\mathbf{r} - \mathbf{r}'|)}{|\mathbf{r} - \mathbf{r}'|} V(\mathbf{r}')\psi_{\mathbf{k}}^{+}(\mathbf{r}') \, d\mathbf{r}'$

Here we have made the identifications

(14.62a) $\langle \mathbf{r} | \varphi \rangle = \varphi_{\mathbf{k}}(\mathbf{r})$

(14.62b) $\langle \mathbf{r} | \psi^{(+)} \rangle = \psi_{\mathbf{k}}^{(+)}(\mathbf{r})$

(14.62c) $\langle \mathbf{r} | \hat{V}\psi^{(+)} \rangle = V(\mathbf{r})\psi^{(+)}(\mathbf{r})$

At large distances from the interaction domain we may write

$$\frac{1}{|\mathbf{r} - \mathbf{r}'|} \simeq \frac{1}{r} + \cdots$$

and

$$k|\mathbf{r} - \mathbf{r}'| = kr\left[1 + \left(\frac{r'}{r}\right)^2 - \frac{2\mathbf{r} \cdot \mathbf{r}'}{r^2}\right]^{1/2}$$

$$\simeq kr\left[1 - \frac{\mathbf{r} \cdot \mathbf{r}'}{r^2} + \cdots\right] = kr - \mathbf{k}' \cdot \mathbf{r}'$$

where we have set

$$\mathbf{k}' \equiv \frac{k\mathbf{r}}{r}$$

Substituting these expansions in (14.61) gives

(14.63) $$\psi_{\mathbf{k}}^{(+)}(\mathbf{r}) = \varphi_{\mathbf{k}}(\mathbf{r}) - \frac{m}{2\pi\hbar^2}\frac{e^{ikr}}{r}\langle\varphi_{\mathbf{k}'}|V|\psi_{\mathbf{k}}^{(+)}\rangle$$

Comaprison with (14.7) gives the following expression for the scattering amplitude:

(14.64) $$f(\theta) = -\frac{m}{2\pi\hbar^2}\langle\varphi_{\mathbf{k}'}|V|\psi_{\mathbf{k}}^{(+)}\rangle$$

In the Born approximation

$$\psi_{\mathbf{k}}^{+} \rightarrow \varphi_{\mathbf{k}}$$

and (14.64) returns our previous finding (14.36). However, one should bear in mind that (14.63) is, more generally, an integral equation for $\psi_{\mathbf{k}}^{(+)}$, solution to which gives a more accurate expression for the scattering amplitude through (14.64).

PROBLEMS

14.9 A beam of photons at a given frequency propagates into a medium of atoms with density n (cm^{-3}). The cross section for absorption of photons at this frequency by the atoms is σ. If J is incident photon flux, then argue that the decrease in J in the distance dx due to absorption is

$$dJ = -\kappa J\, dx$$

where

$$\kappa = n\sigma$$

14.10 A monochromatic beam of photons at frequency $\nu = 10^{14}$ Hz and intensity 1.4 keV/cm^2 s is incident on a gas of atoms of density $n = 10^{18}$ cm^{-3}. At the given frequency the radiation is very near resonance of the atoms. The related transition dipole moment of atoms in the gas has magnitude $0.4a_0e$. At what distance (cm) into the gas will the intensity of the beam be e^{-1} times its starting value? (*Hint:* Use results of the preceding problem.)

14.11 Working in the coordinate representation, employ the Lipmann–Schwinger equation (14.60) to derive its coordinate representation (14.61).

Answer

Let us label

$$\hat{G}_{\pm} \equiv \lim_{\varepsilon \to 0} G^{(\pm)}$$

The Lippmann–Schwinger equation (14.60) may then be written

$$|\psi^{(\pm)}\rangle = |\varphi\rangle + \hat{G}_{\pm}\hat{V}|\psi^{(\pm)}\rangle$$

To obtain the coordinate representation of this equation, we operate on it from the left with $\langle \mathbf{r}|$ to obtain [see (14.62)]

$$\psi_{\mathbf{k}}^{(\pm)}(\mathbf{r}) = \varphi_{\mathbf{k}}(\mathbf{r}) + \langle \mathbf{r}|\hat{G}_{\pm}\hat{V}|\psi^{(\pm)}\rangle \equiv \varphi_{\mathbf{k}}(\mathbf{r}) + I_{\pm}$$

which serves to define the interaction term $I_{\pm}$. Developing this term, we obtain[1]

$$I_{\pm} = \int d\mathbf{r}' \int d\overline{\mathbf{k}}\langle \mathbf{r}|\overline{\mathbf{k}}\rangle\langle \overline{\mathbf{k}}|\hat{G}_{\pm}|\mathbf{r}'\rangle\langle \mathbf{r}'|\hat{V}|\psi^{(\pm)}\rangle$$

We recall[2]

$$\langle \mathbf{r}|\mathbf{k}\rangle = \frac{1}{(2\pi)^{3/4}}e^{i\mathbf{k}\cdot\mathbf{r}}$$

and

$$\hat{H}_0|\mathbf{k}\rangle = \frac{\hbar^2 k^2}{2m}|k\rangle$$

whence

$$\langle \overline{\mathbf{k}}|\hat{G}_{\pm}|\mathbf{r}'\rangle = \lim \frac{1}{E - (\hbar^2 \overline{k}^2/2m) \pm i\varepsilon}\langle \overline{\mathbf{k}}|\mathbf{r}'\rangle$$

Thus we obtain

$$I_{\pm} = \frac{1}{(2\pi)^3}\int d\mathbf{r}' \int d\overline{\mathbf{k}} \frac{e^{i\overline{\mathbf{k}}\cdot(\mathbf{r}-\mathbf{r}')}}{E - (\hbar^2 \overline{k}^2/2m) \pm i\varepsilon}\langle \mathbf{r}'|\hat{V}|\psi^{(\pm)}\rangle$$

Consider the $\overline{\mathbf{k}}$ integration,

$$\Phi_{\pm} \equiv \int d\overline{\mathbf{k}} \frac{e^{i\overline{\mathbf{k}}\cdot\Delta\mathbf{r}}}{R_{\pm}}$$

---

[1] Here we employ the spectral resolution of unity (see Problem 11.1).

[2] Note that this form gives the proper normalization

$$\int d\mathbf{k}\, \langle \mathbf{r}|\mathbf{k}\rangle\langle \mathbf{k}|\mathbf{r}'\rangle = \delta(\mathbf{r} - \mathbf{r}')$$

See (C.11).

where

$$\Delta \mathbf{r} \equiv \mathbf{r} - \mathbf{r}'$$

$$R_{\pm} = E - \frac{\hbar^2 \bar{k}^2}{2m} \pm i\varepsilon$$

With

$$\int d\bar{\mathbf{k}} = 2\pi \int_0^\infty \bar{k}^2 \, d\bar{k} \int_{-1}^1 d\mu$$

$$\mu \equiv \cos \theta = \frac{\bar{\mathbf{k}} \cdot \Delta \mathbf{r}}{|\bar{\mathbf{k}} \cdot \Delta \mathbf{r}|}$$

integration over μ gives

$$\int_{-1}^1 d\mu \, \exp(i\bar{k} \, \Delta r \mu) = \frac{1}{i\bar{k} \, \Delta r} [\exp(i\bar{k} \, \Delta r) - \exp(-i\bar{k} \, \Delta r)]$$

As $\bar{k}/R_{\pm}$ is an odd function of $\bar{k}$, we find

$$\Phi_{\pm} = 2\pi \int_0^\infty d\bar{k} \, \bar{k}^2 \frac{1}{i\bar{k} \, \Delta r} \frac{[\exp(i\bar{k} \, \Delta r) - \exp(-i\bar{k} \, \Delta r)]}{R_{\pm}}$$

$$= \frac{2\pi}{i \, \Delta r} \int_{-\infty}^\infty \frac{d\bar{k} \, \bar{k} e^{i\bar{k}\Delta r}}{R_{\pm}}$$

Next we set

$$E = \frac{\hbar^2 k^2}{2m}$$

which, by conservation of energy, is the same as the free-particle energy of the incident wave, $\varphi_\mathbf{k}(\mathbf{r})$. We obtain

$$\frac{2m}{\hbar^2} R_{\pm} = \left(k^2 - \bar{k}^2 \pm i \frac{2m\varepsilon}{\hbar^2} \right)$$

$$= (k - \bar{k} \pm i\bar{\varepsilon})(k + \bar{k} \pm i\bar{\varepsilon})$$

where $\bar{\varepsilon} \equiv m\varepsilon/\hbar^2 k$. Thus

$$\frac{\bar{k}}{R_{\pm}} = -\frac{2m}{\hbar^2} \left(\frac{1/2}{\bar{k} - k \mp i\bar{\varepsilon}} + \frac{1/2}{\bar{k} + k \pm i\bar{\varepsilon}} \right)$$

We are now prepared to integrate over $\bar{k}$ by contour integration. As the integrand contains the factor $\exp i\bar{k} \, \Delta r$, it must be closed in the domain Im $\bar{k} > 0$, that is, the upper half $\bar{k}$ plane. For Φ_+ there is a pole at

$$\bar{k} = k + i\bar{\varepsilon}$$

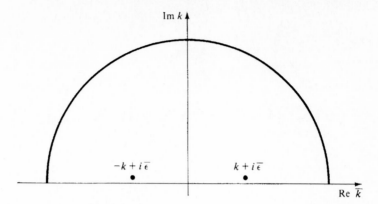

FIGURE 14.12 The contour for the integration in $\Phi_{\pm}$. The pole at $k + i\varepsilon'$ contributes to $\psi^{(+)}$ and the one at $-k + i\varepsilon'$ contributes to $\psi^{(-)}$.

whereas for Φ_- there is a pole at

$$\bar{k} = -k + i\bar{\varepsilon}$$

See Fig. 14.12.

Passing to the limit $\bar{\varepsilon} \to 0$, we obtain

$$\Phi_{\pm} = -\frac{2\pi}{i \, \Delta r} \frac{2m}{\hbar^2} 2\pi i \left(\frac{1}{2}\right) e^{\pm ik\Delta r}$$

$$= \frac{4\pi^2 m}{\hbar^2} \frac{e^{\pm ik\Delta r}}{\Delta r}$$

whence

$$I_+ = -\frac{1}{(2\pi)^3} \frac{4\pi^2 m}{\hbar^2} \int d\mathbf{r}' \frac{e^{ik\Delta r}}{\Delta r} \langle \mathbf{r}' | \hat{V} | \psi^{(+)} \rangle$$

which when inserted in our starting relation

$$\psi_{\mathbf{k}}^{(+)} = \varphi_{\mathbf{k}}(r) + I_+$$

returns (14.61).

14.12 Is (14.61) a valid relation for inelastic collisions—for example, an ionizing collision in which an electron is emitted from an atom due to, say, electron scattering?

Answer

As the energy of the incident electron is not conserved (it loses energy in releasing the bound electron), one cannot equate energy of the scattered electron to its incident free-particle value, and the derivation in the preceding problem is invalid.

15

RELATIVISTIC QUANTUM MECHANICS

In this chapter we present a description of relativistic quantum mechanics. The chapter begins with a review of basic relativistic notions and continues with derivation of the Klein–Gordon equation relevant to relativistic bosons. Incorporating elements of this equation, the Dirac equation appropriate to fermions is obtained. Plane-wave solutions to this equation for a free particle are constructed, components of which are shown to correspond to "spin up" and "spin down" wavefunctions. The energy spectrum of these solutions is noted to have a forbidden gap and an infinite sea of negative values. These properties imply the existence of positrons. The Dirac equation is then shown to imply the correct magnetic dipole moment of the electron. The chapter continues with a derivation of the four-dimensional spin operator in the Dirac formalism and closes with a brief introduction to the covariant formulation of relativistic quantum mechanics.

15.1 PRELIMINARY REMARKS

Postulates, World Lines, and the Light Cone

The two postulates of special relativity are as follows:

1. The laws of physics are invariant under inertial transformations.
2. The speed of light is independent of the motion of the source.

The first postulate states that the result of an experiment in a given inertial frame of reference is independent of the constant translational motion of the system as a

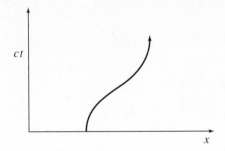

FIGURE 15.1   **The world line of a particle moving in one direction from rest.**

whole. An inertial frame is one in which a mass at rest experiences no force. Thus there is no absolute frame in the universe with respect to which motion of an arbitrary inertial frame is uniquely defined. Only relative motion between frames is meaningful.

Concerning the second postulate, consider a light source fixed in a frame S. The frame moves relative to the observer in a frame S' with speed v. The observer measures the speed of light c, independent of the speed v. This situation is evidently equivalent to one in which S' moves relative to S with speed v. Thus the speed of light is independent of the motion of the receiver, as well as that of the source. This conclusion is alien to our intuitive picture of either wave or particle motion.

Einstein defined an *event* as a point in space-time coordinates. The locus of events of a particle is called the *world line* of the particle. For one-dimensional motion, the world line of a particle is a curve in (x, ct) space (see Fig. 15.1).

Of particular interest in the study of relativity is the concept of the *light cone*. This is the world line of the leading edge of a light wave stemming from a source switched on at a given instant at a given location. The notion of past and future may be defined with respect to the light cone (see Fig. 15.2).

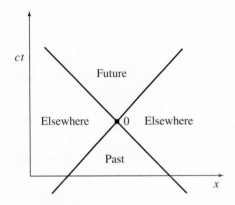

FIGURE 15.2   **Light pulse is initiated at 0. Points in the domain marked "Elsewhere" cannot be reached from 0 at speeds less than or equal to c.**

Four-Vectors

An event at a given location $\mathbf{x}$ and time t may be described by the four-dimensional vector

$$(15.1) \qquad \bar{x} = (\mathbf{x}, ict)$$

which is called a *four-vector*. The momentum four-vector is given by

$$(15.2) \qquad \bar{p} = \left(\mathbf{p}, \frac{iE}{c} \right)$$

Here we have written

$$(15.3) \qquad E = \gamma mc^2 = mc^2 + T$$

for the total energy of a particle (in the absence of potential) of *rest mass m* with kinetic energy T. The parameter γ is written for

$$(15.4) \qquad \gamma^2 \equiv \frac{1}{1 - \beta^2}, \qquad \beta \equiv \frac{v}{c}$$

Note that γ increases monotonically from 1 at $\beta = 0$ to ∞ at $\beta = 1$. The relativistic momentum three-vector is given by

$$(15.5) \qquad \mathbf{p} = \gamma m \mathbf{v}, \qquad \mathbf{v} = \frac{d\mathbf{x}}{dt}$$

where t is time measured in the lab frame.

Let us write p_μ for the components of the four-vector $\bar{p}$. Thus

$$(15.6) \qquad p_4 = \frac{iE}{c}, \qquad -p_4^2 c^2 = E^2 = p^2 c^2 + m^2 c^4$$

In this notation we may write

$$(15.7) \qquad p_\mu = m u_\mu$$

where u_μ is the velocity four-vector,

$$(15.8) \qquad u_\mu = \frac{p_\mu}{m} = \frac{dx_\mu}{d\tau} \equiv \dot{x}_\mu$$

and τ denotes *proper time*. This is the time measured on a clock attached to the moving particle (discussed further below).

Three other important four-vectors are

(15.9a) $$\bar{\mathbf{J}} = (\mathbf{J}, ic\rho)$$

(15.9b) $$\bar{\mathbf{A}} = (\mathbf{A}, i\Phi)$$

(15.9c) $$\bar{\mathbf{k}} = \left(\mathbf{k}, \frac{i\omega}{c}\right)$$

In these expressions, $\mathbf{J}$ is current density, ρ is charge density, $\mathbf{A}$ is vector potential, Φ is scalar potential, $\mathbf{k}$ is a wave vector, and ω is frequency.

Lorentz Transformation

Consider that a frame S' moves at constant speed v in the z direction with respect to a frame S, as shown in Fig. 15.3. It is readily shown from the two postulates above that, if A_μ is a four-vector in S, an observer in S' observes the components

(15.10) $$A'_\mu = L_{\mu\nu} A_\nu$$

where

(15.11) $$L = \begin{pmatrix} 1 & 0 & 0 & 0 \\ 0 & 1 & 0 & 0 \\ 0 & 0 & \gamma & i\gamma\beta \\ 0 & 0 & -i\gamma\beta & \gamma \end{pmatrix}$$

[In (15.10), we have employed the "Einstein convention" in which repeated indices are summed (from 1 to 4).] Note that L is orthogonal; that is,

(15.12) $$\tilde{L} = L^{-1}$$

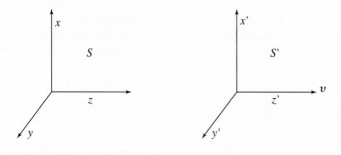

FIGURE 15.3 The frame S' moves at velocity v with respect to the frame S.

where, we recall, $\tilde{L}$ is the transpose of L. Note further that

(15.13) $$\det L = 1$$

(see Problem 11.96). Thus L effects a rotation in complex four-dimensional space. Consequently, the "length" of a four-vector is preserved under a *Lorentz transformation*. That is,

(15.14a) $$A'_\mu A'_\mu = A_\mu A_\mu$$

Let us show this formally:

$$A'_\mu A'_\mu = L_{\mu\nu} A_\nu L_{\mu\lambda} A_\lambda = L_{\mu\nu} L_{\mu\lambda} A_\nu A_\lambda = L^{-1}_{\nu\mu} L_{\mu\lambda} A_\nu A_\lambda = \delta_{\nu\lambda} A_\nu A_\lambda = A_\lambda A_\lambda$$

For example, consider the length of the momentum four-vector,

(15.14b) $\quad p_\mu p_\mu = \gamma^2 m^2 v^2 - \gamma^2 m^2 c^2 = -m^2 \gamma^2 c^2 (1 - \beta^2) = -m^2 c^2$

This returns the useful relation

(15.15) $$E^2 = c^2 p^2 + m^2 c^4$$

For $\bar{\mathbf{x}}$, we find

(15.16) $$x_\mu x_\mu = x^2 - c^2 t^2$$

Such entities, which remain invariant under a Lorentz transformation, are called *Lorentz invariants*. Note in particular that the Lorentz invariant $x_\mu x_\mu$ as given by (15.16) is a reiteration of the second postulate stated above that the speed of light is c in all frames.

The Lorentz matrices (15.11) have the following group property.

(15.17) $$\hat{L}(\beta_1)\hat{L}(\beta_2) = \hat{L}(\beta_{12})$$

where

(15.18) $$\beta_{12} = \frac{\beta_1 + \beta_2}{1 + \beta_1 \beta_2}$$

Note that (15.18) precludes the speed of any object from exceeding c. Consider that one of the frames moves with relative speed c, so that, for example, $\beta_1 = 1$. Then (15.18) returns the value $\beta_{12} = 1$. Furthermore, $L(0) = I$, the identity operator, and $L^{-1} = \tilde{L}$. The Lorentz transformations comprise a group called the *Lorentz group*.[1]

---

[1] Equations written in tensor notation, such as (15.8), (15.14b), and (15.16), are valid in all coordinate frames. Equations written in this form are said to be *covariant*. See Section 15.5.

Length Contraction, Time Dilation, and Proper Time

With relative interframe motion again confined to the z direction, Lorentz transformation of the event four-vector gives

(15.19)
$$\begin{pmatrix} z' \\ ict' \end{pmatrix} = \gamma \begin{pmatrix} 1 & i\beta \\ -i\beta & 1 \end{pmatrix} \begin{pmatrix} z \\ ict \end{pmatrix}$$

We find

(15.20a)
$$z' = \gamma(z - \beta ct)$$

(15.20b)
$$ict' = i\gamma(-\beta z + ct)$$

Let a rod of given length $\Delta z'$ lie fixed in a frame that we label S'. The frame S' moves with speed βc relative to the frame S. At a given instant, the length of the rod is measured in the S frame. This means that the locations of the ends of the rods are measured simultaneously in S. Calling the length so measured $\Delta z = z_b - z_a$, with $t_b = t_a$, (15.20a) gives

(15.21)
$$\Delta z = \frac{1}{\gamma} \Delta z' \leq \Delta z'$$

The rod moving past the frame S is measured to be shortened by an observer in S. This is the phenomenon of *length contraction*.

   Consider next a clock that is at a fixed location in the moving S' frame ($z_b' = z_a'$). To find the manner in which intervals τ' on this clock are observed in S, we write the inverse of (15.20b):

$$ict = i\gamma(\beta z' + ct')$$

Choosing $z' = 0$ gives

(15.22)
$$t = \gamma t' \geq t'$$

Thus an observer in S concludes that intervals on his clock, t, are longer than those on the S' clock or that the S' clock is "running slow." This is the phenomenon of time dilation.

   An important parameter in relativity is that of *proper time*. As noted above, the proper time of a particle is the time measured on a clock that moves with the particle. Thus, if t denotes time in the lab frame and τ proper time, we write

(15.23)
$$dt = \gamma d\tau$$

Comparison with (15.22) reveals that we have identified proper time with the single clock fixed in S'. The square differential element, $(d\tau)^2$ is a Lorentz invariant. (See Problem 15.18.)

15.2   KLEIN–GORDON EQUATION

A relativistic wave equation may be obtained from (15.15) (in which we recall that **p** is a three-vector) together with the substitutions

(15.24)
$$E \rightarrow i\hbar \frac{\partial}{\partial t}, \qquad \mathbf{p} \rightarrow -i\hbar \boldsymbol{\nabla}$$

There results

(15.25)
$$-\hbar^2 \frac{\partial^2 \psi}{\partial t^2} = -\hbar^2 c^2 \nabla^2 \psi + m^2 c^4 \psi$$

This equation is called the *Klein–Gordon equation* and is relevant to the quantum dynamics of spinless bosons such as, for example, π mesons. As developed for the nonrelativistic case (see Section 7.5), expressions for probability and current densities relevant to (15.25) follow from the continuity equation

(15.26)
$$\frac{\partial p}{\partial t} + \boldsymbol{\nabla} \cdot \mathbf{J} = 0$$

To obtain the image of this equation in the present development, we multiply (15.25) from the left by ψ^* and the complex conjugate of this equation from the left by ψ. Subtracting the resulting two equations gives[1]

(15.27a)
$$\rho(\mathbf{r}, t) = \frac{i\hbar}{2mc^2} \left(\psi^* \frac{\partial \psi}{\partial t} - \psi \frac{\partial \psi^*}{\partial t} \right)$$

(15.27b)
$$J(\mathbf{r}, t) = \frac{\hbar}{2im} (\psi^* \boldsymbol{\nabla} \psi - \psi \boldsymbol{\nabla} \psi^*)$$

Electromagnetic Potentials

The Klein–Gordon equation for a particle of charge q in an electromagnetic field with potentials $\mathbf{A}(\mathbf{r}, t)$, $\boldsymbol{\Phi}(\mathbf{r}, t)$ is obtained as follows. With (15.2) and (15.9) it is noted that $(\mathbf{A}, i\Phi)$ and $(\mathbf{p}, iE/c)$ are both four-vectors. Thus (15.15) may be generalized to read

(15.28)
$$(E - q\Phi)^2 = (c\mathbf{p} - q\mathbf{A})^2 + m^2 c^4$$

Substituting the operator equivalents (15.24) for E and **p** into the preceding equation gives the generalized Klein–Gordon equation for a charged particle in an electromagnetic field. [In this equation and in the following, unless otherwise specified, H as well as E and **p**, as given by (15.24), are operators.]

---

[1] This form of ρ may be negative so that it is not a valid representation of position probability density.

We wish to apply (15.28) to the case of charged particle in a static field $\Phi(\mathbf{r})$. There results

$$(15.29) \quad \left(-\hbar^2\frac{\partial^2}{\partial t^2} - 2i\hbar q\Phi\frac{\partial}{\partial t} + q^2\Phi^2\right)\psi = (-\hbar^2c^2\nabla^2 + m^2c^4)\psi$$

Substituting

$$(15.30) \quad \psi(\mathbf{r}, t) = u(\mathbf{r})e^{-iEt/\hbar}$$

into the preceding equation gives the time-independent equation

$$(15.31) \quad (-\hbar^2c^2\nabla^2 + m^2c^4)u = (E - q\Phi)^2u$$

where E is now a c number.

We wish to ascertain the manner in which this Klein–Gordon equation reduces to the time-independent nonrelativistic Schrödinger equation. Toward these ends we set $E = mc^2 + E'$ and assume that $E' \ll mc^2$ and $\| q\Phi \| \ll mc^2$. Inserting these inequalities in (15.31) returns the time-independent nonrelativistic Schrödinger equation

$$(15.32) \quad \left(-\frac{\hbar^2}{2m}\nabla^2 + q\Phi\right)u = E'u$$

See Problem 15.3.

15.3 DIRAC EQUATION [1]

Suppose one attempts to construct a relativistic wave equation from the time-dependent Schrödinger equation

$$(15.33) \quad i\hbar\frac{\partial}{\partial t}\psi(\mathbf{r}, t) = H\psi(\mathbf{r}, t)$$

Again referring to (15.15) and substituting $\mathbf{p} = -i\hbar\nabla$ into (15.33) we find that the resulting expression is not symmetric in space and time derivatives and therefore is not relativistically invariant. Dirac circumvented this inconsistency by constructing a Hamiltonian linear in space derivatives. The simplest such form that maintains Hermiticity is given by

$$(15.34) \quad H = c\boldsymbol{\alpha} \cdot \mathbf{p} + \beta mc^2$$

where the Hermitian, dimensionless parameters $\boldsymbol{\alpha}$ and β are to be determined and $\mathbf{p}$ denotes the three-dimensional momentum vector. [The parameter β in (15.34)

[1] P. A. M. Dirac, *Proc. Roy. Soc.* (London) **A117**, 610 (1928).

should not be confused with β in (15.4).] Substituting the latter expression into (15.33) gives

(15.35a)
$$(E - c\boldsymbol{\alpha} \cdot \mathbf{p} - \beta mc^2)\psi(\mathbf{r}, t) = 0$$

which, with replacements (15.24), gives

(15.35b)
$$\left(i\hbar \frac{\partial}{\partial t} + i\hbar c\boldsymbol{\alpha} \cdot \boldsymbol{\nabla} - \beta mc^2 \right)\psi(\mathbf{r}, t) = 0$$

This latter equation is the Dirac equation for a free particle. As the particle is free, the parameters $\boldsymbol{\alpha}$ and β must be independent of $\mathbf{r}$, t, $\mathbf{p}$ and E (and commute with all these variables). Otherwise the Hamiltonian would contain space-time dependent energies that would give rise to forces.

The structure of the parameters $\boldsymbol{\alpha}$ and β may be obtained from the requirement that any solution to (15.35b) is likewise a solution of the free-particle Klein–Gordon equation (15.25), corresponding to the fundamental energy equation (15.16). (The converse is not true.) Multiplying (15.35a) from the left by $(E + \boldsymbol{\alpha} \cdot \mathbf{p} + \beta mc^2)$ there results

(15.36) $\{E^2 - c^2[\alpha_x^2 p_x^2 + \alpha_y^2 p_y^2 + \alpha_z^2 p_z^2 + (\alpha_x\alpha_y + \alpha_y\alpha_x)p_x p_y + (\alpha_y\alpha_z + \alpha_z\alpha_y)p_y p_z$

$$+ (\alpha_z\alpha_x + \alpha_x\alpha_z)p_z p_x] - m^2 c^4 \beta^2 - mc^3[(\alpha_x\beta + \beta\alpha_x)p_x$$

$$+ (\alpha_y\beta + \beta\alpha_y)p_y + (\alpha_z\beta + \beta\alpha_z)p_z]\}\psi(\mathbf{r}, t) = 0$$

This equation agrees with (15.25), providing

$$\alpha_x^2 = \alpha_y^2 = \alpha_z^2 = \beta^2 = I$$

(15.37)
$$[\alpha_x, \alpha_y]_+ = [\alpha_y, \alpha_z]_+ = [\alpha_z, \alpha_x]_+ = 0$$

$$[\alpha_x, \beta]_+ = [\alpha_y, \beta]_+ = [\alpha_z, \beta]_+ = 0$$

where I denotes the 4×4 identity matrix and

$$[A, B]_+ \equiv AB + BA$$

denotes the *anticommutator*. It follows that the parameter β and the components of $\boldsymbol{\alpha}$ are operators. As H given by (15.34) is Hermitian, each of four parameters $(\boldsymbol{\alpha}, \beta)$ must likewise be Hermitian. The smallest dimension for the operators $(\boldsymbol{\alpha}, \beta)$ to satisfy the anticommutator relations (15.37) and maintain Hermiticity is four.[1] As these operators appear in the Dirac equation, it follows that the Dirac wavefunction, ψ, must likewise be four-dimensional.

As noted in the first equation of (15.37) the squares of all four operators $(\boldsymbol{\alpha}, \beta)$ are unity. Each operator has eigenvalue ± 1. Choosing β as a diagonal matrix, we obtain

---

[1] For further discussion, see L. I. Schiff, *Quantum Mechanics*, 3d ed., McGraw-Hill, New York, 1968, Section 52.

$$\text{(15.38)} \qquad \beta = \begin{pmatrix} I & 0 \\ 0 & -I \end{pmatrix}$$

where I is the 2×2 identity matrix. The anticommutator properties (15.37) are obeyed by the Pauli spin matrices (Problem 11.50). The simplest forms that obey the relation (15.37) in agreement with (15.38) and Hermiticity requirements are given by

$$\text{(15.39)} \quad \alpha_x = \begin{pmatrix} 0 & \sigma_x \\ \sigma_x & 0 \end{pmatrix}, \qquad \alpha_y = \begin{pmatrix} 0 & \sigma_y \\ \sigma_y & 0 \end{pmatrix}, \qquad \alpha_z = \begin{pmatrix} 0 & \sigma_z \\ \sigma_z & 0 \end{pmatrix}$$

The forms (15.38, 15.39) may be more concisely written[1]

$$\text{(15.40)} \qquad \beta = \begin{pmatrix} I & 0 \\ 0 & -I \end{pmatrix}, \qquad \boldsymbol{\alpha} = \begin{pmatrix} 0 & \boldsymbol{\sigma} \\ \boldsymbol{\sigma} & 0 \end{pmatrix}$$

As the $\boldsymbol{\alpha}$ matrices are composed of Pauli spin operators, they obey commutation relations parallel to (11.99), namely,

$$\text{(15.41a)} \qquad [\alpha_x, \alpha_y] = 2i\alpha_z, \text{ etc.}$$

With (15.37) there results

$$\text{(15.41b)} \qquad \alpha_x \alpha_y = i\alpha_z, \text{ etc.}$$

As noted above, $\psi(\mathbf{r}, t)$ is a four-component column matrix

$$\psi(\mathbf{r}, t) = \begin{bmatrix} \psi_1(\mathbf{r}, t) \\ \vdots \\ \psi_4(\mathbf{r}, t) \end{bmatrix}$$

The Dirac equation (15.35b) is then equivalent to four coupled linear homogeneous first-order partial differential equations for the components of ψ.

Density and Current Expressions

A continuity equation may be obtained from the Dirac equation (15.35b) by multiplying it on the left by $\psi^\dagger$ and multiplying the Hermitian adjoint equation

$$-i\hbar \frac{\partial \psi^\dagger}{\partial t} - i\hbar(\boldsymbol{\nabla}\psi^\dagger) \cdot \boldsymbol{\alpha} - \psi^\dagger \beta mc^2 = 0$$

---

[1] Note that with (15.39) the relativistic free-particle Hamiltonian (15.34) is spin-dependent through the inner product term, $\boldsymbol{\alpha} \cdot \mathbf{p}$. (Recall related discussion in Section 11.7.)

on the right by ψ and subtracting the two equations. There results

(15.42)
$$\frac{\partial}{\partial t}\psi^{\dagger}\psi + \boldsymbol{\nabla} \cdot c\psi^{\dagger}\boldsymbol{\alpha}\psi = 0$$

It follows that probability and current densities for relativistic fermions are given by

(15.43)
$$\rho = \psi^{\dagger}\psi, \qquad \mathbf{J} = c\psi^{\dagger}\boldsymbol{\alpha}\psi$$

This form of ρ is nonnegative and may be consistently identified as position probability density. The form of $\mathbf{J}$ given by (15.43) may be shown to reduce to (7.107) in the nonrelativistic limit. (See Problem 15.6.)

Plane-Wave Solutions

Plane-wave solutions to the free-particle Dirac equation (15.35b) are given by

(15.44a)
$$\psi(\mathbf{r}, t) = \bar{a} \exp i(\mathbf{k} \cdot \mathbf{r} - \omega t)$$

(15.44b)
$$E = \hbar\omega$$

(15.44c)
$$\mathbf{p} = \hbar\mathbf{k}$$

where $\bar{a}$ is a four-component column vector. The components $\bar{a}_j$, $j = 1, \cdots, 4$, of $\bar{a}$ as well as the parameters E and $\mathbf{p}$ are c numbers. As in the nonrelativistic counterpart of (15.44a), $\mathbf{k}$ and E are specified in this plane wave. Substitution of (15.44) into (15.35b) gives the following matrix equation for the column vector $\bar{a}$.

(15.45a)
$$(\hbar c\mathbf{k} \cdot \boldsymbol{\alpha} - E + \beta mc^2)\bar{a} = 0$$

which in turn yields the following set of four algebraic equations for the $\bar{a}$ coefficients.

(15.45b)
$$(E - mc^2)\bar{a}_1 - cp_z\bar{a}_3 - c(p_x - ip_y)\bar{a}_4 = 0$$
$$(E - mc^2)\bar{a}_2 - c(p_x + ip_y)\bar{a}_3 + cp_z\bar{a}_4 = 0$$
$$(E + mc^2)\bar{a}_3 - cp_z\bar{a}_1 - c(p_x - ip_y)\bar{a}_2 = 0$$
$$(E + mc^2)\bar{a}_4 - c(p_x + ip_y)\bar{a}_1 + cp_z\bar{a}_2 = 0$$

These equations have a nontrivial solution only if the determinant of the coefficient matrix vanishes. There results

(15.46)
$$(E^2 - m^2c^4 - c^2p^2)^2 = 0$$

which is noted to be in agreement with the fundamental relativistic energy relation (15.15).

Explicit forms of the $\bar{a}_j$ coefficients correspond to choices of the sign of the energy. For each such choice there are two linearly independent solutions. In the following, the four sets of coefficients are written as column vectors and are labeled $\psi_{\uparrow\downarrow}^{(\pm)}$, corresponding to $(+)$ or $(-)$ energy and "spin up" $\uparrow$ and "spin down" $\downarrow$. (Motivation for this spin notation is given below.)

$$E_+ = +(c^2 p^2 + m^2 c^4)^{1/2}$$

(15.47a) $\quad \psi_\uparrow^{(+)} = \begin{bmatrix} 1 \\ 0 \\ cp_z/\Delta_+ \\ c(p_x + ip_y)/\Delta_+ \end{bmatrix}$, $\quad \psi_\downarrow^{(+)} = \begin{bmatrix} 0 \\ 1 \\ c(p_x - ip_y)/\Delta_+ \\ -cp_z/\Delta_+ \end{bmatrix}$

$$E_- = -(c^2 p^2 + m^2 c^4)^{1/2}$$

(15.47b) $\quad \psi_\uparrow^{(-)} = \begin{bmatrix} cp_z/\Delta_- \\ c(p_x + ip_y)/\Delta_- \\ 1 \\ 0 \end{bmatrix}$, $\quad \psi_\downarrow^{(-)} = \begin{bmatrix} c(p_x - ip_y)/\Delta_- \\ -cp_z/\Delta_- \\ 0 \\ 1 \end{bmatrix}$

where

(15.47c) $$\Delta_\pm \equiv (E_\pm \pm mc^2)$$

With N_p representing the normalization coefficient of these eigenstates, we write for, say, $\psi_\uparrow^{(+)}$

(15.48a) $\quad \langle \psi_\uparrow^{(+)} | \psi_\uparrow^{(+)} \rangle = N_p{}^2 \left[1, 0, \dfrac{cp_z}{\Delta_+}, \dfrac{c(p_x - ip_y)}{\Delta_+} \right] \begin{bmatrix} 1 \\ 0 \\ cp_z/\Delta_+ \\ c(p_x + ip_y)/\Delta_+ \end{bmatrix} = 1$

It follows that

(15.48b) $$N_p = \left[1 + \frac{c^2 p^2}{\Delta_+^2} \right]^{-1/2}$$

(See Problem 15.8.)

Relation to Electron Spin

To demonstrate the consistency of spin-up, spin-down notation of the plane-wave states (15.47) we note the following. Nonrelativistic energy of a free particle is positive. It follows that the $\psi^{(+)}$ functions relate to the nonrelativistic limit. This limit is described by the condition $v/c \to 0$. The two lower entries of $\psi_{\uparrow\downarrow}^{(+)}$ behave as $cp/\Delta_+ \sim v/c \to 0$. It follows that in this nonrelativistic limit one may write

(15.49)

$$\psi_{\uparrow}^{(+)}(\mathbf{r}, t) \sim \begin{pmatrix} 1 \\ 0 \end{pmatrix} \exp i(\mathbf{k} \cdot \mathbf{r} - \omega t)$$

$$\psi_{\downarrow}^{(+)}(\mathbf{r}, t) \sim \begin{pmatrix} 0 \\ 1 \end{pmatrix} \exp i(\mathbf{k} \cdot \mathbf{r} - \omega t)$$

These plane waves were encountered previously in Section 11.7 relevant to the propagating of free spin $\frac{1}{2}$ electrons with "spin up" and "spin down," respectively. We may conclude that the notation of (15.47) is consistent.

Electrons and Positrons

Eigenenergies of the plane-wave solutions (15.47) are given by

(15.50)

$$E_+ = +(c^2p^2 + m^2c^4)^{1/2} \geq mc^2$$

$$E_- = -(c^2p^2 + m^2c^4)^{1/2} \leq -mc^2$$

Thus, the energy spectrum of a free particle in the Dirac theory is continuous and ranges from $-\infty$ to $+\infty$ except for a forbidden gap of width

$$2mc^2 = 1.02 \text{ MeV}$$

about the value $E = 0$. The negative-energy states are assumed to be completely filled. Because of the exclusion principle there cannot be any transitions of negative-energy electrons. (See Fig. 15.4.) A manner of interacting with this "Fermi sea" of

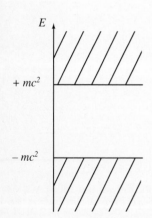

FIGURE 15.4   An energy gap of width $2mc^2 = 1$ MeV exists in the Dirac theory of the electron.

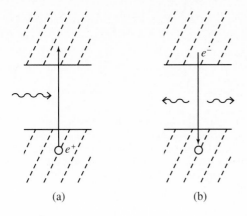

FIGURE 15.5 (a) Electron-positron creation through photon absorption. (To ensure conservation of energy and momentum, this event must occur in the presence of a third body, which carries away momentum.) (b) Electron-positron annihilation and photon emission. This is a composite transition in which the electron first emits a photon and then annihilates with a positron with emission of a second photon. (See Fig. 15.6.)

negative-energy electrons is through photon interaction. Consider a photon of energy $E_\omega > 2mc^2$ that excites an electron in the Fermi sea to positive energy. In so doing, a "hole" is left in the Fermi sea. This hole permits a second negative-energy electron to make a transition to it, leaving a hole elsewhere in the Fermi sea. As in semiconductor physics (Section 12.9), this hole acts as a positive charge and is called a positron.[1] Thus, absorption of a photon in the Fermi sea of energy in excess of $2mc^2$ leads to the creation of an electron-positron pair. Now suppose a vacancy exists in the Fermi sea. A positive-energy electron can fall to this hole and fill the Fermi sea. The decay energy of this electron-positron annihilation is carried away in two photons. (See Fig. 15.5.)

A revealing diagrammatic description of these events may be obtained from the plane-wave form (15.44). A *charge conjugation* operation on the wavefunction of an electron changes it to that of a positron. With the preceding discussion we note that such transformation is effected by the change $E \rightarrow -E$, which gives

$$\psi_{\text{pos}}(\mathbf{r}, t) = a^{(+)} \exp i(\mathbf{k} \cdot \mathbf{r} + \omega t)$$

The difference between the phase of this function and that of the original electron is the change $t \rightarrow -t$. This observation was employed by R. P. Feynman to suggest that in a space-time diagram depicting interacting electrons and positrons, it is consistent to view the positron trajectory moving backward in time. An example of this type of diagram is given in Fig. 15.6 for electron-positron annihilation which, as noted above, is a composite event.[2]

---

[1] The positron was first observed by Carl Anderson in a cloud-chamber experiment [*Phys. Rev.* **41**, 405 (1932)]. (The closely allied *positronium atom* is discussed in Problems 12.32 and 13.56.)

[2] For further discussion, see E. G. Harris, *A Pedestrian Approach to Quantum Field Theory,* Wiley-Interscience, New York, 1972.

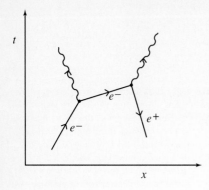

FIGURE 15.6   Feynman diagram for electron and positron annihilation. Time, t, in this graph is nondimensional. The two vertices in this diagram indicate that the interaction is a composite event. Note that the positron is depicted as propagating backward in time.

15.4   ELECTRON MAGNETIC MOMENT

Two-Component Column Wavefunctions

The Dirac equation for a particle of charge q in an electromagnetic field described by the vector and scalar potentials, $A(\mathbf{r})$ and $\Phi(\mathbf{r})$ (where $\mathbf{B} = \mathbf{\nabla} \times \mathbf{A}$ and $\mathbf{E} = -\mathbf{\nabla}\Phi$) stems from generalization of the Hamiltonian (15.34).

$$(15.51) \qquad H = c\boldsymbol{\alpha} \cdot \left(\mathbf{p} - \frac{q}{c}\mathbf{A}\right) + \beta mc^2 + q\Phi(\mathbf{r})$$

The related time-independent Dirac equation is given by

$$(15.52) \qquad \left[c\boldsymbol{\alpha} \cdot \left(\mathbf{p} - \frac{q}{c}\mathbf{A}\right) + \beta mc^2 + q\Phi\right]\psi = E\psi$$

where

$$(15.53) \qquad \psi = \begin{pmatrix} \psi^{(1)} \\ \psi^{(2)} \\ \psi^{(3)} \\ \psi^{(4)} \end{pmatrix}$$

(and, say, $\boldsymbol{\alpha} \cdot \mathbf{p}$ is a sum of 4×4 matrices). The off-diagonal, symmetric property of $\boldsymbol{\alpha}$ (15.40) leads to separation of (15.52) into the coupled equations

$$(15.54a) \qquad \left[c\left(\mathbf{p} - \frac{q}{c}\mathbf{A}\right) \cdot \boldsymbol{\sigma}\right]V + (mc^2 + q\Phi)W = EW$$

$$(15.54b) \qquad \left[c\left(\mathbf{p} - \frac{q}{c}\mathbf{A}\right) \cdot \boldsymbol{\sigma}\right]W - (mc^2 - q\Phi)V = EV$$

where

(15.55)
$$W = \begin{bmatrix} \psi^{(1)} \\ \psi^{(2)} \end{bmatrix}, \qquad V = \begin{bmatrix} \psi^{(3)} \\ \psi^{(4)} \end{bmatrix}$$

(and, say, $\mathbf{p} \cdot \boldsymbol{\sigma}$ is a sum of 2×2 matrices). Solving for V in (15.54b) in terms of W gives

(15.56)
$$V = \left[\frac{c(\mathbf{p} - (q/c)\mathbf{A}) \cdot \boldsymbol{\sigma}}{E - q\Phi + mc^2} \right] W$$

Inserting this expression into (15.54a) gives

(15.57)
$$c^2 \left[\left(\mathbf{p} - \frac{q}{c}\mathbf{A} \right) \cdot \boldsymbol{\sigma}(E - q\Phi + mc^2)^{-1} \left(\mathbf{p} - \frac{q}{c}\mathbf{A} \right) \cdot \boldsymbol{\sigma} \right] W + (mc^2 + q\Phi)W = EW$$

Nonrelativistic Limit

As noted from (15.56), in the nonrelativistic limit ($v/c \to 0$), $\|V\|$ is negligible compared with $\|W\|$. In this same limit, as noted above (15.32), $E' = E - mc^2$ and $\|q\Phi\|$ are far smaller than the rest energy mc^2. This observation permits the expansion

(15.58)
$$(E - q\Phi + mc^2)^{-1} = (E' - q\Phi + 2mc^2)^{-1} = \frac{1}{2mc^2} \left(1 - \frac{E' - q\Phi}{2mc^2} + \cdots \right)$$

Keeping the leading term in the preceding expression, (15.57) reduces to

(15.59)
$$\left\{ \frac{([\mathbf{p} - (q/c)\mathbf{A}] \cdot \boldsymbol{\sigma})^2}{2m} + q\Phi \right\} W = E'W$$

To further simplify this equation we recall the vector operation relations described in Problem 11.101, with which (15.59) reduces to

(15.60)
$$\left[\frac{1}{2m} \left(\mathbf{p} - \frac{q}{c}\mathbf{A} \right)^2 - \frac{q}{mc}\mathbf{S} \cdot \mathbf{B} + q\Phi \right] W = E'W$$

where we have set

(15.61)
$$\mathbf{S} = \frac{\hbar}{2} \boldsymbol{\sigma}$$

Equation (15.60) is the nonrelativistic time-independent Schrödinger equation of a particle of charge q and spin $\mathbf{S}$ in a magnetic field $\mathbf{B}$ and scalar field Φ. It implies a spin dipole moment of value

$$(15.62) \qquad \boldsymbol{\mu} = \frac{q\mathbf{S}}{mc}$$

Setting $q = |e|$, the preceding expression for $\boldsymbol{\mu}$ agrees with measured values [see (11.86)].

Spin Operator in Dirac Formalism

The 4×4 matrix representation of the Pauli spin operator in Dirac analysis is given by the form

$$(15.63) \qquad \boldsymbol{\sigma}' = \begin{pmatrix} \boldsymbol{\sigma} & 0 \\ 0 & \boldsymbol{\sigma} \end{pmatrix}$$

where $\boldsymbol{\sigma}$ denotes the Pauli spin matrices. To demonstrate this property we work in the Heisenberg picture and the Dirac Hamiltonian of a particle in a central field $V(r)$.

$$(15.64) \qquad H = c\boldsymbol{\alpha} \cdot \mathbf{p} + \beta mc^2 + V(r)$$

We note first that orbital angular momentum $\mathbf{L} = \mathbf{r} \times \mathbf{p}$ is not a constant of motion for the Hamiltonian (15.64). Thus, as $\mathbf{L}$ commutes with $V(r)$, we write

$$(15.65) \qquad \frac{d\mathbf{L}}{dt} = -\frac{i}{\hbar}[\mathbf{L}, H] = -\frac{ic}{\hbar}[\mathbf{L}, \boldsymbol{\alpha} \cdot \mathbf{p}]$$

With the relation (Problem 15.9)

$$(15.66) \qquad [\mathbf{L}, \boldsymbol{\alpha} \cdot \mathbf{p}] = i\hbar\boldsymbol{\alpha} \times \mathbf{p}$$

(15.65) reduces to

$$(15.67) \qquad \frac{d\mathbf{L}}{dt} = c\boldsymbol{\alpha} \times \mathbf{p} \neq 0$$

We may conclude that orbital angular momentum is not a constant of the motion for the Dirac Hamiltonian (15.64). To discover the angular momentum constant we write

$$(15.68) \qquad \mathbf{J} = \mathbf{L} + \frac{\hbar}{2}\boldsymbol{\Omega}$$

which obeys standard angular momentum commutation rules. If $d\mathbf{J}/dt = 0$, then $\mathbf{J}$ may be identified with total angular momentum and $(\hbar/2)\,\boldsymbol{\Omega} = \bar{\mathbf{S}}$, with intrinsic spin angu- lar momentum. With (15.67) we note that one must find

(15.69)
$$\frac{d\boldsymbol{\Omega}}{dt} = -\frac{2c}{\hbar}\,\boldsymbol{\alpha} \times \mathbf{p}$$

Consider the matrix $\boldsymbol{\alpha} \times \boldsymbol{\alpha}$ whose time derivative is given by

(15.70) $$\frac{d}{dt}(\boldsymbol{\alpha} \times \boldsymbol{\alpha}) = -\frac{i}{\hbar}[(\boldsymbol{\alpha} \times \boldsymbol{\alpha}), H] = -\frac{ic}{\hbar}[(\boldsymbol{\alpha} \times \boldsymbol{\alpha}), \boldsymbol{\alpha}\cdot\mathbf{p}] = -\frac{4ic}{\hbar}\,\boldsymbol{\alpha} \times \mathbf{p}$$

(see Problem 15.10). It follows that

(15.71a)
$$\boldsymbol{\Omega} = -\frac{i}{2}\,\boldsymbol{\alpha} \times \boldsymbol{\alpha}$$

and

(15.71b)
$$\bar{\mathbf{S}} = \frac{\hbar}{2}\,\boldsymbol{\Omega}$$

Note that

$$\frac{d\mathbf{J}}{dt} = \frac{d\mathbf{L}}{dt} + \frac{\hbar}{2}\frac{d\boldsymbol{\Omega}}{dt} = c\boldsymbol{\alpha} \times \mathbf{p} - c\boldsymbol{\alpha} \times \mathbf{p} = 0$$

We may conclude that the consistent expression for the spin operator in the Dirac formalism is given by (15.71). Explicit components of the 4×4 matrix $\bar{\mathbf{S}}$ are obtained from the rules (15.37) and (15.41). Thus, for example,

$$\bar{S}_z = -\frac{i\hbar}{4}(\alpha_x\alpha_y - \alpha_y\alpha_x) = -\frac{i\hbar}{2}\alpha_x\alpha_y = \frac{\hbar}{2}\alpha_z$$

Similar results pertain to $\bar{S}_x$, $\bar{S}_y$. These findings are summarized by the form

(15.72)
$$\bar{\mathbf{S}} = \frac{\hbar}{2}\begin{pmatrix} \boldsymbol{\sigma} & 0 \\ 0 & \boldsymbol{\sigma} \end{pmatrix} = \frac{\hbar}{2}\boldsymbol{\sigma}'$$

which is the appropriate four-dimensional spin operator in the Dirac formalism.

Note in particular that the Dirac equation for a free particle (15.35b) does not include the diagonal spin operator $\bar{\mathbf{S}}$, but the off-diagonal symmetric $\boldsymbol{\alpha}$ operator (15.40). As noted above, the off-diagonal, symmetric nature of the operator $\boldsymbol{\alpha}$ generates separation of the Dirac equation into coupled equations for a pair of two-component column-vector wavefunctions (15.54) in which the standard 2×2 Pauli spin matrices emerge.

15.5 COVARIANT DESCRIPTION

In this concluding section we present a brief introduction to a covariant description of relativistic quantum mechanics.

The Dirac Equation

To obtain a covariant Dirac equation we introduce the γ matrices

(15.73)
$$\gamma_k = -i\beta\alpha_k, \qquad k = 1, 2, 3$$
$$\gamma_4 = \beta, \qquad \mu \equiv mc/\hbar$$

[The parameter γ denoting these matrices should not be confused with γ defined in (15.4).] The Dirac equation (15.35b) may then be written in the covariant form (with the Einstein summation convention assumed)

(15.74)
$$\left(\gamma_\nu \frac{\partial}{\partial x_\nu} + \mu\right)\Psi = 0$$

[To demonstrate the equivalence of this equation and the Dirac equation (15.35b), multiply (15.74) from the left with β and recall $\beta^2 = I$.] To examine the Lorentz invariance of (15.74), we introduce the transformation

(15.75)
$$\Psi' = \Lambda\Psi$$

where Λ is a unitary 4×4 matrix (Problem 15.19) that transforms the four-column-vector wavefunction Ψ in the S frame to the wavefunction Ψ' in the S' frame. The γ matrices remain unchanged in this transformation. We note further that

(15.76)
$$\frac{\partial}{\partial x'_\nu} = \frac{\partial}{\partial x_\kappa}\frac{\partial x_\kappa}{\partial x'_\nu} = L_{\nu\kappa}\frac{\partial}{\partial x_\kappa}$$

where $L_{\nu\kappa}$ is the Lorentz transformation (15.10). Our demonstration of the Lorentz invariance of the Dirac equation starts by writing (15.74) in the S' frame.

(15.77)
$$\left(\gamma_\nu \frac{\partial}{\partial x'_\nu} + \mu\right)\Psi' = \left(\gamma_\nu L_{\nu\kappa}\frac{\partial}{\partial x_\kappa} + \mu\right)\Lambda\Psi = 0$$

Multiplying this equation from the left with Λ^{-1} gives

(15.78)
$$\left(\Lambda^{-1}\gamma_\nu \Lambda L_{\nu\kappa}\frac{\partial}{\partial x_\kappa} + \mu\right)\Psi = 0$$

Setting

(15.79)
$$\Lambda^{-1}\gamma_\nu \Lambda L_{\nu\kappa} = \gamma_\kappa$$

in (15.78) returns the Dirac equation (15.74) in the S frame, thereby establishing the Lorentz invariance of the equation. Note that (15.79) indicates the manner in which the γ matrices transform under a Lorentz transformation.

The γ matrices (15.73) play an important role in relativistic quantum mechanics. We note the following anticommutation relations

$$(15.80) \qquad\qquad [\gamma_\lambda,\ \gamma_\nu]_+ = 2\delta_{\lambda\nu}I, \qquad \gamma_\nu^{\ 2} = I$$

where I denotes the 4×4 identity matrix.

Covariant Conservation Equations

The covariant structure of the Klein–Gordon equation (15.25) is given by

$$(15.81a) \qquad\qquad (p_\nu^{\ 2} + m^2c^2)\Psi = 0$$

where p_ν represents the quantum four-momentum vector

$$(15.81b) \qquad\qquad p_\nu = i\hbar\frac{\partial}{\partial x_\nu}$$

The corresponding form of (15.81a) for a particle of charge q in an electromagnetic field is given by

$$(15.82) \qquad\qquad \left[\left(p_\nu - \frac{q}{c}A_\nu\right)^2 + m^2c^2\right]\Psi = 0$$

The covariant form of the continuity equation that follows from the Klein–Gordon equation (15.82) is given by

$$(15.83a) \qquad\qquad \frac{\partial J_\nu}{\partial x_\nu} = 0$$

$$(15.83b) \qquad\qquad J_\nu = \frac{\hbar}{2mi}\left(\Psi*\frac{\partial\Psi}{\partial x_\nu} - \Psi\frac{\partial\Psi*}{\partial x_\nu}\right) - \frac{e}{mc}A_\nu\Psi*\Psi$$

This four-current vector is in accord with the basic form (15.9a). In the limit that $A_\nu = 0$, (15.83b) returns the two relations (15.27).

Let us derive the covariant form of the continuity equation which follows from the Dirac equation (15.74). First consider the Hermitian adjoint of (15.74) given by

$$(15.84a) \qquad\qquad \frac{\partial}{\partial x_k}\Psi^\dagger\gamma_k - \frac{\partial}{\partial x_4}\Psi^\dagger\gamma_4 + \mu\Psi^\dagger = 0$$

Here we have recalled that x_4 is imaginary. Multiplying the latter equation from the right by γ_4 and employing (15.80) gives

(15.84b)
$$\frac{\partial}{\partial x_k}\overline{\Psi}\gamma_k + \frac{\partial}{\partial x_4}\overline{\Psi}\gamma_4 - \mu\overline{\Psi} = \frac{\partial}{\partial x_\alpha}\overline{\Psi}\gamma_\alpha - \mu\overline{\Psi} = 0$$

where

$$\overline{\Psi} \equiv \Psi^\dagger \gamma_4$$

is labeled, the "Dirac conjugate" of Ψ. Multiplying the right equation in (15.84b) by Ψ and adding the resulting equation to that obtained by multiplying the Dirac equation (15.74) by $\overline{\Psi}$ gives an equation which when multiplied by the constant ic gives the normal form of the covariant continuity equation appropriate to fermions.

(15.85)
$$\frac{\partial}{\partial x_\nu}(ic\overline{\Psi}\gamma_\nu\Psi) = 0$$

This equation implies the four-current

(15.86a)
$$J_\nu = ic\overline{\Psi}\gamma_\nu\Psi$$

with related position probability density

(15.86b)
$$\rho = \overline{\Psi}\gamma_4\Psi = \Psi^\dagger\Psi$$

and current density

(15.86c)
$$J_k = ic\overline{\Psi}\gamma_k\Psi = c\overline{\Psi}\alpha_k\Psi$$

or, equivalently,

(15.86d)
$$\mathbf{J} = c\overline{\Psi}\boldsymbol{\alpha}\Psi$$

Equations (15.84b through d) indicate that the covariant continuity equation (15.85) is equivalent to the noncovariant continuity equation (15.42).

Additional discussion on covariant relativistic quantum mechanics may be found in the cited references[1] and in Problems 15.19 et seq.

[1] J. D. Bjorken and S. D. Drell, *Relativistic Quantum Mechanics,* McGraw-Hill, New York, 1964, Section 2.4; V. B. Berestetskii, E. M. Lifshitz, and L. P. Pitaevskii, *Relativistic Quantum Theory,* Pergamon, New York, 1979, Chapter III; S. Flügge, *Practical Quantum Mechanics,* Springer, New York, 1974, Problems 192, 193, 199; A. S. Davydov, *Quantum Mechanics,* 2d ed., Pergamon, New York, 1965, Section VIII, 61; F. Constantinescu and E. Magyari, *Problems in Quantum Mechanics,* Pergamon, New York, 1971, Chapter XII.

PROBLEMS

15.1   (a)   Evaluate the squared lengths of each of the three four-vectors (15.9).

(b)   What form do each of these squared lengths assume under a Lorentz transformation to a primed frame?

(c)   What is the name given to the squared length of a four-vector?

15.2   Show that the Klein–Gordon equation (15.25) has a plane-wave solution,

$$\psi(\mathbf{r}, t) = \exp i(\mathbf{k} \cdot \mathbf{r} - \omega t)$$

where

$$(\hbar\omega)^2 = \hbar^2 c^2 k^2 + m^2 c^4$$

15.3   Obtain the first-order relativistic correction to the time-independent Schrödinger equation (15.32) from the Klein–Gordon equation (15.31).

15.4   What is the dimension of the relativistic wavefunction, $\psi(\mathbf{r}, t)$, for a spin $\frac{3}{2}$ particle?

15.5   What are the probability and current densities $(\rho, \mathbf{J})$ for the plane-wave state, $\psi_1^{(-)}$?

15.6   Show that the relativistic expression of current density $\mathbf{J}$, (15.43), reduces to (7.107) in the nonrelativistic limit, $v/c \to 0$.

15.7   Show that the four free-particle eigenstates (15.47) of the Dirac equation are orthogonal.

15.8   Evaluate the normalization constant, $N_p^{(-)}$, corresponding to the normalization, $\langle \psi_1^{(-)} | \psi_1^{(-)} \rangle = 1$.

15.9   Demonstrate the validity of the commutator relation (15.66).

15.10   Demonstrate the validity of the third equality in (15.70).

15.11   Show that the two equations (15.54) for W and V are equivalent to the single Dirac equation, (15.52).

15.12   Evaluate components of the 4×4 matrix, $\bar{\mathbf{S}}_x$, explicitly from (15.71), where $\bar{\mathbf{S}}$ is the spin operator in the Dirac formalism. Show that your result agrees with the general form (15.72).

15.13   (a) Write down the wavefunction for a spin $\frac{1}{2}$ free particle propagating in the z direction with "spin up" and positive energy.

(b)   Employing results of Problem 15.12 and part (a) of this problem, evaluate the expectation, $\langle \psi_1^{(+)} | \bar{\mathbf{S}}_x \psi_1^{(+)} \rangle$, for the particle in this state.

15.14   In Section 15.1 we encountered "lab time" and "proper time." Which of these is the time appearing in the Klein–Gordon equation (15.25) and the Dirac equation (15.35b), respectively?

15.15   Consider the Feynman diagram of Fig. 15.6. Suppose the time axis is labeled ct, where c is the speed of light. Which components of the diagram should be changed and in what manner should these changes be made? Explain your answer.

15.16 Introduce the two- and four-dimensional unit matrices, I_2 and I_4, respectively.

(a) Employing I_4, rewrite the Dirac equations (15.45a) and (15.52) and the Hamiltonian (15.51) in a more self-contained manner.

(b) Employing I_2, rewrite (15.54a, b) in a more self-contained manner.

15.17 (a) Rewrite the set of equations (15.45b) as a 4×4 matrix times the $\bar{a}$ column vector.

(b) Evaluate the determinant of this matrix and verify (15.46).

15.18 Show that $(d\tau)^2$ is Lorentz invariant.

Answer

We may set

$$(d\tau)^2 = -\frac{1}{c^2} \, dx_\nu \, dx_\nu$$

which is evidently a Lorentz invariant. Let us evaluate the right side of this equation in a frame S' in which the particle is at rest. In this frame components dx'_ν are $(0, 0, 0, icdt')$, in which case $(d\tau)^2 = (dt')^2$, and again we find that τ is the time measured on a clock moving with the particle.

15.19 Argue that the Λ operator introduced in (15.75) is unitary and describe the geometrical significance of this property.

Answer

The Lorentz transformation rotates the four-vector, x, in 4-space. It follows that in the transformation $\Lambda\psi = \psi'(x')$, x' is rotated from x. Recalling the discussion in Problem 9.17, we may conclude that for an increment of rotation in 4-space, $\delta\boldsymbol{\phi}$, one may write $\Lambda = \exp i \, \delta\boldsymbol{\phi} \cdot \mathbf{G}$, where $\mathbf{G}$ is the generator of the rotation. It follows that Λ is unitary.

15.20 Verify explicitly that $\alpha_x \alpha_y = -i\alpha_z$.

15.21 Establish the following properties of the γ matrices:

(a) $\gamma_\nu^2 = I$

(b) $\gamma_\nu \gamma_\nu = 4I$

(c) $\gamma_\lambda \gamma_\nu \gamma_\lambda = -2\gamma_\nu$

(d) $\gamma_\lambda \gamma_\nu \gamma_\kappa \gamma_\lambda = 4\delta_{\kappa\nu}$

(e) $\gamma_\kappa \gamma_\nu \gamma_\lambda \gamma_\mu \gamma_\kappa = -2\gamma_\mu \gamma_\lambda \gamma_\nu$

(f) $\mathrm{Tr}(\gamma_\kappa \gamma_\nu) = 4\delta_{\kappa\nu}$

15.22 It is possible to construct 16 linearly independent 4×4 products of the γ_α matrices that span the space of 4×4 matrices. These are given by the following forms:

$$\Gamma^{(1)} \equiv I, \qquad \Gamma^{(2)}_\nu \equiv \gamma_\nu, \qquad \Gamma^{(3)}_{\nu\beta} \equiv i\gamma_\nu \gamma_\beta$$

$$\Gamma^{(4)} \equiv \gamma_1 \gamma_2 \gamma_3 \gamma_4 \equiv \gamma_5, \qquad \Gamma^{(5)}_\nu \equiv \gamma_5 \gamma_\nu$$

These matrices play an important role in construction of Lorentz covariants. In the following problems, $\Gamma^{(n)}$ denotes any of the preceding Γ matrices.

(a) Argue that the space of 4×4 matrices is spanned by 16 independent 4×4 matrices.

(b) What is the matrix representation of γ_5?

(c) Show that $(\Gamma^{(n)})^2 = \pm I$

(d) Show that except for $\Gamma^{(1)}$, for each $\Gamma^{(n)}$ there exists a $\Gamma^{(m)}$ such that

$$[\Gamma^{(n)}, \Gamma^{(m)}]_+ = 0$$

(e) With part (d) show that (again except for $\Gamma^{(1)}$), $\operatorname{Tr} \Gamma^{(n)} = 0$.

(f) Show that the product of any two elements of the set of Γ matrices is proportional to a third element of the set:

$$\Gamma^{(n)}\Gamma^{(m)} = a\Gamma^{(q)}$$

where $a = \pm 1$ or $\pm i$, and $a = +1$ if and only if $n = m$.

(g) If $\Gamma^{(n)} \neq \Gamma^{(1)}$, there exists $\Gamma^{(\kappa)}$ such that $\Gamma^{(\kappa)}\Gamma^{(n)}\Gamma^{(\kappa)} = -\Gamma^{(n)}$.

(h) Show that $[\gamma_\alpha, \gamma_5]_+ = 0$.

(i) Show that if $\Gamma^{(n)} \neq \Gamma^{(1)}$, $\operatorname{Tr} \Gamma^{(n)} = 0$.

(j) Suppose a set of numbers a_n exists such that

$$\sum_n a_n \Gamma^{(n)} = 0$$

Show that there is at least one value of a_n that vanishes, thereby establishing the linear independence of the $\Gamma^{(n)}$ matrices.

Answers (partial)

(a) The indices of a 4×4 matrix each run from 1 to 4. It follows that 16 independent 4×4 matrices span the space of such matrices. As part (g) establishes that the 16 $\Gamma^{(n)}$ matrices are independent, it follows that they are a basis for this space.

(b) $\gamma_5 = \begin{pmatrix} 0 & I \\ -I & 0 \end{pmatrix}$

(c) This property follows from (15.80).

(g) Multiply both sides by $\Gamma^{(\kappa)}$.

(i) $\operatorname{Tr} \Gamma^{(n)} = \operatorname{Tr} \Gamma^{(n)}(\Gamma^{(m)})^2 = -\operatorname{Tr} \Gamma^{(m)}\Gamma^{(n)}\Gamma^{(m)}$
$$= +\operatorname{Tr} \Gamma^{(m)}\Gamma^{(n)}\Gamma^{(m)} = 0$$

The second equality follows from part (d). The third equality follows from $\operatorname{Tr} AB = \operatorname{Tr} BA$.

(j) Multiply the given equation by $\Gamma^{(n)} \neq \Gamma^{(1)}$, take the trace, and use part (i). This gives $a_n = 0$. Now multiply the given equation by $\Gamma^{(1)}$ to find $a_1 = 0$.

LIST OF SYMBOLS

a_0	Bohr radius
a	Scattering length, edge length
$\hat{a},\ \hat{a}^\dagger$	Annihilation and creation operators
$\mathbf{A}$	Vector potential
$Ai(x)$	Airy function
$A_{lk},\ B_{lk}$	Einstein A and B coefficients
$b(\mathbf{k})$	Momentum probability amplitude
$\mathscr{B}$	Magnetic field (also $\mathscr{B}$ in figures)
$\mathfrak{B}$	Basis
c	Speed of light
$c_k(t)$	Transition probability amplitude
$C_{m_1 m_2}$	Clebsch–Gordon coefficient
$\mathfrak{D}$	Coefficient matrix
$\hat{\mathscr{D}}$	Displacement operator
$\hat{D}$	Derivative operator $(\partial/\partial x)$
$\mathbf{d}$	Electric dipole moment
DC	Denotes constant in time
$\mathscr{E}$	Electric field (also $\mathscr{E}$ in figures)
$\mathbf{e}$	Unit vector
e	Charge
$\mathbb{E}$	Photoelectric energy
$\mathbb{E}$	Perturbation energy
E	Energy
E_F	Fermi energy
$\mathscr{E}$	Electric field
$f_{nn'}$	Oscillator strengths
$f(\theta)$	Scattering amplitude
g	Acceleration due to gravity; Landé g factor
$g(E),\ g(\nu)$	Density of states
$\overline{g}$	Degeneracy, density of states
H	Hamiltonian
H'	Perturbation Hamiltonian

$\mathfrak{H}$	Hilbert space
$\mathcal{H}_n(\xi)$	Hermite polynomial
$\mathbb{H}$	r-dependent perturbation Hamiltonian
$h, \hbar$	Planck's constant
$\hat{I}$	Identity operator
j	Total angular momentum quantum number
$\mathbf{J}$	Spin, orbital, or total angular momentum
$\mathbb{J}$	Current density
k_B	Boltzmann's constant
k, κ	Wavenumbers
K^2	Eigenvector of $\hat{J}^2/\hbar^2$
l	Orbital angular momentum quantum number
$\mathbf{L}$	Orbital angular momentum
$\mathcal{L}$	Orbital-angular-momentum-term notation symbol
L	Edge length
$L_p^q(x)$	Laguerre polynomial
m	Mass
m^*	Effective mass
M	Mass of center of mass
$\mathcal{N}_E, \mathcal{N}_k$	Number of states in an energy band
N	Total number of particles
$\mathbf{p}$	Momentum
$\mathcal{p}_r$	Unsymmetrized radial momentum
p_r	Radial momentum
P_{12}	Two-particle joint probability density
P	$P\,d\mathbf{r}$ is probability
P_r	$P_r\,dr$ is probability
$\bar{P}$	Nondimensional P_r
P	Radiated power
$\mathcal{P}$	Momentum of center of mass
P	Permutation operator
$|P|$	Order of permutation
$\mathbb{P}$	Parity operator
P_{lk}	Transition probability
$\bar{P}_{lk}$	Probability relevant to transition to an energy band
$P_l(\cos\theta)$	Legendre polynomial
$P_l^m(\cos\theta)$	Associated Legendre polynomial
P_n^{QM}, P_n^{CL}	Quantum and classical probability densities
q	Charge
q	Coordinate

r	Integral in WKB analysis
$\mathbf{r}$	Radius vector
$\mathscr{R}$	Radius to center of mass
R	Reflection coefficient
R	Radial wavefunction
$\overline{R}$	Nondimensionalized radial wavefunction
$\mathbb{R}$	Rydberg constant
$\hat{R}_{\delta\phi}$	Rotation operator
$\mathbf{S}$	Spin angular momentum
$\hat{\mathbf{S}}$	Transfer matrix
s	Spin angular momentum quantum number
T	Transmission coefficient
T	Temperature
T_C	Critical temperature
T	Kinetic energy
u	Radial wavefunction ($u = rR$)
$u(\nu)$	Energy density per frequency interval
U	Energy density
$\mathscr{U}$	Column vector
v_F	Fermi velocity
V	Potential energy
$\overline{V}$	Average potential
$\mathscr{V}$	Column vector
w_{lk}	Transition probability rate
$\overline{w}_{lk}$	Probability rate for transition to an energy band
W_{lk}	Total atomic transition rate
$\hat{\mathfrak{X}}$	Exchange operator
$Y_l^m(\theta, \phi)$	Spherical harmonic
α	Fine-structure constant
α	Polarizability
α	4×4 matrix
β	Harmonic oscillator wavenumber
β	Speed nondimensionalized with respect to the speed of light
β	4×4 matrix
α, β	Spin eigenstates
ε	Energy
ϵ	Mass ratio, dielectric constant
η	Integral in WKB analysis
Γ	State of a system; 4×4 matrix
γ	4×4 matrix
$\Theta_l^m(\theta, \phi)$	Eigenfunction of $\hat{L}^2$
K	Spring constant

λ	Wavelength
λ	Parameter of smallness
λ_n and λ_p	Characteristic lengths in *p-n* junction analysis
Λ	Angular momentum parameter; scattering centers per unit area
μ	Reduced mass
μ	Chemical potential
$\boldsymbol{\mu}$	Magnetic moment
ν	Frequency
ξ	Nondimensional displacement
ξ	Spin state
ξ, η	Nondimensional wavenumbers
ρ	Nondimensional radius in hydrogen wavefunction
$\hat{\rho}$	Density matrix
$\rho(x)$	Particle density
σ	Stefan–Boltzmann constant
σ	Total scattering cross section
$\hat{\boldsymbol{\sigma}}$	Pauli spin operator
$d\sigma$	Differential scattering cross section
φ	Time-independent wavefunction
Φ	Work function
Φ	Electric potential
$\Phi_m(\phi)$	Eigenfunction of $\hat{L}_z$
χ_S, χ_A	Symmetric and antisymmetric wavefunctions
ψ	Wavefunction
ω	Angular frequency
Ω	Solid angle, spin
$\Omega, \Omega/2$	Cyclotron and Larmor frequencies

Units

A	Ampere
cm, μm, m	Centimeter, micron, meter
C	Coulomb
s, ms, μs, ns	Second, millisecond, microsecond, nanosecond
V, eV, meV, keV, MeV	Volt, electron volt, milli electron volt, kilo electron volt, mega electron volt
W, kW, MW	Watt, kilowatt, megawatt

APPENDIXES

APPENDIX A

ADDITIONAL REMARKS ON THE $\hat{x}$ AND $\hat{p}$ REPRESENTATIONS

Let $|x'\rangle$ represent an eigenstate of $\hat{x}$. Then

(A.1)
$$\hat{x}|x'\rangle = x'|x'\rangle$$

These eigenstates obey the orthonormality condition

(A.2)
$$\langle x'|x\rangle = \delta(x - x')$$

The matrix elements of $\hat{x}$ are then given by

(A.3)
$$\langle x|\hat{x}|x'\rangle = x'\langle x|x'\rangle = x'\delta(x - x')$$

This is a continuous matrix with nonzero entries only on the diagonal $x = x'$.

As remarked in the text, summations over continuous matrices are replaced by integrations. For example, the multiplication of the matrix $\hat{x}$ by the column state vector $|\psi(x)\rangle$ gives

$$\int dx'\,|\psi(x')\rangle\langle x'|\hat{x}|x\rangle = x \int dx'\,|\psi(x')\rangle\,\delta(x' - x) = x|\psi(x)\rangle$$

In the x representation, $\hat{x}$ operating on a state has the effect of multiplying the state by the scalar x.

The projection of $|\psi\rangle$ onto the basis vector $|x'\rangle$ is the coordinate representation of $|\psi\rangle$.

(A.4)
$$\langle x'|\psi\rangle = \int \langle x'|x\rangle\langle x|\psi\rangle \, dx = \psi(x')$$

Here we have employed the spectral resolution of unity,

(A.5)
$$\hat{I} = \int |x\rangle \, dx\langle x|$$

Note in particular that the coordinate representation of $|x\rangle$ is the delta function, $\delta(x - x')$, as given by (A.2). This identification permits one to write the eigenvalue equation for $\hat{x}$ in the form (3.26).

If $|p\rangle$ represents an eigenstate $\hat{p}$, then

(A.6)
$$\hat{p}|p'\rangle = p'|p'\rangle$$

The matrix of $\hat{p}$ in the coordinate representation is given by

(A.7)
$$\langle x|\hat{p}|x'\rangle = -i\hbar \frac{\partial}{\partial x}\delta(x - x')$$

This relation allows us to obtain an explicit form for the transfer matrix $\langle x|p\rangle$.

(A.8)
$$p\langle x|p\rangle = \langle x|\hat{p}|p\rangle = \int dx'\langle x|\hat{p}|x'\rangle\langle x'|p\rangle$$

$$= -i\hbar \int dx' \frac{\partial}{\partial x}\delta(x - x')\langle x'|p\rangle$$

$$= -i\hbar \frac{\partial}{\partial x}\langle x|p\rangle$$

The solution to this differential equation is

(A.9)
$$\langle x|p\rangle = \frac{1}{\sqrt{2\pi\hbar}}e^{ipx/\hbar}$$

The normalization ensures the unitarity of the continuous matrix $\langle x|p\rangle$. To see this, we first recall the condition for unitarity,

(A.10)
$$\int_{-\infty}^{\infty} \langle p|x\rangle^*\langle p|x'\rangle\, dp = \delta(x - x')$$

With the representation (A.9) for $\langle x|p\rangle$ and using the property $\langle p|x\rangle^* = \langle x|p\rangle$, we find that

(A.11) $$\text{LHS(A.10)} = \int_{-\infty}^{\infty} \langle x|p\rangle\langle p|x'\rangle\, dp = \frac{1}{2\pi\hbar}\int_{-\infty}^{\infty} e^{ipx/\hbar}e^{-ipx'/\hbar}\, dp$$

Setting $p/\hbar \equiv y$ reduces the right-hand side of the latter equation to

(A.12)
$$\frac{1}{2\pi}\int_{-\infty}^{\infty} e^{iy(x-x')}\, dy = \delta(x - x')$$

which establishes the unitarity of $\langle x|p\rangle$. Note that the projection (A.9) gives either the coordinate representation of the eigenstates of p or the momentum representation of the eigenstates of $\hat{x}$.

Let us see how the form (A.9) allows one to reconstruct the matrix for $\hat{p}$ as given by (A.7). In the p representation, we have

(A.13)
$$\langle p|\hat{p}|p'\rangle = p'\,\delta(p - p')$$

Using (A.9) together with the last equation gives

$$(A.14) \quad \langle x|\hat{p}|x'\rangle = \frac{1}{2\pi\hbar} \int_{-\infty}^{\infty} \int_{-\infty}^{\infty} dp\, dp'\, p'\, \delta(p - p')e^{ipx/\hbar}e^{-ip'x'/\hbar}$$

$$= \frac{1}{2\pi\hbar} \int_{-\infty}^{\infty} dp\, pe^{ip(x-x')/\hbar}$$

$$= -i\hbar \frac{\partial}{\partial x} \frac{1}{2\pi\hbar} \int_{-\infty}^{\infty} e^{ip(x-x')/\hbar}\, dp$$

$$= -i\hbar \frac{\partial}{\partial x} \delta(x - x')$$

which agrees with (A.7). We may use this relation to calculate the coordinate representation of $\hat{p}|\psi\rangle$.

$$(A.15) \quad \langle x|\hat{p}|\psi\rangle = \int_{-\infty}^{\infty} dx'\langle x|\hat{p}|x'\rangle\langle x'|\psi\rangle = -i\hbar \int_{-\infty}^{\infty} dx' \frac{\partial}{\partial x}\delta(x - x')\psi(x')$$

$$= -i\hbar \frac{\partial}{\partial x}\psi(x)$$

This has the same effect as simply operating on the state ψ with the differential operator $-i\hbar\partial/\partial x$.

As a simple example of these concepts,[1] consider the problem of finding the matrix of $(\hat{x}\hat{p} - \hat{p}\hat{x})$ in the x representation. Let us first examine the term

$$\langle x|\hat{x}\hat{p}|x'\rangle = \int\int\int_{-\infty}^{\infty} dx''\, dp'\, dp\, \langle x|\hat{x}|x''\rangle\langle x''|p'\rangle\langle p'|\hat{p}|p\rangle\langle p|x'\rangle$$

$$= \frac{1}{2\pi\hbar} \int\int_{-\infty}^{\infty} dx''\, dp\, x''\, \delta(x - x'')pe^{ip(x''-x')/\hbar}$$

$$= \frac{x}{2\pi\hbar} \int_{-\infty}^{\infty} dp\, pe^{ip(x-x')/\hbar}$$

$$= \frac{x}{2\pi\hbar}\left(-i\hbar\frac{\partial}{\partial x}\right)\int_{-\infty}^{\infty} dp\, e^{ip(x-x')/\hbar}$$

$$= -i\hbar x \frac{\partial}{\partial x}\delta(x - x')$$

---

[1] For further development of these topics, see W. Louisell, *Radiation and Noise in Quantum Electronics*, McGraw-Hill, New York, 1964.

In like manner we find that

$$-\langle x| \, \hat{p}\hat{x} \, |x'\rangle = i\hbar x' \frac{\partial}{\partial x} \delta(x - x')$$

Combining these results gives

(A.16) $\langle x|\hat{x}\hat{p} - \hat{p}\hat{x}|x'\rangle = -i\hbar(x - x') \dfrac{\partial}{\partial x} \delta(x - x') = +i\hbar \, \delta(x - x')$

In concluding this discussion we note the following. Suppose that a complete set of commuting observables is diagonalized by the ket vectors $|\xi\rangle$. Then the coordinate and momentum representations of these states are $\langle x|\xi\rangle$ and $\langle p|\xi\rangle$, respectively. For example, consider the eigenvectors $|n\rangle$ that simultaneously diagonalize the number operator $\hat{N} = \hat{a}^\dagger\hat{a}$ and the Hamiltonian $\hat{H} = \hbar\omega_0(\hat{N} + \frac{1}{2})$, appropriate to the harmonic oscillator (Section 7.2).

(A.17)
$$\hat{H}|n\rangle = \hbar\omega_0\left(n + \frac{1}{2}\right)|n\rangle$$

$$\hat{N}|n\rangle = n|n\rangle$$

No information is revealed by these equations other than the fact that $|n\rangle$ is an eigenvector of $\hat{H}$ and $\hat{N}$ with respective eigenvalues as shown. If, for example, one wishes the coordinate representation of these states, one must form the projections $\langle x|n\rangle$. These are the weighted Hermite polynomials (7.58).

In a similar vein the coordinate representations of the eigenvectors $|lm\rangle$ of the operators $\hat{L}^2$ and $\hat{L}_z$ are the projections $\langle\theta\phi|lm\rangle$ [i.e., the spherical harmonics, $Y_l^m(\theta, \phi)$].

As a further case of this formalism, consider the following example. We wish to show that the coordinate representation of the Schrödinger equation in abstract ket space

$$i\hbar \frac{\partial |\psi\rangle}{\partial t} = \hat{H}|\psi\rangle$$

is the standard relation (3.45).

Without loss in generality we work in one dimension. Multiplying the given relation from the left by $\langle x|$, we obtain

$$i\hbar \frac{\partial}{\partial t}\langle x|\psi\rangle = \langle x|\hat{H}|\psi\rangle = \int dx' \langle x|\hat{H}|x'\rangle\langle x'|\psi\rangle = \int dx' \, \delta(x - x')\hat{H}(x')\langle x'|\psi\rangle$$

where $\hat{H}(x)$ denotes $\hat{H}$ in the coordinate representation. The further identification

$$\langle x|\psi\rangle = \psi(x)$$

gives the desired relation.

APPENDIX B

SPIN AND STATISTICS

In this appendix we wish to offer a brief elementary outline of the argument connecting spin and the exclusion principle. As described in Chapter 12, particles with integral spin do not obey the exclusion principle, whereas those with half-integral spin do obey the exclusion principle.

The particle quality of a field may be described in second quantization, wherein, in accord with Problems 13.37 and 13.38, the state of the system is written $|n_1, n_2, \ldots \rangle$. In this notation n_i represents the number of particles in the ith state.

There are two prescriptions for the quantization of a field. The first is given by the Jordan–Wigner anticommutation rules (Problem 7.32),

(B.1)
$$\{\hat{a}_n, \hat{a}_m\} = \{\hat{a}_n^\dagger, \hat{a}_m^\dagger\} = 0$$
$$\{\hat{a}_n, \hat{a}_m^\dagger\} = \delta_{nm}$$

The second is given by the Bose commutation rules,

(B.2)
$$[\hat{a}_n, \hat{a}_m] = [\hat{a}_n^\dagger, \hat{a}_m^\dagger] = 0$$
$$[\hat{a}_n, \hat{a}_m^\dagger] = \delta_{nm}$$

As established in Problem 7.32, particles such as electrons, which obey the Jordan–Wigner anticommutation rules (B.1), exist in accordance with the exclusion principle. Number eigenvalues n_i are either 0 or 1 ($n_i^2 = n_i$). Particles, such as photons, which obey the Bose commutation rules (B.2) do not adhere to the exclusion principle.

In his argument relating spin and exclusion, Pauli[1] imposed the following requirements on physical systems.

1. Let A be an observable pertaining to the space–time point $\mathbf{r}_1, t_1$, and let B be an observable at $\mathbf{r}_2, t_2$. Consequently, if $|\mathbf{r}_1 - \mathbf{r}_2|/|t_1 - t_2| > c$, then A and B commute. The rationale behind this stipulation is as follows. In that these space–time points are separated by speeds greater than that of light, relativity (or *causality*) specifies that measurement of A can in no way interfere with measurement of B. Equivalently, we may say that A and B commute.

[1] W. Pauli, *Phys. Rev.* **58**, 716 (1940). See also R. Streater and A. S. Wightman, *PCT, Spin and Statistics, and All That*, W. A. Benjamin, New York, 1964.

2. The total (relativistic) energy of the system is greater than or equal to zero. What Pauli then showed is that

   (a) Quantization of integral spin fields according to Jordan–Wigner anti-commutation rules (B.1), corresponding to exclusion, violates the first postulate.

   (b) Quantization of half-integral spin fields according to Bose commutation quantization rules (B.2) violates the second postulate.

The distinction between half-integral spin fields and integral spin fields enters the argument through the manner in which these fields transform under a Lorentz transformation. The Lorentz transformation relates observation of properties (fields, mass, length, etc.) in one inertial frame to observation in another inertial frame of these same properties. The corresponding matrix is orthogonal (see Table 11.1) and represents a rotation in four-dimensional space. A somewhat similar distinction between integral and half-integral spin states evidenced under ordinary rotation of axes in 3-space, such as described in the discussion on the rotation operator in Section 11.5, is found to persist under Lorentz transformation of spin fields.[1]

Statistics

The property that particles have of either obeying or not obeying the exclusion principle has direct consequence in the distributions in energy that aggregates of particles have in equilibrium at a temperature T. Thus fermions (particles with half-integral spin) satisfy Fermi–Dirac statistics. A collection of such noninteracting particles at the temperature T has the energy distribution

$$f_{\mathrm{FD}} = \frac{1}{e^{(E_i - \mu)/k_B T} + 1}$$

This expression gives the average number of particles per state at the energy E_i. The parameter μ represents chemical potential, which at 0 K reduces to the Fermi energy, E_F. At this temperature, no states of energy greater than E_F are occupied (see Fig. 2.5).

Bosons (particles with integral spin) satisfy Bose–Einstein statistics. A collection of noninteracting bosons at the temperature T has the energy distribution

$$f_{\mathrm{BE}} = \frac{1}{e^{(E_i - \mu)/k_B T} - 1}$$

Here again, f represents the average number of particles per state at the energy E_i and μ is written for the chemical potential. This distribution appears in the Planck radiation formula (2.3) relevant to a photon gas in equilibrium at the temperature T, for which case $\mu = 0$.

---

[1] For further discussion of the distinction between integral spin and half-integral spin fields, see H. Yilmaz, *The Theory of Relativity and the Principles of Modern Physics,* Blaisdell, New York, 1965.

APPENDIX C

REPRESENTATIONS OF THE DELTA FUNCTION[1]

Cartesian Coordinates

$$(C.1) \qquad 2\pi\delta(x - x') = \int_{-\infty}^{\infty} e^{ik(x-x')}\, dk$$

$$(C.2) \qquad \pi\delta(x - x') = \int_{0}^{\infty} \cos k(x - x')\, dk$$

$$(C.3)[2] \qquad 2\pi\delta(x - x') = \sum_{-\infty}^{\infty} \exp\left[in(x - x')\right]$$

$$(C.4)[2] \qquad 2\pi\delta(x - x') = 1 + \sum_{1}^{\infty} 2\cos n(x - x')$$

$$(C.5) \qquad \pi\delta(x - x') = \lim_{\eta \to \infty} \frac{\sin \eta(x - x')}{x - x'}$$

$$(C.6) \qquad \delta(x - x') = \lim_{\epsilon \to 0} \frac{e^{-(x-x')^2/\epsilon^2}}{\epsilon\sqrt{\pi}}$$

$$(C.7) \qquad \pi\delta(x - x') = \lim_{\eta \to \infty} \frac{1 - \cos \eta(x - x')}{\eta(x - x')^2}$$

$$(C.8) \qquad \pi\delta(x - x') = \lim_{\eta \to \infty} \frac{2\sin^2\left[\eta(x - x')/2\right]}{\eta(x - x')^2}$$

$$(C.9) \qquad \pi\delta(x - x') = \lim_{\epsilon \to 0} \frac{\epsilon}{(x - x')^2 + \epsilon^2} = \lim_{\epsilon \to 0} \mathrm{Im} \frac{1}{(x - x') - i\epsilon}$$

Let $\mathcal{H}_n(x)$ be the nth-order Hermite polynomial. Then

$$(C.10) \qquad \delta(x - x') = \sum_{n=0}^{\infty} \frac{1}{\sqrt{\pi}\, 2^n n!} \exp -\left(\frac{x^2 + x'^2}{2}\right) \mathcal{H}_n(x)\mathcal{H}_n(x')$$

[1] Additional properties of the delta function may be found in Problem 3.6.

[2] Domain of validity: $x' - \pi \le x \le x' + \pi$.

853

All the above representations obey the normalization

$$\int_{-\infty}^{\infty} \delta(x - x')\, dx' = 1$$

In three dimensions, with $\mathbf{r} = (x, y, z)$, one has

$$\delta(\mathbf{r} - \mathbf{r}') = \frac{1}{(2\pi)^3} \int\!\!\!\int\!\!\!\int_{-\infty}^{\infty} e^{i\mathbf{k}\cdot(\mathbf{r}-\mathbf{r}')}\, dk_x\, dk_y\, dk_z$$

(C.11)

$$\int\!\!\!\int\!\!\!\int_{-\infty}^{\infty} \delta(\mathbf{r} - \mathbf{r}')\, dx'\, dy'\, dz' = 1$$

The Unit Step Function

Let $S(x)$ denote the unit step function:

$$S(x) = 1, \qquad x \geq 0$$
$$S(x) = 0, \qquad x < 0$$

Then

$$\frac{dS(x)}{dx} = \delta(x),$$

Note also that

$$\frac{d\,|x|}{dx} = S(x) - S(-x)$$

It follows that

$$\frac{d^2|x|}{dx^2} = 2\delta(x)$$

Spherical Coordinates

Let $P_l(\mu)$ be the lth-order Legendre polynomial. Then

$$\delta(\mu - \mu') = \sum_{l=0}^{\infty} \frac{2l + 1}{2} P_l(\mu) P_l(\mu')$$

(C.12)

$$\int_{-1}^{1} \delta(\mu - \mu')\, d\mu' = 1$$

The delta function over solid angle may be written in terms of the $Y_l^m(\theta, \phi)$ spherical harmonics.

(C.13) $\quad \delta(\mathbf{\Omega} - \mathbf{\Omega}') = \dfrac{\delta(\theta - \theta')\, \delta(\phi - \phi')}{\sin \theta} = \displaystyle\sum_{l=0}^{\infty} \sum_{m=-l}^{l} [Y_l^m(\theta, \phi)]^* Y_l^m(\theta', \phi')$

The directional coordinates of $\mathbf{\Omega}$ are θ and ϕ. Normalizations are given by

$$\int_0^{\pi} \delta(\theta - \theta')\, d\theta' = 1, \qquad \int_0^{2\pi} \delta(\phi - \phi')\, d\phi' = 1, \qquad \int_{4\pi} d(\mathbf{\Omega} - \mathbf{\Omega}')\, d\mathbf{\Omega} = 1$$

$$d\mathbf{\Omega} = \sin \theta\, d\theta\, d\phi$$

In three dimensions one obtains the representation

(C.14) $\quad \delta(\mathbf{r} - \mathbf{r}') = \delta(\mathbf{\Omega} - \mathbf{\Omega}') \dfrac{\delta(r - r')}{r^2}$

$$= \dfrac{2}{\pi} \sum_l \sum_m [Y_l^m(\theta, \phi)]^* Y_l^m(\theta', \phi') \int_0^{\infty} j_l(kr)\, j_l(kr')k^2\, dk$$

where $j_l(kr)$ is the lth-order spherical Bessel function.

(C.15) $\quad \displaystyle\int_0^{\infty} j_l(kr)\, j_l(kr')k^2\, dk = \dfrac{\pi}{2r^2}\, \delta(r - r'), \qquad \int_0^{\infty} \delta(r - r')\, dr' = 1$

We note also the differential relations

(C.16) $\qquad\qquad (\nabla^2 + k^2) \dfrac{e^{ikr}}{r} = (\nabla^2 + k^2) \dfrac{\cos kr}{r}$

$$= -4\pi \delta(\mathbf{r})$$

Cylindrical Coordinates

Let $J_m(x)$ be the mth integral-order Bessel function. Then

(C.17)
$$\dfrac{\delta(\rho - \rho')}{\rho} = \int_0^{\infty} J_m(k\rho)\, J_m(k\rho')k\, dk$$

$$\int_0^{\infty} \delta(\rho - \rho')\, d\rho = 1$$

With $k_j \rho_0$ denoting the zeros of $J_0(x)$, that is,

$$J_0(k_j \rho_0) = 0$$

one has the representation

(C.18)
$$\pi \rho_0{}^2 \delta(\rho) = \sum_{j=1}^{\infty} \frac{J_0(k_j \rho)}{[J_1(k_j \rho_0)]^2}$$

$$\int_0^{\rho_0} 2\pi \, \delta(\rho) \rho \, d\rho = 1$$

Three other important normalizations of $J_m(x)$ are

(C.19)
$$k \int_0^{\infty} J_m(k\rho) \, d\rho = 1$$

(C.20)
$$\int_0^{\infty} \frac{J_m(k\rho)}{\rho} \, d\rho = \frac{1}{m} \qquad (m > 0)$$

(C.21)
$$\int_0^{\rho_0} J_0(k_j \rho) J_0(k_l \rho) \, d\rho = \frac{1}{2} \rho_0{}^2 J_1{}^2(k_j \rho_0) \delta_{jl}$$

APPENDIX D

DIFFERENTIAL VECTOR RELATIONS

The line element in orthogonal coordinates is

(D.1) $$ds^2 = (h_1\,dx_1)^2 + (h_2\,dx_2)^2 + (h_3\,dx_3)^2$$

The ∇ operations are

(D.2) $$\nabla\psi = \frac{1}{h_1}\frac{\partial\psi}{\partial x_1}\mathbf{a}_1 + \frac{1}{h_2}\frac{\partial\psi}{\partial x_2}\mathbf{a}_2 + \frac{1}{h_3}\frac{\partial\psi}{\partial x_2}\mathbf{a}_3$$

(D.3) $$\nabla\cdot\mathbf{A} = \frac{1}{h_1 h_2 h_3}\left[\frac{\partial}{\partial x_1}(h_2 h_3 A_1) + \frac{\partial}{\partial x_2}(h_1 h_3 A_2) + \frac{\partial}{\partial x_3}(h_1 h_2 A_3)\right]$$

(D.4) $$\nabla\times\mathbf{A} = \frac{1}{h_2 h_3}\left[\frac{\partial}{\partial x_2}(h_3 A_3) - \frac{\partial}{\partial x_3}(h_2 A_2)\right]\mathbf{a}_1$$

$$+ \frac{1}{h_1 h_3}\left[\frac{\partial}{\partial x_3}(h_1 A_1) - \frac{\partial}{\partial x_1}(h_3 A_3)\right]\mathbf{a}_2$$

$$+ \frac{1}{h_1 h_2}\left[\frac{\partial}{\partial x_1}(h_2 A_2) - \frac{\partial}{\partial x_2}(h_1 A_1)\right]\mathbf{a}_3$$

(D.5) $$\nabla^2\psi = \frac{1}{h_1 h_2 h_3}\left[\frac{\partial}{\partial x_1}\left(\frac{h_3 h_2}{h_1}\frac{\partial\psi}{\partial x_1}\right) + \frac{\partial}{\partial x_2}\left(\frac{h_3 h_1}{h_2}\frac{\partial\psi}{\partial x_2}\right) + \frac{\partial}{\partial x_3}\left(\frac{h_1 h_2}{h_3}\frac{\partial\psi}{\partial x_3}\right)\right]$$

Cartesian coordinates:

$$h_1 = 1, \quad h_2 = 1, \quad h_3 = 1; \qquad x_1 = x, \quad x_2 = y, \quad x_3 = z$$

Cylindrical coordinates:

$$h_1 = 1, \quad h_2 = \rho, \quad h_3 = 1; \qquad x_1 = \rho, \quad x_2 = \phi, \quad x_3 = z$$

Spherical coordinates:

$$h_1 = 1, \quad h_2 = r\sin\theta, \quad h_3 = r; \qquad x_1 = r, \quad x_2 = \phi, \quad x_3 = \theta$$

Differential operator relations in three coordinate frames

Cartesian Coordinates	Cylindrical Coordinates	Spherical Coordinates
$dx,\, dy,\, dz$	$d\rho,\, \rho\, d\phi,\, dz$	$dr,\, r\, d\theta,\, r\sin\theta\, d\phi$

Unit vector and elementary cross relations

Cartesian	Cylindrical	Spherical
$\mathbf{a}_x,\, \mathbf{a}_y,\, \mathbf{a}_z$	$\mathbf{a}_\rho,\, \mathbf{a}_\phi,\, \mathbf{a}_z$	$\mathbf{a}_r,\, \mathbf{a}_\theta,\, \mathbf{a}_\phi$
	$x = \rho\cos\phi$	$x = r\sin\theta\cos\phi$
	$y = \rho\sin\phi$	$y = r\sin\theta\sin\phi$
	$z = z$	$z = r\cos\theta$
	$\rho^2 = x^2 + y^2$	$r^2 = x^2 + y^2 + z^2$
	$\tan\phi = y/x$	$\cos\theta = z/\sqrt{x^2+y^2+z^2}$
	$z = z$	$\phi = \tan^{-1}(y/x)$

$$\mathbf{a}_\rho = \cos\phi\, \mathbf{a}_x + \sin\phi\, \mathbf{a}_y$$
$$\mathbf{a}_r = \sin\theta\cos\phi\, \mathbf{a}_x + \sin\theta\sin\phi\, \mathbf{a}_y + \cos\theta\, \mathbf{a}_z$$

$$\mathbf{a}_\phi = -\sin\phi\, \mathbf{a}_x + \cos\phi\, \mathbf{a}_y$$
$$\mathbf{a}_\theta = \cos\theta\cos\phi\, \mathbf{a}_x + \cos\theta\sin\phi\, \mathbf{a}_y - \sin\theta\, \mathbf{a}_z$$

$$\mathbf{a}_z = \mathbf{a}_z$$
$$\mathbf{a}_\phi = -\sin\phi\, \mathbf{a}_x + \cos\phi\, \mathbf{a}_y$$

$$\mathbf{a}_x = \cos\phi\, \mathbf{a}_\rho - \sin\phi\, \mathbf{a}_\phi$$
$$\mathbf{a}_x = \sin\theta\cos\phi\, \mathbf{a}_r + \cos\theta\cos\phi\, \mathbf{a}_\theta - \sin\phi\, \mathbf{a}_\phi$$

$$\mathbf{a}_y = \sin\phi\, \mathbf{a}_\rho + \cos\phi\, \mathbf{a}_\phi$$
$$\mathbf{a}_y = \sin\theta\sin\phi\, \mathbf{a}_r + \cos\theta\sin\phi\, \mathbf{a}_\theta + \cos\phi\, \mathbf{a}_\phi$$

$$\mathbf{a}_z = \mathbf{a}_z$$
$$\mathbf{a}_z = \cos\theta\, \mathbf{a}_r - \sin\theta\, \mathbf{a}_\theta$$

The gradient of ψ, $\nabla\psi$

Cartesian	Cylindrical	Spherical
$(\nabla\psi)_x = \dfrac{\partial\psi}{\partial x}$	$(\nabla\psi)_\rho = \dfrac{\partial\psi}{\partial\rho}$	$(\nabla\psi)_r = \dfrac{\partial\psi}{\partial r}$
$(\nabla\psi)_y = \dfrac{\partial\psi}{\partial y}$	$(\nabla\psi)_\phi = \dfrac{1}{\rho}\dfrac{\partial\psi}{\partial\phi}$	$(\nabla\psi)_\theta = \dfrac{1}{r}\dfrac{\partial\psi}{\partial\theta}$
$(\nabla\psi)_z = \dfrac{\partial\psi}{\partial z}$	$(\nabla\psi)_z = \dfrac{\partial\psi}{\partial z}$	$(\nabla\psi)_\phi = \dfrac{1}{r\sin\theta}\dfrac{\partial\psi}{\partial\phi}$

The divergence of $\mathbf{A}$, $\nabla\cdot\mathbf{A}$

Cartesian:
$$\frac{\partial A_x}{\partial x} + \frac{\partial A_y}{\partial y} + \frac{\partial A_z}{\partial z}$$

Cylindrical:
$$\frac{1}{\rho}\frac{\partial(\rho A_\rho)}{\partial\rho} + \frac{1}{\rho}\frac{\partial A_\phi}{\partial\phi} + \frac{\partial A_z}{\partial z}$$

Spherical:
$$\frac{1}{r^2}\frac{\partial(r^2 A_r)}{\partial r} + \frac{1}{r\sin\theta}\frac{\partial(\sin\theta\, A_\theta)}{\partial\theta} + \frac{1}{r\sin\theta}\frac{\partial A_\phi}{\partial\phi}$$

The Laplacian of ψ, $\nabla^2\psi$

Cartesian:
$$\frac{\partial^2\psi}{\partial x^2} + \frac{\partial^2\psi}{\partial y^2} + \frac{\partial^2\psi}{\partial z^2}$$

Cylindrical:
$$\frac{1}{\rho}\frac{\partial}{\partial\rho}\left(\rho\frac{\partial\psi}{\partial\rho}\right) + \frac{1}{\rho^2}\frac{\partial^2\psi}{\partial\phi^2} + \frac{\partial^2\psi}{\partial z^2}$$

Spherical:
$$\frac{1}{r^2}\frac{\partial}{\partial r}\left(r^2\frac{\partial\psi}{\partial r}\right) + \frac{1}{r^2\sin\theta}\frac{\partial}{\partial\theta}\left(\sin\theta\frac{\partial\psi}{\partial\theta}\right) + \frac{1}{r^2\sin^2\theta}\frac{\partial^2\psi}{\partial\phi^2}$$

The curl of $\mathbf{A}$, $\nabla\times\mathbf{A}$

Cartesian:
$$(\nabla\times\mathbf{A})_x = \left(\frac{\partial A_z}{\partial y} - \frac{\partial A_y}{\partial z}\right)$$
$$(\nabla\times\mathbf{A})_y = \left(\frac{\partial A_x}{\partial z} - \frac{\partial A_z}{\partial x}\right)$$
$$(\nabla\times\mathbf{A})_z = \left(\frac{\partial A_y}{\partial x} - \frac{\partial A_x}{\partial y}\right)$$

Cylindrical:
$$(\nabla\times\mathbf{A})_\rho = \left(\frac{1}{\rho}\frac{\partial A_z}{\partial\phi} - \frac{\partial A_\phi}{\partial z}\right)$$
$$(\nabla\times\mathbf{A})_\phi = \left(\frac{\partial A_\rho}{\partial z} - \frac{\partial A_z}{\partial\rho}\right)$$
$$(\nabla\times\mathbf{A})_z = \frac{1}{\rho}\left(\frac{\partial(\rho A_\phi)}{\partial\rho} - \frac{\partial A_\rho}{\partial\phi}\right)$$

Spherical:
$$(\nabla\times\mathbf{A})_r = \frac{1}{r\sin\theta}\left(\frac{\partial(\sin\theta\, A_\phi)}{\partial\theta} - \frac{\partial A_\theta}{\partial\phi}\right)$$
$$(\nabla\times\mathbf{A})_\theta = \frac{1}{r\sin\theta}\frac{\partial A_r}{\partial\phi} - \frac{1}{r}\frac{\partial(r A_\phi)}{\partial r}$$
$$(\nabla\times\mathbf{A})_\phi = \frac{1}{r}\left(\frac{\partial(r A_\theta)}{\partial r} - \frac{\partial A_r}{\partial\theta}\right)$$

Formulas of Vector Analysis and Vector Calculus

(D.6) $\nabla \cdot (\mathbf{A} \times \mathbf{B}) = \mathbf{B} \cdot \nabla \times \mathbf{A} - \mathbf{A} \cdot \nabla \times \mathbf{B}$

(D.7) $\nabla(\mathbf{A} \cdot \mathbf{B}) = (\mathbf{A} \cdot \nabla)\mathbf{B} + (\mathbf{B} \cdot \nabla)\mathbf{A} + \mathbf{A} \times (\nabla \times \mathbf{B}) + \mathbf{B} \times (\nabla \times \mathbf{A})$

(D.8) $\nabla \cdot (\nabla \times \mathbf{A}) = 0$

(D.9) $\nabla \times (\nabla \times \mathbf{A}) = \nabla(\nabla \cdot \mathbf{A}) - (\nabla \cdot \nabla)\mathbf{A}$

(D.10) $\nabla \times (\mathbf{A} \times \mathbf{B}) = (\mathbf{B} \cdot \nabla)\mathbf{A} + \mathbf{A}(\nabla \cdot \mathbf{B}) - (\mathbf{A} \cdot \nabla)\mathbf{B} - \mathbf{B}(\nabla \cdot \mathbf{A})$

(D.11) $\mathbf{A} \times (\mathbf{B} \times \mathbf{C}) = \mathbf{B}(\mathbf{A} \cdot \mathbf{C}) - \mathbf{C}(\mathbf{A} \cdot \mathbf{B})$

(D.12) $\mathbf{A} \times (\mathbf{B} \times \mathbf{C}) + \mathbf{B} \times (\mathbf{C} \times \mathbf{A}) + \mathbf{C} \times (\mathbf{A} \times \mathbf{B}) = 0$

(D.13)
$$\mathbf{A} \cdot \mathbf{B} \times \mathbf{C} = \begin{vmatrix} A_x & A_y & A_z \\ B_x & B_y & B_z \\ C_x & C_y & C_z \end{vmatrix}$$

(D.14) $\mathbf{A} \cdot \mathbf{B} \times \mathbf{C} = \mathbf{B} \cdot \mathbf{C} \times \mathbf{A} = \mathbf{C} \cdot \mathbf{A} \times \mathbf{B}$

(D.15) $(\mathbf{A} \times \mathbf{B}) \cdot (\mathbf{C} \times \mathbf{D}) = (\mathbf{A} \cdot \mathbf{C})(\mathbf{B} \cdot \mathbf{D}) - (\mathbf{A} \cdot \mathbf{D})(\mathbf{B} \cdot \mathbf{C})$

(D.16) $(\mathbf{A} \times \mathbf{B}) \times (\mathbf{C} \times \mathbf{D}) = (\mathbf{D} \cdot \mathbf{A} \times \mathbf{B})\mathbf{C} - (\mathbf{C} \cdot \mathbf{A} \times \mathbf{B})\mathbf{D}$

(D.17) $\nabla(\phi\psi) = \phi\nabla\psi + \psi\nabla\phi$

(D.18) $\nabla \cdot (\phi\mathbf{A}) = \mathbf{A} \cdot \nabla\phi + \phi\nabla \cdot \mathbf{A}$

(D.19) $\nabla \times (\phi\mathbf{A}) = \phi\nabla \times \mathbf{A} - \mathbf{A} \times \nabla\phi$

If $\mathbf{r}$ is the radius vector from the origin and $\mathbf{A}$ is a constant vector, then

(D.20) $\nabla \cdot \mathbf{r} = 3$

(D.21) $\nabla \times \mathbf{r} = 0$

(D.22) $\nabla r = \mathbf{r}/r$

(D.23) $\nabla r^{-1} = -\mathbf{r}/r^3$

(D.24) $\nabla \cdot (\mathbf{r}r^{-3}) = -\nabla^2 r^{-1} = 4\pi\delta(\mathbf{r})$

(D.25) $\nabla \cdot (\mathbf{A}r^{-1}) = \mathbf{A} \cdot (\nabla r^{-1}) = -(\mathbf{A} \cdot \mathbf{r})r^{-3}$

(D.26) $\nabla \times [\mathbf{A} \times (\mathbf{r}/r^3)] = -\nabla(\mathbf{A} \cdot \mathbf{r}/r^3)$ for $r \neq 0$

(D.27) $\nabla^2\mathbf{A}r^{-1} = \mathbf{A}\nabla^2 r^{-1} = 0$ for $r \neq 0$

(D.28) $(\nabla^2 + k^2)\dfrac{e^{ikr}}{r} = (\nabla^2 + k^2)\dfrac{\cos kr}{r} = -4\pi\delta(\mathbf{r})$

Properties of Integrals over Vector Functions

The Line Integral

(D.29)
$$\int_{\mathbf{r}_1}^{\mathbf{r}_2} \mathbf{A} \cdot d\mathbf{l} = -\int_{\mathbf{r}_2}^{\mathbf{r}_1} \mathbf{A} \cdot d\mathbf{l}$$

If $\mathbf{A} = \nabla\phi$, or $\nabla \times \mathbf{A} = 0$, then the line integral

(D.30)
$$\int_{r_1}^{r_2} \mathbf{A} \cdot d\mathbf{l} = \phi(\mathbf{r}_2) - \phi(\mathbf{r}_1)$$

is independent of the path of integration from $\mathbf{r}_1$ to $\mathbf{r}_2$. If the path of integration is a closed curve, then

(D.31)
$$\oint \mathbf{A} \cdot d\mathbf{l} = \oint \nabla\phi \cdot d\mathbf{l} = 0$$

Gauss's Theorem Let S be a surface bounding a region of volume V. Then,

(D.32)
$$\int_V \nabla \cdot \mathbf{A} \, dV = \int_S \mathbf{A} \cdot d\mathbf{S}$$

Stokes's Theorem Let S be an open surface bounded by the closed, nonintersecting curve C. Then

$$\int_C \mathbf{A} \cdot d\mathbf{l} = \int_S (\nabla \times \mathbf{A}) \cdot d\mathbf{S}$$

Green's First Identity

(D.33)
$$\int_V (\phi\nabla^2\psi + \nabla\phi \cdot \nabla\psi) \, dV = \int_S (\phi\nabla\psi) \cdot d\mathbf{S}$$

Green's Second Identity

(D.34)
$$\int_V (\phi\nabla^2\psi - \psi\nabla^2\phi) \, dV = \int_S (\phi\nabla\psi - \psi\nabla\phi) \cdot d\mathbf{S}$$

Other Identities

(D.35)
$$\int_V \nabla \times \mathbf{A} \, dV = \int_S d\mathbf{S} \times \mathbf{A}$$

(D.36)
$$\int_C \phi \, d\mathbf{l} = \int_S d\mathbf{S} \times \nabla\phi$$

APPENDIX E

PHYSICAL CONSTANTS AND EQUIVALENCE RELATIONS

Velocity of light in vacuum	c	2.9979×10^8 m/s
		2.9979×10^{10} cm/s
Planck's constant	h	6.6261×10^{-34} J s
		6.6261×10^{-27} erg s
		4.1357×10^{-15} eV s
	$\hbar$	1.0546×10^{-34} J s
		1.0546×10^{-27} erg s
		6.5821×10^{-16} eV s
Avogadro's number	N_0	6.0221×10^{23} atoms/mole
Boltzmann's constant	k_B	1.3807×10^{-23} J/K
		1.3807×10^{-16} erg/K
		8.6174×10^{-5} eV/K
Room temperature		300 K = 0.02585 eV
Gas constant	$R = N_0 k_B$	8.3145 J/mole K
		8.3145×10^7 erg/mole K
		1.9870 cal/mole K
Volume of 1 mol of perfect gas, at normal temperature and pressure		22.421 liters
Electron charge	e	1.6022×10^{-19} C
		4.8032×10^{-10} esu
Electron rest mass	m	9.1094×10^{-31} kg
		9.1094×10^{-28} g
		0.511 MeV
Electron classical radius	$r_0 = e^2/mc^2$	2.818×10^{-13} cm
Electron magnetic moment	μ_e	$1.001 \mu_b$
Fine-structure constant	$\alpha = e^2/\hbar c$	$7.297 \times 10^{-3} = 1/137.04$
Compton wavelength	$\lambda_C = h/mc$	2.426×10^{-10} cm
	$\lambda_C = \hbar/mc$	3.862×10^{-11} cm
Lyman α		0.1216 μm

Gravitational constant	G	6.672×10^{-8} dyne-cm^2/g^2
Proton rest mass	M_p	1.6726×10^{-27} kg
		1.6726×10^{-24} g
		1.0073 amu
		938.27 MeV
Neutron rest mass	M_n	1.675×10^{-27} kg
		1.675×10^{-24} g
		939.57 MeV
Bohr magneton	$\mu_b = \dfrac{e\hbar}{2mc}$	9.27×10^{-21} erg gauss^{-1}
Ratio of proton mass to electron mass	$\dfrac{M_p}{m}$	1836.1
Charge-to-mass ratio of electron	$\dfrac{e}{m}$	1.7588×10^{11} C/kg
		5.2730×10^{17} esu/g
Stephan–Boltzmann constant	$\sigma = \left(\dfrac{\pi^2}{60}\right)\left(\dfrac{k_B{}^4}{\hbar^3 c^2}\right)$	5.6697×10^{-5} erg cm^{-2} s^{-1} K^{-4}
Rydberg constant	$\mathbb{R} = \dfrac{me^4}{2\hbar^2} = \dfrac{\hbar^2}{2ma_0{}^2}$	109,737.32 cm^{-1}
	$= \dfrac{e^2}{2a_0} = \dfrac{\alpha^2 mc^2}{2}$	13.6 eV
Bohr radius	$a_0 = \dfrac{\hbar^2}{me^2}$	0.52918 Å
Triple point of water		273.16 K
Atomic mass unit	1 amu	1.6605×10^{-24} g
		931.5 MeV
		Mass of C$^{12} \times \dfrac{1}{12}$
Wien's displacement law constant	$\lambda_{\max} T$	0.290 cm K

Useful Conversion Constants and Units

Constants, MKS units	ϵ_0	$10^7/4\pi c^2$ F/m
		$= 8.8542 \times 10^{-12}$ F/m
	μ_0	$4\pi \times 10^{-7}$ H/m
$1/4\pi\epsilon_0$		9×10^9 m/F
$1/\epsilon_0\mu_0$		c^2

$$\sqrt{\frac{\mu_0}{\epsilon_0}}$$

	$120\pi \ \Omega = 377 \ \Omega$
Energy in wavenumbers	$E = hc\lambda^{-1}$
	$hc = 1.240 \times 10^{-4}$ eV cm
	$= 1.24$ eV μm
Useful relations	$\hbar/m = 1.16$ cm^2/s
	$cm/e^2 = 118$ s/cm^2
1 electron volt eV	10^{-6} MeV
	1.6022×10^{-12} erg
	1.6022×10^{-19} J $= [e, C]$ J
	3.829×10^{-20} Cal
	$11{,}605$ K
1 coulomb	0.1 abcoulomb (emu)
	$(c/10)$ statcoulomb (esu) $\gg 1$ esu
1 emu	1 esu/c
1 V	$\frac{1}{300}$ statvolts $= \frac{1}{300}$ erg/esu
1 dyne	10^{-5} N
1 erg	10^{-7} J
E (eV)	$1.24/\lambda$ (in μm)
1 W/m^2 = 1 T	10^4 gauss
	$(1$ gauss cm$^2 = 1$ T m$^2)$
1 Cal	4.186 J
	4.186×10^7 ergs
$k_B N_0$	2 Cal/K

$$\left[\frac{e\mathbf{B}}{mc}\right]_{\text{esu}} \rightarrow \left[\frac{e\mathbf{B}}{m}\right]_{\text{emu}}$$

$$\left(\mathbf{p} - \frac{e}{c}\mathbf{A}\right)^2_{\text{esu}} \rightarrow (\mathbf{p} - e\mathbf{A})^2_{\text{emu}}$$

Coulomb's Law

cgs	MKS (SI)
$$F = \frac{q_1 q_2}{\varepsilon(\text{cgs})r^2}$$	$$F = \frac{q_1 q_2}{4\pi\varepsilon(\text{MKS})r^2}$$
$q_1 = q_2 = 1$ esu, at 1-cm separation, in vacuum, gives 1 dyne of force	$q_1 = q_2 = \sqrt{4\pi\varepsilon_0}$ C, at 1-m separation, in vacuum, gives 1 N of force
$$\varepsilon(\text{cgs}) = \frac{\varepsilon(\text{MKS})}{\varepsilon_0(\text{MKS})} \equiv \varepsilon_{\text{relative}}$$	
$\varepsilon(\text{cgs})_{\text{vacuum}} = 1$	$\varepsilon(\text{MKS})_{\text{vacuum}} = \varepsilon_0$

INDEX

accidental, 334, 336p
exchange, 330
of hydrogen energies, 463, 475p
quantum, 660p
symmetry, 334, 339
degenerate material, 666
degenerate perturbation theory, 713, 718
degrees of freedom, 4
De Haas–van Alphen effect, 446
delta function (*see* Dirac delta function)
density of states, 61p, 65p, 746p
in various dimensions, 349ff.
density operator, 559
and equation of motion, 561, 760p
depletion-zone approximation, 674
depletion-zone width, 679
d'Espagnat, B., 566, 581
determinism, 53
deuterium molecule, 398p, 644, 648p
nuclear component wavefunction for, 647
rotational modes of, 645
deuteron, 299p, 456p
ground state, 689ff.
DeWitt, B., 75
diagonalization, 495, 517p
of submatrix, 714
diagonal matrices, 496, 510p
diamond, 329
Dicke, R. H., 564, 607
diffraction, 48, 50, 57
diffusion, quantum mechanical, 173p
dipole radiation, 479
Dirac delta function, 76ff., 169
in Cartesian coordinates, 853
in cylindrical coordinates, 855
in spherical coordinates, 854
Dirac equation, 170, 823
Dirac notation, 99ff.
dispersion relations, 171p
distribution function, 650
donor-type impurity, 661
Doppler shift, 654, 656p
Drell, S. D., 836
drift velocity (*see* velocity, drift)
dumbbell, rigid (*see* rotator, rigid)

effective density of states, 666
effective mass, 306
effective potential, 451, 452
Ehrenfest's principle, 60, 175, 192p

Einstein, photoelectric theory of, 36ff.
Einstein *A* and *B* coefficients, 741
Einstein–Podolsky–Rosen paradox, 60, 566p, 574
Einstein relations, 675
Eisenbud, L., 693
electron emission (*see* cold emission)
electronic distributions in atoms, table of, 622
electron-positron creation, 829
electron radius, classical, 154p
energy bands, 309, 318p, 629p
energy gaps, 308ff., 666, 668, 728, 729
energy operator, 72, 110, 512
ensemble, 118
average, 118, 560
entropy, 64p, 649
equipartition hypothesis, 62p
Ewen, H. I., 700
exchange binding, 641
exchange energy, 637
exchange operator, 553, 615, 619p
exchange symmetry, 616
excitations in liquid helium, 655
exclusion principle, 38, 617, 621, 632, 649
expectation value, 78ff.
time development of, 173ff.
extended states, 308

Fermi, E., 739
Fermi–Dirac distribution, 664, 852
Fermi energy, 38, 65p, 328
of copper, 65p
extrinsic, 669, 671, 673
intrinsic, 668, 673
of sodium, 65p
Fermion, 617
Fermi sea, 38, 828
Fermi's golden rule (*see* golden rule)
ferromagnetism (*see* Heisenberg model for)
Feshbach, H., 693, 730
Feynman diagram, 830
Feynman, R., 82, 538, 655
path integral formulation, 16, 281
fine structure (*see* hydrogen atom)
fine-structure constant, 43p, 53p, 605, 770
Fleischmann, H. H., 427
Flügge, S., 769, 836
Fock, V., 751, 780
forbidden domain, 25
central potential of, 451, 452